ECOLOGY, GENETICS, AND EVOLUTION OF METAPOPULATIONS

ECOLOGY, GENETICS, AND EVOLUTION OF METAPOPULATIONS

Edited by

Ilkka Hanski

Metapopulation Research Group
Department of Ecology and Systematics
University of Helsinki, Finland

Oscar E. Gaggiotti

Metapopulation Research Group
Department of Ecology and Systematics
University of Helsinki, Finland

ELSEVIER
ACADEMIC
PRESS

Amsterdam Boston Heidelberg London New York Oxford
Paris San Diego San Francisco Singapore Sydney Tokyo

Senior Editor, Life Sciences:	Charles Crumly
Editorial Assistant:	Kelly Sonnack
Senior Project Manager:	Angela G. Dooley
Senior Marketing Manager:	Linda Beattie
Cover Design:	Eric DeCicco
Composition:	Integra Software Services Pvt. Ltd, Pondicherry, India
Printer:	The Maple-Vail Manufacturing Group
Cover printer:	Phoenix

Elsevier Academic Press
200 Wheeler Road, Burlington, MA 01803, USA
525 B Street, Suite 1900, San Diego, California 92101-4495, USA
84 Theobald's Road, London WC1X 8RR, UK

This book is printed on acid-free paper.

Cover illustration: Model-predicted probability density of individual locations in a heterogeneous landscape (Ovaskainen 2004, Ecology 85, 242–257) and an empirical example of extinction threshold (Hanski and Ovaskainen 2000, Nature 404, 755–758).

Library of Congress Cataloging-in-Publication Data:
Application submitted

British Library Cataloguing in Publication Data
A catalogue record for this book is available from the British Library

ISBN: 0-12-323448-4

For all information on all Academic Press publications
visit our Web site at www.books.elsevier.com

Printed in the United States of America
04 05 06 07 08 9 8 7 6 5 4 3 2 1

CONTENTS

CONTRIBUTORS

Numbers in parentheses indicate the pages on which the authors' contributions begin.

Jon Aars (515) NERC Molecular Genetics in Ecology Initiative, Aberdeen Population Ecology Research Unit, School of Biological Sciences, University of Aberdeen, Aberdeen AB24 2TZ, Scotland; current address: Norwegian Polar Institute, Polarmiljøsenteret, N-9296 Tromsø, Norway

Janis Antonovics (471) Biology Department, University of Virginia, Charlottesville, Virginia 22904

Ottar N. Bjørnstad (415) Departments of Entomology and Biology, Penn State University, University Park, Pennsylvania 16802

Benjamin M. Bolker (45) Zoology Department, University of Florida, Gainesville, Florida 32611

Cajo J.F. ter Braak (105) Biometrics, Wageningen University and Research Centre, Box 100, NL-6700 AC Wageningen, The Netherlands

Mar Cabeza (541) Metapopulation Research Group, Department of Ecology and Systematics, University of Helsinki, FIN-00014 Helsinki, Finland

Jean Clobert (307) Laboratoire d'Ecologie, Université Pierre et Marie Curie, Bâtiment A, 75252 Parix cedex 05, France

O. Eriksson (447) Department of Botany, Stockholm University, SE-106 91 Stockholm, Sweden

Rampal S. Etienne (105) Community and Conservation Ecology Group, University of Groningen, Box 14, NL-9750 AA Haren, The Netherlands

Oscar E. Gaggiotti (3, 337, 367) Génomique de Populations et Biodiversité, LECA-CNRS UMR 5553 Université Joseph Fourier, F-38041 Grenoble Cedex 9, France

Sergey Gavrilets (275) Department of Ecology and Evolutionary Biology, Department of Mathematics, University of Tennessee, Knoxville, Tennessee 37996

Charles J. Goodnight (201) Department of Biology, University of Vermont, Marsh Life Sciences Building, Burlington, Vermont 05405

Bryan T. Grenfell (415) Department of Zoology, University of Cambridge, Downing Street, Cambridge CB2 3EJ, England

Ilkka Hanski (3, 73, 337, 489) Metapopulation Research Group, Department of Ecology and Systematics, University of Helsinki, FIN-00014 Helsinki, Finland

E.E. Holmes (565) National Marine Fisheries Service, Northwest Fisheries Science Center, Seattle, Washington 98112

Rolf Anker Ims (307) Institute of Biology, University of Tromsø, N-9037 Tromsø, Norway

Tadeusz J. Kawecki (387) Unit for Ecology and Evolution, Department of Biology, University of Fribourg, CH-1700 Fribourg, Switzerland

Matt J. Keeling (415) Maths Institute and Department of Biological Sciences, University of Warwick, Coventry, CV4 7AL, England

Xavier Lambin (515) Aberdeen Population Ecology Research Unit, School of Biological Sciences, University of Aberdeen, Aberdeen AB24 2TZ, Scotland

Mathew A. Leibold (133) Department of Ecology and Evolution, University of Chicago, Chicago, Illinois 60637

Thomas E. Miller (133) Department of Biological Science, Florida State University, Tallahassee, Florida 32306

Atte Moilanen (541) Metapopulation Research Group, Department of Ecology and Systematics, University of Helsinki, FIN-00014 Helsinki, Finland

Isabelle Olivieri (229) Institut des Sciences de l'Evolution UMR5554, Université Montpellier II, Place Eugène Bataillon, 34095 Montpellier cedex 5, France

N.J. Ouborg (447) Department of Molecular Ecology, University of Nijmegen, Toernooiveld 1, 6525 Ed Nijmegen, The Netherlands

Otso Ovaskainen (73) Metapopulation Research Group, Department of Ecology and Systematics, University of Helsinki, FIN-00014 Helsinki, Finland

Stuart B. Piertney (515) NERC Molecular Genetics in Ecology Initiative, Aberdeen Population Ecology Research Unit, School of Biological Sciences, University of Aberdeen, Aberdeen AB24 2TZ, Scotland

Hugh P. Possingham (541) Departments of Zoology and Mathematics, The University of Queensland, St Lucia QLD 4072, Australia

Ophélie Ronce (229) Institut des Sciences de l'Evolution UMR5554, Université Montpellier II, Place Eugène Bataillon, 34095 Montpellier cedex 5, France

François Rousset (307) Institut des Sciences de l'Evolution, UMR5554 Université Montpellier, 34095 Montpellier cedex 5, France

B. Semmens (565) Zoology Department, University of Washington, Seattle, Washington 98195

Sandra Telfer (515) Small Animal Infectious Disease Group, Leahurst, University of Liverpool, Neston CH64 7TE, England

Chris D. Thomas (489) Department of Biology, University of Leeds, Leeds LS2 9JT, England

Claire C. Vos (105) Alterra Green World Research, Box 47, NL-6700 AA Wageningen, The Netherlands

Michael J. Wade (259) Department of Biology, Indiana University, Bloomington, Indiana 47405

John Wakeley (175) Biological Laboratories, Harvard University, Cambridge, Massachusetts 02138

Michael C. Whitlock (153) Department of Zoology, University of British Columbia, Vancouver, BC V6T 1 Z4, Canada

Kimberly A. With (23) Division of Biology, Kansas State University, Manhattan, Kansas 66506

PREFACE

Over the past 15 years, metapopulation biology has developed from a set of ideas, simple models, and a limited number of case studies to an essential part of population biology. Some areas of metapopulation biology continue to flourish with bold new visions and attempts to clarify them with models, but other areas have already become consolidated into a solid body of theory and have been thoroughly investigated empirically. Progress has been so great that the contents of this volume bear only superficial resemblance to the contents of the predecessor, *Metapopulation Biology* (Hanski and Gilpin, 1997), to say nothing about the first edited volume in this series, *Metapopulation Dynamics* (Gilpin and Hanski, 1991).

In this volume we have achieved, for the first time, an equal coverage of metapopulation ecology, metapopulation genetics, and evolutionary metapopulation biology. There is no complete parity, however. Metapopulation ecology, which was at the stage of conceptual development and budding empirical studies 15 years ago, has by now turned to a well-established discipline with substantial impact on practical conservation. In contrast, metapopulation genetic and evolutionary studies are at an earlier stage, with less well-developed integration of theoretical and empirical work. But such integration is undoubtedly coming, and it is hoped that this volume will stimulate further development in this direction.

All the chapters in this volume are entirely new, nothing has been copied from *Metapopulation Biology*. The previous volume includes contributions that are well worth reading even today, but we did not include them here in the interest of giving space to a new set of authors and chapters, and also because the previous volume is still available. One important similarity remains. This is an edited volume in which we have not forced the same approach in treatment of the subject matter in all the chapters. Some chapters are primarily or even entirely theoretical, whereas others are based on empirical research. Some chapters present an overview of one slice of metapopulation biology, whereas others are focused more narrowly on new developments. There are up-to-date reviews of all areas of metapopulation biology. We are confident that there is not a single population biologist in this world who would find nothing new in this volume, nor are there many who would find all the chapters easy bed-time reading. But we trust that most of our readers will appreciate the diversity and the challenge, and will be inspired by at least some of the visions, comprehensive empirical studies, and modeling efforts described in this volume. Our aim was to produce a volume that serves both as a reference for researchers and as a text for advanced students in ecology, genetics, evolutionary biology, and conservation biology. The emphasis is on integration across disciplines. Several chapters are relevant for conservationists in setting the stage for new applications. It is hoped that graduate students will find material in this volume for innovative Ph.D. projects.

We are grateful to a large number of colleagues who provided truly helpful reviews of particular chapters: Miguel Araújo, Frederic Austerlitz, Hans Baveco, Peter Beerli, Thomas Berendonck, Ben Bolker, Cajo ter Braak, Mark Burgman, Jeremy Burdon, Dennis Couvet, Michael Doebeli, Stephen Ellner, Rampal Etienne, Patrick Foley, Robert Freckleton, Sylvain Gandon, Gisela Garcia, Nicholas Gotelli, Mikko Heino, Jessica Hellmann, Eric Imbert, Rolf Ims, Pär K. Ingvarsson, Kevin Laland, Xavier Lambin, Russ Lande, Martin Lascoux, Richard Law, Michel Loreau, Michael McCarthy, Juha Merilä, Atte Moilanen, Allen J. Moore, Isabelle Olivieri, Otso Ovaskainen, John Pannell, Craig Primmer, Jonathan Pritchard, Chris Ray, Steven Riley, Ilik Saccheri, Mikko J. Sillanpää, Jonathan Silvertown, Peter Smouse, Per Sjögren-Gulve, Chris Thomas, Xavier Vekemans, Jana Verboom, and Franjo Weissing. We thank Marjo Saastamoinen and Tapio Gustafsson for indispensable secretarial help. Chuck Crumly from Academic Press had trust in this volume from our very first correspondence, and Kelly Sonnack, Angela Dooley, Michael Sugarman and Eric DeCicco at Academic Press made our task as editors as easy as possible. Finally, our thanks to all the authors for showing great enthusiasm and keeping deadlines.

Ilkka Hanski
Oscar Gaggiotti
April 2003, Helsinki

ACKNOWLEDGMENTS

CHAPTER 1

We thank Rolf Ims and Chris Thomas for comments on the chapter. Supported by the Academy of Finland (Centre of Excellence Programme 2000–2005).

CHAPTER 2

I thank Hans Baveco, Ilkka Hanski, Greg Schrott, Per Sjögren-Gulve, and Jana Verboom for their comments on the chapter. My research on the effects of landscape structure and dynamics on extinction risk for spatially structured populations has been supported by past grants from the National Science Foundation and, most recently, by the U.S. Environmental Protection Agency (R829090).

CHAPTER 3

I thank the Isaac Newton Institute for supporting a workshop on scaling in biological systems where some of these ideas were developed and Toshinori Okuyama and Graeme Cumming for useful discussions.

CHAPTER 4

We thank Ben Bolker, Cajo van ter Braak, Rampal Etienne, and Karin Frank for comments on the chapter. Supported by the Academy of Finland (Centre of Excellence Programme 2000–2005).

CHAPTER 5

We thank Ilkka Hanski, Atte Moilanen, and Otso Ovaskainen for comments on the manuscript, Bob O'Hara and Hannu Toivonen for providing details about their Gibbs sampling algorithm, and Ton Stumpel, Wim Nieuwenhuizen, and RAVON (The Dutch organization for research on reptiles, amphibians, and fishes) for kindly allowing us to use their data.

CHAPTER 6

MAL gratefully acknowledges the Working Group on Metacommunities at the National Center for Ecological Analysis and Synthesis and the National Science Foundation (DEB 9815799). TEM acknowledges support of the National Science Foundation (DEB 0083617 and DEB 0091776).

CHAPTER 7

This work was funded by the Natural Science and Engineering Research Council (Canada). Many thanks to Cort Griswold, Oscar Gaggiotti, and two anonymous reviewers for their useful comments on the manuscript version of this chapter.

CHAPTER 8

I thank Oscar Gaggiotti and Ilkka Hanski for inviting me to contribute to this volume. I am also thankful for helpful discussions with Simon Tavaré and Mark Beaumont concerning the summary–statistic approach to computational inference, with Dick Lewontin about the tests of neutrality, and with Jon Wilkins concerning the dynamics of continuously distributed populations. Finally, I am thankful to Peter Beerli for comments on a previous version of the manuscript. This work was supported by a Career Award (DEB-0133760) from the National Science Foundation.

CHAPTER 10

We thank Frank Shaw, Ruth Shaw, Sylvain Gandon, François Rousset, Oscar Gaggiotti, Ilkka Hanski, and two anonymous referees for helpful comments on the chapter. Sylvain Gandon and Mikko Heino kindly provided the material for Fig. 10.1. This work was supported by the European Union program "Plant Dispersal," Contract EVK2-CT1999–00246, and the French Ministry of Research through the Action Concertée Incitative "Jeune chercheur." This is publication ISEM 2003-049 of the Institut des Sciences de l'Evolution de Montpellier.

CHAPTER 11

I thank my laboratory group, Jeff Demuth, Jake Moorad, Meaghan Saur-Jacobs, Tim Linksvayer, and Troy Wood, for their comments on his work and grant support from NSF DEB-0212582 and NIH GM65414-01A1.

CHAPTER 12

I am very grateful to Randal Acton, Mike Finger, Janko Gravner, Michael Saum, and Michael Vose who greatly contributed at different stages of this project. I thank Frans Jacobs, Oscar Gaggiotti, and two anonymous reviewers for very useful comments on the manuscript. Supported by National Institutes of Health Grant GM56693 and by National Science Foundation Grant DEB-0111613.

CHAPTER 13

We are very grateful to O. Ronce, M. Heino, I. Olivieri, I. Hanski, and O. Gaggiotti for very helpful comments on an earlier version of this manuscript. Part of this research (J. Clobert) has been financed by the European Research Training Network ModLife (Modern Life-History Theory and its Application to the Management of Natural Resources), funded through the Human Potential Programme of the European Commission (Contract HPRN-CT-2000-00051).

CHAPTER 14

We thank Steinar Engen, Patrick Foley, and Russ Lande for comments on the chapter. Supported by the Academy of Finland (Centre of Excellence Programme 2000-2005).

CHAPTER 15

Richard A. Nichols, Steve P. Brooks, Bill Amos, and John Harwood contributed at different stages to the research discussed in this chapter. I thank Jonathan Pritchard, Mikko J. Sillanpää, and Peter E. Smouse for useful comments on the manuscript. Supported by the Academy of Finland (Centre of Excellence Programme 2000-2005).

CHAPTER 16

This work has been supported by the Swiss National Science Foundation.

CHAPTER 17

The authors acknowledge the support of the Royal Society (MJ), the Wellcome Trust (BTG), and the UK BBSRC (Biology and Biotechnology Research Council). We also thank David Earn and Pej Rohani for discussions.

CHAPTER 18

We are grateful to S. Cousins for letting us use unpublished data, to M. Soons and P. Vergeer for giving us preprints of their papers, to J. van Groenendael and K. Oosterhuis for discussion and general support, and to E. Imbert, R.P. Freckleton, and O. Gaggiotti for providing useful comments on earlier versions of this manuscript. This is a publication from the EU-TRANSPLANT project (EVK2-1999–00042).

CHAPTER 19

I especially thank Doug Taylor and Michael Hood, for without their encouragement and help over the past five years, this work would not have continued. Helen Alexander, Francois Felber, and Don Stratton were largely responsible for getting this study underway. I also thank Joe Abrams for much of the data analysis and the maximum likelihood estimation, and Cristina Rabaglia for analyzing the weather data. This research has also been made possible by the generous efforts of many people who each year have voluntarily given their time and energy to coming to Mountain Lake Biological Station, getting up at 5:00 A.M., and scouring the countryside for *Silene*. I thank them all and beg for understanding (and reminders) for names that have been inadvertently omitted: Joe Abrams, Sonia Altizer, Gretchen Arnold, Arjen Biere, Amy Blair, Molly Brooke, Steve Burckhalter, Julie Carlin, Sherri Church, Anita Davelos, Sandra Davies, Lynda Delph, Kyle Dexter, Mike Duthie, Stacie Emery, Annette Golonka, Richard Golumkiewitz, John Hammond, Miriam Heuhsen, Sarah Hyland, Par Ingvarrson, Andrew Jarosz, Sarah Joiner, Britt Koskella, Anna-Liisa Laine, Kathy Lemmon, Emily Lyons, Arlan Maltby, Dave McCauley, Alice McDonald, Kara O'Keefe, Matt Olson, Peter Oudemans, Wendy Palen, Jessica Partain, Amy Pedersen, Todd Preuninger, Sarah Ribstein, Chris Richards, Elizabeth Richardson, Bernie Roche, Laura Rose, Katherine Ross, Meghan Saur, Chad Shaw, David Smith, Dexter Sowell, Pete Thrall, Peter van Tienderen, Henry Wilbur, and Lorne Wolfe.

CHAPTER 20

We are grateful to the large number of people who contributed to the research described in this chapter and to the funding agencies that made it possible. Rob Wilson and Otso Ovaskainen kindly helped produce analyses and/or figures included in this chapter, and Jess Hellmann and Xavier Lambin made helpful comments on the manuscript.

CHAPTER 21

Sandra Telfer and Stuart Piertney were supported by NERC. We acknowledge Balmoral and Assynt Estates for allowing us to conduct our water vole studies, as well as financial support from the People's Trust for Endangered Species. We thank Ilkka Hanski, Oscar Gaggiotti, Rolf Anker Ims, and particularly an anonymous referee for exceptionally insightful comments.

CHAPTER 22

We thank Professor C.D. Thomas for providing Creuddyn Peninsula butterfly data and M. Araújo, M. Burgman, I. Hanski, and A. van Teeffelen for useful comments on the manuscript. This study was supported by the Academy of Finland, Grants 71516 and 74125 to A.M. and M.C., respectively, and a grant from the Finnish Cultural Foundation to M.C.

CHAPTER 23

The graduate students in the Winter 2000 seminar on Spatial Ecology and Conservation Biology at the University of Washington were the initial impetus of this work. Special thanks are given to Amanda Stanley who did many of the simulations that eventually led to the realization of strong patterns within the dynamics of stochastic metapopulations. Jon Hoekstra (NMFS) provided us with data on the Puget Sound chinook ESU. Discussions with Michelle McClure, John McElhany, and Mary Ruckleshaus (leads of the NMFS Technical Recovery Teams for Pacific NW salmon) have been instrumental at various stages of this work. Thanks to an anonymous reviewer whose detailed review and critiques led to many improvements in the presentation of the theory.

Part I

PERSPECTIVES ON SPATIAL DYNAMICS

1. METAPOPULATION BIOLOGY: PAST, PRESENT, AND FUTURE

Ilkka Hanski and **Oscar E. Gaggiotti**

1.1 INTRODUCTION

The term *metapopulation* stems from the general notion of the hierarchical structure of nature. Just like the term population is needed to describe an assemblage of interacting individuals, it seems apt to have a term for an assemblage of spatially delimited local populations that are coupled by some degree of migration — the metapopulation (Levins, 1970). It is conceptually attractive, and helpful for the study of population biology, to explicitly consider the sequence of entities from individuals to local populations to metapopulations. Theoretical studies are greatly facilitated by the view of landscapes as networks of habitat patches inhabited by local populations. And it is not just theory: there are innumerable species that definitely have such a spatial population structure in some landscapes, and continuing habitat loss and fragmentation force ever greater numbers of species to conform to a metapopulation structure. Other species have more continuous spatial distributions in less distinctly patchy environments, but even for these species and for some purposes the metapopulation view of nature can be helpful.

A *metapopulation approach* refers to research or management that, in one form or another, adopts the view that local populations, which the metapopulations consist of, are discrete (or relatively discrete) entities in space and that these local populations interact via migration and gene flow. Classic metapopulation

0-12-323448-4

dynamics in the sense pioneered by Levins (1969, 1970) focus on the processes of local extinction and recolonization in the same manner as population dynamics are concerned with births and deaths of individuals. However, such population turnover is not a necessary condition for the metapopulation approach to be useful, nor a characteristic feature of all species that are structured, in some landscapes, into spatially discontinuous local populations. Important questions need to be asked about the interaction of permanent local populations, for instance in the context of source–sink dynamics (Chapter 16).

Metapopulation biology represents one way of explicitly putting population biology into a spatial context. The basic tenet of *spatial ecology*, which includes metapopulation ecology as well as other approximations (see alter), is that the spatial positions of individuals and populations matter, in the sense of influencing the growth rate and dynamics of populations and metapopulations and their competitive, predator–prey and other interactions. Likewise, we may use the term *spatial population biology* to emphasize the influence of the spatial positions of individuals and populations on their genetic and evolutionary dynamics as well as their ecological dynamics.

That spatial positions matter is a trivial observation for biologists working on plants and other sessile organisms. Thus Harper (1977) entitled one of the five main sections of his *Population Biology of Plants* as "The effects of neighbours." It has been less obvious that spatial positions of individuals matter in the case of mobile animals, which may form more or less random-mating (panmictic) populations. However, from the point of view of ecological interactions, spatial positions often do matter even in mobile animals. One example is the large number of insect species with mobile adults but immobile larval stages. Larvae do most of the interactions and so the spatial distribution of larvae matters greatly to single-species (de Jong, 1979), competitive (Hanski, 1981, 1990a), and predator–prey dynamics (Hassell, 1978, 2000). Indeed, from the 1970s onward, the spatial aggregation of interacting individuals has been one of the most important themes in population dynamics. These types of within-population spatial structures also have evolutionary consequences, which have been investigated by Levins (1970), Boorman and Levitt (1973), Cohen and Eshel (1976), Wilson (Wilson, 1980; Wilson et al., 1992; Mitteldorf and Wilson, 2000), and others. Interestingly, the population genetic modeling of continuously distributed populations initiated by Wright (1940, 1943, 1946) and Malecot (1948) faced difficulties precisely because of the spatial aggregation of individuals (Felsenstein, 1975). Much progress has been made in this area in the last decade using Monte Carlo simulations, spatial autocorrelation methods, and lattice models (Eppenson and Allard, 1989, 1993a,b, 1995; Rousset, 2000).

Taking the population structure in which reproduction is panmictic but ecological interactions are localized, as described earlier, as the starting point, there are two ways of moving to the domain of metapopulation dynamics. First, widespread dispersal may not occur in every generation, in which case patches of microhabitat harbor not just single-generation assemblages of interacting individuals, but multigeneration local populations. Insects living in decaying wood provide good examples, ranging from those that disperse completely in each generation to species that form local populations in particular (large) trunks for tens or even hundreds of generations (Fig. 1.1). The decisive factor is

Fig. 1.1 Oak woodland in Sweden where a long-term study has examined the metapopulation biology of the beetle *Osmoderma eremita*, with long-lasting local populations inhabiting individual oak trees (Ranius, 2000; Ranius and Jansson, 2000). Photograph by Jonas Hedin.

simply the longevity of the microhabitat in relation to the life span of individuals, underscoring the more generally valid point that metapopulation dynamics are typically determined as much, or more, by the structure and dynamics of the physical environment as by the properties of the species. In the population genetics literature, the sort of situation represented by insect populations inhabiting long-lasting microhabitats has been examined under the rubric of the haystack model (Maynard Smith, 1964; Bulmer and Taylor, 1981).

The second way of moving to the metapopulation domain from panmictic local populations is simply by expanding the spatial scale: most organisms have limited dispersal powers, hence there is a spatial scale at which most interactions, including mating, occur "within populations," whereas at larger spatial scales, these local populations are connected by migration and gene flow. It is especially natural to turn to the metapopulation approach if the environment is physically fragmented into pieces of habitat that may support local populations. Metapopulation biology recognizes that many, if not most, ecological, genetic, and evolutionary processes occur at spatial scales that are greater than the scale within which most individuals disperse. Hence there is spatial structure at the metapopulation scale that should not be ignored. Moving to still larger spatial scales, to the geographical ranges of species, brings in other processes that are beyond the metapopulation concept and domain.

We emphasize the significance of metapopulation processes rather than spatial structures. It is tempting to attempt to classify different kinds of spatial population structures (Harrison, 1991, 1994), and some terminology is needed for communication, but the danger is that we impose an order to nature that is not there. Landscapes are all different, hence there must be a huge diversity of "metapopulation structures." Focusing on the processes — migration, gene flow, spatially correlated dynamics, local extinction, genetic drift, local adaptation, and so forth — circumvents the need to infer processes from patterns where this is not necessary (in many cases there is, however, valuable information in patterns that should not be ignored; see Wiegand et al., 2003). By emphasizing the metapopulation approach, we also underscore the point that this is only one approach and not always the most appropriate one (Section 1.3). There is little doubt that spatially localized interactions and movements influence the ecological, genetic, and evolutionary dynamics of the vast majority of species. It is another question which particular approach is the most effective in uncovering the biological consequences of spatially localized interactions and movements for both research and management.

1.2 METAPOPULATION BIOLOGY: PAST TRENDS IN THE LITERATURE

The history of research in metapopulation biology has been narrated by Hanski and Simberloff (1997) and Hanski (1999b). Rather than repeating it here, we will examine that history in light of the number of citations to relevant key words. Such a systematically "documented history" of metapopulation biology goes back to the 1970s. We used the BIOSIS database, which yielded 1087 citations to the key word *metapopulation* in the title of a paper or in its abstract (years 1970–2001). To get a fair idea of the temporal patterns in the number of citations, we divided the yearly totals by the pooled number of citations in the database in that year, a measure of the total volume of the literature.

Thus measured, the number of citations to *metapopulation* has increased more or less linearly since 1990 (Fig. 1.2), with only a few earlier citations, even if the metapopulation concept itself was introduced already in 1970 (Levins, 1970). Some inaccuracy is due to less thorough coverage of the literature in the database in the 1970s than later on, but this does not change the broad picture. One can think about several reasons for the 20-yr time lag in the wider use of the metapopulation concept, which is in sharp contrast to the early success of the island biogeographic theory of MacArthur and Wilson (1963, 1967), published only a few years prior to Levins's (1969, 1970) metapopulation idea and model (Hanski, 1996). First, MacArthur and Wilson published their theory in a leading journal for population biology and as a high-profile monograph, whereas Levins's papers were published in less illustrious journals. Second, MacArthur and Wilson were purposely in the business of turning a page in the history of biogeography, whereas Levins's (1969) immediate goal was more modest, to construct a model to examine alternative strategies of pest eradication. Third, MacArthur and Wilson were widely respected scientists, whereas Levins was a hero for a more limited number of people. Fourth, and what may be really important, the island theory became associated with the species–area relationship, enhancing the theory's popularity because ecologists could use it in their research (whether this application of the theory made a lasting contribution is another matter). There was no similar opportunity to do empirical work that would be similarly linked with Levins's models — a situation that was to change only in the 1990s with further development of the theory (Section 1.3). Finally, the heightened awareness of the dire biological consequences of habitat loss and fragmentation from the late 1980s onward has

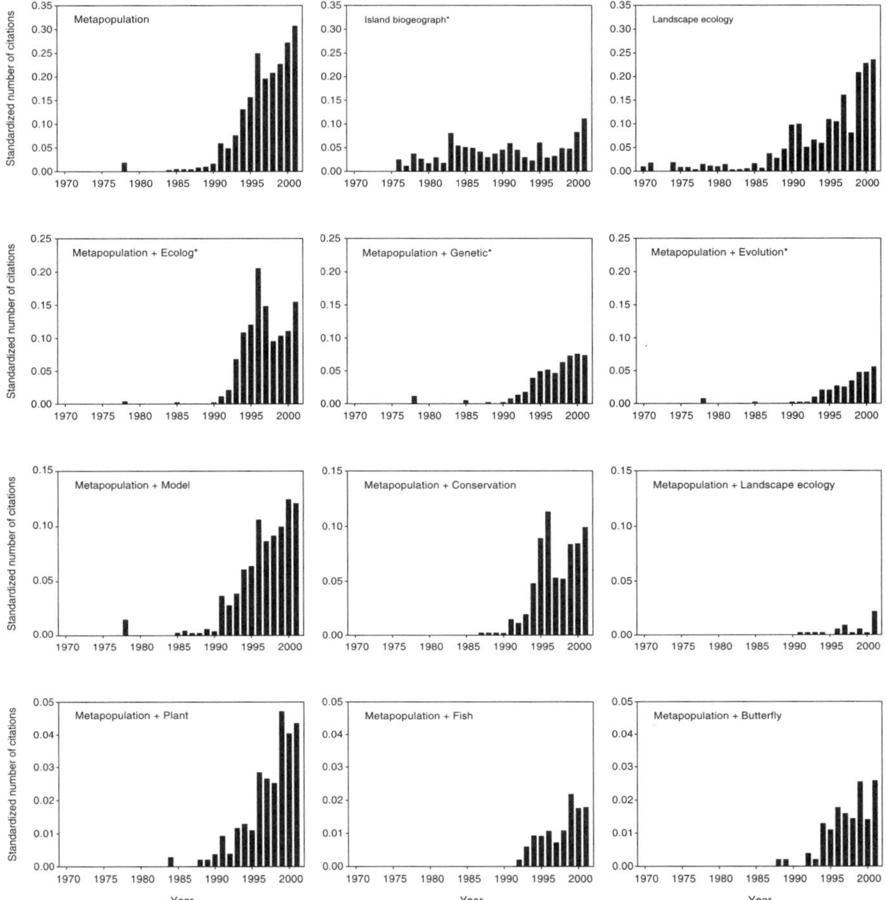

Fig. 1.2 Number of citations in the BIOSIS database to the key words indicated in the panels divided by the total number of citations in a particular year (to control for the increasing total volume of literature over the years). Note that the scale on the vertical axis is different in different rows of panels. See text for discussion.

practically forced an interest in metapopulation biology, making the rediscovery of Levins's early work inevitable.

The top row in Fig. 1.2 gives the number of citations to the key words *landscape ecology* and *island biogeograph** as well as to the key word *metapopulation* (biogeograph* includes all words starting with "biogeograph," such as "biogeography" and "biogeographic"). The temporal patterns show intriguing differences. *Landscape ecology* was established in the literature in the beginning of the period considered, in 1970, but for the next 15 years the frequency of citations remained at a constantly low level. A distinct growth phase began around 1985, and definitely earlier than in the case of *metapopulation*. At present, *metapopulation* is cited somewhat more frequently than *landscape ecology*. *Island biogeograph** has appeared in the literature since the mid-1970s and the frequency has remained high until the present, with ups and downs. Perhaps surprisingly, the standardized number of citations to *island biogeograph** was higher in 2001 than ever before — 34 years since the classic monograph by MacArthur and Wilson (1967) established the modern era in ecological biogeography. It is noteworthy that the peaks in the time series for *landscape ecology* and *island biogeograph** agree rather closely since the late 1980s, suggesting that many papers refer to both key words.

Next we examined combinations of key words, including *metapopulation* and something else. The second row in Fig. 1.2 compares the three subdisciplines *ecolog**, *genetic**, and *evolution**. In all cases, the first papers were published in 1978 and 1985. Most of these papers were in fact listed in all three searches and include Gill's (1978a,b) papers on the metapopulation ecology of the red-spotted newt, its migration rate, and effective population size; Couvet et al.'s (1985) study on the population genetics in spatially structured populations; and Fix's (1985) theoretical study of the evolution of altruism. Since 1990, *ecolog** has accumulated many more citations than *genetic** or *evolution**. The temporal patterns appear to indicate that while *ecolog** has not been growing systematically since 1994, *genetic** has been growing until the late 1990s and the number of citations to *evolution** appears still to be growing. These trends are consistent with our general perception of shifting research interests, as well as with the change in the contents of the three volumes on metapopulation biology (Gilpin and Hanski, 1991; Hanski and Gilpin, 1997; present volume). A somewhat different interpretation of the figures for *ecolog** associates the peak in the number of citations in 1996–1997 to the publication of the previous metapopulation volume, which appeared in the year 1996 (Hanski and Gilpin, 1997). In any case, it is apparent that the number of citations to *ecolog** has increased again since 1996.

The next row in Fig. 1.2 gives some further comparisons. Theory has maintained its position well over the years (key word *model*), although evidence also indicates that empirical work has been catching up to theoretical studies in recent years. This is shown by a significant declining trend in the ratio of citations to *metapopulation* + *model* over *metapopulation* (yearly counts for 1990 until 2001, the 1990 count also including all the previous papers; linear regression, $F = 7.76$, $P = 0.02$). Of course, many of the papers referring to *model* might not be theoretical papers, and part of the continuing increase in *model* papers is due to an increase in genetic and evolutionary metapopulation studies. *Conservation* combined with *metapopulation* has increased steadily for the past decade, with the exception of a striking peak in 1995–1996, paralleling (although not exactly matching) the corresponding peak for *ecolog**. The very low frequency of citations to *metapopulation* + *landscape ecology* is not surprising in the light of the continuing separation of these two disciplines that seemingly have so much in common (more about this in the next section). Let us hope that the relatively large number of citations to *metapopulation* + *landscape ecology* scored for 2001 represents the beginning of a new era!

Finally, the last row in Fig. 1.2 examines three taxa, plants, fishes, and butterflies, all of which show the same increasing trend as *metapopulation* itself. The pooled number of citations to *metapopulation* + "*taxon*" for the years 1996 to 2000 is as follows for the following taxa: *bird*, 22; *mammal*, 85; *fish*, 38; *butterfly*, 49; and *plant*, 94. These overall figures are somewhat misleading, however. For instance, there are many more "hard core" metapopulation papers on butterflies than on birds and mammals, undoubtedly because the metapopulation approach is particularly applicable to many butterflies (Chapter 20; Hanski, 1999, Ehrlich and Hanski, 2004). This is also reflected in the type of the very first papers in the database for the taxa shown in Fig. 1.2. For butterflies the pioneering study is Harrison et al. (1988) on the mainland-island metapopulation structure in the Bay checkerspot butterfly (*Euphydryas editha*) in California, whereas for fishes and plants the first papers are, respectively, Hanzelova and Spakulova's (1992) essentially biometric study and Ellstrand et al.'s (1984) notion of an inflorescence as a metapopulation.

1.3 AN OVERVIEW OF CURRENT RESEARCH

This section outlines some noteworthy recent developments in metapopulation ecology, genetics, and evolutionary studies as well as their integration. This section refers extensively to the remaining chapters in this volume. Although the motivation for research typically stems from past scientific discoveries and perceived opportunities for further discoveries, the ongoing loss, alteration, and fragmentation of natural habitats are widely viewed as other important reasons for conducting research in metapopulation biology.

Ecology

The metapopulation approach is conceptually closely related to the dynamic theory of island biogeography of MacArthur and Wilson (1967). Most importantly, both theories advocate the same "island perspective," whether the islands are true islands or habitat islands, and both theories are concerned with local extinctions and recolonizations, although this is not an exclusive interest in metapopulation biology, as pointed out earlier. The apparent difference in the focus of the island theory on communities and of metapopulation theories on single species is not a fundamental difference, as long as one assumes independent dynamics in the species that comprise the community (as the basic island model does). The similarity between the island biogeographic model and the classic metapopulation model is underscored by the spatially realistic metapopulation theory (Hanski, 2001a; Hanski and Ovaskainen, 2003; Chapter 4; see later), which adds the effects of habitat patch area and isolation on extinctions and colonizations into the classic metapopulation theory. In fact, we can now see that Levins's metapopulation model and MacArthur and Wilson's island model are two special cases of a more comprehensive model (Hanski, 2001a). One advantage of the metapopulation theory over the island theory is that the former but not the latter allows each species to have its own patch network in the same landscape, reflecting differences in the habitat selection of the species. In any case, it is intriguing that the island theory and metapopulation theory have been widely considered as representing two different paradigms in conservation biology (see discussion in Hanski and Simberloff, 1997).

The island theory and metapopulation theory are not the only approaches to spatial ecology. Figure 1.3 gives a simple classification of three main approaches. The key issue is what is assumed about the structure of the environment. In one extreme, labeled as the theoretical ecology approach, the

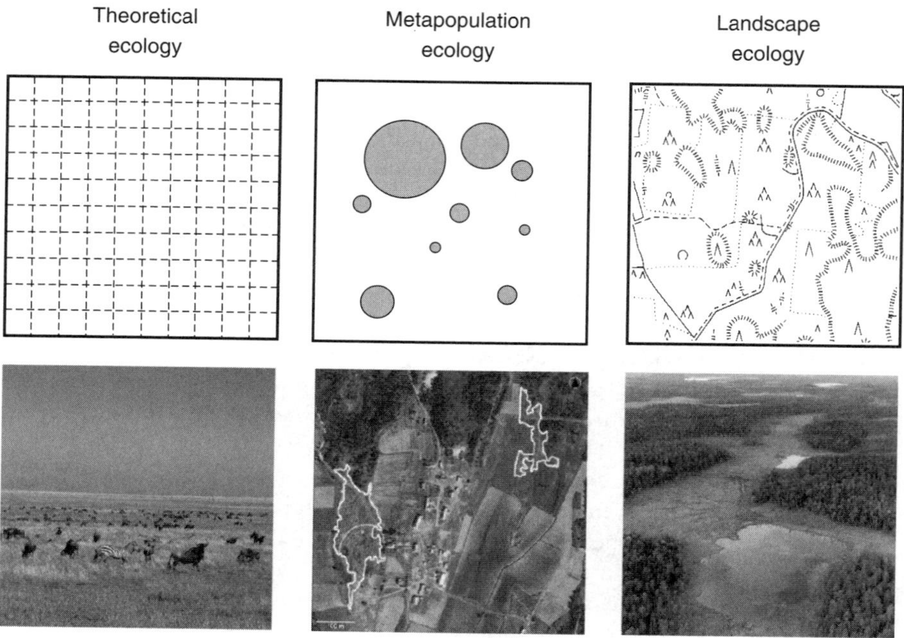

Fig. 1.3 Three approaches to spatial ecology: theoretical approaches assuming homogeneous environment, metapopulation approach, and landscape ecology approach based on a detailed description of the landscape structure (from Hanski, 1998b). Photographs give examples of the three situations to which the three approaches are applicable: uniform grassland; map of the Åland Islands in Southwest Finland with habitat patches (dry meadows) suitable for the Glanville fritillary butterfly delimited; and a mixture of forested landscape with open wetland areas.

common assumption is that the environment is completely homogeneous. Here the primary aim of research is to elucidate the consequences of spatially restricted interactions and/or migration of individuals to the dynamics and spatial structures of populations. Chapter 3 describes at length this approach to spatial ecology. The mathematical tools commonly employed include lattice-based models, such as interacting particle systems, cellular automata and coupled-map lattices, spatial moment equations, and partial differential equations, as well as simulations. Recent work on "neutral" theories of community structure (Bell, 2000; Hubbell, 2001) also fit in this category, although these models deal with evolutionary as well as ecological dynamics (Chapter 6). The assumption of homogeneous space facilitates the study of population processes as opposed to the heterogeneous landscape in creating and maintaining spatial variation in population densities, but this assumption also practically eliminates the possibility of testing model predictions. As suggested in Chapter 3, the models studied by theoretical ecologists are strategic models designed to investigate general principles rather than tactical models designed to answer specific questions about specific populations. Nonetheless, even the general theory has to be related to the real world. It is hence important that recent modeling studies in this framework have attempted to relax the assumption of homogeneous space. For instance, Murrell and Law (2000) have used the method of moments to model the dynamics of carabid beetles in heterogeneous landscapes with three different classes of land type, woodland, agricultural land, and urban areas, and Keeling (2000b) has applied the method of moments to single-species and predator–prey dynamics in coupled local populations in a metapopulation (for further discussion, see Chapter 3).

In the other extreme depicted in Fig. 1.3, which is represented by much of landscape ecology, the starting point is just the opposite, a detailed description of the often complex structure of real landscapes. Chapter 2 presents an overview of landscape ecology as far as it is concerned with population processes. Given the complex description of the landscape structure and the emphasis on individual movements in much of landscape ecology (Schippers et al., 1996; Pither and Taylor, 1998; Haddad, 1999a; Bunn et al., 2000; Jonsen and Taylor, 2000; Byers, 2001), it is not surprising that the prevalent modeling tool has been individual-based simulation (With and Crist, 1995; With and King, 1999b; Hill and Caswell, 1999; Fahrig, 2002; Chapter 2). As seen from Fig. 1.3, we view the metapopulation approach as occupying the middle ground in this classification: the environment is assumed to consist of discrete patches of suitable habitat for the focal species, usually ignoring the shape of these patches, surrounded by the landscape matrix that is not suitable for reproduction but through which individuals may migrate. These assumptions can be somewhat relaxed without compromising the possibility of developing metapopulation theory. For example, one may allow for matrix heterogeneity by calculating effective patch connectivities, and one may replace real patch areas by effective areas allowing for spatial variation in habitat patch quality (Moilanen and Hanski, 1998; Hanski, 1999b). What still remains intact is the core assumption of discrete local populations inhabiting discrete patches of habitat.

In terms of theory in metapopulation ecology, our admittedly partial perspective inclines us to emphasize the significance of *the spatially realistic metapopulation theory* (SMT). The core mathematical models in this theory are *stochastic patch occupancy models* (SPOM). SPOMs assume a network of habitat patches, which have only two possible states, occupied by the focal species or empty. If there are n patches in the network, the metapopulation has 2^n possible states, which is such a large number for large n that a rigorous mathematical analysis is not possible and some simplification is called for. One simplification is to assume a *homogeneous SPOM*, with identical habitat patches, which allows a rigorous analysis of even the stochastic model (this is the familiar "island model"). Another simplification is to resort to deterministic models that ignore spatial correlations in the pattern of patch occupancy and variability due to a finite number of patches in the network. The Levins model makes both simplifying assumptions at the same time — it is a deterministic approximation of a homogeneous SPOM. What we now know is that rigorous theory can be constructed by making just one of the simplifying assumptions. SMT is obtained by combining a *heterogeneous SPOM*, in which patches have different extinction and colonization probabilities, with assumptions as to how the structure of the landscape influences these probabilities (or rates in the case of continuous-time models). Chapter 4 describes SPOMs and the spatially realistic metapopulation theory in detail.

The spatially realistic metapopulation theory makes a contribution toward a unification of research in population biology in several fronts (Hanski and Ovaskainen, 2003). First, as already pointed out, the island theory and the classic metapopulation theory are two special cases of SMT (Hanski, 2001a). Second, SMT contributes to the unification of metapopulation ecology and landscape ecology with its explicit focus on the influence of the structural features of the landscape on population processes. As Chapter 2 shows, some landscape ecologists have worked toward the same goal from their own tradition. These developments suggest that the merging of the two fields of metapopulation ecology and landscape ecology is finally starting to take place. Third, SMT is mathematically closely related to matrix population models (Caswell, 2001) for age-structured and size-structured populations, which have also been employed in the study of source–sink metapopulations (Chapter 16). Fourth, SMT shares common theoretical underpinnings with epidemiological theory (Grenfell and Harwood, 1997; Ovaskainen and Grenfell, 2003). Fifth, a great advantage of the models stemming from SMT is that they can be parameterized rigorously with data on the dynamics and pattern of habitat patch occupancy. Chapter 5 presents a review of the methods of parameter estimation and how the models can be applied to real metapopulations. Chapter 22 employs SMT to combine spatial dynamics with reserve site selection algorithms to incorporate the concept of population persistence into reserve selection procedures. The close linking of theory to empirical research that SMT facilitates is somewhat analogous to the link between the dynamic theory of island biogeography and empirical research on the species–area relationship in the 1970s. The difference, however, is that metapopulation models can be parameterized rigorously with empirical data, whereas just documenting the species–area relationship is not sufficient to parameterize, nor to test, the island biogeographic model. The reason for the success of the metapopulation models in this respect is that they are typically applied to metapopulations with many and often small local populations with a measurable rate of population turnover. Data available hence relate to spatial dynamics as well as to the consequent spatial patterns of habitat occupancy. This is in contrast with past research on the island theory and species–area relationship, which was largely restricted, due to a low rate of population turnover on large islands, to analyses of spatial patterns rather than of processes.

Spatially realistic metapopulation theory is focused on the actual spatial structure of metapopulations, in the sense of specifying the probabilities with which particular habitat patches in a fragmented landscape are occupied. Another class of structured metapopulation models considers the distribution of local population sizes but ignores the actual spatial structure by assuming that all local populations are equally connected (Hanski, 1985; Hastings and Wolin, 1989; Hastings, 1991; Gyllenberg and Hanski, 1992; Gyllenberg et al., 1997). These models are particularly concerned with the influence of emigration and immigration on local dynamics in the metapopulation context and are, in this respect, akin to source–sink models (Chapter 16). The aforementioned modeling studies assume an infinite number of local populations with deterministic local dynamics. Lande et al. (1998) developed another class of models structured by local population size for finite metapopulations with stochastic local dynamics. The most interesting new phenomenon predicted by population size-structured metapopulation models is the possibility of alternative stable equilibria in metapopulation size, one of which corresponds to metapopulation extinction, the other one to a positive and possibly large metapopulation size

(Hanski, 1985; Gyllenberg and Hanski, 1992; Hanski and Gyllenberg, 1993). The processes that lead to alternative stable equilibria are the rescue effect, reduced rate of local extinction due to immigration, and the Allee effect, which increases the rate of successful colonization per capita with increasing immigration rate. These processes can be added to SPOMs only in a nonmechanistic manner (Ovaskainen and Hanski, 2001), as SPOMs are concerned with habitat patch occupancy, not with numbers of individuals. Another great advantage of the metapopulation models structured by the actual size of local populations is the possibility to extend the analysis to evolutionary issues. For instance, Ronce et al. (2000b), Metz and Gyllenberg (2001), and Gyllenberg et al. (2002) studied the evolution of migration rate with population size-structured metapopulation models [see also Heino and Hanski (2001) for a spatially realistic model and Chapter 10 for comprehensive discussion].

Ecological models of metapopulation dynamics tend to make simple assumptions about migration. Emigration is typically assumed to be density independent, and migrating individuals are assumed to follow a correlated random walk or some less mechanistic simple assumption is made about the behavior of migrants. Chapter 13 presents a thorough review of what is known about migration at the level of individual behavior. Not surprisingly, there is no strong support for the simple assumptions made in most models. In contrast, migration is seen as a complex behavior involving a series of decisions that often depend on the state (condition) of individuals and their interactions with other individuals. In particular, migration is often density dependent, although both positive and negative density dependence is commonly reported (Chapter 13). Positively density-dependent emigration and negatively density-dependent immigration are expected to enhance the growth rate of the metapopulation, increasing the range of conditions under which the metapopulation is viable (Saether et al., 1999). These effects occur because the pattern of migration will influence the strength of the rescue effect and the probability of successful colonization. In brief, it is clear that migrants in most species have more sophisticated behavior than assumed by most models. What is not clear, however, is when would it be necessary to (greatly) complicate the models by including many behavioral details, and indeed to what extent should the models be modified. Turning from rigorous mathematical models to simulations just for the sake of adding some "realism" is not necessarily warranted. What is needed is a family of models incorporating different amounts of detail. No systematic study of this type has yet been conducted on migration and metapopulation dynamics.

Ecologists working with population viability analysis tend to prefer individual-based (Possingham and Noble, 1991; Akçakaya and Ferson, 1992; Lacy, 1993, 2000; Akçakaya, 2000a) or population-based (Sjögren-Gulve and Ray, 1996) simulation models. The advantage of these models is that any processes and mechanisms that the researcher may wish to add to the model can be added readily. The disadvantage is that general insights are difficult to extract from complex simulations. Furthermore, it is practically impossible to estimate rigorously the often large number of parameters and to test the structural model assumptions; the modeling results are thus of questionable value for management. The best use of these models, as perhaps of any population models, for conservation and management is to contrast alternative scenarios that differ in only a small number of factors (Hanski, 1997a; Ralls and Taylor, 1997; Beissinger and Westphal, 1998; Akçakaya and Sjögren-Gulve, 2000). One would hope that the result of such comparisons is relatively insensitive to the many uncertainties in parameter values and even in the structure of the model itself. The interested reader is referred to many chapters in two edited volumes (Sjögren-Gulve and Ebenhard, 2000; Beissinger and McCullough, 2001). Our emphasis in this volume is on SPOMs for the reasons that much progress has been made in recent years in developing both the theory (Chapter 4) and applications to real metapopulations (Chapters 5, 20, and 22; see also Dreschler et al., 2003).

To present a balanced view about the standing of the metapopulation approach in ecology, it is appropriate to acknowledge the critical opinions that have been voiced about its general significance. Harrison (1991, 1994; Hastings and Harrison, 1994; Harrison and Taylor, 1997; Harrison and Bruna, 1999) has suggested repeatedly that the occurrence of species "in the balance between the extinction and recolonization of populations is an improbable condition" (Harrison, 1994, p. 115). To some extent, Harrison's concerns are answered by the spatially realistic metapopulation theory, which relaxes many of the simplifying assumptions of the nonspatial homogeneous patch occupancy models, such as the Levins model, and which shows how realistic variation in habitat patch areas and connectivities can be incorporated into models. Another line of response is provided by the scores of empirical studies that demonstrate the operation, in practice, of metapopulation dynamics with a frequent turnover of local populations in systems that lack large and permanent "mainland" populations. Chapter 20 assesses the performance of the metapopulations approach in dynamic (nonequilibrium) landscapes, where Harrison's criticisms initially seem most relevant. In fact, the models perform well in the situations examined and can be used to gain valuable insights about the long-term behavior of metapopulations.

Research on European butterflies, in particular, has produced much empirical evidence for metapopulation processes in shaping not only the ecological dynamics (Thomas, 1994b; Thomas and Hanski, 1997; Hanski, 1999b; C.D. Thomas et al., 2002), but also genetic (Saccheri et al., 1998; Nieminen et al., 2001; Scmitt and Seitz, 2002) and evolutionary dynamics (Kuussaari et al., 2000; Hanski and Singer, 2001; Heino and Hanski, 2001; Thomas et al., 2001; Hill et al., 2002) of butterflies. Chapter 20 in this volume and a volume on the biology of checkerspot butterflies (Ehrlich and Hanski, 2004) present two overviews covering much of this research. Butterflies possess several traits that make them a convenient model group of species for metapopulation research: specific host plant and habitat requirements, meaning that many landscapes are highly fragmented for butterflies; small body size allowing the presence of local breeding populations in relatively small habitat patches; and high population growth rate but also great sensitivity to environmental conditions, leading to high population turnover (Murphy et al., 1990). Additional advantages that butterflies offer include the facility of estimating population sizes and migration rates with mark–release–recapture methods and the often great distinction between the suitable habitat and the landscape matrix. It may remain a matter of opinion as to how representative, and representative of what, the many butterfly studies are, but minimally we expect that butterflies fairly represent a large number of specialized insect species.

Chapters 18, 19, and 21 discuss plant, plant–pathogen, and small mammal metapopulation dynamics, respectively. The metapopulation dynamics of many plants are influenced by the seed bank and very long-lived adult individuals, which complicate empirical studies greatly but also mean that certain phenomena, such as long transients in the dynamics of species in changing environments, are especially important in plants (Chapter 18). The basic issue of delimiting suitable but occupied habitat is often very difficult in the case of plants because plant species typically compete for space and hence a single-species approach is likely to be inadequate. Chapter 19 on plant–pathogen metapopulation dynamics is focused on a two-species interaction, which additionally involves a coupling of ecological and evolutionary dynamics that may be responsible for long-term trends in metapopulation sizes. Metapopulation dynamics in small mammals (Chapter 21) may also often involve more than one species, for instance, a specialist predator driving some of the population turnover in the prey species, potentially leading to spatially correlated patterns of habitat occupancy. More generally, there is a clear need for more studies on metacommunities — assemblages of interacting metapopulations. Chapter 6 reviews the conceptual framework and current research on metacommunities. Not surprisingly, webs of direct and indirect interactions in communities, combined with webs of spatially connected populations, complicate matters greatly, and we may need several different theoretical frameworks to cover the full range of possibilities that arise in metacommunity dynamics.

Returning to the criticism against the general significance of the metapopulation approach, Fahrig (1997, 1998, 2001, 2002) has suggested repeatedly that the persistence of species in (increasingly) fragmented landscapes is little affected by habitat fragmentation as such, but rather what matters is the total area of the (remaining) habitat. In other words, in Fahrig's opinion, the spatial configuration of the habitat makes little difference. If this were generally the case, much of the contents of this volume would be superfluous. We, however, consider that Fahrig's conclusions are too far-fetched. Considering the plane depicted in Fig. 1.4, defined by the proportion of the suitable habitat in the landscape and the migration range of the focal species, habitat fragmentation may indeed be of little significance in most parts of this plane. However, a huge number of species/landscape combinations crowd the lower-left corner of Fig. 1.4: highly fragmented landscapes, in which only a small fraction of the total area is covered by the suitable habitat; and relatively poorly dispersing species at the scale of interest. Furthermore, as we all know, human-caused habitat loss and fragmentation continuously push further combinations of species and landscapes to this corner in Fig. 1.4, where the spatial configuration of the remaining habitat should not be ignored. The metapopulation theory (Chapter 4) is helpful in delineating the parts of the shaded square in Fig. 1.4 that allow long-term metapopulation persistence from those parts that lead to metapopulation extinction.

Genetics

Metapopulation genetic studies have their roots in Sewall Wright's island model of population structure (Wright, 1931), which assumes distinct local populations (colonies, demes) connected by migration and gene flow. In this

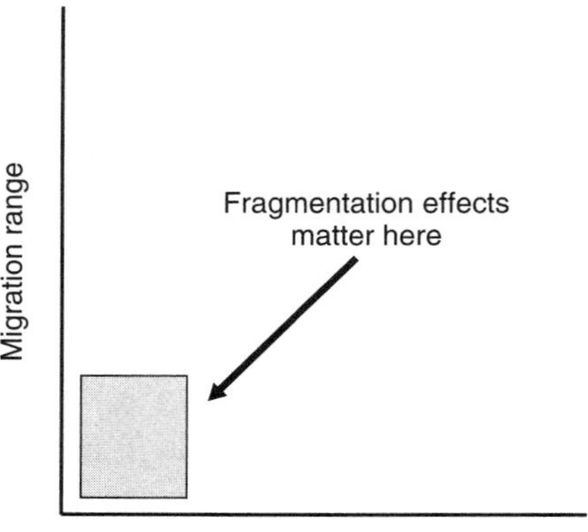

Fig. 1.4 Habitat fragmentation (the spatial configuration of the remaining habitat) matters on the abundance and persistence of species in landscapes where the suitable habitat covers only a small fraction of the total landscape area and the migration range of the focal species is limited.

classic model, all local populations are identical (same size) and equally connected (constant migration rate), which are also the assumptions of the ecological Levins model (a deterministic approximation of a homogeneous SPOM, see earlier discussion). However, while the latter was focused on population turnover, Wright's island model assumed permanent local populations. The first formal application of the classic ecological metapopulation concept in the domain of population genetics was a generalization of the island model to cover the case where local populations would go extinct and new ones were established (Slatkin, 1977). The extension of these ideas to stepping-stone models followed shortly afterward (Maruyama and Kimura, 1980). The pioneering work of Slatkin (1977) was followed by studies by Wade and McCauley (1988) and Whitlock and McCauley (1990). The aim of all these investigations was to clarify the effects that extinctions and recolonizations have on the genetic structure of metapopulations, that is, the partitioning of genetic variability within and between local populations. Just like in classic ecological metapopulation models, the effect of local dynamics on genetic structure was ignored to facilitate the study of factors such as the extinction rate and the genetic composition of the groups of individuals that establish new populations.

The effect of local dynamics on metapopulation genetic structure has been addressed in a series of papers published by Whitlock (1992a), Gaggiotti and Smouse (1996), Gaggiotti (1996), and Ingvarsson (1997). These studies demonstrate that the interaction between local dynamics and migration patterns can have important consequences for the genetic structure of metapopulations. In metapopulations of the Levins type, with all local populations having the same carrying capacity, fluctuations in local population size and/or migration rate increase genetic differentiation among populations (Whitlock, 1992a). Slow population growth following colonization has a similar effect when the migration rate is constant (Ingvarsson, 1997). In the case of source–sink metapopulations, the degree of genetic differentiation among sources and sinks, among sinks, and the level of genetic variability maintained by sink populations is largely determined by the variance in propagule size. The lower the variance, the higher the degree of genetic differentiation and the lower the level of genetic variability maintained by sink populations (Gaggiotti and Smouse, 1996; Gaggiotti, 1996).

All the theoretical studies mentioned so far have been concerned with the genetic structure of selectively neutral genes, which is the most thoroughly studied subject in metapopulation genetics. Following the publication of comprehensive analyses by Whitlock and Barton (1997) and Rousset (1999a,b), which extended the results of the previous studies to models that cover a wide variety of metapopulation scenarios, theoretical research into the genetic structure of metapopulations has diminished. Presently, the most active area in metapopulation genetics is concerned with selected genes and quantitative genetic variation (Chapters 7 and 9). This recent work has added important new processes such as inbreeding, heterosis, and mutation accumulation into the metapopulation approach and its application to conservation and management of endangered species (see Chapter 7). To some extent, further advance in this area is hampered by the substantial lack of knowledge that exists about the rates and effects of spontaneous mutations (discussed in Chapter 14). Indeed, the fact that we have reached a point where further progress in a specific area of metapopulation biology requires the resolution of a fundamental issue in such an established discipline as genetics is an indication of how fast the field has progressed.

Studies reviewed in Chapter 9 have extended classic quantitative genetics theory to metapopulations. The classic theory was concerned with measuring the response to selection and largely ignored epistatic interactions, whereas the more recent metapopulation quantitative genetics theory is concerned with measuring differentiation among populations and emphasizes the importance of epistatic interactions (Chapter 9). This shift in emphasis has uncovered new mechanisms for speciation and is a good example of how a focus on metapopulations can shed new light onto key evolutionary problems.

Another important recent development in metapopulation biology is the extension of the coalescent approach (Kingman, 1982a; reviewed by Fu and Li, 1997) to cover metapopulation scenarios (Wakeley and Aliacar, 2001). The coalescent approach represented a big leap forward for population genetics because it provides a theoretical framework to make inferences about past events based on a genetic sample representing the present population. The essence of the coalescent theory is to start with a sample and to move backward in time to identify events that occurred in the past since the most recent common ancestor of the sample. Chapter 8 provides an overview of the coalescent process in the metapopulation context and describes ways in which it can be used to make statistical inferences. Although current work in this area is highly theoretical, it will lead to useful applications such as the development of statistical approaches for the analysis of molecular data aimed at making inferences about metapopulation processes. This in turn will facilitate the integration of theoretical and empirical work as well as the demographic and genetic approaches to metapopulation biology.

Evolution

Application of the metapopulation approach in the domain of evolutionary biology has been motivated mainly by three broad issues: the shifting balance theory (SBT) (Wright, 1931, 1940), the evolution of migration rate, and the evolution of species' ranges. An important controversy in evolutionary biology deals with two opposing views of adaptation (e.g., Coyne et al., 1997, 2000; Wade and Goodnight, 1998; Goodnight and Wade, 2000). One view, called the Fisherian view, advocates that the bulk of adaptive evolution results from Darwinian mass selection. The other view maintains that adaptation cannot be explained by selection alone and that stochastic processes such as genetic drift often play an important role. Sewall Wright has been the main advocate of this latter view and he formalized it in his shifting balance theory.

The shifting balance theory is based on the idea that species are subdivided into many local populations (demes) that are weakly connected by migration. The small size of the local populations would allow genetic drift to overwhelm the effects of natural selection and take the populations to the domain of attraction of new adaptive peaks (phase I). Individual selection could then move the population toward the new peak itself (phase II), at which point selection among the local populations would act to pull the entire species (metapopulation) toward the new adaptive peak (phase III). At the time of the publication of the predecessor to this volume (Hanski and Gilpin, 1997), the SBT was imperfectly understood and largely untested (Barton and Whitlock, 1997). However, a large number of theoretical studies have provided new insight into the feasibility of the genetic mechanisms underlying the SBT. These studies have also uncovered many alternative forms of evolution in "adaptive landscapes" that are theoretically and empirically better supported than the SBT (Whitlock and Phillips, 2000). Much of this work was influenced by or even based on the metapopulation paradigm. This body of literature and its connection to some recent theories are discussed in Chapters 9, 11, and 12. A particularly brilliant example of how evolution in metapopulations differs from evolution in large panmictic populations is provided by the recent studies of indirect genetic effects (IGEs, Chapter 11). IGEs are genetically based environmental influences that are generated whenever the phenotype of one individual acts as an environment for another (Moore et al., 1997). IGEs create causal pathways between the genes on individuals and the phenotypes of other related or unrelated individuals permitting the coevolution of phenotype and context that is unique to metapopulations (Chapter 11). Another important advance in the evolutionary studies of metapopulations is the recently developed theory of "holey adaptive landscapes" (Chapter 12). This theory provides a genetically explicit approach for the study of the dynamics of speciation and diversification in spatially explicit systems.

Evolution of the migration rate is a well-studied topic in evolutionary ecology, but use of the metapopulation paradigm has shed new light onto the selective pressures created by population turnover (Olivieri and Gouyon, 1997). For example, several studies reviewed in Chapter 10 have shown that under some circumstances, migration is a nonmonotonic function of the extinction rate, with high extinction rates leading to reduced migration propensity, contrary to the prevailing view. The challenge now is to find out what actually happens in real metapopulations. Heino and Hanski's (2001) modeling study of evolution of the migration rate in checkerspot butterflies demonstrated the possibility of a reduced migration rate with an increasing extinction rate, but they concluded that this would not occur under conditions met in natural metapopulations of the butterflies. Other studies have made a start in developing a more general framework of life history evolution in metapopulations, including traits other than migration rate and interactions among different traits. This research is reviewed in Chapter 10.

Another evolutionary problem that has benefited from application of the metapopulation approach is the evolution of species' ranges. The basic question being asked here is: Why do populations at the range margin not adapt to their local conditions and then spread outward (Kirkpatrick and Barton, 1997)? One answer to this question is that peripheral populations receive migrants from the center of the species' range. These immigrants will be well adapted to the conditions at the range center but not to conditions at the periphery and, therefore, the genes that they bring hinder adaptation at the periphery (Mayr, 1963). Thus, peripheral populations are forced into the role of demographic sinks, preventing the range from expanding outward (Kirkpatrick and Barton, 1997). An appropriate conceptual framework used to study the interplay between migration and selection in peripheral populations is the source–sink metapopulation framework (e.g., Holt and Gaines, 1992). The usual approach in this context has been to consider the conditions that would allow the increase of a rare allele with antagonistic effects on fitness in two habitats (Holt and Gomulkiewicz, 1997; Gomulkiewicz et al., 1999; Kawecki, 2000; Kawecki and Holt, 2002). Use of the source–sink metapopulation approach has led to an important general conclusion about sink populations: the parameter that governs the rate of spread of the beneficial mutation is the absolute fitness of the mutant, not its relative fitness, as is the case in populations of constant size (Holt and Gomulkiewicz, 1997). Use of the source–sink metapopulation concept has also shed new light on the evolutionary consequences of asymmetric migration in heterogeneous landscapes (Ronce and Kirkpatrick, 2001; Kawecki and Holt, 2002). These studies are described in detail in Chapter 16.

Integration across Disciplines and Applications

A clear indication of the maturity that the field of metapopulation biology has reached is the appearance of increasing numbers of studies that attempt to integrate many or even all of the main subdisciplines covered by the broader field of population biology. The integration of ecology and genetics has been in the minds of population biologists for a long time. As early as 1931, Sewall Wright attempted the integration of ecological and population genetic processes through his shifting balance theory, as described earlier, with the aim of demonstrating that evolution could proceed rapidly in spatially structured populations. In the years that followed, most of the work that included both ecological and genetic considerations was empirical and did not explicitly attempt such integration. However, the importance of such integration was widely accepted as attested by the conceptual paper published in 1960 by

L. C. Birch. Although using slightly different terms, Birch (1960) referred to many of the problems that are the current focus of metapopulation biology, such as feedback between population dynamics and genetic variability, the importance of sink populations, and so forth, and provided numerous references to the empirical work available at that time. As an aside, it is worth noting that L. C. Birch also made a lasting contribution to early development of the ecological metapopulation ideas by his textbook with H. G. Andrewartha (Andrewartha and Birch, 1954). The first step toward a more formal integration of the two disciplines can be traced back to MacArthur (1962), who analyzed a selection model in which population regulation plays a central role. Subsequent studies continued to explore the way in which population dynamics affects natural selection (e.g., Anderson, 1971; Asmussen, 1979, 1983a,b), but left most other questions unexplored.

New impetus for the integration of the two disciplines came with the realization that human impact is the primary cause of species extinctions in many landscapes and that extinctions are taking place at an alarming rate. Just over a decade ago, little was known about the interaction of demographic, ecologic, and genetic factors in extinction, and Lande (1988) urged population biologists to address this fundamental but difficult problem. Much progress has occurred since Lande's key contribution, and despite its short history, metapopulation biology has facilitated substantial progress in this area. Chapters 13 to 16 cover many of the key contributions of metapopulation biology toward the integration of population biology. For this integration to be truly successful, we need to extend it also to the domain of empirical research. Current developments in the field of statistical genetics provide new tools that will help accomplish this goal. Of particular importance is the development of powerful multilocus genotype methods to make inferences about the origin (natal populations) of migrating individuals (e.g., Smouse et al., 1990a; Rannala and Mountain, 1997; Pritchard et al., 2000; Dawson and Belkhir, 2001). These methods, when implemented under the hierarchical Bayesian framework, can be used to combine genetic, demographic, and environmental data in a single statistical model (e.g., Gaggiotti et al., 2002). This approach, in turn, provides a way of testing hypotheses about the demographic and environmental factors that control metapopulation processes. These very recent developments are covered in Chapter 15. There are already good examples of studies that have employed the metapopulation approach to integrate ecology, genetics, and/or evolution, including studies on host–pathogen interaction (Chapter 19), butterflies (Chapter 20), and small mammals (Chapter 21).

As mentioned earlier, the renewed interest in the metapopulation concept was fostered by its potential application to the field of conservation biology, and it is now clear that the initial expectations were well founded. The design of reserve networks (Chapter 22) is a good example of a problem that needs to be addressed using the metapopulation approach. Another important example is the extension of population viability analysis (PVA) to fragmented populations. In the past, most PVA methodologies either took no account of spatial structure or did so in ways that have unrealistic data requirements. Chapter 23 presents a practical approach that considers spatial population structure and can be parameterized using available data. This chapter describes how such a model can be used in the management of endangered species using the contentious Columbia basin salmon stocks as an example. Practical applications of the metapopulation concept have gone beyond the domain of conservation biology and now include epidemiological studies of infectious diseases in humans and domestic animals. Chapter 17 explores how metapopulation theory at a variety of scales can help understand epidemiological dynamics and how this newly gained insight can be used in the design of efficient vaccination programs.

1.4 CHALLENGES FOR THE NEAR FUTURE

It is generally difficult (and often unnecessary) to try to predict the course that research in a particular field will take even over a short period of a few years. A truly novel discovery may radically change the way we think about a particular issue; new modeling tools are introduced, allowing researchers to tackle questions that previously could be studied only via cumbersome simulations; and new methods of field study may open up possibilities that we could only dream about in the past. One good example is the study of migration and gene flow, which has benefited greatly from new statistical models of both demographic and genetic data and the combination of the two, as well as of the high-resolution genetic markers that have become recently available. Chapters 15 and 21 illustrate the power of these new tools.

We anticipate that the integration of ecological, genetic, and evolutionary studies will continue in the near future. Metapopulation biology is well placed to make ground-breaking contributions here. Theoretical challenges start from the need to combine currently distinct ecological modeling approaches, such as stochastic patch occupancy models, spatial moment equations, and metapopulation models structured by local population size. Adding realistic description of landscape structure into genetic and evolutionary models is another challenge. The new statistical methods that integrate genetic, demographic, and environmental data (Chapter 15) offer a route to merging ecology and genetics but also the possibility of linking theory ever more closely with empirical research. Few of these methods are currently widely available, but we expect that many will be developed further in the near feature. Somewhat more specific research tasks include the need to better understand the interactive effects of populations' age/stage–structure and their spatial structure on the maintenance of genetic variability, genetic clines, inbreeding depression, and so forth (Mills and Smouse, 1994; Gaggiotti et al., 1997; Gaggiotti and Vetter, 1999). To what extent can the metapopulation approach be developed to address such large-scale issues as determination of species' range boundaries and their responses to global changes (Chapter 20), and indeed the global extinction risk of species? We have already commented on the relative lack of studies on metacommunities.

Combining ecology and genetics in the metapopulation context is needed for conservation and epidemiology. Chapter 22 takes an important step forward in adding spatial dynamics to existing reserve site selection procedures. We imagine that including genetics in the same package would be worth the effort. Research on plant–pathogen metapopulation dynamics

(Chapter 19) shows the way forward for epidemiology (Chapter 17) in including genetic and evolutionary issues into the demographic framework. Finally, ever since Lande's (1988) key contribution, ecology and genetics have been integral parts of conservation biology. Opinions have shifted over the years on their relative importance (Chapter 15). The coming years may demonstrate that asking about the "relative importance" has been a somewhat misleading (although necessary) question, as often the real question is about interactions. That being said, we should not lose perspective on the kinds of threats that operate at present, of which habitat loss and fragmentation are the most important ones. The immediate adverse effects of habitat loss and fragmentation are largely ecological, and it remains a major challenge for metapopulation biologists to develop predictive models and robust understanding of this key issue to be able to provide solid scientific advice to the society.

2. METAPOPULATION DYNAMICS: PERSPECTIVES FROM LANDSCAPE ECOLOGY

Kimberly A. With

2.1 INTRODUCTION

It is no coincidence that the current biodiversity crisis occurs at a time when landscapes are being transformed faster than ever before in human history (With, 2004). Many conservation issues are ultimately human land-use issues (Wiens, 2002), which is why the discipline of landscape ecology has become increasingly relevant for the management and conservation of biological diversity (e.g., Gutzwiller, 2002). Processes that operate at broader spatial scales likely influence the occurrence and persistence of an organism at a local scale, and thus a landscape perspective is ultimately required for assessing species' extinction risk.

Such acknowledgment of the importance of landscape ecology for conservation reinforces the common misconception that landscape ecology is concerned solely with broad spatial scales, however. In the present context, this would entail understanding metapopulation dynamics at a "landscape scale" (e.g., Rushton et al., 1997). Apart from the usual broad-scale anthropocentric definition of landscape, a landscape is defined more appropriately as a "spatially heterogeneous area" (Turner and Gardner, 1991) that is scaled relative

0-12-323448-4

to the process or organism of interest (Wiens, 1989). By this rendering, metapopulation dynamics can then be studied in fragmented landscapes that range in scale from that encompassing bacteria and protozoan communities (Burkey, 1997) to spotted owls (*Strix occidentalis*; Gutiérrez and Harrison, 1996). The landscape thus provides a spatial context for understanding processes contributing to metapopulation dynamics and persistence in fragmented landscapes.

Although metapopulation theory is the current paradigm for the conservation of spatially structured populations in fragmented landscapes (Hanski and Simberloff, 1997), landscape ecology provides an additional perspective and suite of approaches that can complement metapopulation theory, particularly in applications that are not handled well by existing theory, such as those involving continuous habitat distributions or recently fragmented landscapes. Metapopulation theory is not applicable to species in landscapes in which the habitat is not distinctly patchy or already fragmented extensively (Moilanen and Hanski, 2001). Nor is the application of metapopulation theory necessarily appropriate for species in recently fragmented systems, given the assumption of equilibrium colonization–extinction dynamics that underlies much of the theory [but see Ovaskainen and Hanski (2002) and Chapter 4 for advances in metapopulation theory involving transient dynamics]. In particular, landscapes fragmented by human land-use activities may represent transient nonequilibrium dynamics in which a formerly continuous population has become subdivided into smaller, more isolated populations. Dispersal among populations is disrupted such that a functional metapopulation is not created; local extinctions are not balanced by recolonization, and consequently, all populations slowly decline to extinction (Hanski and Simberloff, 1997). Thus, a declining population may superficially resemble a metapopulation in structure, but not function like one. Spatial subdivision is a necessary, but not sufficient, condition for metapopulation dynamics.

Metapopulation theory has nevertheless drawn attention to the importance of landscape structure and dispersal for maintaining population persistence (Wiens, 1996). Indeed, the effect of patch structure on dispersal and colonization success is a unifying theme in both metapopulation theory and landscape ecology (Wiens, 1997). Colonization success is not simply a function of the distance between patches, but also depends on the nature of the intervening habitat or land-use matrix through which organisms disperse, which determines the "effective isolation" of patches (Ricketts, 2001). Incorporation of the more complex mosaic structure of real landscapes into metapopulation models has been viewed as the main promise of landscape ecology for metapopulation theory (e.g., Hanski and Simberloff, 1997; Wiens, 1997).

In a couple of earlier reviews, Wiens (1996, 1997) identified several landscape ecological concepts that are relevant to metapopulation ecology and which emphasize the dual importance of dispersal and heterogeneous landscape mosaics for understanding metapopulation dynamics: (1) landscape connectivity, which emerges as the interaction of individual movement with landscape pattern, is important for metapopulation persistence; (2) the landscape matrix matters for metapopulation dynamics because it affects dispersal and thus colonization success; (3) landscapes are heterogeneous mosaics of habitats and land uses, such that habitat quality varies across the landscape, setting the stage for source–sink population dynamics (Chapter 16); and (4) landscape dynamics may affect, or even generate, metapopulation dynamics. The latter represents landscape heterogeneity in time as well as space.

In addition to these potential contributions of landscape ecology to metapopulation ecology, a more fundamental principle emerges from the definition of landscape ecology itself. Landscape ecology is the study of the effect of spatial pattern on ecological process (Turner, 1989). From this it follows that adopting a landscape ecological perspective to metapopulation dynamics entails understanding how spatial pattern, such as habitat fragmentation or heterogeneity, affects processes that contribute to the dynamics of spatially structured populations. This might involve, for example, understanding the relative effects of habitat fragmentation on dispersal (colonization) and demography on metapopulation persistence. This expanded perspective of landscape ecology is adopted in this chapter.

The objectives in this chapter are thus to (1) demonstrate what a landscape ecological perspective can contribute toward understanding metapopulation dynamics, beyond the usual suggestions that landscape ecology offers a broader scale perspective or more spatially complex rendering of landscape structure; (2) discuss how landscape structure is expected, or has been demonstrated, to affect various processes (dispersal, demography) that affect metapopulation persistence and thus extinction risk; (3) assess the implications of adopting a landscape ecological perspective for management and conservation; and (4) identify theoretical and empirical research needs that would help contribute to the further development of this "exciting scientific synthesis" between metapopulation biology and landscape ecology (see Hanski and Gilpin, 1991).

2.2 CONTRIBUTIONS OF LANDSCAPE ECOLOGY TO UNDERSTANDING METAPOPULATION DYNAMICS

This section addresses how landscape structure affects, or is expected to affect, the dynamics of metapopulations. This includes a discussion on issues pertaining to landscape connectivity, landscape connectivity and dispersal thresholds, the relative importance of dispersal for metapopulation persistence, landscape effects on demography and extinction risk, the source–sink potential of landscapes, extinction risk in dynamic landscapes, and the relative effects of habitat loss and fragmentation on metapopulation persistence.

Landscape Connectivity Issues: Patch-Based vs Landscape-Based Measures

Habitat connectivity is a central theme in both landscape ecology and metapopulation ecology (Hanski, 1999a; Tischendorf and Fahrig, 2000a). Connectivity refers to the ability of organisms to access habitat, which affects colonization rates and thus metapopulation persistence on the landscape (e.g., Gonzalez et al., 1998; Kindvall, 1999). The emphasis in metapopulation ecology, however, has been on deriving *patch-based measures* related to the proximity and area of neighboring patches, which quantify the accessibility of habitat patches to an individual on the landscape (Hanski, 1999a; Moilanen and Hanski, 2001). An overall measure of patch connectivity for the landscape can be obtained as a weighted average of patch isolation, which then gives the amount of habitat accessible to a random individual on the landscape (Hanski, 1999a). Overall patch connectivity may give an indication of landscape connectivity, but the latter is not formally derived mathematically from such patch-based measures (Tischendorf and Fahrig, 2001). Patch-based connectivity measures are best applied to extensively fragmented or distinctly patchy landscapes and are less applicable to more continuous habitat distributions (Moilanen and Hanski, 2001).

In contrast, landscape ecologists have focused on deriving measures of overall *landscape connectivity*. Habitat connectivity is thus being assessed at different scales — patch based vs landscape based — in these two disciplines. Landscape connectivity is defined as the degree to which various habitat types facilitate movement across the landscape (Taylor et al., 1993; With et al., 1997; Tischendorf and Fahrig, 2000a) and can thus be assessed for continuous habitat distributions and heterogeneous landscapes (Schippers et al., 1996; With et al., 1997). Landscape connectivity can be quantified in a number of ways, such as by the use of percolation theory and its neutral landscape derivatives (Gardner et al., 1987; With, 1997, 2002; With and King, 1997), graph theory (Urban and Keitt, 2001), and various other approaches (e.g., Schumaker, 1996; Tischendorf and Fahrig, 2000a,b). Although a full rendering of how landscape connectivity can be quantified lies beyond the scope of this chapter, the common theme underlying all of these approaches is how the movement behavior of organisms interacts with the patch structure of landscapes. Landscape connectivity thus emerges as a species-specific response to landscape structure based on factors such as the species' habitat affinities, gap-crossing abilities, movement rates, response to patch boundaries, and differential mortality through elements of the landscape (Wiens et al., 1993; Dale et al., 1994; With, 1997; With et al., 1997; Tischendorf and Fahrig, 2001; Vos et al., 2001).

Landscape connectivity is important for understanding the emergence of spatial structure in populations, which in turn is expected to have implications for the persistence and dynamics of metapopulations. As an example of how species-specific responses to heterogeneity affect landscape connectivity and population distributions, With and Crist (1995) used habitat-specific rates of movement in an individual-based simulation model inspired by percolation theory to predict the distributional patterns of two acridid grasshopper species in a heterogeneous landscape within the shortgrass steppe of the North American Great Plains. The largest species (*Xanthippus corallipes*) moved rapidly through the grass matrix (65% of the landscape), suggesting that the overall landscape was highly connected from the standpoint of this species. Its reduced rate of movement in the remaining third of the landscape resulted in the observed patchy population distribution, consistent with model expectations that good dispersers should exhibit patchy distributions when the landscape contained ≤35% preferred habitat because their high mobility allows individuals to locate and aggregate within the preferred habitat (assuming that individuals reduce their rates of movement and exhibit greater residence times in preferred habitats). In contrast, the lower mobility of the smaller species (*Psoloessa delicatula*) prevented large numbers of individuals from locating and aggregating within its preferred habitat, which constituted a minor (8%) component of the landscape. Because of its relatively greater rates of movement through other grassland habitats, this species was expected to be distributed randomly across the landscape, which was consistent with its observed distribution in the field.

Landscape Connectivity Issues: Data Requirements

Patch-based connectivity measures that form the basis of metapopulation theory have the distinct advantage of ease of model parameterization (Hanski, 1999b; Hanski et al., 2000; Chapter 5). Consistent with metapopulation theory's abstraction of landscape structure as discrete patches embedded in an ecologically neutral matrix, patchwise measures of connectivity have often been based on simplistic measures such as nearest-neighbor distances (Moilanen and Nieminen, 2002). More sophisticated measures of patch connectivity have been developed, however, which incorporate patch-area effects on emigration and immigration rates and species-specific dispersal distances (Moilanen and Hanski, 2001; Vos et al., 2001; Chapter 4). As the aforementioned grasshopper example illustrates, however, the connectivity of habitat patches is not just a simple function of the distance between patches. Because the intervening matrix may determine the effective isolation of patches (Ricketts, 2001), and thus overall landscape connectivity (With et al., 1997), explicit consideration of how the complex mosaic structure of heterogeneous landscapes affects colonization success and population extinction risk has typically been viewed as one of the most important contributions that landscape ecology has to offer metapopulation ecology (Hanski and Simberloff, 1997; Wiens, 1997).

The connectivity of heterogeneous landscapes is not easily captured by a simple index or landscape metric unfortunately, but is commonly tackled with an individual-based simulation modeling approach. Empirical data on habitat-specific movement parameters or residence times within different elements of the landscape are used to parameterize a rule- or vector-based movement model to simulate dispersal across a heterogeneous landscape map. Landscape connectivity is then inferred by extrapolating habitat-specific rates of movement, and perhaps other information (if available) about behavior at habitat edges (Lidicker and Koenig, 1996) or mortality risk while dispersing through the different elements of the landscape, to determine whether individuals are able to colonize a suitable habitat successfully. Some presumed correlate of a connected landscape, such as dispersal success, degree of population aggregation (With and Crist, 1995; With et al., 1997), or population connectivity (Schippers et al., 1996), is then used as an indirect measure of landscape connectivity.

Admittedly, quantifying the resistance of different habitat types to movement is a challenge in practice. Direct observation of individual movement responses to landscape structure is time intensive and is necessarily limited in temporal (and therefore spatial) extent, although this is bound to change with the increasing availability of satellite-tracking devices that permit the near-continuous monitoring of individuals. As an alternative, investigators typically rely upon mark–recapture techniques to derive estimates of interpatch movements in different matrix types (e.g., Pither and Taylor, 1998; Ricketts, 2001) or make inferences about matrix resistance based on observed patterns of patch occupancy in different landscape contexts (e.g., Moilanen and Hanski, 1998). For example, mark–recapture data were used to quantify the resistance of different matrix types to butterfly movement within a naturally heterogeneous landscape located in an alpine valley of the Colorado Rocky Mountains (Ricketts, 2001). Movement through coniferous forests was 3–12 times less likely than movement through willow thickets for the majority of these meadow butterfly species. Meadows separated by coniferous forest are thus effectively more isolated than meadows separated by a similar distance, but embedded in a willow matrix. Because of the difficulty and cost of attempting to reconnect habitat fragments with corridors or stepping stones to enhance dispersal and thus colonization success, it might be more practical to reduce the effective isolation of patches by altering management practices in the surrounding matrix (Ricketts, 2001). Corridors need not be linear features of the landscape, but can occur or be created through the juxtaposition of certain matrix types, such as different habitats or land uses, which serve to funnel individuals among habitat patches (Gustafson and Gardner, 1996; Vos et al., 2002).

Landscape Connectivity Issues: Asymmetrical Connectivity Among Populations

Connectivity in heterogeneous landscapes may thus be difficult to identify from a simple analysis of landscape structure. Nor is the transfer of individuals among patches necessarily symmetrical in a heterogeneous landscape matrix. Gustafson and Gardner (1996) developed an individual-based simulation model to explore how altering landscape heterogeneity affected dispersal success for a generic organism among fragments of deciduous forest in several agricultural landscapes located in the midwestern United States. Emigration and immigration rates among forest fragments often were not symmetrical, leading Gustafson and Gardner (1996) to speculate that such asymmetrical transfers may be the rule rather than the exception in heterogeneous landscapes. If true, this presents a problem for patchwise connectivity measures in which colonization probabilities are based on interpatch distances that ignore matrix effects. To overcome this problem, distances between patches would need to be specified by direction and weighted by the resistance of the intervening matrix habitat to movement (i.e., $d_{ij} \neq d_{ji}$; see Urban and Keitt, 2001).

The potential for asymmetrical connectivity among populations of European badgers (*Meles meles*) was indicated by a GIS-based random walk model applied to a landscape in the central part of The Netherlands (Schippers et al., 1996). Urban areas, canals, and motorways either created barriers to movement or increased mortality such that asymmetries in the connectivity of populations emerged. Asymmetrical connectivity has also been found among populations of the critically endangered Iberian lynx (*Lynx pardinus*) in a human-dominated and therefore extensively fragmented region in southwestern Spain (Ferreras, 2001). Although two source populations had similar emigration rates, they differed in their connectivity to outlying populations because of differences in the matrix type surrounding each source. One of the sources was embedded in an agricultural matrix, which lynx avoided, and were instead funneled along a narrow corridor of a more suitable habitat to populations in the south and west. The other source population was located in the southwestern region, which consisted mainly of Mediterranean scrubland and tree plantations, and most dispersing individuals from this population settled in this region instead of dispersing toward the northeast. Thus, there is an asymmetrical transfer of individuals that occurs among populations: individuals tend to disperse from the northeast to the southwest, but not in the opposite direction. Such asymmetries could lead to a reduction in the effective connectivity, and thus metapopulation capacity (see Hanski and Ovaskainen, 2000), of the landscape for the species.

Landscape Connectivity Issues: Thresholds in Connectivity

Because of the importance of landscape connectivity for evaluating the structure and dynamics of metapopulations, it would be advantageous to identify when landscapes become disconnected and thus when metapopulation processes such as colonization rates are likely to be disrupted. Both patch connectivity measures (Hanski, 1999a) and measures of landscape connectivity (With, 2002) predict critical thresholds in habitat connectivity, where the habitat network becomes abruptly disconnected at a critical level of remaining habitat. In percolation-based approaches, for example, landscape connectivity is assessed by whether a single habitat cluster (the *percolating cluster*) spans the landscape. Landscape connectivity is disrupted when the critical habitat nodes forming the "backbone" of the percolating cluster are destroyed, which abruptly breaks the percolating cluster into two or more fragments. The critical level of habitat at which landscape connectivity becomes disrupted (*percolation threshold*) depends on a number of assumptions regarding species-specific movement attributes (gap-crossing abilities, movement rates through different habitat types, matrix mortality) and the representation and configuration of the landscape itself (grid geometry, degree of habitat fragmentation) (for a review, see With, 2002). The issue, however, is whether thresholds in landscape connectivity, or measures of landscape connectivity more generally, relate to processes such as colonization success and local extinction rates, which are important for predicting metapopulation persistence on landscapes. In other words, is landscape connectivity both a necessary and a sufficient condition for metapopulation persistence (e.g., With, 1999)?

Landscape Connectivity Thresholds and Dispersal Success

Dispersal is the "glue" that keeps metapopulations together (Hansson, 1991), and thus colonization success is deemed crucial to metapopulation persistence. Clearly there should be some relationship between landscape connectivity and dispersal (colonization) success: dispersal success is expected to be higher in landscapes with a high degree of connectivity. What is less clear, however, is whether thresholds in landscape connectivity should necessarily coincide with thresholds in dispersal or colonization success.

To address this, With and King (1999a) quantified dispersal success on a series of landscapes with complex (fractal) habitat distributions that represented a gradient of fragmentation severity (Fig. 2.1a). Dispersal success was defined as the proportion of independent dispersers that successfully located a suitable habitat patch (cell). Consider that if dispersal is truly random, such that dispersal occurs to a random point on the landscape, then the underlying spatial pattern of the landscape is unimportant for predicting dispersal success

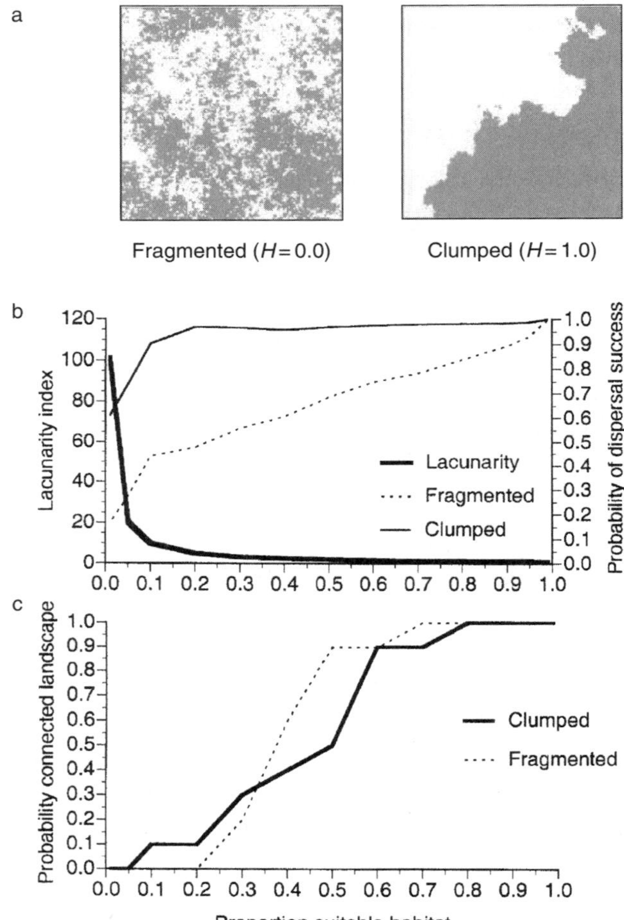

Fig. 2.1 (a) Examples of fractal landscape patterns illustrating extremes in fragmentation severity (*H*, spatial autocorrelation of habitat). (b) Dispersal success declines precipitously below 20% habitat (dispersal threshold), coinciding with the lacunarity threshold of landscape structure, which is a landscape-wide measure of interpatch distances. The lacunarity index is not affected by the landscape pattern for analyses conducted at the finest scale (1 × 1 grid cell), shown here for clarity of presentation. The lacunarity curve at other scales is qualitatively similar, but lacunarity indices tend to be higher in clumped fractal landscapes than in fragmented ones due to the greater variability in gap sizes. (c) Percolation thresholds, a patch-based assessment of landscape connectivity, do not coincide with dispersal thresholds (assuming a 12-cell dispersal neighborhood). Modified from With and King (1999a).

and only the fraction of habitat (h) and number of dispersal steps (m; equivalently the "dispersal neighborhood" or "dispersal ability" of the species) affect dispersal success as

$$\Pr(success) = 1 - (1 - h)^m. \tag{2.1}$$

Equation (2.1) represents the mean-field approximation. If dispersal is instead constrained to movement through adjacent cells (but still random in direction) to force individuals to interact with the patch structure of the landscape, then dispersal success on fractal landscapes can no longer be derived from first principles and may lack a closed-form solution. Thus, dispersal success on fractal landscapes had to be obtained through numerical simulations (With and King, 1999a).

As expected, dispersal success declined with decreasing habitat and increasing fragmentation of the landscape, but the rate of this decline accelerated once the amount of habitat fell below 10–20% (Fig. 2.1b). In other words, dispersal success exhibited a threshold response to habitat amount. This dispersal threshold did not coincide with percolation thresholds used to quantify landscape connectivity, even after allowing for a larger dispersal neighborhood to define habitat connectivity, in which individuals could move through cells of nonhabitat in their search for a suitable habitat site (as in the simulation; Fig. 2.1c). Intuitively, dispersal success is expected to decline as patches become smaller and more isolated because the disperser ends up spending much of its time in the matrix where mortality may be greater. Lacunarity analysis quantifies the "gap structure" of landscapes and is related to the variance-to-mean ratio of the distances among patches on the landscape (Plotnick et al., 1993). The higher the lacunarity index, the greater the variability in distances among patches. Lacunarity is not merely the inverse of some measure of patch structure, such as the fractal dimension of the landscape, however; it can resolve differences in landscape pattern that may be obscured by patch-based measures (Plotnick et al., 1993). Landscape lacunarity exhibited a strong threshold effect around 20% habitat; interpatch distances became greater and more variable when habitat fell below this critical level (With and King, 1999a; Fig. 2.1b). Thresholds in dispersal success thus coincide with lacunarity thresholds rather than percolation thresholds of landscape connectivity.

Empirical tests of percolation theory have been performed in the field with insects moving across experimental "microlandscapes" in which habitat (grass sod) was arrayed as either a random or a fractal distribution (Wiens et al., 1997; McIntyre and Wiens, 1999; With et al., 1999). Although it is not clear whether parameters that describe movement pathways should exhibit threshold behavior, let alone coincide with percolation thresholds in landscape connectivity (With et al., 1999), tenebrionid beetles (*Eleodes obsoleta*) nevertheless exhibited threshold behavior in several movement parameters when grass cover fell below 20% (Wiens et al., 1997). This is in the domain of lacunarity thresholds (With and King, 1999), suggesting that landscape measures of gap structure may ultimately be better predictors of dispersal success than landscape measures of patch structure. This reiterates one of the main tenets of metapopulation theory, that patch isolation measures (and therefore patch-based connectivity measures that incorporate interpatch distances) are a strong correlate of colonization success, at least in extensively fragmented landscapes that meet the metapopulation ideal of habitat patches embedded in an ecologically neutral matrix. As discussed previously, the relationship between landscape structure and colonization success is more complicated in heterogeneous landscapes, where patch isolation may be less important than the quality of the matrix habitat through which the organism disperses. For example, large-scale forestry in Sweden resulted in extensive ditching to drain clear-cut areas, which created an inhospitable matrix that prevented pool frogs (*Rana lessonae*) from colonizing breeding ponds, irrespective of their proximity to an occupied pond (Sjögren-Gulve and Ray, 1996).

The lack of concordance between percolation thresholds and dispersal success has led investigators to invent other, seemingly more-relevant measures of landscape structure for predicting dispersal or colonization success (e.g., Schumaker, 1996; Tischendorf and Fahrig, 2000a). The problem is not the measure used to quantify landscape connectivity, however, but with the scale at which habitat connectivity is assessed relative to the scale of dispersal. Landscape connectivity relates to the potential of organisms to traverse the entire landscape, whereas dispersal or colonization success pertains only to the likelihood that a dispersing organism will successfully find a suitable habitat patch (or cell in a grid-based landscape). Although the two are related, assessments of landscape connectivity and dispersal success are ultimately performed at different scales. Individual movement is constrained in the latter assessment (success is scored for individuals that locate suitable habitat within a dispersal neighborhood, at which point colonization occurs and individuals are assumed to stop moving), but not in the former where the emphasis is on the ability of individuals to move across the entire landscape (whether the organism actually does or not). Thus, the grain of movement may be the same — how individuals move within or between habitat types or cells — but the spatial extent of movement is different.

Habitat connectivity is obviously important for colonization success at some scale. The challenge is to identify what scale is appropriate for predicting colonization success in a given species, however. This involves adopting a species' perspective of habitat connectivity (Wiens and Milne, 1989; With, 1994; Pearson et al., 1996; Vos et al., 2001). Although this has been done using percolation-based neutral landscape models (see With, 2002), a related approach involves the use of graph theory. In graph theory, the grid structure of the landscape is represented as a graph in which habitat patches (vertices or nodes) are connected across varying distances (lines or edges) (Urban and Keitt, 2001). The graph representation permits a process-based measure of connectivity for individual patches as well as the entire landscape. Overall connectivity of the graph (i.e., landscape) is simply assessed in terms of whether each node is connected to some other node. Although there might be several ways to connect the various nodes of the graph to form a *spanning tree*, the one with the shortest length is termed the *minimum spanning tree*. There is a critical threshold distance at which the graph becomes disconnected, reminiscent of the percolation threshold of landscape connectivity for grid-based landscapes (Urban and Keitt, 2001). Using a graph-theoretic approach, van Langevelde (2000) identified different scales of connectivity and related this to colonization patterns of the European nuthatch (*Sitta europaea*) occupying woodlots within fragmented landscapes of The Netherlands (Fig. 2.2a). Patch occupancy patterns of nuthatches were correlated with a critical threshold distance of 2.4–3 km (Fig. 2.2b), such that woodlots located >3 km from a neighboring forest patch were unlikely to be colonized by dispersing nuthatches (van Langevelde, 2000). The extinction of local nuthatch populations is related to both the connectivity and the size of forested patches (Verboom et al., 1991) and underscores again the importance of habitat connectivity — at some scale — for population persistence (e.g., Fahrig and Merriam, 1985). The metapopulation dynamics of nuthatches within this fragmented landscape have also been assessed using an incidence function model (Ter Braak et al., 1998).

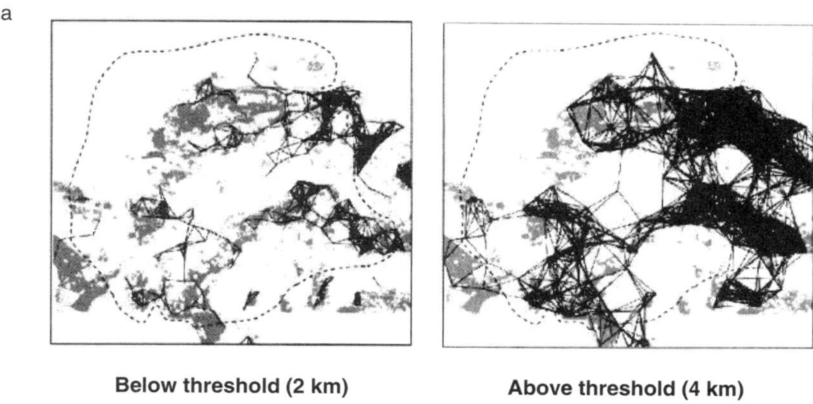

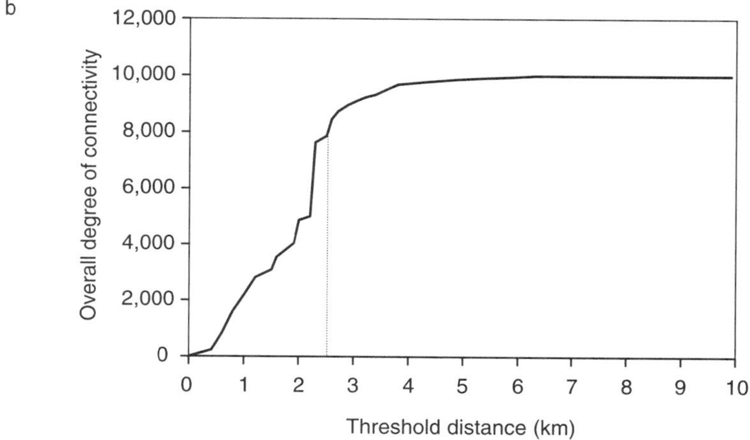

Fig. 2.2 (a) A graph theoretic analysis of habitat connectivity for European nuthatches (*Sitta europaea*) in an agricultural landscape based on an analysis assuming a dispersal distance of 2 km (just below the landscape connectivity threshold; b) and a dispersal distance of 4 km (above the landscape connectivity threshold; b). Gray areas are habitat fragments, and black lines indicate connections among patches based on the indicated dispersal distance. (b) Landscape connectivity exhibits a threshold in this landscape at about 2.5 km. Patches farther than 3 km apart were unlikely to be colonized by nuthatches. Modified from van Langevelde (2000).

Relative Importance of Dispersal for Metapopulation Persistence

Studies on how landscape structure and matrix heterogeneity affect dispersal (colonization) assume that the fine-scale movements of individuals translate into broader patterns of population distribution (e.g., Turchin, 1991; With and Crist, 1996; With et al., 1997), which in turn may have consequences for metapopulation persistence on the landscape. How important is dispersal for predicting metapopulation persistence? Dispersal is a key component of most spatially explicit population models, which serve as the main arsenal in the landscape ecological approach to predicting metapopulation persistence in fragmented landscapes, particularly in evaluating the consequences of different scenarios of land-use change on extinction risk for species of conservation concern (Dunning et al., 1995). As mentioned previously, it is difficult to obtain species- and habitat-specific information on dispersal, which may result in errors in the estimation of dispersal success. Such estimation errors may propagate in spatially explicit models and affect estimates of species' extinction risk (Ruckleshaus et al., 1997), although the magnitude of these errors may initially have been overestimated (Mooij and DeAngeles, 1999).

Landscape structure is not always important for predicting dispersal success, however. Using an individual-based model of dispersal on neutral landscape models, King and With (2002) found that the mean-field approximation [Eq. (2.1)] was sufficient for predicting dispersal success when >40% of the landscape was suitable; below this level, specifics related to dispersal behavior and landscape pattern became more important. Given that many species of conservation concern occur in landscapes with substantially <40% suitable habitat, however, it is likely that landscape structure — the configuration and heterogeneity of land-cover types — will generally be important for predicting dispersal success.

Although dispersal (colonization) success is considered an important process necessary for metapopulation persistence, demographic factors that affect extinction risk may actually be more important for some species (South, 1999; With and King, 1999b). This is especially true for good dispersers, such as birds, where landscape structure has a greater effect on reproductive output through edge effects than on immigration rates (e.g., Donovan et al., 1995b; Dooley and Bowers, 1998). The effect of landscape structure on demography and extinction risk on metapopulation persistence is explored in the next section.

Landscape Effects on Demography and Extinction Risk

Habitat loss and fragmentation pose the greatest threats to biodiversity (Wilcove et al., 1998) and are the inevitable consequence of the transformation of landscapes by humans (With, 2004). Beyond the sheer magnitude and rate of this transformation lies a more insidious problem: the effects of habitat loss and fragmentation on population viability are not linear. Habitat loss may precipitate a sudden and rapid decline in the probability of metapopulation persistence (i.e., a threshold). Using a demographic model founded on Levins' (1969) classic metapopulation model, Lande (1987) first defined *extinction thresholds* for territorial vertebrates as a function of their demographic potentials (k, a composite parameter derived algebraically from life-history parameters such as net lifetime reproductive output, R'_o, and dispersal ability, m). The extinction threshold is the critical level of habitat (h_c) at which the population no longer occurs on the landscape (patch occupancy, $p^* = 0$) and is defined mathematically as $h_c = 1 - k$. The decline in patch occupancy accelerates past a certain reduction in habitat, such that the approach to the extinction threshold (h_c) is usually nonlinear.

Lande's (1987) model was in the tradition of the classical metapopulation model, which only assumes the existence of habitat patch structure (i.e., it is spatially implicit). Habitat is assumed to be distributed randomly across the landscape or, alternatively, is randomly accessible by dispersing individuals. Bascompte and Solé (1996) developed a spatially explicit realization of this model using grid-based landscapes with random habitat distributions and found that extinction thresholds generally occurred at about the same level (h_c) as in Lande's (1987) spatially implicit model, but the decline in patch occupancy occurred faster, resulting in steeper thresholds.

The effect of habitat fragmentation on extinction thresholds has been explored using fractal landscapes, which generate complex landscape patterns across a gradient of fragmentation severity (e.g., Fig. 2.1a; With and King, 1999b; Hill and Caswell, 1999). Because of the complexities of how species interact with fractal landscape patterns, it was necessary to parse the demographic potential (k) into its constituent parameters (R'_o and m) to evaluate the relative effects of these life-history parameters on extinction. Different combinations of R'_o and m may give rise to the same demographic potential (k), but have very different consequences for metapopulation persistence on the landscape in terms of their extinction thresholds. On fractal landscapes, reproductive output (R'_o) had a much greater effect on population persistence (h_c) than dispersal ability (m), which is the opposite of what was found in Lande's (1987) model, which assumes (in essence) a random landscape (i.e., compare the rate at which h_c declines as a function of increasing R'_o as opposed to increasing m in fractal landscapes, relative to the rate at which those same parameters decline in the random landscape, Fig. 2.3). Enhancing reproductive output, such as through the conservation of high-quality habitats or supplementation of nesting habitat, may thus have a greater effect than enhancing dispersal success, by the maintenance or restoration of habitat connectivity, on mitigating extinction risk.

This is not to say that landscape structure had no effect on population persistence, however. Populations in landscapes that were not fragmented ($H = 1.0$) were generally able to persist throughout almost the entire range of habitat availability (i.e., $h_c \leq 0.1$; $H = 1.0$, Fig. 2.3). Reducing fragmentation and maintaining habitat connectivity thus mitigate extinction risk, as expected. In fact, species with low demographic potentials, due to a combination of low reproductive output (R'_o) and poor dispersal ability (m), generally went extinct sooner on fragmented landscapes ($H = 0.0$) than predicted by Lande's (1987) model. Because many species of conservation concern have these combined traits of low fecundity and poor dispersal ability, such species may be at a greater risk of extinction from habitat loss and fragmentation than previously suspected.

The problem of how habitat fragmentation affects extinction thresholds has also been tackled by evaluating the *metapopulation capacity* of the landscape. The metapopulation capacity is basically the sum of the relative contribution of

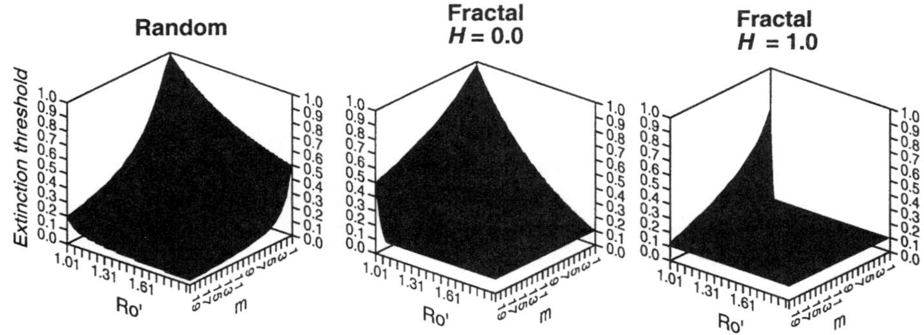

Fig. 2.3 Fragmentation effects on extinction thresholds for species with different demographic potentials (the combined effects of reproductive output, R'_0, and dispersal ability, m). Random landscapes are the most extensively fragmented; fractal landscapes, $H = 1.0$ are the least fragmented (i.e., clumped; Fig. 2.1a). Data from With and King (1999b).

individual patches (their "value") to metapopulation persistence based on their size and degree of connectivity to other habitat fragments in the landscape (Hanski and Ovaskainen, 2000; Chapter 4). This is a spatially realistic extension of Levin's (1969) model in which landscape structure (patch area and isolation) is allowed to affect the metapopulation processes of colonization and extinction. For example, the probability of extinction is calculated as a function of the inverse of patch area because extinction is more likely in small patches than in large ones (see Chapter 4 for the mathematical details of this model). For a given landscape, the metapopulation capacity increases with the dispersal range of the species because dispersal enhances connectivity and thus patch colonization rates. For a given species, landscapes can be ranked according to their capacity to support viable metapopulations. A landscape is capable of supporting a viable metapopulation if its metapopulation capacity (analogous to the fraction of habitat in Lande's model) exceeds a threshold determined by the "metapopulation potential" of the species (the ratio of the species' extinction to colonization rates, analogous to the demographic potential in Lande's model). Landscape fragmentation, created by the random destruction of habitat, resulted in a decline in the metapopulation capacity of the landscape that was roughly proportional to the amount of habitat lost. Destruction of habitat in large blocks caused the metapopulation capacity to decline slower than the loss of habitat, and therefore was less detrimental to metapopulation persistence. Landscape structure thus affects extinction thresholds and metapopulation persistence, which is consistent with the findings of other spatially realistic metapopulation models (Hill and Caswell, 1999; With and King, 1999b).

Although these theoretical investigations demonstrate that some critical level of habitat is required for metapopulation persistence, what empirical evidence is there to support the existence of extinction thresholds? Although extinction thresholds have been quantified mathematically for various species based on available demographic information and estimates of the fraction of suitable habitat in the landscape (e.g., Lande, 1988; Carlson, 2000), extinction thresholds have been identified empirically as an abrupt decline in the occupancy of habitat patches across a series of landscapes that vary in the amount of habitat. For example, the endangered Glanville fritillary butterfly (*Melitaea cinxia*) exhibited a threshold response to declines in the metapopulation capacity of patch networks distributed among the Åland Islands in southwest Finland (Hanski and Ovaskainen, 2000; Chapter 4). Many birds may not exhibit a threshold response to the amount of habitat, however, particularly if they are migratory and exist regionally due to coupled source and sink landscape dynamics (With and King, 2001). Such species may be able to occupy all remaining habitat fragments even in extensively fragmented landscapes, as was found for most neotropical migratory songbirds across a landscape gradient of increasing agricultural dominance in southern Ontario, Canada (Villard et al., 1999). Threshold responses to a reduction in forest cover were generally absent for most species, except for two species (ovenbird, *Seiurus aurocapillus*; black-and-white warbler, *Mniotilta varia*) that were not found in landscapes with <10% mature forest.

For many species, such as neotropical migrants, the efficiency of patch occupancy does not decline with habitat loss and fragmentation. Many plant species may also maintain constant patch occupancy despite a reduction in suitable habitat, due to life-history strategies such as seed dormancy (i.e., species escape in time as well as space; Chapter 18). Although the extinction threshold is defined as the fraction of all sites that are suitable but not occupied on a landscape, Eriksson and Kiviniemi (1999) performed a modified calculation in which the "quasi-equilibrium" threshold was obtained as only the fraction of suitable sites that were not occupied (h'_c). By this measure, 44% (8/18) of the grassland plants they evaluated in southeastern Sweden were existing below the extinction threshold, which means that although such species were currently found in these landscapes, the amount of suitable habitat was not sufficient to permit the long-term persistence of these species; species exhibited a lagged response to habitat loss. Thus, the identification of extinction thresholds, based on site occupancy and availability of suitable habitat, may not always be sufficient for evaluating population persistence and extinction risk. In such cases, an analysis of lagged population responses to landscape change is required (see *Extinction Risk in Dynamic Landscapes*).

The Source–Sink Potential of Landscapes

Spatial heterogeneity, resulting from differences in the size, shape, and quality of habitats comprising the landscape, affects species' demographic rates. Reproductive success may be maximized, or survivorship may be minimized, in a particular habitat. Habitat-specific survivorship and reproductive success set the stage for source–sink dynamics (Pulliam, 1988) in which population growth rates are positive (birth rates exceed death rates) in some patches (sources) but are negative in others (sinks). The relative amount of source and sink habitat on a landscape may thus affect persistence of the metapopulation at the landscape scale (Pulliam and Danielson, 1991; Donovan et al., 1995b; Chapter 16).

A landscape perspective is ultimately required to assess source–sink dynamics and to evaluate how changes in landscape structure, such as from habitat fragmentation or land-use change, may affect these dynamics and thus metapopulation persistence. Unfortunately, most of the previous efforts to model source–sink dynamics have been spatially implicit (e.g., Pulliam and Danielson, 1991), including those that have attempted to determine the effects of habitat fragmentation on the source–sink status of populations (e.g., Donovan et al., 1995a; but see Ritchie, 1997). Although demographic rates vary spatially in such models, they are usually fixed input parameters that are independent of landscape structure. In birds, for example, landscape structure is known to affect reproductive output in many species due to higher *edge effects* in fragmented landscapes in which nesting success is lower in habitat fragments because of greater nest predation or brood parasitism (such as by the brown-headed cowbird, *Molothrus ater*, in North America) along fragment edges (Donovan et al., 1995b, 1997). Thus, reproductive output (a demographic rate) is spatially dependent and varies as a function of patch size and shape, being reduced in fragments dominated by edge and maximized in large patches of contiguous habitat.

With and King (2001) devised a functional relationship between patch structure and reproductive success for neotropical migratory songbirds as part of a spatially structured demographic model developed to assess the source–sink potential of fragmented landscapes. Reproductive success declines as a function of increasing edge (i.e., negative edge effects). For example, some species were "edge sensitive" and exhibited a steep decline in reproductive output in small or irregularly shaped patches that were dominated by edge (edge index → 1.0; Fig. 2.4a). The demographic consequences of landscape structure were assessed as the expected number of female offspring produced per female, per patch, for all individuals across the entire landscape (b_L). A simple two-stage life table combining fecundity (b_L) and survivorship (juvenile, s_o and adult, s) was then constructed for each species (Fig. 2.4a) in a given landscape (e.g., Fig. 2.1a). From the life table, we calculated the finite rate of increase for the entire landscape population (λ_L) as the solution to the characteristic equation (Lande, 1988):

$$\lambda_{L\alpha} - s\lambda L^{\alpha-1} - b_L l_\alpha = 0 \qquad (2.2)$$

for $\lambda \geq 1$ and $0 < s < 1$, where l_α is survivorship at the age of first breeding, s is the annual probability of survivorship for breeding adults (>1 yr), and b_L is derived from the population across the entire landscape. The landscape population was stable when $\lambda_L = 1.0$, declining when $\lambda_L < 1$ and increasing when $\lambda_L > 1$. The annual rate of change in the size of the metapopulation (%/yr) is $(\lambda_L - 1.0) * 100$. Thus, this modeling approach treats a demographic rate (b) as a spatially dependent variable; it is a model output rather than a fixed model parameter as in traditional demographic models. Furthermore, this approach extends the concept of source–sink populations from the scale of patches to the entire landscape, such that the potential of a given landscape to function as a population source or sink is ultimately assessed.

For species with low edge-sensitivity (Fig. 2.4a), landscapes supported viable metapopulations and had the potential to function as sources across a wide range of available habitat (Fig. 2.4b; With and King, 2001). Fragmented landscapes (random) could not support viable metapopulations of this species, however, and functioned as sinks ($\lambda_L < 1.0$) when habitat fell below 30% (Fig. 2.4b). The situation was bleaker for species with high edge-sensitivity, which had a difficult time persisting in landscapes with <50% habitat even when the landscape was managed to preserve large tracts of contiguous habitat ($H = 1.0$, Fig. 2.4c).

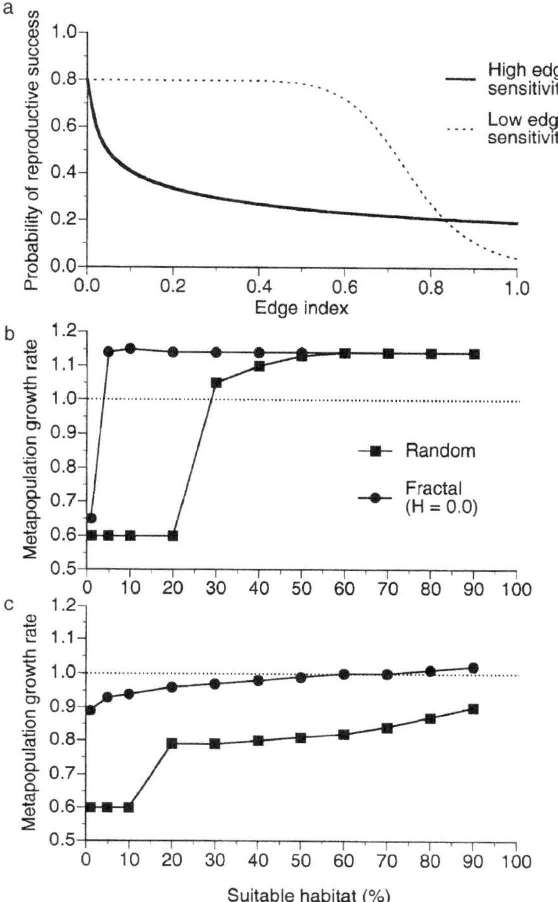

Fig. 2.4 (a) Degree of edge sensitivity — the decline in reproductive success as a function of increasing edge — for a couple of generic migratory songbirds. (b) Effect of fragmentation (random = maximum fragmentation) on a species with low edge-sensitivity. (c) Effect of fragmentation on a species with high edge-sensitivity. Modified from With and King (2001).

This spatially structured avian demographic model was parameterized for the Henslow's sparrow (*Ammodramus henslowii*) in a heavily managed landscape of north-central Kentucky in the eastern United States (King et al., 2000). Henslow's sparrow is an area-sensitive species that requires large tracts of dense tallgrass prairie for nesting (patch sizes ≥30 ha; Herkert, 1994) and is thus a species of conservation concern given that <1% of the historical tallgrass prairie remains throughout the Great Plains of North America (Knopf and Sampson, 1994). Only 0.6% of the managed landscape was suitable breeding habitat, which was fragmented and consisted of many small patches barely large enough to support a breeding pair, let alone meet the area requirements of this species (the largest patch was 51 ha). The finite rate of increase for the Henslow's sparrow metapopulation in this landscape was $\lambda_L = 0.86$. Thus, the Henslow's sparrow was declining at an annual rate of 14%/yr such that the landscape was a sink for this species. Reproductive output, b, must be increased by about 1.5 times its current level to restore the Henslow's sparrow to a stable or increasing metapopulation, which would require an increase in the landscape-wide nesting success from 39 to 58%. To reverse the current decline in this metapopulation, a land manager should thus minimize disturbances that contribute to habitat fragmentation of grassland habitat, thereby increasing patch sizes and minimizing edge effects that decrease reproductive output.

Although this spatially structured demographic model focuses on how patch geometry for a single habitat type affects reproductive success, this approach can be extended easily to heterogeneous landscapes in which reproductive success or survivorship additionally varies as a function of habitat quality.

Extinction Risk in Dynamic Landscapes

The effects of landscape structure on the metapopulation processes discussed thus far have assumed a static landscape in which the amount, suitability, and configuration of habitat patches remain unchanged on the landscape. Even studies that have explored the effects of habitat loss and fragmentation on metapopulation persistence have been conducted on a series of static landscapes representing a gradient of habitat availability and fragmentation severity (e.g., Bascompte and Solé, 1996; Hill and Caswell, 1999; With and King, 1999b, 2001), which assumes that these landscapes all lie on a particular trajectory of landscape change. Real landscapes are not static, however, especially given the current rate at which most landscapes are being transformed by human land-use activities. Different trajectories of land-use change could generate similar landscape patterns, but have very different consequences for the dynamics and persistence of metapopulations on these landscapes.

The rate of landscape change is an important component of landscape structure that affects extinction risk (e.g., Fahrig, 1992; Keymer et al., 2000) and which may be responsible for generating metapopulation dynamics, particularly in ephemeral or successional habitat (Hanski, 1999a; Johnson, 2000). Patch demographics, such as the life span of a patch, drive the dynamics of the metapopulation in these systems. For a given species, there is a critical rate of patch turnover in which the landscape changes too fast relative to the scale of the extinction–colonization process to permit metapopulation persistence (Keymer et al., 2000). Metapopulation extinction is thus predicted to occur more frequently in dynamic than in static landscape scenarios.

These recent theoretical treatments of metapopulations on dynamic landscapes have been concerned primarily with systems in which there is a constant rate of habitat turnover (but see Hanski and Ovaskainen, 2002; Ovaskainen and Hanski, 2002; Chapter 4). Many landscapes are subjected to chronic habitat loss and fragmentation, however, in which habitat that has been destroyed is not restored. To address this latter scenario, Schrott, With and King (unpublished) extended the spatially structured avian demographic model of With and King (2001) to a dynamic landscape context in which habitat was destroyed at various rates (0.5, 1 and 5%/yr) until the landscape was entirely denuded. The most surprising result of this analysis was that the metapopulation appeared to persist across a greater range of habitat destruction when habitat was lost rapidly (5%/yr) than when it was destroyed slowly (0.5%/yr); in other words, extinction appeared to occur sooner in landscapes subjected to lower rates of disturbance (Fig. 2.5a). This paradox is resolved by considering the decline in metapopulation growth rates as a function of time (Fig. 2.5b), which demonstrates that populations on landscapes subjected to rapid rates of habitat loss (5%/yr) will go extinct within 20 yr (the time to total landscape denudation), whereas populations in landscapes subjected to lower rates of habitat loss can persist for up to three times as long. The apparent prolonged persistence of metapopulations in landscapes undergoing rapid change results from a lagged response by the species. The generic migratory songbird being modeled in this study had a life span of 8 yr, such that at a habitat loss rate of 5%/yr, total denudation of the landscape would occur in a little over two generations. The landscape is changing more rapidly than the demographic potential of the species, and thus declines in the metapopulation growth rate (λ_L) lag behind the rate of habitat loss. This "extinction debt" has been demonstrated to be

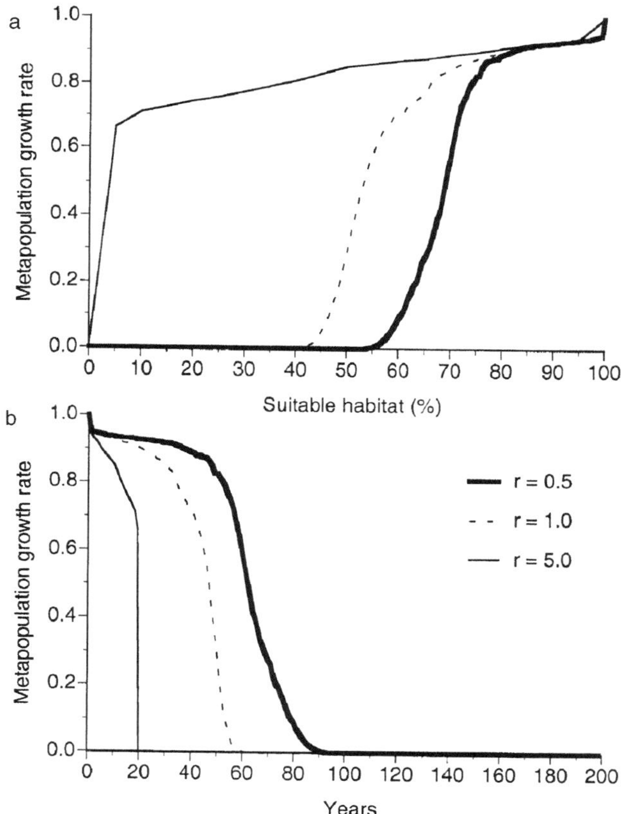

Fig. 2.5 (a) Effect of the rate of habitat destruction (*r*, % habitat destroyed/yr) on metapopulation persistence for a species with intermediate edge sensitivity in moderately fragmented ($H = 0.5$) dynamic landscapes. Because of the initial conditions of the model, in which the metapopulation growth rates are stabilized ($\lambda_L = 1.0$), the landscape population can only decline as habitat is lost. (b) Time to extinction for the same species in dynamic landscapes undergoing moderate fragmentation ($H = 0.5$) at different rates. From Schrott, With and King (unpublished).

especially great for metapopulations that are close to their extinction thresholds following habitat loss and fragmentation (Hanski and Ovaskainen, 2002).

These results emphasize the importance of understanding the historical forces shaping landscape patterns, such as the rate of land-use change or habitat destruction, for assessing species' extinction risk. Two landscapes with identical metrics could have achieved their present state via vastly different trajectories of landscape change, with different consequences for metapopulation persistence (Fig. 2.5a). Conventional landscape metrics thus cannot be used to evaluate extinction risk for metapopulations in dynamic landscapes. Time lags in species' responses to landscape change may also create relict distributions in which the species' occurrence is better explained by a historical landscape configuration than the current one ("ghosts of landscape past"; Nagelkerke et al., 2002). For example, the distribution of carabid beetles (*Abax parallelepipedus*) in a hedgerow network within an agriculturally dominated landscape of France was better explained by the well-connected hedgerow distribution from 40 yr ago than the current network (Petit and Burel, 1998). From a conservation and land-management standpoint, the potential for lagged population responses is especially disquieting because the effects of landscape change may go unnoticed for long periods of time, such that the window of opportunity for affecting a recovery may close before the problem is realized and action is taken. Alternatively, such lagged effects may buy the necessary time in which to implement conservation and restoration measures before the species goes extinct. This assumes that the problem can be recognized in time, which argues for the importance of performing theoretical and empirical analyses of the effect of dynamic landscape change on extinction risk.

Relative Effects of Habitat Loss and Fragmentation on Metapopulation Persistence

A number of empirical and theoretical investigations have attempted to assess the relative importance of the amount of habitat versus the degree of habitat fragmentation on species occurrence and extinction thresholds (e.g., McGarigal and McComb, 1995; Trzcinski et al., 1999; Villard et al., 1999; Fahrig, 1997, 2002; Flather and Bevers, 2002). Part of the difficulty in evaluating the relative effects of these two components of landscape structure, however, has been differences in how extinction thresholds are defined, which is a consequence of the specific modeling approach or measure of population viability used to assess extinction risk. Depending on the modeling construct, extinction thresholds have been defined variously as the critical level of habitat at which the population (1) is unlikely to persist for (or after) some specified amount of time (e.g., duration of run in individual-based simulation models; Fahrig, 1997; Flather and Bevers, 2001), (2) no longer occupies any of the available habitat (metapopulation models, where $p^* = 0$; Lande, 1987; Bascompte and Solé, 1996; With and King, 1999b), or (3) is no longer stable, as assessed by some demographic index, such as net reproductive rate (R_o) or the finite rate of population growth (λ) (spatially structured demographic models, where R_o or $\lambda < 1.0$; With and King, 2001). Some measures of population viability may be more sensitive to fragmentation effects, however (Flather and Bevers, 2002); for example, patch-based measures of population viability (patch occupancy, p^*, of metapopulation models) may be more sensitive to fragmentation effects than probabilities of population persistence derived from individual-based simulation models. Thus, the debate over the relative effects of habitat loss and fragmentation on metapopulation persistence is muddled by the specific modeling approach employed in theoretical studies (Flather and Bevers, 2002), the corresponding measure of population viability used, and the taxonomic group being assessed in the case of empirical studies. For example, birds may not be the best test of the relative importance of fragmentation effects, as many species are efficient at occupying available habitat even when habitat is rare (e.g., Villard et al., 1999).

Nevertheless, habitat amount is generally the best predictor of metapopulation persistence, although fragmentation — the explicit arrangement of patches in space — becomes increasingly important below the extinction threshold (Fahrig, 1997; Flather and Bevers, 2002). As pointed out in the previous section, however, the discussion of the relative effects of habitat loss and fragmentation on metapopulation persistence is moot in the absence of information on landscape dynamics. The rate of patch turnover or habitat destruction has been shown to have a more profound effect on extinction thresholds than either the amount or the fragmentation of habitat (Keymer et al., 2000; Schrott, With and King (unpublished).

2.3 SUMMARY

Although landscape ecology and metapopulation ecology share a common goal in predicting the persistence of spatially structured populations in fragmented landscapes, they differ in scope and academic tradition, which is reflected in the different approaches typically employed by ecologists from the two disciplines. Landscape ecology is an interdisciplinary field that arose from European traditions of regional geography and vegetation science, and which combined the spatial approach of the geographer with the functional approach of the ecologist (Turner et al., 2001). As a consequence, there has been greater emphasis placed on remotely sensed data, geographical information systems, and spatial statistics to generate, display, and analyze complex landscape patterns within landscape ecology, as opposed to the more abstract representation of landscapes inspired by the patch-based ecological theory adopted by metapopulation ecology. In North America, the systems ecology background of many "first-generation" landscape ecologists contributed to the widespread use of computer simulation models to tackle problems related to the effect of land-use change on resource management, an application that additionally could take advantage of GIS. Subsequently, this led to the development of spatially explicit population models, the main tool of the landscape ecologist for assessing population viability in fragmented landscapes. In contrast, metapopulation ecologists hail from a background in population ecology that is rich in mathematical theory and thus tend to approach the problem of population persistence in fragmented landscapes analytically rather than numerically.

Despite its diverse disciplinary breadth, landscape ecology is fundamentally concerned with the effects of spatial pattern on ecological processes (Turner, 1989), at whatever scale spatial heterogeneity emerges. In the context of metapopulation dynamics, any study that incorporates the effect of spatial pattern, such as habitat fragmentation or matrix heterogeneity, on processes contributing to metapopulation persistence is, by definition, adopting a landscape ecological perspective. Thus, it could be argued that the "exciting scientific synthesis" between landscape ecology and metapopulation theory that was envisioned by Hanski and Gilpin (1991) a decade ago is well underway. This is evident in current metapopulation theory that utilizes spatially explicit (or realistic) models (e.g., Hanski, 1999b) and by the use of analytical approaches from metapopulation theory in landscape ecology (e.g., With and King, 1999b).

The goal of this chapter has been to demonstrate that landscape ecology has more to offer metapopulation ecology than just the incorporation of a broader spatial scale or landscape heterogeneity to existing metapopulation theory by providing a spatial context for understanding processes contributing to the dynamics and persistence of metapopulations. Nevertheless, there are still several areas at this nexus of landscape ecology and metapopulation biology that are in need of further research: First, spatially structured demographic models require further development and testing. Demographic factors, such as reproductive output and survivorship, may be spatially dependent. Estimates of metapopulation viability that ignore this spatial dependency may give erroneous and overly optimistic assessments of a species' status on a landscape (With and King, 2001). Second, matrix effects on dispersal success and extinction risk need to be evaluated further. Research has demonstrated the potential importance of managing the matrix for enhancing colonization success and reducing extinction risk, and thus we need to move beyond assessment of mere patch size and isolation effects on metapopulation persistence (e.g., Fleishman et al., 2002), particularly in understanding asymmetrical flows among populations and evaluating the source–sink potential of landscapes for species of conservation concern. Third, the effect of landscape dynamics on extinction thresholds deserves greater attention (e.g., Keymer et al., 2000; Schrott, With and King (unpublished)). Chronic habitat loss and fragmentation increase the potential for lagged responses to landscape change, which may produce an extinction debt (Hanski and Ovaskainen, 2002), such that the status and future viability of metapopulations may not be well predicted by current landscape patterns. Fourth, metapopulation viability analysis needs to be extended to a broader, regional scale. A "metalandscape modeling approach" is particularly important for assessing dynamics among source–sink landscape populations (e.g., Donovan et al., 1995a). Finally, empirical work to address these issues, analogous to how metapopulation theory has been applied to real metapopulations in metapopulation ecology (Chapters 4 and 5), should be a research priority in landscape ecology. Otherwise, continued progress toward this developing synthesis between landscape ecology and metapopulation ecology will slow.

3. CONTINUOUS-SPACE MODELS FOR POPULATION DYNAMICS

Benjamin M. Bolker

3.1 INTRODUCTION

Metapopulation ecologists have explored the dynamics of biological populations that are discontinuous both spatially (because they live in patchy habitats) and temporally (because local populations frequently go extinct).The success of metapopulation theory comes both from the ecological importance of such populations (e.g., populations in fragmented landscapes) and from the simplicity of metapopulation theory. Dealing with population dynamics in more general continuous landscapes — those that are neither completely patchy nor completely homogeneous — requires some compromise between generality and tractability, which has in turn divided ecologists into two camps. *Landscape ecologists* have focused on large-scale patterns of population distributions in heterogeneous environments, generally considering the effects of *exogenous* heterogeneity imposed by the abiotic or biotic environment with less concern for the *endogenous* heterogeneity caused by interactions within the populations. In contrast, *spatial ecologists* have focused on smaller scale patterns of population distribution or expansion and have been more interested in endogenous than in exogenous heterogeneity. Landscape

  0-12-323448-4

ecologists favor *tactical* models (Nisbet and Gurney, 1982), designed to answer specific questions about specific populations (e.g., distributions of extinction times under different management scenarios), whereas spatial ecologists generally prefer *strategic* or "toy" models, which are more abstract and are designed to explore general principles of spatial dynamics (see Chapter 1). Landscape ecologists use computational tools such as geographic information systems (GIS) and individual-based models, which are flexible enough to incorporate realistic patterns of environmental heterogeneity and individual behavior, whereas spatial ecologists use simpler models, such as partial differential equations (PDEs) and interacting particle systems (IPSs), which may lead to more general results and are easier to analyze. (As discussed in Chapters 1 and 4, metapopulation models are used in the entire range from tactical to strategic modeling.)

This chapter gives a (brief!) overview of models for ecological dynamics in continuous space, focusing on spatial rather than landscape ecology. In particular, it explores *spatial moment equations*, a relatively new framework for analyzing spatial dynamics in terms of mean population densities and spatial covariances. The chapter gives a sample derivation of a set of spatial moment equations; summarizes various applications of moment equations for single-species and community dynamics; contrasts the strengths and weaknesses of the approach with other frameworks, such as PDEs and IPSs; and discusses future directions and potential of spatial moment equations. Although this chapter concentrates on the effects of endogenous rather than exogenous variability, it also describes some strategies for incorporating both kinds of heterogeneity and bridging the gap between spatial and landscape ecology. Spatial moment equations are a powerful tool for this task; other advantages (not to leave the reader in suspense for too long) include preservation of the spatial and stochastic character of ecological systems; analytical tractability; and simple connections to field data on individual dispersal and performance and to well-established spatial statistical measures.

3.2 OVERVIEW OF CONTINUOUS-SPACE MODELS

The full range of continuous-space models for ecological systems is far too large to review properly in a short book chapter, but this section gives abbreviated descriptions that serve at least to put spatial moment equations in perspective. For each category discussed earlier, the text describes the basic approach; gives some examples of how it has been used to study ecological dynamics; discusses how models of this type have been used to explore the combination of endogenous and exogenous variability; and gives some starting points in the literature for further exploration.

Table 3.1 categorizes mathematical models for population dynamics in continuous space. Most of these models are discussed in the following sections. Any model with discrete individuals (type = I) can be considered an individual-based model (although the following section focuses on relatively complex, flexible IBMs), whereas models with continuous individuals are covered in the section on continuum models. All of the discrete-space models in Table 3.1 have regularly arranged sites or patches and so fall under the rubric of lattice models (type = L)

TABLE 3.1 Partial Enumeration of Continuous-Space Models

Space	Time	Population	Random	Model	Type[a]
Discrete	Discrete	Discrete	Deterministic	Cellular automaton (CA)	I,L
Discrete	Discrete	Discrete	Stochastic	Probabilistic CA = stochastic CA	I,L
Discrete	Discrete	Continuous	Either	Coupled-map lattice	L
Discrete	Continuous	Discrete	Stochastic	Interacting particle system ≈pair approximations	I,L
Continuous	Discrete	Continuous	Either	Integrodifference equation	C
Continuous	Either	Discrete	Stochastic	Spatial point process ≈spatial moment equations	I
Continuous	Continuous	Continuous	Deterministic	Integrodifferential equation (IDE), partial differential equation (PDE) = reaction–diffusion equation	C
Continuous	Continuous	Continuous	Stochastic	Stochastic IDE, PDE	C

[a] "Type" (I, individual-based model; L, lattice model; C, continuum model) gives a reference to the section(s) that describe different classes of models — some models fall in more than one class. As discussed in the text, "continuum" refers to continuous population densities rather than continuous space.

(as opposed to the "continuous-space" models, where continuity is used in the narrow mathematical sense rather than the broader sense used elsewhere in the chapter). Pair approximations and spatial moment equations, discussed in some detail later, are two classes of approximations to stochastic spatial models; both eliminate the explicitly stochastic, discrete-individual nature of their parent models but preserve some of the important properties of discrete stochastic spatial dynamics (Durrett and Levin, 1994). Other approximations and limits also lead to connections between different classes of models: for example, interacting particle systems converge to PDEs in the "hydrodynamic" limit where individual movement is on a rapid timescale relative to intra- and interspecific interactions (Durrett and Neuhauser, 1994).

Individual-Based Models

By definition, an individual-based model tracks the fates of all of the individual organisms within an ecological community. IBMs need not be spatial, but their flexibility appeals to spatial and landscape ecologists. In addition, keeping track of unique individuals is often the simplest way to model spatial pattern in populations. Spatial IBMs are one kind of *spatially explicit population models* (SEPMs): models that take individuals as discrete individuals occupying a continuous, usually two-dimensional landscape. If individual locations are described as discrete, infinitesimal points (rather than disks with a finite radius, for example), then the model is a *spatial point process* model, which will form the basis for the spatial moment approximations later in this chapter. Spatial IBMs allow dynamics to be stochastic and are usually computational rather than analytical. IBMs are widely used in wildlife and conservation ecology, where researchers have questions about organisms moving and interacting on very specific landscapes (Wiegand et al., 1998). They are convenient because they can incorporate a wide range of individual behaviors and landscape structures. The downside of these models is that they may be computationally intensive, data hungry, and relatively hard to generalize (Wennergren et al., 1995; Ruckelshaus et al., 1997; Mooij and Deangelis, 1999; South, 1999).

In addition to a large variety of tactical simulation models (Turner et al., 1994; Deangelis et al., 1998; Johnson et al., 1998; Werner et al., 2001; Mullon et al., 2002; Woodbury et al., 2002), spatial IBMs have been used to study the endogenous spatial dynamics of predator–prey and larger trophic communities and to compare their results with those from other model formulations (McCauley et al., 1993; Wilson et al., 1993; Keitt, 1997; Schmitz and Booth, 1997; Wilson, 1998; Donalson and Nisbet, 1999). Forest models are a well-developed subset of spatial IBMs that are individually based but only sometimes fully spatially explicit; they usually track the locations of individual trees only to within the nearest "gap," an area of 100 m^2 or so that represents the crown size of an adult canopy tree, although more recent models have localized individuals to points in the plane (Shugart, 1984; Urban et al., 1991; Pacala et al., 1993; Chave, 1999; Phillips et al., 2003). Spatial forest models are used both tactically and strategically.

The flexibility of spatial IBMs lends itself to modeling the effects on population movement and dynamics of different configurations of barriers and habitat types in the landscape (Jonsen and Taylor, 2000; Cumming, 2002). While strategic spatial IBMs focus on endogenous heterogeneity, tactical spatial IBMs often incorporate exogenous heterogeneity — frequently derived from aerial or satellite images of a particular landscape (Caspersen et al., 1999; Murrell and Law, 2000; Mooij et al., 2002).

Further reading. The classic book by DeAngelis and Gross (1992) gives a good introduction to IBMs in general (although most of the material does not emphasize spatial models), whereas Grimm and co-workers give more recent reviews and highlight some of the major directions in IBM research (Grimm, 1999; Grimm et al., 1999).

Lattice Models

Lattice models represent a continuous landscape as a regular (usually) square lattice. Each lattice cell may contain a single individual; a population; or individuals or populations of multiple species, depending on the model. Lattice cells change their state (individuals are born, grow, or die, or are eaten by predators, or sites are taken over by populations of other species) deterministically or stochastically according to rules based on the local density of prey, predators, or competitors. *Interacting particle systems* are stochastic lattice models that run in continuous time. *Stochastic* or *probabilistic cellular automata* model stochastic changes in individual site occupancy in discrete time, whereas *coupled-map lattices* model continuous (deterministic or stochastic) populations in discrete time. Lattice models are another form of SEPM that overlaps with spatial IBMs. A lattice model where single individuals occupy cells is equivalent to a spatial IBM with no within-species variation (so that the only unique property of an individual is its spatial location). However, lattice models (and IPSs in particular) have their own literature, which focuses on general theoretical questions such as competitive coexistence and the connections between IPS and partial differential questions. In contrast to the situation for complex spatial IBMs, there is a well-developed mathematical theory that can be used to construct formal proofs of persistence or extinction in IPS (Durrett, 1988; Bramson et al., 1991; Durrett, 1992; Durrett and Neuhauser, 1994). IPS and stochastic cellular automata have been used to explore persistence of single-species populations (Bascompte and Solé, 1997); competitive interactions (Harada and Iwasa, 1994; Schwinning and Parsons, 1996; Takenaka et al., 1997); predator–prey and parasitoid–host interactions (Hassell et al., 1991; Comins et al., 1992; Wilson et al., 1993; Cuddington and Yodzis, 2000; Hosseini, 2003); and dynamics and evolution of epidemics (Satō et al., 1994; Rand et al., 1995; Boots and Sasaki, 2000, 2002).

Some researchers have used lattice models to explore the effects of environmental heterogeneity, particularly the viability of single-species populations in degraded or fragmented habitats. Bascompte and Solé (1997) and Hiebeler (2000) have both used IPS to (1) understand how population densities vary in response to the amount and pattern of habitat destruction and (2) compare the results to the predictions of patch occupancy models under similar scenarios (Gotelli, 1991; Tilman et al., 1994; Lavorel and Chesson, 1995).

Further reading. Durrett and Levin's classic paper on "The Importance of Being Discrete (and Spatial)" (1994) lucidly compares the dynamics and coexistence properties of nonspatial models, continuum models (see below), and IPS.

Pair Approximation

Pair approximation, an approximate method for analyzing IPS that focuses on the joint occupancy probability of neighboring pairs of cells, has been an important complement to more rigorous analytical techniques (Tainaka, 1988, 1994; Harada and Iwasa, 1994; Harada et al., 1995; Nakamaru et al., 1997; Takenaka et al., 1997; Iwasa, 2000; Satō and Iwasa, 2000; Boots and Sasaki, 2000). Pair approximation makes the assumption of conditional independence. Pairs of neighbors are assumed to be independent (the conditioning is on the presence of a focal individual) so that the probability of a particular configuration of a focal individual (e.g., a predator) and two neighbors (e.g., two different prey individuals) is the product of the probabilities of each pair of neighbors. Although a great deal of foundational work in IPSs has been done on simple models, such as the contact process and the biased voter model (the IPS analogues of logistic growth and simple competition, respectively), pair approximations allow rapid construction and analysis of models dealing with larger communities or more complex biological rules. In addition to basic pair approximation on the lattice with a single, fixed interaction scale, variants of pair approximation have been developed to handle more general spaces, such as networks (Keeling et al., 1997; van Baalen and Rand, 1998; Rand, 1999; Keeling, 1999a,b; van Baalen, 2000); to approximate the spatial dynamics of expanding populations (Ellner et al., 1998); and to model systems with multiple scales of movement and interaction (Ellner, 2001).

Further reading. Chapters by van Baalen (2000) and Rand (1999) give the clearest general explanation of pair approximation derivations (they both discuss network models, which are more general than lattices); chapters by Satō and Iwasa (2000) and Iwasa (2000) are also useful.

Continuum Models

The "continuum" in the definition of *continuum models* refers primarily to the representation of local populations as continuous densities rather than discrete numbers, although the most common continuum models (partial differential equation and integrodifferential equations) also operate in continuous space and time. Continuum models are more common in strategic than in tactical applications. The simplest *reaction–diffusion* models allow only completely local reaction terms (predator–prey interactions, competition, etc.) that are formulated similarly to their nonspatial analogues coupled with movement by local diffusion; more detailed integrodifferential models allow individuals to move or interact nonlocally according to movement or interaction kernels. Continuum models can also be defined in discrete time, in which case they become integrodifference equations.

The continuous-population assumption eliminates demographic stochasticity from continuum models, although the effects of demographic stochasticity can be reintroduced either by assuming Poisson or multinomial statistical variation in the local population (Durrett and Levin, 1994; Grünbaum, 1994) or by adding a stochastic term to create a stochastic PDE or integral equation (Lande et al., 1999; Engen et al., 2002). Continuum models have a proud history in ecology; before computational models were practical, they were the only way to model dynamics in continuous space, and they still benefit from a strong theoretical foundation. Continuum models are particularly powerful for modeling spread and wavefront dynamics of genes, diseases, and populations (Metz and van den Bosch, 1995), but they are also useful for studying equilibrium states and complex spatiotemporal dynamics (Holmes et al., 1994; Klausmeier, 1999; Briscoe et al., 2002). Although many ecologists perceive them as oversimplified, PDEs equations can incorporate fairly sophisticated models of behavior (Kareiva and Odell, 1987; Grünbaum, 1994; Turchin, 1998; Grünbaum, 1998; Moorcroft et al., 1999).

Historically, most continuum models have explored the effects of endogenous variability only. One exception is the large body of work on the distribution of animals following different foraging and dispersal strategies in heterogeneous environments (Grünbaum, 1994); classical work on population persistence and competition in the presence of environmental features, such as resource patches and gradients (Kierstead and Slobodkin, 1953; Pacala and Roughgarden, 1982); and a series of papers by Roughgarden on the shape and scale of spatial distributions of organisms in the presence of particular environmental heterogeneities (Roughgarden, 1974, 1977, 1978; Sasaki, 1997). [Klausmeier (1999) shows how the combination of endogenous dynamics with a smooth slope can generate the heterogeneous pattern of "tiger bush" in semi-arid landscapes.] More recently, Lande et al., (1999) and Engen et al., (2002) have developed a series of approximate and exact stochastic PDE models for the spatial synchrony of population fluctuation that extend Roughgarden's work. Results from this approach of modeling noise-driven correlations on an exogenously variable landscape (also used by Snyder and Chesson, 2002) have begun to converge on some of the results of spatial moment equations described later.

Further reading. Continuum models are presented in Okubo's book on diffusion models (Okubo, 1980) and, more recently, by Shigesada and Kawasaki (1997) [the edited volume by Okubo and Levin (2002) is more up to date than Okubo (1980), but gives more of an overview of developments in ecological diffusion models than an introduction]; Turchin (1998) does a particularly good job connecting continuum models with data on animal movement. Books by Murray (1990) and Edelstein-Keshet (1988) cover the use of PDEs in mathematical biology more generally, but do have useful ecological material.

3.3 SPATIAL MOMENT EQUATIONS

How do spatial moment equations (SMEs) fit into this assemblage of models and tools? The basic structure behind spatial moment equations is that of spatial point processes in a (possibly heterogeneous) two-dimensional space. As pair approximations do for lattice models, SMEs attempt to reduce a spatial point process to a set of equations for the dynamics of population densities and spatial pattern in the system. SMEs capture the spatial properties of an ecological system by tracking the first two spatial moments, the mean densities of different species and the spatial correlations among individuals of all species, which are analogous to the mean and variance of a nonspatial distribution. Like continuum models, the actual representations of spatial pattern are continuous, differentiable functions. While moment equations are not technically stochastic — they represent average behavior over space or across ensembles of similar ecological arenas — they do preserve some of the important effects of stochasticity in finite populations that are lost from typical continuum models.

Many nonspatial models use similar methods to track the properties of population distributions. Moment generating functions (Bailey, 1964) can be used to analyze the entire infinite series of moments of a distribution, but are difficult to apply to nonlinear processes, let alone nonlinear spatial processes. Nonspatial moment equations that use some rule to approximate higher moments (including setting them constant, or to zero) are common in ecology, epidemiology, and genetics (Dobson and Hudson, 1992; Turelli and Barton, 1994; Grenfell et al., 1995; Dushoff, 1999; Cornell et al., 2000). For example, Anderson and May (1985) modeled macroparasite dynamics by assuming that the number of parasites per host has a negative binomial distribution with a fixed overdispersion parameter so that the variance can be calculated for any given value of the mean.

The next section derives the spatial moment equations for the simplest parasitoid–prey system, equivalent to Lotka–Volterra predator–prey equations with a predator efficiency of one (predator bith rate equal to predation rate). The parasitoid moves by random jumps, while the prey is sessile and disperses only as a seed or propagule. When the parasitoid is close enough to the prey, it turns the prey into a single new parasitoid. This system, rather than the analogous equations for competition or epidemics, has been chosen for novelty; derivations of competitive and epidemic systems appear elsewhere (Bolker and Pacala, 1999; Bolker, 1999). Moment equations have been used to model predator–prey metapopulations (Keeling, 2000; Keeling et al., 2002) and spatial epidemics without replenishment (Bolker, 1999; Brown, 2001), and pair approximations have been used for more general epidemics (Satō et al., 1994), spatial moment equations for predator–prey systems have not been published (M. Desai, unpublished manuscript). The derivation is sketched, leaving out the straightforward but scary-looking algebra; this approach makes things slightly harder for the serious reader who wants to follow the algebra line by line, but it allows the discussion to focus on the important steps in the derivation, pointing out where alternative assumptions or procedures are possible. [Similar derivations can be found in papers by Bolker and Pacala (1999) and by Dieckmann and Law (2000).]

Further reading. Other introductions to SMEs can be found in chapters by Bolker et al. and Dieckmann et al. in Dieckmann et al. (2000) and in other papers by Law, Dieckmann, and Murrell (Law and Dieckmann, 2000; Law et al., 2001, 2003).

Write down Stochastic Rules

We start by writing down the expected change in the occupancy of a small patch of size ω located at position $\mathbf{x}$ over a small time Δt, in a landscape of patches Ω. In this system, prey (V) reproduce at a constant *per capita* rate f_V (leading to a Poisson-distributed number of offspring per unit time) and disperse them in a random direction with a distance given by the dispersal kernel $\mathcal{D}_V$. [Kernels are continuous functions that determine the rate or probability with which an individual will interact with another site or individual at a given distance. Dispersal kernels represent probability distributions and so must integrate to 1 over all possible locations; for analytical convenience, we also scale interaction kernels so they integrate to 1. Kernels are typically symmetric, and to avoid interactions at infinite distances, kernels must approach zero as distance becomes large. We typically use simple decreasing functions such as the Laplacian (back-to-back exponential) or Gaussian distribution, but other shapes are feasible.]

Prey die at a density-independent *per capita* rate μ_V, and intraspecific competition leads to additional mortality at a rate proportional to the neighborhood density, measured as the sum of $\alpha\mathcal{U}_{VV}(|\mathbf{y} - \mathbf{x}|)$ for all conspecific neighbors, where α is the overall strength of competition and $\mathcal{U}_{VV}(\rho)$ is the relative strength of competition at a distance ρ. [In the standard nonspatial Lotka–Volterra model, specifying birth and density-independent death rates separately would be redundant, but in stochastic and spatial models the ratio of birth to death rates as well as the difference between birth and death rates is important (Bolker and Pacala 1999).] Self-limitation of the prey may seem to be an unnecessary complication, but it is actually required to keep the spatial moment equations from blowing up (M. Desai, unpublished manuscript).

Parasitoids (P) die at a density-independent *per capita* rate μ_P, and at rate m_P they jump from their current location in a random direction with a dispersal kernel $\mathcal{D}_P$. Parasitism occurs as a function of parasitoid–prey distance, with a rate proportional to $\gamma\mathcal{U}_{PV}(|\mathbf{y} - \mathbf{x}|)$ [again, γ is the overall rate and $\mathcal{U}_{PV}(\rho)$ governs the relative predation rate at different distances]. When predation occurs, the prey turn into newborn parasitoids.

The processes described here can be written formally as stochastic rates:

Event	Change	Rate
Prey birth	$V(\mathbf{x}) \rightarrow V(\mathbf{x}) + 1$	$\sum_{\mathbf{y} \in \Omega} f_V V(\mathbf{y}) D_V (\lvert\mathbf{y} - \mathbf{x}\rvert)\omega$
Prey death (density independent)	$V(\mathbf{x}) \rightarrow V(\mathbf{x}) - 1$	$\mu_V V(\mathbf{x})$
Prey death (density dependent)	$V(\mathbf{x}) \rightarrow V(\mathbf{x}) - 1$	$V(\mathbf{x}) \sum_{\mathbf{y} \in \Omega} \alpha V(\mathbf{y}) U_{VV}(\lvert\mathbf{y} - \mathbf{x}\rvert)$
Parasitoid movement	$P(\mathbf{x}) \rightarrow P(\mathbf{x}) - 1$ $P(\mathbf{y}) \rightarrow P(\mathbf{y}) + 1$	$mp\ \mathrm{P}(\mathbf{x})\ D_P\ (\lvert\mathbf{y} - \mathbf{x}\rvert)\omega$
Parasitoid death	$P(\mathbf{x}) \rightarrow P(\mathbf{x}) - 1$	$\mu_P P(\mathbf{x})$
Parasitism	$V(\mathbf{x}) \rightarrow V(\mathbf{x}) - 1$ $P(\mathbf{x}) \rightarrow P(\mathbf{x}) + 1$	$V(\mathbf{x}) \sum_{\mathbf{y} \in \Omega} \gamma \mathrm{P}(\mathbf{y})\ U_{PV}(\lvert\mathbf{y} - \mathbf{x}\rvert)$

The summation limits represent the set of locations of all patches in the landscape. All events that involve the probability of a propagule or an individual landing in a particular patch (the prey birth rate and parasitoid movement rates in this case) contain a patch size (ω) term because the probability of an individual landing in a patch is proportional to patch area. Interactions that involve two individuals, such as the parasitism term, do not because the corresponding expressions sum over discrete individuals rather than areas. When we take the continuous-space limit later ($\omega \rightarrow 0$), it will be important to have these terms correct.

Alternatives

This derivation is heuristic, starting from small, discrete patches evolving in discrete time and then taking straightforward limits to reach the continuous-space, continuous-time equations. This procedure makes it easier to see when and why (removable) singularities occur in the equations. It is also possible to construct a rigorous derivation starting from infinitesimal space and time increments rather than small but finite ones and using probability measures, but we have always found it easier and clearer to proceed heuristically. Mathematically inclined readers can see Barton et al. (2002) for a more rigorous approach to stochastic population dynamics in continuous space in a population–genetic context.

The predation process just described, where predators turn their prey into a single newborn predator, models a parasitoid–host interaction where only one egg can emerge from a host. This kind of parasitoid model is the simplest to analyze because only two spatial locations are involved in the interaction. If we wanted to allow more than one newborn predator per host, we could define a natal dispersal kernel according to which the newborn predators jumped away from the host, but we would then have to keep track of three spatial locations (predator, prey, and newborn) at a time; to calculate the expected change in predator–predator covariances, we would need to estimate the probability of a four-point configuration (predator, prey, newborn, and neighbor predator). We could mitigate this problem by turning the prey into an incubating prey, which would be represented as a different class of individuals in the equations and which could then be immune from superinfection. At a constant rate, the incubating prey transforms into one or more newborn predators — separating the time of birth from the time of predation eliminates the need to keep track of as many locations (Fig. 3.1).

Similar problems, and similar solutions, arise in predator–prey models. If newborn predators disperse away either from the location of the prey or from the location of their parent, several spatial locations have to be tracked. Another possibility, analogous to incubating prey, is to say that predation turns a hungry predator into a sated predator. The sated predator would "decay" at a constant rate to a pair of predators, a parent and a newborn; if sated predators cannot kill prey, this model structure would also give rise to a handling time. Multiple stages of satiation could make the time lag between predation and predator reproduction gamma distributed (Keeling and Grenfell, 1998); predator efficiency less than one could be handled by requiring that predators experience multiple prey events or by assuming there is a probability less than one that a satiated predator reproduces when it becomes hungry again.

These problems of definition are implicit in all ecological models, from simple nonspatial models to SEPMs. Lattice models often make slightly artificial rules to avoid having multiple individuals in the same lattice cell (Durrett and Levin, 2000); in nonspatial models, these details can often be accounted for by setting up proper functional and numerical responses (Keeling et al., 2000; Cuddington and Yodzis, 2002).

The parasitoids in this model are stupid, neglecting any local cues that could guide them to higher prey densities. We could introduce adaptive foraging rules by letting the probability of movement, or the length of moves,

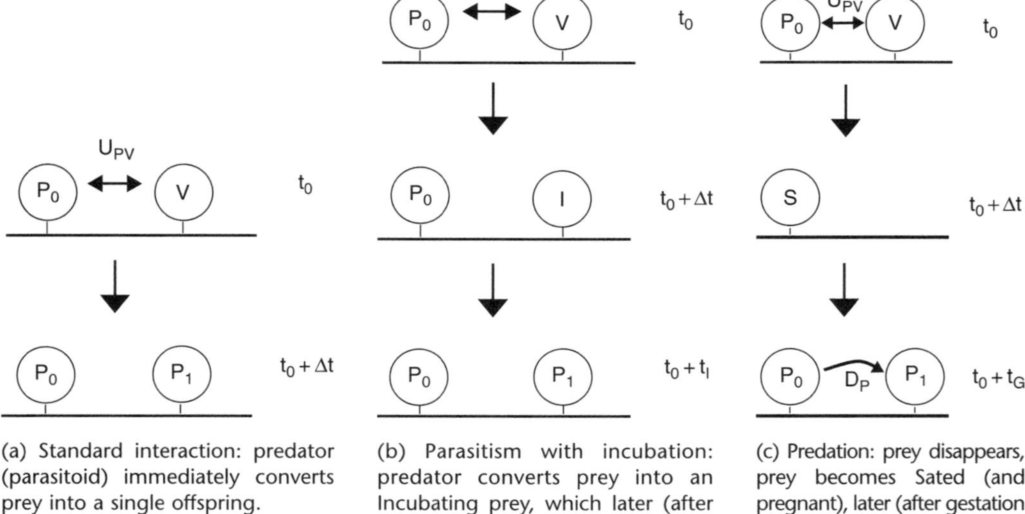

Fig. 3.1 Predator–prey transition rules. Predators are P_0, P_1; prey are V; incubating prey (e.g., containing parasitoid eggs or larvae) are I; and sated, pregnant predators are S. Δt denotes an instantaneous change, whereas other time lags t_I, t_G denote events that occur after an exponentially distributed lag time.

decrease in response to increasing local prey density. This behavioral response would result in an aggregation of parasitoids in areas of high prey density, mitigating the prey depletion that occurs in parasitoid neighborhoods. Another common foraging rule, allowing the directions of successive moves to become less correlated in response to favorable habitat (Turchin, 1998), could be incorporated in the point process model but is harder to preserve in moment equations.

We could also add effects of environmental heterogeneity simply by making some of the parameters be functions of location. Making death or movement rates spatially heterogeneous is simpler (although perhaps less interesting) than making predation or birth rate spatially heterogeneous because making the predation rate heterogeneous induces a triple spatial interaction in the process, which will turn into a four-way interaction when the covariance is calculated.

Compute Expected Changes in Singlet and Pair Densities

For each mean density, the derivation of the spatial moment equations requires two averages or expectations. The first average is the expected change in occupancy over a short time Δt in a patch of size ω for a given starting configuration of neighbors; this is the standard expected change calculated in deriving most ecological models and is denoted by an overbar. In general, this expected rate of change is

$$\overline{\Delta F(\mathbf{x})} = \Sigma(\text{rate of event } i)[\text{change in } F(\mathbf{x}) \text{ if event } i \text{ occurs}]\Delta t$$

(throughout this derivation, F, G, and H refer to arbitrary state variables; $\mathbf{w}$, $\mathbf{x}$, $\mathbf{y}$, and $\mathbf{z}$ are arbitrary locations). Once we have taken this first average to calculate the expected amount of change that will occur at a particular location starting from a particular configuration, we need to average all possible configurations across the ensemble to find the expected amount of change that will occur at the location starting from all configurations with particular mean densities and correlations. This second expectation, denoted by angle brackets $\langle\cdot\rangle$, reduces the information on the right-hand side of the equation from all of the information about the configuration to just the expected patch densities and joint patch densities (doublets).

The same procedure is applied to calculate the average expected rate of change of doublets, e.g., $\langle\overline{\Delta F(\mathbf{x})G(\mathbf{y})}\rangle$, the rate of change of pairs with prey in a patch at point $\mathbf{x}$ and parasitoid in a patch at point $\mathbf{y}$; these rates will involve doublets and triplets.

Alternatives

For simplicity, we assume at this point that Δt is short so that we can drop terms where two events occur at once [which happen with probability proportional to $(\Delta t)^2$]. Retaining finite time steps is also possible [leading eventually to an integrodifference equation for the changes in spatial covariances (Lewis, 2000)], although as usual with discrete-time models, it requires more care in defining the possible orders of events. In addition, retaining double events will lead to fourth-moment terms in expressions for covariances, which will have to be dealt with by some extended form of moment closure (see below).

Assume Spatial Homogeneity

By assuming spatial homogeneity (specifically homogeneity and isotropy or translational and rotational invariance), we can make all singlet expectations into global averages, $\langle F(\mathbf{x})\rangle = \langle F\rangle$. Similarly, joint products ($\langle F(\mathbf{x})G(\mathbf{y})\rangle$) depend only on the distance between two points: $\langle FG\rangle$ ($|\mathbf{x} - \mathbf{y}|$). Assuming that space is homogeneous considerably simplifies the algebra, allowing us to collapse various terms (e.g., $\langle(\Delta F(\mathbf{x}))\, F(\mathbf{y}) + F(\mathbf{x})(\Delta F(\mathbf{y}))\rangle = 2\,\langle \Delta F(\mathbf{x}) F(\mathbf{y})\rangle$). In addition, it simplifies the interpretation of the expectations across ensembles introduced above, which we can now take as averages across space in a single realization rather than as averages across ensembles.

Alternatives

Certain kinds of spatial heterogeneity are easier to incorporate than others. As long as we assume that the probability of events occurring at a point depends only on the occupancy of that point and of its neighborhood, rather than on its absolute position, we can still express the dynamics in terms of moments — covariances between population densities and environmental parameter distributions such as mortality or predation rate.

At the price of some algebraic complexity, we can relax the assumption of spatial homogeneity still further. For example, Lewis (2000) and Lewis and Pacala (2000) analyzed the mean and covariance structure of an invading population by preserving information about the distance of points from the wave front. As another example, nonisotropy (directionality) in some processes could be studied by allowing joint product expectations to depend on direction: $\langle F(\mathbf{x})G(\mathbf{y})\rangle = \langle FG\rangle$ ($|\mathbf{x} - \mathbf{y}|$, θ). Allowing for some degree of absolute (rather than relative) spatial heterogeneity also opens the possibility of modeling covariance dynamics as a function of position along a gradient or distance from an edge. (Edge effects are not easy to incorporate in moment equations, which typically explore dynamics in a hypothetical infinite landscape. Modeling dynamics as a function of distance from edge in a one-dimensional habitat or from an infinitely long, straight edge in a two-dimensional habitat might be feasible, but edge effects in a finite two-dimensional landscape would probably only be calculable numerically.)

Product Expansion/Convert Joint Products to Covariances

This step first expands joint and triple products in terms of means, covariances, and third moments and then subtracts $\overline{\Delta\langle F\rangle\langle G\rangle}$ (the expected change in the product of the mean densities) from joint-product equations to turn them into covariance equations instead of joint-product equations. It is useful to explicitly write out *singular* or "delta function" terms that take into account the possibility that the two individuals share the same patch (when we take the continuous-space limit below, this means that there is really only one individual present). The definition of the covariance is thus

$$\langle FG\rangle(|\mathbf{x} - \mathbf{y}|) = \langle F\rangle\langle G\rangle + C_{FG}(|\mathbf{x} - \mathbf{y}|) + \delta_{\mathbf{xy}}\sigma_{FG}. \tag{3.1}$$

We can expand the triple product $\langle FGH\rangle$ $(\mathbf{x}, \mathbf{y}, \mathbf{z})$ similarly, arriving at a (fairly ugly) formula containing products of the mean densities ($\langle F\rangle$, $\langle G\rangle$, $\langle H\rangle$) and all of the covariances, as well as the third moment $M3_{FGH}(\mathbf{x}, \mathbf{y}, \mathbf{z})$. [We specify third central moments $M3$ by all three locations involved in the triple: with spatial homogeneity and isotropy they are fully determined once we know two sides and an angle (so we could quote them as $M3_{FGH}(|\mathbf{x} - \mathbf{y}|, |\mathbf{x} - \mathbf{z}|, \theta_{\mathbf{xy}-\mathbf{xz}})$), but this notation is simpler.] In addition, the expansion contains a series of singular third-moment terms that result when two or more points in the triangle coincide. For example, a term $\langle H\rangle\sigma_{FG} + C_{H,FG}(|\mathbf{x} - \mathbf{z}|)$ results when F and G are in the same patch: σ_{FG} denotes the covariance of F and G within the same patch (at distance zero), and $C_{H,FG}$ denotes the covariance of H with a patch containing both F and G. When we scale to continuous space below, and patches become too small to contain more than one individual, the singular terms will express the case where one individual plays two roles at once, e.g., as both neighbor and competitor. The within-patch covariance σ_{FG} will equal zero for $F \neq G$ and $\langle F\rangle$ for $F = G$, which we can express by writing it as $\delta_{FG}\langle F\rangle$ (or $\delta_{FG}\langle G\rangle$); $C_{F,GH}$ will similarly become $\delta_{GH}C_{FG}$. One can sometimes justify dropping the third-moment singular terms for analytical simplicity in the limit of long-range interactions (Bolker and Pacala, 1999).

The only delicate part of these expansion is that we have to remember to drop cases where nothing really happens: for example, if the event "parasitoid moves from $\mathbf{x}$ to $\mathbf{y}$" involves the same points $\mathbf{x} = \mathbf{y}$, then no move really takes place and we should omit the singular term.

Alternatives

The covariance is only one way of expressing spatial dependence. We could also leave all expressions written in terms of joint densities $\langle FG\rangle$ or we could write the spatial dynamics in terms of *conditional densities*, *scaled covariances*, *correlations*, or *multiplicative covariances* (neglecting singular terms):

Conditional density	$\langle F\mid G\rangle = \frac{\langle FG\rangle}{\langle G\rangle}$
Scaled covariance	$SC_{FG} = \frac{C_{FG}}{\langle G\rangle} = \frac{\langle FG\rangle - \langle F\rangle\langle G\rangle}{\langle G\rangle} = \langle F\mid G\rangle - \langle F\rangle$
Correlation	$\hat{C}_{FG} = \frac{C_{FG}}{\langle F\rangle\langle G\rangle} = \frac{\langle FG\rangle - \langle F\rangle\langle G\rangle}{\langle F\rangle\langle G\rangle} = \frac{\langle FG\rangle}{\langle F\rangle\langle G\rangle} - 1$
Multiplicative covariance	$\frac{\langle FG\rangle}{\langle F\rangle\langle G\rangle} = \hat{C}_{FG} + 1$

These forms are all equivalent, but have advantages in different situations: for example, using conditional densities makes it easy to take the limit where one species is invading at low densities.

Moment Closure

Moment closure is the key step in the derivation, but also (potentially) one of the simplest. One way or another, we have to deal with the presence of third-moment terms, which incorporate the information about the spatial pattern that cannot be fully captured by pairwise approximations. The simplest solution is *power-1 closure*, which assumes that the probability of a given triangular configuration (e.g., type F at location $\mathbf{x}$, type G at location $\mathbf{y}$, type H at location $\mathbf{z}$) is described adequately by taking the probability of pairs and assuming that the third point in the triangle is independent. This corresponds to the relationship $\langle FGH \rangle = \langle FG \rangle \langle H \rangle + \langle GH \rangle \langle F \rangle + \langle FH \rangle \langle G \rangle - 2 \langle F \rangle \langle G \rangle \langle H \rangle$. If we expand this relationship in terms of central moments, we find happily that the power-1 assumption corresponds to $M3_{FGH}(\mathbf{x}, \mathbf{y}, \mathbf{z}) = 0$, so we simply drop third-moment terms.

Alternatives

Different moment closures are possible (see Box 3.1). In some cases, it may actually be simpler to substitute the relationships between noncentral moments earlier in the derivation: the author's habit of waiting until this point comes from using power-1 closures, which are simplest to apply at this step by setting central moments to zero.

Criteria for choosing a moment closure are analytical simplicity, stability, and numerical accuracy. Different closures may work better for different kinds of problems, for example, in competitive communities vs predator–prey or epidemic systems. There is even a difference between different regions of phase space within the same model. For example, power-1 closure is unstable for calculating invasion rates in a simple epidemic model — in the limit of small densities of infectives, the equations simply blow up — but actually predicts equilibrium densities more accurately than the power-2 closure (Brown, 2001). Power-1 closures are unstable, but analytically tractable (Bolker and Pacala, 1999); power-2 closures may be the best overall choice for stability, although power-3 closures perform better under some conditions (Law et al., 2003). Filipe has discussed a variety of more complex, more accurate closure schemes; while these are discussed in a lattice context, they could easily be adapted to continuous-space models (Filipe, 1999; Filipe and Gibson, 2001; Filipe and Maule, 2003). (These closures are intended, and are probably more useful, for cases where accurate numeric solutions are required rather than for cases where one will attempt to analyze the moment equations.) Hybrid closures, which scale between two different closures in different regions of phase space (e.g., when infection is rare or common), have been used in a nonspatial context by Dushoff (1999). The choice of closures is still more of an art than a science. At present, recommendations are (1) use power-1 closures for analytically tractable results, except possibly in invasion cases; (2) use power-2 closures otherwise; and (3) explore the accuracy of a particular closure with simulations and always check key results.

Take Continuum Limits

Now we let the patch size ω and the time step Δt both become infinitesimally small. Prior to taking the limits we rescale the mean densities $\langle F \rangle$ to densities $f = \langle F \rangle / \omega$, and the covariances C_{FG} to covariance densities $c_{FG} \equiv C_{FG}/\omega^2$.

BOX 3.1 Verbal, Analytic, and Graphical Descriptions of Different Moment Closures

Each moment closure breaks the probability of a triangular configuration $\langle FGH \rangle$ down in a different way: while power-1 assumes that each point is independent of the remaining pairs, power-2 and power-3 deal with pairs only. Power-2 considers pairs (sides) two at a time, while power-3 considers all three sides simultaneously. For each geometric assumption, we can write out the relationship between the triple and the pairs and points (noncentral moments), expand it in terms of central moments, and solve it for $M3$. Dieckmann and Law (2000) give more details on the consistency conditions that must be met by any moment closure rule.

Closure	Noncentral moments ($\langle FGH \rangle$)	Central moments ($M3$)
Power-1	$\langle FG \rangle \langle H \rangle + \langle GH \rangle \langle F \rangle + \langle FH \rangle \langle H \rangle - 2 \langle F \rangle \langle G \rangle \langle H \rangle$	0
Power-2 (asymmetric)	$\dfrac{\langle FG \rangle \langle FH \rangle}{\langle F \rangle}$	$\dfrac{C_{FG}(\lvert \mathbf{y} - \mathbf{x} \rvert)\, C_{FH}(\mathbf{y} - \mathbf{z})}{\langle F \rangle} + \langle F \rangle\, C_{GH}(\lvert \mathbf{y} - \mathbf{z} \rvert)$
Power-3	$\dfrac{\langle FG \rangle \langle FH \rangle \langle GH \rangle}{\langle F \rangle \langle G \rangle \langle H \rangle}$	$(C_{FG}(\lvert \mathbf{y} - \mathbf{x} \rvert)\, C_{FH}(\lvert \mathbf{x} - \mathbf{z} \rvert)\, C_{GH}(\lvert \mathbf{y} - \mathbf{z} \rvert)$ $+ C_{FG}(\lvert \mathbf{x} - \mathbf{y} \rvert)\, C_{FH}(\lvert \mathbf{x} - \mathbf{z} \rvert) \langle G \rangle \langle H \rangle$ $+ C_{FG}(\lvert \mathbf{x} - \mathbf{y} \rvert)\, C_{GH}(\lvert \mathbf{y} - \mathbf{z} \rvert) \langle F \rangle \langle H \rangle$ $+ C_{FH}(\lvert \mathbf{x} - \mathbf{z} \rvert)\, C_{GH}(\lvert \mathbf{y} - \mathbf{z} \rvert) \langle F \rangle \langle G \rangle)$ $/ (\langle F \rangle \langle G \rangle \langle H \rangle)$

The power-2 asymmetric closure shown here multiplies only one pair of sides ($\langle FG \rangle$ and $\langle FH \rangle$). It can be made symmetric by including terms for the other two pairs of sides, but it is useful in the asymmetric form when a triangle naturally contains a focal individual, which breaks the symmetry (Murrell and Law, 2000). Symmetric power-2 closures require a slightly counterintuitive correction term in order to make them consistent (Dieckmann and Law, 2000). Power-2 closure is analogous to the usual pair approximation used in lattice models; it represents an assumption of independence between neighbors of a focal individual.

In the graphical representation below, lines between circles indicate dependence; juxtaposition indicates multiplication (combining independent probabilities).

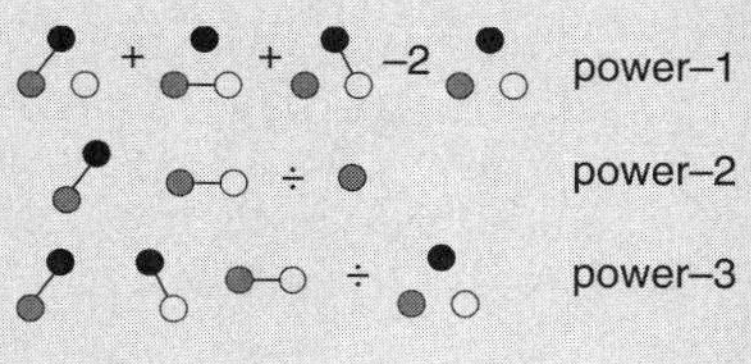

+ + −2 power–1

÷ power–2

÷ power–3

In this limit, summations become integrals and delta functions become Dirac delta functions, but all Dirac delta functions are found inside integrals and can be extracted by the rule that $\int \delta_{\mathbf{x}}\, d\mathbf{x} = 1$. We can also simplify variance terms (σ_F^2) because when there can only be zero or one individual per patch, $\langle FF \rangle = \langle F \rangle = 0$ or 1 and (σ_F^2) = $\langle F \rangle$ (this is equivalent to assuming Bernoulli statistics within each patch).

Alternative

If preferred, we can retain finite patch sizes, although in this case we may have to write out separate equations for the time evolution of the within-patch variances σ_F^2 (Bjørnstad and Bolker, 2000) and the within-patch covariances σ_{FG}.

Simplify Delta Functions and Convolutions

The last step is just algebraic tidying and involves no new assumptions or conceptual leaps. We take delta functions and average values out of integrals (all kernels $\mathcal{K}$ are normalized so that $\int\mathcal{K}(\mathbf{y})\ d\mathbf{y} = 1$); write $|\mathbf{y} - \mathbf{x}| = r$; and write terms of the form $\int\ \mathcal{K}(|\mathbf{z} - \mathbf{x}|)c_{FG}\ (|\mathbf{z} - \mathbf{y}|)\ d\mathbf{z}$ as *convolutions*, $(\mathcal{K}{*}c_{FG})(|\mathbf{x} - \mathbf{y}|) = (\mathcal{K}{*}c_{FG})(r)$ (Bolker et al., 2000). We can also rewrite the *average covariances* that appear in the mean equations, $\int\mathcal{K}(|\mathbf{y} - \mathbf{x}|)c_{FG}(|\mathbf{y} - \mathbf{x}|)d\mathbf{y}$, as $\bar{c}_{\mathcal{K},FG}$ (we can $\mathcal{K}$ omit in the subscript when it is clear from context which kernel is being used as a weighting term).

All the algebra results in the following mean equations for the parasitoid model:

$$\begin{aligned} \frac{dv}{dt} &= f_V v - \mu_V v - \alpha(v^2 + \bar{c}_{\mathcal{U}_{VV},\, vv}) - \gamma(vp + \bar{c}_{\mathcal{U}_{PV},\, vp}) \\ \frac{dp}{dt} &= \gamma(vp + \bar{c}_{\mathcal{U}_{PV},\, vp}) - \mu_P p \end{aligned} \tag{3.2}$$

These are the covariance equations:

$$\begin{aligned} \frac{\partial c_{VV}(r)}{\partial t} = 2[&f_V D_V(r) v + f_V(D_V{*}c_{VV})(r) - \mu_V c_{VV}(r) \\ &- \alpha((\mathcal{U}_{VV} * c_{VV})(r) + v^2\mathcal{U}_{VV}(r) + \mathcal{U}_{VV}(r)c_{VV} + c_{VV}(r)) \\ &- \gamma(v(\mathcal{U}_{PV}{*}c_{PV})(r) + pc_{VV}(r)) \end{aligned} \tag{3.3}$$

$$\begin{aligned} \frac{\partial c_{VP}(r)}{\partial t} = &f_v(\mathcal{D}_V * c_{VP})(\mathrm{r}) - \mu_v c_{VP}(r) - \gamma(pc_{VP}(r) + v(\mathcal{U}_{PV} * c_{PP})(r) + \mathcal{U}_{PV}(r)vp) \\ &+ \gamma(v(\mathcal{U}_{PV} * c_{VP})(r) + pc_{VV}(r)) - \mu_P c_{VP}(r) \\ &- mc_{VP}(r) + m(\mathcal{D}_P * c_{VP})(r) \\ &- \alpha(vc_{VP}(r) + (\mathcal{U}_{VV} * c_{VP})(r)) \end{aligned} \tag{3.4}$$

$$\begin{aligned} \frac{\partial c_{PP}(r)}{\partial t} = 2[&\gamma(pc_{VP}(r) + v(\mathcal{U}_{PV} * c_{PP})(r) + \mathcal{U}_{PV}(r)vp) - \mu_P c_{PP}(r) \\ &- mc_{PP}(r) + m(\mathcal{D}_P * c_{PP})(r)] \end{aligned} \tag{3.5}$$

3.4 NUMERICAL RESULTS

We can evaluate the spatial dynamics of the parasitoid–prey system stochastically, by running an individual-based simulation, or deterministically (and approximately), by integrating the moment equations numerically. As suggested previously, it is important when beginning to work with moment equations to check their accuracy (possibly for a variety of different closures) over a wide range of conditions. Because this chapter focuses on the derivation and meaning of moment equations, the following section shows only a brief overview of the numerical results from the parasitoid–prey system. Starting from the monoculture equilibrium of the prey (running only the equations for V and C_{VV} to equilibrium) and then introducing a small density of parasitoids, the system shows damped oscillations to an equilibrium (Fig. 3.2). (In fact, because moment equations represent the dynamics of the average neighborhood across an infinitely

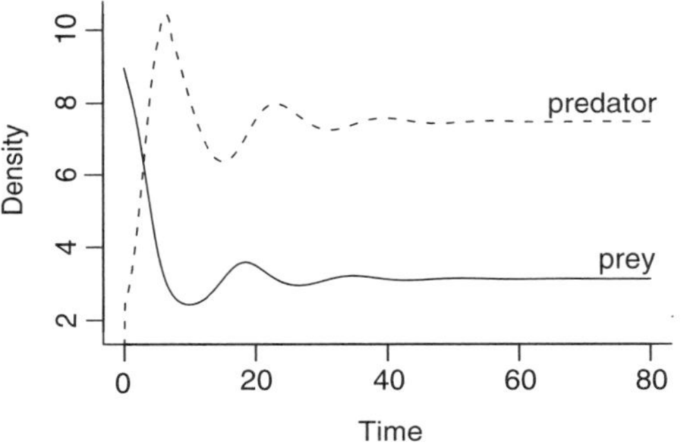

(a) Parasitoid and prey mean densities as a function of time.

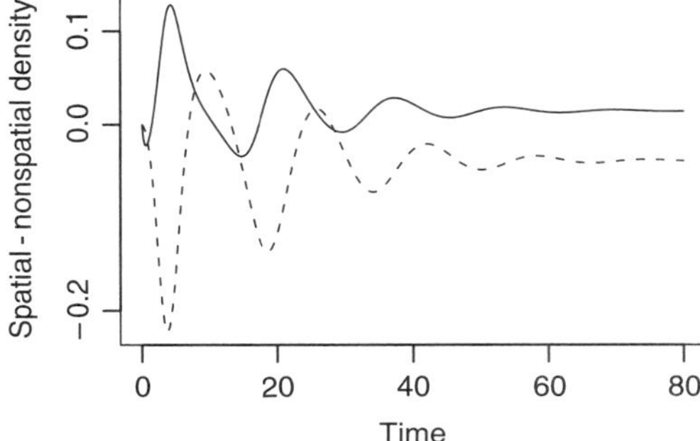

(b) Difference between mean densities in spatial and non-spatial models.

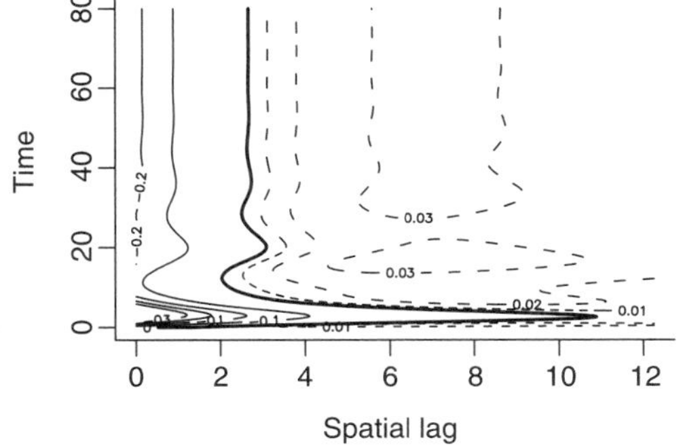

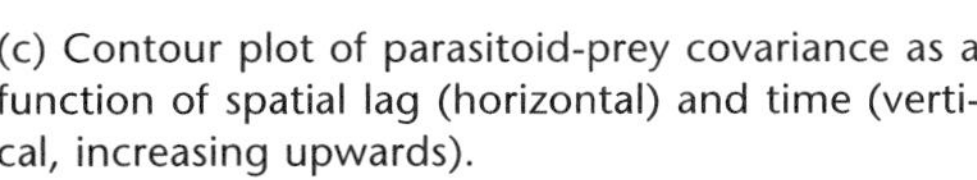

(c) Contour plot of parasitoid-prey covariance as a function of spatial lag (horizontal) and time (vertical, increasing upwards).

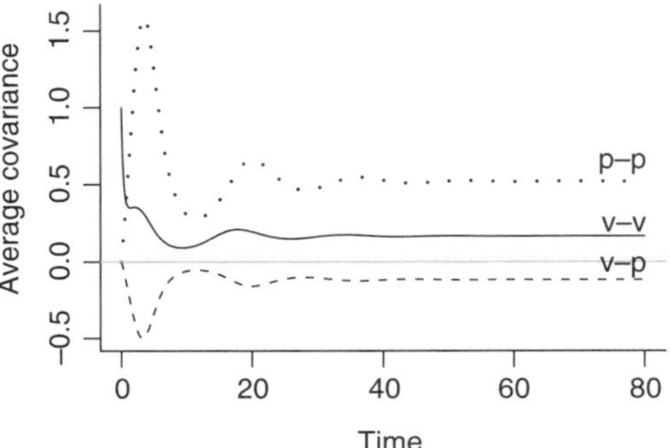

(d) Average covariances ($\bar{c}_{VV} = \int \mathcal{U}_{VV} c_{VV}\, dr$, $\bar{c}_{VP} = \int \mathcal{U}_{PV} c_{VP}\, dr$, $\bar{c}_{PP} = \int \mathcal{U}_{PV} c_{PP}\, dr$) as a function of time.

Fig. 3.2 Numerical solutions of parasitoid–prey moment equations. Parameters (nonspatial): $f_V = 2.0$, $fv = 1.0$, $\alpha = 0.1$, $\mu_P = 0.2$, $\gamma = 0.1$. Parasitoid movement rate $m_p = 1.0$. All kernels exponential (symmetric exponential), $\mathcal{K} = 2/m \exp(-|r|/m)$, with prey dispersal and competition scale $m_{D_V} = m_{\mathcal{U}_{VV}} = 0.2$, parasitoid dispersal and parasitism scale $m_{\mathcal{U}_{PV}} = m_{\mathcal{D}_P} = 1.0$. Solutions by the Euler method with total spatial length 25, $\Delta x = 0.25$, total time = 80, $\Delta t = 0.025$.

extended stochastic system, they should always converge to an equilibrium: without global interactions, there is no way that parasitoid–prey populations can stay in synchrony across an infinite landscape.) For the parameters chosen in Fig. 3.2, the dynamics of the mean densities in the spatial model (Fig. 3.2a) are nearly identical to the dynamics obtained from the analogous nonspatial model (Fig. 3.2b). However, the covariances (Fig. 3.2c and 3.2d) show interesting patterns: the parasitoid–prey covariance at short spatial lags quickly becomes negative as parasitoids deplete the prey in their immediate neighborhood, driving the average parasitoid prey covariance negative (Fig. 3.2d). This zone of depletion grows rapidly up to about $t = 5$ with a maximum correlation length of 12 units and then shrinks back to a length scale of approximately 3 units. The average covariance between parasitoids and prey ($\bar{c}_{VP} = \int \mathcal{U}_{VP}(r) c_{VP}(r) dr$), which describes the net effect of spatial segregation on the parasitism rate, is always negative (Fig. 3.2d): the spatial structure always reduces the effective parasitism rate. The average intraspecific covariance of prey($\bar{c}_{VV} = \int \mathcal{U}_{VV}(r)\ c_{VV}(r)\ dr$) is always positive — prey experience intraspecific crowding that raises their effective density — but drops as parasitoids reduce the prey density. Parasitoids also experience intraspecific aggregation ($\bar{c}_{PP} = \int \mathcal{U}_{VP}(r)\ c_{PP}(r)\ dr > 0$), but this aggregation has no direct effect on the dynamics of the mean densities.

What about the effects of (for example) changing parasitoid scales? Figure 3.3 compares the results of simulations with numerical integration of the moment equations for a broad range of parasitoid scales (parasitoid dispersal scale and parasitism scale were set equal in all cases). For large parasitoid scales, the moment equations are relatively accurate; the prey density converges on the expected density in the nonspatial case; and the parasitoid density is slightly below the nonspatial density. In this case, the only effect of space is to enhance intraspecific competition among the prey. As in many simple parasitoid–prey models, the lowered productivity of the prey base is reflected in a lowered density of the parasitoids rather than in the prey density itself.

As the parasitoid scale decreases, the intraspecific aggregation of the prey increases (from $\bar{c}_{VV} \approx 0.2$ to ≈ 1). The much larger effect, however, is the large increase in magnitude of the spatial segregation between parasitoids and prey ($\bar{c}_{V\,P}$). In effect, this spatial segregation lowers the efficiency of the parasitoids by making it harder for them to find prey nearby. The simultaneous increase in both parasitoid and prey densities with increasing spatial segregation may seem surprising. However, we can understand it by seeing that in the nonspatial analogue of the current model, decreasing parasitoid effectiveness increases both parasitoid and prey density; limiting parasitoid effectiveness increases the productivity of the prey base, which counterintuitively feeds through to increase the parasitoid density as well. (If we were modeling the evolutionary dynamics of this system, we would probably see that the evolutionary stable strategy for parasitoids of unlimited movement and interaction lowers the average density of the population.) By looking at Eq. (3.2), we see that we can encapsulate all of the effects of spatial structure by rescaling the parasitoid effectiveness to $\gamma' = \gamma\,[1 - \bar{c}_{VP}/(vp)]$ and the prey competition to $\alpha' = \alpha\,[1 - \bar{c}_{VV}/(v^2)]$; in general, $\gamma' < \gamma$ and $\alpha' > \alpha$. The mean equations tell us that parasitoid density will begin to drop if γ' falls below a critical value $(f_V - \mu_V)/(2\alpha\mu_P)$, but for the

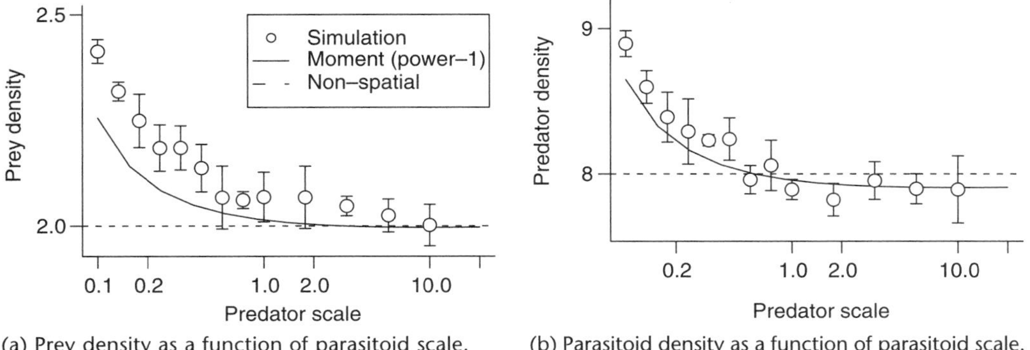

(a) Prey density as a function of parasitoid scale.

(b) Parasitoid density as a function of parasitoid scale.

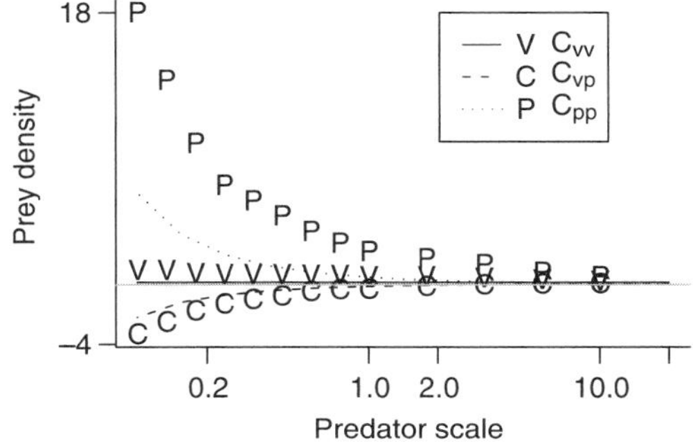

(c) Average covariance as a function of parasitoid scale. Letters show simulation results, lines show moment equation results.

Fig. 3.3 Comparisons of simulations with numerical integration of moment equations. Parameters as in Fig. 3.2, but with varying (equal) scales of parasitism ($m_{\mathcal{U}_{PV}}$) and parasitoid dispersal ($m_{\mathcal{D}_P}$).

parameter values given earlier that would require $\gamma' < 0.4\gamma$ — the effects of spatial structure seen here decrease γ' only to a minimum of 0.84γ.

How accurate are the moment equations at capturing the observed changes in the covariances? In this case, the simple power-1 closure that we have captures the qualitative changes in the means and covariances, although it underpredicts all of the effects of spatial structure. It does the worst job predicting parasitoid clustering (luckily, parasitoid–parasitoid covariance is the one covariance term that has no direct effects on mean density). Other symmetric closures (power-2 and power-3; not shown) give very similar results to the power-1 closure, but an asymmetric power-2 closure fails badly, predicting zero parasitoid–prey covariance. In general, we expect the moment equations to work well (but spatial structure to be irrelevant) for large scales of movement or interaction and to fail at extremely short scales. We can define a *neighborhood size* based on the kernel shape and the scale parameter (Bolker, 1999). For the exponential kernel used here, the neighborhood size equals $4m$. In the past, we have found that moment equations give accurate predictions of mean densities when individuals average $\approx$ 10–100 discrete neighbors within their interaction neighborhood (Bolker and Pacala, 1997), although qualitative results may apply for lower numbers.

Predictions of prey density are reasonably accurate (Fig. 3b) for scales in the range 0.5–2. Because prey density is ≈ 2, this range of scales corresponds to a neighborhood population of 4 to 16, slightly below the expected range. Similarly, for parasitoid density scales between 0.1 and 1.0 work, which for a density of ≈ 8 gives a range of 3.2 to 32 neighbors.

A full analysis of this system would explore the effects of other changes in relative interaction scales; for example, does changing parasitoid movement and parasitism scales independently have any interesting effects? [If one sets all scales (prey dispersal, parasitoid movement, parasitism) equal and varies them simultaneously, the effect of prey aggregation dominates by lowering prey productivity and hence lowering equilibrium parasitoid densities.] One could also look for simplified cases where some analysis is possible; as shown in Bolker and Pacala (1999), one can find the equilibria of moment equations analytically if the problem is restricted to one or two different scales or if some processes are given infinite (global) scales. In the parasitoid–prey case, one might get quite simple results if prey competition were global and parasitoids did not move, being able only to eat prey that settled within their parasitism neighborhood. Ultimately, strategies for simplifying and analyzing the equations, and connecting the results to the more general results from numerical solutions and simulations, depend on the ecological questions of interest.

3.5 APPLICATIONS

Spatial moment equations have yet to see widespread use in ecology, although a few groups of researchers have used them to answer a variety of questions in spatial ecology. This section reviews briefly the (small) literature of applications of spatial moment equations, making connections to nonspatial or noncontinuous models where appropriate.

Single-Species Dynamics

The simplest possible application of moment equations is to populations of a single species; this system is called the *spatial logistic* model and is analogous to both the classical nonspatial logistic equation and the contact process on the lattice (Law et al., 2003). The model defines the fecundity and density-independent mortality rates of individuals, the strength of density dependence, and the kernels for dispersal and competition. Competition can lower the fecundity of parents with increasing neighborhood density; lower establishment probability of offspring with increasing density of the neighborhood where they land after dispersing; or increase mortality probability of adults. The results are simple: spatial structure can be either aggregated (positive average covariance) or even (negative average covariance). Aggregation occurs when individuals have short dispersal scales relative to competition scales and low intrinsic reproductive numbers (fecundity divided by density-independent mortality or expected lifetime reproduction in the absence of competition). The overall strength of spatial effects is largest when neighborhood size, the number of neighbors with which an individual interacts, is small: the neighborhood size is proportional to population density (determined approximately by the nonspatial carrying capacity). In the absence of interspecific competition or environmental heterogeneity, the intrinsic rate of reproduction (fecundity minus density-independent mortality) and the spatial scale of dispersal merely set the temporal and spatial scales of the population without having any qualitative effect on population dynamics (Bolker and Pacala, 1997).

Heterogeneous environments widen the range of possibilities. Heterogeneity could affect fecundity, mortality, competition, or the probability of dispersal (or the scale of dispersal and competition, although this would lead to some thorny technical problems). Models with heterogeneous fecundity and mortality connect directly with applied questions of population viability in degraded and fragmented landscapes, as well as with basic questions about the evolution of dispersal (Rousset and Gandon, 2002). Models with variation in dispersal are connected more closely to classical questions about foraging and aggregation in heterogeneous landscapes, but are also relevant to conservation. Either heterogeneous mortality or heterogeneous movement probabilities can lead to individuals aggregating in a good (low-mortality, low-movement) habitat (Murrell and Law, 2000; Bolker, 2003). This habitat association can shield individuals from the worst effects of habitat degradation, but with sufficiently short dispersal scales the negative effects of intraspecific competition overwhelm the advantages of habitat association [and the moment equations break down; they do not capture the percolation phenomena that can lead to sudden extinction in some continuous-space models (Bascompte and Solé, 1997)].

Finally, one can use single-species moment equations to predict invasions and to begin to understand the feedbacks among environmental heterogeneity, endogenous heterogeneity, and invasion speed at the edge of a spreading wave. Various pieces of the puzzle are present, but they have not been put together into a single picture. The only work with spatial moment equations analyzes the patchiness of invading populations and shows that patchiness and local crowding slow invasion speeds below what would be expected from a continuous population (Lewis, 2000). Other work has been done with lattice models (Ellner et al., 1998) and diffusion equations (Shigesada and Kawasaki, 1997) to explore the effects of endogenous and exogenous variability, respectively, but these phenomena have not been fully integrated into the elegant literature on dispersal and invasion speeds (Skellam, 1951; Mollison, 1991; Clark et al., 1998).

Interspecific Competition

Spatial dynamics and spatial models have been explored extensively in research on competitive coexistence, particularly in plant communities (Tilman, 1994). The natural history of weedy and early successional species clearly suggests that variations in spatial strategies could maintain diversity in environments where there is insufficient variation in resource availability for niche separation to maintain diversity (Tilman and Pacala, 1993). Spatial coexistence can be interpreted as the exploitation of endogenous and exogenous spatial covariance in resource availability. The basic mechanisms of spatial coexistence have been explored in metapopulation (Levins and Culver, 1971; Tilman, 1994; Pacala and Rees, 1998; Chesson, 2000a), lattice (Silvertown et al., 1992; Holmes and Wilson, 1998), and continuous-space models (Gopasalmy, 1977a,b); more recent work has applied the tools of pair (Harada and Iwasa, 1994) and spatial moment equations (Bolker and Pacala, 1999; Bolker et al., 2000; Dieckmann and Law, 2000; Bolker et al., 2003). Chesson (2000a,b) has developed an analytic framework to quantify and test the processes by which coexisting species partition exogenous temporal and spatial variability. His framework largely neglects the dynamics of endogenous variation, although it has recently begun to incorporate continuous space; the results on equilibrium correlations developed by Snyder and Chesson (2003) correspond to the results one would get by deriving moment equations but neglecting the singular terms in the covariances. Finally, the evolutionary dynamics of dispersal and spatial coexistence remain largely unexplored (Hovestadt et al., 2001; Rousset and Gandon, 2002). A full accounting of spatial coexistence will have to incorporate both endogenous and exogenous variation and to recognize that spatial traits such as dispersal represent just a subset of all plant competitive and life history traits and that the coevolution of these traits needs to be considered as a whole (Rees, 1996).

Predator–Prey

The spatial dynamics of predator–prey interactions are currently an area of active research, with many theoreticians trying to understand whether differences between spatial and nonspatial models in stability and equilibrium densities are caused by diffusion limitation or other spatial modifications of the predation process (McCauley et al., 1993; Cuddington and Yodzis, 2002; Hosseini, 2003). Keeling et al. (2002) used moment equations (for a metapopulation system) to understand the broad conditions under which spatial structure stabilizes or destabilizes a predator–prey model.

The usual approaches to understanding predator–prey and other complex ecologies in continuous space make use of simulation models and a variety of scaling rules (Rand and Wilson, 1995; Pascual and Levin, 1999) to identify the relevant scales at which these ecologies are best understood. Looking at the problem from a very different perspective, spatial moment equations have a different set of advantages and disadvantages. Moment equations typically look at the average densities and covariance densities of an entire system and, as a result, have some difficulty capturing local stability or instability. Capturing the effects of finite system size, which might be another way to address the problem, is similarly difficult because of the algebraic complexity of moment equations with spatial inhomogeneity. In addition, some of the nonlinearities and resulting coherent structures (e.g., spirals) that arise in typical predator–prey models may be poorly represented by second-order moment closures (Tainaka, 1994).

However, spatial moment equations also offer unique advantages. Other approaches have explained the changes in predator–prey dynamics from nonspatial expectations as a result of endogenous patchiness, but have not tried to predict the endogenous patchiness itself. By calculating the shape and scale of equilibrium covariances within and between species, we can understand how different spatial and demographic parameters determine the endogenous structure of predator–prey systems in continuous space and, by extension, the spatial dynamics of predator–prey systems (M. Desai, unpublished manuscript). Furthermore, spatial moment equations can naturally be extended to include exogenous variation in predation risk in a more general form than the patchy structure implied by metapopulation models.

Epidemics

Epidemics are a subtype of a predator–prey system that is of obvious practical and theoretical interest. Spatial models of epidemics go back to the pandemic models of Kermack and McKendrick (1927); since then, epidemics have been studied in metapopulation, lattice, and continuum models. Stochastic spatial models have always presented a challenge, however, and until recently, computational simulation models have been the main method of exploration (Duryea et al., 1999). Moment equations and their discrete pair approximation analogues have proved to be powerful tools for analyzing spatial epidemic models (Satō et al., 1994; Keeling et al., 1997; Rand, 1999; Keeling, 1999a; Boots and Sasaki, 2000, 2002). Many of the analyses have used social contact networks as the underlying spatial structure, which is probably more appropriate for human diseases than a two-dimensional plane. For diseases of plants and wildlife, however, lattice or continuum models may be best. Some of the remaining challenges in using spatial moment equations for epidemiology are adapting some of the sophisticated lattice closures developed by Filipe and Gibson (2001) to the continuous-space case; incorporating exogenous spatial heterogeneity in suceptibility and infectivity; and combining disease transmission with other ecological processes such as competition to explore the effects of disease in a broader ecological setting.

3.6 OUTLOOK

Continuous-space models represent a different approach to spatial ecology than the metapopulation models discussed throughout most of this book; they are relevant for populations living in different habitats from the spatially disconnected, temporally persistent patches that are the main concern of metapopulation ecology. This chapter has reviewed continuous-space models with a strong focus on spatial moment equations, a relatively new analytical framework for reducing simple individual-based models to a set of integral equations for the densities and spatial covariances of interacting populations.

Why would one choose continuous-space models (continuum, lattice, or spatial IBM) over patch occupancy or metapopulation models? There are two basic reasons: first, one might be studying an organism or community that inhabits a continuous habitat, where the patchy-habitat assumption is simply unrealistic. Second, one might be asking explicit questions about the scale and pattern of spatial structure that have no meaning in a discontinuous spatial model. Tactical questions and landscapes incorporating specific features such as barriers and complex boundaries between habitat types would favor spatial IBMs, whereas strategic questions and simple landscapes with more continuous variation would favor lattices or continuum models.

Given a continuous-space model, how does one choose among the different options? As described earlier, spatial IBMs are flexible and realistic but hard to generalize, whereas lattice and continuum models are more general and tractable but make different kinds of unrealistic assumptions. Lattice models assume a particular, unrealistic spatial structure that complicates the connection with measured, individual-level data [except in cases where data are gathered at the level of lattice cells; Silvertown et al. (1994)]. Continuum models eliminate demographic stochasticity (although it can be reintroduced by adding noise with the appropriate scaling properties) and tend to reduce the complexity of individual movement rules to diffusion and advection terms. Nevertheless, both of these model types can be extremely useful. Both can be solved numerically by standard packages (lattice models are much simpler computationally) and both are some what tractable analytically, although most nonlinear spatial models of any type can only be analyzed approximately or for special cases such as invasion conditions.

The spatial moment equations emphasized in this chapter represent a different, relatively new set of compromises. Spatial moment equations have two major disadvantages. First, because they depend on a heuristic approximation of the full spatial configuration of the system, they may fail under certain conditions — at present, one must always verify the results of spatial moment equations with simulations for at least a few sets of parameters. Second, the detailed derivation and analysis of moment equations can be daunting for the average ecologist or evolutionary biologist, although some automated tools have been developed to help the process. Spatial moment equations do have strong counterbalancing advantages, however. Because their parameters are expressed directly in terms of individual behavior, moment equations are useful for studying the connections between individual behavior and spatial dynamics. Because their state variables (spatial correlations) preserve information about the shape and scale of spatial pattern, moment equations are useful for understanding the causes and effects of spatial pattern. Spatial moment equations can preserve some aspects of demographic stochasticity and of landscape heterogeneity, although they cannot typically handle movement orientation by individuals or landscape structures, such as edges and barriers, and they have less flexibility than spatial IBMs for incorporating details of individual state; at present, these state dynamics can only be included by adding new types to the model (Fig. 3.1). Spatial moment equations can be analyzed, although often only by simplifying the models considerably. Nevertheless, spatial moment equations can form a bridge between complex spatial IBMs and simple, analytically tractable models.

Like metapopulation models, moment equation approaches can be used to explore a broad range of ecological interactions. Also, like metapopulation models, moment equations work best in communities with a particular kind of underlying spatial geometry. In the case of moment equations, these are communities where both population densities and environmental characteristics change relatively smoothly; where the fate of an individual is predictable (at least on average) from the characteristics of its neighborhood; and where that neighborhood is large enough to encompass at least a dozen or so interacting organisms. Moment equations can predict the shapes and scales of population patterns generated by the interaction between environmental variability and endogenous population interactions, and these predictions can sometimes be handled analytically. Equally important, moment equations have a natural connection with data that are collected at an individual scale, with individual-based computational models, and ultimately with the life and death of individuals in ecological communities.

Part II

METAPOPULATION ECOLOGY

4. METAPOPULATION DYNAMICS IN HIGHLY FRAGMENTED LANDSCAPES

Otso Ovaskainen and **Ilkka Hanski**

4.1 INTRODUCTION

The ecological dynamics of metapopulations have many facets, which can hardly be all studied in any single investigation, nor analyzed with a single metapopulation model. Ecologists have constructed particular models to investigate source–sink dynamics (Pulliam, 1988; Paradis, 1995; Smith et al., 1996; Walters, 2001; Hels, 2002; Chapter 16), the influence of immigration and emigration on the type of local population dynamics (Gyllenberg et al., 1993; Rohani et al., 1996; Ruxton et al., 1997; Saether et al., 1999; Nachman, 2000), models of habitat selection influencing metapopulation dynamics (Ray et al., 1991; Doncaster, 2000; Etienne, 2000; Danchin et al., 2001), and so forth. Models of individual movement behavior in heterogeneous landscapes (Hanski et al., 2000; Ricketts, 2001; Gobeil and Villard, 2002; Ovaskainen and Cornell, 2003) provide building blocks for metapopulation models, and single-species metapopulation models in turn can be extended to model the dynamics of metacommunities (Wilson, 1992; Holyoak, 2000; Klausmeier, 2001; Holt, 2002). The purpose of the models ranges from attempts to clarify general principles, as described by Bolker in

0-12-323448-4

Chapter 3 for spatial continuum models, to spatially extended population viability analysis (Sjögren-Gulve and Hanski, 2000; Akcakaya, 2000; Lindenmayer et al., 2001; Ferreras et al., 2001).

This chapter focuses on models that are in the hard core of ecological metapopulation theory: stochastic patch occupancy models (SPOM) and their deterministic approximations, including the mother of all ecological metapopulation models, the Levins model. There are several reasons for devoting an entire chapter to these models. The classic metapopulation idea of a "population" of local populations is captured, in its bare essence, by SPOMs. In a broader biological perspective, SPOMs belong to "island models," which have played an important role not only in population and community ecology (MacArthur and Wilson, 1963, 1967), but also in population genetics (Wright, 1931; Slatkin, 1977; Wade and McCauley, 1988; Whitlock and McCauley, 1990) and evolutionary studies (Wright, 1931, 1940). These models often allow a rigorous mathematical analysis, an obvious advantage for theory. However, there is yet another reason for focusing on SPOMs in this chapter, a reason that complements the advantages of the models for theoreticians: SPOMs are good models for real metapopulations living in highly fragmented landscapes, to the extent that they can be parameterized with empirical data (Chapter 5) and turned into tools that hold substantial promise for conservation, landscape planning, and management.

Figure 4.1 gives an example of a highly fragmented landscape, in fact two representations of it. The map on the left shows the 56 habitat patches (dry meadows) suitable for reproduction by the Glanville fritillary butterfly (*Melitaea cinxia*) in one part of the Åland Islands in Southwest Finland (Hanski, 1999b). The long-term project on the Glanville fritillary has played an

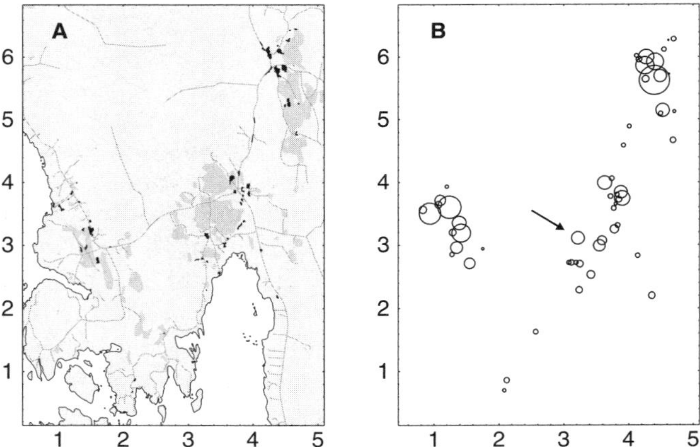

Fig. 4.1 A network of 56 habitat patches used as an example throughout this chapter. This patch network is part of a larger network of habitat patches inhabited by the Glanville fritillary butterfly (*Melitaea cinxia*) in the Åland Islands in Southwest Finland (Hanski, 1999b; Nieminen et al., 2004). Panel A shows a map of the real landscape, whereas panel B shows the simplified view (network of circular habitat patches) assumed by the spatially realistic metapopulation theory. The areas of the circles in B are proportional to the sizes of the habitat patches, but they have not been drawn to scale. The patch indicated by an arrow is analyzed in Fig. 4.7.

important role in stimulating research on SPOMs, and we will use the results of our empirical studies of this butterfly as a focal example in this chapter. The rest of the landscape in Fig. 4.1 mostly consists of cultivated fields and forests. The characteristic feature of "highly fragmented landscapes" is that the pooled area of suitable habitat for the species of interest is limited to say less than 10% of the total area of the landscape, and the habitat occurs in discrete patches, none of which is very large. The map on the right (Fig. 4.1) shows the simplified view of the same patch network in which we only distinguish between the breeding habitat and the remaining matrix and in which the actual shapes of the patches are approximated by circles.

How common are "highly fragmented landscapes?" Nobody knows the answer for a comprehensive sample of species, but there are reasons to suspect that a large fraction of invertebrates and increasing numbers of other species occur in highly fragmented landscapes. Community ecologists have documented repeatedly that most species in most communities are locally rare (Whittaker, 1975; May, 1975; Gaston, 1994). While there are many reasons for local rarity, one likely reason is a limited amount of habitat. Furthermore, locally rare species tend to be rare also in the surrounding landscape (Hanski, 1998b), suggesting that for such species the landscape is often highly fragmented. For the Glanville fritillary in the Åland Islands it has been relatively straightforward to document the degree of fragmentation empirically — ca. 4000 fragments covering 0.6% of the total landscape (Nieminen et al., 2004) — but for most other species it is not. Murphy et al. (1990) have argued that species with small body sizes, high rates of population increase, short generation times, and specific resource requirements are predisposed to have a metapopulation structure. A large fraction of insects possess these attributes. There are also very large numbers of species that live in discrete "minor habitats" (Elton, 1949), such as decaying tree trunks, which are not permanent but change in time. Furthermore, large numbers of parasites inhabit a highly fragmented landscape once we realize that, from their perspective, host individuals are habitat patches (Grenfell and Harwood, 1997). The metapopulation theory developed here applies also to these systems, with due attention to the dynamics of the landscape. Finally, as we all know, the expanding human enterprise typically leads to the loss and fragmentation of many habitat types in most landscapes, increasing the frequency of species whose habitat is highly fragmented.

Stochastic Patch Occupancy Models

Stochastic patch occupancy models are based on two major simplifying assumptions. The first assumption relates to the structure of the landscape, which is assumed to consist of discrete patches of breeding habitat surrounded by the matrix as shown in Fig. 4.1 — the landscape is highly fragmented. The second major assumption concerns the description of population dynamics; SPOMs recognize only two possible states for each habitat patch, which can be either occupied by the focal species or unoccupied. Thus the sizes and structures of local populations are not explicitly accounted for. While these simplifications restrict the range of problems to which SPOMs may be applied, they make the models both tractable for rigorous mathematical analysis

(Ovaskainen and Hanski, 2001, 2002, 2003a; Ovaskainen, 2003) and turn them into effective tools that are used increasingly in empirical studies (Moilanen, 1999, 2000; Hanski, 1999b; Hanski and Ovaskainen, 2000; Ter Braak and Etienne, 2003; Chapter 5).

The original "island model" approach to modeling spatially structured dynamics in population biology assumes that all the patches are identical. In the case of SPOMs, this assumption also means that all the patches and all the local populations occupying these patches are equally connected to each other. We call models making this assumption *homogeneous* SPOMs, which are described in Section 4.2. Assuming identical patches obviously simplifies the models greatly, which has been the overriding justification for this assumption. One might also ask what could be gained by assuming spatial variation in the properties of the patches — would such more complex models provide any novel insight to the processes that SPOMs are used to study? The answer depends on what our aim is. The greatest advantage of employing *heterogeneous* SPOMs with variation in patch properties stems from the opportunity to meaningfully apply them to real metapopulations living in highly fragmented landscapes, either for the purposes of management and conservation or for the purpose of critically testing model predictions for advancing scientific understanding. A core idea in this research is to make structural model assumptions about how the key landscape features, such as habitat patch areas, qualities, and spatial locations, will influence the two key processes of classic metapopulation dynamics; local extinction and recolonization. We have called the combination of heterogeneous SPOMs (assuming dissimilar patches and hence patch-specific transition probabilities) with these structural model assumptions as the *spatially realistic metapopulation theory* (SMT; Hanski, 2001b; Hanski and Ovaskainen, 2003). The assumptions relating population processes to landscape structure are discussed in Section 4.3, where we also describe the two most-studied models based on this theory: the spatially realistic Levins model and the incidence function model.

The stochastic theory of heterogenous SPOMs poses mathematical problems, which relate to the fact that a metapopulation living in a large patch network with n patches and described by the presence or absence of the species in these patches has a huge number of possible states, 2^n. We describe two approaches to overcome these problems. First, for many purposes it is sufficient to use a deterministic approximation of the stochastic model; the deterministic theory is described in Section 4.4. Second, Ovaskainen (2002a) has described an effective approach for analyzing the stochastic model for a heterogeneous patch network with the help of an appropriate stochastic model for a homogeneous network, which latter model can be analyzed mathematically. This approach, described in Section 4.5, is akin to the idea of effective population size in population genetics: construct an unstructured model that behaves similarly to a structured model with respect to some model features of interest. Following the description of deterministic and stochastic theories of SPOMs, Section 4.6. presents some comparisons with other approaches to modeling metapopulation dynamics, particularly focused on one of the key issues in ecological metapopulation dynamics, the extinction threshold for long-term persistence. Section 4.7 discusses the current state of the theory and its applications, and Section 4.8 outlines the broader significance of the spatially realistic metapopulation theory, emphasizing the contributions that it makes toward greater unification of research in population biology.

This chapter is mostly about theory and models, although our purpose is not just to describe a mathematically rigorous body of theory but also to extract biologically relevant messages that stem from SMT. We illustrate many of the results with examples based on the habitat patch network for the Glanville fritillary butterfly depicted in Fig. 4.1. We will not review the literature on the application of the models to real metapopulations, which is covered in Chapter 5 (for more ecologically oriented reviews, see Hanski 1999b, 2001b). The kind of models examined here are employed in some of the analyses presented in Chapters 20–22 in this volume.

4.2 PATCH OCCUPANCY MODELS: HOMOGENEOUS PATCH NETWORKS

Mathematically, SPOMs are defined as Markov chains (discrete-time models) or Markov processes (continuous-time models). However, most of the existing theory does not relate to the full stochastic models, but to their deterministic approximations, which, mathematically speaking, account only for the drift in the dynamics while ignoring the variance due to stochastic fluctuations. This section considers models that assume identical habitat patches. We start with the familiar Levins model, which is helpful in highlighting the basic concepts and qualitative theory behind SPOMs.

Levins Metapopulation Model

Levins (1969, 1970) assumed that the number of habitat patches in a patch network is infinite, which allowed him to formulate a patch occupancy model directly from the deterministic viewpoint. However, we will view here the Levins model as the deterministic mean-field approximation of the stochastic logistic model (Box 4.1), which assumes a finite network of n identical and equally connected patches, each one of which may be either occupied or empty. If a patch is occupied, it is assumed to go extinct at a fixed rate $E = e$. The colonization rate of an empty patch is assumed to depend on the fraction of occupied patches as $C = ck/n$, where c is a colonization rate parameter and $0 \leq k \leq n$ is the number of occupied patches. The reasoning behind this expression for the colonization rate is that k occupied patches are assumed to produce emigrants that disperse randomly to any of the habitat patches (both occupied and empty). As described in Box 4.1, these transition rates from empty to occupied patches and vice versa define the stochastic logistic model as a Markov process.

If the number of patches is large, the stochastic logistic model may be approximated by a diffusion process. To do this, we denote by $p(t) = k(t)/n$ the fraction of occupied patches at time t. The diffusion process is determined by the mean (or drift) $\mu(p) = E[dp]/dt$ and the variance $\sigma^2(p) = E[dp^2]/dt$ of the infinitesimal rate of change (Karlin and Taylor, 1981), which are given,

BOX 4.1 The Stochastic Logistic Model

The stochastic logistic model has been studied in a variety of contexts, including population biology, chemistry, and sociology, and the dynamic behavior of the model is well understood (Kryscio and Lefèvre, 1989; Jacquez and Simon, 1993; Ovaskainen, 2001). We will describe the model here as a SPOM describing the dynamics of a species inhabiting a network of n identical habitat patches.

The stochastic logistic model is defined as a Markov process, which assumes that an occupied patch turns empty at a fixed rate $E = e$ and that an empty patch turns occupied at a rate $C = ck/n$, where c is a colonization rate parameter and $0 \leq k \leq n$ is the number of occupied patches. To put the model into a mathematical setting, $k(t)$ is the random variable giving the number of occupied patches at time t, and $\mathbf{q}(t)$ is the probability distribution describing the state of the metapopulation, with the component $q_i(t)$ being the probability that the system is in state $k(t) = i$ $(i = 0, \ldots, n)$ at time t. By standard theory of Markov processes (e.g., Grimmet and Stirzaker, 2001), the probability distribution $\mathbf{q}$ evolves according to the forward equation $d\mathbf{q}(t)/dt = \mathbf{q}(t)\mathbf{P}$, where $\mathbf{P}$ is the $(n + 1)^*(n + 1)$ matrix

$$\mathbf{P} = \begin{pmatrix} -d_0 & c_0 & 0 & \cdots & 0 \\ e_1 & -d_1 & c_1 & \cdots & 0 \\ 0 & e_2 & -d_2 & \cdots & 0 \\ \cdots & \cdots & \cdots & \cdots & \cdots \\ 0 & 0 & 0 & \cdots & -d_n \end{pmatrix} \quad (1)$$

In Eq. (1), the diagonal elements are defined as $d_i = e_i + c_i$, and $c_i = ci(n - i)/n$ and $e_i = ei$ give total (instead of patch-specific) colonization and extinction rates, assuming that presently i of the n patches are occupied.

In the stochastic logistic model, eventual metapopulation extinction is certain, and the biologically most fundamental quantity that may be derived from the model is the time that the metapopulation is expected to persist. Assuming that the initial state of the metapopulation is drawn from the quasistationary distribution (Box 4.2), the mean time to extinction may be approximated by (Andersson and Djehiche, 1998; Ovaskainen, 2001)

$$T = \frac{1}{e}\sqrt{\frac{2\pi}{n}}\frac{\exp(-np^*)}{(1 - p^*)^{n-1}p^{*2}}, \quad (2)$$

where p^* is the deterministic equilibrium value for the fraction of occupied patches given by $p^* = 1 - e/c$. Equation (2) shows that the time to extinction increases exponentially with the number of habitat patches (assuming that p^* is kept fixed). Before extinction, the process approaches a quasistationary distribution π (Box 4.2), which may be approximated by a normal distribution with mean p^* and variance $(1-p^*)/n$ (Ovaskainen, 2001).

for the stochastic logistic model, as (Saether et al., 1999; Ovaskainen, 2002a)

$$\mu(p) = cp(1 - p) - ep, \quad (4.1)$$

$$\sigma^2(p) = \frac{cp(1 - p) + ep}{n}. \quad (4.2)$$

The Levins model is obtained from the diffusion approximation by assuming that the number of patches is so large that stochastic fluctuations may be ignored. In this case, the model reduces to the familiar deterministic differential equation

$$\frac{dp}{dt} = \mu(p) = cp(1 - p) - ep. \tag{4.3}$$

Figure 4.2 compares the behavior of the stochastic logistic model with the deterministic Levins model. Realizations of the stochastic model fluctuate around the deterministic prediction, and the deterministic equilibrium state p^* corresponds to the peak in the quasistationary distribution of the stochastic model (for the definition of quasistationarity see Box 4.2). A more throughout analysis of the relationship between the deterministic Levins model and the stochastic logistic model has been presented by Ovaskainen (2001), Etienne (2002b), and Etienne and Nagelkerke (2002).

Lande (1987, 1988b) extended the Levins model to account for habitat loss by assuming that a fraction $1 - h$ of the habitat patches are permanently destroyed, and thus only the fraction h remains suitable for colonization. We assume that the migrating individuals are unable to discriminate between suitable and unsuitable patches, hence they attempt to colonize the latter in proportion to their number. The colonization rate of empty patches $(1 - p)$ is thereby reduced to cph, and the model is given as

$$\frac{dp}{dt} = cph(1 - p) - ep, \tag{4.4}$$

where p is the fraction of occupied patches out of the suitable ones. The key prediction made by Lande's model is that the metapopulation will persist (there is a nontrivial equilibrium state) if and only if the threshold condition

$$h > \frac{e}{c} \tag{4.5}$$

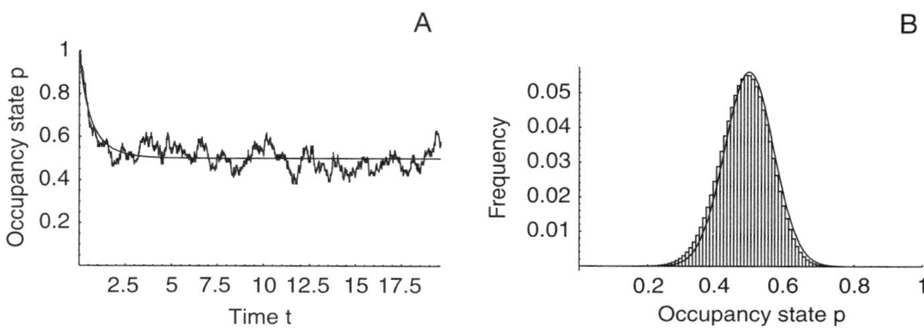

Fig. 4.2 Comparison between the stochastic logistic model and the Levins model. (A) A single simulation realization of the stochastic logistic model compared to the deterministic prediction of the Levins model. (B) The exact quasistationary distribution π and its normal distribution [mean p^*, variance $(1 - p^*)/n$] approximation. Parameter values $e = 1$, $c = 2$, $n = 100$.

BOX 4.2 Quasistationary Distributions and Extinction Times

Mathematically, SPOMs are defined as Markov chains or Markov processes, for which a large amount of general theory is available (e.g., Grimmet and Stirzaker, 2001). We describe briefly here part of the theory that is relevant in the analysis of SPOMS.

As described in Section 4.3, the state of a general heterogeneous SPOM is specified by the random variable $\mathbf{O} = \{O_i\}_{i=1}^n$, where the component O_i specifies whether patch i is empty ($O_i = 0$) or occupied ($O_i = 1$). We assume that these 2^n states are ordered in some manner with metapopulation extinction ($O_i = 0$ for all i) being the first state. Let $1 \leq Q(t) \leq 2^n$ denote the random variable describing the state of the metapopulation at time t and let $\mathbf{q}(t) = \{q_i(t)\}_{i=1}^{2^n}$ denote the probability distribution defined by $q_i(t) = \text{P}[Q(t) = i]$. By standard theory, the probability distribution $\mathbf{q}$ evolves according to the forward equation either as $d\mathbf{q}(t)/dt = \mathbf{q}(t)\mathbf{P}$ (for Markov processes) or as $\mathbf{q}(t+1) = \mathbf{q}(t)\mathbf{P}$ (for Markov chains). The $2^n * 2^n$ matrix $\mathbf{P}$ is called the generator matrix (Markov processes) or the transition matrix (Markov chains) and is composed of transition rates (or probabilities) from any occupancy state to any other occupancy state (Grimmet and Stirzaker, 2001).

If a metapopulation happens to go extinct, there are no occupied patches to produce migrants, and thus empty patches cannot be recolonized. Mathematically, metapopulation extinction is an absorbing state, which the process will eventually reach with probability 1, and thus the stationary distribution of the Markov process (or chain) is concentrated on q_1. Although the stationary distribution is uninformative, it is still meaningful to study the limiting behavior of SPOMs. This may be done in terms of the quasistationary distribution π, which is defined as the limiting distribution conditioned on nonextinction. In practical terms, a metapopulation that succeeds in persisting for a long time converges toward the quasistationary distribution. More precisely, we first condition the probability distribution $\mathbf{q}(t)$ on nonextinction as

$$m_i(t) = \text{Pr}(Q(t) = i | Q(t) \neq 1) = \frac{q_i(t)}{1 - q_1(t)} \tag{1}$$

after which the quasistationary distribution is defined as the limiting distribution, $\pi = \lim_{t\to\infty} \mathbf{m}(t)$. The quasistationary distribution π exists and is unique, provided that the patch network is irreducible, meaning that any patch is able to colonize (possibly through intermediate colonizations) any other patch. Technically, the quasistationary distribution π may be derived as the left leading eigenvector of matrix P_0, which is obtained by deleting the first row and the first column from $\mathbf{P}$. Furthermore, drawing the initial state from the quasistationary distribution, the expected time to metapopulation extinction is given by

$$T = \begin{cases} -1/\rho & \text{for Markov processes,} \\ 1/(1-\rho) & \text{for Markov chains,} \end{cases} \tag{2}$$

where ρ is the leading eigenvalue of $\mathbf{P}_0$ (Darroch and Seneta, 1965, 1967).

is satisfied. In other words, the species is expected to persist only if the amount of habitat (h) exceeds a threshold value ($\delta = e/c$), which is set by the properties of the species. In the original Levins model, the threshold condition is given by $c > e$, as in that model $h = 1$. Based on Eq. (4.5), we may conclude that the long-term persistence of a species in a fragmented landscape is facilitated by increasing amount of suitable habitat (large h), a small risk of local extinction (small e), and a good colonization ability (large c).

The most fundamental message from these simple models for ecology and conservation is that a species may go deterministically extinct even though some suitable habitat remains at the landscape level, but exactly how much habitat is needed for long-term persistence, and how could we estimate this value? The answer given by Lande's model is as follows. First, let us denote by p^* the fraction of occupied patches in an initial situation in which a fraction h_0 of the landscape consists of suitable habitat. Assuming that the species is at its population-dynamic equilibrium, we may solve the value of the species parameter $\delta = e/c$ from Eq. (4.4) as $\delta = h_0(1 - p^*)$, after which the threshold condition [Eq. (4.5)] may be rewritten as $h > h_0(1 - p^*)$. This observation leads to the seemingly very useful result, dubbed the Levins rule (Hanski et al., 1996), that the minimum amount of habitat necessary for long-term persistence can be estimated by just recording the amount of empty habitat while the species is still common.

Carlson (2000) applied the Levins rule to an endangered bird species in Sweden and Finland, the white-backed woodpecker (*Dendrocopos leucotos*). Using data from the Bialowieza National Park in Poland, where the species is still common, he estimated that $h_0 = 0.66$ and $p^* = 0.81$, from which the extinction threshold is estimated as 0.13. This result is consistent with the amount of suitable habitat remaining in Sweden ($h < 0.12$) and in Finland ($h = 0.08$), where the populations have declined severely. The result is encouraging, but we consider that it is based on such a simplified theory and simplified description of the landscape that the Levins rule has primarily pedagogic value. There are several factors that are not included in the model but which are likely to influence metapopulation dynamics and hence the threshold condition in practice, including the rescue effect (Hanski et al., 1996; Gyllenberg and Hanski, 1997). Furthermore, as the models do not take into account variation in the properties of the patches, any estimates of h and p^* are rough approximations at best. The same comments apply to the stochastic logistic model and the Levins and Lande models in general: they provide qualitative insight to classic metapopulation dynamics, but they have limited value for a quantitative metapopulation analysis. To be fair, they were not meant to do that. To follow our interest in developing predictive metapopulation models for highly fragmented landscapes, we next turn to heterogeneous SPOMs, which may be used more readily in the study of real metapopulations in real landscapes. It will turn out, however, that the behavior of the spatially heterogeneous models may often be best understood by studying their homogeneous counterparts, which leads to resurrection of the Levins model via a new interpretation of the model parameters.

4.3 PATCH OCCUPANCY MODELS: HETEROGENEOUS PATCH NETWORKS

The fundamental difference between homogeneous and heterogeneous SPOMs is that in the latter the colonization and extinction probabilities (or rates) are different for different habitat patches. The state of a metapopulation living in a heterogeneous network of n patches, such as shown in Fig. 4.1, is described by the vector $\mathbf{O} = \{O_i\}_{i=1}^{n}$, where the component O_i

specifies whether patch i is empty ($O_i = 0$) or occupied ($O_i = 1$). The metapopulation model may be specified either in continuous time, in which case the model is mathematically a Markov process, or in discrete time, in which case the model becomes a Markov chain. We denote the colonization rate (for Markov processes) or the colonization probability (for Markov chains) of an empty patch i by $C_i = C_i(\mathbf{O})$ and the extinction rate (or probability) of an occupied patch by $E_i = E_i(\mathbf{O})$.

Before turning to specific SPOMs for heterogeneous networks, it is worth noting that the qualitative theory presented in Box 4.1 for the stochastic logistic model applies almost independently of the functional forms for the colonization and extinction processes C_i and E_i. Metapopulation extinction is an absorbing state, which the process will eventually reach with probability one. Before the inevitable extinction, the process converges toward a quasistationary distribution π, which is given as the subdominant eigenvector of a stochastic transition matrix $\mathbf{P}$ (see Box 4.2 for details). This quasistationary distribution, often referred to as "(stochastic) metapopulation equilibrium" (Hanski, 1999b) or as the "established phase" of the metapopulation (Frank and Wissell, 2002; Grimm and Wissel, 2004), is of great importance for ecological applications of the theory. It relates directly to the average size of the metapopulation and is needed for determining the extinction risk of a metapopulation that has already persisted for some time. As the quasistationary distribution does not account for transient dynamics, however, it is not sufficient for the description of the dynamics of a newly established metapopulation nor a metapopulation that has been perturbed recently. In such cases, one needs to find out also the probability of reaching the quasistationary distribution, which influences the extinction risk of the metapopulation within a given time horizon (Verboom et al., 1991a; Stephan, 1993; Ovaskainen and Hanski, 2002; Grimm and Wissel, 2004).

We will next formulate deterministic approximations of the full stochastic SPOM. To do this, we denote by $\mathbf{p} = \{p_i\}_{i=1}^{n}$ a vector with the component i giving the probability that patch i is occupied. A deterministic version of the SPOM may be obtained by replacing the vector of patch occupancies $\mathbf{O}$ by the vector of occupancy probabilities $\mathbf{p}$ (Ovaskainen and Hanski, 2001),

$$\left.\begin{array}{r} \dfrac{dp_i(t)}{dt} \\ p_i(t+1) - p_i(t) \end{array}\right\} = C_i(\mathbf{p}(t))(1 - p_i(t)) - E_i(\mathbf{p}(t))p_i(t). \tag{4.6}$$

In this equation, the upper formula relates to continuous-time models and the lower one to discrete-time models. If the colonization and extinction rates (probabilities) depend on the occupancy state in a linear fashion, Eq. (4.6) may be derived from the stochastic model in the same way as the Levins model was derived from the stochastic logistic model. Note that for nonlinear models, Eq. (4.6) does not necessarily correspond exactly to the drift term, although in practice it often gives a good approximation.

We will put flesh to the skeleton of SPOMs by introducing two examples of biologically reasonable models, which have been dubbed the spatially realistic Levins model (Hanski and Ovaskainen, 2000) and the incidence function model (Hanski, 1994a). Before doing that, we have to add a critical component to the models, the set of assumptions that relate the colonization and extinction rates to the structure of the fragmented landscape.

Spatially Realistic Metapopulation Theory

A particular SPOM is defined by describing how colonization and extinction rates (or probabilities) depend on the structure of the landscape and on the present occupancy pattern. The dependence of population turnover on habitat fragment areas and spatial locations, and possibly on other quantities, such as habitat quality, places the metapopulation theory explicitly in a spatial context. As one may use here landscape measures that describe the structure of real fragmented landscapes, we have called the combination of heterogeneous SPOMs and the assumptions mapping population turnover to landscape structure the spatially realistic metapopulation theory (Hanski, 2001b; Hanski and Ovaskainen, 2003).

The starting point is a description of the landscape structure as depicted in Fig. 4.1. Thus the network is assumed to consist of n circular habitat patches. We denote by A_i the area of patch i, and by d_{ij} the distance between the centroids of patches i and j. The extinction and colonization rates are now defined as functions of A_i and d_{ij}, using some specific arguments, of which the following sections give two examples. In most cases it is sensible to assume that a decreasing patch area increases the extinction rate and decreases the contribution of the respective population to connectivity of other populations because local population sizes tend to decrease with decreasing patch area. Likewise it is sensible to assume that the probability of an empty patch becoming colonized increases with increasing connectivity to existing populations. Figure 4.3 gives an example for the well-studied Glanville fritillary butterfly.

We will note in passing that SMT unites the classic metapopulation theory (CMT) based on the pioneering models by Levins (1969, 1970) and the dynamic theory of island biogeography (DTIB) of MacArthur and Wilson (1963, 1967) (Hanski, 2001b). CMT and DTIB are obviously related because the expected number of species on an island or in a habitat fragment (the

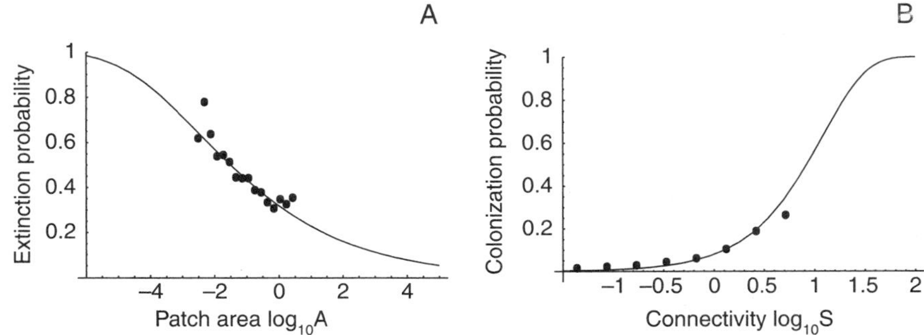

Fig. 4.3 The dependence of extinction probability on patch size (A) and the dependence of colonization probability on connectivity (B). Dots represent data collected for nine successive generations of the Glanville fritillary from a network of ca. 4000 habitat patches in the Åland Islands (Nieminen et al., 2004). Lines depict maximum likelihood estimates based on the entire data set. Parameter values $e = 0.38$, $c = 0.083$, $\alpha = 0.84$, $\zeta_{ex} = 0.17$, $\zeta_{em} = 0.07$ and $\zeta_{im} = 0.30$.

dynamic variable in DTIB) is given by the sum of probabilities of different species occurring in that fragment (the dynamic variable in CMT). Assuming a mainland pool of R identical and independent species, the basic DTIB is obtained by multiplying both sides of Eq. (4.6) by R and assuming the same colonization and extinction rate parameters for all the species. However, assuming a set of identical and equally connected habitat patches as in CMT, Eq. (4.6) gives the rate of change in the probability of occupancy for each fragment, which is also the rate of change in the fraction of occupied fragments.

Spatially Realistic Levins Model (SRLM)

Our first example is the spatially realistic Levins model, which is defined by the following general assumptions about the colonization and extinction rates:

$$\begin{cases} C_i = \sum_{j \neq i} c_{ij} O_j, \\ E_i = e_i, \end{cases} \tag{4.7}$$

where c_{ij} denotes the contribution that patch j (when occupied) makes to the colonization rate of patch i. In other words, the colonization rate of patch i is a linear function of the occupancies of potential source patches, whereas the extinction rate is independent of the occupancies of the other patches. Equation (4.7) is structurally similar to the stochastic logistic model. If the patches are identical and equally connected, SRLM reduces to the stochastic logistic model. Our next task is to make specific assumptions as to how colonization and extinction rates depend on the structure of the patch network.

Local Extinction

Assuming that all the patches are of equal quality, it is reasonable to assume that the carrying capacity of a patch is proportional to its area. Thus $K_i = k\ A_i$ gives the number of individuals in patch i when the local population is at its carrying capacity, and k denotes the density of individuals. If there is variation in density from patch to patch due to differences in patch quality, this can be straightforwardly taken into account in the model, as long as one has estimates of patch-specific density. In order to estimate the extinction rate of a local population, we need to make assumptions on the type of local dynamics. One possibility would be to assume that the local population behaves according to the stochastic logistic model, now viewed as a model for the number of individuals in a local population. In this case, the rate of local extinction would essentially decrease exponentially with local carrying capacity, $Ext_i \propto e^{-K_i}$ (Box 4.1). The exponential scaling arises from demographic stochasticity, which is of major concern in very small populations but not in larger ones. Typically, the extinction risk of all but the smallest populations is dominated by environmental stochasticity (Chapter 14). Adding environmental stochasticity to models for local population dynamics leads to the much milder power-law scaling (Lande, 1993; Foley, 1994)

$$Ext_i = \frac{\widetilde{e}}{K_i^{\zeta_{\mathrm{ex}}}} = \frac{e}{A_i^{\zeta_{\mathrm{ex}}}}, \tag{4.8}$$

where e is an extinction rate parameter. The scaling factor $\zeta_{\rm ex}$ may be interpreted as a measure of the strength of environmental stochasticity, with a low value of $\zeta_{\rm ex}$ indicating that environmental stochasticity is of major importance in determining the extinction risk (Hanski, 1998a).

Colonization

To estimate the colonization rate of an empty patch, we make the simplifying assumption that all the occupied patches are at their local carrying capacity (or that the sizes of the local populations are proportional to the respective carrying capacities). We assume that each of the K_i individuals produces propagules in a continuous fashion with a rate s and that the propagules are randomly (and independently of each other) redistributed according to a radially symmetric dispersal kernel $f(r)$ (Box 4.3). Assuming that each propagule has a constant probability q of successful colonization, it follows that the colonization rate of an empty patch i is given by

$$C_i(\mathbf{O}) = \sum_{j \neq i} K_j s q g_{ij} O_j = cA_i \sum_{j \neq i} = A_j f(d_{ij}) O_j = \sum_{j \neq i} c_{ij} O_j, \qquad (4.9)$$

where $\gamma_{ij} = f(d_{ij})A_i$ is the fraction of propagules produced by an individual in patch j that are expected to migrate to patch i, and the colonization rate parameter c is defined as $c = ksq$. What this formula does is to collate the contributions of all source populations (currently occupied patches) to the colonization rate of the focal patch; the contribution of source patch j depends on the size of patch j and its distance from the focal patch i as specified by the dispersal kernel.

Generally, both the source and the target patch areas may influence individual movements and hence the contribution of patch j to the colonization of patch i in a nonlinear manner (Hanski et al., 2000; Moilanen and Nieminen, 2002; Ovaskainen and Cornell, 2003). Assuming a power-law relationship for both emigration and immigration, Eq. (4.9) may be generalized to the form $c_{ij} = cA_i^{\zeta_{im}} A_j^{\zeta_{em}} f(d_{ij})$ (Ovaskainen, 2002b). Values of the scaling factors ζ_{im} and ζ_{em} depend on the biological processes determining immigration and emigration rates and may either be estimated from empirical data or be derived from submodels for immigration and emigration.

The deterministic mean-field model is obtained by replacing the vector of patch occupancies by the vector of occupancy probabilities (Gyllenberg and Hanski, 1997). For example, Hanski and Ovaskainen (2000) assumed a simple continuous-time deterministic SRLM given by Eq. (4.6) with $E_i = e/A_i$ and $C_i = c\Sigma A_j \exp(-\alpha d_{ij}) p_j$, where we have assumed that $\zeta_{ex} = \zeta_{em} = 1$, an exponential dispersal kernel (the normalization constant is included in the parameter c), and no effect of target patch size on colonization ($\zeta_{im} = 0$). The exponential dispersal kernel used here is phenomenological (does not have a mechanistic explanation) but it typically fits well with observations (e.g., Conrad et al., 1999; Sutherland et al., 2000; Byrom, 2002; Griffith and Forseth, 2002). Note that it differs from the dispersal kernel introduced in Eq. (1) in Box 4.3 by the factor r. As seen in Section 4.4, the deterministic model yields many useful results about metapopulation dynamics.

BOX 4.3 Dispersal Kernels

In order to construct a SPOM, one needs to know the fraction g_{ij} of propagules produced by an individual in patch j that are expected to migrate to patch i. The fraction g_{ij} is often described with the help of a dispersal kernel D, which specifies how propagules are redistributed in a landscape. More precisely, assuming that the propagules are produced in the origin, the fraction of propagules that disperse to a region $X \subseteq R^2$ is assumed to be given by $\int_X D(x,y)dA$, where dA denotes integration with respect to area. As the total number of propagules should be conserved, we always assume that the dispersal kernel is normalized as $\int_{R^2} D(x,y)dA = 1$. In many cases, it is reasonable to assume that the dispersal kernel is radially symmetric, in which case it may be denoted by $f(r) = D(r,0)$, and the fraction of propagules that disperse to distance r with $r_1 < r < r_2$ is given by $2\pi\int_{r_1}^{r_2} f(r)rdr$. For example, if one assumes that propagules move toward a fixed (but random) direction until they settle with a fixed rate, one has

$$f(r) = \frac{\alpha}{2\pi r}e^{-\alpha r}, \tag{1}$$

where the parameter α depends on the ratio of the speed of the individuals and the rate at which they settle. For another example, if one assumes that propagules move according to random walk until they settle with a fixed rate δ, one has

$$f(r) = \frac{\delta}{2\pi a}K_0\left(\sqrt{\frac{\delta}{a}}r\right), \tag{2}$$

where K_0 is a modified Bessel function of the second kind and a is the diffusion coefficient that may be calculated from the parameters of the random walk (Turchin, 1998).

Assuming that the landscape is highly fragmented, i.e., that the patches are small with respect to the dispersal distances, it follows that g_{ij} may by approximated by

$$g_{ij} = f(d_{ij})A_i. \tag{3}$$

Deriving g_{ij} from a dispersal kernel D is best justified for passive propagule dispersal, whereas dispersal strategies with active search behavior may not correspond exactly to any dispersal kernel. For example, if the propagules move according to random walk but bias their movement toward the habitat patches, the patches are, in a sense, competing for the migrants. In such a case, the dispersal rate between any two patches does not depend just on the areas of and the distance between the two patches, but also on the spatial configuration of the entire patch network (Ovaskainen and Cornell, 2003). Further, strong heterogeneity of the dispersal habitat (e.g., dispersal corridors or barriers) may prevent the use of dispersal kernels, although in general they provide a reasonable first approximation.

The SRLM may also be formulated as a discrete-time model by assuming that the time step of the model is sufficiently small. In this case, general expressions for the colonization and extinction probabilities are given by

$$\begin{cases} C_i = 1 - \exp\left(-\sum_{j \neq i} c_{ij} O_j\right), \\ E_i = 1 - \exp(-e_i). \end{cases} \tag{4.10}$$

As continuous-time SRLM is structurally simple and thus analytically tractable, we use it later for developing much of the mathematical theory. However, we will connect the theory to data using the discrete-time SRLM, as the latter is more appropriate for the Glanville fritillary butterfly. For the rest of this chapter, we will call the model 4.10 with parameters estimated for the Glanville fritillary butterfly (Fig. 4.3) as the Glanville fritillary model, whereas SRLM will always mean the continuous-time version of the spatially realistic Levins model.

Incidence Function Model (IFM)

The incidence function model is a discrete-time SPOM that was introduced by Hanski (1994a) and has since been applied widely in empirical studies (Hanski, 1999b, 2001b; Chapter 5). An extended version (Ovaskainen, 2002a) of the incidence function model is defined as

$$\begin{cases} C_i = \dfrac{S_i^z}{S_i^z + 1/c}, \\ E_i = \min\left(\dfrac{e}{A_i^{\zeta_{ex}}}, 1\right)(1 - C_i)^r \end{cases} \tag{4.11}$$

where

$$S_i = A_i^{\zeta_{im}} \times \sum_{j \neq i} A_i^{\zeta_{em}} {}_i f(d_{ij}) O_j \tag{4.12}$$

is the metapopulation dynamic connectivity of patch i to extant local populations. In comparison with the linear SRLM, the IFM is more complex, as it has two structural parameters. First, parameter z relates to the assumptions behind the colonization process, values greater than $z = 1$ reflecting the presence of an Allee effect. In the original version of the IFM, Hanski (1994a) assumed a relatively strong Allee effect ($z = 2$), which could follow from the interaction of immigrants at colonization in a sexually reproducing species. Second, parameter r measures the strength of a rescue effect, describing a reduced extinction risk due to immigrants enhancing the local population size. Hanski (1994a) assumed that $r = 1$. As with the SRLM, the deterministic approximation is obtained by replacing the vector of patch occupancies with a vector of occupancy probabilities in Eq. (4.12).

4.4 DETERMINISTIC THEORY

This section focuses on three ecologically interesting issues that may be addressed by the deterministic theory. First, we examine what kind of threshold conditions SPOMs predict for metapopulation persistence and use these conditions to classify SPOMs into three qualitatively distinct classes. Second, we ask about the "values" of individual habitat patches in the sense of the contributions that the patches make to metapopulation dynamics and persistence.

Third, we explore transient dynamics, the specific aim being to quantify the length of time it takes for a metapopulation to return to equilibrium following a perturbation.

Metapopulation Capacity

The most fundamental question that may be addressed by a patch occupancy model is whether a given species is expected to persist in the long term in a given fragmented landscape. In stochastic models, where eventual extinction is always certain, it is natural to assess the risk of extinction by analyzing the time it takes for the metapopulation to go extinct. In the deterministic framework, a metapopulation may be classified as viable if the model possesses a stable nontrivial equilibrium state $\mathbf{p}^*$, toward which the deterministic model converges asymptotically. The deterministic perspective is useful especially for large metapopulations for which the role of stochastic fluctuations is not as pronounced as for small metapopulations.

In the Levins–Lande model, the deterministic threshold condition for persistence is given by $h > \delta$, where h is the amount of suitable habitat and $\delta = e/c$ is a parameter defined by the properties of the species (Section 4.2). Hanski and Ovaskainen (2000) extended this result to the spatially realistic Levins model, in which the threshold condition for deterministic persistence is given by

$$\lambda_M > \delta \tag{4.13}$$

Here λ_M is called the metapopulation capacity of the fragmented landscape, and $\delta = e/c$ is a species parameter as in the original Levins model. Comparing Eqs. (4.5) and (4.13), we note that in the spatially realistic model, the metapopulation capacity (λ_M) plays the role of the amount of habitat (h) in the nonspatial model. However, λ_M also takes into account how well the patches are connected to each other. Mathematically, the metapopulation capacity λ_M is given as the leading eigenvalue of the $n*n$ matrix $\mathbf{M}$ with elements $m_{ii} = 0$ and

$$m_{ij} = \delta c_{ij}/e_i = A_i^{\zeta_{ex}+\zeta_{im}} A_j^{\zeta_{em}} f(d_{ij}). \tag{4.14}$$

The element m_{ij} gives the contribution that patch j makes to the colonization rate of patch i, multiplied by the expected lifetime of patch i. m_{ij} thus measures the fraction of time that patch i would be occupied if patch j were the only source of immigrants (Hanski and Ovaskainen, 2000; Ovaskainen and Hanski, 2001).

The analogy between λ_M and h extends beyond the threshold condition. In the Levins–Lande model, the equilibrium fraction of occupied patches is given by $p^* = 1 - \delta/h$, whereas in the SRLM it is given by $\mathrm{p}_\lambda^* = 1 - \delta/\lambda_M$, where p_λ^* is a weighted fraction of occupied patches, $p_\lambda = \Sigma_i W_i p_i$, the weights W_i being defined by patch values (see later for details).

Figure 4.4 gives an empirical example of the extinction threshold using the Glanville fritillary model. Due to the nonlinear structure of the discrete-time model, it is not possible to extract a species parameter δ from the other model parameters, and consequently the threshold condition for persistence is given in this model as $\lambda_M > 1$ (Ovaskainen and Hanski, 2001). Although there is a

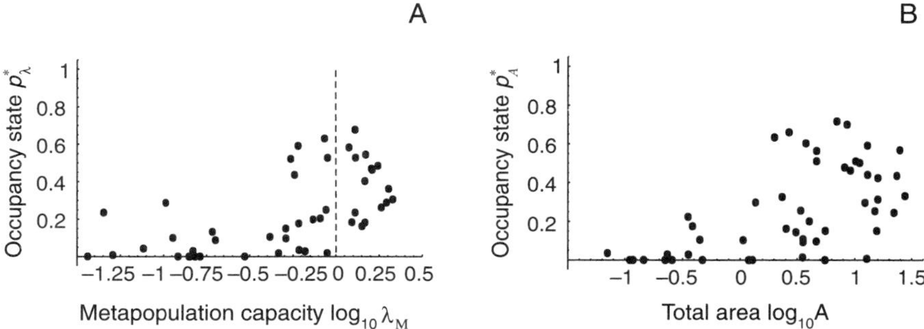

Fig. 4.4 An empirical example of the treshold condition for persistence using the Glanville fritillary model with parameters from Fig. 4.3. Each dot depicts 1 of 59 semi-independent patch networks like the one illustrated in Fig. 4.1. The horizontal axis shows the metapopulation capacity (A) or the total area (B) of the network, and the vertical axis shows the weighted average of the observed probability of patch occupancy during the 9-year study (weights patch values in A and patch areas in B). The dashed line in A depicts the threshold for persistence ($\lambda_M = 1$).

lot of scatter, much of it due to a variation in patch quality and to regional stochasticity, it is evident that the metapopulation capacity is able to rank the networks in terms of the presence or absence of the species more accurately than the total area of habitat in the patch network (Fig. 4.4B). Note that many of the patch networks that fall below the threshold condition $\lambda_M = 1$ have, however, had substantial metapopulations during the study period. The explanation for this result is twofold. First, the different networks are not completely independent of each other, and thus a large metapopulation may keep a nearby small patch network occupied, although the small network would alone be below the threshold condition for long-term persistence. Second, the parameter estimates (Fig. 4.3) are based solely on the observed annual transitions, and they thus ignore the occupancy state observed in the first year (see Chapter 5). As there has been a general decline in the occurrence of the butterfly during the study period (which may, however, be only temporary; I. Hanski, unpublished results), the parameter estimates are likely to be biased in the direction of overestimating the critical amount of habitat needed for metapopulation persistence (Moilanen, 2000). Correcting for the bias would move the vertical line in Fig. 4.4A somewhat toward the left.

Classification of SPOMs

The most fundamental classification of metapopulations in the deterministic context is into those that are beyond the extinction threshold and persist (have a stable nontrivial equilibrium state $\mathbf{p}^* > 0$) and those that are below the extinction threshold and go extinct. A more refined classification considers in addition whether the trivial equilibrium state $\mathbf{p}^* = 0$ corresponding to metapopulation extinction is stable or unstable. If it is unstable, a small metapopulation has a deterministic tendency to grow, and thus the metapopulation may be expected to be able to invade an empty patch network successfully. The threshold condition for successful invasion may be written as $\lambda_I > \delta$, where λ_I is called the invasion capacity of the network (Ovaskainen and Hanski, 2001). In the SRLM, the metapopulation capacity λ_M coincides with the invasion capacity λ_I, but this is not the case in general. Three types of situations may be distinguished based on the relationship between metapopulation and invasion capacities (Fig. 4.5; Ovaskainen and Hanski, 2001). Models for which $\lambda_I = \lambda_M$ are called Levins-type models. Models for which $\lambda_I < \lambda_M$ possess a metapopulation level Allee effect, meaning that although a metapopulation could persist in a network, a single small local population cannot invade an empty network. A metapopulation level Allee effect leads to multiple equilibria in metapopulation dynamics. In mechanistic terms, this may be caused by an Allee effect in the colonization process, by a rescue effect in the extinction process, or by a combination of the two (Ovaskainen and Hanski, 2001). While the SRLM belongs to Levins-type models, the IFM is an example of a model with a strong metapopulation level Allee effect (Fig. 4.5).

Multiple equilibria in metapopulation dynamics are difficult to test because this requires data for several independent patch networks. The long-term study of the Glanville fritillary in the Åland Islands has produced the most convincing example so far (Hanski et al., 1995a), including a demonstration of the rescue effect on local extinction (Hanski, 1999b). A signature of multiple equilibria is a bimodal distribution of patch occupancy frequencies (or, more properly, of p_λ in the case of heterogeneous networks). Putative examples of bimodal "core-satellite" distributions (Hanski, 1982) have been described for a wide range of taxa (Hanski, 1999b). The fundamental message

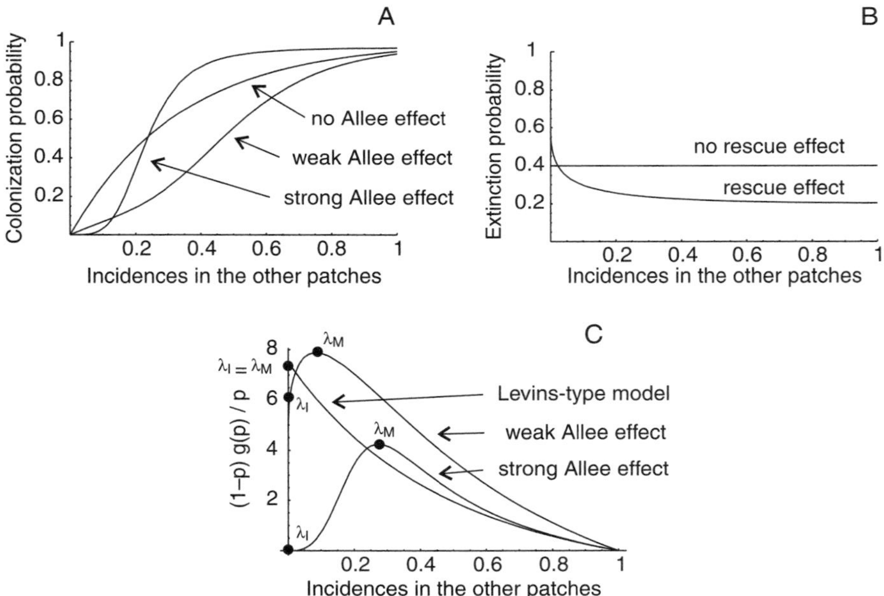

Fig. 4.5 One-dimensional illustration of three qualitatively different metapopulation models. (A) The colonization probability of patch *i* with no Allee effect, with a weak Allee effect, and with a strong Allee effect. (B) Extinction probability of patch *i* with and without a rescue effect. (C) The principal map $(1 - p)g(p)/p$ for Levins-type models and for models possessing a weak or a strong Allee effect. See Ovaskainen and Hanski (2001) for discussion and for the definition of the principal map g (modified from Ovaskainen and Hanski, 2001).

for conservation is that a metapopulation may crash unexpectedly to extinction from a state of commonness. This happens when the metapopulation crosses the unstable internal equilibrium due to perturbation or when the positive internal equilibrium is lost due to an environmental change.

Patch Values

While λ_M characterizes the capacity of an entire patch network to support a viable metapopulation, one might often wish to assess the contributions that particular patches make to the persistence of a metapopulation. For example, in metapopulation management, limited resources may force one to decide which of two patches is more valuable and should be conserved (Lindenmayer and Possingham, 1996a,b; Moilanen and Cabeza, 2002; Chapter 22). In the context of metapopulation dynamics, the value of a habitat patch depends not only on the size and the quality of the patch, but also on its connectivity to the remaining network. Ovaskainen and Hanski (2003a) examined the concept of "patch value" by quantifying the contributions that particular patches make to metapopulation dynamics. They concluded that the value of a particular patch cannot be assessed properly without specifying what exactly is meant by a "contribution to metapopulation dynamics." Table 4.1 lists four biologically meaningful alternatives.

We will here highlight one particular measure, W_i, termed the dynamic value of the patch. More precisely, W_i is defined as the long-term contribution that patch i makes to the colonization events in the network. To define $\mathbf{W} = \{W_i\}_{i=1}^{n}$, we first take into account b_{ij}, the direct contribution that patch j makes to a colonization event in patch i. Assuming that the metapopulation is at equilibrium, it is natural to define b_{ij} as (Ovaskainen, 2002a)

$$b_{ij} = k_i p_j^* \frac{dCol(\mathbf{p}^* + \varepsilon \mathbf{e}_j)}{d\varepsilon} \tag{4.15}$$

TABLE 4.1 Four Measures Used to Characterize Patch Values

Target quantity	Perturbation measures	Dynamic measures	Appropriate for
Metapopulation capacity λ_M	$V_i^S = \frac{d\lambda_M}{dA_i}$ $V_i^L = \lambda^M - \lambda_M^i$	—	Rare species
Colonization events	—	W_i (see text)	Common species
Metapopulation size S	$U_i^S = \frac{dS}{dA_i}$ $U_i^L = S - S^i$	—	Common species
Time to extinction T	$t_i^s = \frac{d \log T}{dA_i}$ $t_i^L = \log(T/T^i)$	—	Rare species

[a] Superscripts S and L refer to small and large perturbations, respectively. Metapopulation size S is defined as $S = \Sigma_i s_i p_i^*$, where s_i is the weight given to patch i. The quantities λ_M^i, S^i, and T^i denote the metapopulation capacity, the metapopulation size, and the time to extinction in a network from which patch i has been removed (modified from Ovaskainen and Hanski, 2003a).

at $\varepsilon = 0$, where $\boldsymbol{e}_j$ is the jth unit vector and the scaling factor k_i is chosen so that $\Sigma_j b_{ij} = 1$. In Eq. (4.15), the term p_j^* measures the fraction of time that patch j may possibly contribute to the colonization rate of patch i, whereas the remaining term measures how sensitive the colonization rate of patch i is to the contribution that patch j makes. For example, in the SRLM, b_{ij} is given as

$$b_{ij} = \frac{p_j^* c_{ij}}{\sum_k p_k^* c_{ik}}. \tag{4.16}$$

While b_{ij} measures the direct contribution that patch j makes to a colonization event in patch i, we would ultimately like to measure the long-term contribution by including the full chain of colonization events through the network. This may be done by raising the matrix **B** to an infinite power, which is equivalent to defining **W** as the left leading eigenvector of matrix $\mathbf{B} = \{b_{ij}\}_{i,j=1}^{n}$ (Ovaskainen and Hanski, 2003a). Doing so, W_j measures the long-term contribution of patch j to colonization events in the entire network and is thus independent of the target patch i. Furthermore, as $\Sigma_j W_j = 1$, W_j represents the relative value of patch j. Figure 4.6 illustrates the behavior of the measure W using the Glanville fritillary model in the network shown in Fig. 4.1. Note that, in this example, the patch values are distributed more evenly than patch areas (Fig. 4.1B). This is not the case in general, but it happens in the Glanville fritillary model as the patch area scaling factor $\zeta = \zeta_{ex} + \zeta_{em} + \zeta_{im}$ is less than 1 (Ovaskainen, 2002b; Ovaskainen and Hanski, 2003a).

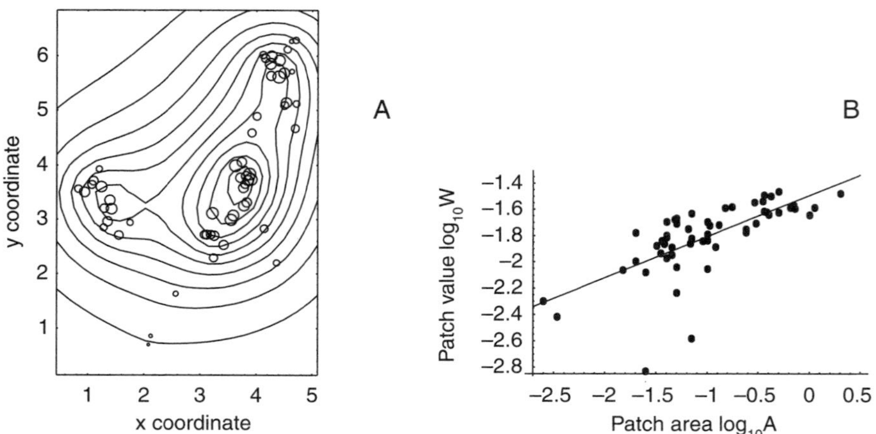

Fig. 4.6 An illustration of the patch value measure **W** in the patch network shown in Fig. 4.1. (A) Sizes of the dots are proportional to the values of the habitat patches, and contour lines indicate the relative value that a hypothetical patch would attain if added to a particular location within the network. (B) Values of the habitat patches with respect to patch areas. The slope of the fitted regression line is 0.31. The figure is based on the Glanville fritillary model with parameters $e = 0.30$ and $c = 0.13$ estimated from data restricted to the network shown in Fig. 4.1 and the remaining structural parameters estimated from the entire metapopulation (Fig. 4.3).

Transient Dynamics

Metapopulation capacity λ_M and patch values refer to the equilibrium state $\mathbf{p}^*$, and they thus relate to the long-term behavior of the metapopulation without any reference to its present state. If the metapopulation happens to be far away from its population dynamic equilibrium, it may be of great importance to be able to say something about the transient dynamics.

For example, consider a species that persists initially very well so that most of the habitat patches would be occupied most of the time. Assume then that due to habitat loss, the situation changes rapidly so that many of the patches are lost from the network. As the metapopulation capacity of the network declines, the metapopulation is expected to move to a lower occupancy state, or it may even go extinct. However, this does not happen instantaneously, and the length of the transient period is often of great interest. A transient may also occur in the opposite direction, as is the case if a species invades an initially empty network. Figure 4.7 illustrates that the length of such transient periods may well be up to 5–10 generations in the Glanville fritillary model.

Ovaskainen and Hanski (2002) investigated the transient time in the SRLM by first demonstrating that the original Levins model is able to approximate the transient behavior of the SRLM and then calculating the transient time for the former. To accomplish the first task, we denote the parameters of the Levins model by $\tilde{c}$ and $\tilde{e}$ so that the model is defined as

$$\frac{dp}{dt} = \tilde{c}p(1 - p) - \tilde{e}p. \tag{4.17}$$

This model approximates the behavior of the SRLM given the following transformations (Ovaskainen and Hanski, 2002). First, one has to interpret the variable p in Eq. (4.17) as $p_\lambda = \Sigma_i W_i p_i$, the weighted fraction of occupied patches. Second, the parameter $\tilde{e}$ is interpreted as the effective extinction rate,

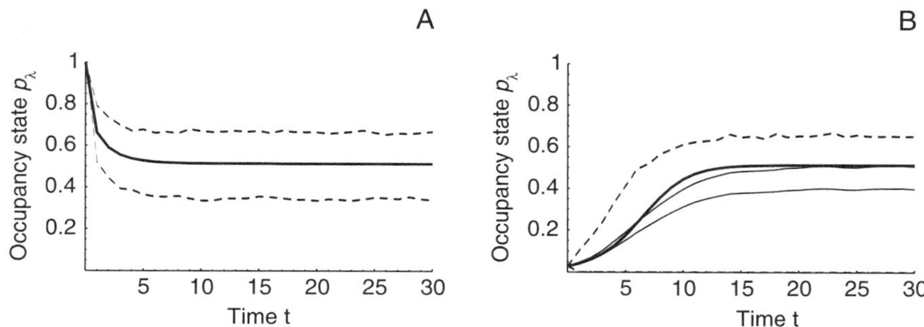

Fig. 4.7 Transient behavior in the Glanville fritillary model in the network shown in Fig. 4.1 with parameter values as in Fig. 4.6. In panel A all the patches are initially assumed to be occupied, whereas in panel B, only the patch indicated by an arrow in Fig. 4.1 is initially occupied. The thick line shows the prediction of the deterministic model. Thin lines show the mean of 1000 simulations, the lower line derived from all replicates and the upper one ignoring such replicates, which have gone extinct by that time. Dashed lines show 95% confidence intervals derived from all simulation replicates.

defined by $\tilde{e} = 1/(\Sigma_i W_i/Ext_i)$. Third, the parameter $\tilde{c}$ is interpreted as the effective colonization rate, defined by $\tilde{c} = \tilde{e}\,\lambda_M/\delta$.

Using the aforementioned transformation of SRLM to the Levins model, the length of the transient period, defined as the time it takes for the metapopulation to return from its initial state to a state close the new equilibrium, may be written as the product of four factors (Ovaskainen and Hanski, 2002). First, the length of the transient period increases with the distance between the present state and the equilibrium state. Second, the length of the transient is longer for species that have slow dynamics (e.g., due to a long life span) than for species with fast dynamics. Third, the length of the transient period is longer in a patch network that has few large patches than in a network with many small patches, as the turnover rate is expected to be slower for larger patches. Fourth and most important, the length of the transient period is expected to be especially long for species that are close to their extinction threshold following perturbation.

The fourth conclusion has the important implication for conservation that many rare species living in recently deteriorated landscapes may be "doomed" to extinction. They still exist because they have not had time to go extinct yet, and the time it takes to go extinct is especially long in the case of species whose long-term persistence is most precarious. The number of species that are predicted to ultimately go extinct due to past habitat loss and fragmentation represents the extinction debt in the community (Tilman et al., 1994; Hanski and Ovaskainen, 2002). The extinction debt is paid either by letting the species go extinct or by improving the quality of the landscape sufficiently for the species that constitute the extinction debt. Hanski and Ovaskainen (2002) discuss an example of extinction debt in beetle species living in boreal forests in Finland.

4.5 STOCHASTIC THEORY

Recall that the analysis of stochastic patch occupancy models is difficult because the size of the state space is 2^n for a heterogeneous network of n patches. This section develops approximation methods that take advantage of possible links between deterministic and stochastic frameworks. In particular, this section shows that patch values may be used to transform a heterogeneous metapopulation to its homogeneous ("ideal") counterpart, which behaves, in some relevant respects, similarly as the original heterogeneous metapopulation.

Effective Metapopulation Size

Let us start with the SRLM. As stated in the previous section, both the equilibrium state and the transient behavior of the deterministic SRLM can be approximated by the one-dimensional Levins model by replacing the original model parameters with the effective colonization rate $\tilde{c}$ and the effective extinction rate $\tilde{e}$. These factors account for the deterministic drift (growth) in the model. The main difference between deterministic and stochastic models is that the latter account for stochastic fluctuations around the mean dynamics, which arise due to the finite size of the network. We may ask whether one could also transform the size of a heterogeneous metapopulation into an effective metapopulation size $\tilde{n}$, which would control for the amount of stochastic fluctuations. To incorporate the latter into the model, we require that the infinitesimal variance σ^2 in Eq. (4.2) be the same for the weighted fraction of occupied patches p_λ in the SRLM and for the simple fraction of occupied patches p in the one-dimensional Levins model. Equating the two variances at the equilibrium state $\mathbf{p}^*$ leads to

$$\tilde{n} = \frac{\tilde{c}p_\lambda^*(1 - p_\lambda^*) + \tilde{e}p_\lambda^*}{\sum_i W_i^2[Col_i(\mathbf{p}^*)(1 - p_i^*) + Ext_i(\mathbf{p}^*)p_i^*]}. \tag{4.18}$$

As expected, the effective number of habitat patches increases with the real number of patches and with decreasing variance in the values of these patches. If all patches are identical, the effective number coincides with the actual number. Transformation of the SRLM is illustrated in Fig. 4.8. The dashed line corresponds to the analytically transformed model, whereas the continuous line corresponds to a numerically fitted model.

In nonlinear models, such as the Glanville fritillary model, or the IFM, it may not be possible to derive analytical expressions for the effective colonization and extinction rates or for the effective metapopulation size. These quantities may still be determined numerically by fitting a structurally similar homogeneous model to the drift and the variance of the heterogeneous model (Ovaskainen, 2002a). Figure 4.9 gives an example for the Glanville fritillary model. The homogeneous model consists of 50 patches, which is somewhat less than the actual number of 56 patches. The difference between the effective number of habitat patches and the actual number of habitat patches is explained by a variation in patch values (Fig. 4.6). In this example, the distribution of patch values is relatively even and hence the difference is not very great.

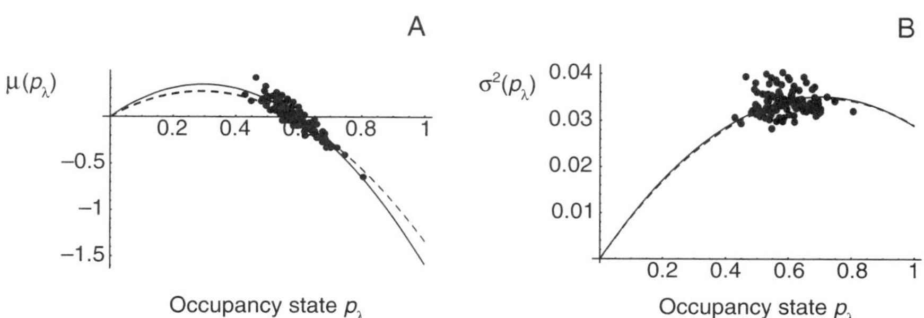

Fig. 4.8 Transformation of the SRLM to the stochastic logistic model in the network shown in Fig. 4.1. Dots represent a sample of 100 randomized occupancy states from the quasistationary distribution obtained by simulation. Lines represent fitted models; the continuous line is based on numerical fitting, and the dashed line is based on the analytical transformation given by Eq. (4.18). The two panels show (A) infinitesimal mean μ and (B) infinitesimal variance σ^2. Parameter values as in Fig. 4.6 except $e = 1$, $c = 0.5$.

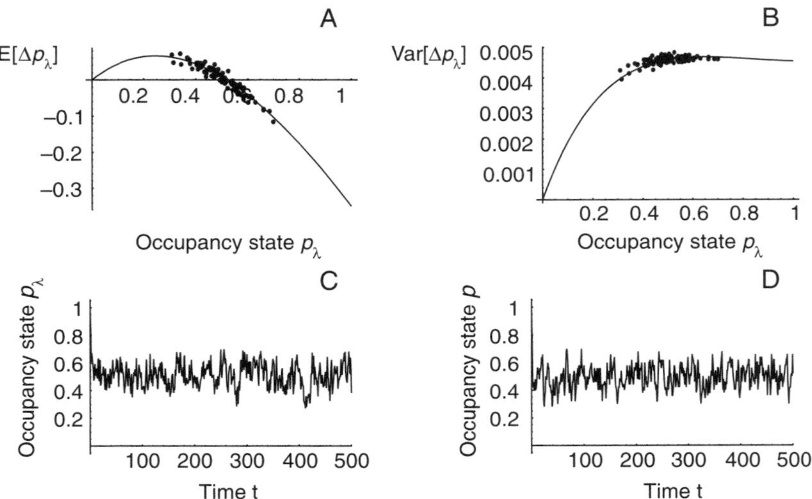

Fig. 4.9 As in Fig. 4.8 but for the Glanville fritillary model with parameter values as in Fig. 4.6. Single realizations of the full heterogeneous model (C) and of the transformed homogeneous model (D).

Spatially Correlated and Temporally Varying Environmental Conditions

Metapopulation models typically assume that the dynamics of local populations are independent of each other and that the environmental conditions remain constant. In reality, both assumptions are violated to a smaller or greater extent (Baars and Van Dijk, 1984; Ims and Steen, 1990; Hanski, 1999b; Lundberg et al., 2000; Nieminen et al., 2004). Spatially correlated and temporally varying environmental conditions present a challenge for metapopulation theory from the viewpoint of both model analysis (Heino et al., 1997; Engen et al., 2002; Ovaskainen, 2002a) and parameter estimation (Thomas, 1991; Hanski and Woiwod, 1993; Bjornstad et al., 1999, Williams and Liebhold, 2000; Peltonen et al., 2002). Viewing the metapopulation as a population of populations, stochasticity in patch occupancy dynamics in a constant environment is analogous to demographic stochasticity, whereas temporal variability in dynamics leads to variability that is analogous to environmental stochasticity. Hanski (1991) has termed these two forms of metapopulation-level stochasticities as extinction–colonization stochasticity and regional stochasticity, respectively. In a finite network of habitat patches, regional stochasticity leads to both spatially correlated and temporally varying parameter values; these two phenomena actually represent the two sides of the same coin.

The effective metapopulation size approach described earlier can be extended to situations in which the parameters vary temporally, allowing one to include regional stochasticity into the model. This is illustrated in Fig. 4.10, which is otherwise identical to Fig. 4.9 but now with the parameter values for the Glanville fritillary butterfly estimated separately for the eight annual transitions present in our dataset. As expected, adding temporal variation to the model increases the variance Var[Δp_λ] (Figs. 4.9B and 4.10B) and thus the amplitude

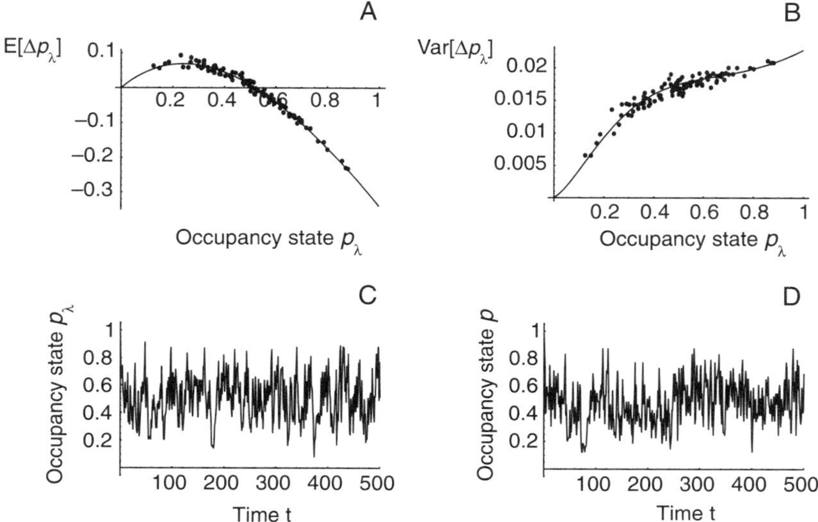

Fig. 4.10 As in Fig. 4.9 but with temporally varying parameters. The parameter distributions $e = (0.24, 0.48, 0.33, 0.34, 0.17, 0.30, 0.53, 0.04)$ and $c = (0.14, 0.10, 0.10, 0.33, 0.08, 0.08, 0.06, 0.22)$ have been estimated separately from the eight annual transitions (see Fig. 4.3), whereas the remaining parameters have been kept fixed.

of stochastic fluctuations (Figs. 4.9C and 4.10C). Figure 4.11 compares the statistical properties of the two models and their homogeneous counterparts. As expected, temporal variation flattens the quasistationary distribution (Fig. 4.11A), but note that it also adds especially low-frequency fluctuations to the behavior of the model (Fig. 4.11B). As illustrated by Figs. 4.9, 4.10, and 4.11, the effective metapopulation size approach gives a very accurate description of the Glanville fritillary model in all the senses investigated here.

The example in Fig. 4.10 demonstrates that temporal variation (spatial correlation) in parameter values increases the extinction risk of a metapopulation by amplifying stochastic fluctuations, which may be restated by observing that temporal variation decreases the effective number of habitat patches.

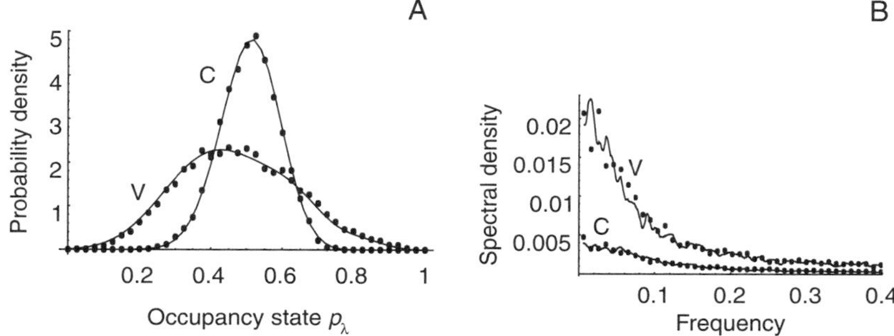

Fig. 4.11 Statistical properties of the Glanville fritillary model in the network shown in Fig. 4.1. (A) Quasistationary distribution and (B) a spectral plot. Lines correspond to the transformed metapopulation model and dots to the original heterogeneous model. The letters C and V refer to models with constant (Fig. 4.9) and temporally varying (Fig. 4.10) parameter values, respectively.

Ovaskainen and Hanski (2003b) made this statement explicit in the context of the stochastic logistic model. They showed that if extinctions and colonizations are correlated, with a correlation coefficient ρ, the effective number of patches is reduced to

$$n_e = \frac{n}{(n-1)\rho + 1}. \tag{4.19}$$

Such a correlation changes the qualitative behavior of the model in several ways. Most importantly, for $\rho > 0$ the mean time to metapopulation extinction does not increase exponentially with the number of habitat patches, as predicted by Eq. (2) in Box 4.1, but it now grows according to the power-law $T \propto n^{1/\rho}$ (Ovaskainen and Hanski, 2003b).

4.6 COMPARISON WITH OTHER MODELING APPROACHES

This section compares stochastic patch occupancy models with other modeling approaches that have been used in metapopulation studies. The other approaches include individual-based simulation models, models structured by population size, lattice models, and spatial moment equations, which were described briefly in Chapters 1 and 3. This section compares the different modeling approaches in the context of extinction thresholds and their relation to landscape structure, which is one of the key issues in metapopulation dynamics for both research and management. To facilitate this comparison, we start by summarizing three ecologically significant messages that have been discussed in this chapter in the context of SPOMs.

- First and most important, there is a critical amount of habitat below which a species is expected to go deterministically extinct. In highly fragmented landscapes, the extinction threshold depends not only on the total amount of habitat, but also on the spatial configuration of the habitat patch network. As characterized by metapopulation capacity, metapopulation persistence is facilitated by increasing connectivity among the habitat patches.
- Second, the contribution that individual patches make to metapopulation persistence depends not only on their area and quality, but also on their position within the network. Well-connected patches make generally a greater contribution than isolated patches.
- Third, metapopulation extinction is a stochastic event, which depends not just on the deterministic drift in metapopulation dynamics, but also on fluctuations around the drift. Stochastic fluctuations increase with a decreasing number of habitat patches, and thus extinction–colonization stochasticity increases the extinction risk, especially in small metapopulations. Furthermore, spatially correlated local dynamics or temporally varying environmental conditions amplify stochastic fluctuations and thus increase the risk of metapopulation extinction.

The first message has been studied extensively in the literature. Fahrig (2002) reviewed a number of lattice-based modeling studies attempting to disentangle the effects of habitat loss (fraction of destroyed habitat) and habitat fragmentation (spatial configuration of the remaining habitat). She categorized the studies into two classes based on the type of models used, either patch occupancy models (colonization–extinction models; Hill and Caswell, 1999; With and King, 1999a) or individual-based simulation models (birth–immigration–death–-emigration models; Flather and Bevers, 2002; Fahrig, 2001). Fahrig (2002) drew three major conclusions, which are all well in line with the results derived from SPOMs. First, all of the modeling studies she considered predicted a threshold amount of available habitat for metapopulation persistence. Second, a smaller total amount of habitat was required for metapopulation persistence if the habitat occurred in clusters. Third, the spatial configuration of the remaining habitat had the strongest effect if only a small fraction of the landscape remained suitable for the species. Similar conclusions were reached by Hiebeler (2000) and Ovaskainen et al. (2002) using a lattice-based patch occupancy model.

A major distinction between patch occupancy models and individual-based simulation models reviewed by Fahrig (2002) was that the latter predicted only a mild effect of habitat fragmentation on extinction threshold, which prediction is in line with much of the empirical literature (Harrison and Bruna, 1999; Fahrig, 2002). Fahrig (2002) argued that the difference between the two types of models occurs because the underlying reasons decreasing metapopulation persistence with increasing amount of habitat loss are different. In patch occupancy models, the colonization rate of empty patches decreases with increasing habitat loss, whereas in individual-based simulation models, an increasing number of individuals spend their time in the landscape matrix, where reproduction is not possible and mortality is assumed to be high. However, as migration mortality is one of the major reasons why a reduced colonization rate with distance is assumed in extinction–colonization dynamics (Hanski, 1999b), the difference between the two causes for the extinction threshold is somewhat superficial. Indeed, one could parameterize a patch occupancy model using data from simulation of an individual-based model, in which case the two modeling approaches should give more or less identical predictions. We suggest that the difference between the two modeling approaches is not fundamental and that the differences observed by Fahrig (2002) were based on quantitative model assumptions. Most importantly, the lattice-based studies are sensitive to spatial scale, especially if single lattice cells (instead of "patches" consisting of a cluster of cells) are assumed to support independent local populations. The simulation studies reviewed by Fahrig (2002) did not include environmental stochasticity at the level of habitat patches, which, in the case of lattice models, would amount to spatially correlated stochasticity in clusters of lattice cells comprising a single habitat patch (Gu et al., 2002). In SPOMs and in reality, the patch area-dependent extinction rate leading to a high rate of extinction of small populations in small patches comprises an important reason why the spatial configuration of the habitat may greatly influence the extinction threshold in highly fragmented landscapes.

Concerning our second message, which relates to the value of individual habitat patches, most of the previous literature has been based on simulation studies. For example, Lindenmayer and Possingham (1996a,b) developed a simulation model to assess the persistence of Leadbeater's possum in south-eastern Australia. In line with our results, they concluded that extinction probability was influenced both by the size and by the spatial arrangement of habitat patches. They found that in some circumstances the probability that a patch is occupied while the metapopulation is extant may be a good measure of its value for metapopulation viability. The habitat patch networks used in their studies varied from a single 300-ha reserve to twelve 25-ha reserves. As expected, metapopulations in networks of many small reserves were vulnerable to extinction due to demographic and environmental stochasticities. Conversely, a metapopulation inhabiting the single large reserve was susceptible to extinction in a catastrophic wildfire, highlighting the need for several more scattered reserves.

Other studies that have addressed patch values include Keitt et al. (1997), who used percolation theory to quantify connectivity at multiple scales and assigned conservation priorities to habitat patches based on their contributions to connectivity. In addition, using sensitivity analysis, Keitt et al. (1997) identified critical "stepping stone" patches that, when removed from the landscape, would cause large changes in connectivity. Moilanen et al. (1998) constructed a somewhat ad hoc measure of patch values while analyzing dynamics in a classic metapopulation of the American pika using the incidence function model. Their measure was an attempt to define the patch value in the sense of our measure W, although Moilanen et al. (1998) considered only a two-generation impact of habitat patches on colonizations. Verboom et al. (2001) emphasized the importance of large "key patches" with a stabilizing role in habitat patch networks. Patch values have also been addressed analytically in the context of metapopulation models structured by the sizes of local populations. Lawton et al. (1994) and Gyllenberg and Hanski (1997) noted that the Levins rule (Section 4.2) may break down due to a variation in habitat patch quality, as destruction of a high-quality patch has a greater impact on metapopulation size than destruction of a low-quality patch.

Our third message, which is concerned with the interplay between stochastic and deterministic dynamics, has been studied by Casagrandi and Gatto (1999, 2002a,b) using metapopulation models structured by local population size. They assumed that the sizes of local populations are negative-binomially distributed to facilitate the model analysis. In line with our results, Casagrandi and Gatto classified the parameter space into four regions, of which one extreme corresponded to deterministic extinction and the other one to metapopulation persistence. Stochasticity was shown to create intermediate zones, in which either demographic stochasticity or environmental catastrophes imposed a substantial risk of extinction, even though the metapopulation was predicted to persist in a deterministic model. Many studies have concluded that the risk of metapopulation extinction increases with increasing regional stochasticity, that is, increasing the scale of environmental variability. For example, Palmqvist and Lundberg (1998) used a coupled map lattice model consisting of local populations with density-dependent dynamics and density-independent migration. They found that the major determinant of the risk of metapopulation extinction is the balance between local population variability and synchrony in local population fluctuations. To what extent regional stochasticity increases the extinction risk of real metapopulations is a quantitative question, which cannot be answered by just demonstrating that there is some degree of spatial correlation in local dynamics.

4.7 CURRENT STATE OF THE THEORY AND APPLICATIONS

Metapopulation theory based on stochastic patch occupancy models has its roots in the Levins model (Levins, 1969), which was originally developed to study alternative pest eradication strategies. In the 1970s, models stemming from the Levins model were constructed to study group selection (Levins, 1970; Levitt, 1978) and competitive (Levins and Culver, 1971; Horn and MacArthur, 1972; Slatkin, 1974; Hanski, 1983) and predator–prey interactions (Maynard Smith, 1974). Patch occupancy models for two or more species were investigated further in the 1990s (Nee and May, 1992; Tilman et al., 1994; Lei and Hanski, 1997; Nee et al., 1997; Holt, 1997). These studies have largely remained at the theory level, although Hanski and Ranta (1983) applied the Levins model extended to three competitors to study the dynamics of *Daphnia* water fleas in rock pools. A few other studies have employed homogeneous SPOMs in the study of actual metapopulations (Lande, 1988b; Kindvall, 1996; Ås et al., 1997; Doncaster and Gustafsson, 1999; Carlson, 2000). However, given the typically marked heterogeneities in real habitat patch networks, these applications are best interpreted as attempts to increase qualitative insight about metapopulation dynamics. Quantitative predictions are best based on heterogeneous SPOMs and the spatially realistic metapopulation theory that has been developed since the first incidence function models (Hanski, 1992, 1994a).

The great advantage of SMT in comparison with much of the other theory in spatial ecology is that the models can be applied to real metapopulations for purposes of research, management, and conservation. The key here is the set of assumptions that these models make on the influence of landscape structure on population processes. SMT prescribes research tasks for ecologists engaged in empirical work in terms of testing model assumptions and predictions. The incidence function model (Hanski, 1994a) has been employed widely in the study of classic metapopulations of insects, frogs, small mammals, and birds (reviews in Hanski, 1999b, 2001b). Statistically rigorous methods have been developed to estimate model parameters of both SPOMs and their mean-field approximations, including maximum likelihood (Moilanen, 1999, 2000) and Bayesian estimation (O'Hara et al., 2002; Ter Braak and Etienne, 2003; Chapter 5).

Although metapopulation theory based on SPOMs has advanced greatly in recent years, the theory is still in a state of active development and major steps forward can be expected to be taken in the near future. We list here some of the challenges for further research. First, it would be interesting to integrate SPOMs with metapopulation models structured by the sizes of local populations (Gyllenberg and Hanski, 1992; Gyllenberg et al., 1997; Casagrandi and Gatto, 1999). Among other things, this would allow one to extend mechanistic models for the rescue effect (Etienne, 2000, 2002b) to heterogeneous patch networks. Second, and related to the former, not accounting for individuals largely prevents the extension of SPOMs to address population genetic and evolutionary questions. However, independently of the ecological research on SPOMs, Whitlock and Barton (1997) have started to develop population genetic theory that takes into account the influence of spatial heterogeneity at the landscape level, and Heino and Hanski (2001) have modeled the influence of landscape structure on the evolution of migration rate (see also Parvinen, 2002; Parvinen et al., 2003). There is much potential in these directions. Third, SPOMs apply best to highly fragmented landscape, although the theory has also been applied to landscapes described as lattices by interpreting the regularly spaced lattice cells as habitat patches (Ovaskainen et al., 2002; Pakkala et al., 2002). In this context in particular, the standard assumption of independent local dynamics becomes very questionable, and ways have to be devised of incorporating spatially correlated dynamics into the model (for one approach with potential for some practical applications, see Gu et al., 2002). Finally, it may be possible to combine SPOMs with other techniques used in spatial ecology, such as spatial moment equations (Chapter 3).

4.8 LINKS WITH LANDSCAPE ECOLOGY, EPIDEMIOLOGY, AND MATRIX POPULATION MODELS

Incorporating the influence of landscape structure on metapopulation processes in the manner that this is done in SMT has led to the realization that more substantial links can be forged between metapopulation ecology and related disciplines than has been perceived previously (Hanski, 2001b; Hanski and Ovaskainen, 2003). In the first place, SMT contributes to the conceptual unification of metapopulation ecology and landscape ecology. Merging of these two fields has been anticipated for a long time (Hanski and Gilpin, 1991), but not much has happened until now, as can be seen from a lack of papers that refer to both disciplines (Chapter 1, Fig. 1.2). Metapopulation ecology and landscape ecology have largely adhered to their own research traditions, to the extent that shared key concepts such as connectivity are used in a different manner (Tischendorf and Fahrig, 2000a; Moilanen and Hanski, 2001). Nonetheless, many landscape ecologists have the same goal in their research as we do, to take account of the influence of the spatial structure of the landscape on population and other processes (Turner et al., 2001). So far, most of these studies have been based on simulations (Chapter 2). SMT represents a coherent body of theory that should go a long way in answering the needs of those landscape ecologists who are interested in population and metapopulation processes.

Second, spatially realistic metapopulation models are closely related to matrix population models for age-structured and size-structured populations (Caswell, 2001). As a matter of fact, much of the mathematical theory is the same (Ovaskainen and Hanski, 2001). Just as traditional matrix models divide populations into age classes, SMT divides metapopulations into individual habitat patches. Important mathematical similarities include the role of the dominant eigenvalue and eigenvector of the respective population matrices in determining population growth rate and persistence. There are also significant differences. For example, in SMT, transitions are possible between any pairs of "classes", whereas in age-structured models, individuals move from one class to another in a predictable manner. Another difference is in the manner in which density dependence typically enters into the models. Age-structured models assume that the effective population density that drives intraspecific competition (e.g., by increasing mortality) is the overall density, calculated as the weighted sum of individuals in the different age classes. This simplification does not apply to spatially structured models, where one needs to know exactly which patches are occupied (local as opposed to global density) to calculate the colonization rate of an empty patch.

Third, SPOMs are closely related to epidemiological models (see Chapter 20 and references therein). The archetypal metapopulation model, the Levins model, is identical to a basic epidemiological model, the susceptible–infected–susceptible (SIS) model. Ovaskainen and Grenfell (2003) utilized the correspondence between metapopulation and epidemiological models in using the patch value measures discussed in Section 4.4 in the study of the effectiveness of various intervention scenarios for sexually transmitted diseases. Further cross-fertilization of these two fields has much potential, including metapopulations inhabiting dynamic landscapes versus disease transmissions in dynamically changing contact networks, the evolution of migration rate and other life history traits versus the evolution of virulence, and predator–prey metapopulation models versus SIR models.

With such links as described earlier developing among metapopulation ecology, landscape ecology, epidemiology, and general population ecology, a new era of exciting research is in the horizon.

5. APPLICATION OF STOCHASTIC PATCH OCCUPANCY MODELS TO REAL METAPOPULATIONS

Rampal S. Etienne, Cajo J.F. ter Braak, and Claire C. Vos

5.1 INTRODUCTION

Consensus has emerged about the generally detrimental effects of habitat loss and habitat fragmentation on the viability of metapopulations (Hanski, 1999). Habitat loss and fragmentation both affect the balance between extinction of local populations and recolonization of empty patches: patch occupancy decreases and the metapopulation is more prone to extinction. Simple statistical models for patch occupancy have been used to ascertain the existence of these detrimental effects in real metapopulations (Van Dorp and Opdam, 1987; Merriam, 1988), but their predictive power to show the magnitude of these effects is very limited, as they do not incorporate the mechanisms underlying metapopulation dynamics (local extinction and colonization). The Levins (1969, 1970) model, the prototype metapopulation model that is based on these mechanisms, is biologically too unrealistic, whereas size-structured and individual-based metapopulation models are mostly too complex to be parameterized with available data. These data almost never consist of accurate

0-12-323448-4

estimates of population sizes. At best, they comprise observations of patches in a patch network being occupied or empty for a single year (snapshot data; Hanski, 1994) or for several, not always consecutive, years (which give information of the population turnover).

There is, however, one class of metapopulation models that can both capture sufficient biological detail and be relatively easily parameterized with these data: the stochastic patch occupancy models (SPOMs, see Chapter 4). Given the current occupancy states of all patches in a patch network, SPOMs predict for each patch the probability that it will be occupied at any time in the future. These probabilities depend on the probabilities of local extinction and colonization, which, in turn, may be constant (resulting in a stochastic Levins model) or depend in a variety of ways on the quantities that are considered relevant, such as patch quality and connectivity (or spatial cohesion; Opdam et al., 2003). Perhaps the simplest such quantities are the area of the patch and the distance between patches, but other quantities are sometimes preferable because area and interpatch distance are not always the best predictor variables (Thomas et al., 2001b). This chapter uses as an example a model for a tree frog (*Hyla arborea*) metapopulation with area and interpatch distance as well two other predictor variables (Vos et al., 2000). For another example, see Hanski et al. (1996).

Application of a SPOM to a real metapopulation consists of four parts. First, patch occupancy data must be collected. Second, the SPOM must be formulated. Here questions such as "what processes do I want to describe and what variables do I need and what processes and variables can I dispense with?" must be answered. Third, given the mathematical representation of the SPOM that results from the second step, the model must be parameterized using the dataset at hand and/or independent information about these parameters. Fourth, predictions can be made with the model by considering different scenarios. The reliability of these predictions must be assessed through uncertainty analysis.

This chapter assumes the dataset and the SPOM to be given and concentrates on the third part by reviewing the various methods that have been developed to parameterize SPOMs using snapshot and/or turnover data. It then focuses on some of the problems still remaining, and their consequences for the fourth part, making model predictions for real metapopulations; these problems are illustrated with the afore-mentioned tree frog example. We show that the model predictions are useful to determine the best conservation strategies for the real metapopulation. For this we compare the distribution pattern in 2002 with the distribution pattern predicted from the model calibrated with data from 1981–1983 and 1986. The chapter ends with a discussion of the four parts in the application of SPOMs and an overview of the insights gained by the application of SPOMs as reported in the literature.

5.2 STOCHASTIC PATCH OCCUPANCY MODELS

Stochastic patch occupancy models have been formulated in discrete time (Day and Possingham, 1995; Hanski, 1994) and continuous time (Verboom et al., 1991; Frank and Wissel, 1998; Etienne, 2002; Etienne and

Nagelkerke, 2002; Chapter 4). Because census data are usually separated by 1 or several years, it seems natural to use a discrete-time model with a time step of 1 year; in this way difficulties with extinction and colonization probabilities not being constant during the year can be avoided. Consider now a patch network of N patches. If a patch i contains a population, this population can go extinct in one time step with probability E_i, and if the patch is empty, it can be colonized with probability C_i (E_i and C_i are denoted by Ext_i and Col_i in Chapter 4, but we adhere to the symbols introduced here for better comparison with the literature cited). Let us denote the state of patch i at time t by $X_i(t)$, which is a binary variable: we have for an occupied patch $X_i(t) = 1$ and for an empty patch $X_i(t) = 0$ (X_i is denoted by O_i in Chapter 4). The total state of the metapopulation at time t, $X(t)$, can then be described by a vector containing N ones and zeros. The colonization probability of patch i at time $t + 1$ usually depends on the state $X(t)$ because this state determines the number of dispersers being sent out that may end up in patch i, but not on earlier states such as $X(t - 1)$. The extinction probability is usually considered to be independent of $X(t)$, but when there is a rescue effect (Brown and Kodric-Brown, 1977) — populations on the brink of extinction are rescued from extinction by immigrants — the extinction probability depends on the colonization probability and thereby on $X(t)$.

The dynamics of the SPOM can thus be described by a single formula, valid for each patch i, that specifies the probability of the state of patch i at time $t + 1$ conditional on the state of the metapopulation at time t:

$$P[X_i(t+1)|X(t)] = \begin{cases} 1 - E_i & \text{if } X_i(t) = 1 \text{ and } X_i(t+1) = 1 \\ E_i & \text{if } X_i(t) = 1 \text{ and } X_i(t+1) = 0 \\ 1 - C_i & \text{if } X_i(t) = 0 \text{ and } X_i(t+1) = 0 \\ C_i & \text{if } X_i(t) = 0 \text{ and } X_i(t+1) = 1 \end{cases} \tag{5.1}$$

in which E_i and C_i are shorthand for $E_i(t)$ and $C_i(t)$, the extinction and colonization probabilities for the transition from time t to time $t + 1$. These may depend on species and landscape characteristics that can be treated either as observed variables or known or unknown parameters in the model.

Because SPOMs possess the Markov property that the probability distribution for the state at time $t + 1$ is completely determined by the state at time t, Markovian theory can be invoked to state various characteristics of the metapopulation. One such characteristic is that the metapopulation, left undisturbed and conditional on the metapopulation not having gone extinct untimely, will settle in a pseudoequilibrium in which the probability that it is in a certain state (e.g., all patches occupied) no longer changes in time. This pseudoequilibrium is also called the quasistationary state (see further Darroch and Seneta, 1965; Gyllenberg and Silvestrov, 1994; Gosselin, 1998; Ovaskainen, 2001; Chapter 4). This is a property used frequently in parameterization methods, as discussed later. We will first give a few examples of SPOMs.

Hanski's Incidence Function Model

In probably the best-known SPOM, described by Hanski (1994) and better known as the incidence function model (IFM), the extinction probability is set proportional to the inverse of patch area:

$$E_i = \min\left(1, \frac{e}{A_i^x}\right) \tag{5.2}$$

where e and x are parameters (e is sometimes written as A_0^x where A_0 is the minimum possible patch area in which a viable population can exist, x is denoted by ζ_{ex} in Chapter 4) and the colonization probability is a saturating function of connectivity measure S:

$$C_i = \frac{S_i^2}{S_i^2 + y^2} \tag{5.3}$$

where y is a parameter and S_i is given by a sum of contributions from all occupied patches weighted by their distance to patch i:

$$S_i(t) = \sum_{j \neq i} X_j(t) A_j \exp(-\alpha d_{ij}) \tag{5.4}$$

where α is a parameter that can be interpreted as the inverse of the mean dispersal distance of the species under consideration. This is Hanski's (1994) model without the rescue effect. Incorporating the rescue effect means multiplying the extinction probability by $1 - C_i$. In later work, Hanski and co-workers added the area of the destination patch A_i to the connectivity and parameters that determine the strength of the contribution of the patches of origin and destination (e.g., Wahlberg et al., 1996; Hanski, 1998a,b; Moilanen and Hanski, 1998; Chapter 4). See Moilanen and Nieminen (2002) for a discussion of different measures of connectivity. Formulas for the extinction and colonization probabilities have some mechanistic basis as pointed out by Hanski (1998a,b).

The model is called IFM because Hanski (1994) derived an incidence function from the extinction and colonization probabilities. For this, he first defined the incidence J_i of patch i as the probability that it is occupied. He then assumed that the metapopulation is in the quasistationary state, which he interpreted as J_i being independent of time. This leads to

$$J_i = J_i(1 - E_i) + (1 - J_i)C_i \Rightarrow J_i = \frac{C_i}{C_i + E_i} \tag{5.5}$$

For example, for the IFM with rescue effect [but without the cutoff at 1 of E_i in Eq. (5.2)], the incidence function becomes

$$J_i = \frac{1}{1 + \dfrac{ey^2}{A_i^x S_i^2}} \tag{5.6}$$

The Extended IFM for Tree Frog Data

Vos et al. (2000) extended the IFM to model a tree frog metapopulation in a pond network in southwest The Netherlands, for which occupancy data have been collected in 1981–1983 and 1986 (see Box 5.1). In their model, the form of the function for colonization probability is relaxed by replacing the exponent of 2 by parameter z (as already suggested by Hanski, 1994) and by allowing the contribution of patch area in the connectivity measure to be determined by parameter b (Wahlberg et al., 1996; b is denoted by ζ_{em} in

BOX 5.1 The Tree Frog Metapopulation in Zealand Flanders, The Netherlands

The tree frog metapopulation exists in an agricultural landscape in southwest The Netherlands (Fig. B5.1). The suitable habitat forms approximately 1.5% of the total landscape cover and is separated by intensively used agricultural fields that are unsuitable for the species. (Semi)natural vegetation can be found on dikes, in coastal sand dunes, and in meadows with cattle-drinking ponds, the aquatic habitat used by the tree frog. The terrestrial habitat of the tree frog consists of shrubs, bushes, and vegetation of high herbs. The tree frog distribution has declined sharply in The Netherlands since the 1950s as a result of intensification of agriculture. Most (semi)natural elements such as cattle-drinking ponds and hedgerows have been cleared.

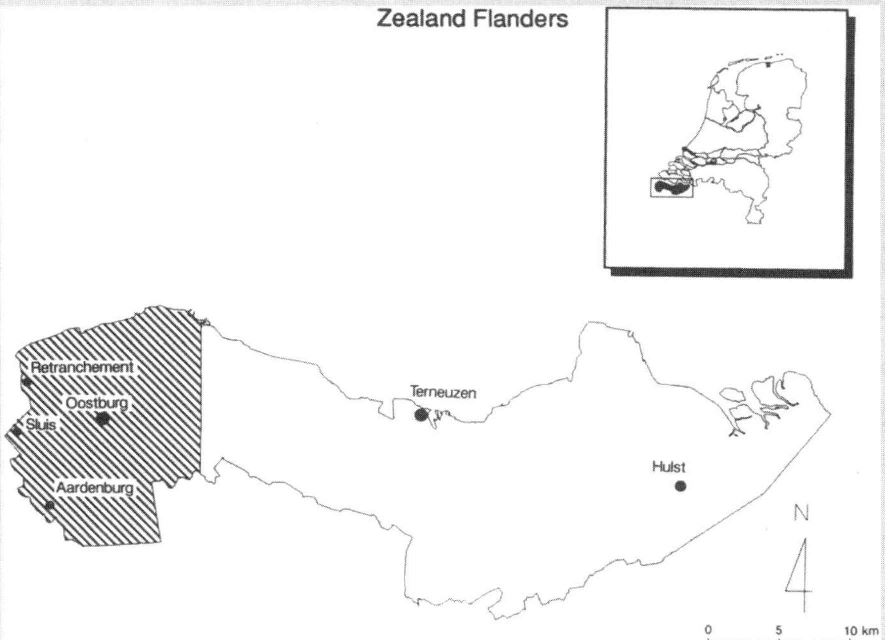

Fig. B5.1 The position of the tree frog study area. Reprinted with permission from Vos et al. (2000).

From 1981 until 1986 the distribution pattern of the tree frog in Zealand Flanders was monitored. Statistical analysis showed that the occupation probability of a pond increased with the number of other ponds and the area of high herbs and bushes in the surrounding (Vos and Stumpel, 1996). Analysis of turnover data from 1981 to 1983 showed that local extinctions were related to spatial features of the landscape: extinction probability decreases with pond size (Vos et al., 2000). Pond size encompasses both pond area and suitable terrestrial habitat within a radius of 250 m around the pond. Also, colonization events were correlated to connectivity as well as habitat quality factors. Comparing dispersal distances with distances between ponds shows that the habitat network is still connected by dispersing individuals (Fig. B5.2). Analysis of dispersal events showed that occupied ponds were preferred over empty

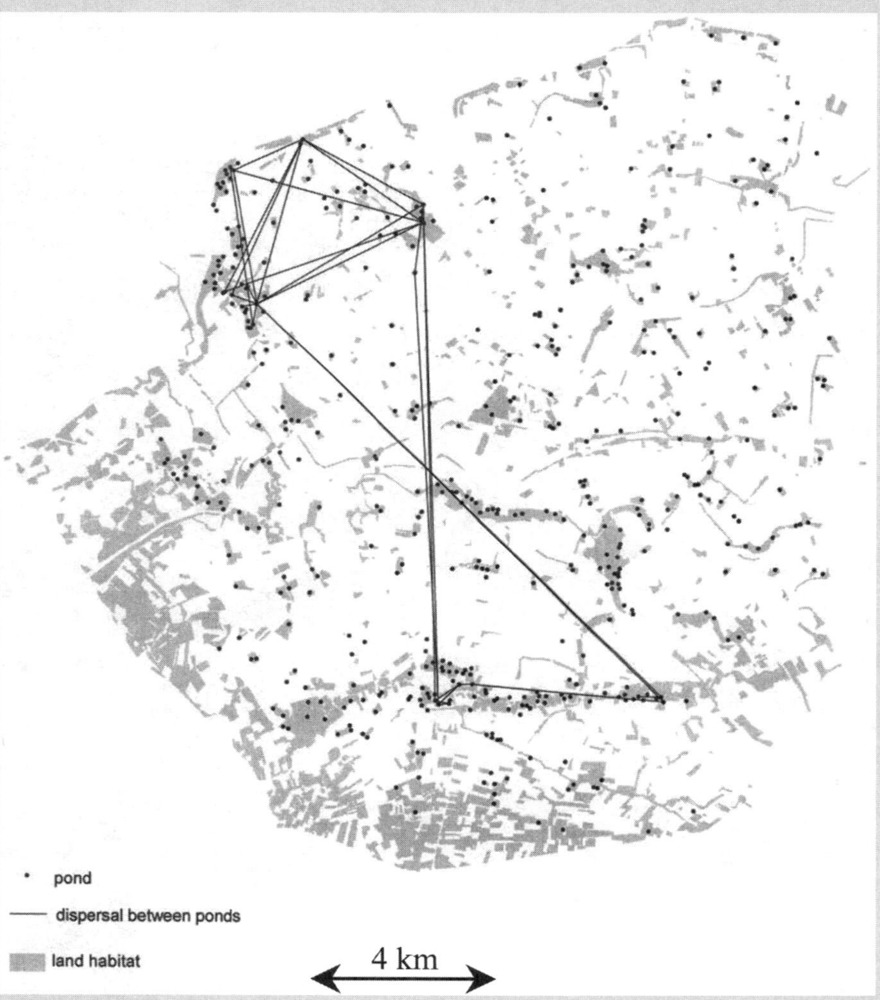

Fig. B5.2 Overview of dispersal events registered by capture–recapture techniques in Zealand Flanders during the period 1981–1989. Reprinted with permission from Vos et al. (2000).

ponds, although many empty ponds were present at close ranges. This implies the importance of conspecific attraction and a rescue effect in this metapopulation pond system.

In the next step toward spatially explicit connectivity measures the permeability of the landscape matrix is incorporated. Results of a radio telemetry study (Vos, 1999), show that moving tree frogs have a high preference for hedgerows, that they avoid arable land, and that pasture takes an intermediate position. Incorporating these characteristics in the connectivity measure of the extended IFM model will improve the model further.

The results seem to justify the application of the metapopulation concept for guidelines for optimal spatial habitat configuration for this species. The protection policy plan for the tree frog in The Netherlands is actually based on metapopulation concepts. Apart from the necessary improvements of habitat quality, the size and connectivity of the habitat networks of the tree frog are being improved (Crombaghs and Lenders, 2001). Results for a tree frog metapopulation in Twente, a region in the east of The Netherlands, already show that the species reacts positively to these restoration measures (Braad, 2000).

Chapter 4). Furthermore, three variables were introduced: B_{ij}, $H_{1,i}$, and $H_{2,i}$. B_{ij} describes the quality of the matrix habitat between patches i and j, and its value was set for all combinations of i and j (e.g., $B_{ij} = 0$ for a total barrier, $B_{ij} = 1$ for no barrier, and $B_{ij} = 0.5$ for a semibarrier). $H_{1,i}$ is the water conductivity, and $H_{2,i}$ is the percentage cover of the water vegetation; both were measured for each pond. Two additional parameters, q_1 and q_2, determine the effect of $H_{1,i}$ and $H_{2,i}$ respectively. The extinction and colonization probabilities are now given by

$$E_i = (1 - C_i)\min\left(1, \frac{eH_{1,i}^{q_1}}{A_i^x}\right) \tag{5.7}$$

and

$$C_i = \frac{S_i^z}{S_i^z + \dfrac{y}{H_{2,i}^{q_2}}} \tag{5.8}$$

with

$$S_i(t) = \sum_{j \neq i} X_j(t) A_j^b B_{ij} \exp(-\alpha d_{ij}) \tag{5.9}$$

(Note that y in this model is not the same as y in Hanski's IFM.) We mention this model, not only as an example of a SPOM with other variables than patch area and interpatch distance, but also because it is used as an illustration of some of the problems in parameter estimation.

5.3 METHODS TO ESTIMATE PARAMETERS OF STOCHASTIC PATCH OCCUPANCY MODELS

Because SPOMs contain few parameters, it is attractive to attempt to estimate them using metapopulation data, that is, data on the occupancy of some or all of the patches in a network for 1 or several years. Several methods have been developed for this purpose. They use only data on transitions between occupancies (extinctions, nonextinctions, colonizations, noncolonizations), data from a snapshot of a single year (which typically shows spatial clusters of occupied patches as the result of the history of extinctions and colonizations in the patch network), or both. This section reviews these methods. Table 5.1 summarizes their main features and shortcomings.

Hanski (1994): Regression of Snapshot Data

In his classic paper, Hanski (1994) not only introduced the IFM, but also showed how it can be parameterized in a simple way using snapshot data of patch occupancies, elaborating ideas presented by Peltonen and Hanski (1991) and Hanski (1992). He proposed nonlinear regression to obtain point estimates of the parameters, with standard errors, by maximizing the logarithmically transformed likelihood, i.e., the loglikelihood,

$$\sum_i \left[X_i \ln J_i + (1 - X_i)\ln(1 - J_i)\right] \qquad (5.10)$$

Note from Eq. (5.6) that in the IFM without rescue effect, e and y form a single composite parameter so that they cannot be estimated separately from snapshot data. Independent information on the minimum sustainable patch size A_0 can help to estimate e and thus separate e and y.

Ter Braak et al. (1998) pointed out that Eq. (5.6) — the IFM with rescue effect and with all patches larger than A_0 — can be linearized by the logit-transformation of J_i giving

$$\log\left(\frac{J_i}{1-J_i}\right) = \beta_0 + \beta_1 \log(A_i) + \beta_2 \log(S_i) \qquad (5.11)$$

with $\beta_0 = -\log(ey^2)$, $\beta_1 = x$, and $\beta_2 = 2$. Therefore, available data being binary values, point estimates for the parameters (with approximate standard errors) can be obtained easily by using logistic regression (McCullagh and Nelder, 1989) of the occupancies X_i on $\log(A_i)$ with offset $2\log(S_i)$ — at least when the values of α and b in the definition of S_i are given. (An offset is a predictor with a regression coefficient of 1.) Although this is mathematically equivalent to Hanski's nonlinear regression, it is computationally much more efficient. It is historically interesting that Eber and Brandl (1994) already performed such logistic regression of patch occupancies without knowing the link to the incidence function model (which had not been published yet); they just used a statistical model. The extended IFM can be fitted to snapshot data using a logistic regression of occupancies X_i on $\log(A_i)$, $\log(S_i)$, $\log(H_{1,i})$ and $\log(H_{2,i})$. We can estimate α and b by logistic regression by calculating

TABLE 5.1 Overview of Features of Existing Methods to Estimate Parameters of SPOMs from Occupancy Data

Method (reference)	Hanski (1994)	Verboom et al. (1991)/Sjögren-Gulve and Ray (1996)	Ter Braak et al. (1998)/ Vos et al. (2000)	Moilanen (1999)	O'Hara et al. (2002)	Ter Braak and Etienne (2003)
What type of information is used?	Spatial structure only	Turnover only	Spatial structure and turnover	Spatial structure and turnover	Turnover only	Spatial structure and turnover
How is quasistationarity incorporated?	Approximation	Not at all (turnover only)	Approximation	Correctly	Not at all (turnover only)	Correctly
Are missing data taken into account?	No (ignored/ considered empty)	No	No (ignored)	Yes	Yes	Yes
What method are parameter estimates based on?	Likelihood optimization (nonlinear regression)/ with rescue effect: logistic regression	Logistic regression	Likelihood optimization (non-linear regression)	Likelihood optimization (MC simulation)	Bayesian MCMC (Gibbs sampling)	Bayesian MCMC (Metropolis-Hastings)
How is uncertainty in estimates specified?	Approximations of standard errors	Approximations of standard errors	Approximations of standard errors	Approximations of confidence intervals	Full uncertainty distribution	Full uncertainty distribution
Is uncertainty analysis of model predictions possible?	No	No	No	No	Yes	Yes
Computational efficiency?	Fast / very fast	Fast	Fast	Slow	Very slow	Slow

the regression over a grid of values of α and b and by locating where the log likelihood [Eq. (5.10)] is maximum.

There are five caveats in the method (Ter Braak et al., 1998; Ter Braak and Etienne, 2003). First, the condition of nonextinction of the metapopulation (required for quasistationarity) is omitted in Eq. (5.5). This is perhaps a valid approximation for large networks that are unlikely to go extinct quickly, but because metapopulation models are often used to explore scenarios intended to conserve threatened metapopulations, this need not always be the case. Second, it is assumed in Eq. (5.5) that the extinction and colonization probabilities of patch i are constant in time. This is not true, for they both depend on the evolving states of the other patches. Third, the connectivity S is calculated using the patch occupancies of the same year and not using the patch occupancies of the previous year as it should according to the model [Eq. (5.4)]; the reason is of course that these are unknown for snapshot data. Fourth, in the log likelihood [Eq. (5.10)], patch occupancies are assumed to be statistically independent. However, this assumption is not met because the patch occupancies are regressed on the connectivity S that is calculated from the same data. Therefore, Eq. (5.10) is at best a pseudo-log likelihood. Fifth, in the derivation of the incidence function Eq. (5.6), the cutoff at 1 of E_i in Eq. (5.2) is neglected. This is warranted only if all areas are larger than A_0.

Missing data (occupancies are unknown for some patches) cannot be dealt with properly in this method. They are basically ignored. It is apparent from Eq. (5.4) that this may affect the connectivity. By ignoring patches of which no data are available, one effectively assumes them to be empty because they do not contribute to connectivity. If the patches were actually occupied, the colonization probability will be underestimated.

Verboom et al. (1991), Sjögren-Gulve and Ray (1996), and Eber and Brandl (1996): Regression of Turnover Data

While Hanski (1994) only considered snapshot data, Verboom et al. (1991), Sjögren-Gulve and Ray (1996), and, less commonly cited, Eber and Brandl (1996) looked only at turnover data [actually, Verboom et al. (1991) also looked at the frequency that a patch is found occupied, see later]. They realized that E_i and C_i in Eq. (5.1) are conditional probabilities that can be fitted to data by logistic regression (McCullagh and Nelder, 1989). To parameterize E_i, they created a dataset with all possible pairs $[X_i(t), X_i(t+1)]$ with $X_i(t) = 1$ and then applied logistic regression with $X_i(t+1)$ as the response variable. Similarly, C_i was parameterized by applying another logistic regression to all possible pairs $[X_i(t), X_i(t+1)]$ with $X_i(t) = 0$ and with $X_i(t+1)$ as the response variable. They thus assumed that the probabilities of extinction and colonization behave logistically:

$$E_i = \frac{1}{1 + \exp(-u_{e,i})}$$
$$C_i = \frac{1}{1 + \exp(-u_{c,i})} \quad (5.12)$$

where u_e and u_c are linear functions of the variables of importance, such as patch area and connectivity, the logarithm of which are treated as explanatory variables in logistic regressions. The parameters in u_e and u_c are fitted from the "extinction dataset" and the "colonization dataset," respectively, giving point estimates and standard errors. They found that extinction is significantly related to patch area (and not connectivity) and colonization to connectivity (and not to area).

Sjögren-Gulve and Ray (1996) used point estimates in subsequent computer simulations of the discrete-time SPOM to predict the long-term trend in occupancy, whereas 5 years earlier Verboom et al. (1991) had missed that opportunity, partly because they adhered to their continuous-time SPOM.

Note that extinction probabilities do not have the logistic form in the IFM [Eq. (5.2)] and they are even linked to the colonization probabilities in the (extended) IFM with a rescue effect [see Eq. (5.7)]. We must therefore resort to nonlinear regression (Vos et al., 2000) where the log likelihood to be maximized is

$$\sum_i [X_i(t)(1-X_i(t+1))\ln E_i + X_i(t)X_i(t+1)\ln(1-E_i)] \\ + \sum_i [(1-X_i(t))X_i(t+1)\ln C_i + (1-X_i(t))(1-X_i(t+1))\ln(1-C_i)] \tag{5.13}$$

Ter Braak et al. (1998) showed that this is equivalent to maximizing the extinction and colonization parts of the log likelihood separately, provided E_i and C_i have different, independent parameters, so giving a formal justification of the approach by Verboom et al. (1991) and Sjögren-Gulve and Ray (1996).

Although this method is theoretically sound (there are no technical difficulties as with snapshot data), using only turnover data has some shortcomings. First, it requires data collection in at least 2, but preferably several consecutive years. Second, while snapshot data are the result of many extinctions and colonizations in the history of the metapopulation and are therefore considered to contain a lot of information, turnover data provide little information if turnover is slow so that data show only a few extinctions and colonizations [most information would be provided if the number of extinctions (cq. colonizations) and the number of nonextinctions (cq. noncolonizations) were equal]. Third, missing data are again ignored, resulting in the same bias as for snapshot data.

Ter Braak et al. (1998) and Vos et al. (2000): Combining the Previous Approaches

To make full use of data, both the historical turnover information contained in the first year of a dataset and the turnover information in the following years should be extracted. Interestingly, Verboom et al. (1991) not only applied logistic regression to turnover events as discussed earlier, but also applied logistic regression to the frequency that a patch is found occupied during the years of survey. For a dataset of only 1 year, the latter is formally equivalent to the snapshot data analysis of Hanski (1994). For data of several years, the method also takes turnover events into account, but extinction and colonization are not separated (a 111000 sequence is considered equal to 101010).

Verboom et al. (1991) must have realized that there is more information in a dataset than only turnover. Eber and Brandl (1994, 1996) certainly realized this, but they did not combine the two approaches.

It was Ter Braak et al. (1998) who mentioned a pragmatic method to combine the two previous approaches. For this, the likelihood of the dataset is partitioned in three parts: the "incidence dataset" from the first year of a dataset, the "extinction dataset," and the "colonization dataset" from the following years. The likelihood to be maximized (see also later) is the sum of Eq. (5.10) and Eq. (5.13), so Hanski's incidence function model [Eq. (5.6)] is used for the first year. This combined approach, first applied in Vos et al. (2000), suffers of course from the same problems as its constituents: it uses a pseudo-likelihood for the first year data, it yields only point estimates of the mean and standard error of the parameters, and cannot handle missing data properly.

Moilanen (1999): Monte Carlo Simulation

Moilanen (1999) provided the first solution to problems involving the pseudo-likelihood and missing data in a new approach based on maximum likelihood estimation using Monte Carlo simulations. Because of the Markov property of the SPOM, the probability of a dataset X of T years, given values for the parameters Θ, can be written as

$$P[X|\Theta]=P[X(1)]P[X(2)|X(1)]\ldots P[X(t+1)|X(t)]\ldots P[X(T)|X(T-1)] \quad (5.14)$$

with, for each year t of this sequence,

$$P[X(t+1)|X(t)]=\prod_i P[X_i(t+1)|X(t)] \quad (5.15)$$

because the states are independent conditional on the state of the system in the previous year. Note the separation between spatial information, contained in $P[X(1)]$, and the turnover information, contained in the remaining conditional probabilities, as noted by Ter Braak et al. (1998). Equation (5.14) is the true likelihood that needs to be maximized.

Instead of using Eq. (5.6), Moilanen (1999) approximated $P[X(1)]$ by Monte Carlo simulation. From an arbitrary state, the IFM is simulated until the quasi-stationary equilibrium is considered to be reached and then for another L time steps to obtain simulated states X_u (u = 1...L). The approximation is then

$$P[X(1)]\approx\frac{1}{L}\sum_{u=1}^{L}P[X(1)|X_u] \quad (5.16)$$

for some large number L. This Monte Carlo approximation derives from the equation (Ter Braak and Etienne, 2003)

$$P[X(1)] = \sum_{k=1}^{K}P[Y_k]P[X(1)|Y_k] \quad (5.17)$$

where the summation is over all possible states Y_k. The probabilities $P[Y_k]$ drop out of Eq. (5.17) because the simulation series X_u is self-weighing, that is, each X_u is generated by the simulation with probability proportional to $P[X_u]$. The reason for the Monte Carlo simulation [Eq. (5.16)] is that K is usually huge so that the summation in Eq. (5.17) cannot be carried out in practice.

Missing years in data can be handled by simulation as well. If, for example, data for year 2 are missing, we require in the likelihood [Eq. (5.14)] the term

$$P[X(3)|X(1)] = \sum_{k=1}^{K} P[X_k(2)|X(1)]P[X(3)|X_k(2)] \tag{5.18}$$

where the summation is over all possible states $X_k(2)$. To approximate Eq. (5.18), the state in year 2 is simulated M times (M being large) from the state in year 1. The Monte Carlo approximation is then [analogously to Eq. (5.16)]

$$P[X(3)|X(1)] \approx \frac{1}{M}\sum_{u=1}^{M} P[X(3)|X_u(2)] \tag{5.19}$$

Although Moilanen (1999) does not mention it explicitly, the same procedure can be applied if, instead of a complete year, only a few occupancies are missing in a single year. Evidently, only the missing data are simulated.

Being a maximum likelihood method, Moilanen's (1999) approach only produces point estimates, although estimates of confidence limits can be computed (but require a lot of computing time).

O'Hara et al. (2002): Bayesian MCMC on Turnover Data

To obtain a full joint probability distribution of the model parameters instead of mere point estimates, the maximum likelihood method must be abandoned. O'Hara et al. (2002) were the first to adopt a Bayesian approach to parameter estimation. The central idea in Bayesian theory (e.g., Gelman et al., 1995) is that our knowledge of the value of a parameter can be represented by a probability distribution and that data containing new information about this parameter can be used to adjust this probability distribution. The probability distributions before and after data have been used to update our knowledge are called prior and posterior probability distributions. Bayes' formula describes how the probability distribution of model parameter Θ is adjusted using data X:

$$P[\Theta|X] = \frac{P[X|\Theta]P[\Theta]}{P[X]} \tag{5.20}$$

The posterior probability distribution $P[\Theta|X]$ can often be approximated through Markov chain Monte Carlo (MCMC) simulation with the Metropolis–Hastings algorithm. The Metropolis–Hastings algorithm consists of the following steps. First, arbitrary values of the model parameters are chosen. New values of the model parameters are then drawn from a probability distribution called the jumping distribution J_u. The form of this jumping distribution is arbitrary, but a smart choice will facilitate calculations and convergence of the simulation to the posterior distribution. These new values, denoted collectively by Θ^*, are now accepted with probability

$$r = \min\left(1, \frac{P[\Theta^*|X]J_u[\Theta^{u-1}|\Theta^*]}{P[\Theta^{u-1}|X]J_u[\Theta^*|\Theta^{u-1}]}\right) \tag{5.21}$$

where Θ^{u-1} represent the previous values of the model parameters. This procedure of drawing new values and accepting or rejecting them is iterated many times (u denotes the iteration number) and creates a Markov chain; acceptance of new parameter values in iteration u only depends on the values accepted in iteration $u - 1$. The set of values Θ^u generated in this way constitutes (a sample from) the posterior distribution $P[\Theta|X]$. The first half of the iterations is often discarded because the simulation needs some time (called the burn-in period) to converge to the stationary distribution of the Markov chain (which has nothing to do with the Markov property of the SPOM).

When Eq. (5.20) is inserted into Eq. (5.21), the probabilities $P[X]$ cancel and only the prior probability distribution $P[\Theta]$ and the probability of data conditional on the model parameters $P[X|\Theta]$ remain. Note that $P[X|\Theta]$ is the likelihood. The prior can be chosen based on prior knowledge of the model parameters, and the likelihood $P[X|\Theta]$ is given by the SPOM itself, that is, Eq. (5.14) with Eqs. (5.15) and (5.1). O'Hara et al. (2002) used only turnover data in which case the state in the first year is considered to be given so that $P[X(1)]$ drops out of Eq. (5.14). For the jumping distribution they used a normal distribution with mean Θ^u and covariance matrix Σ (to be chosen arbitrarily, but a smart choice speeds up convergence).

As in Moilanen's (1999) approach, the problem of missing data can be tackled by simulating them, but in a different way. In the Bayesian context, missing data are in fact treated as parameters; the MCMC thus also yields posterior probability distributions for these missing data. The Metropolis–Hastings algorithm allows alternate sampling of (sets of) parameters, that is, they do not need to be drawn from a single joint jumping distribution simultaneously. It is most convenient to sample the set of model parameters and the set of missing data in turn. O'Hara et al. (2002) chose to sample missing data for each patch in each year separately (see Box 5.2 on Gibbs sampling).

As O'Hara et al. (2002) only considered turnover data, they could not use all information in a dataset. This problem was solved by Ter Braak and Etienne (2003).

Ter Braak and Etienne (2003): Bayesian MCMC on the Full Dataset

While O'Hara et al. (2002) were working on their Bayesian analysis of turnover data, Ter Braak and Etienne (2003) were also developing a Bayesian method. This method turned out to generalize the approach by O'Hara et al. (2002) on two main points.

First, Ter Braak and Etienne (2003) were able to also exploit the information in the first year. Their idea was to extend data with L missing preyears, with L a large number, and to choose arbitrary fixed states for the year $-L$. The likelihood of extended data, given the chosen states in year $-L$, is simply a product of $L + T - 1$ transition probabilities [compare Eq. (5.14) with the

BOX 5.2 Simulating Missing Patch Data One at a Time by Gibbs Sampling

When some patch states are unknown (i.e., with missing data), the expression for the likelihood [Eq. (5.14)] cannot be calculated, even if $P[X(1)]$ were known, because some transition probabilities of Eq. (5.1) are then unknown. The trick is to fill in missing data from the correct conditional distribution. O'Hara et al. (2002) achieved this by simulating each missing patch state in turn, starting from initial guessed states and initial model parameters. After each missing state is simulated once (or more than once), new model parameters are proposed [and accepted with the acceptance probability of Eq. (5.21)]. With the then current model parameters, each missing state is simulated again. Next, new model parameters are proposed and so on until convergence (i.e., when the distribution of the model parameters and simulated states does not change any more).

To simulate a single missing patch, we need the probability p_i that patch i at time t is occupied, given all other patch states, denoted by X_{-i}. This probability can be calculated by

$$p_i = P[X_i(t) = 1|X_{-i}] = \frac{f_i}{1 + f_i}, \tag{1}$$

with f_i the odds ratio (McCullagh and Nelder, 1989)

$$f_i = \frac{P[X_i(t) = 1|X_{-i}]}{P[X_i(t) = 0|X_{-i}]} = \frac{P[X_i(t) = 1, X_{-i}]}{P[X_i(t) = 0, X_{-i}]}, \tag{2}$$

The second equality in Eq. (2) follows from the rule for conditional probability, $P[A|B] = P[A,B]/P[B]$, and by noting that $P[X_{-i}]$ drops out. By applying Eqs. (5.14) and (5.15) to the numerator and denominator of Eq. (2) and observing that all terms cancel except those involving years $t - 1$, t and $t + 1$ we obtain

$$f_i = \frac{P[X_i(t) = 1|X(t-1)]P[X(t+1)|X_i(t) = 1, X_{-i}(t)]}{P[X_i(t) = 0|X(t-1)]P[X(t+1)|X_i(t) = 0, X_{-i}(t)]}. \tag{3}$$

In this method, a 1 is filled in for the unknown state with probability p_i and a 0 with probability $1 - p_i$. Simulating the missing values in this way is known as Gibbs sampling because we sample from the exact conditional distribution. The acceptance probability then equals 1. Equation (3) is not cheap to calculate, as the second term in both the numerator and the denominator of f_i involves the multiplication of $N + 1$ transition probabilities, as is evident from Eq. (5.15). If many patch states are missing for a particular year, it is computationally more efficient to simulate them jointly using the Metropolis–Hastings algorithm of Box 5.3.

If a state is missing in the last year, the corresponding term can simply be removed from the likelihood because the other states do not depend on it. Equivalently, the missing state is simulated as the other missing states but with p_i as defined by Eq. (1).

O'Hara et al. (2002, personal communication) used Metropolis–Hastings to each missing patch in turn with proposals derived from Eq. (5.1), i.e., without conditioning on $X(t + 1)$. This is less efficient than Gibbs sampling because Eq. (3) needs to be calculated for the acceptance probability.

first term dropped]. There are very many missing values in the extended data, but apart from that, there is nothing to prevent a standard Bayesian analysis with MCMC simulation. The validity of choosing arbitrary fixed states for year $-L$ is guaranteed by the Markov property of a SPOM that the probability of the system being in a certain state does not depend on the state of the system in the infinite past. In formula,

$$P[X(1)] = \lim_{L\to\infty} P[X(1)]|P[X(-L)] \tag{5.22}$$

for any state $X(-L)$. Thus, it is wise to take L large. As Eq. (5.17), Eq. (5.22) requires the assumption of quasistationarity.

Second, instead of sampling missing data for each patch in each year separately, Ter Braak and Etienne (2003) chose to sample missing data for all patches in a single year simultaneously, thus only alternating between years (see Box 5.3). This turned out to be much faster than the method of O'Hara et al. (2002).

BOX 5.3 Simulating Missing Patch Data by the Metropolis–Hastings Algorithm

In the Metropolis–Hasting algorithm of Ter Braak and Etienne (2003), missing patch data for a particular year t are simulated jointly by proposing values for the missing states (proposals) and accepting the proposals with an acceptance probability r [Eq. (5.21) with Θ and X interchanged]. Years are updated in turn.

If the metapopulation is believed to have low turnover probabilities (low E_i and C_i), it is easy to generate sensible proposals for all missings in year t, given the current states in the year before and after. If the states of a patch in the years before and after the missing value are the same, propose this state with high probability, say 0.99; if the states differ, choose "1" with probability ½. This rule is applied to all missing patches in year t, yielding an N vector of proposed states denoted by $X^*(t)$ to distinguish it from the current state. Of course, nonmissing data are not simulated, so that for these patches the proposed state and the current state are identical. The so-generated proposal $X^*(t)$ is accepted with acceptance probability r and $X(t)$ is retained if the proposal is not accepted. To calculate the acceptance probability r, we need the ratio of the proposal distributions and the ratio of the likelihoods of $X^*(t)$ and $X(t)$ [compare Eq. (5.21)].

Because proposed patch states are generated independently, the proposal distribution J_u is the binomial probability

$$J_u[X^*(t)|X(t)] = \prod_{i=1}^{N} p_i^{X_i^*(t)}(1 - p_i)^{(1-X_i^*(t))} \tag{1}$$

with p_i the probability in the proposal scheme that a 1 is filled in, i.e., p_i is 0.99, 0.01, or 0.5. The proposal distribution $J_u[X(t)|X^*(t)]$ is obtained as in Eq. (1) with $X(t)$ and $X^*(t)$ interchanged. The ratio of the probability (likelihood) of $X^*(t)$ over that of $X(t)$, given the states of all patch states in other years, is

$$f = \frac{P[X^*(t)|X_{-t}]}{P[X(t)|X_{-t}]} = \frac{P[X^*(t),X_{-t}]}{P[X(t),X_{-t}]} = \frac{P[X^*(t)|X(t-1)]P[X(t+1)|X^*(t)]}{P[X(t)|X(t-1)]P[X(t+1)|X(t)]} \tag{2}$$

and depends only on the states in the neighboring years $t - 1$ and $t + 1$ (see Box 5.2). In Eq. (2) X_{-t} denotes the states of all patches in all years except year t. The acceptance ratio r is

$$r = f \frac{J_u[X(t)|X^*(t)]}{J_u[X^*(t)|X(t)]} \tag{3}$$

In the proposal scheme of Ter Braak and Etienne (2003), the $\{p_i\}$ are not as simple as below Eq. (1). Instead, p_i is calculated as in Eq. (1) of Box 5.2 but with the odds-ratio f_i simplified to

$$f_i = \frac{P[X_i^*(t) = 1|X(t-1)]P[X_i(t+1)|X_i^*(t) = 1, X_{-i}(t)]}{P[X_i^*(t) = 0|X(t-1)]P[X_i(t+1)|X_i^*(t) = 0, X_{-i}(t)]}, \tag{4}$$

which involves four instead of $2(N + 1)$ transition probabilities and is thus inexpensive to calculate. The proposal $X^*(t)$ is not generated with the correct conditional distribution, but that does not matter because the theoretical validity of the method hinges on the correct acceptance probability r, which is calculated via Eqs. (1)–(3). Note that Eq. (1) depends on $X(t)$ because p_i depends on $X(t)$ via f_i in Eq. (4).

It is instructive to see which proposals are generated when E_i and C_i would be constant over time (i.e., when the connectivity of a patch remains the same over time). If $X(t - 1) = 0$ and $X(t + 1) = 0$, then $f_i = C_iE_i/(1 - C_i)^2$, which will be a small value if E_i and/or C_i is small so that p_i is also close to 0 so that with only a small probability a 1 is proposed in year t. Similar considerations show that if $X(t - 1) = 1$ and $X(t + 1) = 1$, p_i is close to 1 so that with a large probability a 1 is proposed in year t. If $X(t - 1) \neq X(t + 1) = 0$, then $f_i = (1 - E_i)/(1 - C_i)$ so that if C_i and E_i are equal or both small, p_i is approximately one-half. The proposal mechanism is thus very similar to the one begun in this Box. It has the advantage of yielding good proposals for more variable metapopulations and does not require a tuning parameter such as the value 0.99 in our initial scheme.

Software to implement the full Bayesian analysis is available in the archives of *Ecology*: http://www.esapubs.org/archive/ecol/E084/005.

5.4 REMAINING PROBLEMS FOR PREDICTIONS: A CASE STUDY

The method of Ter Braak and Etienne (2003) uses the spatial information of the first year of a dataset, reflecting the history of the metapopulation, as well as the turnover information in the following years, it can handle missing data, and it provides a joint probability distribution of the model parameters. Because missing data are treated as parameters that need to be estimated, a probability distribution of the occupancy at the missing data points is also provided. This is very convenient if one wants to make predictions, as shown later. Furthermore, the Bayesian approach allows for many extensions of the method, such as misclassifications of occupancies in the dataset (Moilanen, 2002) and temporal stochasticity (regional stochasticity in the terminology of Hanski, 1991). We refer to the discussion for their possible consequences (see also Ter Braak and Etienne, 2003). However, in addition to technical difficulties of the MCMC approach [When has the MCMC converged? How many preyears (L) should be added? Is there perhaps an even more efficient algorithm than the one presented by Ter Braak and Etienne (2003)?], there are (at least) three fundamental difficulties that need to be considered before calibrated SPOMs can be used for prediction.

First, Ter Braak et al. (2003) invoked the assumption of quasistationarity to be able to use occupancy datasets to their fullest extent. The question is whether this assumption is warranted and what impact it has. Second, even if the model parameters are precisely known, predictions of occupancy, turnover, or the time to extinction will be uncertain due to the inherent stochasticity of the model (termed extinction–colonization stochasticity in Hanski, 1991). If the contribution of the inherent stochasticity to the total uncertainty in model predictions is large, data collection for more than a few years may be rather fruitless. Third, in making predictions it is possible to take into account any planned changes in network structure (habitat creation, barrier removal). Indeed, comparing the outcome of several scenarios is often the main purpose (e.g., Wahlberg et al. 2002a). However, unpredictable changes in the landscape (habitat turnover, unknown management plans) may make reliable predictions rather difficult. These problems are discussed in order using the tree frog SPOM mentioned earlier on the tree frog dataset and on a simulated dataset for the same network with the same pattern of missing values as the real dataset.

The Quasistationarity Assumption

Assuming quasistationarity allowed Hanski (1994) to estimate model parameters from a single year of occupancy data only, and it allowed Ter Braak et al. (1998) and Ter Braak and Etienne (2003) to extract information in a dataset about the history of the metapopulation, as reflected by the first year of a dataset. However, how do we know that the metapopulation is in the quasistationary state or not? To answer this question, we need to know more about the history of the metapopulation (Moilanen, 2000), or we can attempt to find the answer using the dataset itself. Assuming that information about the past history is usually unavailable, it seems worthwhile to explore the dataset for information about the presence or absence of quasistationarity. If we can find such information, we also avoid the objection that assuming quasistationarity makes predictions worthless because it is a self-fulfilling prophecy.

We start by comparing parameter estimates as well as predictions for the complete dataset assuming quasistationarity (which is denoted by QS6), the first year only dataset assuming quasistationarity (QS1), and the turnover dataset (TO). Figures 5.1 and 5.2 show cumulative posterior distributions for the tree frog SPOM and dataset. Figure 5.1 shows that posterior distributions of the model parameters for QS1 are wider than those for QS6 and TO, which are quite alike (but note the differences for x and q_2). This does not necessarily mean that the predictions must also be different because the model parameters may be highly correlated. Indeed, Fig. 5.2A shows that the variance in distributions of the occupancy after 100 years is similar for QS1 and TO, indicating that there is nearly as much information about the occupancy in QS1 as in TO. Combining them in QS6 gives a slightly narrower distribution (i.e., steeper cumulative distribution), as shown in Fig. 5.2A.

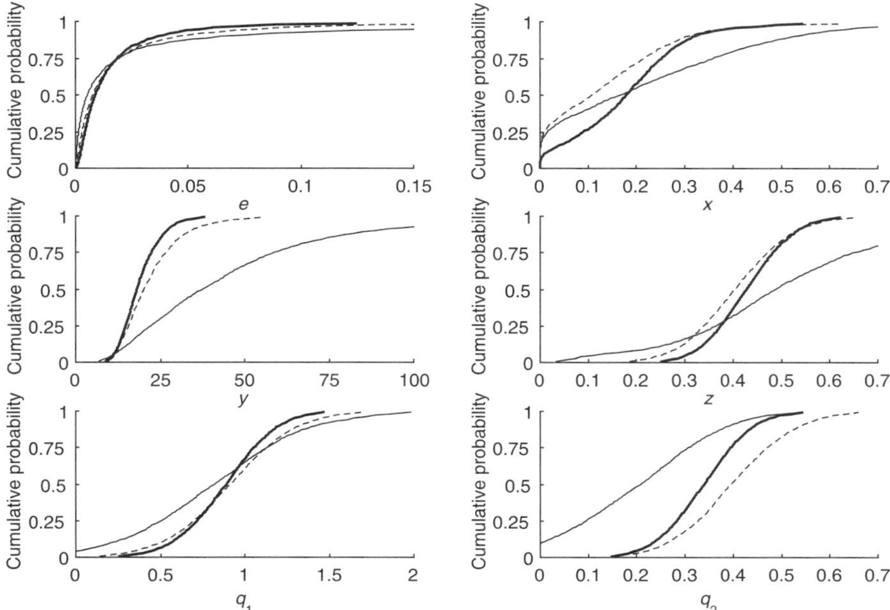

Fig. 5.1 Cumulative posterior distributions of model parameters estimated from the tree frog dataset using QS6 (thick curve), QS1 (thin curve), and TO (dotted curve). Results are based on 500,000 iterations after a burn-in of 500,000 iterations (QS6, TO) or 2,500,000 iterations (QS1, slower convergence). The number of preyears was $L = 25$.

The variances may be similar, the medians are different, with QS6 (0.129) being intermediate between QS1 (0.0941) and TO (0.158). The occupancies in the 4 data years (1981–1983, 1986) are 0.111, 0.111, 0.141, and 0.147, so TO seems to reflect this upward trend. This is not accidental. Moilanen (2000) ascertained that parameter estimation based on a few years of simulated

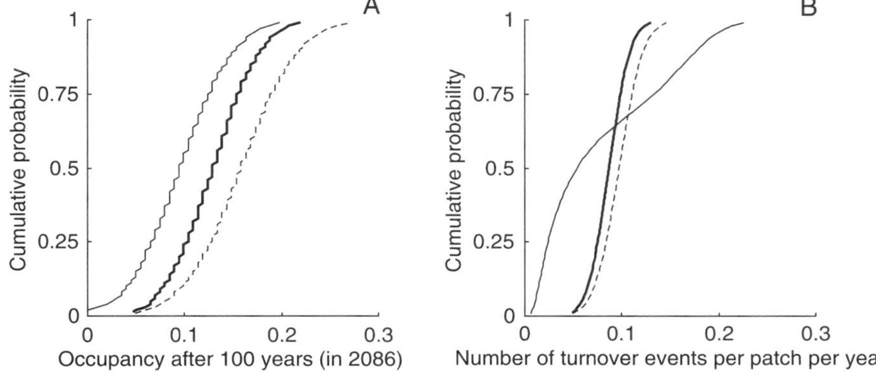

Fig. 5.2 Cumulative probability distributions of predictions for the patch occupancy of the tree frog metapopulation after 100 years (A) and turnover within these 100 years (B) obtained from simulations with model parameters estimated using QS6 (thick curve), QS1 (thin curve), and TO (dotted curve), the posteriors of which are shown in Fig. 5.1.

turnover data often leads to the prediction of a substantially higher or lower occupancy, even if the simulated SPOM was in the quasistationary state. It remains to be seen whether this change in occupancy is still noteworthy when the full posterior distribution is known.

To illustrate this change in occupancy, we simulated the tree frog system using the median values of the QS6 posteriors in Fig. 5.1 for 1000 time steps to reach the quasistationary state and for 6 additional time steps to generate a dataset similar to the real dataset (see Fig. 5.3). We then estimated the model parameters for QS6, QS1, and TO. Figures 5.4 and 5.5 show posterior distributions for the model parameters and the corresponding predictions of patch occupancy and turnover. These figures are similar to Figs. 5.1 and 5.2, respectively, except that TO and QS1 have traded places in Fig. 5.5; medians are 0.158 (QS1), 0.139 (QS6), and 0.124 (TO). Again, TO reflects the trend in data, which are downward in this case (0.158, 0.153, 0.133, 0.129). Nevertheless, because the posteriors of the occupancies for QS1 and TO largely overlap, the assumption of quasistationarity is not refuted by the dataset. Therefore, the use of QS6, which contains more data than TO, appears to be warranted. This warrant becomes even stronger when we calculate the median occupancy in the simulations after 100 iterations (Fig. 5.3), which is 0.139, precisely the median value for QS6. Obviously, for the real dataset we cannot perform this check, but there too we find a large overlap of the posteriors for QS1 and TO, so the assumption of quasistationarity cannot be refuted, and using QS6 seems the best choice. If the posteriors have little overlap, we can interpret this as a sign that the metapopulation may not be in the quasistationary state, and we should perhaps refrain from using QS6 and use TO instead.

Until now, we have only looked at the predictions of the occupancy after 100 years. Predictions of the turnover in 100 years, pictured in Figs. 5.2B and

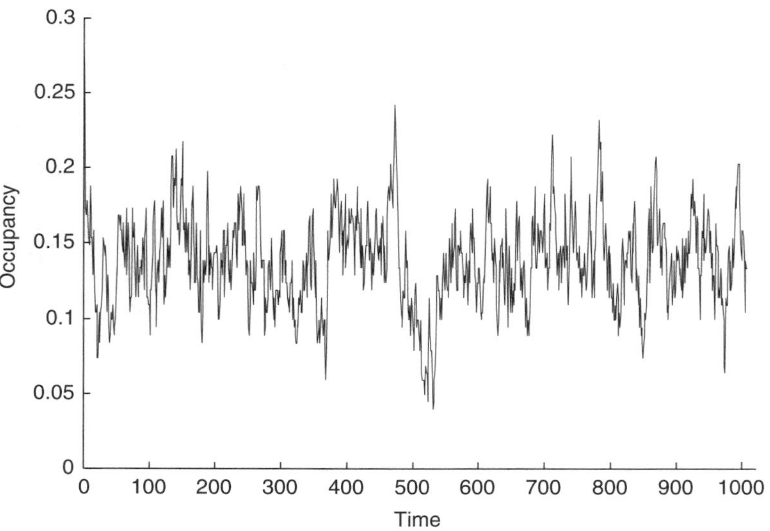

Fig. 5.3 Occupancy in a simulation of the tree frog system. The last 6 years are used to generate a dataset.

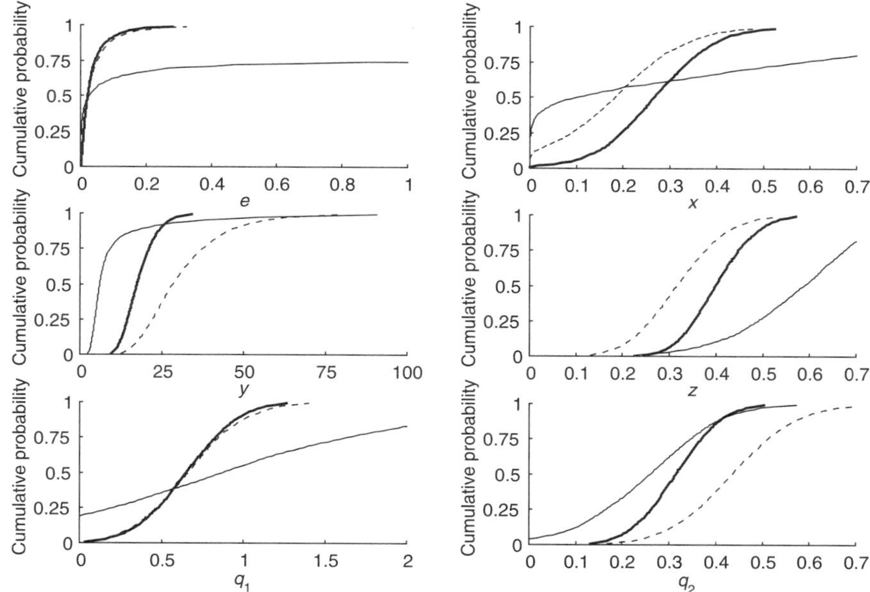

Fig. 5.4 Cumulative posterior distributions of model parameters estimated from the simulated tree frog dataset using QS6 (thick curve), QS1 (thin curve), and TO (dotted curve). These results are based on 500,000 iterations after a burn-in of 500,000.

5.5B, show that QS1 contains little information about turnover. Common sense tells us that there is no real information about turnover in snapshot data, i.e., in QS1, and that any found information must be an artifact of the model. In preliminary MCMC simulations, we sometimes found posteriors for QS1 containing some information about turnover. It turned out that this disappeared when we carried out more iterations or increased the value of L. Hence, if the posteriors for QS1 contain information about turnover, this should be

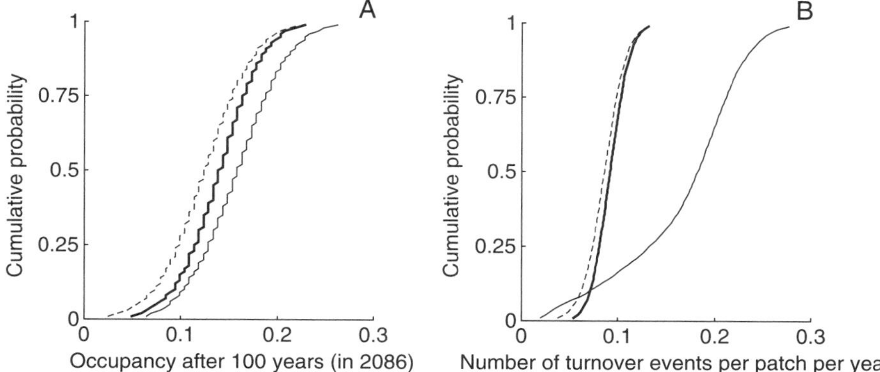

Fig. 5.5 Cumulative probability distributions of predictions for the patch occupancy of the simulated tree frog metapopulation after 100 years (A) and the turnover within these 100 years (B) obtained from simulations with model parameters estimated using QS6 (thick curve), QS1 (thin curve), and TO (dotted curve), the posteriors of which are shown in Fig. 5.4.

interpreted as a sign that the MCMC simulations have not yet converged or that L is not taken large enough.

As already mentioned earlier, differences in the posteriors of the model parameters do not necessarily entail differences in model predictions because parameters may be highly correlated. We remarked that in the IFM with rescue effect, parameters e and y cannot be distinguished by snapshot data; they appear as a product in Eq. (5.6). Although this model is mathematically not completely exact, this high correlation between e and y is still to be expected for QS1. Because the extended IFM of Vos et al. (2000) contains additional parameters q_1 and q_2, we need to correct for these parameters to observe this correlation. The appropriate transformation, relating Eq. (5.7) to Eq. (5.2) and Eq. (5.8) to Eq. (5.3), is

$$\begin{aligned} \log e' &= \log e + (q_1 - \overline{q_1})\overline{\log H_1} \\ \log y' &= \log y + (q_2 - \overline{q_2})\overline{\log H_2} \end{aligned} \tag{5.23}$$

Figure 5.6 confirms that a high correlation exists between e' and y'. It also shows the influence of the cutoff in Eq. (5.7): when e' becomes too large, the extinction probability is unity and the correlation with y' disappears.

Inherent Stochasticity

Evidently, the more data available, the better the predictions, as reflected in steeper cumulative posteriors (Figs. 5.2 and 5.5). These cumulative posteriors will, however, never be vertical regardless of the amount of data available because the model is stochastic. This inherent stochasticity may contribute much more to the total uncertainty than the uncertainty in the model parameters, which would make further data collection rather useless. Hence, the question arises how much data are needed. Figure 5.7 shows predictions for occupancy and turnover for the simulated tree frog system when the model

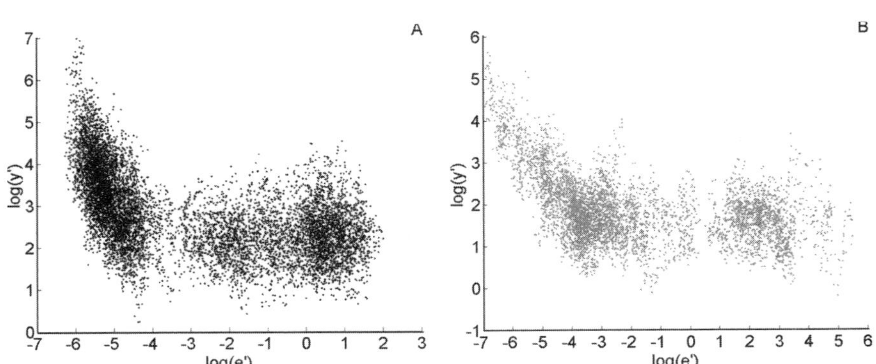

Fig. 5.6 Relationship between log(e') and log(y'), the transformations of log(e) and log(y), for QS1 of the real dataset (A) and the simulated dataset (B). Note the correlation between the two for log(e') <-4 and the near constancy of log(y') for larger values.

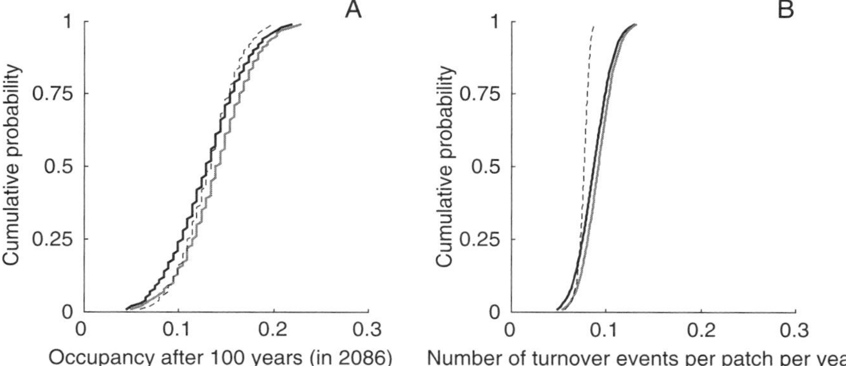

Fig. 5.7 Cumulative probability distributions of predictions for the patch occupancy of the tree frog metapopulation after 100 years (A) and the turnover within these 100 years (B) obtained from simulations based on model parameters estimated using QS6 of real data (black curve), simulated data (gray curve), and based on the model parameters used to generate the simulated dataset (dotted curve).

parameters are known exactly compared to the predictions based on QS6, shown earlier in Fig. 5.2 (real dataset) and Fig. 5.5 (simulated dataset). As far as occupancy is concerned, not much would be gained if more data became available. However, more data would convey more information about turnover. This is important if the metapopulation is at risk of extinction. A metapopulation with high turnover will go extinct much faster than a metapopulation with low turnover and the same occupancy (i.e., the same basic reproduction number R_0 or metapopulation capacity; Ovaskainen and Hanski, 2001). Hence, Fig. 5.7B suggests that data overestimated the turnover and thus underestimated the metapopulation extinction time, that is, the true time to metapopulation extinction is larger than the model parameterized with available data predicts.

Changes in Environment

The aforementioned model predictions all assumed that the system remained unaltered. To weigh several management options (e.g., using decision theory, Possingham, 1996, 1997; Possingham et al., 2001), one needs to be able to change system properties by, for example, introducing patches, increasing patch quality, or reducing the effective interpatch distance by building corridors. Because these changes do not affect the model parameters, their effect can be studied easily with model simulations.

Problems enter, however, when unknown changes need to be dealt with. In our tree frog example, a large proportion of the ponds have disappeared and been created since 1986 because of changes in land use and succession. Precise data are unavailable. We only know that in the year 2002, 23 patches are occupied (in comparison, the number of occupied patches in 1981, 1982, 1983, and 1986 is 22, 22, 28, and 29, respectively) of which just 10 patches existed (and 8 occupied) in 1981–1986. So there are at least 13 new patches, but perhaps many more that are empty. This makes it impossible to compare the predictions of the model and the occupancy found in 2002 quantitatively, let alone that data for 2002 can be combined with 1981–1986 data to calibrate the model even better. Qualitative comparison is, however, still possible. Figure 5.8A shows the spatial occupancy pattern in 1986 and the location of the 23 occupied patches in 2002.

In 1986, eight pond clusters were occupied. In 2002 the metapopulation retreated to only three of these clusters (clusters I, II, and III). It is apparent that the weakest clusters have become extinct in 2002. These are the relatively small and isolated clusters (clusters V, VI , VII, and the one-pond cluster VIII) and a cluster with only one occupied pond in 1986 (cluster IV). Let us examine whether the model is able to predict this spatial pattern. Figure 5.8B shows the probability of patches being occupied in 2002 based on simulations starting with the pattern in 1986 and using posteriors obtained from the real dataset assuming quasistationarity (QS6). The eight clusters (except for cluster VIII) are the only parts of the metapopulation where ponds with a probability of being occupied of more than 0.25 occur, which is consistent with the 2002 data. Clusters I and III, which remained occupied in 2002, are the only clusters that contain ponds with a probability of more than 0.50 or even 0.75 of being occupied. Thus the model is able to locate the parts of the metapopulation with the highest survival potential. The fact that clusters V, VI, VII, and

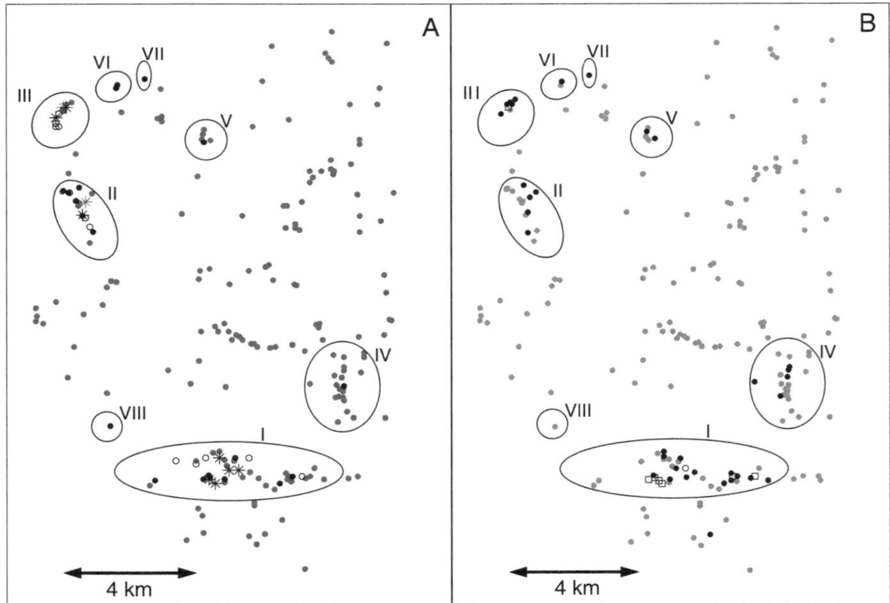

Fig. 5.8 Occupancy patterns for the tree frog system. (A) Comparison of 1986 and 2002: patches empty in 1986 (gray dots), patches occupied in 1986 (black dots), patches occupied in 1986 and 2002 (black asterisks), patches empty in 1986 but occupied in 2002 (gray asterisks), and patches occupied in 2002 that did not exist in 1986 (open circles). (B) Probability P of occupancy in 2002 as predicted from simulation starting in 1986 and using the model parameters of QS6: $0 \leq P < 0.25$ (gray dots), $0.25 \leq P < 0.5$ (black dots), $0.5 \leq P < 0.75$ (open squares), and $0.75 \leq P \leq 1$ (open circle).

VIII have become extinct is in accordance with the model predictions, as in these clusters only one or two ponds with a probability of more than 0.25 were present (the probability of occupancy of the pond in cluster VIII is even smaller than 0.25). However, in contrast to data, the model does not necessarily predict the retreat of the tree frog from cluster IV, as four ponds in this cluster have occupancy probabilities larger than 0.25. At least one occupied pond would have been expected. It is apparent, however, that cluster IV shows the best potential for recolonization, as the cluster is relatively large and has many ponds with a high probability of occupancy. To increase colonization probability, the spatial cohesion with cluster I should be improved. This is in accordance with the actual protection policy in the region (Crombaghs and Lenders, 2001). Thus, the model is a useful tool to determine the optimal spatial strategy to conserve the tree frog metapopulation.

5.5 DISCUSSION

SPOMs were introduced as metapopulation models that combine simplicity with sufficient realism (particularly pertaining to spatial structure), such that they are ideal for assessing the current condition of real metapopulations and for predicting the effect that human intervention has on metapopulation persistence. Hence, it is interesting to see what the literature teaches us about the four parts, mentioned earlier, of applying SPOMs to real metapopulations, that is, to see what metapopulation data have been used, what SPOMs have been developed, what parameterization method has been used, and whether the SPOMs were useful and successful in their predictions.

Perhaps the best-documented metapopulation is the Glanville fritillary (*Melitaea cinxia*) metapopulation on the Åland Islands in southwest Finland (Hanski, 1994; Hanski et al., 1996; Moilanen and Hanski, 1998). There are some 4000 habitat patches, forming several subnetworks of which one 50-patch network has been studied in more detail. It has been modeled with the IFM and parameterized with the incidence method of Hanski (1994). Because the model describes data very well, the two variables of the IFM, area and interpatch distance, have been declared as the main explanatory factors. Other factors do not appear to play a significant role (Moilanen and Hanski 1998). Eber and Brandl (1996), studying the tephritid fly *Urophora cardui* in northeastern Bavaria, came to the same conclusion based on their logistic regression on turnover data, although distance did not matter much because distances between patches were smaller than the average dispersal distance. However, several authors have suggested that it is not generally true. For example, Sjögren-Gulve and Ray (1996), applying their logistic regression method to a dataset of a pool frog (*Rana lessonae*) metapopulation on the Baltic coast of east central Sweden, considered additional variables (mean water temperature, presence of forestry) that were necessary to obtain a good fit. Fleishman et al. (2002), applying logistic regression to an Appache silverspot butterfly (*Speyeria nokomis apacheana*) metapopulation in the Toiyable Range, Nevada, even found that patch area and isolation are not important at all. Other factors that are better indicators of habitat quality than area had much more explanatory power. Lindenmayer et al. (1999) used the IFM to model metapopulations of four marsupials in Buccleuch State Forest, New South Wales, Australia: common brushtail possum (*Trichosurus vulpecula*), mountain brushtail possum (*Trichosurus caninus*), common ringtail possum (*Pseudocheirus peregrinus*), and greater glider (*Petauroides volans*). They parameterized the model with Hanski's (1994) incidence method. For two of these four species, the parameterization produced unreliable results, which Lindenmayer et al. (1999) blamed on the absence of patch quality in the model. These case studies demonstrate that the important variables to be put in the model should be determined on a case by case basis, using expert ecological knowledge and statistical techniques.

While in these examples model structure is altered by additional variables, model structure can also be affected by different processes. Several authors report interesting consequences of their data analysis for model structure in this respect. Crone et al. (2001) applied likelihood maximization on snapshot data of a field vole (*Microtus agrestis*) metapopulation on 71 islands in the Tvärminne archipelago in the Gulf of Finland and concluded that colonization and extinction are highly correlated: when a population is on the brink of extinction, the voles disperse *en masse* to other islands. Although the source population then goes extinct, other populations can benefit from immigration, the rescue effect (Brown and Kodric-Brown, 1977). Peltonen and Hanski (1991) compared three species of shrews (*Sorex araneus, Sorex caecutiens, and Sorex minutus*) on islands in three Finish lakes and on the mainland. They found, using a predecessor of the Hanski (1994) approach, that the colonization rates of these species were identical, but with different contributions from the probability of arrival and the probability of settlement. Hence, the similarity of the colonization rates is perhaps coincidental, in which case these different contributions should really be modeled explicitly. Wahlberg et al. (2002), who employed Moilanen's (1999) Monte Carlo simulation method on snapshot data of the marsh fritillary butterfly (*Euphydryas aurinia*) in 114 patches in the Joutseno region of southern Finland, made yet another important discovery: if the landscape is very dynamic itself (patches appear and disappear), the model must represent this. If patch dynamics are due to succession or due to metapopulation behavior of the patches themselves [e.g., if the patches are host plants to a moth, (Nieminen, 1996) or host butterflies to a parasitoid (Lei and Hanski, 1997)], this can be modeled in a fairly straightforward way, but if, as in our case study, patches disappear because of human activities, any model may be as good as any other.

Moilanen (1999) raised perhaps the most important point about model structure. Temporal stochasticity (regional stochasticity in Hanski, 1991) in model parameters, i.e., for each transition between states there are different values for the model parameters, may determine whether a metapopulation is expected to go extinct or not, so it is crucial whether it should be incorporated in the model. It is relatively straightforward to do so: Moilanen (1999) did this by multiplying patch area by a normally distributed factor with variance σ^2, but it may also be done by putting a certain white noise ε with variance σ_ε^2 on the model parameters. The latter is advocated by Carlsson and Kindvall (2001), who used Moilanen's (1999) approach on the grasshopper *Stauroderus scalaris* in 158 patches on Öland Island, southeast Sweden, and concluded that it is not mechanistic enough to generate the patterns they found. It is also straightforward to estimate the corresponding parameters (σ or σ_ϵ) in the Monte Carlo simulation and Bayesian approaches. However, the estimates may not be very accurate, given the available data. Many years of data are needed to provide information on the variability of the model parameters. Because the transitions between years are then treated relatively independently, each transition is likely to yield a different trend (Moilanen, 2000), potentially leading to an overestimate of temporal stochasticity. This bias may again be reduced by the assumption of quasistationarity. Now we are back at the problem mentioned in the case study: does this assumption apply? Because of temporal stochasticity, it may no longer be able to tell from data whether quasistationarity applies or not, and perhaps we have no other choice than to resort to independent information to decide about quasistationarity (Moilanen, 2000). Further study is evidently needed.

Things may even get worse when data themselves contain errors. Moilanen (2002) mentioned three errors that can be made in data collection. First, patch areas can be estimated wrongly. Second, unknown habitat patches may be located within or around the study area. Third, occupancy observations may be false: truly occupied patches may have been observed as being empty (false zeros). The other possibility is that truly empty patches are observed as occupied (false ones). This may happen if observation of a few individuals is interpreted as an entire population. Moilanen (2002) argued that wrongly estimated patch areas can seriously affect extinction risk, whereas missing patches cause an overestimation of migration distances and colonization ability of the species. False zeros can have a severe effect on both extinction and colonization components, and hence on metapopulation persistence. Like temporal stochasticity, such data errors can be dealt with in the Monte Carlo simulation and Bayesian parameterization methods. Moilanen (2002) gives explicit formulae for the case of false zeros, but again, the parameter estimates will lose some reliability. The great advantage of the Bayesian approach is that this loss of reliability will show up in posterior distributions of the parameters and it can be translated immediately into uncertainties about the predictions. However, no ready-made computer programs exist as yet for Bayesian analysis of these extended models.

Considering all these allegations at the information richness of data, we have to face the important question of how much data (how many patches and how many years) are needed for reliable predictions. Ter Braak et al. (1998) showed by simulation that the curves for extinction and colonization [Eqs. (5.2) and (5.3)] are discomfortingly variable when estimated from 2 years of data with 50 patches, thus mimicking data and analysis of Hanski (1994). Tyre et al. (2001) conducted maximum likelihood regression on snapshot data of the hydrobiid snail *Fonscochlea zeidleri* in only 9 patches in Bopeechee Springs, Australia, with only 3 patches occupied. Their objective was to compare parameter estimates for two sets of snapshot data: before and after human-driven impact (trampling by stock and water extraction). They concluded that the method has too little power to detect this impact. The reason they mentioned is that the number of patches is too small. However, the reason may also be that the number of years is too small. Thomas et al. (2002a) applied maximum likelihood regression on snapshot data of the silver-studded blue butterfly (*Plebejus argus*) in 33 patches on the Creuddyn Peninsula in north Wales. They considered four snapshots spread fairly evenly over a period of 27 years and compared results obtained from one, two, three, or four snapshots. Their message was that predictions based on short-term studies may considerably underestimate turnover and caution must therefore be taken when extrapolating to long-term dynamics. They found themselves supported by the only other long-term study known, by Moilanen et al. (1998), on four snapshots in 20 years of the 66 patch metapopulation of the American pika (*Ochotona princeps*) in Bodie, California. Although both studies used Hanski's (1994) incidence approach on snapshot data and therefore ignored temporal correlation among these snapshots, results of the American pika study were recovered by Moilanen (1999) with his Monte Carlo simulation method that does appropriately deal with temporal correlation. Hence, the warning must be taken seriously. It is interesting, however, to note that it is opposite to Moilanen's (2000) warning that short-term data may indicate a false trend. Still, both advocate the use of many years of data.

These results justify the question of whether there is some rule of thumb about the minimum number of patches and the number of years needed. This question is not easy to answer for (at least) five reasons. First, there is obviously a trade-off between the number of patches and the number of years needed. Second, consecutive years may contain a different amount of information than an equal number of years with missing years in between. Third, it matters whether the assumption of quasistationarity is made. Fourth, the information content of datasets with the same number of patches and years can vary widely. Fifth, the answer depends heavily on what prediction is required to be reliable. For example, our results in the case study (202 patches, 3 consecutive years, 1 additional year after 2 missing years) show that both QS1 (202 patches, 1 year, assumption of quasistationarity) and TO (202 patches, 4 transitions) give fairly reliable predictions of the patch occupancy, but turnover is badly predicted by QS1. Perhaps we are only capable of stating a rule of thumb about what datasets will not give reliable predictions: networks with less than 25 patches and less than 3 years of data will most probably be hopeless to parameterize. At the same time, we note that the only way to be sure is to simulate data you expect to collect and to carry out the parameter estimation with uncertainty analysis of the predictions for a number of simulated datasets. The Bayesian approach is most suited for this purpose, as demonstrated in O'Hara et al. (2002).

Once sufficiently good data have been found to parameterize the model, we are ready for one of the main purposes of developing the model: prediction, not only of what would happen when the system is left to itself, but particularly to assess the impact of conservation strategies. For example, O'Hara et al. (2002) explored the effects of reducing the area of all patches by some factor. Cabeza and colleagues (Chapter 22) used a SPOM for optimal selection of sites to conserve. In our case study, we studied the consequences of inserting patches near the viable subnetworks. Whatever the scenarios to be studied, we urge the use of uncertainty analysis because it is the only way in which different scenarios can be compared properly.

6. FROM METAPOPULATIONS TO METACOMMUNITIES

Mathew A. Leibold and Thomas E. Miller

6.1 INTRODUCTION

Metapopulation thinking (Hanski and Gilpin, 1997, Chapter 1) has led to a remarkable change in the way that population ecologists view population dynamics but is only now beginning to have similar effects on how ecologists view community dynamics. In population biology, the shift in scale from considering local population dynamics to population persistence in a network of connected habitat patches has enabled ecologists to understand the relative importance of different factors, such as habitat quality, species interactions, and migration in a scale-dependent context. One of the earliest and perhaps most obvious extensions of metapopulation theory was to pairs of interacting species (Nee et al., 1997; Hanski, 1999), such as competitors (Levins and Culver, 1971; Horn and MacArthur, 1972; Hastings, 1980) or predator and prey (Caswell, 1978). The development from one- to two-species metapopulation models is important in that it adds a significant level of complexity and realism: the colonization and extinction rates of the focal species become functions of both habitat characteristics and the presence of the other species (Holt, 1993, 1997). Although continued work on pairs of species is important, a bigger challenge is addressing patterns in community structure involving more complex systems. To date, work on metacommunities with this higher level of complexity is sparse and somewhat simplistic, but important progress

  0-12-323448-4

has been made toward understanding patterns of biodiversity at different scales and associated phenomena at the population, community, and ecosystem levels.

It is at this level of complexity that the perspective of metacommunities becomes substantially different from just a simple extension of metapopulation ecology (Wilson, 1992; Holt, 1993). We use the term metacommunity to refer to sets of communities (sets of species that interact with each other) occurring at discrete sites, linked by migration (Gilpin and Hanski, 1991). The crucial elements are multiple potentially interacting species, multiple sites at which such interactions might occur, and migration by individuals of at least some of the species to link species interactions among sites. In its simplest form, a metacommunity may consist of networks of discrete habitats such as islands that are linked by migration of some or all species among the habitats. This kind of metacommunity will have a structure that emerges from the dynamics of the interconnected habitats. However, as with metapopulations, our definition can fit a broader array of situations, including mainland–island metacommunities, where migration from the mainland may largely determine community patterns on islands. Nevertheless, this chapter focuses on metacommunities as independent entities with their own emergent structure.

A community perspective creates an important distinction between metapopulation and metacommunity views. The most basic issue in metapopulation theory is to address what determines the persistence of a metapopulation in a system of connected habitat patches (Chapter 4), whereas the most basic issue in metacommunity studies is to address what regulates coexistence of multiple species in such a system. Excluding the case of exact ecological equivalence among species (Hubbell, 2001), long-term coexistence of species requires that they show some trade-off among important aspects of their biology at some scale (see Chesson, 2000). Theoretical work to date in metacommunities shows that dynamics involving trade-offs are surprisingly constrained for complex reasons. We also examine how our current empirical knowledge is related to the theory, particularly in relation to the role of different possible trade-offs that might regulate metacommunities.

6.2 FOUR BASIC PERSPECTIVES ON METACOMMUNITIES

We can identify four basic perspectives on metacommunities that have generally set the context for metacommunity thinking to date. These simplistic frameworks for metacommunities differ from one another in the degree of spatial heterogeneity among patches and in the degree of connectedness among the patches relative to the migration rate of the component species (Table 6.1). This section also relates these perspectives to the traits of the component species and how these traits covary to produce trade-offs at different scales.

The Patch-Dynamic Perspective

This view is the one that corresponds most closely to a multispecies version of "classic" metapopulations. Patches are generally seen as homogeneous, with local populations subject to local extinctions. Colonization of patches by new

TABLE 6.1 Four Simplified Perspectives on Metacommunities Arranged According to whether Migration Is Sufficient to Alter Local Population Abundances and whether Local Sites or Patches Are Heterogeneous

		Migration sufficient to alter local abundances	
		No	**Yes**
Homogeneous patches	No	Species sorting	Mass effect
	Yes	Patch dynamics	Neutral models

species is thought to occur on a long time scale (through low migration rates), so local population sizes are not regulated by immigration or emigration. Persistence of species in the metacommunity thus depends on how colonization of the species can be balanced against extinctions due either to interactions with other species or to disturbance events that occur at the scale of entire patches (Hanski, 1999). Local communities are frequently "undersaturated" in terms of species number and/or open to invasions by other members of the metacommunity. If disturbance rates are sufficiently high, some species may persist in such metacommunities as "fugitive species" despite being poor competitors. To do so, they must also have sufficiently high migration rates to allow them to coexist in the metacommunity with superior competitors (Hastings, 1980).

The patch-dynamic perspective is perhaps the earliest view of a metacommunity and is closely connected to ideas about fugitive or pioneer species and trade-offs between colonization and competitive abilities. Early theoretical work demonstrated that patch dynamics can contribute to species coexistence (Caswell, 1978; Yodzis, 1978). Recent theoretical work related to this perspective includes studies by Tilman (1994), Kinzig et al. (1999), Yu and Wilson (2001), Levine and Rees (2002), and Holt (2002). The primary result of these studies is that although a trade-off between competitive and colonization abilities can allow for coexistence of species (and potentially many of them), the range of parameters that allow coexistence by this mechanism is surprisingly limited. The patch-dynamic perspective is also associated with certain types of ecosystems: the concept of habitat patches connected by migration underlies our understanding of hard-substrate intertidal systems (Paine and Levin, 1981), the role of gaps in tropical forests (Connell, 1978; Denslow, 1980), and effects of small mammals on certain plant communities (Platt and Weis, 1977; Goldberg and Gross, 1988; Huntley, 1991).

The Species-Sorting Perspective

If connected habitat patches are heterogeneous in some important character, such that different species are favored in different patches, then increased migration can lead to a "sorting" or matching of species with their favored habitats. This species-sorting view focuses on the role of patch-type heterogeneity and examines how species' distributions among patches are related to their relative abilities to exist and interact successfully interact with other species in the larger metacommunity (Clements, 1916; Gleason, 1917; Tilman, 1980). A number of issues in ecology take on different interpretations when viewed in a metacommunity framework. Some of these are highlighted. For species sorting to occur, the migration of species must be sufficient to distribute species among patches but also be insufficient to affect local population abundances. In this view, metacommunity structure emerges as a simple cumulative outcome of all local interactions, and colonization–extinction dynamics are only important for determining the transient dynamics of the system until it reaches a steady state. The steady-state behavior of the metacommunity, including the distribution of species among patches, is thus unrelated to rates of colonization and extinction. Under species sorting, immigration may have little effect on local, within-patch diversity but will be important for determining species identity within patches. Work on this view includes that by Tilman and Pacala (1993), Holt (1993, 1997), and Leibold (1996, 1998).

The "Mass Effects" Perspective

In this view, patch-type heterogeneity is also important, but migration is sufficiently high that local population abundances of component species are affected by net emigration (producing "source" populations) and net immigration (yielding "sink populations"). Under these conditions, metacommunity structure is affected in two ways. First, species that might be expected to do well in a closed community (with no emigration or immigration) can be driven to extinction from local patches where they are competitively superior by strongly subsidized "sink populations" supported by immigration from other patches. Second, such competitively superior species can have reduced competitive abilities in metacommunities because of the emigration of individuals (equivalent to reduced local birth rates) as well. The overall result is that mass effects can increase local diversity (Mouquet and Loreau, 2002) by such "subsidized" population dynamics. The issue is a bit more complicated, however, because such strong subsidy can require very strong emigration that can itself make local populations vulnerable to extinction. Thus these mass effects can reduce the diversity over broader spatial scales (the metacommunity as a whole) and therefore reduce local diversity despite strong mass effects (Mouquet and Loreau, 2002). Research work on this perspective includes that by Holt (1993), Amarasekare and Nisbet (2002), and Mouquet and Loreau (2002).

The Neutral Perspective

Of course, all of this theory presumes that species differ from one another in important ways, including their responses to local conditions (patch-type specialization and/or competitive superiority) and to the landscape structure (migration and background extinction rates). One possible view is that species do not differ in any important respect in any of these ways (Hubbell, 1997, 2001). Although unlikely to be universally true (see Zhang and Lin, 1997; Yu et al., 1998), this view does highlight how important stochastic processes can be in structuring metacommunities and may be important in explaining some important aspects of metacommunities, including abundance distributions and incidence functions (Hubbell, 2001; see Chave et al., 2002).

A Missing Synthesis

Obviously, all of these perspectives are simplifications, and real metacommunities are likely to be structured by some mixture of the processes that are highlighted by each of them. The issue is made more complex because species in real communities may interact with each other despite having fundamentally different migration rates. For example, a predator may have a migration rate sufficiently high that it has a metapopulation structure and significant local population regulation due to migration, while it interacts with prey that have much lower migration rates and in which local population regulation is completely determined by local processes [see Holt (1997) and van Nouhuys and Hanski (2002) for an empirical example]. To date, there is little understanding of how to synthesize these various perspectives into a general framework, even though there is some recognition that such a synthesis is important (Holt, 1997). Surprisingly little work is done at the interface of the species-sorting and patch-dynamics perspectives (but see Holt, 1997; Levine and Rees, 2002; Shurin et al., 2003). Work by Amarasekare and Nisbet (2001) and by Mouquet and Loreau (2002) examines the interface between species sorting and mass effects. It is as yet unclear how to integrate these various perspectives with the stochastic processes described by Hubbell (2001), although work by Chave et al. (2002) indicates that the effects can be subtle.

6.3 THEORETICAL ISSUES ABOUT METACOMMUNITIES

The Importance of Trade-Offs

It is helpful to start thinking about trade-offs from a more conventional basis that ignores metacommunities. Traditional ecological theory about closed communities has drawn attention to trade-offs involving relative biotic abilities. Within a local site, this view involves conventional mechanisms of resource partitioning and associated ecological traits such as susceptibility to predators and so forth. At the local scale, the nature of these trade-offs will determine which species combinations are possible, what the stability of the system might be, and how the community responds to environmental change. Such differences in responses and impacts on resources and other fitness-related factors (this suite of traits is subsumed into a general term we call "biotic ability") can also lead to habitat (patch-type) specialization (involving coexistence at the larger regional scale even if there is no local coexistence). We can think of this process as one of "species sorting" that regulates how well matched species are to their local environments. Although ecologists are familiar with this conventional approach to niche dynamics, they do not always recognize that trade-offs can also be important in regulating metacommunity dynamics. This chapter reviews theory and empirical work that show how important such relative habitat specialization is in regulating metacommunity dynamics even when other important processes are also operating.

We can also approach questions about trade-offs from the perspective of metapopulation theory based on single species. This theory draws attention to the tension between colonization ability and susceptibility to stochastic extinction in regulating the persistence of populations in patchy landscapes where a variety of strategies can allow species to persist in a given landscape (Chapter 4). Similarly, metapopulation theory oriented toward small numbers of interacting species has shown that trade-offs involving colonization ability and biotic ability within patches can also be important (Hastings, 1980; Tilman, 1994; Taneyhill, 2000; but see Klausmeier, 2001; Yu and Wilson, 2001; Levine and Rees, 2002). These dynamics affect species sorting in two opposite ways: increased competitive ability can be related to increased specialization, but high rates of migration may decrease the degree to which local sites are occupied by species that are best matched to local conditions (see Mouquet et al., 2002). Thus connectivity in a metacommunity can change conditions from those that favor habitat specialization and coexistence to those that favor species with traits related to overall biotic ability in the metacommunity as a whole (including dispersal as well as average extinction rates and biotic ability over all patches).

The aforementioned considerations suggest that three broad categories of species differences are all potentially important in regulating the distribution of species in metacommunities: patch specialization by means of biotic ability, migration among patches, and likelihood of local extinctions. These admittedly broad categories could probably be subdivided further and lead to even more possible traits; for example, the process of migration could be dissected into an emigration and an immigration component (Chapter 13).

If we now consider multiple patch types where these species differences can have various effects, then we can consider traits and trade-offs that may be related to these species differences. For example, expression of each of the three categories in two different types of patches produces six traits or ways that species can differ that influence their persistence and coexistence:

1. Biotic ability in patch-type 1
2. Biotic ability in patch-type 2
3. Migration ability into patch-type 1
4. Migration ability into patch-type 2
5. Susceptibility to stochastic extinction in patch-type 1
6. Susceptibility to stochastic extinction in patch-type 2

It is important to understand that any biotic ability that can lead to extinctions (traits 1 and 2) in many metacommunity models is substantially distinct from susceptibility to stochastic extinctions (traits 5 and 6). The latter is viewed as being independent of biotic interactions, due either to stochastic effects acting on the demography of small populations or (perhaps just as commonly) to environmental change within local sites, including disturbances that can affect large populations as well.

Taken in pairwise combinations, this list of six traits can lead to 15 possible trade-offs between species, and more of them involve niche axes that operate at the larger regional scale. This perspective results in many more mechanisms that can regulate coexistence and metacommunity dynamics than are considered in local-scale models. To date, metapopulation theory has addressed only some of these possibilities with any degree of effort.

Current work on metacommunities reveals important interdependent constraints on each of these possibilities. Two features are important. First, the behavior of the system in relation to these traits can be complex because the behavior of metacommunities may vary in response to these trade-offs in ways that are not always consistent. For example, the colonization-rate–competitive-ability trade-off can be important in allowing species to coexist in a metacommunity when local populations are subject to stochastic extinctions, thus allowing competitively subordinate species to coexist with competitive dominant species. Under this scenario, the constraint for coexistence in the metacommunity is that the colonization rate of the subordinate exceed some critical value that allows it to exist as a fugitive species in the metacommunity (i.e., that the migration rate time scale be much shorter than that at which species interactions lead to extinction). The colonization-rate–competitive-ability trade-off can also allow species to coexist when colonization rates are high enough to alter local abundances (through mass effects). However, under this scenario, colonization by the subordinate species must be low enough to prevent it from subsidizing large sink populations in patches occupied by competitive dominants (Amarasekare and Nisbet, 2001).

Second, models based on one mechanism can also be strongly altered by the presence of mechanisms involving other trade-offs. For example, Law and Leibold (in press) have found that a metacommunity of species with nontransitive assembly rules (species A invades and excludes species B, species B invades and excludes species C, and species C invades and excludes species A) shows cycles in the frequency of patch types (a metacommunity becomes dominated, in turn, by sites containing species A, species B, species C, and then species A again). In the absence of background stochastic extinctions, the amplitude of these cycles in patch occupancy varies in a neutral manner (such as the neutral limit cycles of the Lotka–Volterra predator–prey models). However, in the presence of such extinctions, the cycles converge to a stable point at which the frequency of occupancy states is determined by the relative colonization and stochastic extinction probabilities of the three species. Thus the outcome of a model of species interactions (under nontransitive assembly) is strongly modified by the presence of patch dynamics (background extinctions unrelated to species interactions).

The Role of Migration: From Patch Dynamics to Species Sorting to Source–Sink Relations

A broader review of the theory on metacommunities is beyond the scope of this chapter (such a review is under preparation by Holyoak, Leibold, and Holt), but two important conclusions can be drawn about the dynamics of metacommunities. First, the role of migration is not simple (Loreau and Mouquet, 1999; Mouquet and Loreau, 2002; Leibold and Norberg, 2003): Fig. 6.1 presents a hypothetical relationship between migration rate and diversity. Clearly, when there is no migration, then the metacommunity perspective is not needed, the dynamics of local communities are entirely independent of one another, and diversity in local communities is expected to be relatively low due to migration limitation. At minimal levels of migration, every species can potentially get to every patch, but the order of colonization and relative competitive abilities become important in determining local community structure. The result is some degree of species sorting in which migration may not enhance diversity but can allow significant species replacement and some specialization. In particular, "fugitive species" are possible

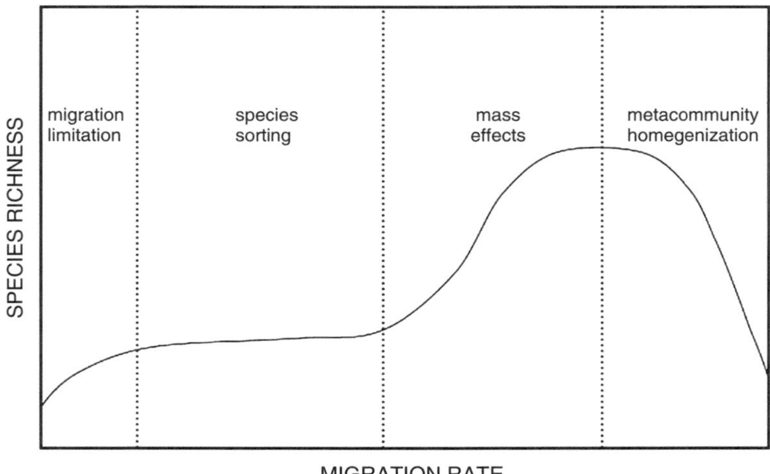

Fig. 6.1 A hypothetical relationship between migration rate and species richness illustrating the major mechanisms by which migration affects community structure.

under these conditions, species that exist in the metacommunity by virtue of temporary existence in patches that have not yet been colonized by patch-type specialists (Hastings, 1980). In contrast, as migration increases until all species can disperse relatively quickly to all patches, the local assembly of communities is much less affected by the order of colonization (because the assembly process can reach its final equilibrium behavior relatively quickly, typically an uninvasible community but also possibly a pattern of repeated heteroclinic cycles), and classic fugitive species are excluded by patch-type specialists. At yet higher rates, migration plays a different role, being high enough to contribute significantly to local growth rates. This allows species that are locally inferior competitors to persist through mass effects (i.e., Mouquet and Loreau, 2002), increasing local diversity (Fig. 6.1). At this point, increased migration may act to decrease diversity of the entire meta-community, with consequential effects on local diversity as well. In the extreme, when migration is so high that local populations are completely mixed, metacommunity diversity (and thus local diversity) collapses into that expected in a single closed community.

The Importance of Habitat Heterogeneity and Patch-Type Specialization

The second important conclusion is that habitat heterogeneity, when it is present, almost always plays an important role in maintaining diversity in metacommunities. Heterogeneity allows habitat specialization, which can provide a local refuge for some specialist species and increase metacommunity (although not local) diversity. At higher rates of migration, such refuges can act as sources that help maintain populations in less favorable patches and increase both local and metacommunity richness (Mouquet and Loreau, 2002). In addition, metacommunity dynamics involving stochastic extinctions at the local scale prevent the occurrence of alternative stable states in local communities in the absence of patch-type heterogeneity (Shurin et al., 2003), eliminating another potentially important mechanism that might maintain high biodiversity in homogeneous patch-type metacommunities. However, in the presence of patch-type heterogeneity, which can prevent exclusion at the metacommunity scale, such alternative stable states can act to increase compositional diversity even among patches of the same type. In both cases, diversity is increased only if patch-type heterogeneity is sufficient to allow the "source" populations that support "sink" populations. Further, even under these conditions, migration is constrained in direct relation to the degree of habitat heterogeneity (Amaresekare and Nisbet, 2001; Mouquet and Loreau, 2002; Levine and Rees, 2002).

Therefore, even if we limit ourselves to considering only the six types of metacommunity-related traits listed earlier, we obtain a complex set of results about metacommunity dynamics and the ways they alter patterns of coexistence and biodiversity. This theoretically rich area of understanding the effects of trade-offs at different spatial scales is only now being addressed, and a number of important results may soon be forthcoming.

6.4 OBSERVATIONAL AND EXPERIMENTAL EVIDENCE ABOUT METACOMMUNITIES

Conventional Analyses of Species Distributions in Relation to Patch-Type Heterogeneity

Empirical work on metacommunities predates the metacommunity concept itself. One of the most basic types of data that describe metacommunities can be summarized as an "incidence matrix," a matrix describing which species are found at each of a variety of sites (Bray and Curtis, 1957; Leibold and Mikkelson, 2002). There are sophisticated ways of analyzing such data, often in relation to null models (Gotelli and Graves, 1996) or environmental gradients (see, e.g., Legendre and Legendre, 1998). Although these data have strong empirical roots and have long been implicitly related to mechanisms important in metacommunity dynamics, the link between incidence matrices and metacommunities has generally not been made explicit.

A good summary of this large body of work is beyond the scope of this chapter, but a few points relevant to the metacommunity concept are highlighted. First, this body of work shows that species associations in local habitats are typically not random (for summaries, see Gotelli et al., 2002 and Leibold and Mikkelson 2002), suggesting that species interactions or similarity of biotic requirements affects species distributions and could affect larger scale (metacommunity) dynamics. Second, environmental variability in abiotic factors (heterogeneity of patch types) is often important in explaining the distribution of species among sites (see, e.g., Bray and Curtis, 1957; Whitaker, 1975). The abundance of individual species or suites of species can be shown to covary significantly with factors such as nitrogen availability or soil moisture. Nevertheless, some spatial effects may be due purely to effects of migration (Borcard et al., 1992; Pinel Alloul et al., 1995).

Important descriptive studies of biodiversity at different scales indicate a complex relationship between local and regional patterns of diversity. For example, Shurin (Shurin, 2000; Shurin et al., 2000; Shurin and Allen, 2001) has shown that commonly observed linear relationships between local and regional diversity were not necessarily associated with uninvasible local communities, but rather could be explained as resulting from a complex assembly process in metacommunities involving multitrophic-level food webs (see also Srivastava, 1999; Mouquet et al., 2003). In another example, Chase and Leibold (2002) showed that local and regional diversity had qualitatively different relationships with productivity; the local relationship was unimodal and the regional relationship increased monotonically (Fig. 6.2). Steiner and Leibold (2003) showed that this situation can result from metacommunity dynamics in which nontransitive invasion cycles (see Morton and Law, 1997) are more common at high productivity. Hubbell (1997, 2001) has also

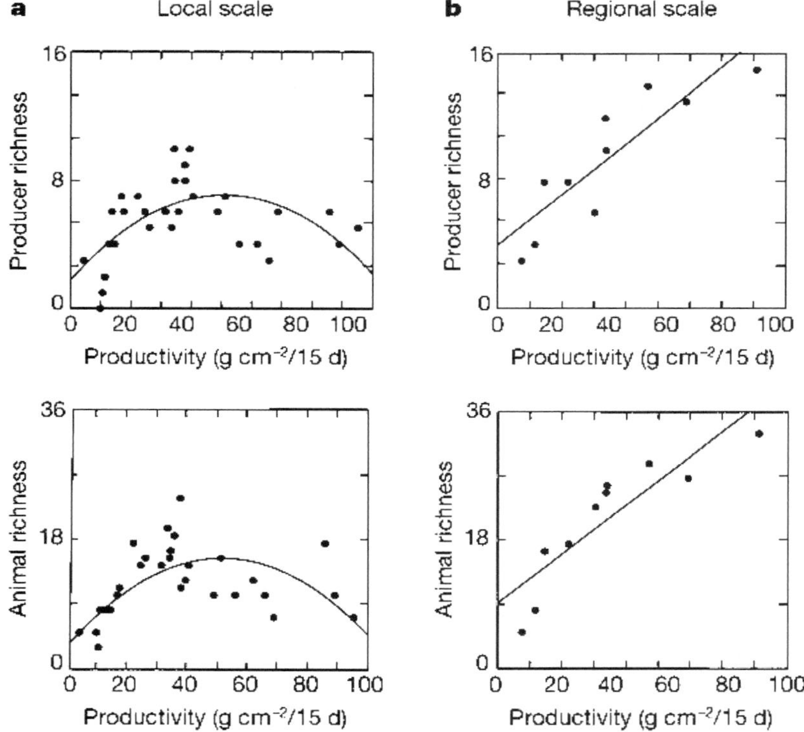

Fig. 6.2 Results from a survey of pond species diversity relative to *in situ* primary productivity at local and regional scales. (Top) Producers (vascular plants and macroalgae). (Bottom) Benthic animals (insects, crustaceans, amphibians, and so on). (**a**) Local, within-pond, species diversity (N = 30). Both relationships are significantly unimodal ($P < 0.05$). The line represents the estimated quadratic function. (**b**) Regional, within-watershed, species diversity. For both producers (regression: $N = 10$, $R^2 = 0.74$, $P < 0.001$) and benthic animals (regression: $N = 10$, $R^2 = 0.75$, $P < 0.001$), regional species diversity was related linearly to primary productivity. From Chase and Leibold (2002).

reinterpreted patterns of diversity using a novel neutral metacommunity perspective in which species have equal effects on one another and the habitat is homogeneous (see also Bell 2000, 2001). The ability of Hubbell's models to predict patterns of relative abundances suggests that many community patterns are created by random "sampling" of larger species pool and cannot be used to imply mechanisms related to species interactions.

Observational Studies with Explicit Metacommunity Approaches

A few studies of unmanipulated systems have specifically used a metacommunity framework for the interpretation of regional patterns. One of the more important of these is work by Cottenie et al. (2001) on zooplankton metacommunity structure in a set of interconnected ponds. They found that individual ponds showed strong habitat heterogeneity related to predator composition; some ponds had dense planktivorous fish populations that may have led to high algal densities through trophic cascades and other ponds had lower planktivorous fish densities leading to higher densities of large grazers and consequent lower algal densities. More importantly, they quantified the amount of migration among heterogeneous ponds to determine whether migration increases diversity through source–sink dynamics. Despite measuring high rates of migration (via stream flow from one pond to the other), they only rarely found evidence for such source–sink dynamics. This work indicates that the strength of species sorting between these different patch types (pond types) is remarkably strong and cannot be affected easily by the homogenizing effects of migration.

Other patterns in natural systems may take on novel significance when interpreted in the context of metacommunity theory, including those related to abundance patterns (Hubbell, 2001; Chave and Leigh, 2002), to similarity and coexistence (see Leibold, 1998), and to ecosystem attributes (see Mouquet et al., 2002; Leibold et al., 1997). Nevertheless, using metacommunity theory to interpret natural patterns of variation is speculative until this body of theory and its principal mechanisms have been evaluated by means of experimental manipulations. A tremendously rich array of work remains to be done, especially work in which experiments are specifically designed to test mechanisms of metacommunity dynamics.

Experimental Work with Metacommunities

A variety of experimental studies have been conducted on metacommunities, going back to Huffaker's (1958) work with predator–prey systems of mites maintained on increasingly complex arrays of oranges. However, relatively few such manipulative studies have been conducted with more than two species in true metacommunities, where a closed system of local communities is linked by migration. Yet such studies are necessary for appropriate tests of factors identified by models as being important for predicting species persistence and diversity in metacommunities. The theoretical work described earlier indicates that metacommunity dynamics are especially likely to result from complex interactions among migration (or isolation), habitat disturbance (or other mechanisms for local stochastic extinctions), patch-type heterogeneity, and the interrelation among species traits that might result in trade-offs.

Migration among local communities is perhaps the hallmark of metacommunities. Several theoretical studies explicitly predict that increasing rates of migration will lead to increasing local diversity until the migration rate is so high that local variation among sites is swamped (Caswell, 1978; Caswell and Cohen, 1991; Loreau and Mouquet, 1999; Mouquet and Loreau, 2002). Increased migration can increase diversity by at least two mechanisms. If species are migration limited, then increased migration may allow species to gain access to communities where they were previously absent (Tilman, 1994; Shurin, 2001; Miller et al., 2002). Alternatively, migration may act to augment or replace local reproduction for some species, allowing them to persist in communities from which they would otherwise be excluded. At extremely high rates of migration, however, both local and regional (metacommunity) diversity will decline, as regionally dominant species exclude others. Several laboratory studies have varied migration rates among local communities, with mixed effects on local diversity. Warren (1996a,b) investigated the role of migration in two laboratory studies of protist metacommunities, enforcing migration at various rates by moving fluid between separate local habitats. In both studies, diversity did increase with increasing migration rate, but the effects were relatively small and depended on the magnitude of other factors, such as disturbance and habitat size (Fig. 6.3). Similarly, Forbes and Chase (2002) varied the degree of connectivity among zooplankton communities in experimental microcosms and found that increased migration had no effect on local diversity but decreased regional diversity through a homogenization of local communities.

Gonzalez and co-workers (1998; Gonzalez and Chaneton, 2002) experimentally varied migration rates in fragmented communities of microarthropods that occur in dense moss growth. They manipulated migration by arranging moss in isolated fragments (very low migration), in fragments connected to larger areas by moss "corridors," and in continuous areas of moss (Fig. 6.4). Lower migration resulted in the loss of rare species, and corridors alleviated the effects of fragmentation greatly. This result indicates that metacommunity dynamics involving colonization–competition trade-offs

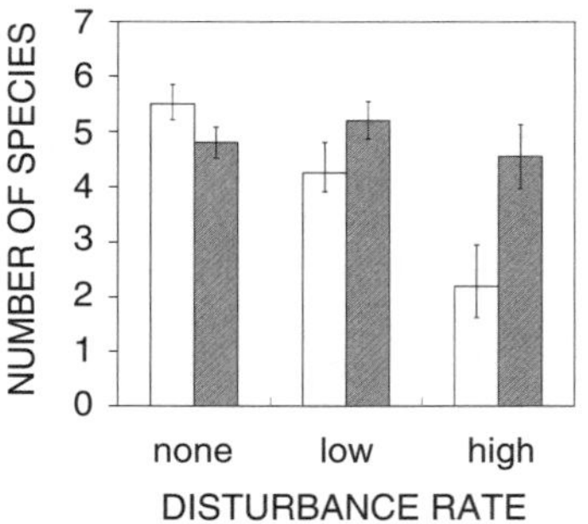

Fig. 6.3 Effects of disturbance and migration rate on local protist species richness in metacommunities consisting of eight experimental 100-ml microcosms. Dispersal took place in 2-ml aliquots of fluid moved at 3-day intervals between either 6 (low dispersal, open bars) or 24 (high dispersal, shaded bars) randomly chosen pairs of microcosms. Disturbance consisted of placing either 1 (low disturbance) or 2 (high disturbance) randomly chosen microcosms in hot water baths every 3 days to kill the protozoa (data adapted from Warren, 1996a).

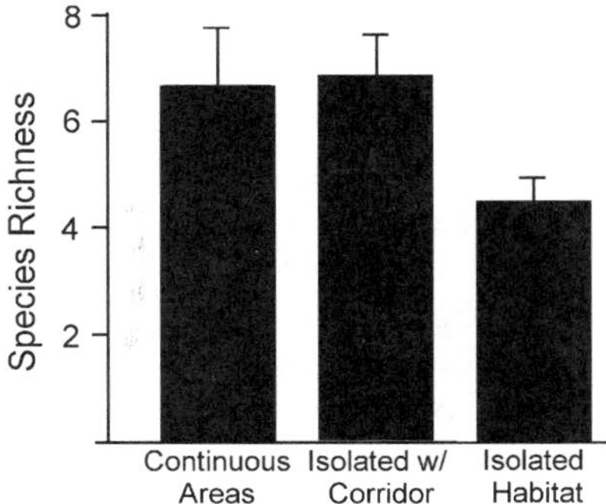

Fig. 6.4 Effect of fragmentation on species richness in a moss invertebrate community. Local communities were either continuous with surrounding moss, isolated but connected with moss corridors, or isolated with corridor. Figure adapted from Gonzalez and Chaneton (2002).

were probably not important. Kneitel and Miller (2003) studied the small communities of protists and rotifers found in the water-filled leaves of pitcher plants. They moved small volumes of water among suites of leaves at three different rates, while also manipulating the presence of predators and the input of resources. Local diversity was highest at intermediate levels of migration, although the presence of predators reduced the effects of migration. Metacommunity dynamics were therefore important in regulating local diversity, but the effects of migration among habitats must be understood in light of local biotic factors.

Disturbance can immediately reduce densities, but it can also create "open" habitats that are available to all species for colonization and may allow fugitive species to persist in metacommunities (Connell, 1978). Disturbance is often considered in continuous communities in which smaller scale disturbances create open patches in a larger community matrix. Although not usually described as doing so, such disturbances create a metacommunity scenario for the community of species that depend on disturbances for persistence in the community. Such species must persist by moving or dispersing propagules to new locally disturbed sites before they are excluded from their current sites. The effects of disturbance in this type of metacommunity should result in a maximum diversity of species at some intermediate rate of local habitat disturbance (Sousa, 1984). For true metacommunities, Warren's (1996a) laboratory protist studies varied the rate of destruction and reestablishment of local communities and showed that disturbance (local patch destruction) resulted only in species loss from the metacommunity, but disturbance was important in determining the effects of migration on protist richness (Fig. 6.3).

Heterogeneity among local habitat patches could be caused by a variation in biotic or abiotic conditions and is predicted to increase metacommunity diversity by increasing the variety of local communities and forming refuges for habitat specialists (Mouquet and Loreau, 2002). Only two studies have directly manipulated heterogeneity within a metacommunity. Forbes and Chase (2002) varied the arrangement of experimental zooplankton communities to influence the probability of moving from similar to dissimilar habitats: arrangement had no effect on local or regional diversity. Miller et al. (2003)

used suites of small laboratory microcosms with protist communities linked by weekly migration. Two types of heterogeneity were maintained: either all local communities composing a given metacommunity were maintained at the same medium resource level or equal numbers of local communities provided low, medium, and high resource levels. Like Forbes and Chase (2002), they found that such variation had no effect on local diversity: species were found to specialize on different resource levels and migration among different resource levels had little effect on establishment. Heterogeneity did lead to higher regional diversity, however, because a greater regional diversity of resources allowed low- and high-resource specialists to persist in appropriate communities.

If we consider all these experimental studies to date, two somewhat surprising findings emerge. First, migration may play less of a role in structuring communities than previously thought. Indeed, some experiments found no effects of migration on local diversity (e.g., Forbes and Chase, 2002) and other studies found only minimal effects. Such results do not preclude the possibility of important effects of migration on community composition through species sorting (e.g., Tilman and Pacala, 1993, Leibold, 1998) or it may be that that experimental migration rates have not been sufficiently high enough to affect diversity through mass effects (Fig. 6.1). Second, a number of other factors, such as trophic interactions, disturbance, and regional heterogeneity, must be incorporated into our understanding of metacommunities; in particular, these factors may have important interactions with migration.

Evidence of the Role of Trade-Offs in Metacommunities

Of greatest interest to us in this chapter are studies that looked explicitly at trade-offs in species traits and how these trade-offs may affect species persistence and coexistence at local and metacommunity scales. To date, remarkably few such studies have been conducted and most deal with a very small number of coexisting species. Yu et al. (2001) investigated factors contributing to the coexistence of two ant genera that specialize on the same plant host. Census and experimental data demonstrated that *Azteca* sp. queens were better dispersers, whereas *Allomerus* cf. *demerarae* colonies were more fecund, presumably making *Allomerus* a better competitor on any given host plant. As a result, species success was a function of interplant distance: at low plant densities, the *Azteca* sp. was dominant, whereas at high densities, the better competitor, *Allomerus*, prevailed. Spatial heterogeneity in host-plant densities allowed regional coexistence due to this trade-off in migration and competitive abilities. This result is intriguing because it highlights a novel possible "niche axis" that might be important in metacommunities, heterogeneity in the local density of patches within a metacommunity (Yu and Wilson, 2001). Lei and Hanski (1998) documented a similar trade-off between competition and colonization in a system of parasitoids attacking the Glanville fritillary butterfly (*Melitaea cinxia*). Again, local density may play a large role in this metacommunity, as changes in host density result in a decline in the proportion of populations attacked by the better parasitoid competitor but have little effect on the proportion of populations attacked by the better colonizer (Fig. 6.5). Working on the community of herbivores on ragwort (*Senecio jacobea*), Harrison et al. (1995) found weak evidence for a colonization–competition trade-off and

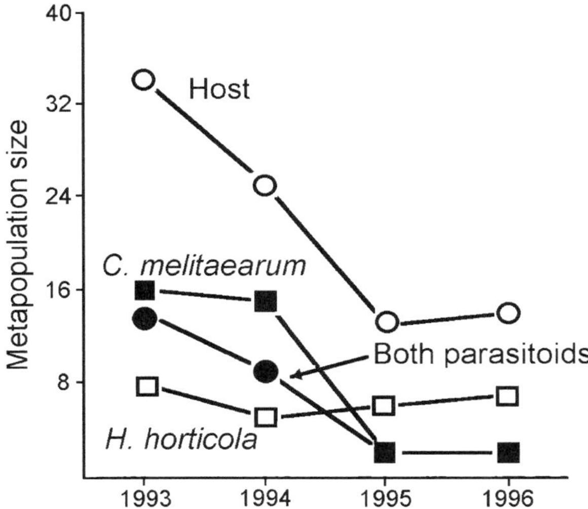

Fig. 6.5 Changes in the abundances of two parasitoids, *Cotesia melitaearum* and *Hyposoter horticola*, and their host, the butterfly *Melitaea cinxia*. As the number of host populations declines, the proportion of populations attacked by the better competitor, *C. melitaearum*, declines, whereas the proportion attacked by the better colonizer, *H. horticola*, is relatively unaffected. From Lei and Hanski (1998).

argued that, even if a trade-off was present, it was not important in regulating metacommunity dynamics. They hypothesized that other mechanisms (possibly related to patch type specialization and species sorting) were most likely allowing for coexistence in this system.

Two studies have more directly tested for the importance of trade-offs that operate at different spatial scales. Amarasekare (2000) studied two egg parasitoids that both use the same host insect. She hypothesized that a trade-off at either of two different scales might explain the coexistence of the two. At the local scale, the inferior competitor might be better able to find unparasitized hosts and thus avoid intraguild predation. At the regional scale, the inferior competitor may have higher migration abilities, allowing it to persist in host patches isolated by distance. Manipulative studies and field surveys were used to demonstrate that the loss of the dominant competitor was related not to host isolation, but to host productivity. Coexistence was therefore best explained by a trade-off between competitive ability and the ability to find unparasitized hosts at the local scale and not by larger scale variation in host abundance. Again, species sorting by local patch-type attributes was surprisingly important in this system, which might at first have appeared to correspond more closely to a metacommunity structured by a colonization–biotic-ability trade-off.

The communities of bacteria and invertebrates found in the water-filled leaves of pitcher plants have the advantage of occurring at discrete scales: within individual leaves (local) and among leaves within a population (regional or metacommunity). Kneitel (2003) proposed to quantify the strengths of trade-offs that may operate at these two different scales. He quantified the growth and colonization rate of protists and rotifers found in the midtrophic-level assemblage of these inquiline communities. To quantify trade-offs, he also determined the competitive ability and tolerance of predation for each of five species using a combination of laboratory and field experiments. Identical experiments were conducted at each of two resource levels, representing different types of local patches. There was little evidence of a trade-off between colonization ability and traits related to species interactions (competitive ability or predator tolerance). Predator tolerance and competitive ability generally showed a strong negative relationship (Fig. 6.6). Further, this negative relationship was strongest between competitive ability at one resource level and predator tolerance at the other, indicating that this trade-off could lead to patch-type specialization. According to Kneitel, these data suggest that species are specialized to do well under different biotic conditions: these species coexist because of a trade-off between competitive ability and predator tolerance that operates at the among-leaf scale.

Clearly, any strong conclusions about how metacommunity dynamics work in natural systems are premature. Few studies have really tackled the difficult issues involved in evaluating complex interactions among migration, patch heterogeneity, and disturbance in naturally existing regional species assemblages. Further, even less has been done to relate any of these results to the pattern of covariation among species traits (or to determine how this covariation results in possible trade-offs). Nevertheless, a few generalizations may be warranted. First, migration can have large effects on local and regional diversity, but its role may depend strongly on other factors. Second, some of the evidence indicates that patch-type heterogeneity and its effect on species sorting are often very strong. This may not always be the case (Hubbell, 2001; Bell, 2000), depending on how evidence from some of these systems is interpreted (see Chave et al., 2002; Chave and Leigh, 2002). Finally, despite a long history

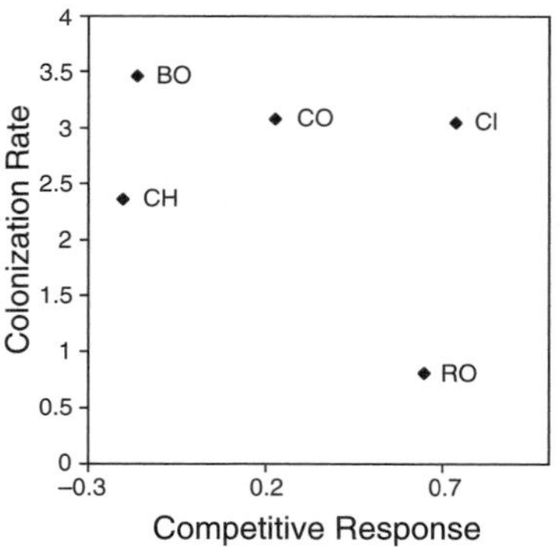

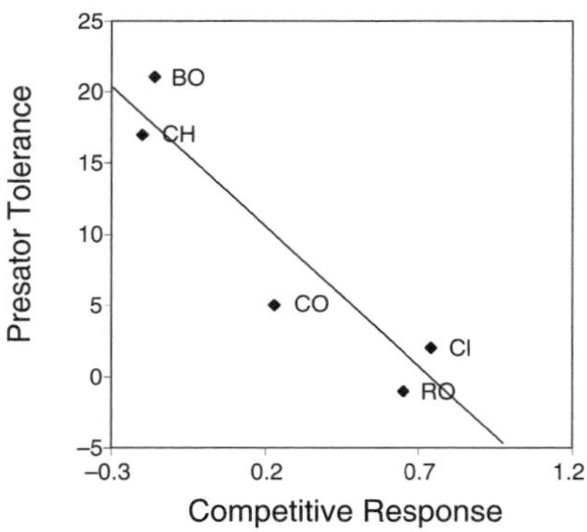

Fig. 6.6 Relationship between the average competitive effect of pitcher plant inquiline species and (a) colonization rate and (b) tolerance to predation. The competitive effect was the average effect of each species in suppressing the growth of competitors in pairwise competition. Predator tolerance is the standardized growth rate of each species in the presence of larvae of the mosquito *Wyeomyia smithii*. The colonization rate was determined by the rate of migration of each species into unoccupied pitchers in the field. Four of the species are protozoans (BO, *Bodo*; CO, *Colpoda*; CI, *Cyclidium*; and CH, chrysomonad sp.); the fifth species is a rotifer (*Habrotochus rosea*) from Kneitel (2003).

in ecological theory, surprisingly little evidence supports the view that a colonization–competition trade-off is important in allowing species coexistence or otherwise influencing diversity. A lack of evidence should not necessarily be interpreted as a general pattern; instead, it should be a call for further study on the role of trade-offs operating at different spatial and temporal scales.

6.5 DISCUSSION

In an intriguing theoretical paper on metacommunities, Wilson (1992) predicted that a surprisingly rich diversity of behaviors, including reciprocal relations between local population dynamics and metacommunity dynamics, the persistence of alternative stable communities in metacommunities, and higher level selection that might affect the nature of ecosystems. Wilson (1992) showed that these phenomena were possible under at least some metacommunity frameworks, but he did not identify the conditions that would influence their occurrence. Work done since then has explored only some of the possibilities, and much remains to be done.

Nonetheless, research in this area is of great potential importance because it is increasingly obvious that many aspects of population biology, community structure, and ecosystem dynamics are influenced by metacommunity processes. Furthermore, numerous environmental issues can be related directly to these processes. Habitat fragmentation, for example, creates patches and affects their isolation (and therefore interpatch migration). Conversely, humans facilitate the movements of other organisms, leading to novel associations of species in metacommunities. Furthermore, humans alter environmental conditions in local patches in a number of ways that alter the local fitness of organisms. Finally, humans also influence extinctions, both on local scales and over larger regional and global scales. These effects are just as likely to alter metacommunity dynamics as they are to affect local ecological processes.

This chapter has tried to emphasize the growth of metacommunity concepts from their origins in metapopulation theory to their current relevance in reevaluating many aspects of community theory. In particular, we argue that although metapopulation theory emphasizes understanding species persistence (Chapter 4), metacommunity approaches are directed toward explaining species coexistence, diversity, and their consequences. Understanding coexistence entails understanding trade-offs in species' responses to different environmental factors, both biotic and abiotic. Although trade-offs have long served as the cornerstone of community theory, metacommunity dynamics draw attention to traits that operate at larger scales and especially to migration and the heterogeneity of local habitats. Among-habitat migration is one of the defining traits of metacommunities and may act frequently to increase not only diversity (if migration rates are greater than extinction rates), but community composition through species sorting. Habitat heterogeneity influences the scale at which species trade-offs are important: homogeneity of habitats may require that trade-offs allow coexistence within local habitats, whereas heterogeneity of habitats expands the scale at which coexistence may occur.

At this stage, we are still learning much about the roles of altered migration, local adaptation, and extinctions in metacommunities. Our knowledge to date of these effects shows that they can be complex and multivaried, but studies have also produced a surprising richness of results that we are only beginning to perceive. Nonetheless, we also think that a couple of generalities are beginning to emerge from both a theoretical and an empirical perspective. Experimental work on migration has suggested its importance, but it may play a larger role in community composition than in community diversity. In addition, habitat heterogeneity appears to be very important for metacommunity dynamics, but very little experimental work has been done in this area. Finally, very few studies of metacommunities have discussed or quantified species trade-offs, a clear call for further experimental work in this area.

A metacommunity perspective, in which scale is incorporated directly into our thinking in the form of local habitats linked by migration, is not a new view for ecologists, but recent theoretical and experimental work in this area demonstrates significant promise. A metacommunity view is clearly important, not just in explaining diversity at the landscape level (something that has been obvious for a long time), but also in constraining the role that disturbance, migration, and heterogeneity can play in regulating biodiversity at local levels. We feel that this approach may also make strong contributions to understanding community composition through understanding traits that allow coexistence at different spatial scales. A rich array of issues awaits both theoretical and empirical investigations.

Part III

METAPOPULATION GENETICS

7. SELECTION AND DRIFT IN METAPOPULATIONS

Hanski and **Gaggiotti,** by **Michael C. Whitlock**

7.1 INTRODUCTION

The distribution of a species over space has many interesting and important evolutionary consequences. All of the basic population genetic forces — drift, selection, migration, mutation, and recombination — act differently in a spatially structured population. Genetic drift can be enhanced or diminished relative to a panmictic population of the same total size. Selection can be more or less effective. Migration is impossible without a spatial context; the consequences of mutations tend to be lowered, and the effective recombination rate is reduced. This chapter reviews some of the effects of population structure, in particular focusing on how selection and drift are changed by the fact that species exist in space. This chapter takes a heuristic and largely nonmathematical look at these issues, trying to express intuitively some recent results in spatial population genetics. This chapter focuses on the dynamics of a single locus, whereas the topics of multilocus selection and quantitative genetics are discussed in Chapters 9, 11, and 12.

One very important summary statistic about the effects of population structure turns out to be one of the oldest: Wright's F_{ST}. There are several ways to define F_{ST}, but they all are standardized measures of the genetic differentiation among populations. Here let us define F_{ST} as the variance in allele frequencies across populations (V_{among}), standardized by the mean allele frequency ($\bar{p}$): $F_{ST} = V_{\text{among}}/\bar{p}(1 - \bar{p})$. F_{ST} has several key features that make

0-12-323448-4

it useful and interesting for the study of evolution in structured populations. First, F_{ST} has the same expectation for all neutral autosomal loci, although even neutral loci can vary substantially around this expectation. Moreover, this expectation is determined by the demographic properties of the species, such as migration rates, local population sizes, and geography F_{ST} can therefore encapsulate a lot of useful information about the demographic history of a species. F_{ST} tends to be larger if local populations are not connected by high rates of migration and/or if local population sizes are small. Finally, F_{ST} is readily measurable from easily obtained data on real populations, and there is already a lot of information about F_{ST} in nature.

There are other useful ways to view the information conveyed by F_{ST} beyond its use as a measure of genetic variance among populations. F_{ST} is also an indication of the amount of relatedness among individuals in the same demes. If F_{ST} is high, then individuals in the same demes are highly related to one another; in other words, they share many alleles. All else being equal, this also pertains to alleles within a diploid individual: if F_{ST} is high, individuals are more likely to be homozygous than would be predicted by Hardy–Weinberg frequencies. These various interpretations and implications of F_{ST} are useful in interpreting the results that follow. It turns out that because F_{ST} represents both the relatedness of individuals within a deme and the excess homozygosity, it is often the only extra parameter needed to describe how population structure changes the pace of evolution.

This chapter reviews the effects of spatial population structure on the amount of genetic drift and the response to selection. The greater part of the chapter then uses these results to discuss basic evolutionary genetic quantities in structured populations, such as the balance point between mutation and selection, mutation load, inbreeding depression, the probability of fixation of new alleles, and other basic quantities. It turns out that these fundamental evolutionary processes are sometimes strongly affected by even a weak population structure.

7.2 GENE FREQUENCY CHANGE IN METAPOPULATIONS

Gene frequency can change in a species by four mechanisms: selection, drift, introgression from other species, and mutation. This section reviews mathematical models that show the effect of population structure in the two more important of these forces, selection and drift.

Genetic Drift

Genetic drift is the change in allele frequency from one generation to the next caused by random sampling of alleles. Genetic drift is nondirectional, meaning that the average change due to drift is zero, but as the population size gets small, the actual change in allele frequency in any given generation can be relatively large.

Effective Population Size

The smaller the effective population size, the more random effects can become important. A key term here is "effective" — the actual amount of genetic drift in a population is determined not only by the actual number of individuals in the population, but also by other factors, such as the distribution of reproductive success in the species. The *effective size*, N_e, of a population is defined as the size of an ideal population, which would be expected to have the same amount of genetic drift as the population in question. An *ideal population* is one in which each of the alleles in the offspring generation have an equal and independent chance of having come from each of the parental alleles. An ideal population would function as though each parent allele contributes an equal and large number of copies to a gamete pool, and then offspring would be formed by random draws from this gamete pool.

Real populations are not ideal though for several reasons. First, and most importantly, in real populations, each individual is not expected to contribute equally to the next generation: some are very fit and have a high reproductive success, whereas others die before even reproducing. This variance in reproductive success tends to reduce the effective population size and therefore increase the rate of genetic drift. Second, in real populations, new individuals are not necessarily formed at random from the available alleles. For example, with inbreeding, individuals are formed with a higher than random chance of having similar alleles at homologous sites. Such inbreeding tends to decrease N_e because each individual effectively carries fewer copies of alleles. Finally, both variation in reproductive success and nonrandom mating can be inherited across generations, and the correlations in reproductive success which result can also affect N_e.

In structured populations, these three factors are even more important. When organisms live in different places, they are likely to experience different conditions, and therefore there is likely to be greater variance in reproductive success than in a single well-mixed population. Population structure causes a kind of inbreeding because locally mating individuals are likely to be related. Finally, if local conditions are correlated positively from one generation to the next, variance in reproductive success will also be correlated among parents and offspring, assuming limited migration.

The effective size of structured populations has been well reviewed by Wang and Caballero (1999).

The Island Model

Describing the effective size of subdivided populations has a long history, beginning with Sewall Wright in 1939. In this paper, Wright derives the effective population size of a species subdivided by an island model, finding it to be

$$N_{e,Island\ Model} = \frac{Nd}{1-F_{ST}}, \tag{7.1}$$

where N is the number of individuals in a deme, d is the number of demes, and F_{ST} is given, for large d at equilibrium, by

$$F_{ST,Island\ Model} \cong \frac{1}{4Nm+1}. \tag{7.2}$$

Here, m is the migration rate among demes. In the island model, each deme contributes a proportion m of its individuals to a migrant pool and then receives the same number of migrants chosen randomly from that migrant pool. It is important to note that these are not random proportions, but that each deme gives and receives exactly Nm individuals to and from the migrant pool each generation, and each deme consists of exactly N individuals. As a result, each deme contributes exactly equally to the next generation. This seemingly innocuous assumption turns out to have fairly important effects on interpreting results obtained from the island model.

If each deme contributes exactly equally to the next generation, then there is no variance in reproductive success among demes. We know from classical population genetics that a lower variance in reproductive success means higher N_e, and in fact this is the case with structured populations as well. Look again at Eq. (7.1). Give that F_{ST} is a quantity that ranges between 0 and 1, the N_e for an island model is always something greater than Nd, in other words greater than the total number of individuals in the metapopulation as a whole. This is because of the assumption that there is no variance among demes in reproductive success.

Relaxing Island Model Assumptions

A more general model of the effective size of structured populations has been derived (Whitlock and Barton, 1997). The general form of the equation for N_e in a species that has reached demographic equilibrium is given by

$$N_e = \frac{\overline{N}d}{\sum_i \frac{N_i w_i^2(1-F_{ST,i})}{\overline{N}d} + 2\sum_i \sum_j \frac{w_i w_j N_i N_j \rho_{ij}}{\overline{N}d}} \tag{7.3}$$

where N_e is shown to be a function of the local population sizes (N_i), the relative contributions of each deme (w_i), the F_{ST} predicted over a set of demes with demographic properties such as deme i ($F_{ST,i}$), and the correlation among demes of allelic identity (ρ_{ij}, which is defined similar to F_{ST}, but instead using covariance of pairs of demes). This equation makes few assumptions about the nature of the spatial subdivision among populations, allowing for variable migration rates over different population pairs, including isolation by distance, local changes in population size, including local extinction, and new population formation via colonization or fission.

While general, Eq. (7.3) is a bit unwieldy for intuitive use. To aid in explaining a few key features of this result, let us use a simplified version of this equation that makes a few more assumptions. If all demes have the same size as each other, but contribute unequally to the next generation via differential migration, then we can write V as the variance among demes in the expected reproductive success of individuals from that deme (i.e., $V = \mathrm{Var}[w_i]$). The effective population size is then

$$N_e = \frac{Nd}{(1 + V)(1 - F_{ST}) + 2NF_{ST}Vd/(d - 1)} \tag{7.4}$$

(Whitlock and Barton, 1997).

Let us examine two extremes using Eq. (7.4). If, as in the traditional island model, the variance among demes in reproductive success is zero, then Eq. (7.4) reduces to Eq. (7.1) [This is true not only for the island model, but for any model for which each deme is equal in size and contributes exactly equally to all other demes, provided that all demes are ultimately reachable by each deme via migration (Nagylaki, 1982). This includes classic stepping stone models.]

At the other extreme though, let us imagine that one deme is extremely successful and produces all of the offspring that fill all d of the demes. In this case, we would intuitively predict that the effective population size of the whole system should be the same as the size of the single successful deme, and Eq. (7.4), with appropriate modification, shows us that this is in fact the case. (In this extreme, the F_{ST} would be zero and the variance among demes of allelic reproductive success would be $d - 1$.) This extreme example tells us that N_e can be much smaller than the census size with population structure.

The truth obviously lies somewhere in the middle. It turns out that the boundary between whether population structure increases or decreases N_e is approximately whether or not demes have greater or less variance in reproductive success than would be expected by a Poisson distribution. In other words, if demic structure acts to increase variance in reproductive success relative to that expected by chance, N_e would be reduced. Only if the effects of population structure are to reduce the variance among demes in reproductive success to less than random would N_e be increased. This is perhaps biologically unlikely, yet this is the requirement for the results from the simple island model to hold qualitatively. In real species, the opposite is likely to be true: different demes are likely to have different amounts of resources, and different demes are likely to experience different levels of other ecological factors that might affect success, such as levels of parasitism, disease, predation, weather fluctuations, and other catastrophes. Realistic ecology implies higher than random variance among demes in reproductive success, and therefore the effective size of a subdivided species is likely to be reduced, perhaps substantially. The island model is not a good descriptor of typical population structure, for this and many other reasons (see Whitlock and McCauley, 1999).

Extinction and Colonization

It will be useful to consider a couple of specific cases that go beyond the simple island model. One aspect of population structure that has attracted some attention is the possibility of local extinction and recolonization (Slatkin, 1977; Maruyama and Kimura, 1980; Whitlock and Barton, 1997). The models considered in these papers are similar: the basic structure is like an island model, except that each deme has some chance per generation of going extinct independently of its genotype frequencies. An equal number of new demes are colonized, either in the same places recently vacated by the extinction events or in other vacant sites, by a small number of individuals. As a major, unrealistic simplification, each new deme then immediately grows back to N individuals, like all other demes.

With local extinction and recolonization, population structure contributes in an obvious way to the variance in reproductive success among demes. Even though this model is based on the island model, even a small rate of extinction is enough to cause the effective population size of the species to be reduced rather than increased. The main reason is perhaps obvious: with extinction and recolonization, some demes have zero reproductive success,

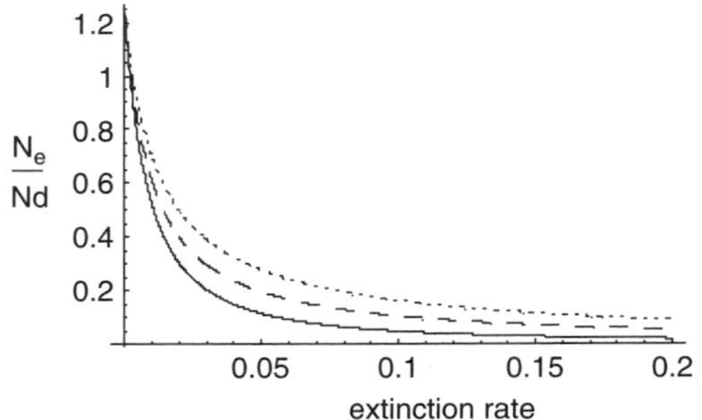

Fig. 7.1 The effective size (displayed as a proportion of the census size) of a metapopulation with local extinction and colonization. Here each deme has 100 individuals, and each new population is founded by four individuals. These colonists have a probability ϕ that they come from the same source population, with $\phi = 0$ in the dotted line, $\phi = 1/2$ in the dashed line, and $\phi = 1$ in the solid line. The migration rate was 0.01 in all examples. As the local extinction rate increases, the effective population size is reduced greatly.

whereas others — those that manage to survive and send colonists to start new demes — have a reproductive jackpot. Thus there is a great deal of variance among demes in reproductive success, which causes the effective size to be reduced. This reduction can be extreme (Fig. 7.1).

Sources and Sinks

In some species, some populations have large amounts of resources, whereas others have so few that they cannot replace themselves without migration (Pulliam, 1988; Dias, 1996; Holt and Gaines, 1992). These so-called "sources" and "sinks," respectively, cause the population dynamics to be different from the island model: demes do not contribute equally to the migrant pool, and therefore there is variance in reproductive success. If the quality of patches of resource is correlated positively over time, then the effect on N_e is even more extreme.

The effects of source–sink structure and correlation over time in patch suitability can be best seen by another extreme example. Imagine that a fraction of demes, say 20%, reside in productive source patches, and the other 80% of demes are what Bob Holt has called "black-hole" sinks — that is, these demes never contribute migrants to other demes and only persist because of migration from source populations. In this case, it is clear that only alleles in individuals in source populations can contribute to future generations and so the only individuals that matter to the evolution of the species are in the source populations. Therefore the N_e of the species should reflect only the effective size of the source populations alone. Thus the N_e of this species should be only 20% of what it would have been with equal migration.

To be more general, we can apply useful results from Nagylaki (1982), who showed that the N_e of a system of populations with a constant migration matrix could be described with the left eigenvector of that matrix. (This assumes a few technical details, such that all demes are ultimately reachable by migration from all other demes, even if it takes multiple steps.) Consider a case where migration is via a migrant pool so all emigrants from all demes are

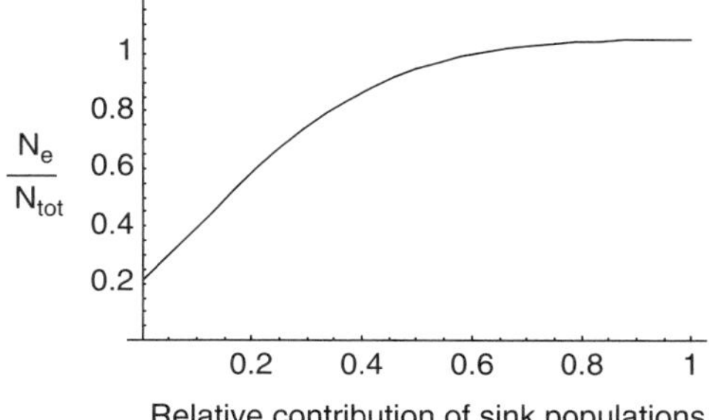

Fig. 7.2 The effective size of a species in which 20% of the demes are sources and the rest are sinks. The *x* axis varies the contribution of the sink populations, expressed as a fraction of the contribution of the sources. Here each 100 demes have 100 individuals, and each receives five immigrants per generation sampled from the migrant pool. As the contribution of sinks reaches zero, the effective size of the system is the same as an island model with only the 20 source populations.

mixed together and then moved on to recipient demes at random with respect to where they originate. Source demes contribute a large number to the migrant pool, whereas "sink" demes contribute a fraction of that number. For simplicity, each deme receives a constant number of immigrants from the migrant pool. This ensures that the F_{ST} among sources and among sinks are approximately equal. Figure 7.2 shows the effective size of these systems as a function of the relative contribution to the migrant pool by sinks. [To make the calculations in Fig. 7.2, Nagylaki's (1982) results were used, accounting for the fact that Nagylaki's definition of N_e differs from the usage here. Nagylaki calculates the N_e that would give the same amount of variance *within* a deme at mutation–migration–drift balance; in other work including in this chapter N_e predicts the amount of variance predicted by the average allele frequency of the species as a whole. The second of these two quantities can be found from the first by dividing by 1-F_{ST}. Details are given in Whitlock (2003).]

Note that with this form of source–sink structure, the effective size of the species is just the effective size of just the source populations when the sinks do not contribute to the future, and it reduces to the island model results when "sinks" contribute equally to sources.

Selection

With good reason, the study of selection in subdivided populations has, in the past, focused on the effects of spatially heterogeneous selection (e.g., Felsenstein, 1976). A great deal of important and interesting evolutionary biology results from variation in selection over space, but population structure, perhaps surprisingly, has a lot of interesting effects even on uniform selection. Arguably, most loci have approximately similar selection in different demes, even though the more obvious and more polymorphic cases may reflect spatially divergent selection. This chapter focuses on this special case in which genotypes have the same relative fitness in each population of the species.

When selection is uniform across populations, it becomes possible to follow the state of the metapopulation by following the mean allele frequency across all local populations, $\overline{q}$. Consider simple selection between two alleles at the same locus, with the fitnesses of the three genotypes given by $1 : 1 + h\,s : 1 + s$. In this case, the change in allele frequency due to selection within each population is a third-order function of q; therefore, to understand how the mean allele frequency would change by selection requires knowing the expected values of q, q^2, and q^3. Fortunately, under most circumstances the dynamics of the expected value of q^3 can be well enough predicted by an understanding of changes in the first two, which reduces the problem to understanding $\overline{q}$ and $E[q^2]$. The expected value of q^2 may seem like an exotic quantity to keep track of, but remember that the variance among demes is derived easily from $\overline{q}$ and $E[q^2]$, and F_{ST} is derived easily from the variance in allele frequency and $\overline{q}$. Thus, a very good understanding of the change in allele frequency across a metapopulation can be obtained by knowing $\overline{q}$ and F_{ST}. Moreover, as long as the selection coefficient is not much greater than the rate of migration into a deme, the F_{ST} predicted from neutral theory works extremely well to predict allele frequency change in structured populations. These conclusions are derived and discussed in greater detail in Whitlock (2002).

[One technical note is necessary: when calculating these quantities, it is essential to weight each individual equally. The usual calculations of F_{ST} weight each local population equally, independent of size. Most models of population structure have assumed equal deme sizes, and therefore they predict the right quantity. Most empirical measures do not measure the appropriate F_{ST} exactly. This may be an important issue in some cases; for example, if smaller demes have higher extinction rates, then the subset of the population with the highest F_{ST}'s would properly be weighted least.]

It will help to look at the equation for the change in mean allele frequency due to selection. From Whitlock (2002), we get

$$\Delta_s\overline{q} \cong \overline{p}\,\overline{q}\,s(1-r)(F_{ST}+(1-F_{ST})(h(1-2\overline{q})+\overline{q})) \tag{7.5}$$

where r is the relatedness of two random individuals competing for resources.

Let us consider the various parts of this equation in turn. First, we see that the response to selection is a function of the mean allele frequencies and the strength of selection $\overline{p}\,\overline{q}\,s$. These are the classic terms that would appear even without population structure: the response to selection is proportional to the allelic variance $\overline{p}\,\overline{q}$ and to the strength of selection.

Next, we find that the response to selection is proportional to one minus the relatedness of competing individuals. This last phrase deserves some explanation. Consider a classic dichotomy introduced by Dempster (1955; see also Christensen, 1975) between local and global competition for resources, i.e., soft versus hard selection. With soft selection, each deme contributes a number of individuals to the next generation (whether via resident individuals or migrants) *independent* of the genotypes of the deme. With hard selection, each deme contributes to the next generation in proportion to its mean fitness determined by its genotype distribution. Under soft selection, individuals are competing locally for resources, and therefore there is competition between relatives. The mean relatedness of individuals from the same deme (without inbreeding within demes) is given by $r \equiv 2F_{ST}/(1 + F_{ST})$. At the other extreme, under hard selection, there is no local competition for resources, and the relatedness of competing individuals is zero. Putting these equations into Eq. (7.5), we find that hard selection is always more effective than soft selection in changing allele frequency. With local competition for resources, if an individual does well because of having a good genotype, it will, through competition, reduce the resources available to other individuals in the same deme. With population structure, these other local individuals are likely to share alleles. Therefore the event that would have boosted the number of copies of this good allele in the next generation (the first individual doing well) is partially counterbalanced by competition against the same genotypes.

Note that for the relatedness term, increasing population structure tends to weaken the response to selection. With soft selection, increasing F_{ST} results in greater relatedness and therefore a lower response to selection, *all else being equal*.

Finally, we see in the last term $(F_{ST} + (1 - F_{ST})(h(1 - 2\bar{q}) + \bar{q}))$ a reflection of the effects of increasing homozygosity on the response to selection in structured populations. As F_{ST} increases, so does the proportion of individuals that are homozygous, even for the same mean allele frequency. Greater homozygosity, for the same q, increases the magnitude of the response to selection. This increase is particularly important if $\bar{q}$ is small and the allele is at least partially recessive ($h < 1/2$). In these cases, with panmixia, most alleles appear as heterozygotes and selection therefore cannot discriminate the recessive alleles. As F_{ST} increases, most of the selection is experienced by alleles in the homozygous state, where the alleles have relatively large effects. Thus, in opposition to the effect of relatedness given earlier through its effects on increasing homozygosity, population structure tends to *increase* the response to selection. For nearly recessive alleles, this boost can be extremely large.

This effect of excess homozygosity has been described much earlier with respect to inbreeding within populations (Ohta and Cockerham, 1974). In fact, with hard selection, there is no distinction between the effects of inbreeding due to population structure and that due to local inbreeding; they enter the response to selection equations in exactly the same way. With soft selection, however, the extra effects of competition among relatives change the relationship between F and response to selection.

The balance between these two effects (competition among relatives and homozygosity) depends on the details. With hard selection, there is no effect of relatedness, and population structure therefore always increases the rate of response to uniform selection. With soft selection, response to selection can be either increased or decreased depending on the dominance coefficient of the locus under selection and F_{ST}. The following section shows examples of both. The effects of population structure on even uniform selection are quite complicated.

With this selection equation available, a variety of results on basic selection become easy to derive. The next few sections of this chapter show some of these results.

7.3 MAINTENANCE OF GENETIC VARIATION IN SUBDIVIDED POPULATIONS

One of the oldest questions in population genetics is "what forces are most important in maintaining genetic variation ?" Population subdivision can affect the maintenance of genetic variation in a variety of ways. This section reviews a few of these briefly, focusing on the case of spatially uniform selection.

Mutation–Selection Balance

Estimates have shown that the genomic rate of mutation to deleterious alleles is reasonably high, ranging from a few per thousand individuals to much greater than one per each new individual (Lynch et al., 1999; Keightley and Eyre-Walker, 2000). Although natural selection operates to reduce the frequency of these deleterious alleles, they are not immediately eliminated completely. As a result, some deleterious alleles are always segregating in populations at a frequency determined by the balance between mutation and selection. Some have argued that levels of standing genetic variance observed in natural populations could be explained largely by this mutation–selection balance.

Mutation is likely not much affected by population structure, but the previous section showed that the efficacy of selection can be affected greatly by subdivision. At mutation–selection balance, the deleterious allele is likely to be rare, which simplifies Eq. (7.5) to

$$\Delta_s\overline{q} \cong \overline{q}s(1 - r)(F_{ST} + (1 - F_{ST})h) \tag{7.6}$$

The equilibrium allele frequency at mutation–selection balance is then given by

$$\hat{\overline{q}} \cong \frac{\mu}{-s(1 - r)(F_{ST} + (1 - F_{ST})h)} \tag{7.7}$$

(Remember that in the way we have defined fitness in this chapter, a deleterious allele has $s < 0$.) For recessive alleles in particular, the frequency of deleterious alleles at mutation–selection balance is much reduced with population structure due to the more effective selection against homozygotes. See Fig. 7.3, for some examples. As a result, the amount of variation maintained by mutation selection balance can be reduced greatly in large metapopulations, depending on the distribution of dominance coefficients. Most current estimates of the mean dominance coefficient of mildly deleterious alleles give answers around $h = 0.1$ (Houle et al., 1997; García-Dorado and Caballero, 2000; Peters et al., 2003), so the reduction in variance can be substantial even for relatively small F_{ST} values.

The predominant model of the genetic mechanism for inbreeding depression claims that inbreeding depression results from deleterious recessive alleles segregating in populations at mutation–selection balance. With population structure, the reduction in mean deleterious allele frequency

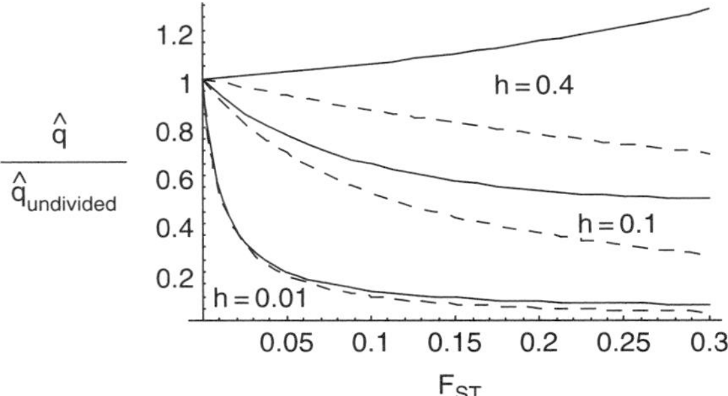

Fig. 7.3 The equilibrium value of the frequency of a deleterious allele can be changed substantially by population structure. Here solid lines indicate pure soft selection and dashed lines indicate pure hard selection. With very recessive alleles, the equilibrium allele frequency is reduced greatly relative to the case in an undivided population (where $\hat{q} \cong -\mu/hs$). Parameter values used for these calculations were $s = -0.1$, $\mu = 10^{-6}$, and the three lines correspond to $h = 0.4$, 0.1, and 0.01 from top to bottom. From Whitlock (2002).

results in a potentially large reduction in the amount of inbreeding depression predicted for a species, even at relatively low F_{ST} values (see Whitlock, 2002).

Balancing Selection

Balancing selection, by definition, occurs when selection acts to increase the frequencies of rare alleles. This can happen with overdominance, negative frequency-dependent selection (where rare alleles are favored because they are rare), or by spatially heterogeneous selection. Each of these are affected by the spatial population structure.

Overdominance

With overdominance, the heterozygote is the most fit genotype. For this section only, let us redefine the fitnesses of the three genotype **AA**, **Aa**, and **aa** as 1-s : 1 : 1-t, such that the fitness of the two homozygote genotypes is reduced by a factor s or t. With overdominance in a large randomly mating population, there is an intermediate equilibrium allele frequency that stably maintains variation in the population as a result of the heterozygote being selected for whenever one or the other of the two alleles becomes too rare.

In structured populations, the extra homozygosity caused by population structure can change the dynamics of the maintenance of variance. Nonrandom mating causes the marginal fitnesses of the two alleles to be determined more by their homozygous effects and less by their effects in heterozygotes. As a result, if the two homozygotes fitnesses are not equal ($s \neq t$), then the allele associated with the fitter homozygote will have a higher frequency than expected under random mating. Mathematically, that frequency is given by

$$\hat{q} \cong \frac{s - tF_{ST}}{(s + t)(1 - F_{ST})}, \tag{7.8}$$

so long as this value is between zero and one, which it need not be (Whitlock, 2002). If F_{ST} is large enough, the expected equilibrium leaves the population fixed for the allele with the most fit homozygote. Thus population structure tends to reduce the amount of variation maintained by overdominance.

Frequency Dependence

In some cases, the fitness function changes with the frequency of alleles in the population or species; this is called frequency-dependent selection. If selection displays negative frequency dependence, then alleles are more fit when rare than when the same allele is common. In this case, selection can act to maintain variation in a population because as alleles get rare (as they would on the path to being lost from the population), their fitness increases and therefore their frequency climbs again.

One of the most studied examples of negative frequency dependence is the self-incompatibility (SI) alleles common to many species of plants. With SI, pollen (or, in some cases, its parent plant) that shares alleles with the maternal plant are not allowed to fertilize ovules. These processes presumably evolved as a mechanism to prevent self-fertilization, but they also prevent unrelated individuals that share alleles from mating. As a result, rare alleles at the SI locus have higher fitness because they are able to mate with more other individuals in the population. All else being equal, the system always favors new alleles being introduced into the population, but real species have limited numbers of SI alleles because of loss due to genetic drift. The smaller the effective population size, the fewer SI alleles maintained at equilibrium.

With population structure, one might imagine that different alleles might be maintained in different populations, thereby increasing the total diversity in the species as a whole. It turns out that this is true for species with very low migration rates between demes, but with realistic, intermediate levels of migration the total number of SI alleles maintained is slightly lower than would be expected with panmixia (Schierup, 1998; Schierup et al., 2000; Muirhead, 2001).

Heterogeneous Selection

It has been known since at least the 1950s that spatially varying selection can maintain genetic variation, especially if there is soft selection (Levene, 1954; Dempster, 1955). The conditions for this are narrower than was commonly thought (Maynard Smith and Hoekstra, 1980), requiring strong, relatively symmetric selection. Felsenstein (1976) and Hedrick (1986; Hedrick et al., 1976) reviewed the theory and empirical evidence for and against the maintenance of genetic variance by heterogeneous selection.

A different form of heterogeneous selection can emerge in populations in which there is already a lot of genetic differentiation among populations. In these cases, epistatic interactions between loci can cause different alleles to be favored locally even when the underlying function describing the relationship between fitness and genotype is uniform across space (see Chapters 9 and 11). This sort of heterogeneous selection depends on there being selectively and epistatically different alleles in different local populations, which becomes important only under extremely restricted gene flow or extreme drift.

One special case of epistasis that may be quite common is that generated on approximately additively interacting alleles that form a phenotype under stabilizing selection. Stabilizing selection causes the fitness effects of alleles to vary depending on whether the sum of the effects of all other alleles in the individual add up to a value above or below the optimum for the trait; hence with stabilizing selection, a population near its optimum will have mainly epistatic variance for fitness associated with that trait (Whitlock et al., 1995). Barton and Whitlock (1997) have shown that with uniform stabilizing selection and low migration, the amount of genetic variance for a trait that can be maintained can be increased substantially as a result of this epistasis. However, this is only likely to be important in species with very high values of F_{ST}, in the range of $F_{ST} > \sim 0.2$.

7.4 ADAPTATION IN SUBDIVIDED POPULATIONS

Population structure can affect the pace of adaptive evolution. We have already discussed the conditions under which the response to selection is increased or decreased with population structure. The subdivision also allows novel patterns of adaptation, such as local adaptation (see Barton, 2001), shifting balance evolution [Wright (1931), but see Coyne et al. (1999) and Whitlock and Phillips (2000)], and more rapid evolution with epistatic interactions (Bryant et al., 1986; Goodnight, 1988; see Chapter 9). More fundamentally though, population structure strongly affects the pace of evolution even for those alleles that are uniformly selected without any complicating interactions with other loci. This section reviews the effects of population structure on the probability of fixation of new mutations.

Probability of Fixation

One of the most remarkable results in population genetics has to be Haldane's (1927) result that a new beneficial allele with heterozygous benefit of hs has only about $2hs$ chance of ultimate fixation. Haldane assumed that the species in question was ideal (i.e., its census size equaled its effective size) and undivided. Even in an infinite population, if a new allele is introduced as only a single copy, the fate of that allele is partially determined by stochastic changes in the numbers of copies of the allele left in each generation. It turns out that by introducing an allele as a single copy (as a rare mutation would likely do), even alleles with moderate selective advantage are more likely to be lost stochastically from the population than fix. Kimura (1964; see also Crow and Kimura, 1970) modified this result to allow for nonideal populations and allowed arbitrary dominance for deleterious alleles as well. He found that the probability of fixation of a beneficial allele is given approximately by $2hsN_e/N$, where N is the census size of the population.

In 1970, Maruyama achieved the first results on the probability of fixation in subdivided populations. He showed that in an island model, the probability of fixation for an additively acting allele was simply s. (For additive alleles, $h = 1/2$, so this result is equivalent to the $2hs$ of Haldane.) Maruyama (1974) and others (Slatkin, 1981; Nagylaki, 1982) extended this result to deal with any model such that each deme contributes exactly equally to the next generation; the probability of fixation with population structure with this restriction remained s. This was viewed by some as an invariant result of population structure; the claim was made that population structure therefore did not affect the probability of fixation of beneficial alleles. However, this conclusion was premature because other models of population structure are possible (and even more reasonable than the island model) and because the effects of dominance were not properly accounted for. The first demonstration that this was not true was a model of extinction and two specific types of recolonization by Barton (1993). In these cases, the probability of fixation was much reduced by population structure relative to the panmictic case.

The probability of fixation in a more general model of structured populations has been found (Whitlock, 2003). Based on Kimura's diffusion equations, this work shows that the probability of fixation can be derived from the equations for drift and response to selection presented earlier in this chapter. Moreover, as long as the strength of selection is lower than the typical immigration rate, the F_{ST} expected for neutral loci can be used in these equations, which expands their usefulness greatly. For dominance coefficients differing from 1/2, the equations cannot be solved directly, but the answers can be obtained with numerical integration. In the interests of space, this chapter will not review the mathematics of the general equations, but will focus on the additive case, as well as an approximation that works very well for beneficial alleles even with arbitrary dominance. More details can be found in Whitlock (2003).

For additive alleles, such that $h = 1/2$, the probability of fixation in structured populations is given by

$$u[q] = \frac{1 - \exp[-2s(1 - F_{ST})N_e q]}{1 - \exp[-2s(1 - F_{ST})N_e]} \tag{7.9}$$

for soft selection and

$$u[q] = \frac{1 - \exp[-2s(1 + F_{ST})N_e q]}{1 - \exp[-2s(1 + F_{ST})N_e]} \tag{7.10}$$

for hard selection, where q is the initial allele frequency of the allele in the metapopulation. If the population starts with a single copy of the new allele, then $q = 1/2N_{tot}$, where N_{tot} is the total size of the metapopulation. These equations look fearsome, but in fact they are quite similar to the equations for the panmictic case derived by Kimura (1964). There are two differences. First, the N_e here is the effective size of a subdivided population, given by Eq. (7.3). Second, the strength of selection s is now modified by a term involving F_{ST}, which reflects the change in the efficacy of selection from population structure.

For beneficial alleles, we can write a simple equation for the probability of fixation of a new mutant, even with arbitrary dominance:

$$u \cong 2s(1 - r)(F_{ST} + (1-F_{ST})h)N_e/N_{tot}. \tag{7.11}$$

Here it is possible to see that this result builds directly on Kimura's. As F_{ST} goes to zero, this approaches the $2hsN_e/N$ given earlier.

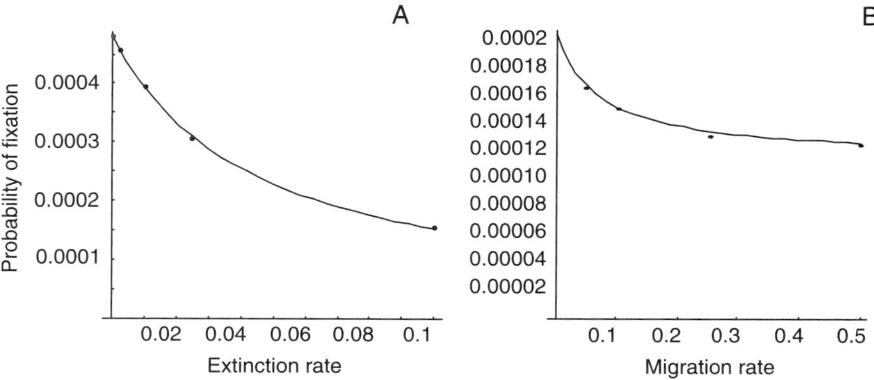

Fig. 7.4 Examples of the fixation probabilities of nearly recessive beneficial alleles ($h = 0.01$) with soft selection. (A) Extinction and recolonization. In this example, the migration rate between populations was 0.05, colonization occurred by four individuals with a probability of common origin of 1/2, $s = 0.002$, and there were 100 demes with 100 diploid individuals each. (Each point represents results from 10^7 simulations, so the standard error ranges from 6.9×10^{-6} on the left to 3.9×10^{-6} on the right.) As the extinction rate increases, the effective population size of the metapopulation decreases, and therefore so does the probability of fixation. (B) A one-dimensional stepping-stone model. With a stepping-stone model, F_{ST} (and therefore N_e) increases as the migration rate drops so the probability of fixation also increases with lower migration. This is particularly true with recessive alleles, which are expressed often in the homozygous state with the concomitant increase in the efficacy of selection. (There are 100 demes with 100 diploid individuals each, $s = 0.0002$ and dots represent 10^6 simulations each.)

These results have been tested by simulation in a wide variety of models of population structure, including the island model, extinction–recolonization, stepping-stone models, and source–sink models. They work remarkably well (see Figs 7.4 and 7.5).

The probability of fixation of beneficial alleles tends to be much reduced with population structure. This is mainly a result of the fact that the effective

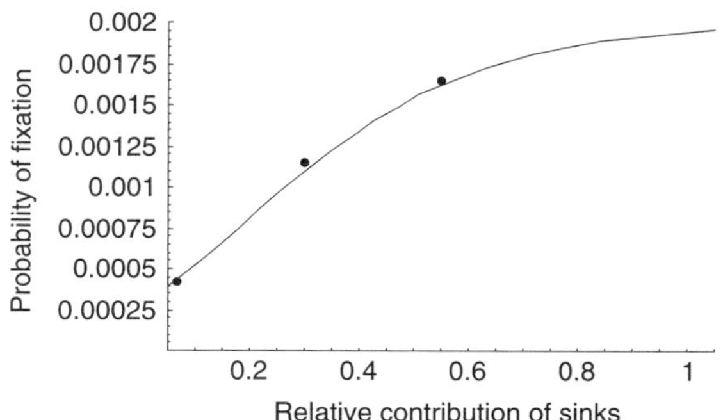

Fig. 7.5 The probability of fixation in a source–sink model. Here there are 100 demes, 20 of which are "sources" and the rest are "sinks". Each deme has 100 individuals, and the immigration rate to the sources is 0.2, whereas it is 0.25 in sinks. Demes exchange migrants by a modified island model, where each sink's contribution to the migrant pool is a fraction of that of each source. As this asymmetry increases, the effective population size is reduced and the probability of fixation of beneficial alleles drops. For these examples, $s = 0.002$ and $h = 1/2$, and dots represent results of 10^7 simulations.

population size is reduced in most models of population structure. The probability of fixation can be increased for some loci, especially for nearly recessive alleles that can be expressed more strongly in structured populations because of increased homozygosity.

Let us return to the island model. As mentioned earlier, the island model is an extreme description of population structure because it allows no variance among populations in reproductive success. For additive alleles, Maruyama and successors found the probability of fixation to be simply s in an island model, the same as in an unstructured population. The more general model predicts that the probability of fixation should be $s\,(1 - F_{ST})\,N_e/N_{tot}$ (because the island model in its basic form as used by Maruyama is also a soft selection model). Remember that the island model has the unusual property of having a larger N_e than census size: $N_e = N_{tot}/(1 - F_{ST})$. Putting this N_e into the probability of fixation equation simplifies it to simply s. The results are consistent; what is more important is that the island model is unrealistic and extreme. Most real species will have $N_e < N_{tot}$, and so most will have lower probabilities of fixation of beneficial alleles than predicted by Maruyama's formula. Probabilities of fixation are not invariant with respect to population subdivision.

Relaxing the assumption of uniform selection has been investigated using the island model by a variety of authors (Barton, 1987; Tachida and Iizuka, 1991; Gavrilets and Gibson, 2002). Population structure tends to increase the probability of fixation relative to that expected by the mean fitness of the alleles across demes. It is not yet known what effect heterogeneous selection would have with a more realistic model of subdivision.

Population structure also substantially affects the time taken for fixation of new alleles (Whitlock, 2003).

7.5 GENETIC LOAD IN SUBDIVIDED POPULATIONS

Genetic load is the reduction in the mean fitness of a population relative to an optimal genotype caused by some particular factor, such as deleterious mutation, genetic drift, and segregation (Crow, 1993). Load is sometimes strongly affected by population structure, as reviewed in this section.

Mutation Load

Mutation load is the reduction in mean fitness caused by recurrent deleterious mutations in a population. Mutation load is usually calculated at mutation–selection balance: that is, it is the mean reduction in fitness associated with an allele frequency predicted by the equilibrium between mutation and selection. In panmictic populations, the load associated with an allele that is not completely recessive is $L = 2\mu$ (where μ is the mutation rate from wild type to deleterious allele; remarkably, this is not a function of the strength of selection against the deleterious allele).

With population structure, load equations become more complicated (Whitlock, 2002):

$$L \cong -(2h(1-F_{ST})+F_{ST})s\hat{\bar{q}} \tag{7.12}$$

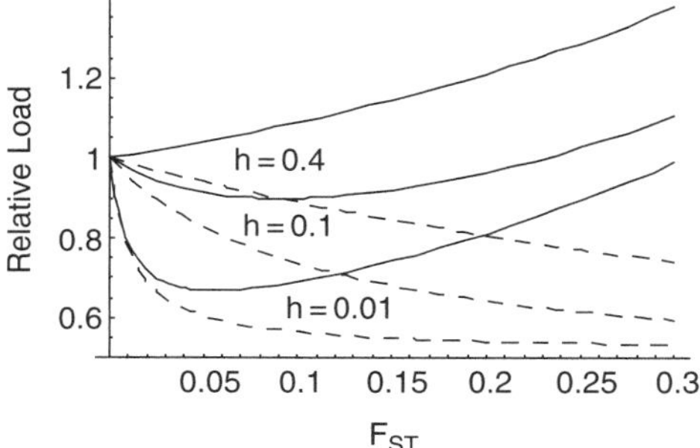

Fig. 7.6 The mutation load in a metapopulation relative to the load at a similar locus in an undivided population (~2μ). For the values of F_{ST} likely to be found within species and relatively small values of the dominance coefficient h, the mutation load can be reduced substantially in a subdivided population. Solid lines show pure soft selection, whereas dashed lines correspond to pure hard selection. Parameters for this example are $s = -0.1$, $\mu = 10^{-6}$, and the three pairs of curves correspond to $h = 0.4$, 0.1, and 0.01 from top to bottom.

where the value of $\hat{\bar{q}}$ is given by Eq. (7.7). Note that s will cancel out when this substitution for $\hat{\bar{q}}$ is made, but load remains a function of the dominance coefficient, unlike the panmictic case. Figure 7.6 shows the change in load as a function of population subdivision. Load is always reduced with hard selection, but with soft selection, load is increased for high values of F_{ST} and near additivity. With nearly recessive alleles, the reduction in load can be nearly 50%.

Segregation Load

Segregation load is the reduction in fitness caused by the inability of a population to be composed entirely of heterozygotes even when these genotypes are the most fit. As such, segregation load requires overdominance. With population structure, there are even fewer heterozygotes in a species than under Hardy–Weinberg conditions so the segregation load would be more pronounced. Using the same notation as in the overdominance section given earlier, the segregation load is expected to be

$$L = \frac{(1 + F_{ST})st}{s + t}, \tag{7.13}$$

which reduces to the segregation load in a panmictic population when $F_{ST} = 0$ (Crow, 1958). Therefore, the segregation load is $(1 + F_{ST})$ times as great in a subdivided population as in an undivided one, as expected by the increased number of homozygotes.

Drift Load

Drift load is the reduction in fitness caused by drift changing allele frequencies away from those favored by selection. An extreme form of drift load results from fixation of deleterious alleles by drift. Drift load has received a lot of attention in the last several years because of the possible mutational meltdown of small endangered populations (Lande, 1994; Lynch et al., 1995a,b).

The rate that deleterious alleles accumulate in a species is a function of the efficacy of selection and of the effective population size; the smaller these two values are the faster drift load will accumulate. We have seen that selection is often more effective in structured populations (although not always), but more importantly, the effective population size tends to be reduced by structure. Because the latter of these two effects turns out numerically to be more important, in most cases, population structure increases the rate of accumulation of deleterious alleles (Higgins and Lynch, 2001; Whitlock, 2003). This is most pronounced in cases with large variance in reproductive success among demes, such as with extinction and recolonization or source–sink models. Figure 7.7 shows that the change in the probability of fixation of deleterious alleles can be reasonably large (two- to three fold), although perhaps in most cases the change is less than a doubling.

Migration Load

If the local population in a deme is well adapted to local conditions and if migrants to this population come from populations adapted to other conditions, then the alleles that come into the population by migration are likely to be poorly adapted to local conditions. The reduction in mean fitness that results is called *migration load*. Migration load increases with increasing differences in the selection coefficients among populations and with migration rate. In some species, migration load is likely to be the most important type of genetic load. Migration load may be key in determining the range limits of species because migration from the species center may prohibit further local adaptation at the margins (Mayr, 1963; Kirkpatrick and Barton, 1997).

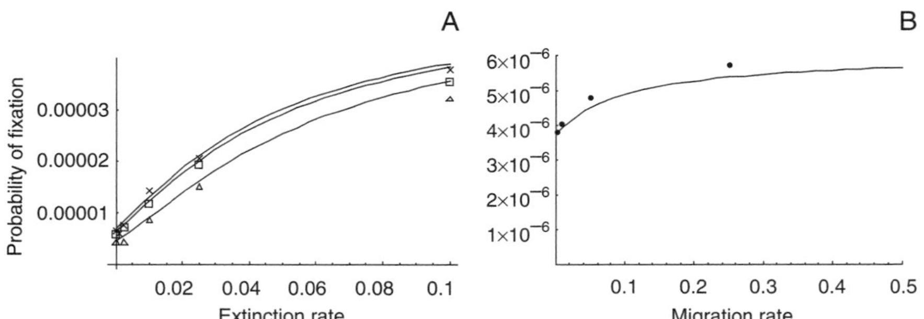

Fig. 7.7 The probability of fixation of deleterious alleles with (A) extinction and colonization or (B) a one-dimensional stepping stone model. (A) The three lines plot, from bottom to top, the predicted probability of fixation for alleles with dominance coefficients of 0.5, 0.1, and 0.01, respectively. The symbols mark simulation results over a minimum of 10^7 replicates each, with the three dominance coefficients represented by triangles, squares, and crosses, respectively. Other parameters used for these examples were $s = -0.0002$, $m = 0.1$, 100 demes of 100 diploid individuals each, and colonization by four individuals with a probability of common origin equal to 1/2. The probability of fixation is increased substantially by the reduction in N_e that accompanies extinction dynamics. (B) The parameters in these examples were $h = 0.01$, $s = -0.0002$ with 100 demes of 100 diploid individuals. Points represent the results of 10^8 simulations.

Local Genetic Load and the Consequences of Migration

In subdivided populations, weakly deleterious alleles can rise by drift to high frequencies within local populations, even if selection is effective at keeping their overall frequency low throughout the species. Crow (1948) proposed that this could be the mechanism for the commonly observed pattern of heterosis, the increase in fitness often observed in hybrids between different populations. We examined this hypothesis using Wright's distribution of allele frequencies for the island model (Whitlock et al., 2000; Ives and Whitlock, 2002) and found that Crow's hypothesis was extremely credible. We referred to the reduction in mean fitness caused by this local increase in the frequency of deleterious alleles *local drift load* and showed that reasonably large values of heterosis were consistent with what is known about mutation rates and population structure. These results have been extended by Morgan (2002) and Glémin (2003). Morgan (2002) showed that

$$\frac{\overline{w}_{hybrid}}{\overline{w}_{local}} = \left(\frac{(1-hs)^2}{1-s}\right)^{nV_{among}} \tag{7.14}$$

where V_{among} is the variance among demes in allele frequency as defined and n is the number of loci. With this we can write a prediction for the heterosis in terms of F_{ST} and $\overline{q}$:

$$heterosis = \frac{\overline{w}_{hybrid}}{\overline{w}_{local}} - 1 = \left(\frac{(1-hs)^2}{1-s}\right)^{nF_{ST}\overline{q}\overline{p}} - 1 \tag{7.15}$$

If the metapopulation itself is relatively large and at equilibrium, then $\overline{p} \cong 1$ and $\overline{q}$ is approximately $\hat{\overline{q}}$ from Eq. (7.7).

Heterosis has an interesting biological consequence. If offspring formed by crosses between demes have selective advantage, then the offspring of migrants will have increased fitness (Ingvarsson and Whitlock, 2000; Morgan, 2002). Thus the genetic effects of migration will be increased relative to the actual observed migration rate. The *effective migration rate* for a neutral locus is approximately

$$m_e = m\, e^{heterosis/\tilde{r}}, \tag{7.16}$$

where $\tilde{r}$ is the harmonic mean recombination rate between the neutral locus and all selected loci (Ingvarsson and Whitlock, 2000). For low values of F_{ST}, the magnification of the effective rate of migration can be severalfold. This can be counterbalanced or reversed by sufficient local adaptation or strong differences among populations in epistatic interactions.

Load in Subdivided Populations, a Summary

Several types of load are affected by population structure. Mutation load tends to decline at equilibrium with structure, and migration load is lowered with lower migration rates, whereas drift load, segregation load, and local drift load tend to increase. Because these different genetic loads are cumulative, the mean fitness of the population with three different types of genetic load is approximately $(1 - L_1)(1 - L_2)(1 - L_3)$. If the loads are small (they are not in general expected to be) then the overall load is approximated by the sum over the types of load. Whether population structure increases or decreases mean fitness on average depends on a large number of circumstances. If habitat conditions vary strongly, then population structure allows local adaptation (in other words, reducing migration load) and this effect can be paramount. However, if migration rates become too small and local population size is low, then local drift load will become very important and essentially the population will suffer from inbreeding depression. Species-level drift load could become important if there is a lot of variance among demes in reproductive success and if the total census size of the species was small (so that the effective size was low), but is likely not very important if the effective size of the species is over about 10,000. Mutation load may be reduced by population structure (at equilibrium), but not by more than a half. In some species, for example, those in which the genomic deleterious mutation rate is high, this could be a major effect; but for species with lower mutation rates, this could be a trivial effect. The balance of the effects of these processes will depend on the specifics of the species.

7.6 CONCLUSIONS AND INCONCLUSIONS

The course of evolution is changed quantitatively and qualitatively by the subdivision of populations over space. All of the population genetic processes that act in unstructured populations are affected, sometimes substantially, and some kinds of evolution are only possible with structured populations. This chapter focused on the former: quantitative changes in evolutionary rates from population subdivision. Even with uniform selection, the rate of genetic drift and the response to selection are changed substantially.

For some of the quantities described in this chapter (e.g., N_e, the probability of fixation of beneficial alleles), results have already been found for a special case of population structure: the island model. The island model is the oldest in population genetics, and it is rightfully the first to turn to when considering new problems because of its simplicity. Unfortunately, the very simplicity that make it appealing also makes it an aberration. The island model assumes that all demes are equal; not only do all demes have the same population size and migration and immigration rates, but more importantly, it implicitly assumes that all demes contribute exactly equally to the next generation. Clearly these conditions do not apply to most (or even all) natural populations, but this would not matter if these assumptions had no effect on our evolutionary predictions. Unfortunately, this assumption of equal reproductive success has a qualitative effect on our predictions, especially for questions that involve effective size. In this subtle but key respect, the island model is an extreme model, and some of the predictions made from the island model are extreme as a result.

Fortunately, it is possible to derive theory that predicts the necessary parameters for other models of population structure. The last couple of decades have seen a lot of development of models, including isolation by distance, local extinction, population size change, variable migration rates, and asymmetric migration. Even more fortunately, the results described in this chapter show that, at least for weak selection, most of the effects of population structure can be described in a few summary statistics, especially F_{ST} and N_e. This is extremely useful because we know a lot about how F_{ST} is changed by various demographic processes and we have the theory to predict the effective size for a broad class of models. F_{ST} in particular has been very well studied, with many empirical studies devoted to measuring it in a wide variety of species and a large number of theoretical models. These include extinction and recolonization (Wade and McCauley, 1988; Whitlock and McCauley, 1990), population fission and fusion (Whitlock, 1994), source–sink models (Gaggiotti, 1996), and stepping-stone models (Kimura and Weiss, 1964). In all of these cases, F_{ST} differs significantly from that predicted by the island model, and in most the effective population size is also substantially different (and usually much less than the census size). Moreover, it is usually straightforward to calculate F_{ST} even for a novel system.

As an aside, the reason that F_{ST} has been measured empirically so often has little to do with its importance to predict the effects of population structure on selection or drift. F_{ST} has been measured usually because of the false hope that it could be used to estimate the number of migrants coming into a population per generation (Whitlock and McCauley, 1999). It is fortunate then that this effort has not been wasted, and it is important not to throw the evolutionary baby out with the estimator bathwater. F_{ST} is an excellent descriptor of the nature of population structure and should be calculated in genetic studies of metapopulations. Unfortunately, the same cannot be said for its properties as an estimator of dispersal.

There are many unresolved questions on evolution on space. We have made some progress in understanding the effects of population structure on response to uniform selection, but we have not yet made similar progress with the heterogeneous selection case. All of the results considered here deal with discrete populations in which organisms are grouped into demes with the space between them empty. Most of the questions presented here have not solved for the spatial case in which individuals are spread continuously over space, a much more challenging topic. These results all assume weak selection, yet some of the most interesting cases involve selection coefficients stronger than migration rates.

We also need many more empirical studies on these topics. This chapter has not reviewed the empirical literature at all, but most of the theory presented here remains untested experimentally. Furthermore, we need better measures of some key parameters. The dominance coefficient has a tendency to cancel out of panmictic calculations, but this is not true for evolution in structured populations; we have very few estimates of the distribution of dominance coefficients. We desperately need more empirical studies of the effective size of structured populations. We also need to develop individual-weighted estimators of F_{ST}, as has been shown to be required by this theory.

The subdivision of a species over space can affect its evolution strongly and in a variety of ways. Because most species in nature are subdivided over space, it behooves us to understand this nearly ubiquitous feature of the natural world.

8. METAPOPULATIONS AND COALESCENT THEORY

John Wakeley

8.1 INTRODUCTION

Coalescent theory, or the study of gene genealogies, provides the framework for empirical molecular population genetics. It is a rapidly moving field that at once draws upon the long history of population genetics theory and responds to the latest advances in biotechnology. The essence of the coalescent is that it models the genealogical history of a sample of genetic data and, via that history, makes predictions about patterns of variation that might be observed among members of the sample. During the development of the coalescent approach between the early 1970s and the early 1980s, there was a switch in viewpoint from the prospective view taken by classical population genetics to a new one that begins with a sample and looks backward in time (Ewens, 1990). The immense practical benefit of this was that it was no longer necessary to describe the properties of an entire population and then imagine sampling from it in order to make predictions about a sample of genetic data: only the direct ancestors of the sample mattered. The aim of this chapter is to describe the basic features of coalescence in unstructured populations, to discuss how this forms a basis for inference about population history, and then to discuss the ways in which metapopulation structure changes these basic features and what, in turn, the prospects are for historical inference in metapopulations. In taking the coalescent approach, this chapter complements those of

  0-12-323448-4

Chapters 7 and 9, which consider classical, forward-time dynamics of genetic variation in a metapopulation and the genetics of quantitative traits in a metapopulation, respectively.

8.2 COALESCENCE IN PANMICTIC POPULATIONS

Although the seeds of genealogical thinking and coalescence date back at least to the work of Ewens (1972) and Karlin and McGregor (1972) on the sampling theory of selectively neutral alleles under the infinite alleles model of mutation, and are obvious in the work of Watterson (1975) on the number of segregating sites in a sample under the infinite sites model of mutation without recombination, it was not until the early 1980s that the familiar ancestral process known as the coalescent was firmly established. Almost simultaneously, Kingman (1982a,b,c) proved the convergence of the ancestral process for a sample to this simpler, pure death process, which he called the n-coalescent, whereas Hudson (1983a) and Tajima (1983) explored many properties of gene genealogies that are of direct interest and use to biologists. The mathematically similar theory of lines of descent was introduced just before this by Griffiths (1980). Tavaré (1984) reviewed these early mathematical developments, and Hudson (1990) and Nordborg (2001) reviewed the broader biological scope of coalescent theory.

This section explores the properties the standard coalescent process that Kingman described and called the n-coalescent (a bit of terminology that never quite caught on, at least among more biological practitioners). Note that terminology is used loosely in general here: for example, ancestral process, coalescent process, and genealogical process are used interchangeably, without reference to any particular model of a population. The standard coalescent involves a number of assumptions in addition to the assumption that the population is well mixed, or panmictic, i.e., mating randomly across the entire range if we are talking about diploid species. It is also assumed that variation is selectively neutral, that the effective size of the population has not changed over time, and that there is no recombination within the locus under study. The choice of a mutation model is rather flexible and typically depends on the type of data that is available or will be gathered. Deviations from each of these assumptions will be considered here and subsequent sections, although the main focus of this chapter is to describe the effects of metapopulation structure on gene genealogies and the coalescent process and to discuss the implications of this for inference.

The Structure of Gene Genealogies

For a wide variety of population models, which differ in terms of important biological properties, such as the distribution of offspring number among members of the population and whether generations are overlapping or discrete, Kingman (1982a,b,c) proved that the ancestral process for a sample of finite size n converges to the coalescent as the population size tends to infinity. In this limit, all of the myriad possible events that could happen to the sample looking back in time a single generation reduce to two: either all items have distinct parents or two members of the sample share a common ancestor. The other possibilities, whose probabilities become negligible as the population size goes to infinity, are those in which more than one of these common ancestor events happens in a single generation. For example, in a small population, two pairs of samples may have common ancestors or more than two members of the sample may share a single common ancestor in the immediately previous generation, and the probability of this cannot be neglected. In a large population, the genealogical history of a sample is simplified greatly because such events are extremely unlikely. The resulting genealogies are easy to describe in words and to model mathematically.

This simple process holds not only for the currently sampled items, but also for the lineages ancestral to them that existed at some time in the past. Thus, the genealogy of a sample under the standard coalescent is simply a series of common ancestor events between pairs of lineages, by which the sample of n items, or lineages, can be traced back to a single common ancestor. An example genealogy is shown in Fig. 8.1.

Times to Common Ancestry

The history of a sample of n items includes exactly $n - 1$ coalescent intervals. These are the times in the history of the sample during which there were n, $n - 1, \ldots, 3, 2$ lineages ancestral to the sample. In Fig. 8.1, T_i is used to denote the time during which there were i ancestral lineages. For a broad class of models of a population — the "exchangeable" models of Cannings (1974) — Kingman showed that these times are independent and distributed exponentially:

$$f_{T_i}(t_i) = \binom{i}{2} e^{-\binom{i}{2} t_i} \tag{8.1}$$

when time is measured in units of G/σ^2 generations, where G is the total number of copies of each genetic locus in the population, and σ^2 is the variance in offspring number. Under the commonly used Wright–Fisher model (Fisher, 1930; Wright, 1931) of a diploid, monoecious organism, G is equal to $2N$,

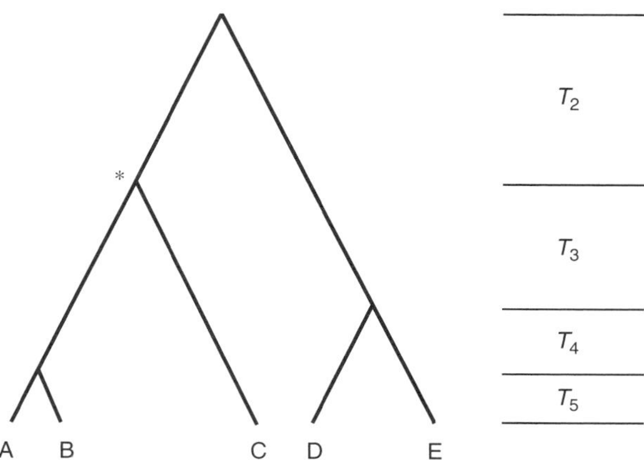

Fig. 8.1 An example genealogy of five items under the standard coalescent.

where N is the number of individuals and σ^2 is equal to 1. Strictly speaking, the coalescent is a model for a haploid population. However, it holds exactly as stated for the diploid Wright–Fisher model, due to the assumptions of random mating and monoecy, and it holds when there are two sexes if N is replaced with the appropriate effective population size (Möhle, 1998b). Thus, on average, genealogies will look something like the example in Fig. 8.1 in which $T_2 > T_3 > T_4 > T_5$, although the expectation is even more skewed than what is shown in the Fig. 8.1, as $E[T_i] = 2/[i(i-1)]$. The variances of these times are quite large as well. Nordborg (2001) displayed several realizations of the coalescent process to illustrate this.

Using only Eq. (8.1) and the fact that T_i and T_j are independent if $i \neq j$, many useful analytical results can be obtained, including the distribution of the time to the most recent common ancestor of the sample (T_{MRCA}) and the distribution of the total length of the genealogy (T_{Total}), i.e., the sum of the lengths of all the branches in the genealogy. Expressions for these two probability functions are complicated, not particularly illuminating, and are not reproduced here (see Tavaré, 1984). From these distributions, or directly from Eq. (8.1), one can obtain the familiar expressions for the expected values of these quantities:

$$E[T_{MRCA}] = 2\left(1 - \frac{1}{n}\right), \tag{8.2}$$

$$E[T_{Total}] = 2\sum_{i=1}^{n-1}\frac{1}{i}. \tag{8.3}$$

As seen later, the second of these determines the expected number of polymorphic, or segregating, sites at a locus when mutations occur according to the infinite sites model (Watterson, 1975).

In addition, it is not too difficult to derive the expected total length of branches in the history that have i descendents in the sample:

$$E[\tau_i] = \frac{2}{i}, \tag{8.4}$$

(Fu, 1995), which for $1 \leq i \leq n-1$ are the individual terms in the expected sum of all branch length [Eq. (8.3)]. In Fig. 8.1, the branch above the asterisk, up to the root of the tree, has three descendents. It is the only branch in that particular genealogy that can contribute to τ_3, whereas all the other branches contribute to either τ_1 or τ_2, and none in that tree can contribute to τ_4. Under the infinite sites model of mutation, τ_i represents the opportunity for the creation of a polymorphic site at which the ancestral base is in $n - i$ copies and the mutant base in i copies in the sample. Thus, Eq. (8.4) is important in making predictions about base frequencies at polymorphic sites.

Branching Pattern of Genealogies

Under the standard coalescent model, every pair of ancestral lineages has an equal chance of being the pair that coalesces at each common ancestor event. In fact, the simple ancestral process and the rate factor $\binom{i}{2}$ in specific follow from the fact that each pair of lineages in the coalescent limit coalesces with rate 1 independently of all other pairs. Thus, every possible random-joining tree is equally likely under the coalescent. In addition, the structure of the genealogy and the coalescent times are independent of one another. The many useful results of the coalescent, some of which are discussed in Section 8.3, follow from these facts together with the specific distributions of coalescence times described earlier. In fact, we have already seen one such result, Eq. (8.4), in which the derivation depends on the random-joining structure of genealogies (see Fu, 1995).

Genealogies and Recombination

If all variation is selectively neutral, and the other assumptions of the standard model are true, then the marginal distribution of the genealogy at any nucleotide site is given by the coalescent. Thus, quantities such as $E[T_{MRCA}]$, $E[T_{Total}]$, and $E[\tau_i]$ that do not depend on the joint distribution of the histories at multiple sites do not in turn depend on the rate of recombination. The joint ancestral process at two or more sites depends critically on recombination. Therfore, quantities that do depend on the joint histories of sites are affected by recombination. For example, the variances of T_{MRCA}, T_{Total}, and τ_i at a locus can be expressed as functions of the covariance in coalescence times at pairs of sites in the sequence, which in turn are functions of the recombination rate between the sites (Hudson, 1983b; Hudson and Kaplan, 1985). McVean (2002) provided a simple genealogical derivation of the correlation in coalescence time for a pair of sites.

Hudson (1983b) and others, including Kaplan and Hudson (1985), Griffiths and Marjoram (1996), and Wiuf and Hein (1999), have studied the coalescent process at a multisite genetic locus. If there is no recombination, then the entire locus follows one genealogy. Recombination events, viewed backward in time, cause the ancestral segments on either side of the recombination breakpoint to be separated onto two different copies of the chromosome. Genealogies under recombination become complicated webs, as ancestral sites travel together for periods of time on the same chromosome and are split up by recombination, possibly coming back together later in coalescent events. However, the genealogy of each site individually remains a simple random-joining tree, with marginal distribution described by the standard coalescent. If the recombination rate is very high, then the genealogy of every site is independent of the genealogy of every other site. The effect on the covariances in coalescence times at pairs of sites in the sequence is predicted from these considerations. It approaches zero as the recombination rate becomes large and the sites' genealogies become independent, and it grows as the recombination rate decreases.

Extensions to the Coalescent

The basic coalescent technology of modeling the genealogical process for a sample of genetic data has been extended in many different directions. Examples include Slatkin and Hudson (1991), who considered changes in population size over time; Notohara (1990), who gave a general mathematical model of coalescence in a geographically structured population; Kaplan et al. (1988), who modeled strong selection, and Krone and Neuhauser (1997), who described a framework for the coalescent with weak selection. Theoretical works tend to treat these phenomena in isolation, but it will often be necessary to include several factors when interpreting data. An example is Kaplan et al. (1991), who used a model that includes balancing selection, recombination, and geographic subdivision to explain the decrease in levels of polymorphism with distance from a selected site in samples of the *Adh* gene in *Drosophila melanogaster* (Kreitman, 1983).

In addition, the standard coalescent has been obtained under a variety of circumstances, such as the case of a two-sex diploid population mentioned earlier, in which it is not obvious at first that such a simple model should hold (Möhle, 1998c). These "robustness" results for the coalescent derive from a lemma of Möhle (1998) on the convergence of discrete Markov chains with two timescales to simpler, continuous time processes. For example, in the case of two sexes, lineages switch back and forth between males and females much faster than they coalesce, and the standard coalescent is obtained, only with a rescaled effective population size that is a function of the number of males and females in the population. The result for genealogies in a metapopulation described in Section 8.3 is based on this kind of separation of timescales.

Mutation and Patterns of Genetic Variation in a Sample

An aligned set of DNA sequences sampled from a population is a potentially rich source of information about the history and current demography of the population. The coalescent, together with a model of mutation, can be used to make predictions about levels and patterns of variation in a sample. The most frequently used mutation model for DNA sequences is the infinite sites model, which assumes that every mutation happens at a previously unmutated site. Thus, the infinite sites model is appropriate when the per-site mutation rate is low. Recombination can, of course, be an important factor in determining patterns of genetic variation, and workers have variously assumed no recombination at all (Watterson, 1975), independent assortment among all sites (Kimura, 1969), or any intermediate level of recombination (Hudson, 1983b). As noted earlier, the importance of modeling recombination will depend on how data are analyzed.

Undoubtedly the most important assumption about mutation in the standard coalescent is that all variation is selectively neutral. Genetic similarities and differences among sampled sequences are a view to past and present demography, such as metapopulation structure, rather than directly the subject of natural selection, although as noted earlier it is possible to extend the coalescent to include selection at a site linked to the locus under study. Due to the way in which time is measured under the coalescent, the appropriate mutation parameter, θ, is similarly scaled. Under the Wright–Fisher model, $\theta = 4Nu$, where u is the rate of neutral mutation per locus copy per generation. Thus θ is equal to twice the average number of mutations introduced into the population each generation. The extra factor of two is due to the historical importance of the notion of heterozygosity, the expected value of which in a randomly mating population is equal to θ.

Because the mutation rate per generation is very low, mutation is modeled accurately as a Poisson process along the branches of genealogy. Specifically, the number of mutations on a lineage, lineages, or entire genealogy of given length t follows the Poisson distribution with parameter $\theta t/2$. Mutations under neutrality, by definition, do not affect the reproductive rates of individuals. Thus, the genealogical process and the mutation process can be treated separately. This allows predictions to be made easily under the coalescent about many measurable aspects of DNA sequence polymorphism. The reason for generating such predictions is of course twofold: it builds our understanding of how the forces that maintain variation work, and the predictions can be used for making inferences about populations.

Predictions about Full Data Patterns

It is possible under the coalescent with infinite sites mutation and no recombination to analytically compute the probability of observing any possible data set, i.e., the full sample of DNA sequences, using a recursive equation (Griffiths and Tavaré, 1995). The method can be extended to more general models that, for instance, include geographic structure (Bahlo and Griffiths, 2000). This analytic method is infeasible except for small samples though because the number of equations that must be solved simultaneously becomes astronomically large for complex data sets of many sequences. However, as Section 8.3 describes, this general approach can be turned into a Monte Carlo method of inference.

A second issue with this recursive method on full data patterns is that it does not lend itself to investigations of how the forces that produce and maintain variation act to shape patterns of genetic variation. This is a general concern rather than a problem with this particular analytic method. While there is obviously a wealth of information in a data set of DNA sequences, it is poorly known which aspects of data contain the bulk of information about each factor of evolution that might have been important in the history of the particular species under study. From the theoretical perspective, another aspect of this problem is that some of the parameters in a complicated historical model might be nonidentifiable (Beaumont et al., 2003). There is essentially just one result in population genetics in which a measure has been shown to contain all of the information about a population parameter. In the case of allelic or haplotypic data from an unstructured, constant-sized population in which all variation is selectively neutral, Ewens (1972) showed that the number of alleles is a sufficient statistic for θ, i.e., that the frequencies of the alleles contain no additional information.

Identifying patterns in data that correspond to particular phenomena and making statements, even approximate ones, about the sufficiency of statistics will likely be a major focus of research in the future given the current trends in inference discussed in Section 8.3. Work under the coalescent has focused on how various deviations from the standard model affect gross summaries of the data such as the expected base frequencies at polymorphic nucleotide sites, which form the basis of the "neutrality" tests of Tajima (1989) and Fu and Li (1993). In the context of subdivided populations, the majority of effort has gone to studies of Wright's (1951) F statistics, most notably the fixation index F_{ST}, even though the significance of F_{ST} in most situations is unclear (Whitlock and McCauley, 1999). In order to untangle the complex current and historical demography of populations, for instance, those exhibiting metapopulation dynamics using summary statistics of DNA sequence data, it will be necessary at a minimum to expand the battery of such measures to include at least as many measures as the number of parameters affecting the population.

Predictions about Summary Measures

The measures of DNA sequence polymorphism that have received the most attention in theoretical studies and the most use in empirical work are the total number of polymorphic (or segregating) sites S, the average number of pairwise differences π, and the number of polymorphic nucleotide sites, η_i, at which the least frequent base is in i copies in the sample of size n. The reason for focusing on these is partly historical. For example, the significance of π comes from the fact that it can be used to estimate the heterozygosity of a diploid population (Tajima, 1993). However, concerns about the efficiency of inferences made from sequence data have also been important. For example, Fu (1994) studied the properties of various possible estimators of θ using linear combinations of the η_i. Again, for subdivided populations, the focus has been on F_{ST}, which can be seen as a simple extension of average pairwise differences to a structured population (Slatkin, 1991; Wilkinson-Herbots, 1998).

Because mutational and genealogical processes can be treated separately under neutrality, predictions about these and other summary measures can be made by conditioning on the genealogy or on some relevant aspect of the genealogy. Conditional on the genealogy, the number of mutations in the history of the sample is, again, Poisson distributed, and under the infinite sites model each mutation produces a polymorphic site. Thus, from the probability density function for the total length of the genealogy, T_{total}, the probability function for the number of segregating sites, S, can be obtained as $\int_0^\infty P\{S = k|t\} f_{T_{total}}(t)dt$. Beyond this, it is difficult to obtain analytical expressions for the probability functions for measures of DNA sequence polymorphism.

However, the derivation of expected values, variances, and covariances of these measures is straightforward. From Eqs. (8.3) and (8.4), and considering the Poisson nature of the mutation process, we have

$$E[S] = \theta \sum_{i=1}^{n-1} \frac{1}{i}. \tag{8.5}$$

(Watterson, 1975) and

$$E[Z_i] = \frac{\theta}{i}, \tag{8.6}$$

(Tajima, 1989; Fu and Li, 1993), where Z_i is the number of polymorphic sites at which the mutant base is in i copies in the sample. Typically, because it is not known which is the mutant and which is the ancestral base, we have

$$E[\eta_i] = \frac{\theta(\frac{1}{i} + \frac{1}{n-i})}{1 + \delta_{i,n-i}} \tag{8.7}$$

for the "folded" site frequency spectrum, that is, when the patterns Z_i and Z_{n-i} are indistinguishable. The $\delta_{i,n-i}$ term in the denominator is equal to one if $i = n - i$ and zero otherwise. It is needed in order to avoid counting Z_i twice in the case where $i = n - i = n/2$. The variances and covariances of the η_i can also be obtained (Fu, 1995). The expected value of the average number of pairwise differences, which can be expressed as a simple linear combination of the mutant base frequencies, $\pi = \Sigma_{i=1}^{n-1} i\,(n - i)\; Z_i/\binom{n}{2}$, is equal to θ (Tajima, 1983), and this is of course identical to the expected value of S when $n = 2$. Tajima (1983) also obtained the variance of π. These and other analytical results have been important in building an understanding about the ancestral process for a sample and in making inferences about populations, both when estimating population parameters and when testing the assumptions of the standard coalescent model.

Making Inferences Using the Coalescent

Because of its close connection to samples of genetic data, the coalescent approach provides a natural framework for inference about the structure and history of populations (see Stephens, 2001). Inferences can of course be made using the classical forward-time approach to population genetics, but in this case it becomes a two-step procedure. First the properties of the entire population are considered and then the process of sampling from the population is modeled and the properties of such samples determined. In some cases, the classical approach may be preferable. For instance, much of the ease and computational efficiency of the coalescent evaporate when weak selection acts on variation (Krone and Neuhauser, 1997; Neuhauser and Krone, 1997). However, the convenience and efficiency of the coalescent approach under neutrality, which stems from the fact that the genealogy of a sample can be modeled without reference to the rest of the population, make the coalescent a very powerful inferential tool.

Analytical Methods

Where analytical results are available, such as those presented in Section 8.3, corresponding inferences can be made. For example, the analytical expression for the probability of observing S segregating sites in a sample of size n can be used to make maximum likelihood estimates of θ under the assumption of no intralocus recombination and infinite sites mutation. However, most analytical methods of inference use the method of moments, i.e., to equate the observed value of a measure of sequence polymorphism with its analytical expectation then to solve for the parameter of interest. This has led to a multitude of estimators of the fundamental parameter, θ, of the population based on S (Watterson, 1975), π (Tajima, 1983), and η_1 or other combinations of the η_i (Fu and Li, 1993; Fu, 1994). Among these, π has a rather undesirable statistical property: it is inconsistent. That is, its variance does not decrease to zero as the sample size tends to infinity (Tajima, 1983; Donnelly and Tavaré, 1995). Therefore, estimates based on the number of segregating sites, S, or on linear combinations of the site frequencies, η_i, are preferable to those made using pairwise differences.

These moment-based estimators are unbiased and easy to implement. They also have the advantage of being applicable regardless of the recombination rate because the relevant expected values do not depend on the rate of recombination. In fact, their accuracy will increase with the rate of recombination, as the sites in the sequence become more and more independent. Of course, it is not satisfactory to make only point estimates of parameters, and recombination must be considered if any statement about the error of these moment-based estimates is to be made. Obtaining analytical results about the variances of S, π, and η_i for arbitrary levels of recombination is not trivial. In fact, it is only for π that such results are available (Hudson, 1987; Pluzhnikov and Donnelly, 1996). In addition, the variance of an estimator is only a useful piece of information when its errors are distributed normally or at least when its distribution is known and is symmetric, which is almost never the case for measures of sequence polymorphism.

Computational Methods

It is straightforward and extremely efficient to simulate genealogical histories (Hudson, 1983b, 1990) because it is not necessary to simulate the entire population, just the sample history. The ease of simulations, together with the desire to know more about how the errors of parameter estimates are distributed, has led to a recent explosion of computational coalescent methods of inference. While the field is still in developement, it does appear that the focus has shifted somewhat over time. The first of these methods used Monte Carlo integration to compute the likelihood of a full data set under the coalescent model. Simulations were used to "integrate" over genealogies by simulating a large number of them and averaging the results. The quotation marks are in recognition of the complexity of genealogies, having continuously distributed branch lengths and discrete tree structures. This is an impossible task if genealogies are simulated using the standard coalescent without reference to data because the overwhelming majority of genealogies make a negligible contribution to the likelihood unless the sample size or the number of polymorphic sites is small.

Two different solutions to this problem were proposed. One was to use the recursive equations for the probability of data under the infinite sites model, discussed in Section 8.3, to define an ancestral Markov chain conditional on data and to sample genealogies from this rather than from the "unconditional" coalescent process (Griffiths and Tavaré, 1994a,b). The probability of data is then the average value of a function computed for each simulated path, i.e., genealogy, through this Markov chain. Because only genealogies that are minimally compatible with data under the infinite sites model are generated, the likelihood can be estimated with relative ease. This method has been extended to cover geographically structured populations, both with migration (Nath and Griffiths, 1996) and without (Nielsen, 1998), and loci that undergo recombination (Griffiths and Marjoram, 1996). In the case of recombination, the straightforward application of this approach is still quite inefficient, and a more optimal scheme that makes better use of importance sampling has been proposed (Fearnhead and Donnelly, 2001).

The other solution to the problem of the enormity of the space of all possible genealogies was to use a Markov chain Monte Carlo (MCMC) method to focus on genealogies that do contribute substantially to the likelihood (Kuhner et al., 1995). The chain is run with a starting genealogy, and each subsequent step involves the proposal of a new genealogy and then its acceptance according to a probability that depends on how new genealogies are proposed and on the relative contributions of the current and the new genealogies to the likelihood. This is an application of Metropolis–Hastings sampling. If the chain is run long enough, then sampling genealogies from it is equivalent to sampling them with respect to their relative contribution to the likelihood of data. Mutation models other than infinite sites can be incorporated easily, which is an advantage of this approach over the one described earlier. This method has also been extended beyond the standard coalescent to include subdivison, both with (Beerli and Felsenstein, 1999) and without (Nielsen and Wakeley, 2001) migration, and to include recombination (Kuhner et al., 2000).

By definition, these full-data likelihood methods extract the greatest possible information from data. However, despite being made feasible by focusing on genealogies relevant to data, they are highly computationally intensive, sometimes prohibitively so. Further, it is unclear whether all of this computation is justified in relation to the questions of statistical sufficiency discussed in Section 8.3. For example, it would be a waste of time to design a full-data method of estimating θ if the data were allele counts under the infinite alleles model, as all the information about θ is contained in the number of alleles, not the frequencies. While little is known about the axes of information content in samples of DNA sequences, it cannot be expected that all of the many facets of polymorphism will contribute equally to inferences about particular parameters. Work is clearly needed in this area, both to build our knowledge and intuition and to aid in the development of better computational techniques of inference.

Partially in response to these concerns about information content, but mostly due to interest in computational feasibility, there is a growing trend to design computational methods of inference using summary measures of polymorphism rather than full data. These methods date back to Fu and Li (1997), Tavaré et al. (1997), and Weiss and von Haeseler (1998), and a more recent example is Beaumont et al. (2003). They use simulated genealogies to compute the probability of observing a set of summary measures that are identical to or sufficiently close to the values observed in data. The advantage of this approach is that a much larger fraction of randomly generated genealogies have a chance of producing the observed data summaries than the fraction that contribute significantly to the likelihood of the full data.

Another recent trend in inference is the growing popularity of the Bayesian approach (see Beaumont et al., 2003). The difference between the likelihood and Bayesian approaches is less in the mechanics of the computational methods than in the interpretation of the output, i.e., as a likelihood surface or as a posterior probability distribution (see also Box 15.1, Chapter 15). In the former case, the large body of statistical theory on the distribution of likelihood ratios, which holds asymptotically as the sample size tends to infinity, is used to construct confidence intervals and test hypotheses. In the latter case, the credible intervals for parameters given data are drawn so that 95%, or some other chosen percentage, of the posterior distribution lies inside the credible interval. The size of the credible interval can depend strongly on the prior distribution of the parameter, which may be viewed as a drawback of the Bayesian approach. In defense of Bayesian methods, it is questionable whether the asymptotic theory of likelihoods is valid except when data are available from a large number of independent loci.

"Neutrality" Tests

The analytical results for S, π, and η_1 (or Z_1), which produced series of unbiased method of moments estimators of θ, have also spawned a series of statistical tests of the standard coalescent model. Tajima's (1989) D and the tests of Fu and Li (1993) are the best known, although a number of others have been proposed (Simonsen et al., 1995). All of these tests are based on the fact that a number different measures of polymorphism, representing different aspects of data, can all be used to estimate the single parameter θ. Under the null model, the expected value of the difference between two such estimates is equal to zero. Deviations from zero, the significance of which are best measured by the simulation scheme of Simonsen et al. (1995), lead to rejection of the standard coalescent. Unfortunately, although the standard coalescent involves a number of assumptions, there is a strong tendency to see these tests as tests of selective neutrality only.

In fact, all of these tests simply measure deviations from the site-frequency distribution predicted by the coalescent. Figure 8.2 displays the expectation for this distribution for a sample of size $n = 10$, with the heights of the bars scaled, by dividing $E[Z_i]$ by $E[S]$ so that they sum to one. The various test statistics detect deviations only in two directions — positive or negative — and if the standard model is rejected there are a number of possible explanations, which include selection and also demographic and/or historical factors (Simonsen et al., 1995). Some statistics assume that information about the ancestral state at each site is known and are thus functions of the mutant base counts, Z_i. Others do not make this assumption, instead assuming that the patterns Z_i and Z_{n-i} are indistinguishable as in Eq. (8.7). The latter, which include Tajima's (1989) D, are positive when the Z_i around $i = n/2$ are inflated relative to Fig. 8.2 and negative when Z_i near either $i = 1$ or $i = n - 1$ are inflated. The statistics that assume the ancestral states are known have the potential to detect differences between inflated Z_i near $i = 1$ and inflated Z_i near $i = n - 1$. As shown in Section 8.3, metapopulation dynamics can produce a wide variety of site frequency distributions, putting the status of these "neutrality" tests in further jeopardy.

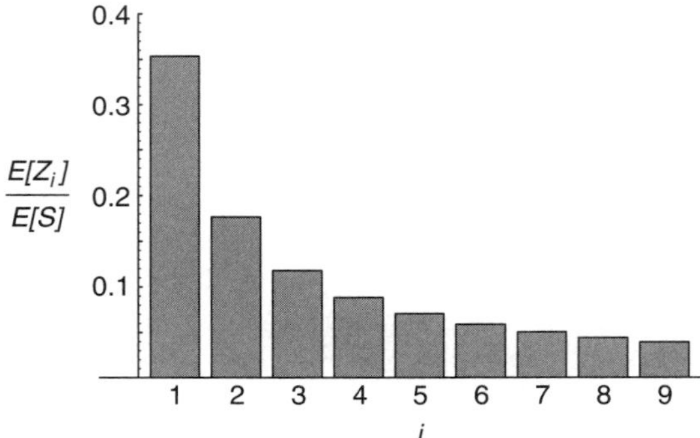

Fig. 8.2 The expected site frequency distribution under the standard coalescent.

8.3 COALESCENCE IN METAPOPULATIONS

The word metapopulation, implying a "population of populations," can be applied very broadly to any geographically structured species, particularly ones that exhibit local extinction and recolonization (Levins, 1968a,b). Hanski and Gaggiotti (Chapter 1) discussed the current metapopulation concept, and other chapters in this book attest to the variety of situations to which the concept has been applied. Because population geneticists had been studying metapopulations for decades before the word metapopulation was introduced, the terminology can be confusing. In particular, "population," "subdivided population," "structured population," and "total population" are often used interchangeably to refer to a metapopulation, and any of "deme" (Gilmour and Gregor, 1939), "subpopulation," and even "population" are used to refer to geographically local populations within a metapopulation. This section uses metapopulation and deme. In addition, except for some brief review of work on general models of population structure, this chapter adopts the common metapopulation notion that the number of demes is not small. Instead of making this assumption implicitly, the ancestral metapopulation process described here exists in the limit as the number of demes grows, in much the same way that the standard coalescent holds in the limit as the population size of an unstructured population approaches infinity.

There is a long history of work on the genetics of structured populations, dating back at least to Wright (1931). Much of this work is relevant to the discussion of metapopulation structure, especially Wright (1940), who early on saw a major potential role for extinction and recolonization in his shifting balance theory of evolution. Slatkin (1977) formulated the basic population genetic model of metapopulation dynamics that is still commonly used and identified the major possible effects of extinction and recolonization on genetic variation: (1) that the turnover of demes and recolonization by small numbers of individuals can decrease overall levels of variation and (2) that the movement of founders across the metapopulation can decrease levels of differentiation among demes in a manner similar to migration. Pannell and Charlesworth (2000) provide an excellent review of these and later works (see also Chapter 7) which have focused on the effects of metapopulation structure on well-known summaries of polymorphism such as Wright's (1951) F_{ST}.

The formal extension of the coalescent approach to the case of population structure occurred only recently, with a general model called the structured coalescent (Notohara, 1990; Nordborg, 1997; Wilkinson-Herbots, 1998). The structured coalescent does not include extinction and recolonization of demes, only migration between them, but it could be reformulated to do so. The backward migration rate, m_{ij}, is defined to be the fraction of deme i that is replaced by migrants from deme j each generation. The structured coalescent exists in the limit as the sizes of demes go to infinity but the scaled backward migration rates, $M_{ij} = 4N_i m_{ij}$, remain finite. Thus, it assumes that migration is a weak force, with a rate roughly comparable to that of genetic drift/coalescence. This is not a weakness of the model. If $M_{ij} = 4N_i m_{ij}$ does not remain finite as the N_i goes to infinity, then migration is a much faster process than drift/coalescence, and the dynamics of the metapopulation converge on those of an unstructured population, both forward (Nagylaki, 1980) and backward

(Notohara, 1993) in time. This is known as the strong migration limit. In practice, the effects of structure are very difficult to detect once the rates M_{ij} are greater than about 10. Nordborg and Krone (2002) studied a structured coalescent in which some of the M_{ij} remain finite whereas others increase without bound and showed that the ancestral process converges to a structured coalescent among the subunits of the metapopulation that have finite M_{ij}.

One of the major influences of Wright (1951) was to focus attention on F_{ST} as a summary measure of metapopulation structure. The connection to coalescent theory was made by Slatkin (1991), who showed that, in the limit of small mutation rate, inbreeding coefficients such as F_{ST} can be expressed in terms of expected pairwise coalescence times. Under the infinite sites mutation model, the expected values of pairwise differences are linear functions of θ so that taking ratios of observed pairwise differences within and between demes provides a way of estimating migration parameters. For example, under the island model of metapopulation structure (Wright, 1931), in which every deme exchanges migrants with every other deme at the same rate and all demes are of the same size, expectations of the average number of pairwise differences within and between demes, π_w and π_b, respectively, are given by

$$E[\pi_w] = \theta \quad \text{and} \quad E[\pi_b] = \theta\left(1 + \frac{1}{M}\right), \tag{8.8}$$

where θ is the scaled mutation parameter for the entire metapopulation. The sole migration parameter of the island model can be estimated as

$$\hat{M} = \frac{\pi_w}{\pi_b - \pi_w} \tag{8.9}$$

although this moment-based estimator is certainly not unbiased. However, the island model, which is a particularly unrealistic model for many organisms, is the only model for which there is a simple connection between F_{ST}, or average pairwise differences, and the parameters of the metapopulation (Whitlock and McCauley, 1999).

Theoretical work on metapopulations has focused on pairwise coalescence times or pairwise differences due to their connection with F_{ST} and their utility in estimating migration rates, but also in no small part due to the fact that analytical results for larger samples under the structured coalescent are difficult to obtain. This is unfortunate because, as noted earlier, estimates made from pairwise differences have relatively poor statistical properties (Tajima, 1983; Donnelly and Tavaré, 1995). In hindsight, the historical focus on F_{ST} and, relatedly, on pairwise sequence comparisons within and between demes may have drawn attention away from the true goal of such work, which is to understand the dynamics of metapopulations and how these shape the patterns of genetic polymorphism. To some degree, this is an unfair statement. The profound importance and utility of Eqs. (8.8) and (8.9), for example, cannot be questioned. At the same time, it is clear that these simple measures of polymorphism are not sufficient to untangle the complicated demography of metapopulations (Pannell and Charlesworth, 1999; Pannell, 2003).

The study of genealogies of samples from a metapopulation will help identify the patterns in DNA sequences or other genetic data that are likely to contain substantial information about the dynamics of the metapopulation. Again, the focus here is on the case of a large number of demes. With the additional assumption that the migrants and/or founders that arrive at a deme could have originated in any one of a large number of source demes, it is possible to describe the general features of these histories, which are expected to hold for many different types of metapopulation structures. Beyond this it will be necessary to make more detailed assumptions about the structure of the metapopulation in order then to explore specific effects on patterns of polymorphism.

Sampling for population genetic studies is typically not done at random across the geographic range of a metapopulation. Instead, a number of samples are taken from a number of different locations, with the geographic distances between samples from the same location being smaller than those between samples from different locations. This is true even for species with apparently continuous ranges and it is forced by local abundance in species composed of more discrete demes. This is a logical approach to the study of geographic structure, but one that is also conditioned by long-standing notions about the importance of F_{ST} and results such as Eq. (8.9), which, again, come directly from the island model of a metapopulation. A consideration of sample genealogies in metapopulations may also aid in the design of better sampling strategies for studying geographic structure. Another noteworthy aspect of samples from a metapopulation composed of a large number of demes or distributed over a very broad range is that many demes or locations will not be sampled at all.

Now consider the locations of the lineages ancestral to the sample at some time in the past. Unlike the locations of the present-day samples, which are under experimental control, these will be determined by the history and dynamics of the metapopulation and, of course, the depth of time considered. Recently in the past, samples from the same place will tend still to be close together and will be relatively likely to coalesce. Lineages from different locations are less likely to share a common ancestor recently. In the more distant past, lineages originally sampled from the same location, if they have not coalesced, will have instead moved, by migration and/or extinction/recolonization, to other locations. In a metapopulation with a large number of demes and in which the number of source demes of migrants and founders is large, these ancient lineages are not likely to be in the same deme nor are they likely to be in any of the originally sampled demes. Their locations will be the result of their random movement across the metapopulation according to the rates of migration and extinction/recolonization between demes. They will tend to accumulate in the parts of the metapopulation that contribute greatly to the migrant pool or that send out an abundance of founders, and they will spend little time in regions that act as "sinks" instead of "sources" (Pulliam, 1988). There will be chances for such ancient ancestral lineages to coalesce, mediated by migration and extinction/recolonization, and it may require a lot of wandering of the lineages across the population before the most recent common ancestor of the entire sample is reached.

Thus, for a broad range of specific metapopulation structures that have a large number of demes in common, sample genealogies should exhibit a recent burst of coalescent events among samples taken from the same locality followed by a more ancient historical process for the remaining ancestral lineages. These have been called the scattering phase and the collecting phase (Wakeley, 1999), and details of them depend on the details of the dynamics of the metapopulation. The next section describes an idealized model of a metapopulation in which this behavior emerges in the limit as the number of demes goes to infinity and for which simple precise descriptions of the scattering and collecting phases are possible. A simulation study of Pannell (2003) showed, among other things, that this twofold structure of genealogies is realized in a special case of this model even if the number of demes is not terribly large. In addition, Ray et al. (2003) found this behavior in a model very different from the one described later. Ray et al. (2003) simulated a metapopulation that expands from a single deme over a two-dimensional grid and showed that genetic signatures of this expansion in a sample from a single deme are strong only if the backward migration/colonization rate of the deme is high. If the migration rate is low, then few lineages will escape the recent burst of scattering-phase coalescent events, the effective number of lineages will be small, and the power to detect the expansion will be low.

The Effect of Metapopulation Structure on Genealogies

The model considered here is essentially the same as the model in Wakeley and Aliacar (2001), although it is unnecessary to assume that the individual demes are large (Lessard and Wakeley, 2003). The model assumes that there are K "regions," each of which resembles the metapopulation described by Slatkin (1977). Each region may have different values of all parameters, and among regions there is some explicit geographic structure. For ease of discussion, consider the case where the demes are large so that the relevant scaled parameters for each such region, or class of demes, are the migration rate $M_i = 4N_im_i$, the extinction/recolonization rate $E_i = 4N_ie_i$, and the propagule size k_i, which is the number of founders of the deme each time it is recolonized after going extinct. Note that k_i is the number of founding gametes rather than diploid individuals, as in Slatkin (1977), and that the parameters M_i and E_i differ by a factor of two from those in Wakeley and Aliacar (2001), in keeping with the scaled migration rate M used earlier in this section. The index i of these parameters ranges from one to K. Figure 8.3 depicts one example of such a metapopulation.

There are D demes total in the metapopulation, and a fraction β_i of these are in class or region i, where $\Sigma_{i=1}^{K}\beta_i = 1$. Thus there are $D\beta_i$ demes of class i so that in the limit $D \to \infty$ considered later, the number of demes in each class also approaches infinity. Looking back in time, when a lineage experiences a migration event or an extinction/recolonization event, it has some probability of having come from a deme in each of the other regions and some probability of coming from a deme in its own region. That is, $m_i = \Sigma_{j=1}^{K} m_{ij}$ and $e_i = \Sigma_{j=1}^{K} e_{ij}$, and these probabilities of movement, given that a migration event or an extinction/recolonization event has occurred, are given by m_{ij}/m_i and e_{ij}/e_i, respectively. Every deme in a region has an equal chance of being the source deme of a migrant/colonist from that region. The only constraints on the structure of movement are that lineages can get from any of these K

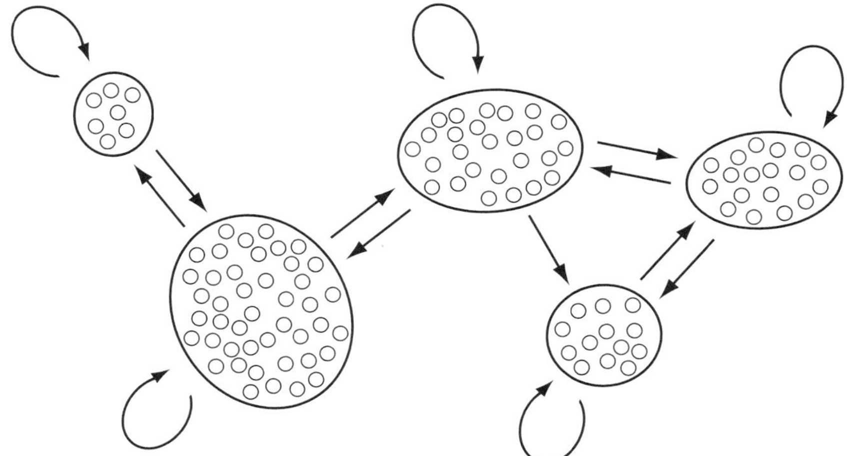

Fig. 8.3 One possible metapopulation that conforms to the model described in the text, in this case with $K = 5$ and with an arbitrarily chosen structure.

regions to any other, given enough time, and that there is a nonzero probability that a lineage will remain in the same region. Even very strongly constrained patterns of movement among regions, such as the one-dimensional stepping-stone model (Kimura and Weiss, 1964), conform to this assumption. A surprising result of this model is that in the limit as D goes to infinity, the details of this aspect of the geographic structure of the metapopulation are obliterated, similarly to the way in which all structure disappears in the strong migration limit (Nagylaki, 1980; Notohara, 1993).

Branching Pattern of Genealogies

Under this model, with the aid of the convergence result of Möhle (1998), it is possible to show that the more ancient part of the history, the collecting phase, converges to Kingman's (1982a,b,c) coalescent as the number of demes goes to infinity, but with a rate, or an effective size, that depends on all the parameters of the model (Wakeley and Aliacar, 2001). This had been found previously in models that include migration but not extinction and recolonization (Wakeley, 1998, 1999, 2001). Simulations imply that the predictions of the model are accurate as long as the number of demes is at least three to four times the sample size (Wakeley, 1998; Pannell, 2003). Surprisingly, a great deal of the geographic structure and dynamics of the metapopulation — the details of movement among regions and the values of M, E, and k for unsampled demes — is manifest only through the single effective size of the metapopulation during the collecting phase. This results from the fact that when the number of demes is very large, the lineages will migrate so many times as to reach a stationary distribution over deme types, determined by the movement matrices for migration and extinction/recolonization, before two of them end up in the same deme and have the chance to coalesce. Overall, then, if time is scaled by the effective size and θ is defined accordingly, all the detailed results of the standard coalescent model, including those discussed in Section 8.2, hold for the these collecting-phase lineages.

In this model, for any sample, the transition from the scattering phase to this coalescent collecting phase occurs as soon as each ancestral lineage is in a separate deme. Thus, the history of a sample taken singly from different demes in a metapopulation of this sort is also described by the standard coalescent. The only evidence of the structure in this "scattered" sample will be in the magnitude of θ, if, for instance, it could be compared to a scattered sample from another metapopulation of the same total size but with different details of structure and thus a different effective size. How many lineages then will enter the collecting phase for other kinds of samples?

This is determined by the random outcome of coalescent, migration, and extinction/recolonization within demes. Assuming that demes are large in size, in a deme of type i that contains j lineages these will occur with relative rates $j(j-1)$, jM_i, and E_i. In the limit as D goes to infinity, the number of demes will be much greater than the number of lineages ancestral to the sample, and migration events will send lineages off to demes that do not contain other ancestral lineages. If the demes are small in size, these rates apply only roughly and it will be possible for multiple migration and coalescent events to occur in a single generation. In either case, both single coalescent events and migration events both bring the sample closer to the transition to the collecting phase by decreasing the number of lineages in the deme by one. If an extinction/recolonization event occurs, whatever lineages remain in the deme will be related through the k_i founders. Even if the deme size is large, it is possible that several common ancestor events will occur in this step because k_i may not be large.

Figure 8.4 shows one possible scattering phase for a sample of size $n = 8$ from a single deme in which a series of three coalescent events and two migration events are followed by an extinction/recolonization event in which the remaining three lineages all coalesce. The result is $n' = 3$ lineages that will enter the collecting phase. These three lineages have different numbers of descendents in the sample, or different "sizes." Because whatever labels we might assign to these collecting-phase lineages are arbitrary — they are exchangeable — we can write $P(n'; a_1, a_2, \ldots, a_n)$ for the probability that there are n' lineages are the end of the scattering phase and among these, a_1 have one descendent in the sample, a_2 have two descendents in the sample, and so on. The possible size configurations are all those that satisfy $\Sigma_{i=1}^{n} ia_i = n$ and of course $\Sigma_{i=1}^{n} a_i = n'$.

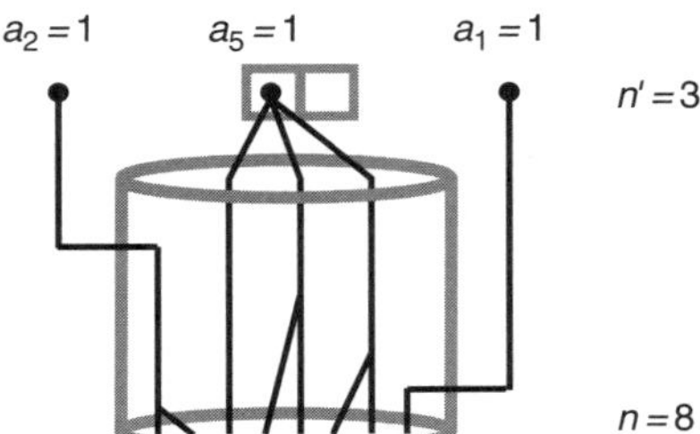

Fig. 8.4 A realization of the scattering phase for a sample from a single deme. The gray cylinder represents the deme back through time and lines represent lineages ancestral to the sample. The two attached boxes represent the $k = 2$ founders of the deme.

Different details of the dynamics within demes will give different distributions, $P(n'; a_1, a_2, \ldots, a_n)$. For example, when the deme size is large and there is no extinction/recolonization, the distribution is identical to Ewen's (1972) distribution, but with infinite alleles mutation replaced by infinite demes migration; the number of alleles becomes the number of collecting-phase lineages (n') and the counts of the allele become the number descendents of these lineages (a_i) in the sample (Wakeley, 1998, 1999). However, if the rate of extinction/recolonization is high and migration is absent, then $P(n'; a_1, a_2, \ldots, a_n)$ will be the result of tossing n balls into k boxes, with a_i being the number of boxes that contain i balls (Wakeley and Aliacar, 2001). Comparable levels of migration and extinction/recolonization combine both these effects, and there are of course many other possibilities, depending on the local dynamics within demes and the sizes of demes.

Finally, in the limit as D goes to infinity, the scattering phase occurs independently within each sampled deme so that the probabilities $P(n'; a_1, a_2, \ldots, a_n)$ are simply multiplied over demes to obtain the overall chance that n' lineages enter the collecting phase, with some distribution of sizes, given that a sample of size n has been sampled among some number of demes. In sum, the topological structure of sample genealogies in a metapopulation will be identical to that in the standard coalescent, except that now the number of (collecting phase) tips of the tree is a random variable and each tip will have an associated stochastic size that can be greater than one and is equal to the number of descendents of that branch in the sample.

Times to Common Ancestry

The collecting phase is a metapopulation-wide process. Its effective size is roughly on the order of the total size of the metapopulation, although low rates of migration can make it larger than this and some types of extinction/recolonization can make it smaller. In some situations it is important to consider these effects on the timescale of the coalescent, for instance, in the context of divergence between two metapopulations or species, where this timescale determines the probability of reciprocal monophyly of samples, among other things (Wakeley, 2000). Here, we have simply defined the parameter θ for the metapopulation-collecting phase, and the importance of its effective size is mostly in comparison to that of the scattering phase. This more recent phase, which occurs within demes, depends on the effective sizes of demes. Thus, the effective size of the collecting phase is about D times larger than that of the scattering phase. In the limit as D goes to infinity, the duration of the scattering phase becomes negligible in comparison to that of the collecting phase.

Clearly, as in the standard coalescent, the genealogy of a sample from a metapopulation contains exactly $n - 1$ coalescent events. Under the limiting process described earlier, which holds for metapopulations with a large number of demes, the first $n - n'$ (scattering-phase) coalescent events have negligible branch lengths, whereas the remaining $n' - 1$ have branch lengths determined by the Kingman's coalescent process. The scattering phase becomes an instantaneous adjustment of the sample size and structure, which can be used to obtain results for times to common ancestry as well as predictions about the level and pattern of polymorphism in the sample. It is no longer possible to write down a formula like 8.1 because these will depend on the realization of n' for the sample. However, by conditioning on the scattering phase, it is possible to derive equations like 8.2 through 8.4, for the expected values of T_{MRCA}, T_{total}, and τ_i. We can write $P[n'|n]$ to denote the probability that n' lineages enter the collecting phase, i.e., without regard to their sizes. Properties of T_{MRCA} and T_{total} depend only on this overall number n', and relatively simple analytic expressions can be obtained in some cases (Wakeley, 1998, 2001). Quantities such as τ_i depend on the sizes of the collecting-phase lineages. In the context of a metapopulation, these frequency measures should be redefined to represent the joint frequencies among sampled demes; (e.g., see Wakeley, 1999).

The Effect of Metapopulation Structure on Genetic Variation in a Sample

The level and pattern of genetic variation in a sample from a metapopulation are determined by the recent history of coalescent, migration, and extinction/recolonization events in the sampled demes and by the more ancient coalescent process that occurs among the remaining lineages. To obtain predictions for polymorphism in the sample, it is only necessary to condition on the outcome of the scattering phase. In some cases this can be done analytically, whereas in others it will be necessary to use simulations. These simulations are nearly as straightforward as simulations of the standard coalescent, the difference being the addition of the scattering phase. First the scattering phase is simulated, and all branches during this period of the history have lengths set to zero. The remaining lineages are then fed into the usual coalescent simulation, for instance, as described in Hudson (1990). The advantage of this in terms of the efficiency of the simulations is in not having to represent all of a large number of demes, only those from which samples have been taken. In addition, convergence of the collecting phase to Kingman's coalescent shows that a lot of time could be wasted trying to represent the myriad details of movement of lineage across the metapopulation during the collecting phase.

There are two main effects of metapopulation structure on patterns of genetic variation, which are represented conveniently and separately in the scattering phase and the collecting phase. First, of course, overall levels of variation are determined by the collecting-phase coalescent process, but the connection of the sample to this more ancient history is mediated by the scattering phase. In particular, the overall levels of polymorphism in a sample are greater when n' is larger and smaller when n' is smaller. In fact, if n' is equal to one, which is possible only if all samples come from the same deme, there will be no variation in the sample. This seems surprising, but it is an understandable consequence of the number of demes being very large and θ being finite: the values of θ for individual demes must be infinitesimal. This is probably appropriate for low mutation rate data such as DNA sequence data. The alternative, that demic θ values are not small, dictates that the metapopulation-wide θ approaches infinity as the number of demes grows, predicting an infinite number of polymorphisms in the sample. The assumption that demic θ values are not small may be appropriate for loci with higher mutation rates, such as microsatellites. In this case, the model would predict an infinite number of microsatellite mutations during the collecting phase, and it would be appropriate to study sample probabilities of identity as in Vitalis and Couvet (2001a,b). Under the assumption of finite total-population θ made here, all genetic variation is ancestral in the sense that it is not due to mutations that happened during the recent (scattering-phase) history of the sampled demes.

Scattered samples, taken singly from some number of demes, will show patterns of polymorphism identical to those in a completely unstructured population, the characteristics of which are reviewed in Section 8.2. To the extent that multiple samples are taken from single demes, the ancestral collecting-phase variation will be partitioned within and among demes' samples. The following discussion essentially assumes a constant value of θ for the metapopulation; changes in parameters are interpreted as different potential properties of the sampled demes.

Full Data Patterns

The connection of the ancestral process for samples from a metapopulation to that of samples from an unstructured population means that the tools of the standard coalescent can be adapted for use here. For example, in principle, it should be straightforward to use the recursive approach of Griffiths and Tavaré (1995), with the recognition that the number of migration and extinction/recolonization events in the history is not fixed. In other words, it will be necessary to account for the stochastic-scattering phase, although in the case of a scattered sample the standard coalescent methods can be used directly. At present, this remains one of several possible areas of future research.

Summary Measures

Predictions for summary measures, such as the number of segregating sites S, the average number of pairwise differences π, and the site-frequency distribution, can be made by modeling the scattering phase. It is possible to make analytical predictions about them by conditioning on the number of lineages n' that remain at the end of the scattering phase. In general, any process that tends to decrease n', such as restricted migration and extinction/recolonization with a small number of founders, will tend to decrease the number of (ancestral) polymorphisms found to be segregating in the sample. Wakeley and Aliacar (2001) showed that larger values of M produce larger average values of S, as do larger values of E if the number of founders, k, is large, whereas increasing E when k is equal to one decreases the average value of S. In addition to effects on the average value of S, effects on the shape of its distribution can be investigated analytically or using simulations.

There are many possible summary measures of sequence polymorphism in addition to S and π, including the site frequencies, Z_i or η_i, and it is hoped that the study of gene genealogies in a metapopulation will aid in the development of new statistics that capture the essential features of the dynamics of the metapopulation. Figure 8.5 shows computer simulation results for site frequencies, Z_i, in a sample from a single deme. In a single-deme sample, these are adequate to describe the frequency spectrum when patterns of linkage among sites are not a concern. Of course, there is likely to be some extra

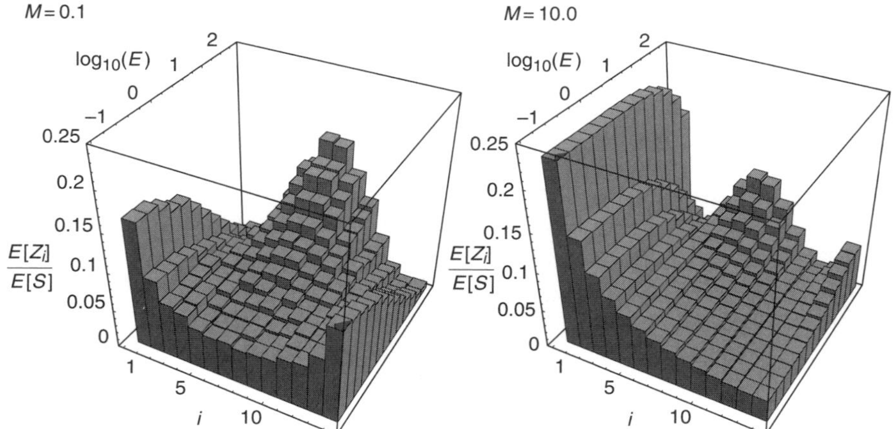

Fig. 8.5 Expected "unfolded" site frequencies, i.e., expected relative counts of the mutant base at polymorphic sites, in a sample of size $n = 15$ from a single deme with propagule size $k = 2$.

information about the metapopulation contained in the joint frequencies at two or more linked polymorphic sites. In samples from multiple demes, there may be information contained in the joint frequencies of alleles among demes. However, the extent of this information in both cases (among loci and among demes) and the potential to construct sufficient or nearly sufficient statistics for metapopulation parameters have not yet been fully explored.

Figure 8.5 demonstrates that the counts of mutant alleles in a sample from a single deme contain substantial information about the three relevant paramters of this model: the values of M, E, and k for the sampled deme. In addition, the overall number of polymorphic sites is likely to constitute the bulk of the information available about θ. In Fig. 8.5, site frequencies are shown as fractions for the total number of polymorphic sites. In the case of low migration ($M = 0.1$), when extinction and recolonization are weak forces, i.e., when E is small, the site-frequency distribution is U shaped. This is similar to the case of alleles under positive Darwinian selection, considered by Fay and Wu (2000) and Kim and Stephan (2000), so we can expect our ability to distinguish between positive selection and migration using single-deme samples to be low. At the other extreme for small M, when extinction/recolonization is a much stronger force than migration, the site-frequency distribution has a mode at the middle frequencies. This is similar to what might be expected in a combined sample from two demes in a metapopulation or if balancing selection were operating between two alleles at a locus, so it is surprising to find it here for a single-deme sample. The explanation is that large E means that the deme from which the sample was drawn is very likely to have experienced a recent extinction/recolonization event. The n lineages were immediately related through k ancestors, given a mode in the site frequency distribution around the expected number of descendents per ancestor, n/k, which in this case is equal to 7.5.

In the case of $M = 10$ (left side of Fig. 8.5), if E is small, the site frequencies are close to those predicted for a sample from a panmictic population shown in Fig. 8.2. As the extinction/recolonization rate increases, a mode again develops around n/k. However, because migration is strong, the panmictic pattern continues to hold and the interaction of these two patterns produces an average site-frequency distribution that has three modes. On the one hand, this means that there is potentially a great deal of information about the parameters of the model even in samples from a single deme. On the other hand, this has rather dire consequences for tests of the standard neutral model discussed in Section 8.2. It looks as if there will always be a *neutral* metapopulation explanation for any significant deviation in these statistics.

A Framework for Inference in Metapopulations

Clearly, there is great potential to adapt the many useful methods that have been developed for the standard coalescent model to the case of a metapopulation. Again, this simply means taking the scattering phase into account, and again this work is in its infancy.

Analytical Methods

As in the standard coalescent, if predictions about summaries of polymorphism do not depend on the joint distribution of genealogical lengths at pairs or groups of sites, then they will be accurate regardless of the rate of recombination. Thus, unbiased method of moments estimators of metapopulation parameters could be devised. Because there are at least four (θ, M, E, k) parameters in the metapopulation model used here, so this will require the proposal of some new summary statistics, tailored to samples from metapopulations. It is clear that simple, commonly used measures, such a pairwise differences within and between demes, will not suffice (Pannell and Charlesworth, 1999; Pannell, 2003).

One example of the possibilities for inference is the analytical method given in Wakeley (1999), which bases inferences on the joint distribution of allele frequencies among demes in a multideme sample. It was assumed that the demes were not subject to extinction and recolonization, only migration, and predictions like those shown in Fig. 8.5 formed the basis of a maximum likelihood method of inference using data from unlinked loci such as RFLP or SNP data. The model also incorporated a change in the effective population size at some time in the past, illustrating the ease with which such complications can be treated when part of the history of the sample is given by the standard coalescent process.

Computational Methods

The development of full data methods such as those of Griffiths and Tavaré (1994a,b) and Kuhner et al. (1995) appears promising because those methods, developed for unstructured populations, can be applied directly once the scattering phase is taken into account. For the case of metapopulations with large numbers of demes, this will be much more efficient than the current methods (Beerli and Felsenstein, 1999; Bahlo and Griffiths, 2000), which require the estimation of migration rates between every possible pair of demes and which so far have assumed that the sampled demes are the only demes in the metapopulation. Still, a full data method for a metapopulation would have to deal with more complicated data and a greater number of potential histories of the sample than a panmictic population requires, so a method of this sort is expected to have a greater number of potential drawbacks than panmictic methods.

One potential solution to this is the development of computational methods that use summary statistics, again following the work on the panmictic coalescent (Fu and Li, 1997; Tavaré et al., 1997; Weiss and von Haeseler, 1998; Beaumont et al., 2003). These should be much more efficient computationally than full data methods and hold the promise of being able to focus on data patterns that contain most of the information about the parameters of the model, assuming that such statistics can be developed. The elucidation of complicated patterns like those displayed in Fig. 8.5 is a step toward this goal. An example of a summary statistic computational method for the sort of model discussed here is the approximate maximum likelihood method for multilocus data given in Wakeley et al. (2001) in which numbers of polymorphisms were used to make inferences about θ, the distribution of those polymorphisms among demes was used to make inferences about migration parameters, and the overall frequency of polymorphisms in the sample was used to make inferences about a possible change in effective size at some time in the past.

8.4 SUMMARY AND CONCLUSIONS

It is important to remember that the results presented in this chapter hold only for large metapopulations, within which the number of possible source demes of migrants and colonists is large. Whether this is justified or not will depend on the species under study (see Chapters 13 through 23). If the number of demes is small, the standard structured coalescent (Notohara, 1990; Nordborg, 1997; Wilkinson-Herbots, 1998) is a more appropriate framework. The finding of Kingman's coalescent as part of the history of any sample from a large metapopulation immediately makes applicable a plethora of theoretical and inferential results and methods. One interesting consequence of this, which might be considered unfortunate, is that many of the details of the dynamics of the metapopulation are folded into a single population parameter: the effective size of the collecting phase, or its mutation-scaled equivalent, θ. This means that many phenomena of biological interest will not produce any observable effect on patterns of genetic polymorphism. However, this is akin to the standard coalescent in which the distribution of offspring numbers among individuals in the metapopulation affects levels and patterns of polymorphism only through the effective mutation parameter, θ. Outweighing this is the fact that by modeling gene genealogies in a metapopulation, we gain intuition about the potential of further theoretical study and the design of more optimal methods of inference. Even the little that is currently known about the complex patterns of genetic variation in samples from a metapopulation, e.g., Fig. 8.5, provides a great deal of hope.

9. METAPOPULATION QUANTITATIVE GENETICS: THE QUANTITATIVE GENETICS OF POPULATION DIFFERENTIATION

Charles J. Goodnight

9.1 INTRODUCTION

The field of quantitative genetics can be traced to Fisher (1930), although the roots of this field can be traced further back to the "biometician" school of evolution, and ultimately to Darwin (Provine, 2001). The original goals of quantitative genetics included explaining and describing the response to directional selection and to providing analytical tools that could be used in the breeding of livestock. With respect to the goal of providing tools for breeders, quantitative genetics has been stunningly successful.

In developing quantitative genetics, Fisher made the assumption that populations were large, unstructured, and mated randomly. These assumptions are inappropriate for metapopulations, which are, by definition, structured and in which the individual demes are frequently small. Fisher's methods remain

0-12-323448-4

valuable in that they predict the response to selection within demes; however, this prediction generally will not provide a useful description of the evolution of the metapopulation as a whole. Of the many reasons that this may be true, such as environmental heterogeneity and multilevel selection, one of the most interesting is the effect of population structure on the underlying effects of genes on the phenotypes. In particular, when there is epistasis, defined as interactions among alleles at different loci, the effect of a gene on the phenotype of an individual is a function not only of the gene, but also of the genetic background in which it is found. What this means is that even when selection acts on the phenotype in the same manner across all demes, the genetic consequences of that selection may be different in different demes. Thus, an allele favored by selection in one deme may be eliminated by the same selective regime in a second deme with a different genetic background.

The goal of metapopulation quantitative genetics is to describe the variation among demes in the phenotypic effect of alleles. Conceptually, it is a question of what is the variance in the phenotypic effect of a particular allele when it is "inserted" into the different demes in a metapopulation? This variance must be corrected for the effects of the overall deme mean, which will necessarily affect the phenotypic variance in the effects of an allele across a metapopulation. Thus, the more correct quantity is the phenotypic variance in the effect of an allele relative to the effects of other alleles at the same locus measured in the same demes. If this variance in the relative allelic effects is zero, or in an experimental situation small and not significant, then allelic effects measured in one deme are indicative of allelic effects (relative to other alleles at the same locus) measured in any deme. However, if the variance in the allelic effects is nonzero, then the allelic effects measured in one deme are not predictive of the relative allelic effects in other demes.

When there is variance in the effects of alleles, phenotypic selection acting uniformly in all demes will become a diversifying selection at the genic level. That is, selection favoring an allele in one deme may lead to a decrease in the frequency of that same allele in a second deme. This potentially leads to a selective restriction of migration between demes, as the offspring of migrants will be of low fitness. This could interfere with ecological and demographic processes, such as the "rescue effect" of migrants. In the extreme, the fitness of the offspring of migrants may be so low that interdemic gene flow is eliminated, effectively turning the different demes into separate species. Finally, it is important to note that variation in the effects of alleles need not be correlated with variation in deme means. Thus, even if two demes have very similar mean phenotypes, they may nevertheless be differentiated for genic effects. Conversely, two demes with very different mean phenotypic values need not be differentiated for genic effects.

This chapter briefly describes traditional or "Fisharian" quantitative genetics and uses this as a framework to discuss some of the modifications of this theory that are necessary when applying this theory to a metapopulation rather than a single panmictic population. It then discusses the interpretation of metapopulation quantitative genetics. In particular, whereas traditional or Fisherian quantitative genetics are naturally related to measuring the response to selection, metapopulation quantitative genetics is more naturally related to the differentiation of populations and, as a consequence, speciation.

9.2 FISHERIAN QUANTITATIVE GENETICS

When he originally developed the field of quantitative genetics, Fisher (1930) used the assumption that traits were determined by a very large (in the limit, infinite) number of loci each with a very small (in the limit infinitesimal) effect. Under this assumption, long-term directional selection would lead to a linear change in the mean phenotype with no discernible change in gene frequency at any given locus. Indeed, at the limit of an infinite number of loci, each with an infinitesimal effect on the phenotype, changes in gene frequency would be infinitesimal as well. More importantly, he assumed that there was no population structure and that populations were very large. Finally, in order for many of the relationships that Fisher described to work, there must be random mating (Falconer, 1985).

Quantitative genetics is built on the idea of partitioning the phenotype into components. If the *i*th individual has a phenotype P_i, then this can be divided into components due to genetics (G_i) and the environment (E_i):

$$P_i = \mu + G_i + E_i \tag{9.1}$$

where μ is the mean of the population. In this partitioning, it is assumed that there are no genotype by environment interactions or correlations. These are incorporated easily, but are not necessary for the topics discussed in this chapter. The genetic component can be further divided into components including the breeding value (additive effects, A), a component that can be attributed to interactions between alleles at the same loci (dominance, D), and interactions between alleles at different loci (epistasis). Epistasis can be further divided into components due to the nature of the particular interaction. For example, two-locus interactions can be divided into additive by additive epistasis (AXA), additive by dominance epistasis (AXD), dominance by additive epistasis (DXA), and dominance by dominance epistasis (DXD) (Table 9.1). Similarly, three locus and higher interactions can, in principle, be added. Thus, the value of the phenotype of the *i*th individual becomes

$$P_i = \mu + A_i + D_i + AXA_i + AXD_i + DXA_i + DXD_i + \ldots + E_i \tag{9.2}$$

This partitioning of the phenotype into components is done statistically using the regression of phenotype on the genetic variance components (Hayman and Mather, 1955; Goodnight, 2000a,b). Understandably, the regression model can potentially becoming quite complicated as more of the genetic effects are included.

It is important to emphasize that this partitioning is a statistical partitioning done by multiple regression. Further, this multiple regression is weighted by the frequency of the different genotypes. As a consequence, when gene frequencies change, the partitioning will also change. Thus, the breeding value of an individual is not only a property of the genes that make up that individual, but also of the population in which it is measured (Falconer and Mackay, 1996).

Variation is necessary if there is to be evolution, and as a result it is the partitioning of the phenotypic variance that is of interest. Because the phenotype has been divided into genetic and environmental components using a least-squares

TABLE 9.1 The Eight Genetic Effects Used in This Chapter[a]

Additive A locus

	A_1A_1	A_1A_2	A_2A_2
B_1B_1	1	0	−1
B_1B_2	1	0	−1
B_2B_2	1	0	−1

Additive B locus

	A_1A_1	A_1A_2	A_2A_2
B_1B_1	1	1	1
B_1B_2	0	0	0
B_2B_2	−1	−1	−1

Dominance A locus

	A_1A_1	A_1A_2	A_2A_2
B_1B_1	−1	1	−1
B_1B_2	−1	1	−1
B_2B_2	−1	1	−1

Dominance B locus

	A_1A_1	A_1A_2	A_2A_2
B_1B_1	−1	−1	−1
B_1B_2	1	1	1
B_2B_2	−1	−1	−1

Additive by additive epistasis

	A_1A_1	A_1A_2	A_2A_2
B_1B_1	1	0	−1
B_1B_2	0	0	0
B_2B_2	−1	0	1

Additive by dominance epistasis

	A_1A_1	A_1A_2	A_2A_2
B_1B_1	1	0	−1
B_1B_2	−1	0	1
B_2B_2	1	0	−1

Dominance by additive epistasis

	A_1A_1	A_1A_2	A_2A_2
B_1B_1	1	−1	1
B_1B_2	0	0	0
B_2B_2	−1	1	−1

Dominance by dominance epistasis

	A_1A_1	A_1A_2	A_2A_2
B_1B_1	−1	1	−1
B_1B_2	1	−1	1
B_2B_2	−1	1	−1

[a] These effects fully describe any two-locus two-allele genetic effects.

regression, these components are statistically independent of each other. As a result, phenotypic variance can be partitioned in exactly the same manner as the individual phenotype (Fisher, 1930; Hayman and Mather, 1955; Falconer and Mackay, 1996; Goodnight, 1998, 2000a,b):

$$\mathrm{V_P = V_A + V_D + V_{AXA} + V_{AXD} + V_{DXA} + V_{DXD} + \ldots + V_E} \qquad (9.3)$$

As with the partitioning of the phenotype of an individual, the components of phenotypic variance are statistical properties of a population. As gene frequencies change, the partitioning among variance components will also change. Under Fisher's assumption of large population size, and many loci each with small effect gene frequencies will not change appreciably, and the variance components will remain approximately constant.

The primary utility of quantitative genetics is that it can be used to predict the response to selection using the standard "breeder's equation" (Falconer and Mackay, 1996):

$$r = h^2 s \qquad (9.4)$$

where r is the response to selection measured as the difference in mean phenotype between parents and offspring, s is the selection differential measured as the difference in mean phenotype between the selected parents and all parents, and h^2 is the heritability. Heritability is the ratio of the additive genetic variance to the phenotypic variance, $\frac{V_A}{V_p}$. It is also a constant of proportionality that "converts" change due to selection within generations into change between generations (Falconer and Mackay, 1996).

For predicting the response to selection, only two variance components and the selection differential are needed. The two variance components are the additive genetic variance and the phenotypic variance. The phenotypic variance is simply the variance in the trait observed in the population. Thus, the additive genetic variance, the genetic variance that can contribute to the resemblance between parents and offspring, is important. In addition, it is often useful to measure the dominance variance both because it is often a large portion of the phenotypic variance and because it is experimentally reasonable to measure. Fisher (1930) felt that the epistatic variance components could be relegated to the environmental variance component. This emphasizes that the "environmental" variance is perhaps better referred to as the "residual" variance, as it is the sum of all of the unmeasured factors, genetic and environmental, contributing to the phenotypic variance.

9.3 GENETIC VARIANCE COMPONENTS IN A TWO-LOCUS TWO-ALLELE SYSTEM

Consider a system with an A locus and a B locus. At each locus there are two alleles, A_1 and A_2 alleles at the A locus, and B_1 and B_2 alleles at the B locus. In a system such as this there are nine possible genotypes. For heuristic purposes, it is convenient to array these as a three by three matrix (Table 9.1). Such a system can be fully described by nine orthogonal equations (Goodnight, 2000a,b). One of these is the mean genotypic value, leaving eight equations to fully describe the genetic effects. Thus, a two-locus two-allele system can always be divided into eight independent genetic effects. Assuming a gene frequency of 0.5 for both alleles at both loci and a population in two–locus Hardy–Weinberg–Castle equilibrium, the eight genetic effects used in this chapter are given in Table 9.1. At a gene frequency of 0.5, these genetic effects are orthogonal; however, if gene frequencies deviate from 0.5 for either locus, they will no longer be independent. In general, a change in gene frequency will tend to shift the forms of genetic variation involving more interaction into forms involving less interaction.

The statistical shift of genetic variation into forms involving less interaction means that periods of small population size will tend to cause dominance and epistasis to diminish and additive genetic variance to increase. Figure 9.1 is a pair of graphs of the additive genetic variance as a function of the Wright's inbreeding coefficient, *F*, with single locus effects (additive and dominance effects, Table 9.1) shown in Fig. 9.1a and digenic epistasis shown in Fig. 9.1b.

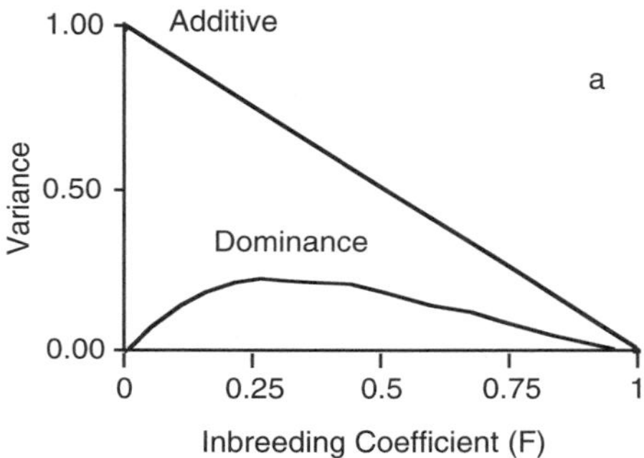

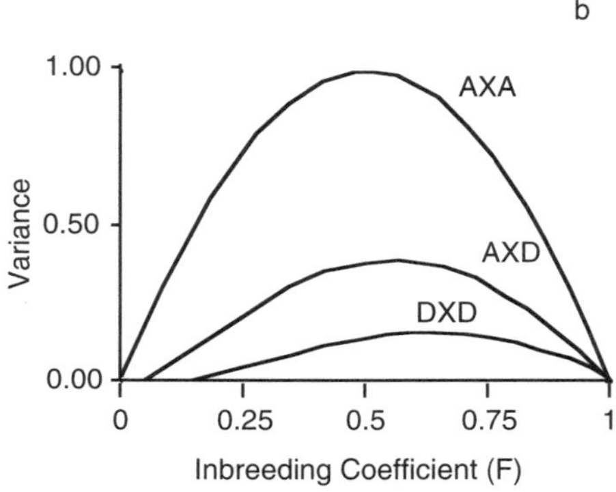

Fig. 9.1 Additive genetic variance in a population as a function of Wright's inbreeding coefficient, *F*. Total genetic variance in the outbred population ($F = 0$) is standardized at 1. (a) Single locus effects, additive effects (Additive) and dominance effects (Dominance), (b) Two locus interactions, additive by additive epistasis (AXA), additive by dominance and dominance by additive epistasis (AXD), and dominance by dominance epistasis (DXD).

When there are only additive effects, the additive genetic variance declines linearly as a function of F (Fig. 9.1a). For dominance (Fig. 9.1a) and all forms of epistasis (Fig. 9.1b), the additive genetic variance increases as a function of F until it reaches a maximum value at an intermediate level of inbreeding before declining. In the absence of gene interaction, additive variance declines in direct proportion to the heterozygosity (equal to $1-F$, where F is Wright's inbreeding coefficient). In the presence of gene interaction, the change in additive genetic variance due to a small population size will depend on the competing processes of the increase due to the conversion of dominance and epistasis to additive variance and the loss due to an overall loss in genetic variation.

If weighted by the genotype frequencies at Hardy–Weinberg–Castle proportions, the genotypic values of each genetic effect in Table 9.1 sums to zero. This allows them to be used as orthogonal contrasts in a linear regression and, as a consequence, provides a simple method for partitioning any set of two-locus two-allele genotypic values into independent genetic effects. To do multiple regression, the nine observed genotypic values are used as dependent variables and the eight genetic effects are used as independent variables. This is an unusual regression because with nine genotypic values there are only eight degrees of freedom. With each of the genetic effects using one degree of freedom, there are no degrees of freedom left for error. Thus, with only nine genotypic values, the genetic variance can be partitioned into genetic variance components, but no estimate of the accuracy of that partitioning is possible. If, however, a set of individuals are scored for both the phenotypic value of the trait and their genotype (possibly using quantitative trait loci), then a regression could be done that would provide an error variance. The genotypic values must be weighted by their genotype frequencies; thus, a regression done at one gene frequency will give a different answer than a regression done at a different gene frequency. Again, this is the basis of conversion of a nonadditive genetic variance into an additive genetic variance (Fig. 9.1). Finally, the regression must be done using sequential or type 1 sums of squares. Typical regression packages use iterative or type 3 sums of squares. While appropriate for standard uses of regression with moderate unbalance (type 1 and type 3 sums of squares give identical results for balanced data sets), type 3 sums of squares give incorrect results for the genotypic regressions that are unbalanced due to changes in gene frequency. An issue with using type 1 sums of squares is that the order in which the independent variables are entered into the regression model will change the results of the regression. In the case of regression on independent genetic variance components, the order in which the variables must be entered is additive locus A and additive locus B (order not important); dominance locus A and dominance locus B (order not important); additive by additive epistasis; additive by dominance and dominance by additive epistasis (order not important); and finally dominance by dominance epistasis (Goodnight, 2000a).

9.4 METAPOPULATION QUANTITATIVE GENETICS

To extend Fisherian quantitative genetics to a metapopulation setting, assume that there is a metapopulation structure consisting of a large set of demes linked by migration. Define the local breeding value of the ith individual in the jth deme to be A_{ij}. In a large randomly mating population, the breeding value is the mean phenotype of the offspring of an individual. In a structured population, it is necessary to identify both the individual, and the deme from which its mates are drawn. The author refers, to A_{ij} as the local breeding value to distinguish it from the breeding value as defined by Fisher (1930). It differs in that an individual has only a single Fisherian breeding value, which is an average across the metapopulation, whereas an individual has a distinct local breeding value in each deme of a metapopulation. The local breeding value is taken as a deviation from the metapopulation mean. Using least-squares partitioning, the breeding value can be divided into components due to an individual effect (A_i), a deme effect (A_j), and an individual by deme interaction (A_{i*j}) (Wade and Goodnight, 1998):

$$A_{ij} = A_i + A_j + A_{i*j} \tag{9.5}$$

Within mean additive genetic variance within demes will be given by

$$\overline{V}_A = \frac{1}{J}\sum_j\left(\sum_i p_{ij}(A_{i\bullet} + A_{i*j})^2 - A_{\bullet j}^2\right) \tag{9.6}$$

where $\overline{V}_A$ is the mean within deme additive genetic variance, p_{ij} is the frequency of the ith genotype in the jth deme, and J is the number of demes. The effect of population size on this equation is worth discussing. First, when the population is unstructured, p_{ij}, the frequency of the ith genotype in the jth population is simply the frequency of the ith genotype. In a structured population, genetic drift will, by random chance, cause frequencies of genotypes to change and some genotypes to disappear.

It is often convenient to focus on allele frequencies rather than genotype frequencies. Although selection acts on entire phenotypes, much of the recent genetical data has taken the form of mapping the chromosomal regions that affect quantitative traits, i.e., quantitative trait loci (QTL). These studies, which are naturally related to a gene level focus rather than a whole phenotype level focus, allow a detailed dissection of two locus interactions that would be difficult or impossible using more traditional quantitative genetic methods. In systems without gene interaction and random mating, the frequencies of individual alleles are adequate for describing the genetic variance. Wright's inbreeding coefficient, F, is a measure of the increase in correlation among alleles in a randomly mating population and thus provides a summary measure of the expected change in allele frequencies. In an additive system, the additive genetic variance is a function of F and decreases as the inbreeding coefficient increases:

$$V_A^* = (1 - f)V_A \tag{9.7}$$

where V_A^* is the additive genetic variance in the derived population. It must be emphasized that this formula applies only in the special case of complete additivity with no dominance or epistasis.

The term $A_{i\bullet}$ should remain approximately constant with inbreeding, as it is an average of breeding value across all demes. This will not be true if there is inbreeding depression or other factors that cause shifts in breeding values associated solely with increased F.

A_{i^*j}, however, is a function of the inbreeding in the population when there is gene interaction. In particular, in an outbred deme ($F = 0$), A_{i^*j} will equal zero. As the inbreeding coefficient increases ($F > 0$), demes will become progressively more differentiated and A_{i^*j} will increase. It is only when there is gene interaction (dominance or epistasis) and inbreeding that A_{i^*j} is nonzero. Figure 9.2 is a graph of the variance in local average effects for the different genetical effects as a function of inbreeding coefficient, F. In the absence of gene interaction, A_{i^*j} is always zero. For this reason, additive effects are not shown. For the additive by dominance (and dominance by additive) gene interaction, the variance in local average effects is shown separately for the "additive" locus (A locus in the additive by dominance interaction) and the "dominance" locus (B locus in the additive by dominance interaction). The "additive" locus refers to the locus with additive genotypic values within the genotype of its interacting pair. For example, for additive by dominance epistasis within the B_1B_1 genotype, genotypic values for the A locus are 1, 0, and –1 for the A_1A_1, A_1A_2, and A_2A_2 genotypes, respectively. In contrast, the "dominant" locus has dominant genotypic values within genotypes of the interacting pair. For example, within the A_1A_1 genotype, the B locus has genotypic values of 1, –1, and 1 for the B_1B_1, B_1B_2, and B_2B_2 genotypes, respectively.

Thus, there are two competing processes that occur as demes differentiate. The loss of genetic variation is reflected in the p_{ij} and has the effect of decreasing the additive genetic variance, whereas when there is gene interaction, the

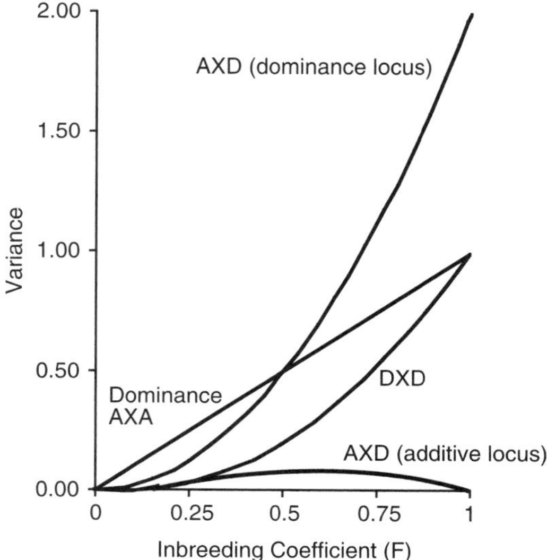

Fig. 9.2 Variance due to the allele by deme interaction, $Var(\alpha_j{}^*kl)$, as a function of inbreeding coefficient, F, for the different forms of genetic effects. The variance in local average effects for additive effects is zero for all values of F. Dominance effects (Dominance), additive by additive epistasis (AXA), additive by dominance and dominance by additive epistasis (AXD), and dominance by dominance epistasis (DXD) are shown. For additive by dominance epistasis, additive and dominance loci are listed separately.

differentiation of populations will tend to cause the A_{i^*j} to increase, which will in turn increase the additive genetic variance. Thus, the question of whether the additive genetic variance will increase or decrease following a population bottleneck depends on the relative magnitude of these two effects. When there are only additive effects (no dominance or epistasis), the A_{i^*j} will equal zero regardless of the level of inbreeding. With gene interaction, A_{i^*j} will generally be nonzero and will increase with increasing F. Goodnight (1988) showed for additive by additive epistasis that if the ratio of VAA/VA > 1/3, a one-generation bottleneck will lead to an increase in the additive genetic variance. Similar results have not been developed for other forms of epistasis.

9.5 METAPOPULATION QUANTITATIVE GENETICS AND POPULATION DIFFERENTIATION

Some of the most important effects of gene interaction appear primarily among demes. Typically, discussions of population differentiation focus on the differentiation of the means of the populations. That is, a common formula for the genetic variance among demes (in an additive system) is (e.g., Hedrick, 2000)

$$V_{\text{bet}} = 2f\, V_A \tag{9.8}$$

where V_{bet} is the variance between demes. Because the additive genetic variance is a portion of the phenotypic variance, V_{bet} is the variance among deme mean phenotypes. However, the following discussion shows that it is also necessary to consider the differentiation of genetic effects. It also show that differentiation for genetic effects may not be related to differentiation of deme means.

The increase in additive genetic variance following bottlenecks in systems with gene interaction must come from somewhere. The additive genetic variance was given earlier in terms of breeding value; however, in a randomly mating population, the breeding value of an individual is the sum of the average effects of the underlying alleles. In the following discussion it will be convenient to focus on average effects. The Fisherian average effect of an allele is defined to be the mean deviation from the population mean of individuals that received that allele from one parent, with the allele received from the other parent having come at random from the population. Using this relationship, the additive genetic variance in a randomly mating population can be shown to be (Falconer and MacKay 1996)

$$V_A = 2 \sum_{k=loci} \sum_{l=alleles} p_{kl}\alpha_{kl}^2 \tag{9.9}$$

where k refers to the summation over all loci affecting the trait, l refers to the summation over all alleles at each of the k loci, p_{kl} is the frequency of the lth allele at the kth locus, and α_{kl} is the average effect of the klth allele. Examination of this equation reveals that there are only two ways to increase the additive genetic variance. One is to increase the effective number of alleles. Although perhaps not obvious from this form of the equation, increasing the number of alleles and decreasing the mean p_{kl} has the effect of increasing the additive genetic variance. However, genetic drift leads to an increase in F and, on average, will decrease the effective number of alleles (defined as $n_e = \frac{1}{F}$) (Crow and Kimura, 1970). As a result, the increase in additive genetic variance does not come from this source. The other possible way for the additive genetic variance to increase is for the average effects to change. This shift in the average effects of alleles is the cause of the individual by deme interaction (A_{i*j}) and causes the increase in the additive genetic variance.

Thus, when the additive genetic variance increases as a result of small population size, this is evidence that there is an individual by deme interaction. What this means is that genes have different effects in different demes in a genetically differentiated metapopulation.

The gene by deme interaction requires a bit more discussion. In moving to a metapopulation setting, it is again necessary to distinguish the Fisherian average effect from local average effects. Local average effects, like local breeding values, are similar to average effects, but defined separately for each deme. Thus, the local average effect of an allele measured in a deme is the mean deviation from the metapopulation mean of individuals that have the allele in question, with the other allele, and all alleles at other loci having come at random from the deme in question (Goodnight, 2000a,b). In theory, measuring this would require substituting an allele into genotypes drawn from the deme in question without modifying the alleles at any other loci. While this ideal is not possible experimentally, two locus local average effects can be reconstructed from QTL data. Unfortunately, whereas in a randomly mating panmictic population the breeding value of an individual is a simple sum of the average effects of the underlying alleles, the local breeding value of an individual is not a simple sum of the underlying local average effects. Factors such as linkage disequilibrium generated by drift in a structured population complicate this summation. Interestingly, this is an example of how reductionist methods that work well in randomly mating populations often fail when there is population structure.

For the purposes of this discussion, local average effects are useful because they help in describing the behavior of the different forms of gene interaction and are useful in interpreting quantitative trait loci data. Using local average effects, the mean additive genetic variance within a single deme in a metapopulation becomes

$$\overline{V}_A = 2 \sum_{j=demes} \left(\sum_{k=loci} \sum_{l=alleles} p_{jkl}\alpha_{jkl}^2 - \left(\sum_k \alpha_{jk\bullet} \right)^2 \right) \tag{9.10}$$

where j is the summation of demes, k is the summation over loci, and l is the summation over alleles at the kth locus. The frequency of the lth allele at the kth locus in the jth deme is p_{jkl}, α_{jkl} is the local average effect of the klth locus in the jth deme, and $\alpha_{jk\bullet}$ is the deme mean local average effect at the kth locus.

As with the local breeding value, the local average effect can be divided into components due to a deme effect, $\alpha_{jk\bullet}$, an allele effect, $\alpha_{\bullet kl}$, and a deme by allele interaction, α_{j*kl}, with the only difference being that effects are locus (k) specific. Using this formulation the mean within deme additive genetic variance becomes

$$\overline{V}_A = 2\sum_{j=demes}\left(\sum_{k=loci}\sum_{l=alleles} p_{jkl}\left(\alpha^2_{\bullet kl} + \alpha^2_{jk^*l}\right) - \left(\sum_k \alpha_{jk\bullet}\right)^2\right) \quad (9.11)$$

It is particularly interesting to examine the among demes variance in local average effects. Conceptually, this is the equivalent of measuring the local average effect of a single allele in each deme and measuring the variance in these within allele local average effects. The (mean) variance in the local average effect of an allele at the kth locus is

$$\begin{aligned} Var(\alpha_{kl}) &= \sum_j p_j(\alpha_{jkl} - \alpha_{\bullet kl})^2 \\ &= \sum_j p_j\left(\alpha^2_{jk\bullet} + \alpha^2_{j^*kl} - \alpha^2_{\bullet kl}\right) \\ &= Var(\alpha_{jk\bullet}) + Var(\alpha_{j^*kl}) \end{aligned} \quad (9.12)$$

The variance due to the deme effect on the kth locus, $Var(\alpha_{jk\bullet})$, can be distinguished from the variance due to the allele by deme interaction $Var(\alpha_{j^*kl})$ by examining the among demes variance in the mean local average effects:

$$\begin{aligned} Var(\alpha_{k\bullet}) &= \sum_j p_j(\alpha_{jk\bullet} - \alpha_{\bullet k\bullet})^2 \\ &= \sum_j p_j\left(\sum_l p_l\alpha_{jkl} - \alpha_{\bullet kl}\right)^2 \end{aligned} \quad (9.13)$$

The allele by deme interaction can then be obtained by subtraction (Goodnight, 2000a).

Variance in the $\alpha_{jk\bullet}$ is not particularly interesting. The mean local average effect is how much on average the mean allele changes the phenotype of an individual. If a deme has a particularly high frequency of alleles conferring a large phenotypic value, the mean local average effect will be negative; if the deme has a low frequency of these alleles, the mean local average effect will be positive. Indeed, in a system without gene interaction, $\alpha_{jk\bullet}$ is directly proportional to the phenotypic variance among demes. More importantly, this component of the local average effects does not drive any differentiation the local average effects of alleles because the local average effects of all alleles at a locus will be affected by the deme effect in the same way. For example, consider a locus affecting body weight. If, when measured in one deme, one allele causes the body weight to be 3 g heavier than a second allele, this 3-g difference will be maintained regardless of in which deme they are measured. The mean mass of individuals will vary as a result of the variation in the $\alpha_{jk\bullet}$; however, the mean difference in weight between individuals carrying different alleles will be a constant.

It is the gene by deme interaction, α_{j^*kl}, that is of more interest. As with the individual by deme interaction, this value equals zero in the absence of gene interactions. When there is dominance and epistasis, it will generally be nonzero (Fig. 9.2). Unlike the deme effect, this interaction does shift the local average effects of alleles relative to each other. In the example given earlier, if the difference in local average effect between a pair of alleles in one deme is 3 g, this will not be predictive of the difference in other demes. In a second deme the difference may be 1 g or the ordering of local average effects may be reversed.

To quantify the extent to which the local average effects of alleles vary among demes together due to $\alpha_{jk\bullet}$ compared to the extent to which they vary among demes independently due to α_{j^*kl}, Goodnight (2000a) suggested using the intraclass correlation in local average effects:

$$cor(\alpha_{kl}) = \frac{Var(\alpha_{k\bullet})}{Var(\alpha_{kl})} \tag{9.14}$$

This intraclass correlation will vary between zero and one. If it is one, this indicates that the local average effects of the alleles at a locus are maintaining their relative ranking in all demes. If the correlation is less than one, it indicates that the local average effects are varying among demes relative to each other. The amount it is less than one indicates the extent to which the local average effects of the alleles are varying independently among demes.

In Fig. 9.3, the intraclass correlation in local average effects as a function of the inbreeding coefficient is shown for all of the types of genetical effects. The "additive" locus (A locus of additive by dominance interaction, Table 9.1) and the "dominance" locus (B locus of additive by dominance interaction, Table 9.1) δ are shown separately for additive by dominance and dominance by additive epistasis. Several points can be made from Fig. 9.3. First, additive gene action is a special case and the only case in which $cor(\alpha_{kl})$ is one. This means that regardless of the level of inbreeding, if one allele confers an advantage over a different allele in one deme, it will maintain that advantage regardless of in what deme it

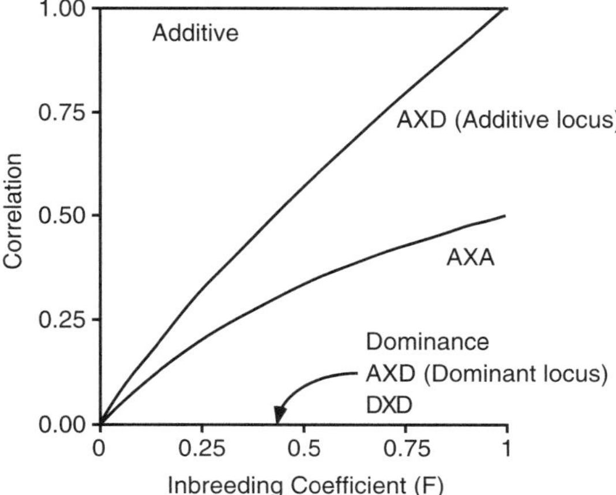

Fig. 9.3 Intraclass correlation in local average effects as a function of inbreeding coefficient for the different forms of genetic effects. Additive effects (Additive), dominance effects (Dominance), additive by additive epistasis (AXA), additive by dominance and dominance by additive epistasis (AXD), and dominance by dominance epistasis (DXD) are shown. For additive by dominance epistasis, additive and dominance loci are listed separately. Additive effects give a correlation of one, which indicates that the values of alleles relative to other alleles at the same locus are a constant. All other effects have correlations less than one (AXA, AXD additive locus) or zero (all interactions involving dominance).

is measured. For the case of the "additive" locus of additive by dominance and dominance by additive epistasis, the correlation approaches one as the inbreeding coefficient (*f*) approaches one; however, this is very different from the additive case where the correlation is fixed at one. Second, all interactions involving dominance, i.e., dominance, "dominance" locus of additive by dominance and dominance by additive epistasis, and dominance by dominance epistasis, have a correlation in local average effects of zero for all inbreeding coefficients. This means that to the extent that the local average effects of alleles vary among demes, they vary independently. Thus, for these pure forms of gene interaction involving dominance, regardless of the inbreeding coefficient, the relative ranking of alleles in one deme is not predictive of the ranking in other demes.

Actual two locus interactions will involve a mixture of different genetical effects. When interacting quantitative trait loci have been measured, a large fraction of the variation is typically additive (Cheverud, 1995; Goodnight, 2000b). Thus, it is unlikely that in most circumstances the extreme of $cor(\alpha_{kl}) = 0$ will often be observed in experimental situations.

When there is dominance or epistasis, the intraclass correlation in local average effects is less than one. This indicates that when there is gene interaction, populations differentiate for local average effects, whereas when there are only additive effects, no differentiation occurs. This can be contrasted to the differentiation of population means (Fig. 9.4) for the different forms of pure genetical effects. Additive effects cannot contribute to the differentiation

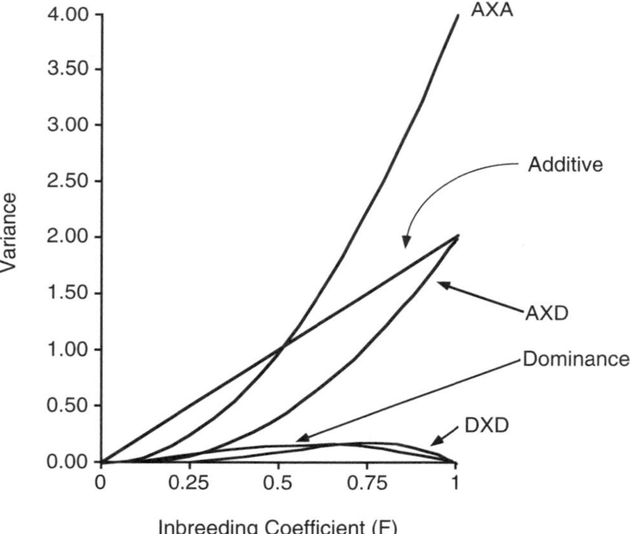

Fig. 9.4 The variance in mean phenotype among demes as a function of inbreeding coefficient for the different forms of genetic effects. Additive effects (Additive), dominance effects (Dominance), additive by additive epistasis (AXA), additive by dominance and dominance by additive epistasis (AXD), and dominance by dominance epistasis (DXD) are shown. Additive effects have a large effect on the phenotypic variance among demes, but no effect in the variance in local average effects. Conversely, dominance effects and DXD epistasis have little effect on the phenotypic variance among demes, but a much larger effect on the variance in local average effects (Fig. 9.2).

of average effects, but can contribute to the differentiation of population means, whereas interactions involving dominance have a large effect on the differentiation of average effects, but little effect on the differentiation of population means.

9.6 MIGRATION

The effects of migration on quantitative genetics parameters in metapopulations have not been well studied. Whitlock et al. (1993) used a descent measure model to show that with additive by additive epistasis the highest level of additive genetic variance is attained at an intermediate level of migration.

The relationship between island model migration and inbreeding coefficient can be used to make some observations concerning the effects of migration on metapopulation quantitative genetic measures. Following Hedrick (2000) under an infinite alleles model in a metapopulation with an infinite number of demes, but a small population size within each deme and island model migration among demes, the change in inbreeding coefficient (F or, in this context, more correctly F_{ST}) is given by

$$F_{(t+1)} = \left[\frac{1}{2N} + \left(1 - \frac{1}{2N}\right)F_t\right](1 - m)^2 \qquad (9.15)$$

where N is the within deme population size and m is the migration rate. A reasonable approximation for the equilibrium value of F is found by setting $F_{(t+1)} = F_{(t)}$ and ignoring terms on the order of m^2. Using this approximation, it can be shown that (Hedrick, 2000)

$$\hat{F} \approx \frac{1}{4Nm + 1} \qquad (9.16)$$

This emphasizes the classical result that it is the number of migrants (Nm) not the per-capita migration rate (m) that determines population differentiation. Solving Eq. (9.16) for Nm gives

$$Nm \approx \frac{1 - F}{4F} \qquad (9.17)$$

Figure 9.5 is a graph of the additive genetic variance (relative to a variance of one in the panmictic population) as a function of the number of migrants per deme for each of the different pure forms of genetical effects listed in Table 9.1. In agreement with intuition, when there are only additive effects, the additive genetic variance increases as a function of the number of migrants and is maximum when mixing is complete. The conversion of dominance to additive genetic variance is maximal at a migration rate of slightly over one migrant every other generation ($Nm = 0.575$). For digenic epistasis, the conversion of epistatic variance to additive variance is maximum at even lower migration rates of between one migrant every four generations and one migrant every six

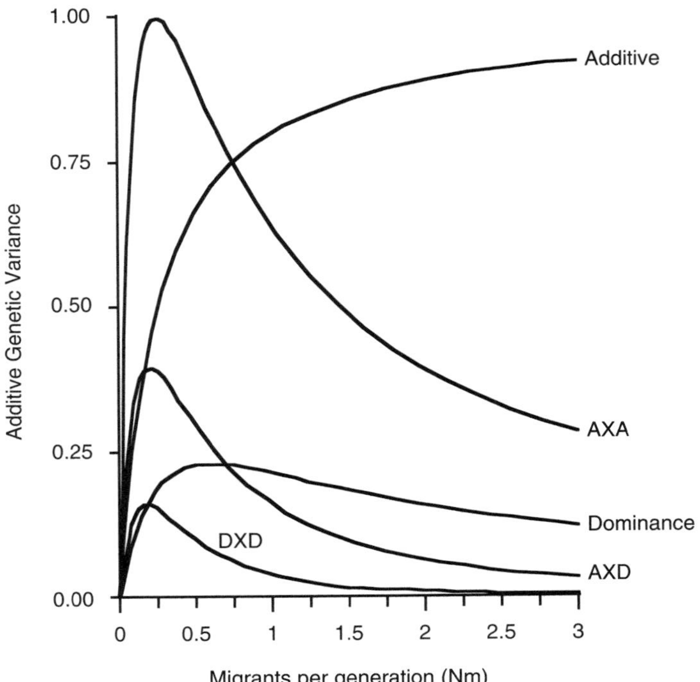

Fig. 9.5 Additive genetic variance within demes as a function of the number of migrants per deme at equilibrium. Total genetic variance in the outbred population ($F = 0$) is standardized at 1. Additive effects (Additive), dominance effects (Dominance), additive by additive epistasis (AXA), additive by dominance and dominance by additive epistasis (AXD), and dominance by dominance epistasis (DXD) are shown. For additive effects, the additive genetic variance increases with the number of migrants. For other genetic effects, the additive genetic variance is maximized between 0.175 and 0.575 migrants per generation.

generations ($Nm = 0.25 - 0.175$). Nevertheless, for several forms of gene interaction, notably dominance and additive by additive epistasis, the conversion of nonadditive variance to additive variance can be substantial, even with migration rates as high as three or more individuals per generation.

It is also interesting to examine population differentiation for average effects as a function of migration rate. Figure 9.6 shows the variance in local average effects as a function of number of migrants per deme for dominance and each of the different forms of epistasis (for additive effects, this value is zero for all migration rates). For the additive by dominance (and dominance by additive) interaction, the "additive" locus (A locus of additive by dominance interaction, Table 9.1) and the "dominant" locus (B locus of additive by dominance interaction, Table 9.1) are plotted separately. From Fig. 9.6 it is apparent that the differentiation of local average effects is very sensitive to the migration rate. For all effects except the additive by dominance additive locus, the differentiation of local average effects declines with migration rate, and with the exceptions of dominance and additive by additive epistasis, the differentiation for average effects is very low with as little as one migrant per generation.

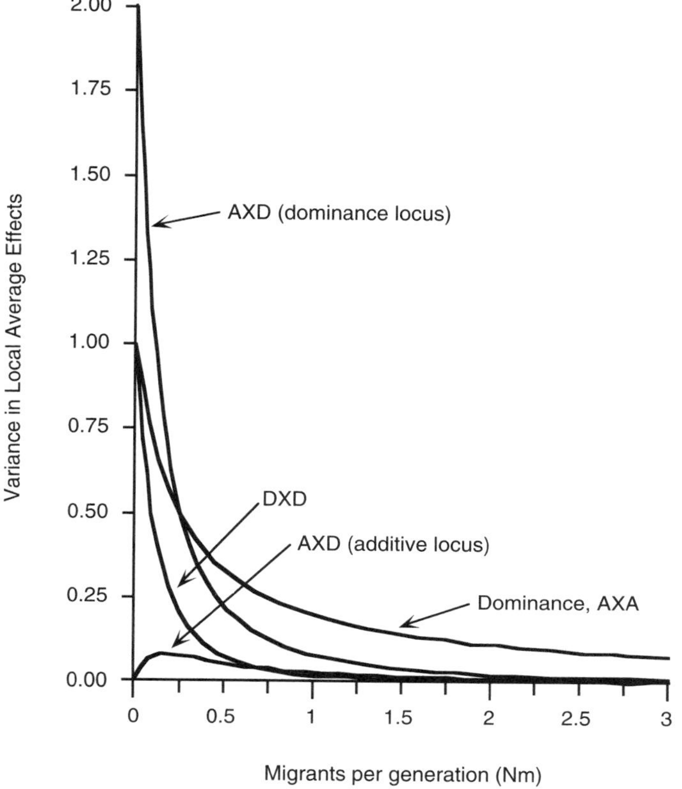

Fig. 9.6 Variance due to the allele by deme interaction, $Var(\alpha_{j^*kl})$, as a function of number of migrants per deme per generation. The variance in local average effects for additive effects is zero for all values of *Nm*. Dominance effects (Dominance), additive by additive epistasis (AXA), additive by dominance and dominance by additive epistasis (AXD), and dominance by dominance epistasis (DXD) are shown. For additive by dominance epistasis, additive and dominant loci are listed separately. The variance in local average effects decreases rapidly as a function of the number of migrants and is very small unless the number of migrants per deme is less than one per generation.

It is important to note, however, that these calculations assume that there is no selection acting. Selection can potentially lower the effective migration rate. Consider the situation of a metapopulation with a migration rate of three individuals per generation and considerable additive by additive epistasis. In this case, the additive genetic variance would be elevated by the conversion of epistasis to additive genetic variance (Fig. 9.5) and the response to selection would reflect this. In addition, there would be some small differentiation of average effects due to the additive by additive epistasis (Fig. 9.6). This indicates that to some degree, directional selection would be driving a differentiation of average effects as it acted on the gene interaction. This potentially leads to a lowering in fitness of the offspring of migrants (Goodnight, 2003), which in turn would lower the effective migration rate. The potential exists for a positive feedback to develop. Thus, the interaction among selection, migration, and drift may generate much more population differentiation than predicted by the apparent migration rate alone. This process has not been explored and deserves more attention.

9.7 EMPIRICAL EXAMPLE: THE METAPOPULATION QUANTITATIVE GENETICS OF A TEOSINTE–MAIZE CROSS

Doebley et al. (1995) identified quantitative trait loci for several traits in a wide cross between teosinte (*Zea mays* ssp. *parviglumis*) and cultivated maize ("corn," *Zea mays* ssp. *mays*). The details of the mapping procedure are described in Doebley et al. (1995; see also Doebley and Stec, 1993). Several pairs of QTL regions were shown to interact epistatically. One pair of epistatically interacting markers is BV302 and UMC107, which are located on different chromosomes. One trait these loci affect is "PEDS," the percentage of cupules with the pedicellate (maize like) spikelet (for a more complete description of this trait, see Doebley et al., 1995).

The genotypic values of PEDS for the different two locus genotypes are shown in Table 9.2. The genotypic values are simply the mean phenotypes of those individuals with the genotype of interest. Using the regression procedure described earlier, the partitioning of genetic variance into components for this trait is shown for three gene frequencies in Table 9.3. This partitioning includes total genetic variance and percentage contributions due to additive effects, dominance effects, and digenic epistasis. At all three gene frequencies the genetical effects are mainly attributable to additive and single locus dominance effects. However, in all cases, digenic epistasis accounts for a substantial proportion (between 20.9 and 39.4%) of the total genetic variance. This is reflected in Fig. 9.7, which is a graph of additive genetic variance due to the BV302 QTL as a function of the frequencies of the BV302 QTL and the UMC107 QTL. Within each gene frequency of the UMC107 locus, the additive variance of the BV302 locus varies in a manner consistent with a standard locus with dominance (e.g., Falconer and Mackay, 1996); however, the shape

TABLE 9.2 Genotypic Values for PEDS, the Percentage of Cupules with the Pedicellate (Maize-like) Spikelet for a Cross between Teosinte (T) and Maize (M)[a]

	UMC107		
BV302	**T/T**	**T/M**	**M/M**
T/T	0	0	0.8
T/M	0.3	0.2	1.3
M/M	1.1	0.4	7.3

[a] Note that crosses were done with teosinte as the seed parent. The teosinte cytoplasm and random sampling of genes at other loci affecting this trait are likely responsible for the low proportion of maize-like cupules even when both loci are from maize (data from Doebley et al., 1995).

TABLE 9.3 Decomposition into Genetic Variance Components for a Pair of Interacting Loci (UMC107 and BV302) Affecting the Percentage of Cupules Lacking the Pedicellate (Maize-like) Spikelet.

Effect	Freq($BV302_{teo.}$) = 0.25 Freq($UMC107_{teo.}$) = 0.25	Freq($BV302_{teo.}$) = 0.5 Freq($UMC107_{teo.}$) = 0.5	Freq($BV302_{teo.}$) = 0.75 Freq($UMC107_{teo.}$) = 0.75
Additive (UMC107)	3.7209 (38.6%)	0.6328 (21.6%)	0.0258 (8.9%)
Additive (BV302)	2.7719 (28.7%)	0.5513 (18.8%)	0.0697 (24.1%)
Dominance (UMC107)	0.8016 (8.3%)	0.4556 (15.5%)	0.0665 (23.0%)
Dominance (BV302)	0.3433 (3.5%)	0.1406 (4.8%)	0.0137 (4.7%)
AXA	1.4400 (14.9%)	0.4556 (15.5%)	0.0089 (3.1%)
AXD	0.3165 (3.3%)	0.3613 (12.3%)	0.0476 (16.4%)
DXA	0.2109 (2.2%)	0.1953 (6.7%)	0.0130 (4.5%)
DXD	0.0445 (0.5%)	0.1406 (4.8%)	0.0445 (15.4%)
Total genetic variance (percentage due to epistasis)	9.6501 (20.9%)	2.9331 (39.3%)	0.2897 (39.4%)

[a] Numbers in parentheses are the percentage of the total genetic variance due to the component. Total genetic variance and percentage due to digenic epistasis are also listed.

of the curve changes dramatically as the gene frequency at the UMC107 QTL changes. Figure 9.8 is a graph of the total variance in digenic epistasis in the population. Note that epistatic variance is greatest at intermediate gene frequencies. Fixation of either locus leads to a loss of epistasis and its conversion to additive and dominance variance at the locus that is still segregating.

When a metapopulation undergoes genetic drift, gene frequencies at both loci in the different demes will change randomly. The expected distribution of gene frequencies can be described using a Markov chain and Wright's inbreeding coefficient (F). Figure 9.9 is a graph of the effect of drift (measured by F) on mean additive genetic variance and the variance in local average effects (corrected for demic effects) for the BV302 QTL. Also shown for comparison is the variance in deme means. Not that due to epistatic interactions with UMC107, the additive genetic variance has a maximum value at an intermediate value of F. The early increase in additive genetic variance is due to a conversion of epistatic variance to additive variance. At higher values of F the effects of fixation within demes at both loci overwhelm the conversion of epistasis to additive variance and the overall additive genetic variance declines until it reaches zero when F equals one. Along with this increase in additive genetic variance there is also an increase in the variance in the local average effects of the BV302 alleles. In the absence of gene interaction, this variance in local average effects would be zero. The variance in local average effects indicates that BV302 alleles in a teosinte–maize metapopulation would have different effects on the PEDS phenotype in different demes.

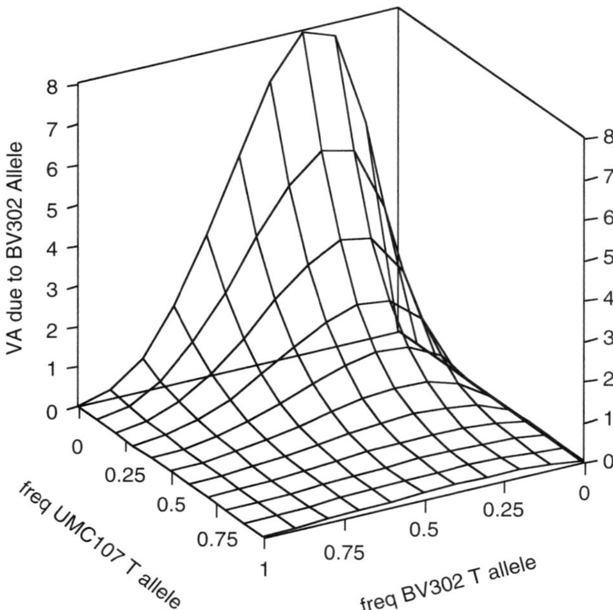

Fig. 9.7 Additive genetic variance due to the BV302 locus as a function of the frequency of BV302 and UMC107 teosinte alleles. Additive genetic variance at this locus is greatest when UMC107 is fixed for the maize allele and BV302 is at an intermediate gene frequency.

The local average effects for the BV302 locus (corrected for deme mean) are shown in Fig. 9.10 as a function of the gene frequencies at the two loci. In this particular case, maize genes at this locus always code for a more maize-like phenotype; thus, the interaction is expressed as a change in scale rather than a reversal in sign. This makes sense given that this interaction has a substantial additive

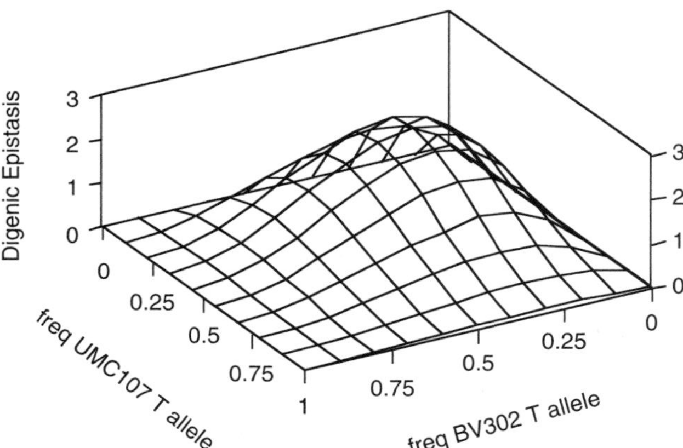

Fig. 9.8. Variance due to digenic epistasis (sum of additive X additive epistasis, additive by dominance epistasis, dominance by additive epistasis, and dominance by dominance epistasis) as a function of gene frequency of BV302 and UMC107 teosinte alleles. Epistatic variance is greatest at an intermediate frequency at both loci.

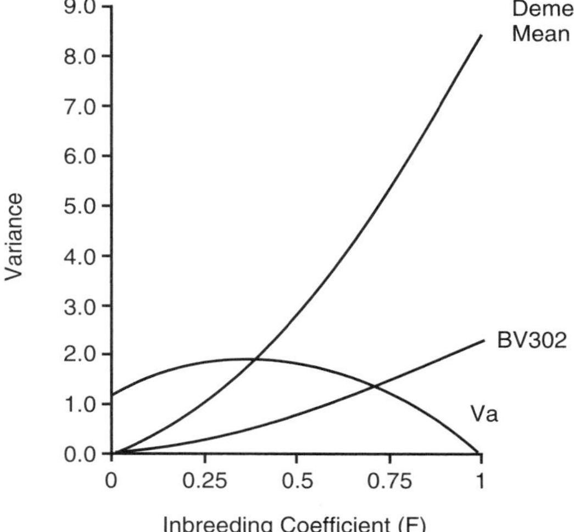

Fig. 9.9 Among deme phenotypic variance (Deme Mean), variance in local average effects for the BV302 locus corrected for demic effects (BV302), and mean within deme additive genetic variance (Va) as a function of an inbreeding coefficient. The additive genetic variance shown is the sum of the effects for the BV302 locus and the UMC107 locus. Maximum additive genetic variance occurs at an intermediate inbreeding coefficient, indicating that a conversion of non-additive variance into additive variance is occurring. This is reflected in the variance among demes, which is much greater than expected from an additive model. A maximum phenotypic variance among demes of 4.74 is possible under an additive model. It is also reflected in the variance in local average effects for the BV302 locus, which would be zero in an additive model.

component at all gene frequencies (Table 9.3). Nevertheless, it is quite apparent that the UMC107 locus has a very large effect on the average effects of BV302 alleles. The consequences of this can be seen by considering the difference between the two BV302 alleles in demes with different frequencies of the UMC107 alleles. In those demes with low frequencies of the UMC107 teosinte allele (and therefore high frequencies of the maize allele), differences between the two BV302 alleles are pronounced. Selection acting on PEDS would effectively distinguish between the two alleles. However, in demes with high frequencies of the UMC107, tesinte allele selection would be much less effective, as the difference between the alleles is much smaller. As an aside, the teosinte ancestor of corn may have had genetical effects at the BV302 locus similar to the back corner in Fig. 9.10. For this set of genotype frequencies, there is almost no difference between the two alleles, and the BV302 "maize" gene would have been nearly neutral with respect to the PEDS phenotype. As the domestication of teosinte progressed and it became more maize like, the frequency of the UMC107 maize allele would have presumably increased. This in turn would have acted to magnify the differences between the BV302 alleles. Thus, this is an interesting case where selection converts formerly neutral variation into large differences that can respond to selection. In this case, selection, rather than using up additive genetic variance, generates new additive genetic variance. These results also suggest why in the past there was considerable debate over the origins of maize (Beadle, 1980). Genes coding for a "maize" phenotype are nearly neutral in a teosinte genetic background, and the pathway for selecting maize from teosinte is not clear.

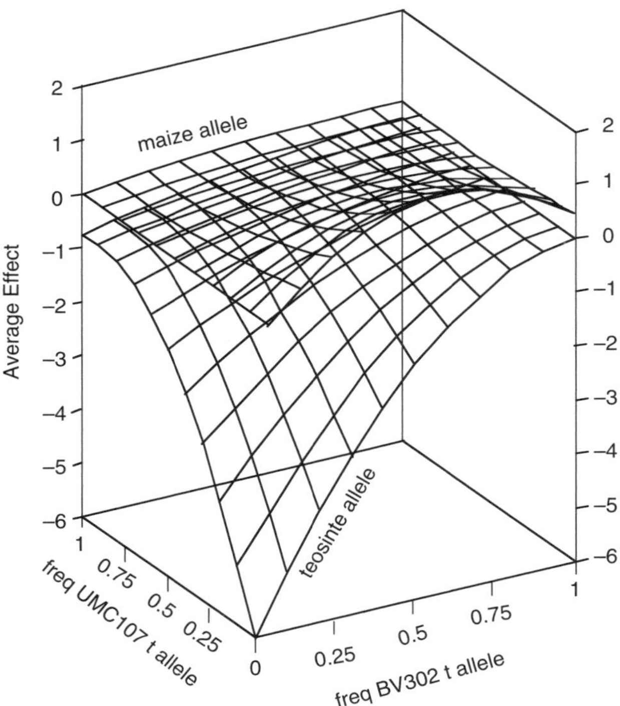

Fig. 9.10 Average effects of the BV302 allele as a function of frequency of BV302 and UMC107 teosinte alleles. These are Fisherian average effects that are taken as a deviation from the local population mean. Average effects vary as a function of both the interacting alleles. There is little difference between the two alleles in a teosinte-like genotype (back edge of graph), but a very large difference exists between alleles in a maize-like genotype (front edge of graph). In a system without gene interactions, planes describing the average effects of the two alleles would be parallel.

9.8 GENE INTERACTION AND SPECIATION

Although there is a substantial literature on speciation (reviewed in Templeton, 1981; Turelli et al., 2001; Schluter, 2001; Kondrashov, 2001; Wu, 2001), a general understanding of the relationship between quantitative genetics or population genetics and speciation has been elusive (Hedrick, 2000). One possible explanation for this is that although the importance of gene interaction in speciation is acknowledged (Muller, 1939; Mayr, 1963; Templeton, 1981; Futuyma, 1986; Orr, 1995), the majority of models that can be applied to the study of speciation assume only additive gene action. For example, in his review article, Wu (2001) focused on "speciation genes," which he identified as genes responsible for differential adaptation primarily to the ecological or sexual environment. The effect of gene interaction on the shift in the local average effects of alleles is a potential genetical mechanism for speciation that does not depend on shifts in the ecological or sexual environment. This is not a new idea; for example, Muller's (1939; Orr, 1995) model provided a qualitative model of how speciation would occur through the accumulation of mutations that caused reproductive incompatibility between two populations. However, these models lack a mechanism other than random mutation for the accumulation of incompatible mutations. The model discussed in this chapter suggests that a possible mechanism for speciation is that genetic drift generates small differences between demes and directional selection works to magnify these differences to the point that the two populations become reproductively isolated. Importantly, directional selection may be uniform at the phenotypic level, but nevertheless diversifying at the genetic level. That is, although selection may favor the same phenotype because of the different genetic backgrounds, different alleles would be favored in the two populations.

To see how genetic drift coupled with directional selection can lead to speciation, consider the example of dominance by additive epistasis (Table 9.1; Fig. 9.11a; Goodnight, 2000b). With dominance by additive epistasis, the A locus is overdominant, neutral, or underdominant, depending on the genotype at the B locus, and the B locus is additive with the favored allele being dependent on the genotype at the A locus.

Consider a metapopulation segregating for the A locus, but fixed for the B_2 allele. This situation corresponds to the bottom row in Fig. 9.11b. In this situation, the A locus is exhibiting simple overdominance with no apparent epistasis. Stabilizing selection will tend to drive the gene frequency at the A locus to approximately 0.5 for both alleles. Note, however, that because demes within the metapopulations are finite and likely small, there will be deviations from this equilibrium gene frequency.

If a B_1 allele is introduced into a deme either by mutation or by migration it will be neutral provided that the gene frequency at the A locus is exactly 0.5 (Fig. 9.11c, middle column). Any deviations from a gene frequency of 0.5 at the A locus will result in directional selection favoring the B_1 allele (Fig. 9.11c, left and right columns). The resulting increase in the frequency of the B_1 allele will weaken the strength of stabilizing selection on the A locus, resulting in neutrality and ultimately disruptive selection at the A locus (Fig. 9.11b, middle and top rows).

This is a positive feedback system, wherein genetic drift at the A locus results in weak directional selection at the B locus. This directional selection on the B locus has the effect of weakening the strength of stabilizing selection at the A locus, which in turn, when coupled with genetic drift, will increase the strength of directional selection on the B locus. Once the frequency of the B_1 allele exceeds 0.5, the A locus will experience disruptive selection and be fixed quickly for either the A_1 or the A_2 allele.

If this process occurs in several demes within the metapopulation some of the demes will, by random chance, become fixed for the A_1 allele, whereas others will become fixed for the A_2 allele. If the strength of destabilizing selection is strong enough, this could be sufficient to cause reproductive isolation and, as a consequence, speciation.

Prior to the introduction of a B_1 allele, there would be no reason to consider the A locus part of an epistatic interaction, nor any reason to consider it as a candidate for a locus that could drive speciation. Indeed, the A locus would appear to be a simple overdominant locus maintained by stabilizing

a

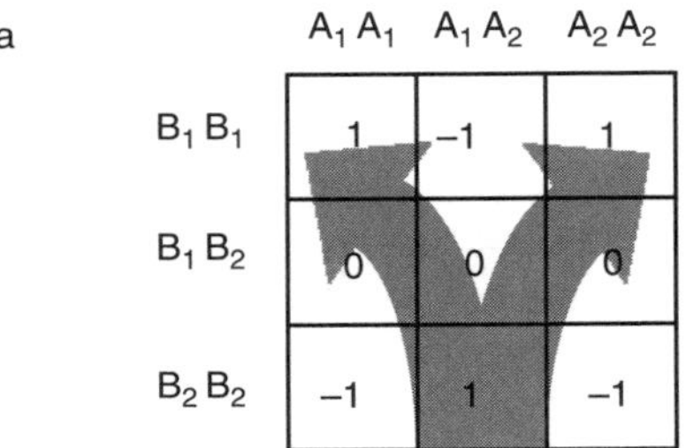

b

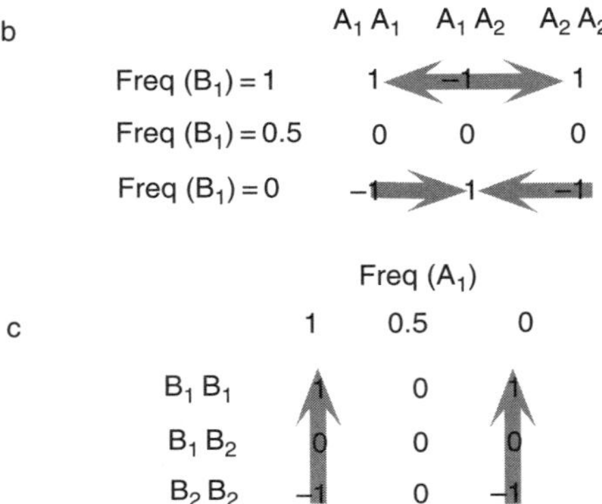

c

Fig. 9.11 The potential role of dominance by additive epistasis in speciation. (a) Genotypic values for dominance by additive epistasis. Gray arrows indicate the change from stabilizing selection to disruptive selection at the A locus that occurs as the frequency of the B_1 allele changes from zero to one. (b) Genotypic values for the three A locus genotypes when the frequency of the B_1 allele is 0, 0.5, and 1. When the B_1 allele is rare, there is stabilizing selection at the A locus, whereas when the B_1 allele is common, there is disruptive selection at the A locus. (c) Genotypic values for the three B locus genotypes when the frequency of the A_1 allele is 0, 0.5, and 1. At a frequency at the A locus of 0.5, the B locus is neutral. However, if frequency at the A locus drifts from 0.5, then the B locus will be under directional selection favoring the B_1 allele. When the B_1 allele is rare, genetic drift at the A locus will interact with directional selection at the B locus, eventually leading to fixation of the B_1 allele and either the A_1 or the A_2 allele. Redrawn from Goodnight (2000a).

selection. Similarly, at the end of the divergence process, the metapopulation will be fixed for the B_1 allele, and again the A locus will appear to be a simple underdominant locus with no evidence that it is part of an epistatic interaction. Thus, just by examining the end points of this process, there would be little indication that an epistatically interacting B locus was involved in the speciation process.

Finally, note that this process involves an interaction between the random process of genetic drift and the deterministic process of directional selection. Random drift begins the process (the B_1 allele is neutral at a gene frequency of 0.5 at the A locus), and directional selection enhances the power of drift and greatly accelerates the drive to the fixation of one of the two alleles at the A locus.

9.9 SUMMARY AND CONCLUSIONS

Fisher developed quantitative genetics to address specific questions about evolution within a single population. Indeed, quantitative genetics in its original formulation was concerned solely with describing the process of adaptation to directional selection acting within populations. The goals of metapopulation quantitative genetics are different from those of Fisherian quantitative genetics. Metapopulation quantitative genetics primarily provides measures of population differentiation rather than focusing on the response to selection that is central to Fisherian quantitative genetics. One of the central observations of metapopulation quantitative genetics is that populations can be differentiated both for population means and for local average effects. The first is well known from studies of genetic drift in additive systems. The second, while it has been observed qualitatively (e.g., Dempster, 1963), had not been quantified in the past. The differentiation of populations for local average effects is of particular interest because it is a measure of what alleles do in different demes. This form of population differentiation, unlike differentiation of population means, is directly related to reproductive isolation and speciation. However, it need not be related to the differentiation of population means. For example, additive effects cause differentiation of population means but no differentiation of local average effects, whereas at fixation ($F = 1$) there is no differentiation of population means due to dominance, but the populations are strongly differentiated for local average effects. Population genetics and Fisherian quantitative genetics have been remarkably successful for developing our understanding of evolution within populations, but interestingly, these disciplines have also been remarkably unsuccessful at developing our understanding of how these within population changes lead to speciation and evolution above the species level. By providing new measures of population differentiation, metapopulation quantitive genetics may shed light on this important subject.

Part IV

EVOLUTIONARY DYNAMICS IN METAPOPULATIONS

10. LIFE HISTORY EVOLUTION IN METAPOPULATIONS

Ophélie Ronce and Isabelle Olivieri

10.1 INTRODUCTION

Life history theory deals with the evolution of those traits that shape an organism's age schedules of birth and death (Calow, 1998). Many biological traits potentially affect the patterns of reproduction and mortality throughout the life cycle. Life history traits therefore constitute a loosely defined set of morphological, developmental, or behavioral characteristics, including, for instance, body size, growth patterns, size and age at maturity, reproductive effort, mating success, number, size, and sex of offspring, and rate of senescence. Despite this diversity of traits, by the early 1990s, life history evolution had grown successfully into a very productive field organized around a few central questions with a very strong unifying theoretical background, grounded in both optimization principles and quantitative genetics (Stearns, 1992; Roff, 1992). Then, the realm of most studies of life history theory was that of a single, large, undisturbed and spatially homogeneous population (see, however, Kawecki and Stearns, 1992; Kawecki, 1993). Through several examples, Olivieri and Gouyon (1997) illustrated how disequilibrium and the spatial structure characteristic of metapopulations might significantly affect the evolution of life history traits, a phenomenon they called "the metapopulation effect." Five years later, despite increasing awareness of the importance of metapopulation structure and dynamics for the demography, genetics, and

0-12-323448-4

conservation of many species, little is understood about how much these characteristics have shaped basic life histories. The present chapter reviews empirical and theoretical studies published since 1997 that have addressed the evolution of life history traits in a metapopulation context. We first comment generally about the development of this field of research since 1997.

First, studies of life history evolution in a metapopulation context remain rare as compared to the blooming of metapopulation demographic studies (see Chapters 4 and 5). Using a literature search engine (ISI Web of Science v04), a search with the combined key words "metapopulation" and "life history evolution" gave only 1 reference matching the query since 1997 compared to, respectively, 962 and 297 when searching with the key words "metapopulation" and "life-history evolution" alone. This result reflects, not so much the scientific production on the subject, but rather the fact that life history evolution in a metapopulation context is poorly identified as a distinct field of research. We restrict our review to empirical or theoretical studies considering intraspecific variation in life history traits in a landscape characterized by spatial structure, local extinction, and dispersal among patches of habitat. Many studies outside this range are related to the field of life history evolution in a metapopulation. For instance, many models of subdivided populations deal with the evolutionary consequences of spatial structure but do not take into account extinction–recolonization dynamics [see, e.g., Gandon (1999) and Pen (2000), respectively, for models of dispersal and reproductive effort evolution]. We refer to such studies when we feel that they point toward relevant and unexplored aspects of life history evolution in a metapopulation. Finally, life history evolution in a metapopulation context is closely related to some aspects of community dynamics [see e.g., Tilman et al., (1997) or the discussion about colonization–competition trade-offs in Chapter 6]. Testing predictions about how metapopulation dynamics affect selection on life histories might actually be achieved more easily by documenting changes in specific composition within a community rather than by studying genetic differences within a species. To limit the scope of the present chapter, we do not incorporate community-based studies in our review but we invite the reader to keep in mind the connection when reading Chapter 6.

Second, the field is largely dominated by theory, with very little empirical research due to obvious practical difficulties. Most empirical evidence of a metapopulation effect on life history evolution in natural systems is indirect. Metapopulation theory predictions have been tested by comparing mean phenotypes among populations that have been founded for different times (Cody and Overton, 1996; Piquot et al., 1998; Hill et al., 1999; Hanski et al., 2002) or by comparing the mean phenotypes among landscapes with different degrees of fragmentation (Thomas et al., 1998; Hill et al., 1999; Hanski et al., 2002). Whether those phenotypic differences are ultimately due to evolutionary change and not to environmental effects is still too rarely investigated (but see Thomas et al., 1998; Hill et al., 1999; Hanski et al., 2002). Artificial metapopulations of short-lived organisms in controlled conditions provide a fascinating opportunity to witness evolutionary change and test metapopulation theory predictions more accurately (Warren, 1996; Buckling et al., 2000), but such projects, though growing in numbers, are still in the process of development (Lavigne et al., 2001). How much artificial metapopulations inform us about the relevance of metapopulation theory for life history evolution in the real world is also open to question. The increasing imbalance between theoretical production and data collection is somehow worrying for the development of the discipline. Despite our efforts to integrate relevant empirical examples, the present review, with its strong focus on theory, reflects this bias.

The third point concerns the way the theoretical part of the field has developed. Life history theory specific to metapopulations has bloomed essentially around questions related to dispersal evolution (we counted more than 40 theoretical papers on dispersal evolution published since 1997). Because a whole chapter of the present volume is devoted to dispersal (Chapter 13, see also Chapter 16), we will not here review exhaustively models of dispersal evolution. Instead, we focus on those studies that help us to understand better how the evolution of dispersal contributes to an organism's general life history strategy. In particular, this is illustrated with studies of variation in dispersal strategies with age and the interaction of dispersal with other life history characters. Comparatively, the evolution, in a metapopulation context, of other more classical life history traits such as life span (Kirchner and Roy, 1999), age at maturity (de Jong et al., 2000), or reproductive effort (Ronce and Olivieri, 1997; Ronce et al., 2000c; Crowley and McLetchie, 2002) has received little attention to date. In particular, the evolution of senescence patterns or, more generally, of age-specific reproductive strategies, while major subjects of classical life history theory, are almost unexplored theoretical questions in the context of a metapopulation (with the exception of the evolution of delayed reproductive strategies such as dormancy and diapause; for a review, see Olivieri, 2001). We suggest reasons why these questions might constitute promising investigation areas. Life history traits other than dispersal also deserve more attention because they may, in some instances, be easier to measure empirically than dispersal and would thus allow more precise tests of the theory.

This chapter is organized by looking for common patterns explaining results obtained in different specific studies. Founding events and small local population size in a metapopulation are two causes of genetic resemblance among neighbors exploiting the same local environment. This chapter illustrates how this genetic structure makes life history evolution in a metapopulation deviate from that expected in a single large panmictic population. Changes in population age structure and density following disturbance and recolonization are major features of life in a metapopulation. Species whose biology is described most adequately using the metapopulation framework also often occur in habitats subject to successional changes. Such variations in selection pressures associated with colonization and succession have deep implications for life history evolution.

10.2 RESEMBLANCE BETWEEN NEIGHBORS

Fragmentation of the habitat is often associated with small population size in remnant patches of habitat. Patches of the now classic example of metapopulation, the Finnish populations of the butterfly *Melitaea cinxia* in the Åland Islands, contain at most a few sib families. Both genetic and demographic stochastic processes take an increasing importance in small populations.

In a system of small and poorly connected populations, genetic drift results in both the loss of genetic diversity within each local population and an increasing variance in allelic frequencies among populations. Similarly, a transient reduction in population size associated with disturbance and/or recolonization by a few founders can leave a significant signature in the genetic composition of populations long after population regrowth (Ingvarsson, 1998; Ingvarsson and Giles, 1999). Extinction–recolonization processes can either attenuate or accelerate the effect of drift on population differentiation, depending on the details of recolonization, dispersal, and the length of the period of transient growth following recolonization (Slatkin, 1977; Wade and McCauley, 1988; Whitlock and McCauley, 1990; Whitlock and Barton, 1997; Ingvarsson, 1997; Pannell and Charlesworth, 1999). How metapopulation dynamics affect the structuring of genetic diversity within and among populations is reviewed in Chapters 7, 8, and 9 (see also Pannell and Charlesworth, 2000). Both founder effects and subsequent genetic drift within local populations have the result that two individuals interacting in the same patch of habitat have a higher probability of sharing alleles than individuals belonging to different patches. Such a genetic structure holds major implications for the evolution of life history traits, which have been explored incompletely and are often neglected. This section illustrates those consequences through three examples concerned, respectively, with the evolution of dispersal, life span, and sex allocation strategies.

Dispersal

Multiple Causes for the Evolution of Dispersal

Dispersal is often considered in a metapopulation context as a risky behavior, compensated by the potential benefit of founding a new population in an empty patch of habitat. Such a view, adopted by early students of dispersal evolution (van Valen, 1971; Gadgil, 1971; Roff, 1975), appeared to be somehow too simple after Hamilton and May (1977) discovered that selection should favor frequent dispersal behavior in subdivided populations even in the absence of empty patches. Further theoretical work by Frank (1986) and Taylor (1988), in particular, allowed a better understanding and quantification of the forces selecting for dispersal in demographically stable but genetically structured populations. Dispersal in such a theoretical context can be seen as an altruistic act by which an individual risks its own fitness to alleviate kin competition within the natal patch. As with any altruistic act, such a behavior is favored as long as the individual fitness cost endured by the disperser is smaller than the inclusive fitness benefit of its departure for its kin. As resources freed by the departure of an individual are shared among all its neighbors, the inclusive fitness benefit will depend on its relatedness to other residents in the natal patch compared to its relatedness with the occupants of its new patch (Gandon and Rousset, 1999).

Dispersal is a complex character with multiple consequences, whose evolution is affected by multiple causes (Clobert et al., 2001). Kin competition avoidance and recolonization of empty patches are not mutually exclusive selective forces acting on dispersal evolution. However, theoretical studies addressing the evolution of dispersal often consider one force or the other as the major explanation for dispersal by either neglecting genetic drift (Olivieri et al., 1995; Holt and McPeek, 1996; Doebeli and Ruxton, 1997; Parvinen, 1999; Ronce et al., 2000b; Mathias et al., 2001; Parvinen, 2002; Kisdi, 2002) or ignoring complex metapopulation dynamics (Ezoe, 1998; Gandon, 1999; Gandon and Rousset, 1999; Hovestadt et al., 2001; Leturque and Rousset, 2002). Such decisions are often linked to trivial technical choices, such as modeling population numbers as a continuous rather than discrete variable or using deterministic rather than stochastic models [see Ronce et al. (2001) for further discussion on this topic]. Individual-based evolutionary simulation models necessarily incorporate the kin selection phenomena associated with drift and genetic structure, although this is not always clearly acknowledged (see, e.g., Travis and Dytham, 1998).

A Kin Selection Model for the Evolution of Dispersal in a Metapopulation

One might wonder about how kin selection and local extinctions interact in a metapopulation and about their relative importance in explaining patterns of dispersal in recurrently disturbed systems. Quite early on, Comins et al. (1980) incorporated the two forces in the same model. An analytical model by Gandon and Michalakis (1999), building on the work of Comins et al. (1980), helped clarify this question (see Table 10.1 for the main assumptions of their model). The evolutionarily stable (ES) dispersal rate, i.e., the fraction of individuals leaving their natal patch before reproduction, can be expressed as a simple function of the extinction frequency, the extramortality or "cost" associated with dispersal, the average relatedness among individuals born in the same patch, and the probability of common origin of immigrants. When immigrants in the same patch have a null probability of common origin (the migrant pool model, see Slatkin, 1977), the ES dispersal rate increases with higher extinction rates and higher within local population relatedness. This happens, respectively, because more empty patches are available for colonization and because kin competition is more intense for philopatric individuals, as was predicted by previous models that have considered kin competition phenomena or the extinction–recolonization dynamics separately. However, more complex patterns emerge due to the interaction of those two forces.

In particular, for a very low probability of surviving migration, the ES dispersal rate can increase with increasing dispersal cost (Fig. 10.1.A), whereas previous models predicted that dispersal should always decrease with increasing cost of dispersal (but see Comins et al., 1980). Gandon and Michalakis (1999) explained this unexpected pattern by a simple kin selection argument. As the dispersal mortality increases, a larger fraction of the individuals competing in the same patch are philopatric (because immigration is very low), which increases the probability of competing with related individuals in the natal patch. In a system with no empty patches (Frank, 1986), the lower inclusive fitness of philopatric individuals is compensated by the increasing difficulty of immigrating into extant populations. The empty patches created by local extinction, as in Gandon and Michalakis (1999), however, represent an extra benefit for dispersers, leading to increasing dispersal rates for very high dispersal costs.

Interactions between kin competition and metapopulation dynamics are also more complex because the average level of relatedness among individuals born

TABLE 10.1 Effect of Fragmentation on Dispersal Evolution: Comparison of Two Models and Empirical Evidence

Models	Landscape characteristics affected by fragmentation	Effect on relatedness within local population	Effect on the number of empty patches	Effect on ES dispersal rate
Gandon and Michalakis (1999) Spatially implicit Infinite island model of dispersal Catastrophic local extinction independent of population size	Dispersal mortality increases	Increases	None	Decreases, then increases
All patches recolonized before disturbance Local carrying capacity reached at foundation	Patch carrying capacity decreases	Increases	None	Increases
Heino and Hanski (2001) Spatially explicit Distance-dependent dispersal model based on focal butterfly species behavior Local extinction due to demographic stochasticity	Dispersal mortality per unit distance increases or distance between patches increases	Increases	Increases	Decreases then increases
Not all patches recolonized Local carrying capacity reached at foundation	Patch carrying capacity decreases	Increases	Increases	Increases

Empirical evidence	Measure for landscape fragmentation	Measured trait	Correlation
Thomas et al. (1998) *Plebejus argus*	Inversely related to heathland area (13 metapopulations)	Total mass Thorax/abdomen allocation	Positive None
Hill et al. (1999) *Hesperia comma*	Mean distance between patches (2 metapopulations)	Thorax/abdomen allocation	Positive
Hanski et al. (2002) *Melitea cinxia*	Connectivity (2 metapopulations)	Dispersal propensity	None

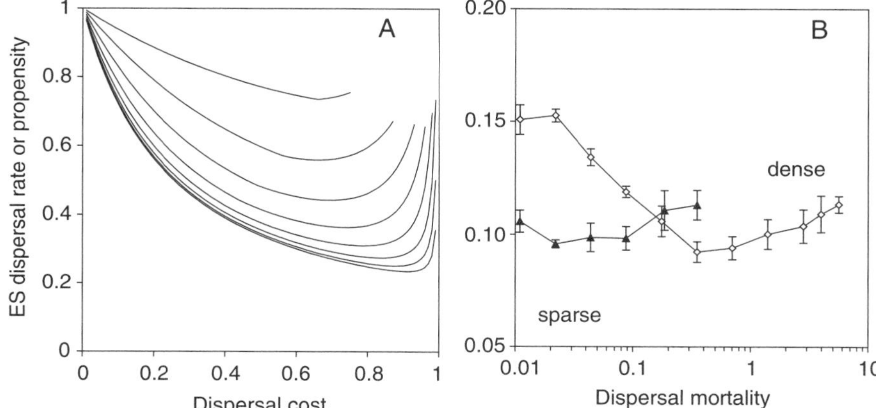

Fig. 10.1 Effect of dispersal cost on the evolution of dispersal. (A) ES dispersal rate as a function of dispersal cost as predicted by a spatially implicit kin selection model for different values of patch carrying capacity: from top to bottom $N = 1, 2, 4, 8, 16, 32, 64$, and 128; generated from the model by Gandon and Michalakis (1999), with different parameter values; here extinction rate $e = 0.2$ and fecundity $n = 5$. (B) ES dispersal propensity as a function of dispersal mortality per unit of distance as predicted by realistic spatially explicit simulations based on the biology of Finnish *Melitea* butterflies for two hypothetical patch networks characterized, respectively, by large interpatch distances (sparse) or short interpatch distances (dense). Redrawn from Heino and Hanski (2001); for original parameters, values see their Fig. 4A.

in the same patch depends on the intensity of founder effects and subsequent drift within local populations. Local population size, fecundity, extinction rate, dispersal cost, and the dispersal rate itself all affect indirectly the evolution of dispersal through their effect on relatedness. Consistent with what has been observed in subdivided populations with no extinction, the ES dispersal rate always decreases with increasing local population size (see Fig. 10.1A) because genetic drift is then weaker and within-local population relatedness decreases. A sensitivity analysis (Gandon and Michalakis, 2001) showed that population extinctions may have a higher qualitative or quantitative impact on dispersal evolution than kin competition or other forces, such as inbreeding avoidance. This is especially true of metapopulations made of local populations of large size, with weak founder effects. In situations with strong genetic spatial structure, kin selection phenomena are likely to play a considerable role in the evolution of patterns of dispersal.

Spatially Explicit Simulations for a Butterfly Metapopulation

Most of the qualitative conclusions of the simple theoretical study of Gandon and Michalakis (1999) have been reproduced by a much more realistic model of dispersal evolution (Heino and Hanski, 2001) based on the biology of Finnish metapopulations of *Melitaea cinxia* and *Melitaea diamina* (see Table 10.1 for a comparison of the assumptions and predictions of the two models). Each local population in these butterfly metapopulations is composed of few sib families, and mating occurs mostly within local populations before dispersal (for a detailed description of the system, see Hanski, 1999). The level of relatedness among individuals born in the same patch is thus likely to be high and kin competition avoidance may be a potent force in dispersal evolution in those species, together with the frequent local extinctions. Both the effect of migration mortality (Fig. 10.1B) and the patch carrying capacity on the evolution of dispersal propensity (see Table 10.1) can be interpreted with the same kin selection arguments as in the analytical model of Gandon and Michalakis (1999). However, an alternative explanation provided by Heino and Hanski (2001) is that, in their simulations, the number of empty patches increases when recolonization becomes more difficult because of higher dispersal cost or when demographic stochasticity increases because of a small local population size (see Table 10.1). More numerous empty patches represent more opportunity for colonization and favor the most dispersing genotypes. Contrary to analytical studies, it is difficult to disentangle the respective role of kin selection phenomena and extinction–recolonization dynamics in simulation results (see, however, the attempts to remove the genetic structure artificially in Heino and Hanski, 2001). Spatially realistic models, designed to fit the biology of a focal species, offer nonetheless the major advantage that they can be compared to data more readily and may help build a more precise test of the theory. The predicted average dispersal propensity for the well-studied *M. diamina* metapopulation was not significantly different from that estimated empirically, which is encouraging even though it does not provide a critical test of the theory. Predictions of the model concerning regional variation in dispersal propensity among different clusters of patches could, in the future, be compared to the observed variation in butterfly behavior among such clusters.

Evolutionary Consequences of Landscape Fragmentation

Predictions of the previous models have deep implications for our understanding of the effect of fragmentation on the evolution of life histories (see Table 10.1). For a long time the evolutionary effects of fragmentation were compared to syndromes of insularity. A rapid loss of dispersal ability of plants in island populations (Cody and Overton, 1996) was expected to also happen in fragmented habitats where the increasing cost of dispersal would select against dispersal. Previous models (Comins et al., 1980; Gandon and Michalakis, 1999; Heino and Hanski, 2001) show that the expected relationship between dispersal propensity and dispersal mortality or distance between patches is more complex than previously imagined (see also Leimar and Norberg, 1997). Another complicating factor is that fragmentation is also associated with habitat loss, and thus with a likely reduction in local population size: previous models showed that a reduction in local population size selects for increased dispersal. How do the effects of increasing dispersal cost and decreasing local population size combine in a fragmented landscape to act on the evolution of dispersal? A simulated habitat change in the *M. diamina* metapopulation (Heino and Hanski, 2001) led to complex predictions depending on the type of change (removal of whole patches, reduction in patch area, or quality). In some instances, the effect of decreasing local population size and increasing distance between patches counteract each other completely so that we would not expect any evolutionary change in dispersal in the new landscape.

Empirical evidence for a correlation between habitat fragmentation and dispersal propensity in natural systems is ambiguous (see Table 10.1 and also Chapter 20). Thomas et al. (1998) found that in *Plebejus argus* butterflies, total mass, potentially associated with flight ability, increased in the most fragmented heathlands. Comparing two metapopulations of the butterfly *Hesperia comma*, Hill et al. (1999) found that the relative size of the thorax was larger in the landscape characterized by larger distances between patches, which suggests that higher fragmentation is associated with higher dispersal. The fact that such patterns were observed in controlled conditions, together with strong family effects, suggests that morphological differences are based genetically and that these characters have indeed evolved (or could evolve in the future) in response to fragmentation. Whether selection on dispersal is the major explanation for such patterns is, however, difficult to establish. Hanski et al. (2002) found no obvious relationship between morphological measurements and estimated dispersal propensity for Glanville fritillary butterflies, *M. cinxia*. Contrary to Hill et al. (1999), they found no differences in migration propensity among metapopulations characterized by different degrees of connectivity.

Life Span

Evolution of Reproductive Effort in Genetically Structured Populations

Does the genetic structure generated by drift and founder effects affect the evolution of life history traits other than dispersal? A model of a subdivided population (Pen, 2000) suggests that the same processes are likely to alter the evolution of reproductive effort in a metapopulation. Pen's (2000) model does not incorporate extinction–recolonization dynamics (see Table 10.2). Adults are sessile and competitively superior to juveniles. They allocate their resources between reproductive and maintenance physiological functions, which generate a trade-off between fecundity and survival. In a genetically structured population, surviving adults prevent the recruitment of related juveniles in the same patch, whereas dispersing juveniles compete with unrelated individuals. A higher allocation to offspring production instead of survival can then be seen as another mechanism for kin competition avoidance. As a result, Pen (2000) found that the ES reproductive effort increases with the increasing level of relatedness among competitors in the same patch. We conjecture that results obtained by Pen (2000) could be generalized to other life cycles in that the existence of a strong genetic structure should favor the allocation of resources to life history stages with the highest dispersal propensity. Because of their indirect effect on within-local population relatedness, a lower dispersal of juveniles, higher dispersal cost, and smaller local population size all select for higher reproductive effort in his model. We reached similar conclusions (see Table 10.2), although for entirely different reasons, in a metapopulation model with local extinctions but no genetic structure (Ronce and Olivieri, 1997). Both models suggest that increasing fragmentation of the habitat would lead to the evolution of a higher reproductive effort and shorter life span for species in which adults disperse less than juveniles. Such a prediction should be checked by examining rigorously the interactions between kin selection and extinction–recolonization dynamics for the evolution of reproductive effort as Gandon and Michalakis (1999) did for dispersal. From an empirical point of view, it would be interesting to compare patterns of allocation to reproductive

TABLE 10.2 Theoretical Predictions for the Correlation Between Life Span and Juvenile Dispersal Rate[a]

Reference	Genetic structure	Extinction–recolonization	Trade-off between survival and fecundity	Evolving trait	Predicted correlation between life span and dispersal
Pen (2000)	Yes	No	Yes	Reproductive effort (RE)	+
Taylor and Irwin (2000)	Yes	No	No	Dispersal	+
Kirchner and Roy (1999)[b]	Yes	Yes No transient population growth after colonization	No	Life span	+
Ronce and Olivieri (1997)	No	Yes Transient population growth after colonization Dynamic number of colonizers	Yes	RE	+
Crowley and McLetchie (2002)	No	Yes Transient population growth after colonization Fixed number of colonizers	Yes	RE jointly with dispersal	–
Ronce *et al.* (2000c)	No	Yes Transient population growth after colonization Dynamic number of colonizers	Yes	RE jointly with dispersal	+/–

[a] In all models, adults are sessile and competitively superior to juveniles.

[b] This model involves interactions with sterilizing pathogens.

and maintenance structures in landscapes of different fragmentation levels, as was done for traits related to dispersal.

Evolution of Life Span in a Metapopulation with Sterilizing Parasites

With a different set of assumptions than in Pen (2000), Kirchner and Roy (1999) showed that resemblance between neighbors generated by founder effects could profoundly modify the evolution of life span in a metapopulation. Contrary to Pen (2000), they assumed no direct trade-off between fecundity and adult survival (see Table 10.2). Their evolutionary scenario involves interactions with sterilizing pathogens. In a large panmictic population, a long life span carries an epidemiological cost, as increasing adult survival increases the prevalence of the pathogen in the population. This happens because infected hosts remain in the population longer. Despite this cost expressed at the scale of the population, longer lived individuals still enjoy a higher fitness than short-lived hosts and alleles increasing longevity invade the population readily. The situation is different in a metapopulation with strong founder effects and little dispersal among extant local populations. Then, the variance in gene frequencies among local populations advantages the short-lived genotypes. When short-lived individuals have mainly short-lived neighbors, the local prevalence of pathogens is low and the average fecundity per individual is much higher than in local populations dominated by long-lived individuals where levels of infection and sterility are both high. Because of the higher offspring production of local populations with a high frequency of short-lived genotypes, new populations in empty patches are founded more often by such genotypes. Selection among local populations favors a short life span, whereas selection within local populations still advantages genotypes with a long life span. The frequency of short-lived individuals in the metapopulation depends critically on the rates of dispersal and local extinction. Note that, differently from other models that have explained the coexistence of different life history strategies in a metapopulation by a competition-colonization trade-off (Olivieri and Gouyon, 1997; Lehman and Tilman, 1997; Jansen and Mulder, 1999), the trade-off here emerges from the genetic structure of the metapopulation, as short-lived individuals indeed enjoy a higher fecundity only when they are surrounded by other short-lived individuals. Predictions of the model could be tested by comparing survival rates of uninfected individuals belonging to *Silene alba* metapopulations (see Chapter 19) differing in the prevalence of their sterilizing parasites *Ustilago violacea* or in their connectivity.

Life History Syndromes: Correlation between Life Span and Dispersal

Starting from Grime (1977), much effort in life history theory and ecology has been devoted to identifying patterns of covariation among species life history traits and relating these patterns to characteristics of their habitat (Southwood, 1988; Taylor et al., 1990; Silvertown and Franco, 1993; Westoby, 1998; Charnov, 2002). This led to the definition of syndromes, such as the syndrome of the "fugitive species" (Hutchinson, 1951), the "ruderal strategy" (Grime, 1977), or the "colonizer syndrome" (Baker and Stebbins, 1965), which associates high dispersal ability, high fecundity, and short life span as a set of coadapted traits symptomatic of unstable habitats. How does the genetic structure of metapopulation affect the evolution of such syndromes? Based on the two previous theoretical studies, one would predict a positive correlation between lifespan and dispersal ability, contrary to the colonizer syndrome (see Table 10.2). This happens because higher dispersal decreases genetic differentiation between local populations, which selects for a longer life span. Conversely, what are the consequences of increased life span for the evolution of dispersal? Despite the potential consequences for life history evolution, our understanding of how complex life cycles with several age classes affect the genetic structure in a metapopulation is still very incomplete (but see Rousset, 1999). Several models of subdivided populations, assuming that only juveniles disperse, have shown that increasing adult survival rates led to a higher level of relatedness among juveniles born in the same patch, favoring the evolution of increasing juvenile dispersal (Taylor and Irwin, 2000; Irwin and Taylor, 2000; Ronce et al., 2000a). They therefore predict again a positive association between life span and dispersal (see Table 10.2).

As such studies do not incorporate extinction–recolonisation dynamics, it is, however, difficult to conclude about the generality of the syndrome (see Section 10.3 for a further discussion of syndromes). Better predictions could also be achieved by letting both life span and dispersal evolve jointly in the same metapopulation. However, theoretical studies addressing this question do not take into account the genetic structure of the metapopulation (see Table 10.2 and Section 10.3). Artificial microcosms using mutant strains of *Caenorhabditis elegans* with modified dispersal and life history traits represent an exciting perspective for testing experimentally the role of kin structure and extinction–recolonization dynamics in the evolution of life history syndromes (Delattre and Felix, 2001; Friedenberg, 2003).

Variation of Allocation Strategies with Age

Iteroparity (i.e., the existence of several reproductive episodes within an organism's life span) can also lead to the evolution of offspring dispersal strategy varying with maternal age (Ronce et al., 1998, 2000a), in part because relatedness with other juveniles born in the same patch increases with the number of times their parent reproduced in the patch. We conjecture that variation in the intensity of kin competition could also alter the evolution of age-specific reproductive effort in such subdivided populations, and thus affect patterns of senescence. How those results generalize to metapopulations, with large disequilibrium in age structure generated by local extinction, is an open question (see Section 10.3 for predictions, not based on genetic resemblance, concerning age-specific allocation strategies).

Sex Allocation

Spatial Structure for Genes Involved in Sex Determination

Similarly, several theoretical studies have investigated how genetic structure and context-dependent fitness could affect the evolution of resource allocation between sexes in a metapopulation. Such studies focus particularly on the case of gynodioecy or of androdioecy, in which hermaphrodites coexist with, respectively, female or male individuals (Pannell, 1997a; McCauley and Taylor, 1997; Couvet et al., 1998; Pannell, 2000; McCauley et al., 2000). Gynodioecious species, such as *Thymus vulgaris* (Dommée et al., 1978) and *Beta maritima* (Boutin et al., 1988), and androdioecious species, such as *Datisca glomerata* (Liston et al., 1990), *Mercuria annua* (Pannell, 1997b), *Schizopepon bryoniafolius* (Akimoto et al., 1999), all occur in recurrently disturbed habitats. In gynodioecious species, gender is often determined by epistatic interactions between nuclear and cytoplasmic factors (Kaul, 1988; Frank, 1989; Charlesworth and Laporte, 1998). High levels of genetic differentiation among patches for both neutral cytoplasmic markers in linkage disequilibrium with cytoplasmic male sterility alleles (Cuguen et al., 1994; Manicacci et al., 1996; McCauley et al., 2000) and nuclear genes involved in sex determination (Manicacci et al., 1997) suggest that founding events and limited gene flow leave a strong signature in gynodioecious metapopulations, such as those of *B. maritima, T. vulgaris*, or *Silene vulgaris*. Sex ratio is, moreover, highly variable among populations in such species (Frank, 1989). Despite these common features shared by many well-studied gynodioecious systems, theoretical studies disagree about the consequences of genetic structure for the evolution of sex ratio in a metapopulation. Some models predict that the frequency of females would be lower in a metapopulation than expected in a large panmictic population (Pannell, 1997a; McCauley and Taylor, 1997; McCauley et al., 2000), whereas others reach the opposite conclusion (Couvet et al., 1998). We here suggest which specific assumptions may be responsible for those discrepancies and show how different models shed light on different mechanisms acting on sex ratio evolution in a genetically structured metapopulation. A list of the models' main assumptions and predictions is given in Table 10.3.

Local Sex Ratio Variation and Pollen Limitation

All four models focus on the evolutionary consequences of the recolonization of disturbed patches by a small number of individuals. Such founder effects generate a large variation in the local sex ratio (see Table 10.3). How does the spatial clustering of female and hermaphrodite genotypes affect the evolution of the sex ratio at the metapopulation scale? In *S. vulgaris*, both the pollination rate of females and the number of viable seeds per fruit born by hermaphrodites were shown to increase with the frequency of hermaprodites in the neighborhood (McCauley and Brock, 1998; McCauley et al., 2000), suggesting that seed production is indeed limited by the availability of outcross pollen in that species. Because of increased pollen competition, the siring success of an hermaphrodite plant is, however, correlated negatively with the local frequency of hermaphrodites (McCauley and Brock, 1998). Using frequency-dependent male and female fitness components as in *S. vulgaris*, McCauley and collaborators (2000; McCauley and Taylor, 1997) modeled the combined effects of pollen limitation and local sex ratio variation in a gynodioecious metapopulation of an annual plant where seed dispersal is global but pollen flow strictly localized. They concluded that the expected frequency of females in the metapopulation was lower than in a large panmictic population and declined with increasing variance in local sex ratio. A higher variance in the local sex ratio at foundation increases the spatial aggregation of females. Most of them are then found in patches with numerous other females and suffer from pollen

TABLE 10.3 Models for the Evolution of Gynodioecy

Reference	Inheritance of sex	Local extinction	Recolonization	Migration	Pollen limitation	Main result
McCauley and Taylor (1997) McCauley et al. (2000a)	Purely nuclear or purely cytoplasmic	All patches, every generation	Sampling variance at foundation	Global seed dispersal No pollen flow	Yes	Less females in a metapopulation
Pannell (1997)	Nuclear	Random Extinction of only female demes	Sampling variance at foundation Demes grow until carrying capacity is reached No drift in extant demes	Stochastic immigration of seeds in extant patches No pollen flow	No	Less females in a metapopulation
Couvet et al. (1998)	Nucleocytoplasmic	Random Extinction of only female demes	Sampling variance at foundation Demes grow until carrying capacity is reached No drift in extant demes	No gene flow after foundation	No	Generally more females in a metapopulation Polymorphism maintained at both cytoplasmic and nuclear loci

limitation. Similar arguments have been used to explain the predicted lower frequency of sex ratio distorters or feminizing parasites in genetically structured animal metapopulations as compared to panmictic populations (Hatcher et al., 2000). Note that this prediction goes in a direction opposite to the classic local mate competition hypothesis, which states that in the presence of a strong spatial genetic structure, higher sperm or pollen competition among related individuals should favor the evolution of female-biased sex ratios (but see de Jong et al., 2002). In the model of McCauley and Taylor (1997), polymorphism could not be maintained at equilibrium at both nuclear and cytoplasmic sex loci. The prediction of lower female frequency, however, held whether the determinism of sex was purely nuclear (McCauley and Taylor, 1997) or purely cytoplasmic (McCauley and Taylor, 1997; McCauley et al., 2000).

Reproductive Assurance and Recolonization

The same conclusion of a lower frequency of females in a gynodioecious metapopulation was reached by Pannell (1997a) in a model with no pollen limitation (see Table 10.3). Pannell (1997a), however, assumed that females fail to found new populations in the absence of hermaphrodites. Successful immigration of female genotypes in a patch can only occur after the arrival of hermaphrodite genotypes. As a result, the frequency of hermaphrodite genotypes in recently founded populations is higher than expected on the simple basis of their frequency in the migrant pool. Hermaphrodites benefit more than females from the relaxed competitive conditions and higher recruitment rates that prevail in recently founded local populations. Stochastic variation in the composition of founders in such a metapopulation therefore tends to favor cosexual hermaphrodites, at the expense of unisexuals, such as females in gynodioecious species, but also males in androdioecious species (Pannell, 1997a). Such an argument bears close connections to Baker's law and the reproductive assurance concept (Pannell and Barrett, 1998).

Intragenomic Conflicts and Founder Effects

Assumptions of Couvet et al. (1998) are very similar to those used by Pannell (1997a), as can be seen in Table 10.3. However, Couvet et al. (1998) reached strikingly different conclusions. For some parameter sets, the predicted female frequency in the metapopulation is lower than expected in a large panmictic population, but, for most of the explored parameter range, the reverse prediction holds. In particular, relatively high frequency of females can be maintained in a metapopulation for parameter values that do not allow the presence of females in a single panmictic population. Such discrepancies are ultimately due to the mode of sex inheritance in the two models. Pannell (1997a) assumed pure nuclear control of sex, whereas Couvet et al. (1998) considered the case where sex is determined by both nuclear and cytoplasmic loci. Contrary to McCauley and Taylor (1997), polymorphism at both types of loci was protected in the metapopulation for a large range of parameters. Why are assumptions about the genetic architecture of sex so important ?

In Couvet et al. (1998), founding events have qualitatively different consequences than envisioned in all previous models in this section. Stochastic variation in the identity of founders not only generates phenotypic correlations among neighbors, but also results in local variation in the mode of sex inheritance. Depending on the diversity of cytoplasmic and nuclear alleles borne by founders of a new population, sex can show no heritable variation locally, have a strict nuclear or cytoplasmic inheritance, or be determined by variation at both types of loci. In local populations with a strict cytoplasmic inheritance of sex, female frequency is predicted to reach a frequency close to 100% ultimately, as soon as females produce more seeds than hermaphrodites (Lewis, 1941). In contrast, in a population in which only nuclear alleles can modify sex expression, females will be progressively eliminated if they produce fewer than twice the number of seeds of an hermaphrodite and, whatever their fecundity advantage, will never exceed 50% of the local population at equilibrium (Lewis, 1941). Such a discrepancy is explained by the fact that cytoplasmic genes are usually transmitted through seeds only, whereas nuclear genes are transmitted through both pollen and seeds.

In Couvet et al. (1998), the initial variation in sex ratio generated by founding events is thus exaggerated by further selection within the established local populations. Because of their higher seed production, local populations with a sex ratio highly biased in favor of females grow faster than other local populations, consistent with the observed larger size of female-dominated patches in natural *T. vulgaris* populations (Manicacci et al., 1996). The combination of within-local population selection favoring a large frequency of females in local populations with a cytoplasmic inheritance of sex and the more dynamic growth of such local populations results in an overall increase in female abundance at the scale of the metapopulation. Note that founding events here affect the evolution of the metapopulation sex ratio not because female fitness is affected by their neighbors phenotype, but because founding events allow cytoplasmic male sterility alleles to escape the control of nuclear genes in some local populations.

Conclusion

Because these different studies have described different facets of the evolutionary consequences of founding events, it is difficult to combine their messages to appreciate their relevance for sex ratio evolution in real metapopulations. In gynodioecious metapopulations, observed female frequency is often much higher than expected in a single panmictic population (Couvet et al., 1990) and is still higher than predicted by the metapopulation model of Couvet et al. (1998) for the same female fecundity advantage. The observed decline in female abundance with time since foundation observed for *T. vulgaris* (Belhassen et al., 1990; see also Olivieri and Gouyon, 1997) is consistent with the predictions of Couvet et al. (1998), but not those of Pannell (1997a). Further theoretical and empirical work is needed to estimate the relative impact of frequency dependent fitness and variation in sex transmission mode for the evolution of sex ratio within metapopulations.

Genetic Resemblance: Conclusion

We find it useful to distinguish two types of genetic resemblance among neighbors: relatedness for genes directly affecting the expression of the trait of interest and relatedness for genes affecting selection on the trait only indirectly.

In the first case, the fact that an individual with a given life history trait is more likely to be surrounded by individuals with the same phenotype will modify selection on that trait whenever selection is frequency dependent. Examples reviewed here have shown that frequency-dependent selection can affect the evolution of a large variety of life history traits. When a life history character has a complex inheritance mode, as sex in some plants and animals, variation of the genetic composition of local populations can result in differences in the transmission of such character, with potentially important consequences for its evolution. Whether intragenomic conflicts affect the evolution of life history characters in addition to sex allocation is, however, open to question (but see the imprinting phenomena and parental conflicts about maternal investment during pregnancy in mammals; Hurst et al., 1996). We have not discussed here the evolutionary consequences of the second type of genetic resemblance, namely that concerning genes affecting fitness but not directly the expression of the life history trait of interest. In particular, the loss of genetic diversity associated with founding events and subsequent drift can result in the local or global fixation of deleterious mutations, which can have dramatic effects on population viability (Saccheri et al., 1998; Nieminen et al., 2001; Higgins and Lynch, 2001). Several theoretical (Fowler and Whitlock, 1999; Whitlock et al., 2000; Whitlock, 2002; Couvet, 2002) and empirical (Haag et al., 2002; Groom and Preuninger, 2000) studies have investigated how metapopulation functioning affects mutational load and inbreeding depression. Very little is known about how inbreeding depression and heterosis indirectly affect the evolution of life history traits in a metapopulation. Gandon (1999) studied how kin competition avoidance and inbreeding avoidance interact to influence the evolution of dispersal in a subdivided population with no local extinction (see also Perrin and Mazalov, 2000). Could similarly the evolution of higher reproductive effort or age-specific reproductive strategies be understood as inbreeding avoidance mechanisms?

10.3 CHANGING LIVING CONDITIONS

The recognition of the changing and ephemeral nature of life is deeply rooted in the metapopulation concept, which acknowledges that, just as individuals, populations do not persist forever. Evolution of many life history traits, such as dispersal, dormancy (Venable and Brown, 1988; Rees, 1994), iteroparity (Rees, 1994), clutch size (Orzack and Tuljapurkar, 2001), or age at maturity (Lytle, 2001) can be understood as adaptations to this fundamental uncertainty. Bet-hedging strategies diminish the risks of genotype extinction by spreading reproduction over several years or several sites. This section focuses on the evolutionary consequences of a different type of variability in a metapopulation: we are interested in changes associated with return to the equilibrium condition within disturbed populations. Founding events not only leave a signal in the genetic composition, but also in the demographic structure of a population. It may be useful to consider the fact that local populations have a history, characterized by a transient period following recolonization, where density, age structure, and genetic diversity change with time, and a quasistationary period where those population variables are

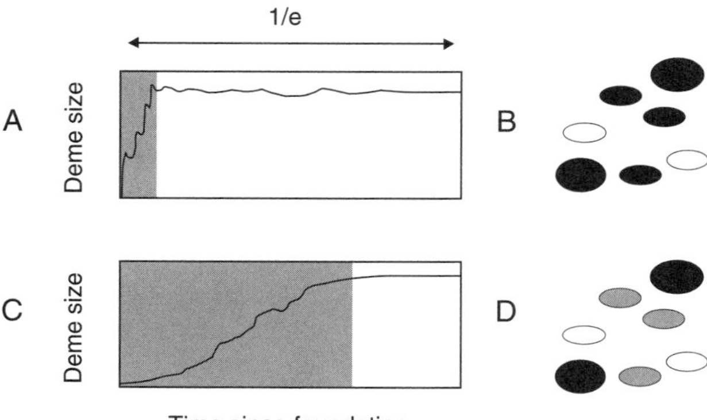

Fig. 10.2 Temporal and spatial variability in population size for hypothetical "fast" (A and B) and "slow" (C and D) species. Metapopulations of the two species are characterized by the same expected persistence time for local populations, equal to $1/e$ if e is the per time probability of extinction. Demes of the fast species go through a short period of transient growth (A: hatched area) but spend most of their lifetime in a quasistationary state; as a result, most of the occupied patches (in black) in the metapopulation at equilibrium are at this stationary density (B). Demes of the slow species spend a large amount of time in the transient phase (C); such temporal dynamics translates into a large amount of spatial variability in density in the metapopulation at equilibrium (D). Density now varies not only among empty (white) and occupied patches (black or hatched), but differs among occupied patches depending on time since their colonization (hatched for demes still growing, black for demes at their stationary density).

approximately constant (see Fig. 10.2). Note that this stationary state will be reached at different times for the different parameters (in particular, genetic equilibrium will be reached long after demographic equilibrium). If local extinctions and recolonization events are not synchronized perfectly among different patches, such temporal changes within populations translate into increased spatial heterogeneity at the scale of the metapopulation (see Fig. 10.2). Populations in different patches of habitat then differ not only on the basis of intrinsic differences in patch quality or stochastic factors, but also because different amounts of time have elapsed since their foundation.

Theoretical studies, starting with the classical Levins (1969) metapopulation model, often assume that the transient period in population dynamics is relatively short as compared to the mean local population life span (see, e.g., Gandon and Michalakis, 1999). Such an assumption fits well the biology of some empirically studied metapopulations (Heino and Hanski, 2001). In that case, we can safely ignore the heterogeneity among patches recolonized at different times and assume that local populations in the landscape have all reached some demographic stationary state (Drechsler and Wissel, 1997). One may, however, wonder about the generality of this assumption: do the majority of species living in a metapopulation spend most of their time in populations at equilibrium or in populations still recovering from the last disturbance? Do metapopulation dynamics select for life history traits that allow the transient period following recolonization to be short? Or is the very concept of stationary demographic state a mere abstraction, a concept difficult to reconcile with the ever changing demographic conditions that prevail, for instance, in successional systems? Answers to those questions are still unclear.

Through several theoretical and empirical examples, this section illustrates how periods of transient dynamics or, more generally, varying ecological conditions with time since foundation can affect the evolution of life histories in a metapopulation. In general, spatial or temporal heterogeneity in selection has three possible consequences: (i) it may facilitate the coexistence of genotypes with different life history strategies in the same metapopulation, (ii) it may lead to the evolution of strategies that represent a compromise between conflicting selection pressures, or (iii) it may promote the evolution of plastic genotypes, which can express different life history strategies depending on their location or at different times in their life. Such consequences are illustrated through the example of the evolution of sex allocation (i), age at maturity (i), reproductive effort (i and ii), and dispersal (ii and iii).

Coexistence

Sex Allocation and Inbreeding

Pannell (2000) suggested that a change in inbreeding levels within local populations following recolonization could allow the invasion of males in a hermaphrodite metapopulation. Androdioecy has always been a puzzle for evolutionary biologists (see also Section 10.2). Because males transmit their genes only through pollen, they must produce and disperse more than twice the quantity of pollen produced by hermaphrodites to be maintained in a population [Charlesworth and Charlesworth (1978), but see Vassiliadis et al., (2000) for an exception]. This condition is less difficult to achieve if hermaphrodites do not allocate their resources fairly between male and female functions and invest more in ovule production. In large populations, hermaphrodites are expected to have a female-biased investment only if they self a large fraction of their ovules. Invasion of male phenotypes in a hermaphrodite population is, however, made more difficult by increasing the level of selfing in such hermaphrodites (Charlesworth, 1984). These paradoxical requirements could explain why androdioecy is so rare. Pannell (2000) claimed that metapopulation dynamics can generate the situation with a high female investment and low selfing rate in hermaphrodites necessary for invasion by males. His argument involves the change in selfing rate that goes along with the change in density following recolonization. Delayed selfing consists in various mechanisms favoring outcrossing in dense populations but allowing self-pollination in the absence of outcross pollen. Such a strategy is thought to be particularly adaptive in metapopulations with recurrent bottlenecks that severely limit the availability of pollen donors as well as that of pollinators. Strong inbreeding in generations following recolonization tends to favor the evolution of increased female allocation in hermaphrodites, whereas the low selfing rate in denser and older populations allows males to invade in those populations. Such a scenario is consistent with the fact that known androdioecious species, both plants and animals, are frequently found in recurrently disturbed habitats, have highly variable sex ratios, and are known or suspected to vary in selfing rates in a density-dependent way (see the review in Pannell, 2000). The verbal model of Pannell (2000) does not allow us to conclude rigorously that coexistence of males with hermaphrodites in such a metapopulation is stable, but suggests that a change in inbreeding levels during transient stages of population dynamics may create some evolutionary window favorable to the establishment of different life history strategies.

Age at Reproduction

Changes in population density during the transient period of growth following recolonization can also affect the evolution of life history strategies because it affects the probability of successful recruitment of juveniles born in populations established for different times. The evolution of age at reproduction in the perennial monocarpic *Carlina vulgaris* provides an illustration (de Jong et al., 2000). In semelparous organisms, age at reproduction represents a compromise between the benefits of delaying reproduction to attain a greater size and thus higher fecundity and the risk of dying before reproduction. Single population models based on this basic trade-off, however, largely overpredict the age of first reproduction observed in natural populations of *C. vulgaris*. In the system studied by de Jong et al. (2000), the rare thistles grow in ephemeral patches at the edge of willow scrub. Local populations are founded by seeds, grow to reach a peak density, and then decline to extinction (Klinkhamer et al., 1996). Recruitment of seeds is highly variable among patches, being high in the period of population growth but very low in the period of decline. Rosette survival is not affected by density, but varies widely between patches within a year. Peaks of density are asynchronized among patches and are probably related to vegetation succession.

de Jong et al. (2000) incorporated these features in a metapopulation model with local extinctions and limited dispersal between patches. The predicted age at maturity for *C. vulgaris* was then closer to that observed in natural populations than the age predicted by the single population model for the same rosette mortality. In some instances, they found that two genotypes with different ages at reproduction could coexist in the same metapopulation. In the period of growth following colonization, plants reproducing early are favored because their progeny benefits from more benign competitive conditions than progeny born later in denser populations. Once safe sites have become scarce, it, however, pays to delay reproduction to reach a higher fecundity and the late-reproducing genotypes are advantaged.

Reproductive Effort

The latter results bear close resemblance to what was obtained about the evolution of reproductive effort in a metapopulation (Ronce and Olivieri, 1997). We indeed similarly found situations where two genotypes with different reproductive effort strategies could coexist in a metapopulation with a transient period of growth, but neither in an unstructured population nor in a metapopulation where the stationary density would be reached immediately upon recolonization. Within each population, genotypes with a higher reproductive effort are selected for during the period of growth because they occupy space more rapidly, but they are replaced progressively by genotypes with a higher investment in survival as soon as juvenile

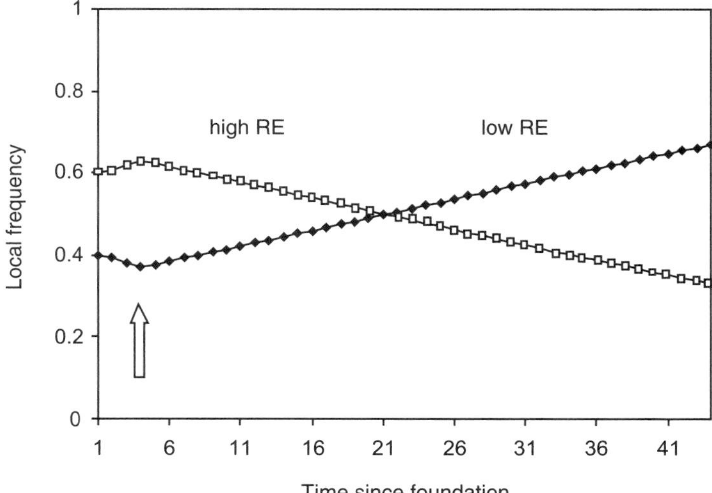

Fig. 10.3 Frequency of two genotypes with different reproductive efforts in populations established for different times. In white, the genotype with the larger reproductive effort (49% of resources allocated to reproduction); in black, the genotype with the lower reproductive effort (41% of resources allocated to reproduction). The arrow indicates the time at which the stationary ceiling density is reached within local populations. Generated from the same model as in Ronce and Olivieri (1997), with different parameters as in their Fig. 9.

recruitment becomes severely limited by adult survival (Fig. 10.3). Such a prediction is consistent with the higher frequency in recently founded populations of genotypes investing in sexual reproduction rather than asexual growth as observed in the macrophyte *Sparganium erectum* along a successional gradient (Piquot et al., 1998).

Cases of coexistence mediated by a competition–colonization trade-off (see Olivieri and Gouyon, 1997) often involve conflicting selection pressures at the level of the metapopulation and at the level of each population [see Olivieri et al. (1995) for dispersal polymorphism and Jansen and Mulder (1999) for reproductive effort]. The present example (Ronce and Olivieri, 1997), together with that of De Jong et al. (2000) about age at maturity, suggests that the succession of antagonistic selection pressures with time within each local population may broaden the range of conditions allowing the maintenance of protected polymorphism in a metapopulation. Empirical evidence on side-blotched lizards, *Uta stansburiana*, similarly suggest that temporal changes in density help maintain polymorphism for clutch size strategies (Sinervo et al., 2000). Orange-throated females producing large clutches of small offspring are favored during the period of population growth, whereas yellow-throated females producing fewer offspring of higher quality are favored at high densities. The previous theoretical models also lead us to think differently about colonization: successful colonization is a complex process, which implies not only arriving in empty patches of habitat, but also an efficient strategy of space occupation once arrived [see also Bolker and Pacala (1999) on the same topic].

Evolutionary Compromises

Length of the Growth Period: Effect of Productivity

Classic life history theory has for long shown that an increase in productivity, such that fecundity increases, leaving survival unchanged, should select for a higher reproductive effort in an exponentially growing population (Charnov and Shaffer, 1973). In a density-regulated population where juveniles establish only in safe microsites freed by the death of an adult, increasing productivity has, however, no effect on the evolution of reproductive effort (see Kisdi and Meszena, 1995) because the number of recruited offspring is then limited by adult mortality and not by productivity. In a metapopulation where populations go through a period of transient growth after recolonization, the evolutionary pattern is strikingly different from that predicted in a single population. The ES reproductive effort increases but then decreases with increasing productivity [Fig. 10.4B, and similar pattern in Ronce and Olivieri (1997) and Ronce et al., (2000c), but see Crowley and McLetchie, (2002), for a different conclusion]. Such a nonmonotonic response reflects the conflict in selection pressures in the metapopulation. Within each growing local population, increasing productivity selects for increasing reproductive effort, but the period of growth during which higher reproductive efforts are favored also becomes shorter as local populations grow faster and reach their stationary state sooner (Fig. 10.4A). At any point of time, the fraction of local populations in the landscape still in the transient phase of growth declines with increasing productivity. Lower reproductive efforts are selected for at the scale of the metapopulation. This example illustrates how changes in demographic parameters affect the evolution of life history traits, not only by changing selection pressures within local populations, but also by changing the very composition of the landscape.

Longer periods of transient growth tend to select for lower dispersal rates in the metapopulation (Olivieri et al., 1995, Ronce et al., 2000b,c; see also Crowley and McLetchie, 2002). In particular, ES dispersal rates decrease with decreasing productivity (see Fig. 10.4B, and also Fig. 6 in Crowley and McLetchie, 2002). The availability of safe sites within recently founded, low-density local populations indeed makes the venture of risking death to reach an empty patch of habitat less worthwhile compared to a situation where almost all occupied patches are fully crowded (Fig. 10.4A). Ellner and Schmida (1981) used a similar argument to explain the rarity of long-range dispersal adaptations in desert floras. Consistently, Wilson et al. (1990) showed that the relative frequency of plant species whose seeds are dispersed by wind or vertebrates increased along a fertility gradient in Australian forests, whereas the frequency of species with ant-dispersed seeds or no special dispersal device decreased.

Coupling between Local and Regional Dynamics: Dispersal and Extinctions

Dispersal has been perceived as an adaptation to habitat instability. The frequency of winged species in insect communities (Southwood, 1962; Denno, 1994) or of winged morphs in populations of the same species (Roff, 1994; Denno et al., 1996) decreases with increasing stability of the habitat. Most

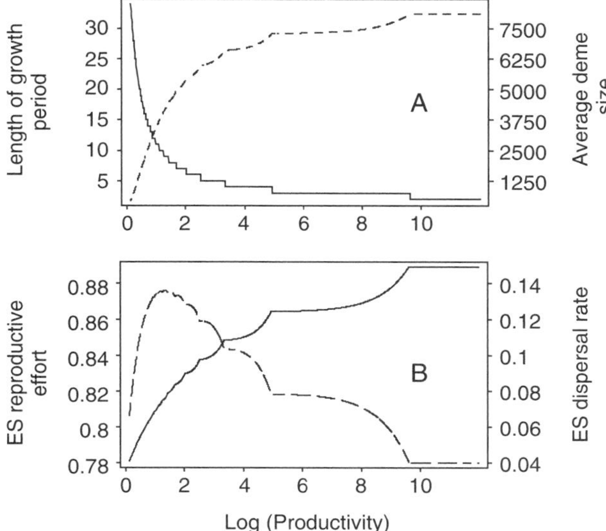

Fig. 10.4 Effect of productivity on the joint evolution of reproductive effort and dispersal in a metapopulation with transient periods of growth after foundation. The model generating the predictions is similar to that of Ronce et al. (2000c) except for two assumptions: here the number of founders is fixed ($k = 1$) and there is no immigration in extant populations (as in Crowley and McLetchie, 2001). Local demes grow exponentially until they reach a ceiling density. The maximal size of local demes is $K = 10^4$, the probability of extinction per generation is $e = 0.1$, the minimal mortality rate per generation is $e^{-\delta}$, with $\delta = 2 - \sqrt{3}$; the maximal fecundity achievable by semelparous individuals in the environment is taken as a measure of productivity. (A) Effect of increasing productivity on population dynamics descriptors: the length (in generations) of the growing period or age at which the ceiling density is reached (continuous) and average density in occupied patches in the metapopulation at equilibrium (dashed). (B) Effect of productivity on evolutionary dynamics: ES reproductive effort, measured as the ratio of the fecundity over the maximal fecundity (dashed), ES dispersal rate (continuous). The discontinuities in curves are due to (i) the ceiling mode of population regulation and (ii) the time discreteness of the model. Note that patterns of evolution in reproductive effort and dispersal correspond to changes in average within-patch density. The assumption of a fixed number of colonizers and absence of immigration does not change qualitatively the patterns of evolution when compared to Fig. 2 in Ronce et al. (2000c).

models predict that ES dispersal rates should increase monotonically with an increasing frequency of local extinction (see, e.g., Gandon and Michalakis, 1999). Such models generally assume that the period of transient growth following recolonization is so short that it can be neglected altogether. By relaxing this assumption, two deterministic models (Ronce et al., 2000b; Parvinen et al., 2003) found that the ES dispersal rate may vary nonmonotonically with the frequency of catastrophic extinctions. On the one hand, more frequent disturbances create more empty patches in the landscape and thus more colonizing opportunities, which selects for increasing dispersal ability. On the other hand, disturbances also affect local dynamics indirectly by reducing the total metapopulation size and thus the average number of immigrants arriving in any given patch. As a result, local populations start off smaller and grow longer before reaching some equilibrium density. The longer growth period and higher frequency of low-density local populations tend to select against frequent dispersal. The conflict between the effects of disturbance at the two spatial scales results in a hump-shaped pattern of dispersal evolution (Ronce et al., 2000b). Note that it is not so much the inclusion of a transient phase of local population growth that created this pattern, but the coupling between local and global dynamics in the metapopulation. For instance, Crowley and McLetchie (2002) assumed a fixed number of founders per new population and no subsequent immigration (see Table 10.2): they found that ES dispersal rates always increased with increasing frequency of local extinction because increasing the fraction of empty patches in the landscape then had no effect on the local dynamics.

It has, however, been questioned whether the decreasing part of the hump-shaped curve relating dispersal to local extinctions could be observed in nature, the most serious objection being that in the presence of demographic stochasticity, metapopulations may not persist sufficiently long in this high disturbance regime to observe the evolution of declining dispersal rates (Heino and Hanski, 2001; Poethke et al., 2003). Our deterministic model (Ronce et al., 2000b) assumes that the global landscape composition affects the local dynamics, but does not incorporate the reverse feedback (but see Parvinen et al., 2003). This might not describe accurately the more general situation where environmental fluctuations combine with demographic stochasticity to lead to extinction (see Heino and Hanski, 2001). Poethke et al. (2003) showed that, in the latter situation, the pattern of covariation between dispersal rates and extinction rates is much more complex than previously envisioned. This happens because the evolving dispersal rate itself then affects the probability of extinction. In particular, increasing environmental variability may select for larger dispersal rates, which in turn reduces the local extinction rate because of the rescue effect (Brown and Kodric-Brown, 1977). Dispersal and local extinctions, both dynamic parameters, may then covary positively or negatively along gradients of environmental variability or dispersal mortality. Poethke et al. (2003) suggested that the two latter parameters would be better predictors of the ES dispersal rates than the frequency of local extinction.

Habitat Templates and Life History Syndromes

Taking into consideration transient population dynamics deeply affects our understanding of life history syndromes. In particular, we showed that predicted patterns of association between reproductive effort and dispersal were strikingly different in "fast species" where local populations reach their equilibrium density immediately upon recolonization and in "slow species" where the length of the growth period varies dynamically with landscape and life history characteristics (Ronce et al., 2000c). First, both changes in dispersal and reproductive effort can affect population dynamics and the length of the transient phase or, more generally, the distribution of population densities in the landscape. Evolution of one trait may thus modify the selection pressures on the other indirectly through changes in demographic dynamics, just as they do through changes in the genetic structure of the metapopulation (see Section 10.2). Two models (Ronce et al., 2000c; Crowley and McLetchie, 2002), however, suggest that such evolutionary interactions may be of minor importance

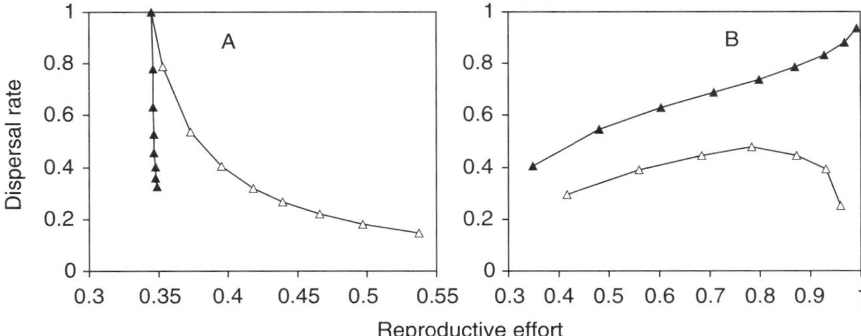

Fig. 10.5 Covariation among dispersal and reproductive effort evolving jointly along environmental gradients. (A) Along a gradient of increasing habitat fragmentation in a fast species (in black) where ceiling density is reached immediately upon recolonization and in a slow species (in white) with variable periods of transient growth; the dispersal cost varies from 0 to 0.9. (B) Along a gradient of increasing disturbance in a fast species (in black) and in a slow species (in white); the local extinction rate varies from 0.1 to 0.9 or 0.6 for, respectively, the fast species or the slow species. Parameters and results for the ES life history traits are as in Fig. 3 and 4 in Ronce et al. (2000c).

(see Table 10.2). Second, because both traits respond in opposite directions to change in the length of the transient period, we may often expect negative correlations between reproductive effort and dispersal along environmental gradients (see Fig. 10.5), contrary to the classical colonizer syndrome (for further discussion see Ronce et al., 2000c). Again, a clearer understanding of such life history syndromes may be achieved by acknowledging the fact that a change in the landscape structure or quality will affect the distribution of population numbers at two different spatial scales. The fraction of occupied patches in the landscape and the fraction of occupied space within occupied patches can be considered two measures of the degree of resource monopolization in the metapopulation, reflecting, respectively, the intensity of global or local competition. Unexpected life history syndromes can emerge because the evolution of long-distance dispersal, contrary to reproductive effort, responds differently to change in the intensity of competition at the global and local scales. While decreased competition at the global scale favors increased allocation to reproduction and dispersal, decreased competition locally selects for more reproduction but less dispersal.

Plasticity

Density-Dependent Life Histories

If the benefits of dispersal vary with local density, as suggested by previous theoretical work, it is quite natural to expect the evolution of dispersal strategies conditional on local density. Theoretical predictions and empirical evidence for density-dependent dispersal strategies are reviewed in detail in Chapter 13. While attention to density-dependent dispersal in a metapopulation has certainly increased in recent years (see, e.g., theoretical work by Jánosi and Scheuring, 1997; Travis et al., 1999; Johst and Brandl, 1997; Metz and Gyllenberg, 2001; Poethke and Hovestadt, 2002), the same conclusion does not hold about the evolution of other plastic life history characteristics. Yet given the conflicting selection pressures acting on reproductive effort or age at maturity in growing and crowded populations, one might expect similarly that such life history decisions may depend on the density of neighbors. Quintana-Ascencio et al. (1998) found that the reproductive effort of transplanted *Hypericum cumilcola* changed with time since fire and the density of conspecifics in patches of Florida rosemary scrub. Variation in mean life histories with population age observed in natural populations is often consistent both with the succession through time of genotypes with different allocation strategies and with the expression of plastic allocation strategies (see, e.g., Houssard and Escarré, 1995).

Time since Foundation and Individual Age

Many population characteristics other than density might change with time since foundation, which consequences for the evolution of conditional dispersal behaviors or more generally conditional life history strategies have been little explored. Drastic changes in the age structure may occur during the transient phase following recolonization, especially in those species where the age structure of migrants does not reflect the stable age structure of local populations (see, e.g., Joly and Grolet, 1996). If competitive abilities or, more generally, demographic parameters, such as fecundities or survival rates, vary among age classes, the disequilibrium in age structure will generate a large variation in recruitment probabilities or expected reproductive success among populations founded for different times, even if they have the same density. Patterns of genetic diversity may change with population age (Whitlock and McCauley, 1990), affecting the level of within local population relatedness compared to genetic resemblance to individuals from other local populations. One might then expect the evolution of variable kin competition avoidance depending on population age. We are, however, aware of no study trying to relate patterns of change through time in the genetic structure of populations to the expression of altruistic behaviors. The probability of extinction of a local population may also depend on the time elapsed since foundation, either through a change in population size and demographic stochasticity or because of negative or positive temporal autocorrelation in disturbance events. In many species either exploiting an ephemeral and nonrenewable resource or subject to successional replacement, the very quality of the habitat deteriorates with time (Valderde and Silvertown, 1998). This led Olivieri and Gouyon (1997) to predict that ES conditional dispersal in a metapopulation should increase with population age. A marginal value argument like that used by Metz and Gyllenberg (2001) and Poethke and Hovestadt (2002) when studying density-dependent dispersal enabled us to predict the ES reaction norm for dispersal as a function of local population age for a species subject to successional replacement, with no constraint on the shape of the reaction norm (O. Ronce, S. Brachet, I. Olivieri, P.-H. Gouyon and J. Clobert, unpublished result). Our analysis does not incorporate kin selection effects. In most cases, we verified the general prediction of Olivieri and Gouyon (1997) of an increasing dispersal rate with time since foundation. In details, the patterns of variation in optimal dispersal rates can be more complex than envisioned in their simple study, with several outbreaks of dispersal during the population lifetime (Fig. 10.6.A). Such oscillations occur when equilibrium density is reached before the stable age structure: the peaks match the oscillations in age structure during the transient period (Fig. 10.6A). A consequence of such conditional dispersal strategy is that high rates of emigration in old populations accelerate the rate of successional replacement and shorten the life span of the population. This population level pattern interestingly parallels what happens in the evolution of senescence at the individual scale.

These highly theoretical considerations do not help us understand how individuals may acquire information about the age of their population and adjust their dispersal strategy accordingly. A candidate cue for population age could be the age of the individual itself in a long-lived species in which populations take a long time before reaching their stable age structure. We indeed showed that in the presence of strong variance in age structure among populations founded for different times, juvenile dispersal strategies conditional on

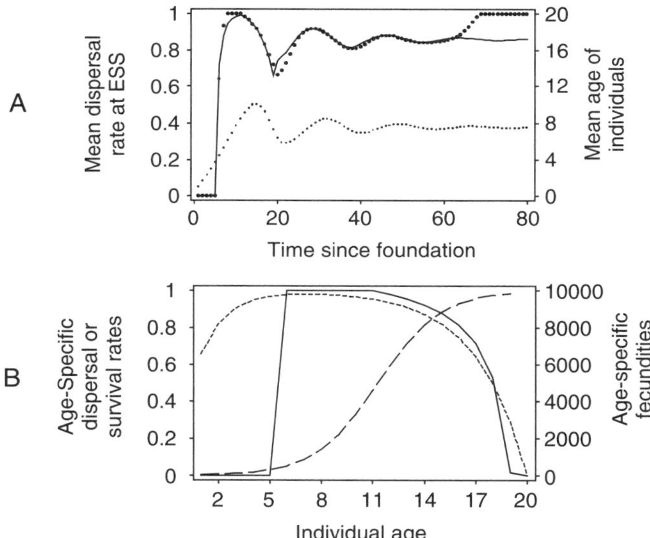

Fig. 10.6 Evolution of conditional dispersal strategies conditional on individual age or deme age. (A) Variation of the mean dispersal rate per local deme as a function of time since foundation for plastic ES dispersal strategies. Large dots: ES dispersal strategy conditional on deme age when individuals have access to complete information about time since foundation. Small dots: mean age of individuals per deme as a function of time since foundation. Continuous line: mean dispersal rate per local deme when dispersal varies with individual age (see B) and individuals have no information about time since foundation. (B) Variation of demographic parameters as a function of individual age. Long dashes: age-specific fecundity. Short dashes: age-specific survival rate. Continuous line: age-specific dispersal rate when individuals do not have access to information about time since foundation (generating the continuous line in A). Demes persist for 80 yr with a probability of local extinction of 0.1 per year. After 80 yr, local demes go extinct and remain uncolonizable until the arrival of the next disturbance (which occurs with a probability of 0.01) to mimic the effect of successional replacement. All demes have a constant size of 200 adults. Only juveniles disperse and establish either in empty patches or within occupied patches in microsites freed by the death of adults. Seventy percent of the dispersing juveniles die during migration.

maternal age could be selected for in a metapopulation (Fig. 10.6.B; O. Ronce, S. Brachet, I. Olivieri, P.-H. Gouyon and J. Clobert, unpublished result). A variation in mean dispersal rates among populations due to the combination of variable age structure and age-specific dispersal rates results in a pattern mimicking the optimal reaction norm that would evolve if individuals had complete information about the time of foundation of their population (Fig. 10.6.A). This example illustrates how metapopulation processes may alter the evolution of age-specific life history strategies. Similar processes might affect the evolution of age-specific reproductive effort or rates of senescence. How much of the classic life history theory would be modified in a metapopulation with a variable age structure?

Change in Living Conditions: Conclusion

Previous authors have found it useful to distinguish between selection pressures acting at the local and global level to understand the evolution of life history traits in a metapopulation (Olivieri and Gouyon, 1997). We have extended this view by showing through several examples how a variation of selection pressures within local populations, ultimately due to the disequilibrium generated by local extinction, could affect the evolution of life history traits. The same theoretical examples suggest that the evolutionary importance of such transient dynamics may depend on the coupling between landscape processes, such as extinction and recolonization, and local growth. Immigration and recolonization of empty patches of habitat are the two events that connect the dynamics within each local population to the regional state of the metapopulation. Although the demographic consequences of this coupling have started to be explored both theoretically and empirically (see, e.g., the relationship between species abundance and distribution as discussed in Hanski, 1999), little is known about its impact on life history evolution. We have presented a few examples where this coupling results in initially counterintuitive patterns of evolution. However, much achievement in metapopulation theory has been realized by uncoupling the local and regional dynamics, relying on the timescale difference between the two processes (Drechsler and Wissel, 1997; see also Chapter 4). Theory based on such assumptions has often led to remarkably robust conclusions (Keeling, 2002). Some authors have also argued that this very uncoupling might define what is a "real" or "classic" metapopulation rather than a simple patchy population (Harrison and Taylor, 1997). We do not share this point of view. Rather, we think that details of the focal species biology determine the degree of coupling between processes at different spatial scales and that any feedback between local and global dynamics should be neglected only with great caution. This is especially true of modeling exercises. For instance, Crowley and McLetchie (2002) consider that the average number of founders per new population is independent of the global demographic state of the metapopulation, while we assume that the number of empty patches in the landscape is independent of the distribution of local population sizes (Ronce et al., 2000b,c). More realistically, both parameters vary dynamically with the number of migrants in the metapopulation: Both models therefore miss part of the real picture. Data are sorely needed to assess, in a diversity of organisms, the extent and importance of the transient phase of population dynamics and the degree of coupling between regional and local dynamics.

10.4 GENERAL CONCLUSION

Perspectives for Theoretical Studies of Life History Evolution in a Metapopulation

The two parts of this chapter unfortunately reflect a major division in theoretical studies of life history evolution in a metapopulation context. Studies that have investigated the effect of genetic structure on life history evolution have generally ignored the complex population dynamics generated by extinction and recolonization and vice versa. Such a theoretical gap is essentially explained by the difficulty to analyze kin selection models with complex demography. Individual-based simulations allow the consideration of both demographic and genetic stochasticity, but it remains difficult to disentangle processes and to identify the mechanisms leading to some evolutionary pattern. Without a mechanistic understanding of such patterns, it may be difficult to determine their generality. Two methods have been proposed to integrate genetic drift and demographic stochasticity in the same analytical framework (Metz and Gyllenberg, 2001; Rousset and Ronce, 2004). The latter bears more resemblance to classic kin selection models, expressing selection gradients as functions of probability of genetic identity between pairs of individuals found in the same or different local populations. The effect of kin selection processes can then be assessed readily by setting all probabilities of identity to zero. The application of such methods is still restricted to a particular class of models (infinite island models with no age structure) and relies on extensive numerical computations. Still, they offer the exciting possibility to better identify and disentangle the effects of demography and genetic structure acting on the evolution of life history traits in a metapopulation. Finally, with the exception of models for dispersal evolution (Heino and Hanski, 2001; Travis and Dytham, 1999), the evolution of life history traits in a metapopulation context has essentially been modeled in a spatially implicit way. We therefore know very little about how the details of spatial configuration of patches may affect the previous models' predictions. A better understanding of this question could be achieved by a careful comparison of spatially explicit simulations and analytical models as we did here with the work of Gandon and Michalakis (1999) and Heino and Hanski (2001).

Most of the theoretical studies reviewed in this chapter make predictions about evolutionary stable life history strategies in a metapopulation, i.e., evolutionary end points, without worrying about realistic amounts of genetic variation maintained in such metapopulations, which eventually fuels evolutionary change. Both genetic drift and demographic asymmetries not only affect the direction of selection, but also constrain the evolutionary response to selection (see Chapters 9 and 16). Although an increasing number of theoretical studies address questions related to the balance among selection, dispersal, and drift in a metapopulation context, little is still known about how constraints on selection affect the evolution of particular life history traits. Boughton (1999) provided a nice empirical example in the checkerspot butterfly *Euphydryas editha*, where extinction–recolonization dynamics and patterns of gene flow severely constrain the evolution of host preference.

Perspectives for Empirical Studies of Life History Evolution in a Metapopulation

As stated in the introduction, the present chapter strongly reflects the bias affecting the field, with a scarcity of empirical studies. Data are needed to assess the validity of model assumptions about the genetic structure or the demography of species thought to have evolved in a metapopulation context. Obtaining the latter type of information can be extremely demanding as it often requires long-term surveys at multiple, and sometimes very large, spatial scales. Such data accumulated on a few systems now allow us to ask evolutionary questions about such metapopulations (as the evolution of dispersal and host preference in *M. cinxia*). Observing evolutionary changes at the scale of a metapopulation is a challenging task. Patterns of variation in life history traits along environmental or successional gradients offer an alternative for testing evolutionary model predictions, but it requires evaluating the relative role of genetic differences and plasticity in the production of such patterns. Proving that the same mechanism as envisioned in the model explained the observed pattern represents another difficulty of this approach, as for instance seed size or flight ability may be selected for reasons not related to dispersal.

Manipulation of natural systems offers a much more powerful test of evolutionary scenarios. For instance, Sinervo et al. (2000) manipulated offspring size in lizards to assess selection on this trait along cycles of low and high density in lizard populations. There is a large amount of heritable variation for life history strategies in their system due to the coexistence of very distinct female reproductive tactics. Sinervo et al. (2000) documented existing patterns of genetic covariation among different life history characters, but also artificially modified the phenotypes of individuals to test selection scenarios. In the same study, they measured the frequency-dependent component of selection acting on offspring size. Combining such approaches with a spatial perspective on such selective processes would provide a fascinating opportunity for empirical studies of life history evolution in a metapopulation. Parasites exploiting individual hosts or groups of hosts have been pointed out as organisms particularly amenable for testing metapopulation theory (Grenfell and Harwood, 1997). Many studies have now documented interspecific or intraspecific variation in parasites life history traits in relation to the structure or dynamics of their hosts populations (see, e.g., Sorci et al., 1997; Morand, 2002; Parker et al., 2003; Thomas et al., 2002). An experimental test of theoretical predictions about life history evolution in metapopulations using host–parasite systems and manipulating the characteristics of the host population (see, e.g., Koella and Agnew, 1999) should be further encouraged.

We conclude by indicating potential directions for empirical investigation of life history evolution in a metapopulation, summarizing questions emerging from our review. First, in order to judge the generality of particular models predictions, one would wish to have a better sense of the importance of transient population dynamics, of the degree of coupling between local and global dynamics, and of the extent of among-local population differentiation for quantitative characters in diverse species. As seen in this review, kin selection and demographic disequilibrium can profoundly modify our expectations concerning life history evolution, but the quantitative importance of such phenomena in real metapopulations is somehow still unclear. Second, if details of the population dynamics and genetic structure matter so much, it casts doubts on the potential of broad scale comparative studies to identify a robust pattern of variation in life history syndromes associated with metapopulation processes (for a discussion of such syndromes, see Poethke et al., 2003; Ronce et al., 2000c). Focusing on well-studied systems, for which such details are known and where quantitative predictions can be achieved, sounds as a more promising alternative. In particular, the present review points toward two main questions deserving more empirical exploration. The first concerns the patterns of evolution for life history traits other than dispersal in relation to landscape fragmentation. Candidate traits include, for instance, reproductive effort, longevity, age at maturity, and clutch size. The second question is related to the evolution of plasticity or, more generally, of conditional life history strategies as adaptations to a variable environment in a metapopulation. Of particular interest is the evolution of senescence patterns or age-specific reproductive strategies for species subject to extinction–recolonization dynamics or successional processes.

11. SELECTION IN METAPOPULATIONS: THE COEVOLUTION OF PHENOTYPE AND CONTEXT

Michael J. Wade

11.1 INTRODUCTION

Most organisms live in metapopulations, interacting locally with conspecifics in ways that affect individual fitness and in genetic contexts that vary from one deme to another (Loveless and Hamrick, 1984; Whitlock, 1992; Kelly, 1996; Hanski and Gilpin, 1997; Ingvarsson, 1999; Wade and Goodnight, 1998). As a result, natural selection in metapopulations *always* differs from that in large, randomly mating populations, whether or not there is an added component of higher level selection acting among demes. This difference between adaptive evolution in metapopulations and that in nonsubdivided populations has been overlooked by evolutionary genetic theory with its tradition of partitioning phenotypic variation into dichotomous genetic and environmental factors (e.g., Falconer and MacKay, 1996). This traditional approach founders whenever some of the factors responsible for variation in individual phenotype or fitness are both genetic *and* environmental at the same time. It is these factors, often called "indirect genetic effects" (IGEs) (Moore et al., 1997;

0-12-323448-4

Wolf et al., 1998, 1999), that create causal pathways between the genes in one individual and the phenotypes expressed by others, even if unrelated, and that permit the coevolution of phenotype and context that is unique to metapopulations. Even individual traits with no additive genetic variance can evolve in metapopulations when the mean of a trait in the social partners (an IGE) evolves (Moore et al., 1998; Agrawal et al., 2001).

An IGE can be the outcome of an overt social interaction between individuals, as in altruism or mutualism, or a less conspicuous competitive interaction, such as competition for sunlight or nutrients in plants. The evolution of interactions among individuals typically has been investigated in the context of optimality and game-theoretic models (e.g., Hamilton, 1964a,b), a conceptual context that does not deal very effectively with genotype-by-environment interactions, epistasis, or indirect genetic effects (Wolf and Wade, 2001; Shuster and Wade, 2003). An IGE is an environmental source of phenotypic variation that evolves itself in response to selection and random genetic drift. Thus, IGEs are fundamental to the coevolution of phenotype and context. Whenever there are IGEs, the phenotype of an individual depends both on its own genotype and on the genotypes of those with which it interacts (Griffing, 1967, 1977, 1989; Moore et al., 1998), permitting coevolution of individual and social context (see Fig. 11.1 and 11.2). The magnitude of an IGE depends on the sign of its effect on the individual's phenotype, the reciprocity of the interactions among individuals, and the genetic correlations with other traits in the individual and in those with which it interacts. However, like the coevolution of traits in different species (Gomulkiewicz et al., 2000; Wade, 2003), genetic correlations are not necessary for IGEs to affect the rate or direction of evolutionary response to local selection.

Whenever different individuals respond differently to the same context, then IGEs introduce a *nonlinear* component to the evolution of interactions. In large, randomly mating and mixing populations, these effects of IGEs are of limited importance because all individuals experience essentially the same average context (Wade and Goodnight, 1998). In metapopulations, however, nonlinear interactions become more important because they can enhance the among-deme variation in the contexts experienced by individuals and are

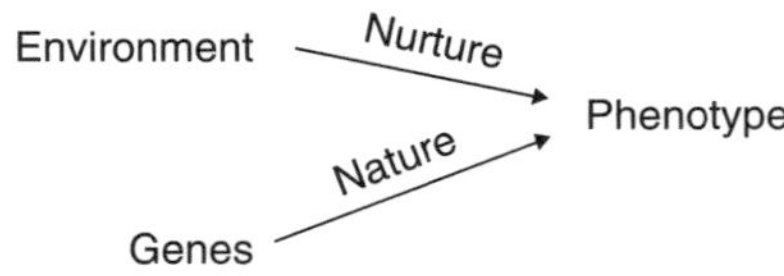

Standard Decomposition of the *Individual* Phenotype:

$$z_I = a_I + e_I$$

Variation in z_I is caused by variation in a_I or in e_I:

$$V(z_I) = V(a_I) + V(e_I)$$

Heritability: $h^2(z_I) = V(a_I)/V(z_I)$

Fig. 11.1 Standard decomposition of an individual phenotype into genetic (a_I) and environmental (e_I) components.

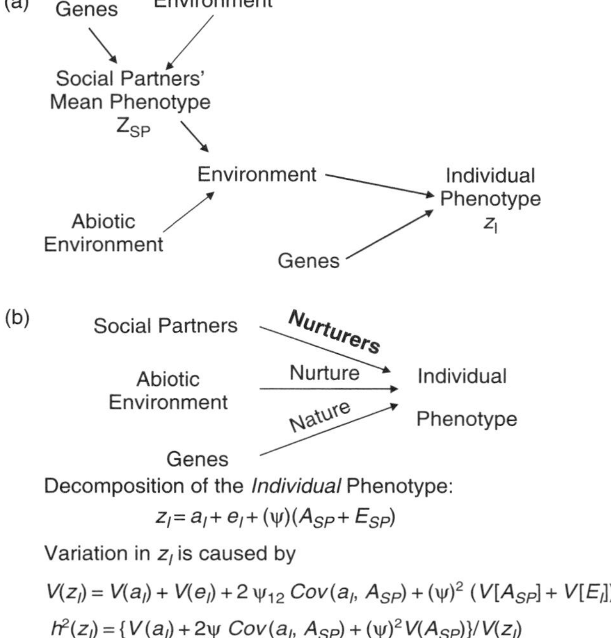

Fig. 11.2 (a)The influence of environment on an individual's phenotype can be partitioned into abiotic and social components (IGEs). Importantly, the social component itself can be decomposed into genetic and environmental components. Because the environment contains genes, the individual phenotype and the social context can coevolve. (**b**) The decomposition of an individual phenotype into genetic (a_I), environmental (e_I), and IGE components ($\Psi_{12}Z_{SP}$). The mean phenotype of the social partner, Z_{SP}, is partitioned into genetic (A_{SP}) and environmental (E_{SP}) components.

more likely to create higher levels of selection (e.g., Aviles et al., 2002). In short, the evolutionary role of IGEs in metapopulations is always greater than it is in large panmictic populations, which are lacking in conspicuous genetic structure (see Fig. 11.4).

In considering the role of IGEs in metapopulations, this chapter shows that it is important to determine (1) how trait evolution is affected by local context, (2) which traits contribute to local context, (3) how trait and context are correlated ecologically and genetically, and (4) whether variation in mean fitness leads to selection among contexts. This chapter emphasizes that the traits affected by context (1) may be the same or different from those traits contributing to context (2) and that these may or may not be genetically correlated. Furthermore, even in the absence of among-deme selection (4), metapopulations permit the coevolution of phenotype and context where the more traditional evolutionary genetic models do not (cf. discussion in Wade, 1996; Wade and Goodnight, 1998; Goodnight and Wade, 2000). Finally, the chapter discusses how these considerations might influence the coevolution of species in metacommunities, where the presence or absence of interacting species contributes to a variation in local context.

11.2 KINDS OF CONTEXTS

It is fundamental in evolutionary genetic theory that the environment affects how a nuclear genotype is expressed as an individual phenotype. The general concept of "environment" includes intraindividual cytoplasmic organelles, such as mitochondria and chloroplasts, as well as endosymbionts such as *Wolbachia* and *Buchnera*, which can interact epistatically with the host nuclear genome and affect expression of the host phenotype. Classical physiological epistasis, based on interactions between genes in the same genome (see Chapter 9), or nonclassical epistasis, based on interactions between genes in the maternal and zygotic genomes (Wolf, 2000; Wade, 2000), are ubiquitous intraindividual genetic contexts, wherein the genetic background at one locus modifies the phenotypic expression of another locus. Variation in these heritable contexts can cause a focal gene to change its Mendelian phenotypic identity from dominant, to recessive, to additive, to neutral, or to overdominant (Wade, 2001; see also Chapter 9). The variance in context caused by additive-by-additive epistasis alone can change a gene's phenotypic effect from major to minor and change the sign of its effect on fitness so that it is a "good" gene in one deme but a "bad" gene in another (Wade, 2002; Goodnight, 1995, 2000). These effects of the intraindividual genetic environment are always more important in metapopulations, where subdivision leads to among-deme variance in context, than in large populations, where panmixia with recombination diminishes the contextual variation experienced by single genes (Fig. 11.3). Indeed, the enduring controversy at the foundation of evolutionary theory between S. Wright and R. A. Fisher (Coyne et al., 1997, 2000; Wade and Goodnight, 1998; Goodnight and Wade, 2000) can be seen as a controversy over the evolutionary significance of such variations in genetic context. The patterns of migration within and among demes and the local population size affect the variance in context.

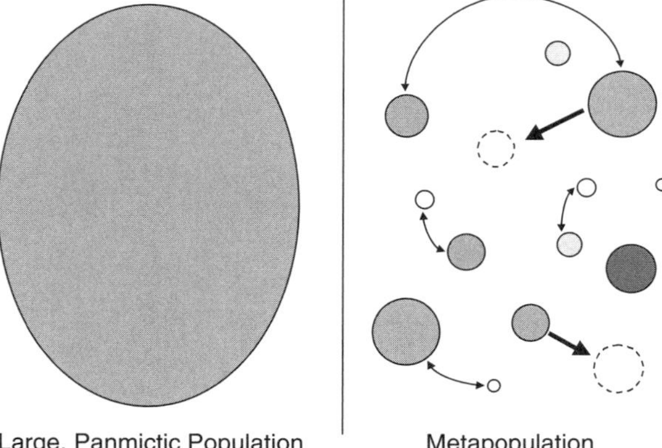

Fig. 11.3 A large, randomly mating "Fisherian" population (left) and a genetically subdivided metapopulation (right). The different size circles in the right panel represent demes of different age (smaller is younger) and different abundance. Demes with dotted borders are local extinctions. Small arrows connecting different demes represent migration, whereas larger arrows indicate colonization events.

The abiotic environment, external to the individual, is the most commonly considered context in relation to phenotypic variation among individuals. It is the reference context for most discussions of adaptation and, in most evolutionary theory, the external environment is treated as temporally and spatially invariant, at least on average. The traditional partitioning of the variation in a phenotype, z_1, into additive genetic, a_1 (with mean A_1) and environmental, e_1, components reflects this viewpoint (Fig. 11.1):

$$z_1 = a_1 + e_1. \tag{11.1}$$

Genotype-by-environment interactions ($G \times E$) are the most prevalent format for the introduction of variation in environmental context into evolutionary models (Schlichting and Pigliucci, 1998). In spatially heterogeneous habitats, geographic variation in parameters of the niche can lead to variation in local selective pressures (Holt and Gaines, 1992). Migration among habitats with $G \times E$ constrains evolution because genotypes adapted to one habitat are moved to other habitats where they are less fit by migration. Habitats where abundance is greatest will tend to dominate evolution in those where abundance is lower if migration is proportional to local abundance (Holt and Gaines, 1992). Under some circumstances, however, 'adaptive plasticity' can evolve as a response to spatial variation. Plasticity plays a central role in our understanding of the evolution of polyphenisms, particularly those involved in an adaptive response to temporal and spatial heterogeneity of the abiotic environment. The standard partitioning of phenotypic variation is expanded to include a term for $G \times E$, c_{axe},

$$z_1 = a_1 + e_1 + c_{1axe}. \tag{11.2}$$

This approach is appropriate for the abiotic environment but not for biotic environments. This formulation permits a genetic response to abiotic environmental variation, but does not allow the reciprocal because the environment, although variable, is not evolutionarily dynamic. Only the genotype changes evolutionarily if fitness varies with context in this way. Thus, Eq. (11.2) cannot adequately account for the context provided by conspecifics or by other species in an ecological community because these ecological contexts (unlike e_1) also have a genetic component. Thus, many ecologically important contexts differ from $G \times E$ because they are evolutionarily dynamic and can coevolve with z_1. Interactions between males and females, between parents and offspring, between age cohorts, between competitors, between predators and prey, or between hosts and pathogens all contribute to individual phenotypic and fitness variation in plants and animals. Notably, these contexts contain genes and, when they vary from deme to deme across a metapopulation, genes and context can coevolve (Thompson, 1994; Wolf et al., 2003; Wade, 2003).

11.3 EVOLVING CONTEXTS

In order to investigate evolving contexts, additional terms must be included in the standard formulation given earlier. Imagine a second trait, z_2, which is sensitive to the local mean value, Z_{1i}, of the first trait. The first trait, z_1, is the "context" and its local mean value influences individual values of phenotype, z_2, in that deme as follows:

$$z_{2i} = a_{2i} + \Psi_{12}Z_{1i} + e_{2i}. \tag{11.3}$$

The second term, $\Psi_{12}Z_{1i}$, is the IGE, where the coefficient, Ψ_{12}, represents the scale or magnitude of the linear effect of Z_{1i} relative to the direct effects whose coefficients have been set to 1. The effects of Z_{1i} can include both linear and nonlinear components (cf. Agrawal et al., 2001). Whenever it contains nonlinear terms, then it is not generally possible to scale the global means, Z_1 and Z_2, to zero without also a temporal and spatial rescaling of the coefficients of all other terms in Eq. (11.3) each generation. Thus, the presence of an evolvable context means that the coefficients of additive genetic effects, such as a_{2i} (here assumed to be 1.0), must be reevaluated from generation to generation instead of remaining constant as is done in standard theory. It is the nonzero values of Ψ_{12} that permit the environment to coevolve (Fig. 11.2).

Substituting Eq. (11.1) into Eq. (11.3), we find that

$$z_{2i} = a_{2i} + \Psi_{12}(A_{1i} + E_{1i}) + e_{2i}. \tag{11.4}$$

This is the effect of IGEs: an individual's phenotypic value, z_2, depends on the genetic and phenotypic value of its local, social context $\Psi_{12}Z_{1i}$ (Fig. 11.2). Furthermore, with genetic subdivision ($F_{ST} > 0$), the average social context itself will vary from deme to deme across a metapopulation. Thus, there can be heritable variation in the additive genetic values of social context from deme to deme, affecting the evolution of traits sensitive to context. If F_{ST} is Wright's measure of metapopulation genetic subdivision and $V(a_j)$ is the variance in a_j in a population without subdivision (i.e., $F_{ST} = 0$), then in the absence of epistasis, the average variance within demes is $(1 - F_{ST})V(a_j)$ and the variance among demes is $2F_{ST}V(a_j)$ (Hartl and Clark, 1997).

When there is additive-by-additive epistasis, $V(a_i \times a_j)$, the average additive genetic variance within demes can be greater than $(1 - F_{ST})V(a_j)$ (Goodnight, 1987, 1990, 1995, 1999) by an amount approximately equal to $4F_{ST}(1 - F_{ST})V(a_i \text{ x } a_j)$ (Whitlock et al., 1993). With nonlinear interactions among genes, the identity of the genes contributing to the local additive variance will vary from deme to deme as will the coefficients of a_1 and a_2 in Eqs. (11.1)–(11.3) (Goodnight, 2000). If the IGE term, $\Psi_{12}Z_{1i}$, is itself partitioned into linear and nonlinear components, then not only will the coefficients change from deme to deme and from generation to generation, but the strictly additively determined phenotypes, z_1 and z_2, will now have an added epistatic (i.e., a_1 x a_2) component.

11.4 SELECTION IN METAPOPULATIONS

To see how selection in metapopulations is different with IGEs, consider selection only within demes. We need to first define the relative fitness of an individual in the ith deme,

$$w_i = k + \beta_{1i}z_{1i} + \beta_{2i}z_{2i} + B_1 Z_{1i} + B_2 Z_{2i} \quad (11.5a)$$

and then substitute the genetic values contributing to the phenotypes,

$$w_i = k + \beta_{1i}a_{1i} + \beta_{2i}(a_{2i} + \Psi_{12}A_{1i}) + B_1 A_{1i} + B_2(A_{2i} + \Psi_{12}A_{1i}). \quad (11.5b)$$

[Note that in Eq. (11.5 b), only the additive genetic components of fitness are expressed and the abiotic environment and epistasis are left out.] Here, k is a constant common to all individuals and it is assumed that the abiotic environment remains constant across generations. The coefficients, β_{1i} and β_{2i}, represent the strength of individual selection within the ith deme on the two phenotypes, z_{1i} and z_{2i}, and the coefficients, B_1 and B_2, represent the strength of interdemic selection acting on the means, Z_1 and Z_2, of each phenotype. Setting both B_1 and B_2 equal to zero eliminates interdemic selection. Note that, considering selection only within the ith deme, the term, $\beta_{2i}\Psi_{12}A_{1i}$, would not vary among individuals and thus would not affect the variance in relative fitness within the deme.

The metapopulation mean values, Z_1 and Z_2, change in proportion to the covariance between the additive genetic components of z_1 and z_2 and relative fitness. Let selection within all demes be the same so that β_{ji} equals β_j. Considering only the genetic effects for trait 1, taking the covariance of z_1 and w_1, we have

$$\Delta Z_1 = (\beta_1 V[a_1] + \beta_2 C[a_1,a_2])(1 - F_{ST}) + 2F_{ST}V(a_1)(B_1 + \beta_2\Psi_{12} + B_2\Psi_{12}) + 2F_{ST}C(a_1,a_2)B_2, \quad (11.6a)$$

where $C(a_1,a_2)$ is the genetic covariance between z_1 and z_2 within a nonsubdivided population. In the absence of interdemic selection (i.e., B_1 and B_2 equal to zero), this reduces to

$$\Delta Z_1 = (\beta_1 V[a_1] + \beta_2 C[a_1,a_2])(1 - F_{ST}) + 2F_{ST}V(a_1)(\beta_2\Psi_{12}). \quad (11.6b)$$

The first two terms of Eq. (11.6 b) are the standard expressions for the change in the mean, Z_1, in a genetically subdivided metapopulation that result from (1) the average effects of direct selection within-demes on phenotype, z_1, and (2) the average effects of indirect selection within-demes on phenotype, z_2. However, the third component is new and results from genetic variation in social context among demes (i.e., $2F_{ST}V[a_1] > 0$). It not only affects the expression of z_2 but also affects selection on z_1. As a result, social context affects its own evolution in a metapopulation through its indirect effect on the context sensitive phenotype, z_2. This effect of context is different from standard indirect selection because it occurs even when the two traits are not genetically correlated (i.e, with $C[a_1,a_2] = 0$). In standard theory, absent a genetic correlation, the evolution of one phenotype would be completely independent of and unconstrained by the evolution of the other. Thus, in a metapopulation with IGEs (i.e., $F_{ST} > 0$ and Ψ_{12} is nonzero) and without a genetic correlation between z_1 and z_2, *individual selection* on z_1 will always be affected by selection on z_2, even in the absence of interdemic selection. This is different from evolution in a large randomly mating and mixing population where a genetic correlation is necessary for the evolution of one phenotype to constrain another.

Selection on z_2, which is sensitive to the social context provided by z_1, is more complicated, even omitting terms in $(\Psi_{12})^2$,

$$\Delta Z_2 = (\beta_2 V[a_2] + \beta_1 C[a_1,a_2])(1 - F_{ST}) + 2F_{ST}\{(\beta_2 + B_2)V[a_2] + \Psi_{12}(\beta_1 + B_1)V[a_1]\} + 2F_{ST}C(a_1,a_2)\{2\beta_2\Psi_{12} + 2\Psi_{12}B_2 + B_1\}, \quad (11.7a)$$

and, in the absence of interdemic selection,

$$\Delta Z_2 = (\beta_2 V[a_2] + \beta_1 C[a_1,a_2])(1 - F_{ST}) + 2F_{ST}(\beta_2 V[a_2] + \Psi_{12}\beta_1 V[a_1]) + 4F_{ST}C(a_1,a_2)\beta_2\Psi_{12}. \quad (11.7b)$$

Clearly, in metapopulations, a variation in genetic context creates new sources of covariance between genes and fitness that affect the response to selection. In the absence of interdemic selection, both Eqs. (11.6b) and (11.7b) have terms with the coefficient $2F_{ST}\Psi_{12}$. Thus, in a metapopulation with IGEs (i.e., $F_{ST} > 0$ and Ψ_{12} is nonzero), *individual selection* will be always be different from what it is in a large randomly mating and mixing population, even in the absence of interdemic selection. This is true as long as Z_1 varies among demes (i.e., $2F_{ST}V[a_1] > 0$), whether or not there is a genetic correlation between social context, z_1, and the sensitive trait, z_2, either within or among demes.

With interdemic selection, as in Eqs. (11.6a) and (11.7a), the response to selection becomes substantially more complex. Note especially that all of the interdemic selection terms have F_{ST} as a coefficient: there can be no interdemic selection in the absence of among-deme genetic variation. Note also that interdemic selection on *either* trait results in a response to interdemic selection in the other trait. This means that whether a trait is a social context trait or is sensitive to a social context trait, interdemic selection on one affects the evolution of the other and does so independent of genetic correlations.

11.5 THE PERVASIVENESS OF EFFECTS OF INDIRECT GENETIC EFFECTS IN METAPOPULATIONS

Local density dependence is an ubiquitous example of an IGE (like z_1 shown earlier) that affects evolution in metapopulations. Many morphological and behavioral traits exhibit a response to local density. The emigration rate in the flour beetle, *Tribolium castaneum*, is a good example of a behavior whose expression, like z_2 given earlier, is influenced both by the genotype of the individual and by the local population density (Craig, 1982). The tendency to emigrate of an individual beetle is determined both by its own genotype and by the local density of conspecifics that it experiences during development (Fig. 11.4). An individual reared in a high-density environment is more prone to emigrate than a genetically similar individual reared in a low-density environment, even when both are tested for emigration tendency at the same intermediate density. That is, the past experience of local density as a larva influences the emigratory tendency of the adult. So, like the theory presented earlier, an individual's emigration tendency (z_2) is sensitive to a genetically

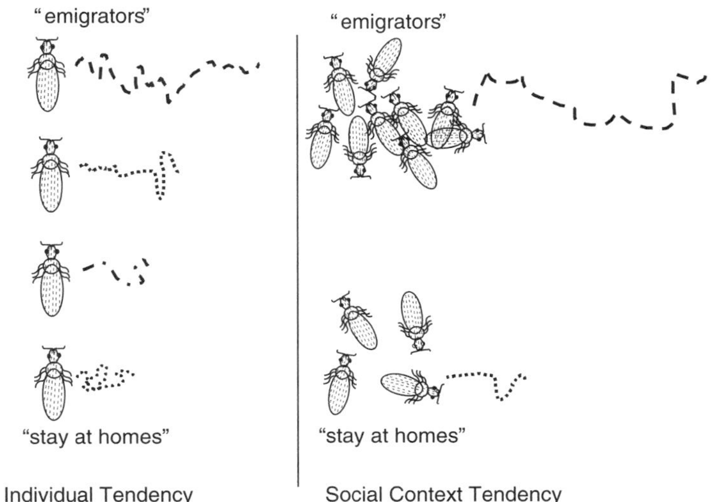

Fig. 11.4 The emigratory tendency of a single flour beetle (*Tribolium castaneum*) is influenced by its own genotype (left) and by the social context represented by local density (right). Because local density has a genetic component, it is an IGE and a potentially coevolving environment. (See text for further discussion.)

variable social context, local density (Z_1). In this species, there is no evidence for a genetic correlation between emigration tendency and fecundity.

When Craig (1982) imposed individual and interdemic selection for increased and decreased emigration rate (B_1 and $\beta_1 \neq 0$ but B_2 and $\beta_2 = 0$), the local density changed as a correlated response to selection on the emigratory rate. The regression coefficient of emigration tendency on reproductive capacity (both expressed as percent deviation of the selected treatment means from unselected control mean) was 0.94 ($p < 0.015$) and it explained 86% of the variation in the emigration rate among treatments. Thus, the change in local density, the IGE, accounted for a proportionally greater amount of the response to selection in an individual's tendency to emigrate than the individual's own genotype! This demonstrates that the coefficient of context, Ψ_{12}, is large relative to the coefficient of the direct additive term. Genetic variation among demes in local density exists in flour beetles and rapidly becomes partitioned among demes in laboratory metapopulations (Wade and McCauley, 1980; Wade, 1982; Wade and Griesemer, 1998). Indeed, the biology of flour beetle metapopulations is even more complex because different genotypes are not only differentially sensitive to density (Wade and McCauley, 1980; Wade and Griesemer, 1998; Wade, 2000), but also differentially contribute to density effects (Lloyd, 1968; Wade, 2000).

Plant leaf area in the cress, *Arabidopsis thaliana* (Goodnight, 1985), is another morphological trait whose expression is sensitive to social context and is representative of a myriad of ubiquitous plant–plant competitive interactions. In *A. thaliana*, an individual's genotype influences its leaf area and so do the leaf areas of neighboring plants, whether genetically related to the focal individual or not. Here, the same trait, leaf area, provides both the direct effect and the social context. In the model given earlier, z_2 would be affected by its own mean, Z_{2i}. Importantly, as a result of competition for sunlight, the direct effect of an individual's own genotype is in conflict with the IGE of its neighbors, i.e., the coefficient of the direct additive term is positive while Ψ_{12} is negative. As a result, artificial individual selection to increase leaf area produces an average decline in leaf area (Goodnight, 1985) as it does in many other domesticated plant species. Without the social effect on leaf area, this would not occur. Plants are not typically regarded as social organisms, like the hymenoptera, and leaf area is not typically considered a social context trait or an IGE.

The results of Goodnight's experiments support the theoretical findings of Griffing (1967, 1977, 1981, 1989). He showed that whenever direct and indirect "associative" effects are of opposite sign, as they tend to be in competitive interactions, individual selection alone cannot maximize the response to selection. His findings are particularly important to plant breeders trying to maximize yield. A large plant has a manifestly greater yield than a small one, but a stand of *N* large plants does not produce *N* times the yield of one large plant, but rather a much reduced total yield due to interplant competition, which unfortunately intensifies with plant size. Griffing (1967, 1977, 1981, 1989) illustrated theoretically how and why interdemic selection could achieve maximal yields whereas individual selection alone could not. Goodnight (1985) demonstrated empirically that artificial interdemic selection alone resulted in a greater response to selection for increased leaf area than artificial individual selection alone of comparable intensity. Indeed, Goodnight showed, as Griffing (1967, 1977, 1981, 1989) predicted, that individual and interdemic selection together would interfere with one another, resulting in a total response to combined selection that was less than the sum of the expected responses to each level of selection acting alone (see also the discussion of this interference in Agrawal et al., 2001). Evolution in metapopulations with IGEs is much more complex, with or without interdemic selection, than the uncritical use of Occam's razor (Williams, 1966) would suggest. Applying Occam's razor to the adaptive increase in plant leaf area, one would get both the mechanism (interdemic selection) and the direction of response to individual selection wrong without considering IGEs, which are not part of standard evolutionary genetic theory.

Muir (1996) capitalized on similar findings and used artificial intergroup selection on sire-families to increase egg laying in domestic chickens, *Gallus gallus*. Egg production in long-term individual selection lines is limited by the practice of maintaining hens in group cages, where competitive interactions have such severe affects on mortality and condition that debeaking is a common practice. (Debeaking is the removal of most of the beak to minimize feather plucking, injuries, and cannibalism among cage mates. It may be done more than once with laying hens. It limits but does not prevent deleterious interactions among birds.) As with plants, a large hen might yield more or larger eggs than a small hen, but a group of *N* large hens produces less than *N* times the yield of a single hen housed alone. That is, social context as well as individual genotypes have profound effects on the number and size of eggs yielded by a group of hens. The behavioral concept of a dominance hierarchy or "pecking order" was also developed in studies of this species. However, the crude attempts to take the peck out of the pecking order by debeaking individual birds (removing trait, z_1) did not eliminate the negative effect of IGE on egg production as effectively as focusing artificial interdemic selection on the IGE.

The response to artificial intergroup selection for increased egg laying in the Muir experiments was spectacular, especially considering that his founding stock was an elite breed, derived by the application of the most effective and efficient individual selection protocols for over 50 generations. In only six generations of interdemic selection, mortality in group cages declined sevenfold, from 68.8 to 8.8%, which is equal to the background mortality of hens housed individually. The negative effects of social context were essentially eliminated in six generations of interdemic selection. Notably, this is a result that could not be achieved by debeaking individual birds. Mean egg number per hen increased more than twofold, from 91 to 237 eggs, in part because of a 16% increase in eggs per hen per day, but also in part because of a doubling of hen longevity. In an individually selected control with single-hen cages, the response to selection was much slower and the setup impossible to implement for large-scale egg production, which requires group cages. The competitive interactions diminished in response to intergroup selection to the point that birds with or without beaks had equivalent survival and debeaking was unnecessary to obtain the increased productivity.

It must be emphasized that competitive interactions, like those associated with leaf area in *A. thaliana*, with density in *T. castaneum*, or with fighting in *G. gallus*, are a common form of IGE. Indeed, intraspecific competition was the essential concept from Malthus that Darwin realized would ensure a struggle for existence, making natural selection inevitable. In Darwin's words, ". . . it at once struck me that under these circumstances favorable variations would tend to be preserved and unfavorable ones to be destroyed" (Darwin, 1876, pp. 119–121). That is, intraspecific competition is an integral component of the conceptual logic of Darwinian evolution by natural selection and one of the most ubiquitous of IGEs.

The effects of IGEs, however, are even more pervasive in metapopulations than the three examples would indicate. Like the role of the beak in competition in *G. gallus* or leaf area in *A. thaliana*, the expression of most social traits involves one or more morphological traits. For example, in the flour beetle, *T. confusum*, egg cannibalism involves the interaction of both larval mandible size and egg size (Teleky, 1980). Large eggs are safe from predation by small larvae simply by virtue of their size relative to the mandible size of early instar larvae. Thus, "egg cannibalism" is an IGE that involves the mandible size of the prospective cannibal, its genetic propensity toward cannibalism, and the mean and distribution of egg sizes in the local environment, which are also influenced by the genes in the laying mothers as well as genes determining the ejaculate quality of their mates. The totality of the interaction is highly nonlinear; in laboratory metapopulations, conspicuous among-deme variation in the level of egg cannibalism arises even in the absence of interdemic selection (Wade, 1978, 1979, 1980).

Neither egg size nor mandible size would be a likely candidate for an IGE. Nevertheless, the social trait of cannibalism involves both of these morphological traits and associates them with fitness via egg viability and via the nutritional fitness advantages that accrue to the cannibal. In this way, many morphological traits become subject to the evolutionary consequences of selection of IGEs in metapopulations. Whether a trait is a social context trait or is sensitive to a social context trait, *interdemic selection* on one trait affects the evolution of the other and does so independent of the existence of genetic correlations. Furthermore, in a metapopulation with IGEs (i.e., $F_{ST} > 0$ and θ_{12} is nonzero), *individual selection* will be always be different from what it is in a large randomly mating and mixing population, even in the absence of interdemic selection.

11.6 EFFECTS OF INDIRECT GENETIC EFFECTS IN METACOMMUNITIES

Whenever the environment contains genes, context can evolve along with the evolutionary response to context. This is true whether the environmental context is other conspecifics (Wolf et al., 1998), as in the examples given earlier, or other species, as in ecological communities (Wolf et al., 2002). Extending the definition of IGEs to include other species is a warranted extension of earlier definitions (Wolf et al., 1998). In the words of Goodnight (1991, p. 343), "Inter-species interactions are different from other genotype–environment interactions because not only can a second species be a significant component of the environment, but it is also an evolving entity that can change as a result of deterministic forces such as natural selection and random forces such as genetic drift."

Keister et al. (1984) modeled the interaction and coevolution of two species interacting randomly in a large panmictic population. They showed how the response to selection on a trait in one species is dependent on the mean value of the context provided by the other species and vice versa. More important for evolution in metapopulations, however, was the finding that random genetic drift affecting two trans-specific, coevolving characters depended on the species with the *smaller* effective population size. Thus, if two coevolving species, A and B, interact and species A is strongly affected by genetic subdivision but species B is not, traits in species B will nevertheless evolve as though species B were as highly genetically subdivided as species A. Thus, the effects of IGEs discussed in the preceding section for traits in one species apply to coevolving traits in other species, even if less genetically subdivided!

A metacommunity can be defined by analogy with metapopulation to be a more or less genetically subdivided collection of interacting species (see also Chapter 6). Because of interspecific IGEs, coevolution in a metacommunity will result in hot spots, where reciprocal coevolution is strong, and cold spots, where it is weak, according to Thompson's (1994) geographic mosaic hypothesis. Empirical evidence for this kind of variation in the outcome of coevolution has been forthcoming from the recent studies of natural metacommunities of toxic newts, genus *Taricha*, and their garter snake predator, *Thamnophis sirtalis* (Geffeney et al., 2002; Brodie et al., 2002). The ecological factors that affect the local abundance of one species are experienced by the other species as among-deme variation in context, which affects the response to evolution of *both* species across the metacommunity.

Just as the mean of a trait without heritable variance can change in response to selection on a contextual trait (Moore et al., 1997), in a metacommunity, the heritability of a trait in species A might depend on the value of context provided by species B. This has been referred to as "community heritability" and defined as the among-community fraction of the genetic variance affecting coevolving traits (Goodnight, 1991; Goodnight and Craig, 1996). Goodnight (1991) created 10 replicate small communities using laboratory populations of two species of flour beetles, *T. castaneum* and *T. confusum*, and allowed these communities to codifferentiate by random genetic drift for 16 generations (Fig. 11.5). After that period, he factorially combined the members of each species from each community to create 100 new two-species communities and replicated each three times (Fig. 11.6). In Fig. 11.6, the 10 shaded diagonal squares represent the codrifting 10 communities of Fig. 11.5. This design is analogous to the standard diallele design used to detect epistasis as a significant interaction between crossed inbred strains (see, e.g., Wade and Griesemer, 1998). In this case, however, the "epistasis" or "interspecific intermixing ability" is due to genetic interactions between species. For each community, Goodnight measured four traits, offspring numbers of each species and adult emigratory rate of each species. In addition to main effects of community of origin for each species, he found significant interactions (i.e., significant interspecific intermixing ability). Some strains of *T. castanuem* were much more productive with particular strains of *T. confusum* and vice versa. This means that community effects on fitness and emigration cannot be decomposed into simple additive effects of the separate species. This has very important implications for ecological models of metapopulations where species' growth rates and emigration rates are often assumed to be constant. Importantly, they are

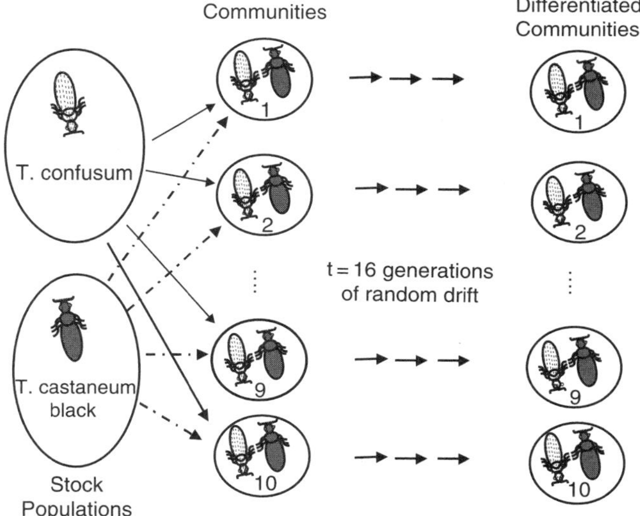

Fig. 11.5 A schematic representation of the experimental design used by Goodnight (1991) to create genetically divergent small communities, each consisting of two species of flour beetles, *Tribolium castaneum* and *T. confusum*. Each of the 10 communities was established by taking groups of 16 adult beetles from laboratory stock populations of each species. These communities were held at constant size, 32 beetles (16 of each species), and were allowed to differentiate by random genetic drift for 16 generations. (See text and Fig. 11.6 for further discussion.)

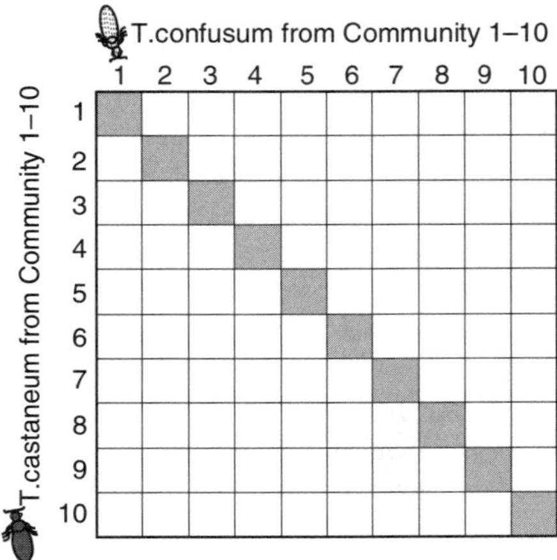

Fig. 11.6 The 10 differentiated communities generated by the protocol depicted in Fig. 11.5 were separated into 10 *Tribolium castaneum* and 10 *T. confusum* lineages, one from each community. These lineages were combined factorially and replicated (three replicate populations per cell). For each of these 100 experimental communities, emigratory rates and population growth rates of both species were measured. Using ANOVA, Goodnight tested for direct genetic effects of *T. castaneum* (row effect) and/or *T. confusum* (column effects) as well as "intermixing ability" (interaction between rows and columns). (See text for a discussion of the findings.)

not constant in two different ways: (1) direct genetic effects, where the genetic strain of the focal species affects its population growth and emigration rate and (2) the indirect genetic effect, where the genetic strain of the competing species affects population growth and emigration rate of the focal species. These data are the clearest evidence that random genetic drift acting in a relatively brief period of time can create among-community heritable variation and that a component of this variation is not observable in single species and is a heritable property of the interaction. This experiment is particularly interesting because the heritable variation developed during a known period of community isolation by random drift.

The only estimate of community heritability per se comes from studies of competitive ability in metacommunities of the flour beetles, *T. castaneum* and *T. confusum*, conducted by Goodnight and Craig (1996). They established single-species metapopulations as well as metacommunities, with both species coexisting together. After a period of subdivision with no artificial selection either within or among demes, they measured the competitive ability of beetles from each type of subdivided population using a method similar to that of the classic ecological studies of Park (1948). Goodnight and Craig (1996) found that metacommunity structure affected two ecological aspects of competitive ability: (1) the heritability for competitive outcome (i.e., the identity of winning species) and (2) the time to extinction of the losing species.

In natural communities, the studies of Brodie et al. (2002) and Geffeney et al. (2002) document geographic covariation between prey toxicity in the newt, *Taricha granulosa*, and predator resistance in the garter snake, *Thamnophis sirtalis*. This kind of variation across natural metacommunities could arise by a combination of drift and natural selection. These studies of natural metacommunities complement those of Goodnight and Craig (1996) with laboratory metacommunities and demonstrate the potential effects of trans-specific IGEs on fundamental ecological processes, such as predation and competition, which shape ecological communities. This important aspect of metapopulation evolution has received relatively little theoretical attention (but see Agrawal et al., 2001; Wade, 2003; Wolf et al., 2003).

11.7 DISCUSSION

Moore et al. (1997) emphasized a critical feature of IGEs, ". . . interacting phenotypes differ from other traits because they are determined in part by an environment that can evolve." They investigated the effects of IGEs in the absence of population genetic structure and showed that IGEs profoundly affect the expected response to selection. In the model given earlier, selection on z_1 produces a change in the mean of trait z_2 because the mean value, Z_{1i}, is a context that affects the phenotypic expression of trait z_2. Just as poor nutrition can induce small body size while good nutrition can induce large body size, changes in the mean value of z_1 produce effects on the phenotype, z_2, even in the absence of genetic variation for z_2. The genetic subdivision of metapopulations (i.e., $F_{ST} > 0$) enhances the influence of IGEs on evolution.

Griffing (1967, 1977, 1981, 1989) showed how one of the most ecologically common IGEs, intraspecific competition, negatively affects the response to individual selection, especially selection to increase the rate of population increase. His theoretical findings have found empirical support in a number of organisms, and similar effects of competition and density have been detected in several studies (cf. review by Goodnight and Stevens, 1997). Because of IGEs and the genetic structure of metapopulations, the mean value of one trait becomes an evolutionarily dynamic component of the environment experienced by another trait. Thus, IGEs in metapopulations alter the outcome of local individual selection and reinforce the view of Goodnight and Wade (2000), "Multilevel selection is far more common in nature than previously believed, and "pure" individual selection is far less common" (p. 322).

In metacommunities, interactions between species create a novel class of trans-specific IGEs, in which the mean value of a trait in one species becomes an evolutionarily dynamic component of the environment experienced by another trait in another species. Because the species with the smaller effective population size determines how random genetic drift affects the coevolving traits, the metapopulation genetic structure experienced by one species will affect the evolution of all the other species with which it interacts, whether or not they have a conspicuous metapopulation structure. Whenever a trait in one species is an ecological context trait in another species, then evolution in both species will be uniquely influenced by metapopulation genetic structure in either species. Furthermore, interdemic selection on one species will affect the coevolution of traits in the other species with which it interacts. Thus, the metapopulation genetic structure of one species, with or without interdemic selection, will have important consequences for its evolution and for the coevolution of many other species in its metacommunity.

12. SPECIATION IN METAPOPULATIONS

Sergey Gavrilets

12.1 INTRODUCTION

Analysis of ecological and evolutionary dynamics in metapopulations, that is, in populations subdivided into a large number of local subpopulations that become extinct and are recolonized from other locations, has been a focus of numerous experimental and theoretical studies (e.g., Hastings and Harrison, 1994; Harrison and Hastings, 1996; Hanski and Gilpin, 1997; Hanski, 1998; this Volume). The main interest of most of the previous work on the evolutionary effects of local extinction and colonization in metapopulations has been mainly on the levels of genetic variation within and between local populations, on the fixation probabilities and fixation times, and on Wright's shifting balance theory (e.g., Wright, 1940; Levins, 1970; Slatkin, 1977, 1981, 1978; Wade, 1978; Lande, 1979, 1984, 1985, 1992; Wade and McCauley, 1988; Whitlock and McCauley, 1990; Barton, 1993; Michalakis and Olivieri, 1993; Whitlock et al., 1993; Le Corre and Kremer, 1998; Pannell and Charlesworth, 1999).

Several recent studies modeled the joint dynamics of speciation, extinction, and colonization in a spatially explicit framework using phenomenological descriptions of speciation. These studies did not consider any underlying genetics and simply postulated that a new species with a certain number of individuals (one or more) emerges with a certain probability out of the ancestral species. A major focus of these studies was on explaining the so-called

0-12-323448-4

species–areas curves. A species–areas curve relates the number of species S found in a region with its area A. These curves are usually described using a power-law relationship

$$S = cA^b \tag{12.1}$$

where c is a constant and b is an exponent, which typically ranges from 0.15 to 0.40 (e.g., Rosenzweig, 1995; Hubbell, 2001).

For example, Bramson et al. (1996) and Durrett and Levin (1996) considered a two-dimensional space divided into square cells. Each cell is occupied by one individual, and each individual is characterized by its "type." Each individual has four "neighbors" (above, below, left, and right). During each unit time interval, the state of each individual changes to that of a randomly chosen neighbor with a small probability δ. This event represents the death of the individual immediately followed by dispersal of an offspring of one of its neighbors into the vacant cell. With a small probability ν the state of each individual changes to a new type not previously present in the system. This event represents the replacement of the individual by its mutant offspring that belongs to a new species. Numerical simulations of this model show that individuals of the same type tend to form clusters in space (Bramson et al., 1996; Durrett and Levin, 1996; Hoelzer, 2001). This dynamic pattern is explained by the simple fact that all individuals of the same species are ancestors of a single mutant individual and therefore are more likely to be found close to each other (provided, of course, the dispersal is limited). The rate of death and replacement δ and the rate of speciation ν control the properties of the resulting stochastic equilibrium via their ratio $\alpha = \nu/\delta$. Using analytical methods, Bramson et al. (1996) and Durrett and Levin (1996) showed that these properties are different at different spatial scales. The characteristic linear dimension is

$$l = \frac{1}{\sqrt{\alpha}} \tag{12.2}$$

For large areas, that is, for squares with area $A > l^2$, the number of species found is given by Eq. (12.1) with $b = 1$ and

$$c = \frac{1}{2\pi} \alpha(\ln \alpha)^2 \tag{12.3}$$

(Bramson et al., 1996). For smaller areas, that is, for squares with area $A < l^2$, the results of Durrett and Levin (1996) suggest that the species–area curve is given approximately by Eq. (12.1) with

$$b = \frac{2 \ln l + \ln(2/\pi)}{2 \ln l} \tag{12.4}$$

and $c = 1$. For example, as α decreases from 10^{-4} to 10^{-12}, the exponent b decreases from 0.283 to 0.174.

The model studied by Bramson et al. (1996) and Durrett and Levin (1996) was actually a simplification of a model proposed and numerically studied by Hubbell (2001, Chapter 6) who allowed for a number (≥ 1) of individuals at each cell (interpreted as a "local community"). In Hubbell's model, each death in a local community resulted in replacement from the local community or from one of the neighboring local communities with probabilities $1 - m$ and m, respectively. Parameter m is interpreted as the probability of migration. Hubbell (2001) demonstrated the linearity of the species–area curves on "intermediate" spatial scales and showed that the slope of these curves increases with decreasing the rate of migration m. He also noticed that the dependence of S on A becomes approximately linear for areas that are (much) smaller than those predicted by the characteristic distance l defined by Eq. (12.2). Assuming no dispersal limitations (i.e., disregarding the spatial structure of the system), Hubbell (2001) numerically compared two versions of the model differing with regard to the number of individuals in the new species. In the "point speciation model" each new species starts with exactly one individual. In the "random fission model" the new species gets a random proportion (sampled from the standard uniform distribution) of individuals of the ancestral species. This latter procedure was supposed to model speciation resulting from a vicariance event. Hubbell noticed dramatic differences in various characteristics of the system between these two speciation scenarios. A variant of the "random fission model" was used by Barraclough and Vogler (2000) in their numerical study of the dynamics of species ranges that did not allow for extinction.

Both Hubbell (2001) and Bramson et al. (1996) and Durrett and Levin (1996) were primarily concerned with the number of species found within a certain area embedded within a much larger area. Their results cannot be used to evaluate the overall number of species in the system (i.e., the overall diversity). This latter question was approached by Allmon et al. (1998) using numerical simulations of finite two-dimensional square-lattice systems in which each square cell was interpreted as a population rather than an individual. [Note that the same interpretation is applicable to the models of Hubbell (2001), Bramson et al. (1996), Durrett and Levin (1996).] Allmon et al. (1998) allowed for the probability of speciation to be dependent on the number of and the distance to other populations of the same species. The main focus of Allmon et al. (1998) was on demonstrating that the overall diversity is maximized at intermediate rates of extinction of local populations.

Pelletier (1999) used numerical simulations to study the species–area curves both for areas embedded within much larger regions and for isolated areas. He used a different modeling framework that explicitly treated the dynamics of local population densities and allowed for density-dependent dispersal of individuals. The latter process was modeled using a diffusion equation. In his simulations, the probability of speciation at a given grid point was set to be inversely proportional to the species abundance at this point. This assumption implied that speciation was most probable in small populations. Pelletier's results demonstrated the linearity of species–area curves on the log–log scale [which implies that Eq. (12.1) was adequate]. Curves corresponding to nested subareas had shallower slopes and were positioned higher than curves corresponding to isolated areas. Pelletier (1999) also demonstrated that in his model the number of new species originating from a given species has a power-law distribution and the time series of extinctions and originations have a $1/f$ power spectrum where f denotes frequency.

Although these earlier approaches are very useful for training our intuition about the process of diversification in metapopulations and for providing a basis for additional numerical and analytical work, the phenomenological treatment of speciation they employ is not satisfactory. Excluding processes such as polyploidy and major chromosomal changes, speciation does not occur instantaneously. Changes in at least several loci are needed for the degree of reproductive isolation or morphological change necessary for assigning an individual or a population to a new species. The probability of this happening instantaneously is extremely small. For example, if 10 genes have to be changed, then using the standard estimates of the mutation rates (10^{-5}–10^{-6}), the probability of instantaneous speciation is on the order of 10^{-50}–10^{-60}. If there are L loci and changes in any K of them will result in strong reproductive isolation (or a significant morphological difference), then the probability of speciation is approximately $\binom{L}{K}10^{-5K}$. Here, $\binom{L}{K}$ is the binomial coefficient (i.e., the number of combinations of K objects chosen from a set of L objects) and it is assumed that the mutation probability is 10^{-5} per locus per generation. For example, with $L = 100$ and $K = 10$, the probability of instantaneous speciation is approximately 10^{-35}. Hubbell's numerical work shows that both the initial number of individuals of the new species and their initial spatial distribution have dramatic effects on various dynamic characteristics of the system. However, the phenomenological approaches cannot provide information on the appropriate values or ranges of these parameters. Hubbell (2001) and others argued convincingly that species–area curves must be derived from the underlying processes of extinction, speciation, and dispersal rather than be postulated to follow a certain statistical distribution. However, in a similar way, the dynamics of speciation must be derived from the underlying microevolutionary processes rather than be postulated to follow a certain statistical distribution (e.g., a "point speciation" model or the "random fission" model).

This chapter attempts to expand these previous approaches by developing a model incorporating multiple genetic loci underlying reproductive isolation and species differences. It starts by describing an approach for modeling speciation that is based on the classical Bateson–Dobzhansky–Muller (BDM) model. Then this approach is incorporated into the metapopulation framework. The resulting model allows one to study the dynamics of both the overall diversity and the genetic structure of the system of populations undergoing radiation.

12.2 MODELING GENETICS OF REPRODUCTIVE ISOLATION

Most of the existing approaches for modeling the genetics of reproductive isolation utilize the idea first expressed explicitly by Bateson (1909), Dobzhansky (1937), and Muller (1942) according to which reproductive isolation is a consequence of "incompatibilities" between different genes and traits. This section starts by describing a simple two-locus two-allele model formalizing this idea. Then it briefly discusses the notions of nearly neutral networks and holey adaptive landscapes that provide a multilocus generalization of the Bateson–Dobzhansky–Muller model. Finally, a specific multilocus model is described, which is used later in studying the dynamics of speciation and diversification within the metapopulation framework.

The Bateson–Dobzhansky–Muller Model

The BDM model makes a very specific assumption about the genetic architecture of reproductive isolation, namely that there are two alleles at different loci that are "incompatible." Growing experimental evidence strongly supports this assumption (Wu and Palopoli, 1994; Orr, 1995; Orr and Orr, 1996; Wu, 2001). Let the alternative alleles at the two loci under consideration be denoted as **A**, **a** and **B**, **b**. The easiest way to illustrate the BDM model is by means of the corresponding adaptive landscapes shown in Fig. 12.1. Figure 12.1 assumes that alleles **a** and **B** are incompatible in the sense that individuals carrying both of them have zero viability (Fig. 12.1a) or that females carrying allele **a** do not mate with males carrying allele **B** (Fig. 12.1b). Speciation occurs if two populations that have initially the same genetic composition end up at the opposite sides of the ridge of high fitness values at genetic states with genotypes **AABB** and **aabb**. There are two possibilities for this to happen. The two populations can start with the same genotype **AAbb** and then evolve in different directions by fixing incompatible alleles (as illustrated by the arrows in Fig. 12.1a) or the populations can start with the same genotype **AABB** (or **aabb**) and then one of them will fix two new alleles (as illustrated by the arrows in Fig. 12.1b). In this model, the populations are not required to cross any adaptive valleys to evolve reproductive isolation, as they simply follow a ridge of high fitness values. The evolution along this ridge can be driven by any of the evolutionary factors, such as mutation, random genetic drift, and natural or sexual selection (Gavrilets, 1999a, 2000).

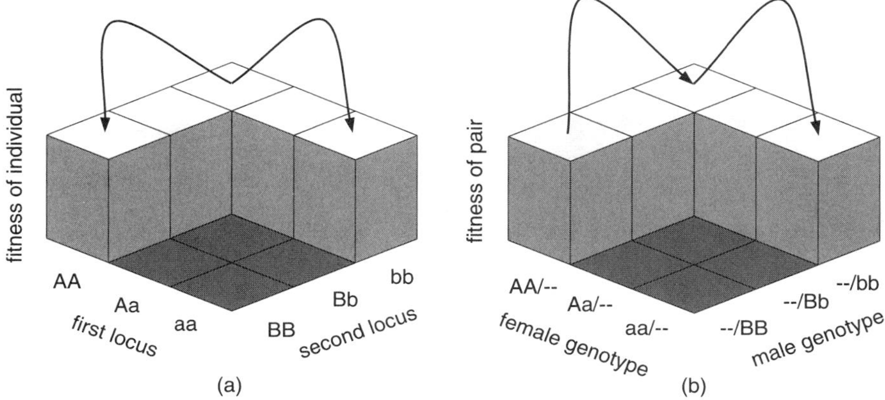

Fig. 12.1 Fitness landscapes in the diploid BDM model. (a) Fitness of an individual. (b) Fitness of a mating pair.

Holey Adaptive Landscapes

How common is the genetic architecture implied by the BDM model illustrated in Fig. 12.1? Starting with Wright (1932) typical adaptive, landscapes are usually imagined as very rough surfaces with many different peaks and valleys (see Fig. 12.2a). Continuous evolution on such landscapes requires crossing adaptive valleys. Wright's metaphor of rugged adaptive landscapes enforces a belief that the ridges of high fitness values, which are explicit in the BDM model, are very improbable. However, analyses have shown that the properties of the three-dimensional geographic landscapes implicit in Wright's metaphor are a rather poor indicator of the properties of adaptive landscapes describing genetic systems with thousands of loci (Gavrilets and Gravner, 1997; Gavrilets, 1997, 2003). The most prominent parts of three-dimensional landscapes are peaks and valleys. In contrast, the most prominent feature of adaptive landscapes of very high dimensionality are extensive *nearly neutral networks*, that is, connected networks of genotypes with very similar fitnesses that expand throughout the genotype space (Gavrilets and Gravner, 1997; Gavrilets, 1997, 2003; Reidys, 1997; Reidys et al., 1997). Among different nearly neutral networks, those with sufficiently high fitnesses are of particular importance as they allow for continuous evolutionary innovations without any significant loss in fitness. An important notion describing such networks is that of "holey adaptive landscapes." A *holey adaptive landscape* is defined as an adaptive landscape where relatively infrequent high-fitness genotypes form a contiguous set that percolates (i.e., expands) throughout the genotype space. An appropriate three-dimensional image of such an adaptive landscape that focuses exclusively on the percolating network of genotypes is a nearly flat surface with many holes representing genotypes that do not belong to the network (see Fig. 12.2b). The smoothness of the surface in Fig. 12.2 reflects close similarity between the fitnesses of the genotypes forming the corresponding nearly neutral network. The "holes" include both lower fitness genotypes ("valleys" and "slopes") and very high fitness genotypes (the "tips" of the adaptive peaks). The BDM model considered earlier provides one of the simplest examples of a holey adaptive landscape. Many more examples are known (Gavrilets and Gravner, 1997; Gavrilets, 1997, 2003).

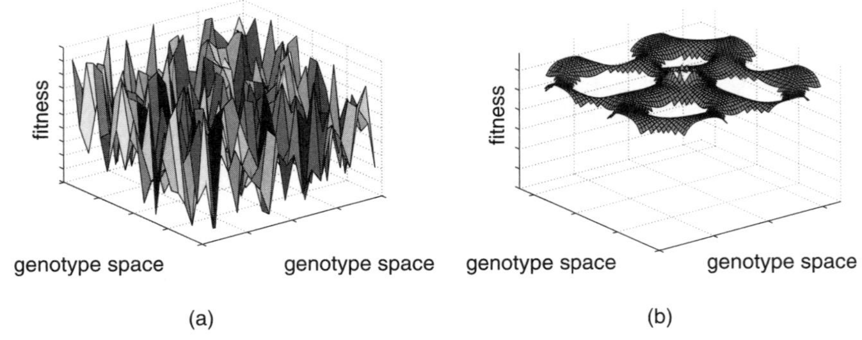

Fig. 12.2 Adaptive landscapes. (a) A rugged adaptive landscape. (b) A holey adaptive landscape.

A Multilocus Generalization of the BDM Model

The BDM model was formulated in terms of only two loci. However, existing data on the genetics of reproductive isolation show that typically there are many different loci underlying reproductive isolation, even at very early stages of divergence (Wu and Palopoli, 1994; Naveira and Masida, 1998; Wu, 2001). When there are many loci, rather than studying the dynamics of speciation *given a specific genetic architecture* (which is generally unknown), it becomes much more fruitful to look at the dynamics of speciation expected "on average." One such approach reduces the complexity of adaptive landscapes underlying reproductive isolation to simpler effects of genetic "incompatibilities" of certain types that arise with certain probabilities (Orr, 1995; Orr and Orr, 1996; Orr and Turelli, 2001).

Let us assume that there are two populations that have diverged in d diallelic loci potentially affecting reproductive isolation. Let us consider all possible combinations of k alleles with each allele taken from a different locus. For example, in the BDM model, $d = 2$ (see Fig. 12.2). If $k = 2$, there are two "parental" combinations, **AB** and **ab**, and two "hybrid" combinations, **Ab** and **aB**. With $d = 5$ loci, there are 38 "hybrid" combinations if $k = 3$ alleles, and 78 such combinations if $k = 4$ alleles. Assume that each "hybrid" combination can be incompatible with probability q. Here, "incompatibility" means epistatic interaction between the alleles potentially resulting in a loss of fitness. Note that in the BDM model it is assumed that combination **aB** is definitely incompatible and that combination **Ab** is definitely compatible. In contrast, in the model just formulated, the overall number of incompatibilities is a stochastic variable.

The next question is how exactly the incompatibilities translate into reproductive isolation between the populations. Following Orr and Orr (1996), let us assume that complete reproductive isolation occurs when C incompatibilities separate the populations. Note that in the BDM model, $C = 1$. In general, one expects that as genetic distance d between the populations increases, the expected number of incompatibilities, I, increases as well. One also expects that the probability $w(d)$ that two populations are not reproductively isolated decreases. Box 12.1 makes this intuition more precise. Results given in Box 12.1 are compatible with the general observation that the degree of reproductive isolation increases with genetic divergence between the parental organisms (Edmands, 2002). It should also be intuitively clear that adaptive landscapes implied by this model are "holey" and, thus, allow for extensive genetic divergence in a (nearly) neutral fashion.

To predict the dynamics of speciation, it will be assumed that the probability of compatibility of two populations that have diverged in d loci is given by the threshold function

$$w(d) = \begin{cases} 1 & \text{for } d < K, \\ 0 & \text{for } d \geq K \end{cases} \tag{12.5}$$

(Gavrilets et al., 1998; Gavrilets, 1999a, 2000). This function implies that genotypes that are different in less than K loci are perfectly compatible, whereas genotypes different in K or more loci are isolated reproductively. The neutral case

BOX 12.1 Properties of a Multilocus BDM Model

Consider two populations that have diverged in d loci. The number of incompatibilities between them is a random variable that follows a Poisson distribution with parameter

$$I = q\binom{d}{k}(2^k - k - 1), \tag{B1}$$

where $\binom{d}{k}$ is the binomial coefficient (Walsh, 2003). Parameter I gives the expected number of incompatibilities. The value of I increases very rapidly ("snowballs") with genetic distance d, approximately as the kth order of genetic distance d. Note that the snowball effect is much more pronounced with larger values of k (Orr, 1995).

Because the number of incompatibilities follows a Poisson distribution, the probability that two genotypes (or populations) at genetic distance d are not isolated reproductively is approximately

$$w(d) = \sum_{i=0}^{C-1} \exp(-I)\frac{I^i}{i!} = \frac{\Gamma(C, I)}{\Gamma(C)} \tag{B2}$$

Here $\Gamma(\cdot,\cdot)$ and $\Gamma(\cdot)$ are the incomplete gamma function and gamma function, respectively (Gradshteyn and Ryzhik, 1994), and I is given by Eq. (B1). The probability that two genotypes at distance d are isolated reproductively is $1 - w(d)$. The average K, variance var(K), and the coefficient of variation CV(K) of the number of substitutions required for speciation (i.e., for complete reproductive isolation) can be found in a straightforward manner using Eq. (B2). These values are

$$K = \nu_k^{-1/k}\frac{\Gamma(C+1+1/k)}{\Gamma(C+1)} \approx \left(\frac{C}{\nu_k}\right)^{1/k}, \tag{B3a}$$

$$\text{var}(K) = \nu_k^{-2/k}\frac{\Gamma(C+1+2/k)\Gamma(C+1) - \Gamma(C+1+1/k)^2}{\Gamma(C+1)^2}, \tag{B3b}$$

$$CV(K) = \sqrt{\frac{\Gamma(C+1+2/k)\Gamma(C+1)}{\Gamma(C+1+1/k)^2} - 1} \approx \frac{1}{k\sqrt{C}} \tag{B3c}$$

where $\nu_k = q\,(2^k - k + 1)/k!$ and the approximations assume that both k and C are not too small (>3). Figure B12.1 shows that as the genetic distance d exceeds the value K, the probability of no complete reproductive isolation undergoes a rapid transition from 1 to 0. This "threshold effect" is especially strong when many complex incompatibilities are required for complete reproductive isolation. The latter feature is also apparent from the fact that the coefficient of variation CV(K) quickly goes to zero as C or k become large.

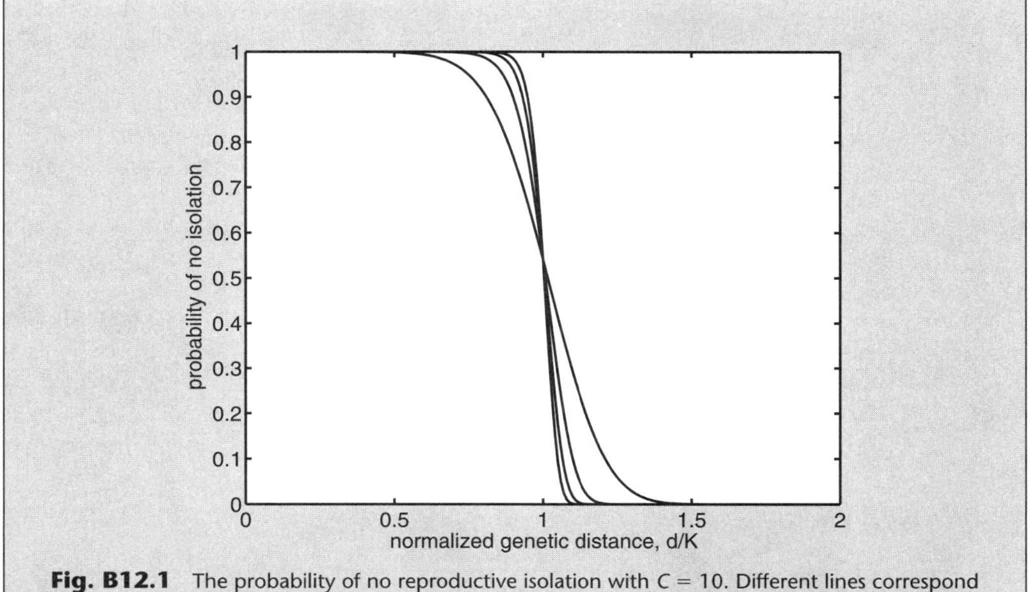

Fig. B12.1 The probability of no reproductive isolation with $C = 10$. Different lines correspond to $k = 2$ (the shallowest), 4, 6, and 8 (the steepest). Note that on the normalized scale d/K, the probability of no reproductive isolation does not depend on q.

(i.e., the case of no reproductive isolation) corresponds to K larger than the number of loci. The function defined by the Eq. (12.5) can be viewed as a limiting case of the function $w(d)$ given by Eq. (12.20) in Box 12.1 when k or C are large, that is, when reproductive isolation is due to many complex incompatibilities. As noted previously, the threshold function of reproductive compatibility was utilized to explore various features of the dynamics of speciation in systems of two stable populations using analytical approximations (Gavrilets, 1999a, 2000) and in one- and two-dimensional stepping-stone systems with stable populations using individual-based simulations (Gavrilets et al., 1998, 2000b).

12.3 DYNAMICS OF DIVERSIFICATION IN DEME-BASED MODELS

Analyzing the dynamics of genetic diversification in systems with a large number of interconnected populations and with individuals characterized by a large number of genes requires simplifying approximations. The approximations made in this study are discussed explicitly later at appropriate places. However, there is one approximation that has to be clarified right away to avoid possible confusion. A straightforward interpretation of the models to be considered next is that they describe *speciation caused by the spatial spread of mutually incompatible nearly neutral genes*. That is, these models are "neutral" (cf. Hubbell, 2001) in the sense that they do not explicitly specify the effects of genetic differences on viability and fertility of individuals (and populations). However, such effects are implicitly present. The underlying picture is that of the evolution along an extensive system of ridges in a holey adaptive landscape with reproductive isolation following as a by-product of sufficient genetic divergence (as in the model discussed in the previous section). The existence of "holes" in the adaptive landscape means that the differences in fitness are present explicitly. However, because of (i) the separation of the time scales (i.e., rapid "adaptive" evolution from a "hole" toward a high-fitness ridge followed by slow "nearly neutral" evolution along the ridge), (ii) the assumption that there are always many possible directions (i.e., ridges) for the evolution of populations, and (iii) the fact that the "choice" of a specific direction is to a large degree random, the dynamics of speciation can be treated as effectively neutral. This "nearly neutral" approach is problematic in situations where reproductive isolation depends on ecological factors that vary between the populations. However, this approach appears to be a good approximation when reproductive isolation is controlled genetically and is not affected by external conditions. The end of this section discusses how adaptation is expected to affect the conclusions reached within the nearly neutral framework.

Model Description

Throughout this section, sexual species are considered with nonoverlapping generations. The main motivation of this section is to develop a mathematical model describing the dynamics of (adaptive) radiation following colonization of a new environment or appearance of a "key innovation." [The word "adaptive" is put in parentheses because adaptation is treated implicitly rather than explicitly.] The whole set of populations in the system is considered a "clade."

Spatial Arrangement

A habitat is considered subdivided into a large but finite number n of discrete "patches" arranged on a line (in the case of one-dimensional systems) or on a two-dimensional square lattice. Each patch can support one population of a species. Each patch can receive colonizers from up to two (in the one-dimensional cases) or up to four (in the two-dimensional cases) neighboring patches. The number of neighboring patches will be smaller for patches at the boundary.

Population State

It is assumed that there is a large number L of possibly linked diallelic loci affecting reproductive isolation or other phenotypic traits (morphological, behavioral, etc.) that differentiate species (genera, families, etc.). Each population is characterized by the genetic sequence of its most common genotype. Note that because L is large, the number of possible sequences is enormous. This chapter neglects within-population genetic variation. This implies that the size of local populations is relatively small and that the rates of mutation and migration are small as well.

System State

The system state is characterized by the set of states of the populations present. One can image a population as a point in the multidimensional genotype space. The clade will be a cloud of points that changes both its structure and location in the genotype space as a consequence of ecological and evolutionary processes.

Fixation of Mutations

At each time step in each population a mutation can be fixed at each locus under consideration with a very small probability μ. The fixation rate per genotype, $\nu = \mu L$, is assumed to be small as well. Following the general framework discussed earlier, it is assumed that mutations are nearly neutral. For neutral mutations, the probability of fixation is equal to the mutation rate (Kimura, 1983). Thus, for haploid species, μ is equal to the probability of mutation per allele, whereas for diploid species, μ is equal to twice the probability of mutation per allele.

Extinction and Recolonization

At each time step, each population may go extinct with a small probability δ. Extinction is followed rapidly by colonization from one of the neighboring patches chosen randomly. Alternatively, one can think of extinction of a local population as being *caused* by successful invasion from one of the neighboring demes. A newly established population grows to the equilibrium size rapidly.

Genetic Clusters and Species

It will be assumed that genetic differences lead to genetic or phenotypic incompatibilities between different populations, for example, as specified in the previous section. Different populations are assigned to different genetic clusters (species, families, genera, etc.) based on the degree of genetic divergence characterized by genetic distance d. Recall that d is defined as the number of genes that differ between two populations. The maximum divergence allowed within a cluster is characterized by parameter K [see Eq. (12.5)]. In the numerical simulations described later, the single linkage clustering technique is used (e.g., Everitt, 1993). This means that two populations separated by a distance d equal to or larger than the corresponding threshold K may potentially still belong to the same cluster if there is another "intermediate" population "linking" them together. For example, if both the genetic distance d_{12} between populations 1 and 2 and the genetic distance d_{23} between populations 2 and 3 are smaller than K, then all three populations 1, 2, and 3 will belong to the same cluster, even if the genetic distance d_{13} between populations 1 and 3 is equal to or larger than K. According to this definition, all populations forming a ring species (e.g., Mayr, 1942, 1963; Wake, 1997; Irwin et al., 2001) would belong to the same species. Note that the case of $K = 1$ is compatible with the previous work discussed in Section 12.1. In this case, each cluster (e.g., species) is defined by a unique sequence of genes (at the set of loci under consideration). The case of $K = 2$ corresponds to the BDM model. Alternatively, if new species result from the accumulation of a number of genetic (morphological) differences, then larger values of K are more appropriate. Genetic clusters corresponding to different values of K can also be interpreted as describing different levels of taxonomic classification. For example, let us specify an increasing sequence $K_1 < K_2 < K_3 < \ldots$. Then, all populations at a genetic distance less than K_1 can be thought of as belonging to the same species, all populations at genetic distances that are larger or equal than K_1 but are smaller than K_2 can be thought of as belonging to different species within the same genus, all populations at genetic distances that are larger or equal than K_2 but are smaller than K_3 can be thought of as belonging to different species and genera within the same family, etc.

Migration into Occupied Patches

It is assumed that migration into occupied demes has no effect on the genetic composition of the resident population even if immigrants are coming from the same species and are genetically compatible and able to mate with the residents. As a working example, a plant metapopulation is envisioned where local demes produce a large number of seeds of which only few germinate. In this case, migrant seeds will have an extremely small probability of germinating unless there is an extinction event eliminating all or most resident plants. In a similar way, if there is frequency-dependent selection against immigrants, then again one can neglect effects of migration (other than bringing colonizers into an empty patch). The assumption of no effects of gene flow is justified only if the rates of immigration are very small or the selection against immigrants is very strong. The effects of gene flow on the model's dynamics will be considered elsewhere (M. Saum and S. Gavrilets, unpublished results).

Dynamic Scenario

Identifying and understanding dynamic features of the process of diversification following colonization of a new environment are key. As an initial condition, it is assumed that all patches are occupied by populations with exactly the same "founder" genotype. This implicitly assumes that the spread of the species across the system of patches from the point of its initial invasion happens on a (much) shorter time scale than that of mutation. This assumption appears to be reasonable. Initial spread is followed by the diversification phase during which the founder species splits into an increasing number of different clusters. Eventually the system reaches a state of stochastic equilibrium in which the number of clusters and their different characteristics fluctuate around certain values. Note that even after reaching this state, the clade keeps evolving, as different species (or clusters) go extinct and their place is taken by new species (clusters). Parameters and different dynamic characteristics of the model to be studied are defined in Box 12.2.

The analytical approximations for the diversity S, the average cluster range R, and the turnover rate T will assume that the system size is sufficiently large so that the effects of the boundaries are negligible. For the model under consideration, the characteristic linear size is

$$l_c = \sqrt{\frac{\delta K}{\nu}} \tag{12.6}$$

(cf. Sawyer 1977a,b, 1979; Bramson et al., 1996; Durrett and Levin, 1996). Patches separated by (spatial) distances much larger than l_c demes are expected to behave largely as independent. Also for systems with the linear dimension l larger than l_c, the effects of borders will be small. [In one-dimensional systems, $l = n$, whereas in in two-dimensional systems, $l = \sqrt{n}$.] This implies that for large systems (with $l > l_c$), the diversity S will increase linearly with the number of patches in the system, whereas the range R will not depend on n. Note that for the parameter values used in the simulations, l_c ranges

BOX 12.2 Model Characteristics

Parameters and dynamic characteristics. The following is a list of the parameters of the model considered in the main text:

- the dimensionality of the system (one dimensional or two dimensional)
- the system size n (i.e., the number of patches in the system)
- the local extinction–recolonization rate δ
- the fixation rate per genotype ν (which is actually the product of the number of loci L and the fixation rate per locus μ)
- the clustering level K

The effects of these parameters on the following characteristics need to be understood:

- the average time to the beginning of radiation, t_b, defined as the average waiting time until the first split of an initially uniform population into at least two clusters
- the average duration of radiation, t_d, defined as the average waiting time from t_b to the time when the number of clusters reaches the (stochastic) equilibrium value for the first time
- the diversity, S, defined as the average number of clusters in the clade at the stochastic equilibrium
- the average cluster range, R, at the stochastic equilibrium, defined as the number of populations that belong to an average cluster
- the average pairwise genetic distance, D_c, between the members of the same cluster
- the cluster diameter, $D_{c,\max}$, defined as the maximum genetic distance between the members of the same cluster
- the turnover rate, T, defined as the number of new clusters emerging per unit of time divided by the standing diversity S
- the clade disparity, D, defined as the average pairwise distance between all populations in the system
- the average genetic distance of the clade from the founder, d_f, defined as the average of pairwise distances between all populations and the species founder

The main text described analytical approximations for S, R, T, and d_f and used numerical simulations both to check the validity of these approximations and to understand the dynamics of other characteristics. The following is a list of parameter values used in numerical simulation:

- number of loci $L = 100$
- fixation rate per locus $\mu = 10^{-6}$, 4×10^{-6}, 16×10^{-6}
- extinction–recolonization rate $\delta = 0.25 \times 10^{-2}$, 10^{-2}, 4×10^{-2}
- clustering levels $K = 1, 2, 4, 8, 16$
- system sizes:
  $-8^2 \times 1$, $16^2 \times 1$, $32^2 \times 1$, $64^2 \times 1$ for one-dimensional systems and
  -8×8, 16×16, 32×32, 64×64 for two-dimensional systems
- the number of runs for each parameter configuration is 40

from 1.25 (for the smallest extinction rate $\delta = 0.0025$, $K = 1$, and the largest fixation rate $\mu = 16 \times 10^{-6}$) to 80 (for the largest extinction rate $\delta = 0.04$, $K = 16$, and the smallest fixation rate $\mu = 10^{-6}$). This suggests that in the one-dimensional numerical examples, the effects of borders will be insignificant except for the smallest system (64×1) with the largest K (=16). In contrast, in the two-dimensional examples, the effects of borders will be important even in the largest system (64×64) if K is large. Unfortunately, increasing the system size is currently impossible because of computation speed considerations.

Transient Dynamics

Figure 12.3 illustrates the transient dynamics of the number of different clusters as well as the clade disparity D and the average distance from the founder d_f. The dynamics of the two latter measures do not seem to depend on spatial dimensionality.

The dynamics of the average distance from the founder depends only on the fixation rate per locus μ and is approximated by

$$d_f(t) = \frac{L}{2}\left[1 - e^{-2\mu t}\right] \tag{12.7}$$

(Gavrilets, 1999b). That is, d_f asymptotically approaches the distance equal to one-half of the maximum possible distance. This implies that after a sufficiently long time the members of the clade will be different from the founder in half of the genes on average. Moreover, the clade can be equally likely found in any part of the genotype space. Equation (12.7) can be used both to check the constancy of the rate of evolution in time and to estimate its value (see Gavrilets, 1999b).

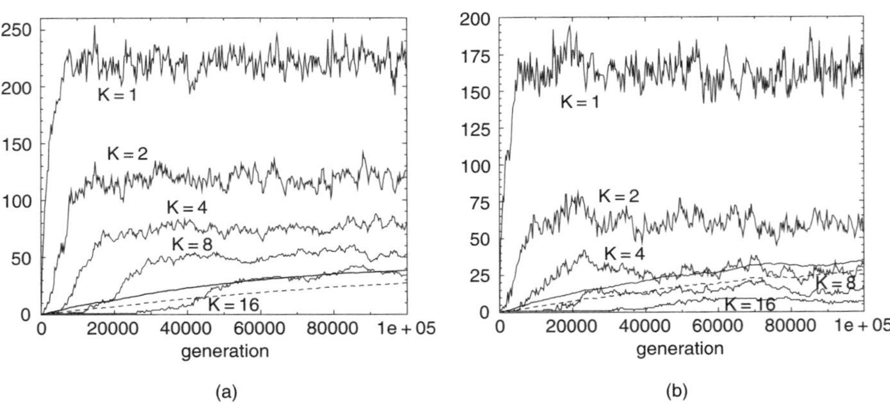

Fig. 12.3 Dynamics of diversity S at five different clustering levels K (marked in the figure), of clade disparity D (bold line), and average distance from the founder d_f (dashed line). Parameters: $\mu = 4 \times 10^{-6}$, $\delta = 0.01$. The statistics are computed every 250th generation. (a) One-dimensional $32^2 \times 1$ system. (b) Two-dimensional 32×32 system. Notice the difference in the scale of the vertical axes.

The initial dynamics of the clade disparity D are similar to that of d_f with the difference that D increases twice as fast [i.e., the exponential term in an analog of Eq. (12.7) has $4\mu t$ rather than $2\mu t$]. A simple explanation of this fact is that while the disparity is computed on the basis of pairs of evolving lineages, in computing d_b one lineage in each pair (i.e., the founder) does not change. For a clade with no spatial structure the dynamics of disparity D are understood (Gavrilets, 1999b). Unfortunately, for spatially explicit systems, neither the equilibrium value of D nor its dynamics on the intermediate time scales are known. In larger systems, approaching an equilibrium for D takes a very long time. Figure 12.3 illustrates the important observation that the diversity S at low taxonomic levels (i.e., at small K) equilibrates faster than the clade disparity D, whereas the equilibration of the diversity at higher taxonomic levels (i.e., at large K) can take a comparable or longer time. In the latter case, very high values of D (relative to its asymptotic equilibrium value) can be observed simultaneously with very low taxonomic diversity. The pattern of elevated disparity early in the history of many clades has been traditionally explained by paleontologists by invoking explanations that postulate temporal changes in the types and/or levels of forces driving divergence (Valentine, 1980; Foote, 1992, 1999; Erwin, 1994; Wagner, 1995; Lupia, 1999). However, both previous work (Gavrilets, 1999b) and models studied here show that these patterns are perfectly compatible with the null model of time homogeneous diversification.

The dynamics of other characteristics depend crucially on the spatial dimensionality of the system. Note that the one-dimensional version of this model was introduced and analyzed in Gavrilets et al. (2000a). Figure 12.4 illustrates the dependence of the time to the beginning of radiation, t_b, and the duration of radiation, t_d, on parameters in more detail. The time to the beginning of radiation increases (approximately exponentially) with K. In biological terms, higher taxonomic groups arise later in the history of the clade. t_b decreases with the fixation rate ν (apparently as $1/\nu$). Increasing ν by a certain factor results in a smaller increase in t_b than the proportional increase in K. The time t_b increases weakly with the extinction/recolonization rate δ and system size n. At $K = 1$, t_b can be approximated as the inverse of the expected number of mutations per clade, that is, $t_b \approx 1/(n\nu)$. The duration of radiation t_d is not very sensitive to K and δ (increasing weakly with both these parameters) but is much more sensitive to ν. It appears that t_d is on the order of $1/\nu$. This feature of the dynamics of radiation is compatible with that for the dynamics of parapatric speciation where the time interval during which the intermediate forms are present has the order of the reciprocal of the mutation rate (Gavrilets, 2000). Numerical simulation also show, as expected, that small systems (with small n) reach stochastic equilibrium faster than large systems (with large n).

Stochastic Equilibrium: One-Dimensional Systems

One can use certain approximations to evaluate cluster ranges, diversity, and turnover rates (Gavrilets et al., 2000a). The following formulas assume that the fixation rate per genotype is much smaller than the extinction/colonization rate ($\nu \ll \delta$).

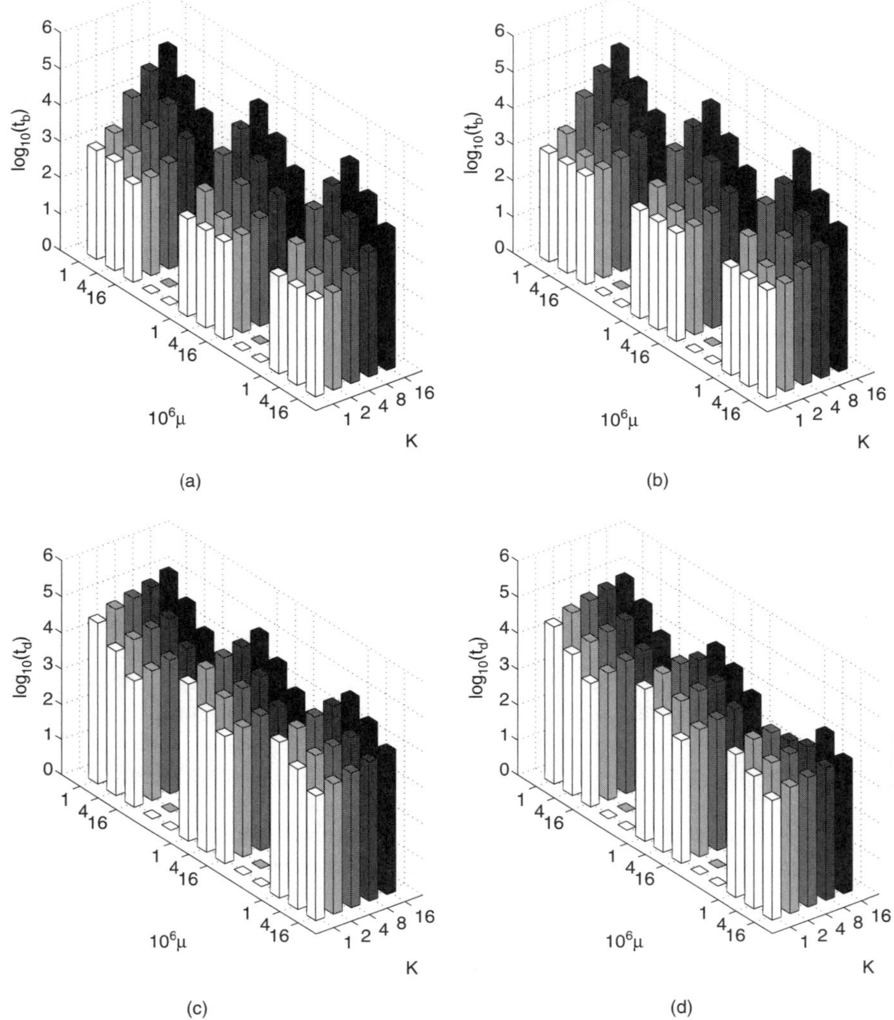

Fig. 12.4 The average waiting time to the beginning of radiation and the average duration of radiation. Left column: $32^2 \times 1$ system. Right column: 32×32 system. First row: the average time to the beginning of radiation t_b. Second row: the average duration of radiation t_d. Within each figure the three sets of bars correspond to $\delta = 0.0025$, 0.01, and 0.04 (from left to right). Average values over 40 runs.

The average range of a cluster can be approximated as

$$R = \sqrt{\frac{\pi\delta}{2\nu}} \frac{(K-1)!}{\Gamma(K-1/2)}, \tag{12.8a}$$

where Γ is the gamma function (Gradshteyn and Ryzhik, 1994). This expression simplifies to

$$R = \sqrt{\frac{\delta}{2\nu}} \tag{12.8b}$$

for $K = 1$ and to

$$R = \sqrt{\frac{\pi\delta K}{2\nu}} \qquad (12.8c)$$

for large K. R is obtained as the inverse of the probability that two populations belong to the same cluster. The aforementioned equations assume that each mutation is unique. A correction can be made to account for backward mutations. In this case, instead of K one needs to use

$$\widetilde{K} = \frac{\ln(1 - \frac{2}{L}K)}{\ln(1 - \frac{2}{L})}, \qquad (12.9)$$

which is the overall expected number of mutations needed to move at genetic distance K from a reference genotype. The aforementioned expression for $\widetilde{K}$, which was found from Eq. (12.7), simplifies to K if the number of loci L is very large.

The average diversity is just $S = n/R$, leading to

$$S = \sqrt{\frac{2\nu}{\pi\delta}} \frac{\Gamma(K-1/2)}{(K-1)!} n. \qquad (12.10)$$

Figure 12.5a illustrates the dependence of S on the parameters of the model observed in simulations. Biological intuition tells one that increasing the rate of fixation of new mutations should increase the rate of speciation, thus increasing the number of species in the system. Decreasing the rate of

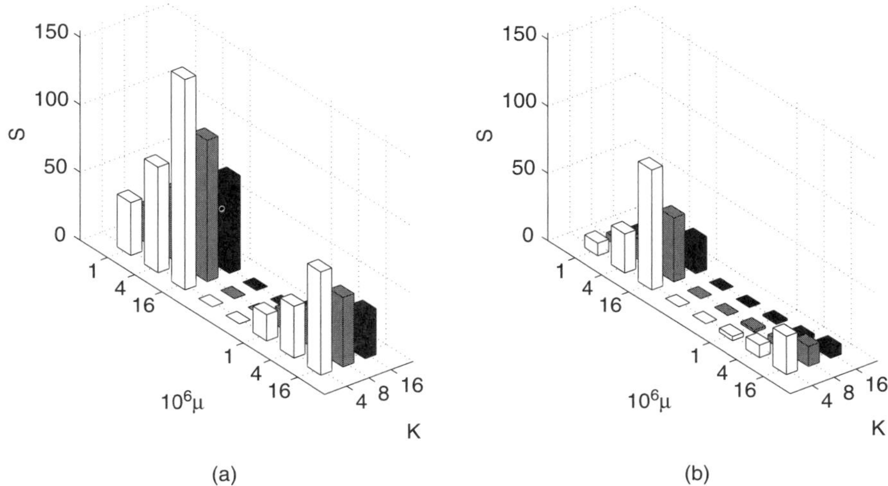

Fig. 12.5 The number of clusters. (a) $32^2 \times 1$ system. (b) 32×32 system. Within each figure the two sets of bars correspond to $\delta = 0.01$ (left) and 0.04 (right). The averages over generations 200,000 through 500,000 and over 40 runs.

extinction–colonization should have a similar effect because larger levels of genetic variation will accumulate in the system. Equations (12.8) and (12.10) support these intuitions. For example, decreasing δ by factor 4 will result in an increase in the number of species by factor 2. These results provide a formal justification for the idea that species can accumulate rapidly after colonizing a new environment if local populations in the novel environment have a reduced probability of extinction (e.g., Mayr, 1963; Allmon et al., 1998; Schluter, 1998, 2000). Similar effects can be achieved by increasing the fixation rate ν.

With $K = 1$, the rate of turnover, T, can be evaluated by dividing the number of new clusters per generation, which is νn, by the standing diversity S, leading to

$$T = \sqrt{\frac{\delta\nu}{2n}}. \tag{12.11a}$$

The consideration of the time that it takes for a typical cluster to go extinct leads to

$$T = \frac{\nu}{K} \tag{12.11b}$$

for large K (Gavrilets et al., 2000a). The turnover rate depends weakly on δ for $K = 1$ and becomes approximately independent of δ for large K. The latter counterintuitive prediction is explained by the fact that the increase in the overall extinction rate of species resulting from an increase in δ is exactly balanced by a decrease in the number of species S maintained in the system.

To check these analytical approximations and to get further insights into the model dynamics, numerical simulations were performed (Gavrilets et al., 2000a). In most cases, Eq. (12.10) underestimates the average number of species by a couple of percents, whereas Eq. (12.11) overestimates the turnover rate by about 5–10%. In the case of the smallest mutation rate, the errors are slightly higher.

Additional simulations were used to analyze the structure of the clade in the genotype space. The left column of Fig. 12.6 illustrates within-cluster average pairwise distance, D_c, and cluster diameter $D_{c,\max}$. Figure 12.6 shows that although there is plenty of within-cluster genetic variation, typical members of a cluster are at distances that are smaller than K. Effects of ν and δ do not seem to be significant. In fact, Fig. 12.6 and similar results not shown here suggest that for one-dimensional systems, roughly $D_c \approx K/4$ and $D_{c,\max} \approx K/2$.

Stochastic Equilibrium: Two-Dimensional Systems

Approximating the average range of clusters R in the two-dimensional case is much more difficult than in the one-dimensional case. If $K = 1$, then the results of Bramson et al. (1996) on the number of species $\hat{R}$ found within a square impeded within a much large area [see Eqs. (12.1) and (12.3)] give

$$\hat{R} = \frac{2\pi\delta}{\nu} \frac{1}{\left[\ln(\delta/\nu)\right]^2}. \tag{12.12a}$$

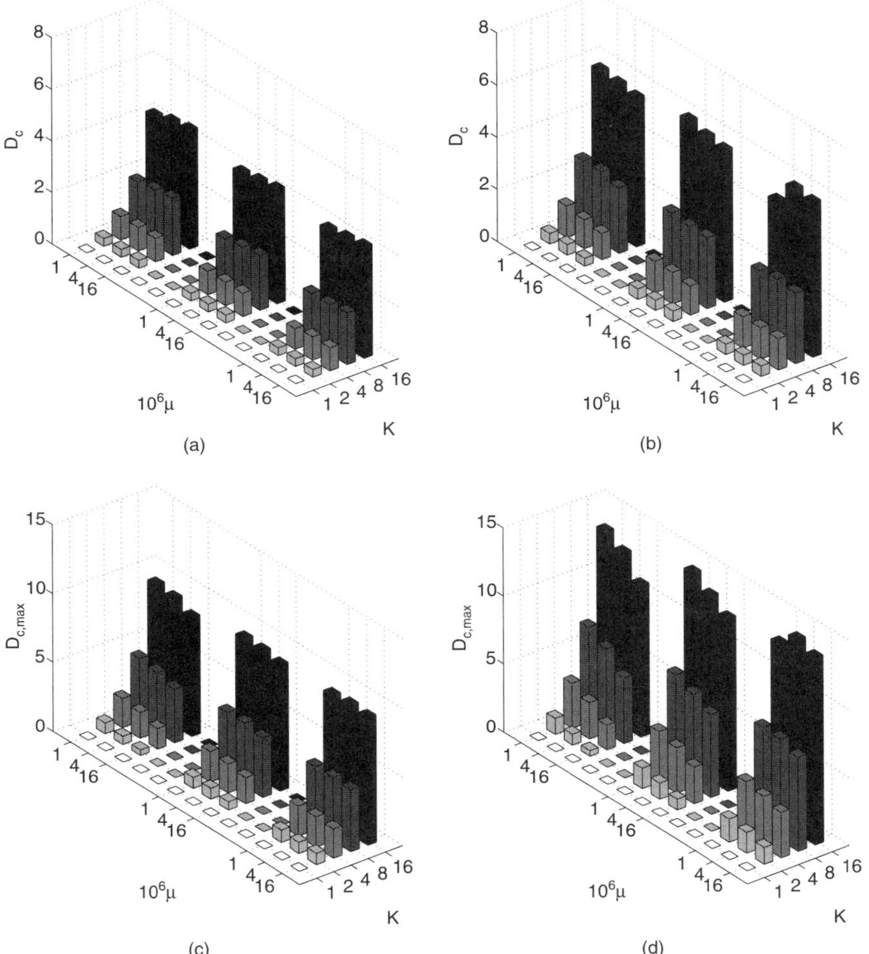

Fig. 12.6 Within-cluster genetic variation. First column: $32^2 \times 1$ system. Second column: 32×32 system. First row: average pairwise distance D_c, Second row: cluster diameter, $D_{c,\max}$. Within each figure the three sets of bars correspond to $\delta = 0.0025$, 0.01, and 0.04 (from left to right). The averages over generations 200,000 through 500,000 and over 40 runs.

This approximation implies that the size of the system is sufficiently large. $\hat{R}$ overestimates R because many species occupying nearby areas may still have few representatives within the sampling square.

One can also estimate the average number of populations $\tilde{R}$ that have the same type as the population from a randomly chosen patch. $\tilde{R}$ is somewhat larger than R. For example, if there are five clusters with 1, 2, 3, 4, and 5 populations, respectively, then $R = (1 + 2 + 3 + 4 + 5)/5 = 3$ but $\tilde{R} = (1^2 + 2^2 + 3^2 + 4^2 + 5^2)/15 = 3.67$. If $K = 1$, then range $\tilde{R}$ can be found by integrating the probability of identity $I(d, \nu)$ of two genotypes found a certain distance apart (as found by Sawyer, 1977a) over all possible spatial positions. This approach leads to

$$\tilde{R} = \frac{\pi\delta}{4\nu} \frac{1}{\ln(\delta/4\nu) + 2\pi\delta}. \tag{12.12b}$$

(see Appendix 12.1). For large K,

$$\tilde{R} = \frac{\pi\delta K}{2\nu}\,\Phi(K,\nu), \tag{12.12c}$$

where $\Phi(K, \nu)$ is a function that depends only weakly on its arguments (see Appendix 12.1).

Simulations were performed to check the validity of the approximations (12.12) for the average range size R. The fit was within 30–50%, which is satisfactory given a number of simplifying assumptions involved and the small size of systems used in numerical simulations.

As before, the average diversity is just $S = n/R$ $\left(\text{or } n/\tilde{R}\right)$. Both analytical approximations and numerical results show that the diversity in two-dimensional systems is (much) lower than in comparable one-dimensional systems. The differences are most apparent when S is relatively small, which happens with large δ and K and small μ. Some data on the number of clusters are summarized in Fig. 12.5b.

With $K = 1$, the turnover rate can be estimated by dividing the number of new clusters, νn, by the standing diversity S, leading to

$$T = \frac{2\pi\delta}{\left[\ln(\delta/\nu)\right]^2} \tag{12.13a}$$

in the case of R as given by Eq. (12.12a) and to

$$T = \frac{\pi\delta}{4} \frac{1}{\ln(\delta/4\nu) + 2\pi\delta}. \tag{12.13b}$$

in the case of $\tilde{R}$ as given by Eq. (12.12b).

To approximate the turnover rate in the case of $K > 1$, an intuitive but not rigorous approach is to consider the average waiting time until a cluster of an average size goes extinct. Because the process is "neutral," the average time to extinction starting with a size R should be the same as the average time t^* to grow to size R starting with a single population. The latter time t^* can be approximated by the solution of equation

$$R = \frac{\pi\delta t^*}{\ln(t^*)}$$

(Kelly, 1977; Sawyer, 1979; Bramson and Griffeath, 1980). Then the average lifetime of a cluster of average size is $2t^*$. The turnover rate T is $1/(2t^*)$, which leads to T being given by a solution of

$$T = \frac{\nu}{K} \frac{1}{\Phi(K,\nu)\ln(1/2T)}. \tag{12.13c}$$

This suggests that T is of the order of ν/K [the logarithmic dependence in Eq. (12.13c) is less important]. The latter expression does not depend on the extinction rate, which is similar to the one-dimensional case.

Simulations were also used to analyze the structure of the clade in the genotype space. The right column of Fig. 12.6 illustrates the average pairwise within-cluster distance, D_c, and cluster diameter $D_{c,\max}$. As in the one-dimensional case, Fig. 12.6 shows that although there is plenty of within-cluster genetic variation, typical members of a cluster are at a distance that is smaller than K. Effects of ν and δ on D_c and $D_{c,\max}$ do not seem to be significant. Typically, populations that belong to the same cluster are also spatially contiguous.

The number of patches in the system n is also of importance. If n is too small, significant diversification will be prevented. For example, consider a case with $\delta = 10^{-2}$, $\mu = 10^{-6}$, and $K = 16$. Then numerical simulations show that small systems with 8×8 patches fail to diversify and have a single cluster present. In contrast, in 16×16 systems, there are on average just over two clusters, whereas in 32×32 systems this number goes to eight.

There is also a number of additional observations valid for both one- and two dimensional systems that follow from numerical simulations similar to those described here and individual-based simulations reported elsewhere (Gavrilets et al., 1998, 2000b).

- Clusters differ in their "life spans." The number of unsuccessful speciation events (i.e., the number of clusters that are very short lived) is much larger than the number of "real" speciation events (i.e., the number of clusters that exist for a long time).
- The distribution of species range size is right skewed on the linear scale and becomes left skewed on the log scale. These properties are similar to those of the species range distributions estimated from real data (e.g., Gaston, 1996, 1998). Species are more likely to break at the center of their range (cf. Gavrilets et al., 1998, 2000b).
- The larger the species range, the more likely it will break. However, because there are not many species with very large range sizes, the species that contribute the largest number of new species are those with intermediate range sizes.

Effects of Adaptation

The aforementioned results are based on models treating the dynamics of diversification as a neutral process. The justification of this approach was given at the beginning of the previous section. An important question is how these results will be affected by adaptation that is expected to take place simultaneously with diversification. One simple qualitative approach is to consider the expected effects of adaptation on the parameters ν and δ (which were assumed to be constant in the previous subsection).

Adaptation to local conditions can be controlled by loci different from the loci underlying reproductive isolation, or the two sets of loci can overlap partially or completely. Let us first assume that the two sets of loci are completely different. Then adaptation to local conditions is expected to result in decreasing the rate of local extinction δ and is not expected to affect the rate of fixation of new alleles ν in the genes underlying reproductive isolation. A consequence of these changes will be an increase in the equilibrium level of species diversity S. The turnover rate T will not be affected or will decrease somewhat.

Next assume that the two sets of loci overlap. Now the process of fixation of mutant alleles will not be neutral anymore. At the early stages of adaptation (and diversification), one expects many possible directions for evolution. However, as the clade as a whole rises higher and gets closer to a ridge of high fitness values in the adaptive landscape, one expects that it will become more and more difficult to find mutations increasing adaptation further. This will result in a decrease in the rate of fixation ν from a high level, expected when mutations are adaptive, to a lower level, expected when mutations are nearly neutral. As before, adaptation is expected to result in decreasing the rate of local extinction δ. The effects of a simultaneous decline in ν and δ on the clade diversity will depend on which parameter has experienced a larger change. A larger change in ν than in δ will result in a drop in the species diversity S. A smaller change in ν than in δ will result in increasing S. In both cases, one expects a decrease in the turnover rate T. These conclusions are preliminary and more concrete modeling work is definitely necessary.

12.4 DISCUSSION AND CONCLUSIONS

The dynamics of speciation and diversification in spatially explicit systems undergoing frequent local extinction and recolonization have been a subject of several recent theoretical studies (Bramson et al., 1996; Durrett and Levin, 1996; Allmon et al., 1998; Pelletier, 1999; Hubbell, 2001). In describing speciation, these studies used heuristic approaches, which postulated that new species emerge with certain probabilities and at certain population densities. Both the probability of speciation and the number of individuals in the new species have been shown to be very important in controlling various aspects of the diversification process. However, the heuristic nature of the approaches used did not allow one to uncover the relationships between these characteristics and microevolutionary processes. The major goal of this study was to develop more general approaches in which speciation is modeled explicitly rather than heuristically.

The approach adapted here is based on a multilocus generalization of the classic two-locus two-allele Bateson–Dobzhansky–Muller model. In the BDM model, reproductive isolation is reduced to a single "incompatibility" of two alleles at two different loci. This incompatibility is manifested in a reduced fitness component such as an individual's viability (in the case of postmating reproductive isolation) or the probability of mating between two parental forms (in the case of premating reproductive isolation). In the BDM model, reproductive isolation evolves as a by-product of genetic divergence, which can be driven by any of the evolutionary factors, such as mutation, random genetic drift, selection for adaptation to local biotic/abiotic environment, and sexual selection. The BDM model provides a way for the (sub)population to avoid any adaptive valleys on its was to a state of (complete) reproductive isolation, as the (sub)population evolves along a ridge of high fitness values in the corresponding holey adaptive landscapes. Although the genetic architecture implied by the BDM model, which results in a ridge of high fitness genotypes, may appear to be rather specific and uncommon, theoretical studies of multidimensional adaptive landscapes strongly suggest that it should be widespread (Gavrilets and Gravner, 1997; Gavrilets, 1997; Reidys, 1997; Reidys et al., 1997). These studies have shown that neutral and nearly neutral divergence along the corresponding holey adaptive landscapes can lead to strong reproductive isolation.

This chapter developed a model representing a straightforward multilocus generalization of the BDM model for the case when complete reproductive isolation requires a number C ($\geq$1) of incompatibilities between sets of k ($\geq$2) loci. The adaptive landscape underlying this model belongs to a class of holey adaptive landscapes. The accumulation of reproductive isolation in the model is characterized by the "threshold effect": as genetic distance between the two parental forms exceeds a certain value, the strength of reproductive isolation undergoes a rapid transition from low to high. This transition is especially rapid if both C and k are large. This property has been used previously to explore various features of the dynamics of speciation in systems of two stable populations (Gavrilets, 1999a, 2000), in one- and two-dimensional stepping-stone systems with stable populations (Gavrilets, et al., 1998, 2000b), and in one-dimensional metapopulations (Gavrilets et al., 2000a). This section consideres both one- and two-dimensional metapopulations, paying special attention to the cluster genetic structure of the diversifying clade. Speciation and diversification were modeled as a continuous process of mutation accumulation accompanied by the generation of new genetic clusters and contractions or expansions of their ranges.

The main motivation was to get a better understanding of the processes following colonization of a new environment or appearance of a new key innovation. This section focuses both on the properties of the transient dynamics of diversification and on the characteristics of the long-term stochastic equilibrium. Currently, empirical data on the dynamical features of (adaptive) diversification are scarce (Schluter, 2000; Section 3.5), with the best data coming from the fossil record (e.g., Valentine 1980; Foote, 1992, 1999; Erwin, 1994; Wagner, 1995; Lupia, 1999). A number of potentially important generalizations have emerged from the analyses described here.

i. The waiting time to the beginning of radiation t_b increases with decreasing the fixation rate per locus μ and increasing the number of genetic changes necessary for speciation K. The local extinction/recolonization rate δ, the dimensionality of the system, and the number of patches n have much smaller effects. In numerical simulations, the waiting time to the beginning of radiation was on the order of 10^3 generations (for $K = 1$ and 2) to 10^5 generations (for large K).
ii. The duration of radiation t_d depends mostly on the fixation rate μ. t_b increases weakly with K and system size n and decreases weakly with δ. The duration of radiation is longer in one-dimensional systems than in two-dimensional systems. In numerical simulations the order of t_b ranges from 10^4 to 10^5 generations.

iii. The transient dynamics of the diversity S (i.e., the number of clusters) and the disparity D (i.e., the average pairwise distance between populations in the clade) are decoupled to a certain degree. At low taxonomic levels (with small K), the diversity increases faster than the disparity, whereas at high taxonomic levels (with large K), the diversity increases slower than the disparity. This observation explains the difference between the patterns of diversification as observed in the fossil record (which are usually summarized at higher taxonomic levels, e.g., Valentine, 1980; Foote, 1992, 1999; Erwin, 1994; Wagner, 1995; Lupia, 1999) and for more recent groups (which are usually summarized at lower taxonomic levels, e.g., Schluter, 2000).

iv. The average genetic distance from the species founder increases monotonically at a constant rate controlled by the fixation rate. Note that Gavrilets (1999b) used this property to develop a method for testing the constancy of the rate of evolution and estimating its rate using morphological data.

v. The clade as a whole keeps changing genetically as it moves along the underlying holey adaptive landscape even after the number of species (or other genetic clusters) has approached an equilibrium level.

vi. The general effects of the model parameters on different equilibrium characteristics of the metapopulation are mostly as suggested by biological intuition. For example, diversity increases with mutation rate and decreases both with the local extinction/recolonization rate and the number of genetic differences required for speciation. However, the model predicts counterintuitively that the turnover rates do not depend (or weakly depend) on extinction rates and are controlled mostly by parameters ν and K. Intuition is a poor guidance as far as the structure of different genetic clusters in the multidimensional genotype space is concerned. Results of numerical simulations show that both the average pairwise distance within cluster and the cluster diameter are mostly controlled by parameter K and are close to its numerical value.

vii. Diversification requires that the overall number of patches in (or spatial area of) the system exceeds a certain minimum value. This effect may have contributed to the fact that in adaptive radiation of the west Indian *Anolis* lizards, within-island speciation occurred only on bigger islands, despite the fact that the degree of spatial heterogeneity does not seem to differ between the islands (Losos, 1998). In very large systems, the overall diversity increases linearly with the number of patches (or area).

viii. The results presented here show profound effects of spatial dimensionality on the dynamics of diversification and significant differences between one-dimensional systems (such as describing rivers, shores of lakes and oceans, and areas at a constant elevation in a mountain range) and two-dimensional systems (such as describing oceans and continental areas).

a. In general, the characteristics of two-dimensional systems are (much) more sensitive to parameter values than those of one-dimensional systems. For example, increasing the local extinction rate δ by a factor 25 will typically decrease the species diversity by the same factor in two-dimensional systems. In contrast, in one-dimensional systems the decrease will only be by factor 5. Therefore, the diversity of one-dimensional systems is expected to be more stable over a long period of time.

b. The diversity in one-dimensional systems is predicted to be (much) higher than that in two-dimensional systems. For example, let there be $L = 100$ loci, the mutation rate per locus per generation be $\mu = 4 \times 10^{-5}$, the local extinction/recolonization rate be $\delta = 10^{-2}$ per deme per generation, and let there be $n = 1024$ local demes. Then if the demes are arranged in a 32×32 square, numerical simulations show that there are, on average, 2.6 genetic clusters at the clustering level corresponding to $K = 16$. In contrast, if the demes are arranged on a line (i.e., in a $32^2 \times 1$ pattern), there are, on average, 17.2 such clusters. These effects may have contributed to the extraordinary divergence of cichlids in the great lakes of Africa, most species of which inhabit the relatively narrow band along the shoreline (e.g., Kornfield and Smith, 2000).

c. Typically, the genetic clusters in one-dimensional systems are denser (i.e., are characterized by smaller values of D and D_{max}) than those in two-dimensional systems.

ix. It has been argued that species can accumulate rapidly after colonizing a new environment if the species in the novel environment have a reduced probability of extinction. This could happen because reduced extinction can extend the lifetime of a lineage, thus increasing its chance to accumulate enough genetic changes to result in reproductive isolation (Mayr, 1963; Allmon, 1992; Schluter, 1998, 2000). The same could also happen after developing a "key innovation" decreasing the extinction rate. The results presented here quantify these arguments. As discussed earlier, decreasing the extinction rate by a certain factor in a two-dimensional metapopulation will increase the equilibrium diversity by approximately the same factor.

The afotementioned conclusions are based on a specific model and certain cautiousness is required when trying to apply them to more general and realistic situations. A number of directions must be pursued in order to evaluate the generality of the results presented here.

Here the spatial arrangement of demes in a two-dimensional system was (unrealistically) symmetric. Allowing for some demes to be unsuitable is expected to increase the possibilities for differentiation and speciation in both one- and two-dimensional systems. However, if space were continuous rather than discrete, these possibilities would be reduced significanly in two-dimensional systems.

To achieve more realism, one needs to account for the effects of migration into occupied patches and the resulting gene flow. Migration is expected to make splitting of the population into different clusters much more difficult. The big question is how the characteristics studied here scale with the migration rate. The common wisdom is that migration rates on the order of one immigrant per population per generation are sufficient to prevent any significant divergence in neutral alleles. One can be tempted to interpret this as proof that speciation will not be possible either. However, this interpretation is not necessarily justified because it does not account for the possibility of large fluctuations in the genetic distances between neighboring populations, which can lead to reproductive isolation (Gavrilets, 2000). Numerical individual-based simulations (with no extinction allowed) show that speciation by random drift and mutation is possible even if migration rates are on the order of several immigrants per population per generation (Gavrilets et al., 1998, 2000b). A simple approach to account for the effects of migration within the framework used here is to adjust the probability of mutation in a locus from μ to

$$\mu_e = \mu + m\mathcal{N},$$

where m is the rate of migration and $\mathcal{N}$ is the number of neighboring populations that have the alternative allele fixed at the locus under consideration. The aforementioned expression utilizes the fact that the probability of fixation of a neutral allele is equal to its frequency. With migration, new alleles are brought in the patch both by mutation (at rate μ) and migration (at rate $m\mathcal{N}$). In this approximation, the only role of migration is to bring in new alleles that are fixed quickly or lost by random genetic drift. For example, if initially both the population under consideration and its four neighbors have allele 0 at the locus under consideration, then the probability that an alternative allele 1 is fixed is $\mu_e = \mu$. However, once this has happened, the probability of switching back to allele 0 is $\mu_e = \mu + 4m$. If the migration rate is larger than the mutation rate per locus, switching back will happen much faster. However, because there are many genes and many populations, the accumulation of enough genetic differences may eventually take place, resulting in the splitting of the system into different clusters.

Also, it is necessary to consider the effects of allowing for multiple populations per demes. A simple approach for doing this is to introduce another threshold genetic distance, say K_{comp} ($> K$), reaching which will allow for coexistence in a deme. If the genetic divergence is below the threshold, the competition between different species prevents their coexistence. In this case the expected evolutionary dynamics will consist of a series of parapatric splits followed by range expansions and an increase in the number of populations per deme after accumulating enough genetic differences.

These two generalizations are discussed elsewhere (M. Saum and S. Gavrilets, unpublished results). It is also important to introduce spatial heterogeneity of selection into the modeling framework. This heterogeneity is expected to affect significantly both the probabilities of fixation and the overall dynamics of diversification (Ohta, 1972; Eldredge, 2003; Gavrilets and Gibson, 2002). Finally, and most importantly, one needs to analyze the effects of adaptation explicitly.

APPENDIX 12.1

Derivation of Eq. (12.12b)

Sawyer's (1977a) Eq. (3.2) describes the probability $I(r, \nu)$ that two genes found a distance r apart are the same type in an infinite allele selectively neutral migration–mutation–random drift model with mutation rate ν:

$$I(r,\nu) \approx \frac{2\,K_0\big(q(x)(2u)^{1/2}\big)}{\ln(1/2u) + 4\pi\big[2N\sigma_1\sigma_2 + C_0\big]},$$

where $K_0(\cdot)$ is the Bessel function of the second type of order zero. In terms of our model, $2N = 1$ (there is a single sequence per colony) and $u = \nu$ (which is the mutation rate per sequence per generation). The probability that a sequence in a given site is substituted by the sequence from a neighboring site is $\delta/4$ (where δ is the probability of site extinction, and factor 1/4 because there are 4 neighbors each of which can colonize the extinct site). Therefore, Sawyer's $\sigma_1^2 = m_1$ [see below his Eq. (5.1)] is equal to $\delta/2$ and, in a similar way, $\sigma_2^2 = m_2 = \delta/2$. Using Sawyer's expression for C_0 given below his Eq. (3.4), $C_0 = (1/4\pi)\ln(\delta/2)$. Finally, $q(x) = \sqrt{(2/\delta)x_1^2 + (2/\delta)x_2^2}$. Thus, we can rewrite the equation for $I(r, \nu)$ as

$$I(r,\nu) \approx \frac{2K_0\left(\sqrt{4\frac{\nu}{\delta}x_1^2 + 4\frac{\nu}{\delta}x_2^2}\right)}{\ln(1/2\nu) + 2\pi\delta + \ln(\delta/2)} = \frac{2K_0\left(\sqrt{y_1^2 + y_2^2}\right)}{\ln(\delta/4\nu) + 2\pi\delta},$$

where $y_1^2 = 4\frac{\nu}{\delta}\,x_1^2$, $y_2^2 = 4\frac{\nu}{\delta}\,x_2^2$. The average range of a cluster is

$$\hat{R} = n \times \frac{\iint I(r,\nu)dx_1dx_2}{n} = \iint I(r,\nu)dx_1dx_2,$$

where n is the number of sites in the system. Then

$$\hat{R} = \frac{2}{\ln(\delta/4\nu) + 2\pi\delta}\iint_0^\infty K_0\left(\sqrt{y_1^2 + y_2^2}\right) dx_1dx_2$$

$$= \frac{\delta}{4\nu}\,\frac{2}{\ln(\delta/4\nu) + 2\pi\delta}\iint_0^\infty K_0\left(\sqrt{y_1^2 + y_2^2}\right) dy_1dy_2.$$

Finally, using polar coordinates r and θ and the fact that $\int_0^\infty K_0(r)rdr = 1$, one finds that

$$\iint_0^\infty K_0\left(\sqrt{y_1^2 + y_2^2}\right)dy_1dy_2 = \int_0^\infty\int_0^{\pi/2} K_0(r)rdrd\theta = 1 \times (\pi/2) = \pi/2,$$

which leads to Eq. (12.12b).

Derivation of Eq. (12.12c)

The expected number $\tilde{R}$ of the populations that belong to the same cluster as the population at patch 0 is estimated. Looking back in time, demes 0 and x can be traced to a single founding deme $\tau_{0,x}$ generations ago. The coalescence time $\tau_{0,x}$ is a random variable. These two demes belong to the same cluster if they have accumulated less than K mutations since time $\tau_{0,x}$ generations ago. With small ν the process of mutation accumulation is approximately Poisson and the expected number of mutations separating demes 0 and x is $2\nu\tau_{0,x}$. The latter equation assumes that each fixation results in a genotype that is completely new to the system. $X(\lambda)$ is used to denote a generic Poisson random variable with parameter λ. Therefore,

$$\widetilde{R} = \sum_x \Pr(\text{populations at 0 and } x \text{ differ by less than } K \text{ substitutions} \tag{12.14a}$$

$$= \sum_x \sum_t \Pr(\tau_{x,0} = t) \Pr(X(\lambda) < K) \tag{12.14b}$$

$$= \sum_t \left[\sum_x \Pr(\tau_{x,0} = t) \right] \Pr(X(\lambda) < K), \tag{12.14c}$$

where $\lambda = 2\nu t$ and the sums are taken over all demes and over all possible coalescence times.

The probability $\Pr(X(\lambda) < K)$ can be written as

$$Pr(X(\lambda) < K) = \sum_{i=0}^{K-1} \frac{e^{-\lambda}\lambda^i}{i!} = \frac{\Gamma(K, \lambda)}{\Gamma(K)} \tag{12.15}$$

(e.g., Gavrilets, 1999a; Gavrilets et al., 2000a).

The sum in the square brackets in Eq. (12.14c) can be approximated by the derivative of $\sum_x \Pr(\tau_{x,0} \leq t)$ with respect to t. The latter sum can be approximated as

$$\sum_x \Pr(\tau_{x,0} \leq t) \approx \frac{\pi\delta t}{\ln(t)} \tag{12.16}$$

for large t (e.g., Kelly, 1977; Sawyer, 1979; Bramson and Griffeath, 1980), leading to an approximation

$$\sum_x \Pr(\tau_{x,0} = t) \approx \frac{\pi\delta}{\ln(t)} \tag{12.17}$$

for large t. Therefore,

$$\hat{R} \approx \sum_{t>1} \frac{\pi\delta}{\ln(t)} \frac{\Gamma(K, 2\nu t)}{\Gamma(K)} \tag{12.18a}$$

$$\approx \frac{\pi\delta K}{2\nu} \sum_{t>1} \frac{1}{\ln(t)} \frac{\Gamma(K, 2\nu t)}{\Gamma(K+1)} 2\nu \tag{12.18b}$$

$$\approx \frac{\pi\delta K}{2\nu} \Phi(K, \nu), \tag{12.18c}$$

where

$$\Phi = \int_{4\nu}^{\infty} \frac{\Gamma(K, \lambda)}{\Gamma(K+1) \ln(\lambda/2\nu)} d\lambda$$

TABLE 12.1 Values of Φ

ν	$K = 2$	$K = 4$	$K = 8$	$K = 16$
10^{-4}	0.122	0.113	0.105	0.098
4×10^{-4}	0.148	0.135	0.124	0.114
16×10^{-4}	0.187	0.167	0.150	0.136

Note that $\lambda = 4\nu$ corresponds to $t = 2$. Table 12.1 was found by evaluating Φ numerically. Table 12.1 shows that the dependence of Φ on its arguments is weak.

Part V

INTEGRATION AND APPLICATIONS

13. CAUSES, MECHANISMS AND CONSEQUENCES OF DISPERSAL

Jean Clobert, Rolf Anker Ims, and **François Rousset**

13.1 INTRODUCTION

The movement (or dispersal) of propagules among suitable patches of habitats is an essential ingredient of metapopulation dynamics. At the birth of the metapopulation concept, Levins (1970) only considered colonization, (i.e., movement to empty patches). However, all patches within a metapopulation are to some extent exchanging individuals due to dispersal, even those which are already occupied. This phenomenon leads to a reenforcement of extant local populations (the rescue effect; Brown and Kodrick-Brown, 1977).

This chapter focuses on condition-dependent dispersal because we feel it is important to take condition dependence into account to make realistic predictions about dispersal evolution and its consequences. In Levins' original model and some subsequent extensions of it, dispersal was mostly considered as a fixed trait (i.e., any individual in the metapopulation had the same probability of dispersing successfully). In more recent developments of metapopulation theory, dispersal has been considered to be function of the density in the patch of departure and other features such as the patch size and the distance between

0-12-323448-4

patches (Hanski, 1999b). In addition to these attempts, dispersal was largely considered to be unconditional of the status of the individual, the potential donor, and recipient patches, as well as the matrix between patches.

Similarly, migration has been mostly considered as a fixed trait in population genetic models (Barton et al., 2002; Chapter 8). There are some exceptions where dispersal is allowed to vary with some aspects of individual condition, such as sex or age (Prout, 1981; Chesser, 1991; Rousset, 1999b), or with the deme the individual is belonging to (Whitlock and Barton, 1997; Rousset, 2004), but these do not consider the evolution of dispersal and its demographic consequences [see Maynard Smith (1966) for an exception].

The origin of dispersal as a behavioral trait at the level of individuals and its population level consequences were not fully recognized before a population ecologist, Charles Krebs et al. demonstrated in 1969, by an elegant experiment, that population dynamic of voles was affected dramatically when individuals were prevented to move freely (the fence or Krebs effect). Krebs himself emphasized this phenomenon's behavioral component by invoking the term spacing behavior (Krebs et al., 1969). Lidicker (1962) also pointed out early on that dispersal was most probably a complex phenomenon by identifying two types of movement: presaturation and saturation dispersal. Somewhat later the field of behavioral ecology emerged. Greenwood (1980, 1983) used a comparative approach based on what was known about dispersal in birds and mammals at this time to conclude that mating system, resources levels, and inbreeding were the forces shaping sex-specific natal and breeding dispersal (see also Greenwood and Harvey, 1982). The evidence that individual departure from its natal site was dependent on local crowding and that individuals were not choosing to settle in a new habitat at random also started to accumulate (Lambin et al., 2001; Kokko et al., 2001). From an evolutionary (and theoretical) viewpoint, many biotic and abiotic factors were identified as potential causes for dispersal evolution (reviewed in Clobert et al., 2001). However, up to recently, there has been no comprehensive consideration of the evolution of a state-dependent dispersal, whereas the available empirical evidence was pleading for such a theory.

In the last two decades, dispersal has been subject to renewed interest. At least five books have been produced on this subject (Stenseth and Lidicker, 1992; Dingle, 1996; Clobert et al., 2001; Woiwood et al., 2001; Bullock et al., 2002), and the number of papers, especially theoretical, has increased markedly. Indeed, understanding why and how animals and plants are moving has become of prime importance, especially if we want to predict what will be the result of habitat fragmentation and global changes.

It emerges from these new bodies of empirical and theoretical studies that previous assumptions about dispersal modeling were far too simple. Then, the question of how much details on a species' dispersal ecology must be known to predict metapopulation dynamics and evolution is still largely unknown. For instance, under which circumstances can we consider dispersal, at least practically, as the kind of random process usually assumed in models of metapopulations? To what extent are dispersal patterns, in term of dispersal distances and rates, molded by the cause of dispersal? To what extent are patch settlement decisions conditional on causes of departure/emigration? Are the effects of dispersal on local patch-specific dispersal proportional to the fraction of individuals leaving and arriving? When can the matrix between habitat patches which organisms must disperse through be considered as neutral? Because every movement might be undertaken for a different reason and may modify both the local conditions of the patch of departure and arrival, it is not intuitively obvious to predict its overall impact on the ecological and evolutionary dynamics of the metapopulation. Indeed, the multiplicity of causes and potential feedback effects demonstrated by some new empirical studies strongly complicates the issue of the role of dispersal in metapopulations.

To make the topic tractable for this chapter, we will restrict ourselves to consider natal and breeding dispersal [see Greenwood and Harvey (1982) for definitions]. We use the term of dispersal rather than migration for reasons given in Clobert et al. (2001), but both terms have been used for describing the same phenomenon throughout this book. Although sometimes important to consider in a metapopulation framework, other kinds of movements (e.g., feeding migrations) are not discussed here. Due to our background there will be an inherent zoological bias in our perspectives on dispersal.

In order to discuss the consequences of dispersal on metapopulation dynamics and evolution, we will first review the potential causes of dispersal evolution. The many factors proposed to promote the evolution of dispersal (Comins et al., 1980) can be grouped in three categories (Fig. 13.1).

1. Habitat-specific factors. All abiotic and biotic factors that are not intrinsic to the organism itself or conspecific individuals. Examples are temperature, food, predators, parasites, and interspecific competitors.
2. Factors related to mate choice (i.e., inbreeding, mating system). Although constituting a special case of social factors (see later), we will

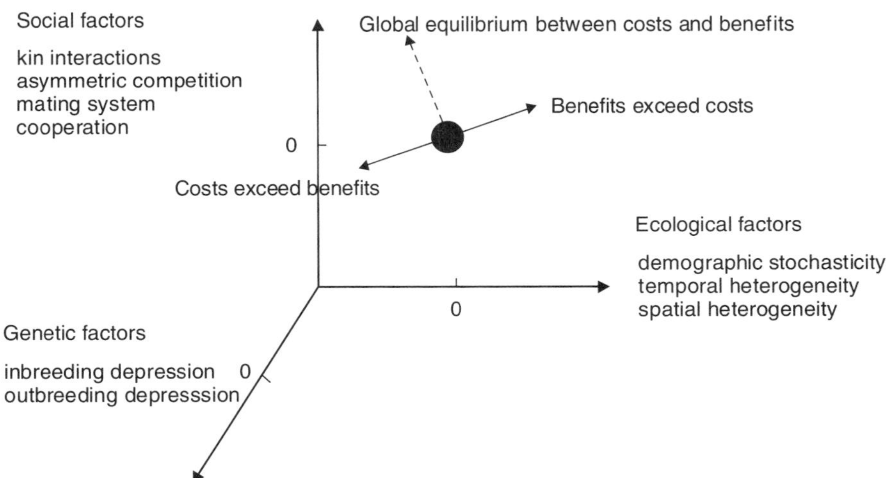

Fig. 13.1 Multiple causes that act on dispersal evolution. To each cause is associated some cost and benefits, the zero on the axes symbolizing the point where costs and benefits balance each other. The evolution of dispersal may result from all these selection pressures acting together, depending on how the movement called dispersal can be considered as one well-defined trait or uncover distinct behaviors under the control of different mechanisms. Modified from Clobert et al. (1994) and J. F. Le Galliard (unpublished results).

treat aspects of mate choice separately, as they have been seen as a major force driving the evolution of dispersal and because it may be strongly involved in the process of local extinctions in metapopulations (Hanski, 1999b; Higgins and Lynch, 2001)

3. Social factors. All types of intraspecific interactions fall into this category. In the broadest sense, such interactions may be treated as density-dependent sources of dispersal. However, it may be useful to distinguish between competitive interactions taking place within and among different age or life cycle stages, sexes, and individuals with different genetic relationship (e.g., kin or nonkin), as they may lead to different dispersal strategies.

We will also consider interactions among these factors and the likely mechanisms by which dispersal decisions (state-dependent dispersal) can be achieved.

We will then try to characterize them in term of their potential effects on movement patterns and see to what extent such movements are likely to end in successful settlement either reinforcing local populations or creating new ones (colonization of empty patches). However, because of the lack of empirical examples and a distinct theory, a discussion of the consequences of considering more realistic dispersal scenarios in a metapopulation context would necessarily be very speculative and will not constitute the main body of the chapter. Indeed, there are still large methodological challenges (reviews in Ims and Yoccoz, 1997; Clobert et al., 2001; Bullock et al., 2002) associated with obtaining reliable empirical information about virtually all aspects of dispersal that are relevant for metapopulations.

13.2 DISPERSAL AND HABITAT SPECIFIC FACTORS

Habitat Heterogeneity in Space and Time

Habitat heterogeneity in space has long been taught not to be sufficient to promote evolution of dispersal (Hastings, 1983; Holt, 1985), especially in source–sink metapopulations (McPeek and Holt, 1992). When dispersal is costly, the inclusion of temporal fluctuations (of which extinction is an extreme case) was found to be necessary to promote the evolution of dispersal (Levin et al., 1984). Indeed, any population limited by the carrying capacity of the habitat has an extinction probability equal to one (Caswell, 2001) due to demographic or environmental stochasticity.

In this context, it seems trivial to predict that genotypes that have the capacity to produce some dispersing offspring will enhance their fitness compared to those that do not. Indeed, dispersing individuals will escape this local certitude of extinction and will colonize other patches or reenforce other populations (rescue effect; Brown and Kodric-Brown, 1977). Rescue effects are usually viewed as something advantageous at all levels in the hierarchy, from the immigrant individual, the recipient population, and to the metapopulation as a whole. However, this is not universally true. For example, dispersing individuals may disrupt the social structure of a local population, which can lead to negative population growth (Gundersen et al, 2002). Furthermore, connecting two landscapes with different locally adapted populations might drive both populations to extinction (Olivieri and Gouyon, 1997). Dispersal may increase the metapopulation extinction probability because the spatial autocorrelation of temporal fluctuations increases (Heino et al., 1997). Moreover, in a source–sink metapopulation (Pulliam, 1988), a too high dispersal rate from the source to the sink may drive the entire system to extinction (Pulliam, 1996; Gundersen et al., 2001). In this case of source–sink dynamics, dispersal should be selected against unless dispersal back to the source is permitted (Watkinson and Sutherland, 1995). In some cases, dispersal is found to be counterselected at high extinction probabilities (Ronce et al., 2000b; Parvinen et al., 2003), although this result has been found to be debatable (Heino and Hanski, 2001).

Spatial heterogeneity by itself has been found to have an impact on evolution of the dispersal rate (McPeek and Holt, 1992). These findings might be understood in the context of habitat selection theory, using the ideal free distribution (IFD) concept derived by Fretwell and Lucas (1970). With a simple model, Fretwell and Lucas (1970) suggested that animals were distributing themselves such as to equalize each individual's contribution to the future gene pool of a population. The IFD concept has been most developed in studies of animal behavior, particularly foraging studies (Kennedy and Gray, 1993; Tregenza, 1995). The theory of IFD postulates that an individual will change foraging habitat only when its realized fitness will be higher by moving than by staying at the same place. Application of the IFD concept to the evolution of dispersal was first touched upon by Levin et al. (1984) and Holt (1985) and was then developed by McPeek and Holt (1992). Using a simple two-patch model with dispersal independent of density, but potentially dependent on habitat quality, McPeek and Holt (1992) found [see also Lemel et al. (1997) for an extension] that when the fecundities at the carrying capacity were equal and carrying capacity differed between habitat patches, the evolutionary stable habitat-dependent dispersal rate was inversely proportional to the carrying capacity. This illustrates the superiority of state-dependent (here habitat) dispersal strategies over state-independent ones. IFD here develops in the sense that long-term reproductive success will be equalized across habitats by selection (Holt and Barfield, 2001; Khaladi et al., 2000; Lebreton et al., 2000; Rousset, 1999a). Moreover, these IFD-based models predict balanced exchanges (equal number of emigrants and immigrants) among patches, and these predictions have been found compatible with some empirical results (Doncaster et al., 1997; Diffendorfer, 1998; although see Rousset, 1999a).

However, the theoretical results have been derived based on stringent assumptions with respect to the environment (no temporal fluctuation, no constraints on dispersal), population dynamic and structure (fixed point equilibrium, number of patches), mode of life cycle (timing and number of dispersal event), and behavioral capacities of the organisms (existence of environmental and social cues, perfect knowledge of the landscape), all of which may be severely violated in the typical settings of metapopulations. Some recent explorations of more realistic models have indeed demonstrated that, in many situations, deviations from IFD distribution are found and that the relationship between habitat-specific dispersal rate and habitat-carrying capacity can be varied (Leturque and Rousset, 2002). In particular, the size of a patch as well as the distance among patches might play an important role in both metapopulation persistence and the evolution of dispersal (Hanski, 1991). Indeed, many models predict complex spatial patterns of patch occupancy or abundance arising from an interaction between spatial constraints on dispersal range (maximum dispersal distance) and specific local population dynamics (see Chapter 3). The kind of emergent large-scale dynamics resulting from increased fragmentation may then feed back on the evolution of dispersal as to decrease or increase the dispersal rate, depending on the balance between forces operating at the local (i.e., within the population) versus a more regional (e.g., metapopulation) scale. Based on such considerations, Olivieri et al. (1995), for instance, predicted a decrease of the proportion of a dispersal genotype with population age.

Individuals' Decisions in Space and Time

Most predictions about the evolution of dispersal in a metapopulation setting rely on the assumption that individuals are moving at random or are following simple density-dependent rules and thus have limited capacities for making dispersal choices.

Much of the empirical information about individual decisions that are conditional on spatial heterogeneity encountered during the different stages of dispersal (departure, transience, and settlement) has been accumulated over the last decade. For example, during transience, individuals are typically not moving at random with respect to spatial characteristics of their environment [see Ims (1995) for a review]. Much experimental evidence shows that habitat corridors facilitate fast and straight-lined movements in several species (Andreassen et al., 1996a; Rosenberg et al., 1997; Aars and Ims, 1999; Haddad, 1999b; Tewksbury et al., 2002). Many ground-dwelling species are following landscape features such as habitat patch boundaries. Consequently, the distance between two patches might be quite different from how a human will perceive it on a map.

Species-specific environmental tropisms will interact strongly with the landscape structure to produce a dispersal pattern probably far from that produced by a random walk (Wiens, 1997). Specifically, the perception of landscape heterogeneity by an organism will strongly depend on the graininess of the landscape in terms of its mobility and assessment of risks/cost per time unit during transience. Thus, a stretch of matrix habitat between two habitat patches will be perceived as less hostile if a species needs only 10 s for crossing it than if 1 h is needed or if it is devoid of predators. This kind of variability can also, to some extent, be found between individuals within a species with a significant temporal component due to changing ambient abiotic (wind, humidity, temperature) or biotic conditions (food resources, predators) (Wiens, 2001). Especially high levels of predation are often thought of as the main obstacle during the transient phase in animals. However, little empirical evidence available that can be used to quantify such a cost during the transient phase of dispersal (Bélichon et al., 1996; Johannesen and Andreassen, 1998; Woodroffe, 2000).

The nature of the transient habitat (i.e., the matrix between habitat patches) has also a potentially strong impact on departure and settlement decisions.

There are now many studies where the matrix surrounding suitable patches of habitats or the distance separating suitable patches has been found to influence dispersal propensity. For example, in a study of the common lizard, the exchange rate between two populations separated by a distance less than a home range diameter (20 m) was decreased from 50% to 0 when open habitat was replaced by forest (Clobert et al., 1994). Rather small gaps in habitat corridors may be sufficient to impede movements significantly in voles (Andreassen et al., 1996b). Reviews on how spatially explicit landscape features, such as patch size, patch edge characteristics, matrix structure, and interpatch distances, affect the rate and direction of dispersal have shown that is it difficult at the present stage to find valid generalizations (e.g., Ims, 1995; Ims and Yoccoz, 1997; Wiens, 2001). The existence of dispersal functions valid as a species-specific fixed trait is most probably a myth. We think, however, that with more relevant empirical information it will be possible to establish dispersal functions that are conditional on a spatially explicit landscape feature.

A lot of empirical studies show that departure and settlement decision depend on habitat quality in terms of food resources, amount of refuges, predators, parasites, and intra- and interspecific competitors (Clobert et al., 2001). Most species studied appear to have a state-dependent dispersal response to changes in habitat quality (particularly well-documented examples are aphids, MacKay and Wellington, 1977; Weisser et al., 1999; reptiles, Massot et al., 2002). The degree of state dependence is, however, likely to vary among the species, and some species might show no variation of their dispersal potential [see Roff and Fairbairn (2001) for a review]. In the present absence of quantitative data available for metaanalyses, we predict that species with the least environment-sensitive dispersal strategies will be highly specialized species (in term of habitat requirement) or species living in habitats varying either in a systematic (seasonal or successional) or a random manner (Ronce et al., 2001). In such cases, there is either no need of information (when environment is changing systematically) or there is no information (randomly varying environments) available at any given time to predict the value that the environment will take later on. The conditions are then met for the evolution of a fixed dispersal rate.

13.3 DISPERSAL, SEX-SPECIFIC INTERACTIONS, AND INBREEDING

Sex-biased dispersal (i.e., either males or females are dispersing in higher proportions) has been observed in a large number species. In addition to their sex, males and females differ in many respects mainly because they are not subject to the same selection pressures. Because of sexual selection, the sexes may differ in their morphology, physiology, and behavior (Gross, 1996). In particular, females are limited by the number of zygotes they can produce and therefore will maximize offspring quality though resource acquisition and/or mate choice, whereas males are more limited by the number of mates to which they have access (Andersson, 1994). The type of mating system (monogamy, polygamy, polyandry) will also constrain the way sexual selection will operate on each sex and, as a by-product, influence their respective investment into resource and mate acquisition.

Based on these arguments, Greenwood (1980) attributed the widespread sex-biased dispersal observed in mammals and birds to the fact that resource-based territoriality was mainly found in one of the sexes (males in birds and females in mammals). Territorial sex was the philopatric sex, as nondispersing offspring of the same sex would have good chances of inheriting its father/mother territory. Advantages for philopatric sex are important: familiarity with the habitat and not much competition to acquire a territory. The opposite sex offspring would also have some advantages in such circumstances, but would have to mate with relatives and thus pay the cost of inbreeding depression.

Although evidence for inbreeding depression is accumulating (e.g., Saccheri et al., 1998; Ebert et al., 2002), its reported impact on dispersal has been mainly correlative (but see Wolff, 1992). Moreover, the hypothesis that inbreeding avoidance is a major determinant of dispersal has been challenged on several grounds. There are other mechanisms to avoid mating with relatives, such as kin recognition. Also, moderate inbreeding may even be advantageous under certain circumstances, such as to avoid breaking coadapted genes complexes and to purge deleterious alleles (although the latter may have little impact on dispersal evolution).

Inbreeding avoidance per se was proven theoretically to be able to promote the evolution of dispersal (Motro, 1991), but objections have been raised based on a set of theoretical considerations by Gandon (1999), Perrin and Malazov (1999), and Perrin and Goudet (2001) about the conditions that promote a sex-biased dispersal. They demonstrated that inbreeding depression is promoting dispersal in one sex only (the best solution being one sex dispersing all, the other remaining philopatric). However, in a vast majority of species, both sexes are dispersing to some degree, suggesting that factors other than inbreeding are important. In addition, inbreeding is not the only force that can promote sex-biased dispersal (Perrin and Goudet, 2001). Therefore, the precise role of inbreeding in generating dispersal movement (including inbred individuals having a tendency to disperse more, see Cheptou et al., 2001) is still a widely open question.

However, it is difficult to imagine that the cost of inbreeding will not be important to consider in a metapopulation context, particularly at colonization or in small and isolated patches. It is, however, possible that situations leading to a potential risk of inbreeding are incidentally avoided by dispersal having evolved for solving other types of individual, especially kin-based interactions.

13.4 DISPERSAL AND SOCIAL FACTORS

Individual Interactions

It was not until very recently that theoreticians demonstrated that demographic stochasticity will favor a density-dependent dispersal (Travis et al., 1999; Poethke and Hovestadt, 2002; Cadet et al., 2003). Interestingly, demographic stochasticity in this case is playing somehow the same role as spatially uncorrelated environmental stochasticity.

Although it was demonstrated quite early on that dispersal was crucial for population regulation [the enclosure experiments by Krebs et al. (1969), and

Boonstra and Krebs (1977), but see Ostfeld (1994)], the mechanisms involved have, until recently, been elusive. For example, to what extent dispersal was negatively or positively density dependent has, until recently, been controversial (Gaines and McClenaghan, 1980; Stenseth, 1983; for a review, see Ims and Hjermann, 2001). These inconsistencies are likely to be the result of confusion at several levels: (1) direct versus delayed effect of density and (2) density as an ultimate cause versus a proximate cue.

Individuals might leave a patch because density is a good descriptor of the current level of intraspecific competition. However, ideally, local crowding has to be compared to the one found in other patches. For instance, as the sign of density dependence is likely to differ in the emigration (usually positive) and immigration process (negative), respectively (e.g., Andreassen and Ims, 2001), the resultant dispersal rate will depend on the spatial covariance in population density in relation to the dispersal range of a given species (Ims and Hjermann, 2001; see later). Moreover, population regulation may impose a negative temporal autocorrelation in population density in which case crowding at time t predicts less crowding at $t + 1$. Thus life stages that can endure temporal crowding may therefore choose to stay in a patch at high densities. Depending on the level of temporal and spatial autocorrelation in densities, one might predict very different outcomes for the way density affects dispersal. Appropriate study designs taking such scaling considerations seriously have rarely been used in studies of density-dependent dispersal.

Another aspect that may complicate the density dependence of dispersal processes is that density not only determines the potential for antagonistic interactions in competition for resources, it may also act as a proximate cue for habitat quality. The idea that the presence of conspecifics can be used as a cue for habitat quality has been proposed by many researchers (Danchin and Wagner, 1997) and was first demonstrated experimentally by Stamps (1988, 1991). Evidence for the fact that the presence of conspecifics and their specific characteristics (e.g., their reproductive success) may influence departure from and arrival to a patch has since accumulated [see Stamps (2001) and Danchin et al. (2001) for reviews]. The use of density of conspecifics as a cue for habitat quality can therefore explain some of the cases where an inverse relationship between dispersal and density has been found (Denno and Peterson, 1995; Lambin et al., 2001; Ims and Hjermann, 2001). In such cases, the quality component of the cue "density" may overshadow the competition component. In the case of the Glanville fritillary butterfly (Kuussaari et al., 1996), where individuals left patches with a low population density, it was suggested that low density may serve as a cue for low mating probability. It may also be that different individual categories are responding differently to density, depending on their position in the competitive hierarchy (their competitive ability) or their life history characteristics (see Gundersen et al., 2002). In that case, the realized density-dependent dispersal rate will be conditional on the demographic structure in a given patch.

Although of prime importance to understanding dispersal processes, density dependence is a complicated issue because it may act both as an ultimate cause and as a proximate cue. Unfortunately, there are few experiments explicitly designed to unravel the effects of population density on departure and settlement processes (for some exceptions; Aars and Ims, 2000; Gaggiotti et al., 2002; Gundersen et al., 2002).

Interactions between Kin

The nature of the interacting individuals, in particular their genetic relatedness, is central to the question of sociality in animal populations and, more specifically, how altruistic behaviors have evolved. Genetically proximate individuals should tend to congregate spatially by means of offspring philopatry and/or delayed dispersal as to avoid misdirect helping and invasion by cheaters (Packer and Pusey, 1997). For example, in the Seychelles warbler (Komdeur et al., 1997), where most of the good available territories are occupied, offspring produced early in the reproductive life of a pair holding a good territory tend to stay to help their parent raise other offspring. Philopatry then enhance the chances of inheriting a high-quality parental territory. In lekking species, genetically related males tend to concentrate in the same leks (Petrie et al., 1999) because females tend to be attracted by big groups of males. However, the intensity of the cooperation is also state dependent. For example, Lambin and Yoccoz (1998) showed that spatial association among kin Townsends voles increased as the population increased (kin cooperation for resource holding). The same pattern has been found in red grouse (Lambin et al., 2001).

In nonsocial species, where competitive interactions dominate over cooperative interactions, genetically similar individuals are expected to avoid situations of competition. Hamilton and May (1977) provided the first theoretical support for dispersal being one way to avoid such interactions. They demonstrated that, in a homogeneous population, kin competition was promoting the evolution of dispersal even in the presence of a high cost to disperse. Further extensions relaxing some of the assumptions of earlier models (Perrin and Goudet, 2001; Gandon and Michalakis, 1999; Ronce et al., 1998) all converge to the same conclusion (Gandon and Michalakis, 2001).

Kin competition can take several forms, (i.e., between parents and offspring, among offspring of opposite or same sex). In some species of jays, dispersers are actively expelled siblings by other siblings from family groups (Strickland, 1991). In some rodents, dispersal increases with the number of siblings in the litter (Ribble, 1992). In the common lizard, mother–offspring competition leads to the dispersal of female offspring (de Fraipont et al., 2000). In the latter case, not only the presence of the mother but also her condition determines the likelihood that her offspring will disperse (Léna et al., 1998). However, most of these cases are observational and can be explained by other factors, such that empirical evidence for any form of kin competition promoting dispersal is still scarce in the literature. Only a few experiments have demonstrated an effect of kin interactions on dispersal (e.g., Lambin, 1994; Léna et al., 1998).

Restricted dispersal is not always a prerequisite of the evolution of altruism, not only because cycles of coevolution between the two traits can lead to a temporary positive relation between dispersal and altruism (Le Galliard and 2003b et al., 2003a), but because individuals might have evolved ways to assess their new social environment in term of genetic relatedness (Hamilton, 1987). For example, a colonial ascidia was found to settle in the vicinity of genetically similar individuals (Grosberg and Quinn, 1986). The best evidence of dispersal being caused by competition among or by attraction toward genetically similar individuals comes from an experiment done on offspring dispersal in the side-blotched lizard *Uta stansburiana* (Sinervo et al., 2003). In this annual species, three male morphs distinct by their throat color during reproduction (orange, blue, and yellow) coexist in a frequency dependent way analogous to a rock–paper–scissor game (Sinervo and Lively, 1996): orange males are very aggressive toward any other males and easily take over females of blue-throated males; yellow males, which look like females, sneak females of orange-throated males; and blue-throated males successfully avoid being sneaked by yellow-throated males. To avoid being sneaked by yellow-throated males, blue-throated males appear to cooperate. After having randomly distributed offspring within the population, young orange-throated males were found 1 yr later to actively avoid each other based on their genetic proximity, whereas genetically proximate blue-throated males were found closer to each other than expected by chance alone (Sinervo and Clobert, 2003). Therefore, it seems that both departure and settlement decisions can be based on kin and, more generally, on the local genetic structure depending on the cost and benefits expected of the interactions within a local population. Empirical evidence is, however, only starting to accumulate. The case of the side-blotched lizard, however, already strongly suggests that individuals have derived direct or indirect ways to assess the level of expected kin-based interactions and condition their dispersal behavior to this information.

How interactions between kin affect dispersal probabilities as opposed to interaction between unrelated individuals is probably a very relevant question in the typical metapopulation setting. Relatedness in local populations may be expected to depend on patch size and isolation, as well as the time since colonization, and is thus a factor by which spatial structure and demography may feed back on the dispersal rate and ultimately on the dynamics of the metapopulation.

13.5 DISPERSAL: A SAME RESPONSE FOR DIFFERENT FACTORS?

For a long time, the goal of many researchers has been to discover *the* ultimate cause of the evolution of dispersal. In the realm of this effort, many factors has been demonstrated theoretically or empirically to promote dispersal (Clobert et al., 2001). Most of these studies have been unifactorial in the sense that the effect of one factor is considered at a time. However, the most common situation in nature is that individuals are affected simultaneously by multiple factors that may be involved in the decision of whether an individual should depart from a patch or not, how far it will move, and eventually where it should settle. How such multiple factors interact to shape the overall dispersal patterns is a question of critical importance to our understanding of the evolution of dispersal as well as for predicting transfer rates between patches in a given ecological setting.

It is only recently that models of the evolution of dispersal considered the action of several factors at the same time. Perrin and Mazalov (1999, 2000) considered the evolution of sex-biased dispersal under the joint effects of inbreeding depression, local mate competition, and local resource competition. They found that different assumptions about the way local competition affects each sex may result in sex biased dispersal or not and that inbreeding depression could enhance biases due to other factors. Similarly, Gandon and Michalakis (2001) examined the respective roles of local extinction, kin competition, and inbreeding on the evolution of dispersal. Extinction seems to be a stronger selective force than the other two, and the interaction among forces leads sometimes to counterintuitive results. For example, the dispersal rate was sometimes found to increase with the cost of dispersal. They also pointed out (see also Perrin and Goudet, 2001) that many factors are covarying because variation in one of them feeds back into the other (especially kin interaction and inbreeding). A few other models have attempted to incorporated ecoevolutionary feedbacks such as those imposed by local (patch-specific) population dynamics on relatedness, probability of inbreeding, and probability of extinction (Doebeli and Ruxton, 1997; Mathias et al., 2001; Heino and Hanski, 2001; Parvinen et al., 2003, Rousset and Ronce, 2004).

These modeling exercises have been fruitful in pointing out the complexity of the interactions among variables describing the social environment and in understanding the respective role of each factor. The models are, however, still too simplistic and neglect many important facets of the problem. In addition, in most models, dispersal is not state dependent, whereas results of these models pledge for the widespread existence of such a state-dependent dispersal. Indeed, variables describing the social environment are both predictors of the future social environment and are influencing each other in a somehow predictable way. Thus temporal autocorrelation in the social environment then sets the ground for the evolution of a sensitivity to cues reflecting it. Finally, factors influencing settlement decisions are not taken into account in models, whereas empirical research on habitat selection has proved that such decisions are influential (Stamps, 2001; Danchin et al., 2001).

Thus dispersal models are still far from offering a good predictive framework with respect to the interplay between multiple determinants of dispersal. However, empirical evidence regarding this issue is also poor, and only a few studies have reported interactive effects between determinants of dispersal. Although many of them are correlational, these studies have recurrently found interactions among factors acting either at the same time and location or at different locations (departure and settlement) and different moments in the life of an individual. For example, in the English grain aphid (Dixon, 1985) and in a collembolan [*Onychiurus armatus*, Bengtsson et al., (1994), see Ims and Hjermann (2001) for further examples], dispersal at one location was triggered by an interaction between habitat quality and population density. However, in an experimental setting, competition with kin and with unrelated congeners has been demonstrated to be additive (Le Galliard and 2003a et al., 2003b). In the collared flycatcher, the decision to leave and to select a patch are both influenced by the average reproductive success in that patch, whereas population density only affects departure not settlement (Doligez et al., 1999, 2002). Interactions often result from factors exerting their influence at different ontogenetic stages (even at the level of grandmother in aphids; MacKay and Wellington, 1977). In a reciprocal transplant experiment on the common lizard, Massot et al. (2002) found that local conditions (e.g., humidity and temperature) at different moments during the embryonic development interact among them as well as with the local conditions at birth to shape the dispersal response of the juvenile.

Although the empirical evidence at present is scarce and needs more experimental studies, multiple state dependence (Ims and Hjermann, 2001) is expected to be common (Massot and Clobert, 2000). Based on our present knowledge, the evolution of dispersal under multiple dependence can be seen in two opposite ways: (1) a slow, progressive building up of factors influencing dispersal starting from a primitive cause. For example, one theoretical perspective is that the minimal model of dispersal evolution is a model with kin competition alone (Perrin and Goudet, 2001; Gandon and Michalakis, 2001; Leturque and Rousset, 2002). Indeed, solving kin competition problems is inherent to most, if not all, organisms because of the obligate spatial cooccurrence, at least for a certain amount of time, between parents and offspring or of offspring. All other forces will come as modifiers of this initial situation and therefore should interact with kin competition. (2) At the opposite, an omnibus response to different unrelated problems with their own controlling pathways. One might think that kin interactions, mate searching, intraspecific competition, and habitat characteristics are all perceived at different spatio-temporal scales (Krebs, 1992; Ims, 1995; Ims and Hjermann, 2001). In this situation, dispersal is just a common response to very different situations and one should expect many additive effects of various factors with only limited interactions among them.

At this stage, no perspective can be discarded. Indeed, although a state-dependent dispersal at the level of single factor seems to be the rule and a fixed dispersal the exception, the few studies looking to the existence of interactions among factors used at departure and/or settlement (Stamps, 2001; Doligez et al., 2002; Le Galliard et al., 2003a) give contradictory results. The study of the mechanisms and cues involved in departure, transience, and settlement might help build a more precise view of dispersal evolution, as well as its expected consequences on the metapopulation dynamic.

13.6 PROXIMATE CONTROL OF DISPERSAL

Genetic Control

Evidence for the genetic control of dispersal is found more easily in organisms that produce offspring with some specialized dispersal morphs. In insects, dispersal morphs may be characterized by a winged morphology (Roff and Fairbairn, 2001), physiology (Clark, 1990), or behavior (Carrière and Roitberg, 1995). Roff and Fairbairn (2001) presented a number of cases where significant heritability has been found for a trait associated with dispersal and/or migration. In plants, seed dimorphism is known in the Asteraceae, and genetic effects have been demonstrated (Venable and Burquez, 1989; Imbert, 2001). Dispersal morphs are known in the naked mole rat but genetic effects have not been described in this case. Several traits must be present simultaneously to facilitate dispersal; those that trigger dispersal and those that subsequently facilitate movements and finally settlement in a new patch. In vertebrates, most studies report a strong correlation in dispersal propensities within families (Massot and Clobert, 2000). Whether a family component of dispersal is due to common genes or environment is undecided in most field studies.

Two studies, however, provide good support for a genetic basis of dispersal. The first one reports that in the rhesus macaque (a monkey), the timing of dispersal (early versus late in life) is conditioned to the presence of a specific allele at a serotonin transporter gene (Trefilov et al., 2000). The second one concerns the side-blotched lizard *U. stansburiana*, where offspring sired by males with different throat color morph (genetically based) have subsequently different dispersal patterns (Sinervo et al., 2003). There is little doubt that dispersal in this case has a genetic basis, although it is far from being clear what are mechanisms of the genetic control. For example, there are few cases where there seems to be a simple genetic mechanism underlying a state-independent dispersal strategy. An exception seems to be the cricket *Gryllus* sp. (Roff and Simons, 1997; Roff et al., 1997). The widespread presence of a state-dependent, environment-driven dispersal in nature is, however, incompatible with a simple, rigid genetic control. On the contrary, state-dependent dispersal suggests a genetic control in terms of dispersal plasticity and heritability of norms of reaction. Reaction norms may explain the apparent contrasting heritabilities for dispersal heritabilities in *Drosophila* spp. (Lefranc, 2001). To our knowledge, no studies have been done on the genetic control of phenotypic plasticity in dispersal-related traits.

Physiological and Behavioral Control

In most organisms, there is ample evidence that many aspects of the environment (for a review, see Ims and Hjermann, 2001) can condition dispersal. It is somehow trivial to say that a state-dependent dispersal will be selected if organisms can predict its expected reproductive success in a spatial setting. Indeed, there is hardly any environmental factor that has never been found to affect dispersal in some species or under some circumstances.

Examples of individuals using concurrent cues for dispersal decisions are numerous and documented for almost any environmental factors (Dixon, 1985; Denno et al., 1991; Denno and Peterson, 1995; Lidicker and Stenseth, 1992; Ims and Hjermann, 2001). It is, however, not clear when and how organisms can do so. Indeed, individuals also use cues at some earlier stage to disperse at a later stage (delayed dispersal). In many species of birds, individuals are often assessing habitat quality the year before they actually leave their breeding area to settle in a new one (Danchin et al., 1998; Doligez et al., 1999). The maintenance of information-gathering systems, especially on the scale of a lifetime, is most probably costly, especially if several mechanisms are needed to acquire, store, and process the necessary information. To reduce these costs, evolution might have driven species to select integrative cues (describing several aspects of the environment) as well as to use existing physiological systems. For example, Danchin et al. (2001) proposed that many animal species evaluate environmental quality through the success of conspecifics. This parameter obviously integrates several dimensions of habitat quality.

In other cases, dispersal-conditioning cues entailed modification of the individuals' internal condition (Nunes et al., 1998) or in the development of the phenotype (O'Riain et al., 1996), which then later on will influence dispersal decisions (for a review, see Ims and Hjermann, 2001). Effects of the maternal and in some cases even the grandmaternal environment, for instance, in terms of food and presence of predators, on the production of winged offspring are particularly well exemplified in aphids (Dixon, 1985; MacKay and

Wellington, 1977). Modifications of the phenotypic traits affecting dispersal propensities are sometimes subtle. For instance, different cues may act on the physiology or behavior (de Fraipont et al., 2000; Meylan et al., 2002) at different moments of individual ontogeny (Ims and Hjerman, 2001; Dufty et al., 2002; Massot et al., 2002). Although there are still relatively few empirical examples of both maternal effects and stage-dependent dispersal cues, they are likely to be quite common given the rapidly accumulating evidence for dispersal being largely state dependent. Combining present and past information on several environmental factors might be done through the organizational and activational effect of hormones (Dufty and Belthoff, 2001), and, indeed, hormone have been demonstrated to trigger dispersal decision at different developmental stages [Dufty et al. (2002) for a review on vertebrates, see Zera and Denno (1997) for insects]. For example, corticosterone during ontogeny has an influence on brain organization and on the distribution, type, and density of hormonal receptors in different part of the body (organizational effect). This will in turn set the behavioral repertoire and hormone-mediated stimuli reaction profile (activational effect) later in life. In other words, variations in several environmental factors might be translated into the modification of one or a few hormones (or other message-carrying substances) during development and/or at adulthood and be the proximate and common mechanism underpinning a majority of dispersal decisions.

The questions now are what sort of state dependence is prevailing in a given species and ecological situation, what are the underlying mechanisms that shape a given response, and which cues for assessing the environment do the organisms use? These depend on many factors, including which environmental cues are available, what are their predictive powers, how organisms cope with different degrees of spatial and temporal variability/predictability (autocorrelation), and the simultaneous actions of different (and even opposing) cues. The fact that varying dispersal rates feed back on the social and competitive environment, for instance, adds to the complexity. The problem of density-dependent dispersal is a good example (see later and Ims and Hjermann, 2001). Most of the species are living in environments where conditions cueing for dispersal exhibit various degrees of autocorrelation. The individuals' abilities of gathering information about the spatial scaling of critical environmental factors are determinant for optimizing their choice (Doligez et al., 2002). Species-specific exploration ranges relative to the spatial scaling of the environment are obviously important in this context. Very little is known about this.

In the present lack of a unified theory of the evolution of state-dependent dispersal, we suggest a preliminary framework that centers around the information acquisition process that has to precede any condition-dependent dispersal event. Indeed, the presence of cues that can be sensed and assessed by the organism is a prerequisite for condition dependence to evolve. Three important aspects are recognized (Fig. 13.2): (1) Information accumulated over an organism lifetime will lead to an increasingly accurate knowledge of the actual situation, (2) the spatial and temporal autocorrelation in environmental factors will determine the reliability of the information gathered, and (3) the value of the information gathered is decided by its relevance to the organism's stage-dependent reproductive values (Ims and Hjermann, 2001; Dufty et al., 2002).

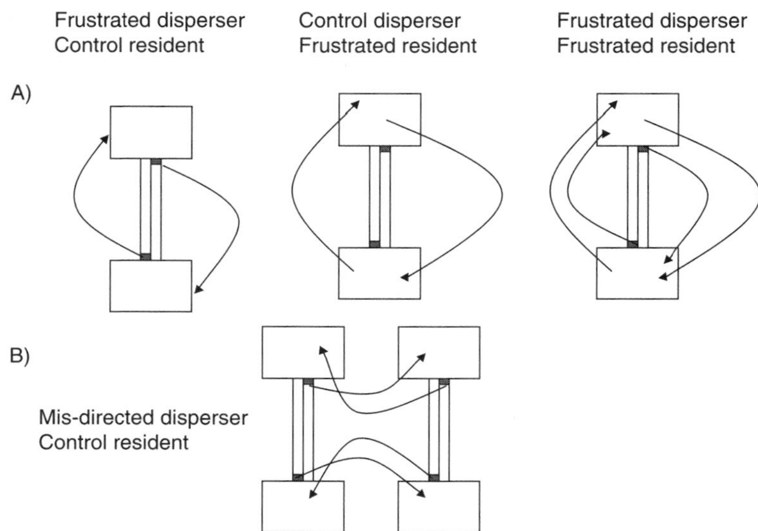

Fig. 13.2 Examples of experiments aiming to study differences in fitness between dispersing and resident individuals. Two patch systems connected by one-way corridors help identify emigration attempts (for an example, see Boudjemadi et al., 1999) by capturing individuals at the end (gray square) on each one-way corridor. This can be further studied by varying population characteristics in the two patches.

Although research along these lines has just started to be carried out, it is tempting to think that the apparent complexity of factors acting on dispersal can be reduced to the study of the action of a few proximate mechanisms. If this is true, it will then make the study of dispersal, including its evolution and (meta)population consequences, easier than it might appear based on the present review.

13.7 DISPERSAL: PHENOTYPIC ADAPTATION, COST, AND BENEFITS

Initial Differences among Dispersers and Philopatrics

There is evidence in many taxa that dispersers often are characterized by a special apparatus, which should enhance their ability to disperse (see Section 13.6). In many cases, the production of such structures (or the proportion of individuals, among the progeny or in the populations, with these specializations) is condition dependent. For example, when food is lacking, some ciliates are able to elongate their body and their flagellae, reaching 10 times the speed of a normal cell (Nelson and DeBault, 1978). In most species, however, such extreme specializations are lacking, and differences between dispersing and nondispersing individuals are often subtle. Indeed, although not yet well documented, dispersers are found to be a nonrandom sample of the population (for a review, see Swingland, 1983). They might slightly differ in morphology, physiology, or behavior (Murren et al., 2001). In gray-sided voles, asocial individuals were predominant among dispersers (Ims, 1990). Sometimes such differences have been taken as evidence for dispersers being inferior individuals in terms of competition for resources or mates at their natal site (Ims and Hjermann, 2001; Stenseth and Lidicker, 1992). It was, however, demonstrated in a few cases that dispersers were not inferior individuals. For juveniles of the commonlizard which dispersed were the biggest individuals (Léna et al., 1998; de Fraipont et al., 2000).

In addition to adaptations related directly to mobility, evolution may have shaped philopatric and dispersing individuals differently according to coadapted sets of traits (similar or not to syndromes found at the interspecies level; see Chapter 10) maximizing their respective fitness. The presence of disperser traits subject to state dependence offers parents a possibility of manipulating offspring dispersal propensity. Although evidence for the parental control of offspring dispersal is only starting to appear (Massot and Clobert, 1995; Meylan et al., 2002), the existence of parental control on other traits is well demonstrated (Mousseau and Fox, 1998). These considerations, which might seem at first glance of little importance, may, however, have a strong impact on two essential aspects of dispersal, there is very little information about: the costs associated with dispersal and the settlement/colonization success.

Individual Level: Cost and Benefits of Dispersal

The transience phase has been presented recurrently as the most costly stage of dispersal. For example, Aars et al. (1999) and Ims and Andreassen (2000) found that dispersal movements in experimental vole metapopulations increased predation risk quite dramatically. The risk associated with the transient stage of dispersal is likely to be species and condition dependent, and other studies have not been able to demonstrate a significant cost of movement (Bélichon et al., 1996). However, it is likely that dispersal in the typical setting of metapopulations with highly fragmented habitats imbedded in a hostile matrix is more costly than in other situations.

Clearly, specific traits of dispersing individuals (as opposed to philopatric individuals) may render them less susceptible to the risks associated with dispersal. Dispersers may be better skilled than other individuals to becoming integrated in a novel population or colonizing an empty patch. For example, Danielson and Gaines (1987) showed that individuals colonizing an empty habitat had a higher growth rate and survival than nondispersing or frustrated individuals (dispersers forced to be resident). It was also demonstrated experimentally that immigrants into a population of *Daphnia* spp. had a higher long-term fitness than local individuals (through heterosis; Ebert et al., 2002). In a review based mostly on observational studies, Bélichon et al.(1996) found that dispersers, when settled in a new population, did not necessarily have lower fitness than residents. However, all these studies have their interpretation complicated by problems of study designs (size of the study area, type of habitats, etc) and nature of the dispersal events.

In the very few experimental studies aimed at comparing the fitness of dispersers and residents, only a restricted number of situations have been explored (see Fig. 13.3) such that a same-ground comparison of the two strategies has seldom been done. Gundersen et al. (2002) were able to compare

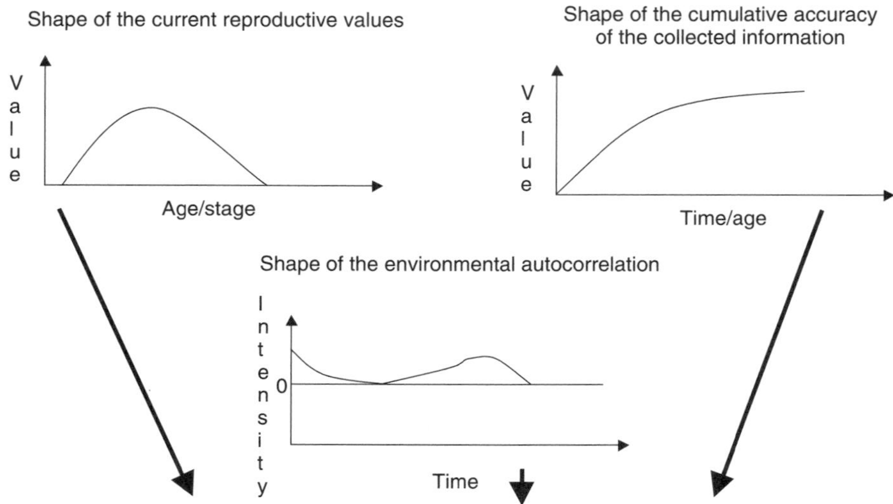

Resulting shape of phenotype sensitivity to environmental cues during individual's lifetime

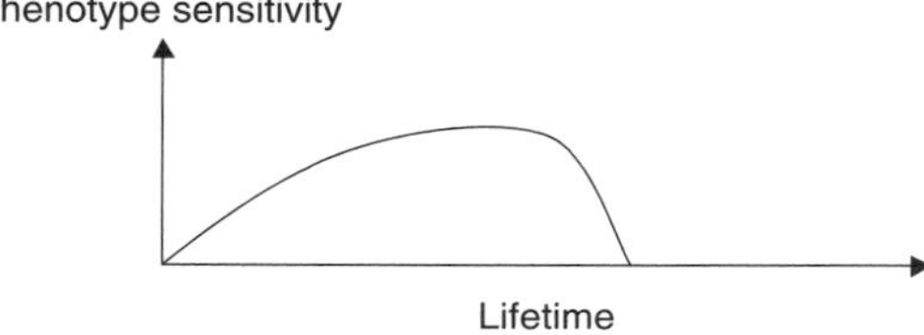

Fig. 13.3 Condition-dependent dispersal. A dispersal decision may be influenced by the present environment in interaction with information of the past history gathered by various means. The way this information will affect the dispersal decision and phenotype should depend on the shape of the current age- or stage-dependent reproductive value, the shape of the cumulative accuracy of the collected information, and the shape of the environmental autocorrelation. The shapes of these curves are not the results of a model, but are just for illustrating the way a dispersal decision could be influenced by cues collected at different times in the development of the phenotype.

settlement successes on experimental patches of same-age subadult voles that had shown different propensities to disperse from their natal patch. Individuals that already had shown some propensity for dispersal had much better success in settling, surviving, and reproducing on patches with a resident population than those who had never attempted to disperse. This experiment also showed that the density-dependent nature of the settlement success was sex specific (Fig. 13.4). Moreover, the cost of forced dispersal in residents (important to consider when patches are suddenly disappearing) may be different than that of forced residency of individual destined to disperse.

The actual cause of dispersal also matters when measuring costs and benefits of dispersal. Indeed, phenotypic adaptations to disperse, and the associated costs and benefits, may be dependent on the cause of dispersal itself. The available evidence of a phenotypic differentiation dependent on the dispersal cause is contradictory. For example, in a study on aphids (Dixon 1985), the disperser phenotype was not specific to the cause of dispersal (most factors studied did induce the production of winged offspring). On the contrary, in the common

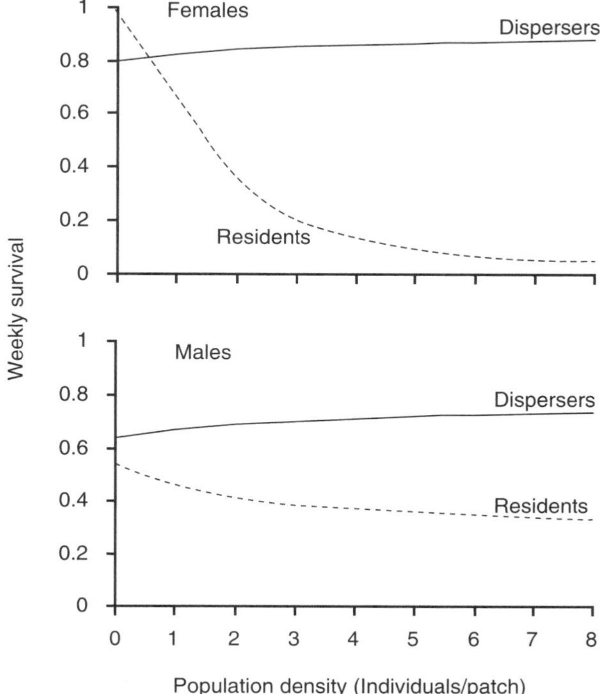

Fig. 13.4 Settlement success. Settlement success measured by the weekly survival rate of root voles introduced experimentally to patches with varying densities of resident animals (modified from Gundersen et al., 2002). Two types of animals were introduced. *Dispersers* were animals that earlier had shown an ability to disperse, whereas *residents* were same-age animals (belonging to the same cohort) that never left their natal patch. Survival probabilities were modeled by logistic regression.

lizard, phenotypic differences between residents and dispersers were only found in the context of a mother–offspring competition (Léna et al., 1998), although (or maybe because) it is predicted that, under kin competition, the evolution of dispersal could be achieved even in the presence of a high cost to disperse (Hamilton and May, 1977; Murren et al., 2001). Therefore, we hypothesize that individuals dispersing for different causes display different skill at colonizing empty patches or integrating already occupied patches.

The aforementioned hypothesis might even be more important to investigate when dispersal distances are considered. Indeed, it has been recurrently proposed that dispersal distances are increasing from socially based dispersal to habitat-based dispersal (Krebs, 1992; Ronce et al., 2001). However, the relationship between different dispersal causes imposing different phenotypic adaptations and dispersal distances may in fact be more complex than previously thought. For instance, consider the possibility that, for a given dispersal cause, a mother (through maternal effects) imposes a specific phenotype to its offspring (such as in the case of kin competition in the common lizard; Léna et al., 1998) so as to enhance their overall efficiency at dispersing. In such a case, it might be predicted that those individuals going to disperse over long distances or show particular ability to colonize empty habitat are individuals with the most adapted phenotype, i.e., for the dispersal cause that produces the best-adapted phenotype to disperse.

Although the individual consequences of philopatry and dispersal are only starting to be considered, the first reported studies clearly demonstrate the importance of such considerations and their potential impact on dispersal cost and settlement success. Overall, they also pointed out a lack of theoretical studies on females strategies producing a dispersal-dependent offspring phenotype.

13.8 DISPERSAL, POPULATION GENETICS, AND GENETIC FEEDBACK MECHANISMS

Local Adaptation

Low dispersal sets the conditions for local adaptation, as predicted repeatedly by models of evolution on a spatial gradient (e.g., Nagylaki, 1975). It follows, therefore, that recurrent recolonizations by immigrants will reduce the potential for local adaptation to abiotic factors in metapopulations. This in turn feeds back on the evolution of dispersal, as local adaptation makes dispersal more costly. In a two-habitat model with environmental fluctuations, the most common outcome of the coevolution of dispersal and habitat specialization was the coexistence of two habitat specialists with low dispersal (Kisdi, 2002), although the relaxation of some model assumptions on habitat patchiness and species dispersal range might change these predictions.

A different outcome may result from host parasite coevolution. Parasites may be locally adapted to their hosts in that they bear virulence alleles that best match resistance allele of their local hosts. This implies that hosts will be locally maladapted to their parasites. Conversely, hosts may be locally adapted to their parasites, whereas parasites may be maladapted. With limited dispersal between populations, different combinations of virulence/resistance alleles may evolve in different populations, with some average tendency for either hosts or parasites to be locally adapted. The partner that is locally adapted is the one that evolves faster in response to changes in the other partner's genotypes. Parasites tend to be advantaged as they often have shorter generation times, but the speed of evolution also depends on the local input of genetic variation through immigration and mutation. Thus the partner with higher dispersal and mutation rates tends to be locally adapted (e.g., Gandon, 2002). Some studies confirm this trend (Dybdhal and Lively, 1996; Davies et al., 1999; Kaltz et al., 1999; Delmotte et al., 1999).

Genetic Diversity

Recurrent recolonizations by a few founders reduce both local and global genetic diversity. The impact of demographic process is commonly described in terms of genetic diversity and of spatial genetic structure. The total genetic diversity of a species is, in principle, determined by its effective size, which may be defined as measuring the rate at which gene lineages in different individuals merge in a common ancestral lineage (asymptotic inbreeding effective size, also known as eigenvalue effective size; Ewens, 1982, Whitlock and

Barton, 1997). As guessed, frequent extinctions and recolonizations reduce this effective size, particularly when recolonization is done by a few number of individuals. This prediction is supported by a series of simple models [see Chapter 7 and Rousset (2003, 2004) for reanalyses and interpretations in a coalescent framework].

Genetic diversity and spatial genetic structure may be viewed as reflecting two different forms of "inbreeding" effects. First, offspring sired by parents coming from a same population may be less fit than those of parents coming from different populations (heterosis). Second, "inbreeding" may reduce genetic variability as well as promote fixation of deleterious mutation in the total population (which does not result in heterosis). The importance of such effects has been investigated by simulation (Whitlock, 2000; Theodorou and Couvet, 2002) and analytical models (Glémin et al., 2003), including some metapopulation models [Whitlock, (2002) and Chapter 7; although see Roze and Rousset, (2003) for alternative analyses of such models]. With sufficiently weak deleterious effects and strong density regulation, inbreeding may have little or no effect on population demography, but otherwise population numbers will be reduced (see, e.g., Saccheri et al., 1998). This reduced size may feedback on effective size and genetic diversity so that the metapopulation becomes less and less fit, eventually leading to its extinction ("mutational meltdown"). Models have shown that meltdown could occur in principle (Lande, 1994; Lynch et al., 1995a) and that it could occur much faster if dispersal is restricted to adjacent populations rather than following an island model (Higgins and Lynch, 2001). However, the value of the key parameters of this process, the genomic rate of deleterious mutations and the distribution of deleterious effects of individual mutations, is still debated (e.g., Keightley and Bataillon, 2000; Chapter 14). In addition, most of these models neglect the genomic rate of beneficial mutations, although its importance has been demonstrated in other cases [models, Poon and Otto (2000); data, Shaw et al. (2002)], also a matter of debate (Keightley and Lynch, 2003; Shaw et al., 2003). Observations of local adaptation, despite low population sizes and low diversity of molecular markers, also raise doubt about such processes (McKay et al., 2001).

Information on population structure may be useful in an appreciation of the importance of competition between relatives in metapopulations. Wright's F_{ST} is here the relatedness parameter weighting kin selection effects relative to the direct fitness effects of an individual on its own number of offspring (see Chapter 10 for an example of kin selection effects in metapopulations). Whether the metapopulation turnover reduces or increases spatial structure, as measured by F_{ST}, depends on whether recolonizers tend to come from the same or from different populations (Wade and McCauley, 1988). Data have been little analyzed in light of these models (see, however, Chapter 15), although estimates of their parameters have been obtained for two fungus beetles (Whitlock, 1992a; Ingvarsson et al., 1997). Cases where immigrants tend to come from the same origin are expected not to be rare because there is a strong family effect of dispersal propensity (Massot and Clobert, 2000).

Gene genealogies in some metapopulation models may be understood as "structured coalescents" in which the coalescence of gene lineages from different demes is described by Kingman's (1982a) coalescent process (Wakeley and Aliacar, 2001; Chapter 8). This suggests that simulation algorithms such as those of Nielsen and Wakeley (2001) or Griffiths and Tavaré (1994a) for maximum likelihood estimation of genetic parameters could be adapted to the metapopulation context.

Methodological Issues

A weakness of most of the aforementioned analytical approaches is that they are based on the island model of dispersal, which is often poorly suited for data analysis (e.g., Hanski, 1997a). At the other extreme, simulation can be used to analyze complex demographic scenarios, including localized dispersal and density dependence (e.g., Barton et al., 2002). Some genetic patterns may be interpretable in terms of "effective density" and "effective dispersal" parameters (Rousset, 1999b). In this case, the large number of parameters of a complex model is reduced to a small number, such as metapopulation size, effective density, and effective dispersal. In principle, complex life cycle and various causes of dispersal could be taken in account by such effective parameters, but several difficulties still impede progress. First, there is no validated and practical method for estimating the effective size of a metapopulation. Available formulas call for more data than are available. Second, the most important demographic parameters for metapopulation processes are not necessarily extracted easily from the reduced set of genetic parameters. Indeed, both effective density and effective dispersal parameters are expected to be complex functions of age structure, age-dependent dispersal and fecundities, and so on (Rousset, 1999b; and case study in Sumner et al., 2001). Furthermore, the effective parameters do not describe well patterns of genetic differentiation at short distances (Rousset, 1999c), yet the latter may be important for quantifying kin competition between neighbors. Thus, it is still unclear what can offer genetic diversity analyses in the absence of detailed demographic observations.

13.9 FEEDBACK BETWEEN METAPOPULATION DYNAMICS AND DISPERSAL

The role of dispersal as a determinant of colonization–extinction dynamics is a main topic in metapopulation biology, which has been explored by numerous studies (see Chapters 4, 14, and 19–22). Naturally, colonization of patches following extinction can only take place as result of dispersal, and dispersal affects extinction probability by increasing it through emigration and decreasing it by immigration (i.e., rescue effect) (Martin et al., 1997; Hanski, 1999b, 2001). Indeed, most metapopulation studies directly or indirectly aim to address dispersal as a driver or a cause of metapopulation dynamics through its effects on colonization and extinction rates. However, the fact that there is potential for a dynamic duality in the dispersal-metapopulation dynamics relation, in the sense that dispersal appears as a consequence of a particular (meta)population dynamics in addition to being a cause, has been less appreciated. In particular, the extent to which there is a feedback between spatio-temporal population dynamics and dispersal parameters (e.g., rate and distance) has been explored to a limited extent. The potential importance of the cause–consequence duality of dispersal in a metapopulation setting is, for instance, evident concerning the relationship between spatial population synchrony and dispersal. The degree of population synchrony, itself an important determinant of metapopulation extinction probability, is expected to increase with increasing dispersal rate (Lande et al., 1999), whereas both the rate and the distance of dispersal are likely to be dependent on the degree of synchrony (Ims and Hjermann, 2001).

Understanding which dispersal cause has the highest impact on metapopulation dynamics is still a largely open question. It is most likely that the answer will depend on species biology. For example, one might imagine that species living in highly disturbed or evolving habitats (high level of extinction) will be dominated by habitat-driven dispersal and what will be most important to consider in a metapopulation framework will be habitat selection, i.e., the dynamics will be driven by the settlement phase of dispersal. In species living in a less disturbed habitat, social characteristics of the local patch are likely to become important, and two types of dispersal, kin and density-dependent dispersal, are to be considered.

Multiple Action of Density

The most obvious way by which population dynamics may feed back on dispersal is through density-dependent dispersal. However, the common assumption of a positive, linear relationship between population density (density as an ultimate and proximate cause) and dispersal (for a review, see Hanski, 1999b) has been questioned by several empirical studies and reviews (Hanski, 2001; Ims and Hjermann, 2001; Lambin et al., 2001; Chapter 21; Section 13.6). Indeed, negative density-dependence has been found frequently. Understanding why such negative density-dependence occurs and what are the consequences for metapopulation dynamics are critically important. With respect to potential proximate causes of negative density-dependent dispersal, one possibility is that density has effects on other traits that are related indirectly to dispersal. For example, in species where the ontogeny of dispersal is linked to puberty and sexual maturation (Dufty and Belthoff, 2001), a density-induced delayed sexual reproduction at high densities will then also result in delayed dispersal, possibly to time periods with lower densities in temporally fluctuating populations. For individual organisms that require a certain amount of stored energy reserves for emigration to be triggered (e.g., Nunes et al., 1997), a high local density resulting in intense resource competition can also affect the rate of emigration negatively, so precise species biology has to be known in order to model the effect of dispersal in a metapopulation.

Unless a high emigration rate from low-density patches is compensated for by a high immigration rate, emigration will be a likely cause of extinction (see Kuusaari et al., 1996; Andreassen and Ims, 2001). During the transient phase, a high density of territorial individuals may impede movements and thus reduce dispersal distance. This second mechanism of negative density dependence (see Section 13.4 for more details) is probably less likely in a typical metapopulation setting in which most of the dispersal trajectory takes place in an empty matrix between suitable patches. Still the effect of a *social fence* (see Hestbeck, 1982) may be relevant for individuals situated near the center of large patches

(Stamps, 1987) and for which within-patch, population-level dispersal precedes between-patch metapopulation level dispersal. Thus the social fence effect opens the possibility that the size of a local population, in addition to its density, can determine the per capita emigration rate. However, in a metapopulation the impact of social fences is most relevant in the immigration stage of dispersal in which settlement success is likely to be related negatively to the density of conspecifics in putative immigration patches. Experimental studies have demonstrated that such a density-dependent immigration rate both tends to rescue small (extinction prone) populations and evens out spatial variance in population density among patches (Aars and Ims, 1999; Gundersen et al., 2001, 2002; Lambin et al., 2001; Lecomte et al., 2003; Chapter 21).

Which Densities?

The kind of dispersal stage-specific response to population density described previously complicates the study of metapopulation dynamics–dispersal interface. For instance, densities in the emigration and all immigration patches within the exploration range of an organism must be mapped and the density dependence of both emigration and immigration probabilities must be estimated to predict the organism's overall transfer probability. Having such information available from a relatively simple, small-scale and transparent experimental model system, Andreassen and Ims (2001) showed that root voles tended to disperse most frequently from relatively low-density patches to patches with even lower population density. In this case it appeared that emigration probability from a given patch could be modeled as a function of density in the emigration patch and the coefficient of variation in density among patches in the metapopulation (Fig. 13.5).

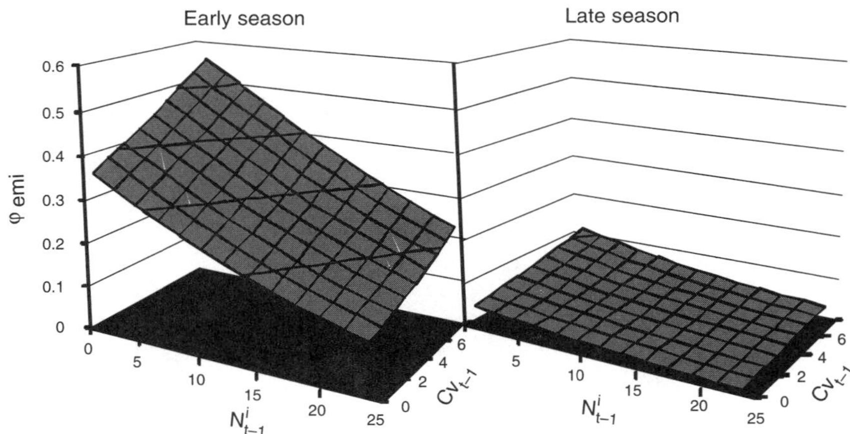

Fig. 13.5 Emigration probability. The emigration probability from patches in experimental metapopulations of root voles (Andreassen et al., 2002) depending on local density, regional density variation, and time in season. φ_{emi} is the specific emigration probability from patch *i*, N^i_{t-1} is the density of animals at time $t-1$ within patch *i*, CV_{t-1} is the coefficient variation between patches in the metapopulation at the same time, and the week denotes the specific time in the breeding season.

However, limited knowledge about the scale and mode of patch and matrix exploration of dispersing individuals (Ims, 1995; Wiens, 2001; Ricketts, 2001; Conradt et al., 2003) represent obstacles for partialling out the effects of density-dependent dispersal in most situations. The exploration stage of dispersal is important to consider because it imposes a feedback between immigration and emigration stages in the dispersal process. In particular, if immigration is negatively density-dependent and there are only high-density patches within the exploratory range, any departed animal will tend to return to its patch of origin. This will markedly decrease colonization of even more distant patches. However, many organisms are not capable of exploring their environment before deciding to disperse, either because they entirely lack this ability or because the nearest patch is out of exploration range. In this case, density-dependent can only act locally and through its influence on emigration probability; i.e., density in potential immigration patches will only affect whether emigration is succeeded by successful immigration (i.e., whether dispersal is *efficient*) and eventually dispersal distance (how far an animal must move before finding a suitable patch; Ronce et al., 2001). Whether the immigration rate can be expected to be lower or higher depending on whether an organism is able to explore patches is unclear. Nevertheless, this point is important because it concerns the impact of dispersal on local dynamics (e.g., the effect of immigration versus other *in situ* demographic processes), the degree of coupling (synchrony) of dynamics between populations, and the possibility for feedback between metapopulation level processes and dispersal. Many more empirical studies on the metapopulation–dispersal interface are urgently needed. Central questions such as what is the sign, strength, functional form, and spatial scale of density-dependent emigration and immigration processes are largely unexplored even for well-studied metapopulations.

Qualitative Effects

In addition to the form of the density-dependent dispersal rate, metapopulation dynamics can also be affected by the nature of the dispersing or philopatric individuals. As discussed in Section 13.7, there is accumulating evidence that dispersers are not a random subset of their population of origin (Bélichon et al., 1996; Murren et al., 2001). Consequently, the success of a propagule and its effectiveness at reenforcing existing populations or colonizing empty patches might be strongly dependent on the "quality" of dispersers. In other words, there may be an interplay between the cause of dispersal and the effect of colonists and immigrants have on the growth rate in previously empty and already occupied patches. For this reason, the specific cause of dispersal may be important to know when considering the consequences of dispersal in a metapopulation setting.

Just as high-quality propagules may enhance population growth in recipient patches and populations, the extinction of donor populations may also be enhanced by the poor quality of the remaining individuals, which may be decreased even further by the potential deleterious effect of inbreeding. If organisms use conspecifics as cues when selecting a new patch, a patch containing individuals in a poor shape is less likely to be rescued by high-quality immigrants because of habitat selection. This may further increasing extinction probabilities (extinction vortex). Thus generally, the interplay between departure and settlement decisions may be of prime importance to better understand colonization–extinction processes.

13.10 DISPERSAL AND SPECIES INTERACTIONS

As discussed earlier, species are likely to influence each other dispersal strategies. Indeed, seeds are dispersed by dispersing seed predators, predators (parasites) have to search for their prey (host), and prey (hosts) try to avoid predation (parasitism) by many ways, including emigration from patches with a high predation (parasitism) pressure. For example, aphids increase the proportion of winged offspring when predation increases (Weisser et al., 1999). Moreover, dispersal was enhanced under high predation rates in experimental metapopulations of voles (Ims and Andreassen, 2000). Dispersal in host–parasite, predator-prey, or multitrophic systems has therefore received increasing attention.

Indeed, the inclusion of a spatial dimension into systems of interacting species has revealed an extension of the domain of coexistence in many cases, but it has also shown an increased complexity of the overall population dynamics (Bernstein et al., 1999; van Baalen and Hochberg, 2001). Of particular interest in a metapopulation context is the opposite action on local population persistence of the degree of information a species has about its environment and the spatial heterogeneity of this environment. In the case of two interacting species, the degree of information leads to an ideal free distribution (IFD) for both species, but to a low probability of coexistence, whereas spatial heterogeneity leads to distributions far from IFD, but to an increased probability of coexistence (van Baalen and Sabelis, 1993). There is a certain analogy here with the effect of spatial autocorrelation of environmental variation (see earlier discussion). Another analogy with the one-species case is that habitat selection by one individual depends on the other individuals' (of all species) decisions. Thus, one expects individuals to show some dynamical responsiveness, which may lead to condition-dependent dispersal in both species. However, constraints on information accessibility and the capacity to disperse typically vary among species. For example, the degree of local adaptation in the predator (parasite) and prey (host) strongly depends on their respective dispersal capacities, as discussed earlier (Hochberg et al., 1992; van Baalen and Hochberg, 2001).

The inclusion of dispersal into models of metapopulation of interacting species has therefore deep consequences on the overall dynamic of the system, but more theoretical and experimental research has to be conducted to measure the actual magnitude of such an effect. In turn, these ecological consequences will feed back on the evolution of dispersal in both interacting species. The extent to which such evolutionary feedbacks will be important depends on the other forces molding dispersal evolution as well as on the species capacity to invest in other mechanisms, such as defensive structures, chemical weapons, or immunity, which prevent them to escape without moving.

13.11 CONCLUSION

Underlying the omnibus term dispersal, there is large set of mechanisms and adaptations, each of which is likely to generate different qualitative and quantitative short- and long-term effects on metapopulation dynamic. We have provided a rather general review, in particular focusing on the many factors promoting the departure of individuals from patches and directing them to arrive at new ones. The various factors may cause different rates of dispersal, different phenotypic profiles in dispersing and resident individuals, and different links between leaving and settlement decisions. Despite the fairly long history of dispersal studies and the renewed interest due to the recent focus on landscape ecology, spatial population dynamics, and metapopulation biology, all of which are research disciplines where dispersal is a key parameter, our present knowledge about dispersal is rather scattered. There is no such thing as an unified theory of dispersal, although dominant causal factors have been proposed. For example, the rate of extinctions has been proposed to be dominant over other causes for dispersal evolution based on theoretical considerations (Gandon and Michalakis, 2001), whereas empirically, density dependence has recurrently been found to be one of the most important proximate factor influencing the dispersal rate (Lambin et al., 2001). Moreover, there is accumulating evidence, both theoretically and empirically, that kin interactions are important to consider in the study of dispersal (Perrin and Goudet, 2001; Leturque and Rousset, 2002; Le Galliard and 2003b). Thus, dispersal can quite certainly be found to respond to almost any potential cause. Indeed, dispersal, in the majority of cases, has been found to be state dependent. Potentially this has profound implications on the role of dispersal in metapopulation dynamics and evolution. Even though recent models of metapopulations take some local factors (i.e., local population dynamic) into account (i.e., through density-dependent dispersal), very few, if any, have an empirical basis for the local factors that are included and how they are modeled, for example, the strengths and shapes of the functional relationships and between dispersal rate and the local factors. Moreover, it is not made clear what is to be considered as local factors (within the influence or exploration scale of an individual) and what are more regional factors and, eventually, how local and regional factors interact. Population density is naturally a prominent factor for most individual dispersal decisions as it can be seen both as an indication of intraspecific competition (density as an ultimate cause) and as a sign of habitat quality (density as proximate cue). The nature of individuals themselves (age, size, and relatedness) will influence the way density is perceived. For this and other reasons, the effect of population density in donor and recipient patches is likely to relate differently to the emigration and immigration probabilities, respectively. That population density both indicates and determines different aspects and processes complicates how density-dependence dispersal should be estimated, interpreted, and modeled. We expect that the question of how dispersal affects the density and correlated descriptors of population characteristics (age structure, sex ratio, genetic makeup) will be an important and rich field for research in the next decade.

Throughout our review on the causes and consequences of dispersal, we have referred to its three stages: emigration, transience, and settlement (i.e., immigration and colonization). While the three stages all are obviously important for the likelihood of a successful transfer of the individual in metapopulations (Ims and Yoccoz, 1997), it seems that the settlement stage is the least appreciated by metapopulation biologists. Dispersers are not at random within the metapopulation. If we except idiosyncratic models (MacDonald and Johnson, 2001), settlement decisions are almost never taken into account in attempts to model the evolution of dispersal and to understand its genetic and population consequences. This is certainly one of the most important weaknesses of the current metapopulation paradigm. There is indeed ample evidence that most species are able to assess the quality of their abiotic and biotic environment and that this assessment serves as a basis for settlement strategies. The knowledge of these strategies and an understanding of environmental and internal cues individuals are using as a basis for a given strategy will certainly prove to be necessary if we want to have a more realistic view of dispersal within metapopulations.

A pervasive theme in empirical studies of dispersal is that dispersers are not a random subset of the individuals in the source population: they may differ in morphology, physiology, or behavior. Whether dispersers have features that make them more successful is important, especially when the cost of transience and the success at settlement are considered. Indeed, although still open to debate, the success of an immigrant when compared to the resident in a resident patch is often found to be different. This "quality" effect does not seem to be associated to all dispersal causes, as dispersers are not found to be less competitive individuals in many cases. For example, the study on the dispersers phenotype in relation to the cause of dispersal suggests that, at least in some species, there are two types of dispersers depending on the settlement conditions that might be important to distinguish: those individuals movements that are completed by settlement in an already occupied patch (reenforcement) and those that end up in an unoccupied patch (colonization). Consider that these two types of dispersal are performed by qualitatively different types of individuals that are preconditioned by the environmental conditions in the patches of departure. Then, in case of a large-scale environmental change, for instance, global warming, metapopulation survival will be enhanced by individuals tending to leave degraded patches and colonize newly suitable patches. In that case, metapopulation survival will be very much dependent on how the conditioning for immigration influences the sensitivity for different settlement cues. It is possible that individuals destined for settlement in empty patches are sensitive to cues (e.g., density and quality of the conspecific individuals) other than those more likely to settle in already occupied patches (e.g., resource levels and abiotic conditions). To which extent these "quality" differences are important to the dynamic of extinction and recolonization has just started to be considered and definitively deserves more studies.

Genetic models bring us a better understanding of who should disperse and when, and these questions are being investigated with increased realism in metapopulations. However, other contributions of genetic studies of metapopulations have been limited (beyond topics that are not specific to metapopulations per se). Dispersal surely feeds back on many aspects of the genetics of metapopulations, but claims about the importance of genetic effects on metapopulation persistence remain highly speculative. Attempts to estimate demographic parameters from genetic structure will face conflicting issues. On the one hand, theory aims to analyze the genetic structure in terms of a few synthetic effective parameters (population size, density, and dispersal rates). On the other hand, there are hopes to uncover complex demography out of these few parameters. There is no evidence, however, that the genetically effective parameters are related in a simple way to parameters that would appear most important for population dynamics or life history evolution in metapopulations. In such a context, we cannot advise putting more effort in studies of genetic structure at the expense of demographic studies, with the important exception of studies of relatedness between competing individuals, as they may help understand many behavioral decisions.

There is no doubt that dispersal is important to metapopulation dynamic and evolution. The extent to which a detailed knowledge of dispersal is necessary to understand and predict metapopulation dynamics and evolution is still a largely open question. We nevertheless suggest that model adjustments to actual metapopulation data should not be used as a demonstration that a more detailed, mechanistic knowledge is unnecessary. Indeed, simple models used for conservation purposes can indeed yield a good fit to the actual dynamics of colonization–extinction observations (Schoener et al., 2003), but often lead to false conclusions with respect to the underlying processes generating these patterns when compared to more realistic models.

14. MECHANISMS OF POPULATION EXTINCTION

Oscar E. Gaggiotti and **Ilkka Hanski**

14.1 INTRODUCTION

Population ecologists have traditionally been concerned with questions about population regulation and the mechanisms that increase population stability (Elton, 1949; Nicholson, 1954, 1957; Milne, 1957, 1962; Andrewartha, 1957; den Boer, 1968; Andrewartha and Birch, 1984; Sinclair, 1989; Hanski, 1990b; Price and Cappuccino, 1995; Turchin, 1995, 2003). Population ecologists tended to study large populations, often of recognized "pest" species, which appeared to exhibit great persistence. In fact, until the early 1960s the predominant view in population ecology considered population extinctions unlikely in the presence of effective population regulation, wide dispersal, and generally large population sizes. This view predominated because little attention was paid to the actual spatial structure of populations (Allee et al., 1949). Notable exceptions were three Australian ecologists who recognized the possibility of small populations with high rate of extinction, although they reached this conclusion for entirely different reasons. Nicholson (1957), the principal architect of the population regulation paradigm, envisioned spatially structured populations and extinctions of small local populations, but principally in the case of host–parasitoid dynamics with strong density dependence leading to oscillations with increasing amplitude and, therefore, to local extinction (Nicholson, 1933). In contrast, Andrewartha and Birch (1954), who were not impressed by the

0-12-323448-4

effectiveness of population regulation in preventing extinctions, developed proto-metapopulation ideas of large-scale persistence of species with ephemeral local populations (see discussion in Hanski, 1999b). One of the more influential studies that gradually changed ecologists' views about the spatial structure and dynamics of populations was Ehrlich's study on the checkerspot butterfly *Euphydryas editha* in California, showing apparently independent dynamics of similar populations over short distances in the absence of obvious density dependence (Ehrlich, 1961, 1965; Singer, 1972). The checkerspot studies are noteworthy for having addressed both ecological and genetic processes in local populations and for having contributed many insights about the processes of population extinction for the last four decades (for a comprehensive review, see Ehrlich and Hanski, 2004).

The perspective in population biology changed greatly in the 1970s in the wake of the emergence of modern conservation biology and its emphasis on questions about reserve design and population viability (Simberloff, 1988; Hanski and Simberloff, 1997). Questions about reserve design stemmed from the dynamic theory of island biogeography (MacArthur and Wilson, 1963, 1967), which, of course, was explicitly concerned with population extinctions. Early analyses of population viability in conservation biology emphasized genetic factors, inbreeding and drift (Foose, 1977; Chesser et al., 1980; Soulé and Wilcox, 1980; Frankel and Soulé, 1981; O'Brien et al., 1983; Schonewald-Cox et al., 1983; Soulé 1986). In the late 1980s, increasing recognition of habitat loss and fragmentation as the main threats to biodiversity (Wilson, 1988, 1989; Reid and Miller, 1989; Groombridge, 1992; Ehrlich and Daily, 1993) contributed to the growth of metapopulation biology (see Fig. 1.2 in Chapter 1), with emphasis on the spatial structure of populations and on the often high rate of extinction of small local populations (Gilpin and Hanski, 1991; Hanski and Gilpin, 1997; Hanski, 1999b).

The relative importance of ecological versus genetic factors in population extinction has been the subject of controversy ever since the birth of modern conservation biology. As already mentioned, conservation biology emerged as a discipline on two foundations, the island theory and the vision of population extinction due to genetic deterioration. Lande's (1988) influential paper reviewed the issue 15 years ago. He concluded that focusing primarily on genetic mechanisms of extinction was misguided and would not provide an adequate basis for understanding the processes underpinning the survival of endangered species. He also stressed the need for a realistic integration of demography and population genetics that would be applicable to species in their natural environments. Following the publication of this paper, a consensus started to form supporting the primary role of ecological factors in extinction. This consensus was later challenged by a series of theoretical studies (see later) of the decrease in fitness due to the accumulation of deleterious mutations ("genetic meltdown"). These analyses suggested that even relatively large populations might go extinct due to genetic deterioration. Undoubtedly, it has been difficult to reach a robust understanding about the mechanisms of population extinction because of the multitude of factors involved and the likely interactions among them, including ecological and genetic factors. Despite these difficulties, there has been substantial progress in this area during the last decade.

Although it is appropriate to emphasize interactions among different kinds of mechanisms influencing population extinction, it is practical to start with a review of particular ecological and genetics factors, which is done in Sections 14.2 and 14.3. One way of integrating the different factors is to relate them to the most important correlate of extinction risk, small population size. A common surrogate of local population size in metapopulation studies is the size of the habitat fragment in which the population occurs. Effects of population size and habitat patch size on extinction risk are reviewed in Section 14.4. The range of significant extinction mechanisms is expanded further when we consider local extinction in the metapopulation context (Section 14.5) and extinction of entire metapopulations (Section 14.6). Some challenges for further research are discussed in Section 14.7.

It is customary in reviews like the present one to make the point that the reasons why populations and species are currently going extinct at a distressingly high rate have primarily to do with loss of habitats and interactions with species that humans have displaced around the globe. This is what Caughley (1994), in an influential paper, called the declining-population paradigm. In contrast, most of the factors reviewed in this chapter belong to Caughley's (1994) small-population paradigm and relate to the ecological and genetic mechanisms that render the persistence of small populations precarious even without any added threats introduced by humans. A major exception is metapopulation theory, which can be employed to elucidate the risk of metapopulation extinction due to habitat loss and fragmentation (examined in Chapter 4; see also Chapter 2 on landscape ecology). It is important to realize that such a distinction can be made, but it is equally important to realize that, to some extent, Caughley's (1994) dichotomy is false (Hedrick et al., 1996; Holsinger, 2000). The dichotomy between small-population and declining-population paradigms is partly false because mechanisms in the two realms interact. This is especially apparent in the context of metapopulation biology, where our interest is focused on species with spatially structured populations, often consisting of many small local populations even if the metapopulation as a whole is large. To properly understand the dynamics and population biology of such species, we need to understand the mechanisms of extinction of the local populations that are often small. The main objectives of this chapter are to provide an update on the status of our understanding of these issues and to outline avenues of future research that could help improve it.

14.2 POPULATION EXTINCTION: ECOLOGICAL FACTORS

Demographic and Environmental Stochasticities

The classic models of population dynamics are deterministic and of little use in the study of population extinction, except in making the trivial but hugely important point that if the population growth rate r is negative, the population will surely, and rather quickly, go extinct. This is important because the human onslaught on the environment introduces changes, such as habitat loss and alteration, and spreading of invasive species, which will make r negative in many populations. In deterministic models without age structure, the time to extinction from initial population size N_0 (which is assumed to be much below the carrying capacity) is given by $-\ln N_0/r$ (Richter-Dyn and Goel, 1972). In contrast, populations with $r > 0$ will not go extinct in simple deterministic models.

Deterministic models are inadequate for real populations because their dynamics are influenced by stochastic effects. It is useful to distinguish between two forms of stochasticity. *Demographic stochasticity* is due to random independent variation in the births and deaths of individuals. *Environmental stochasticity*, in contrast, is generated by random effects affecting all individuals in the population similarly. The label "environmental" signifies that the effects are caused by the shared environment of the individuals in the same population, such as adverse weather effects increasing mortality. These are the exogenous factors of population ecologists (Turchin, 2003).

In line with the two forms of stochasticity maintaining fluctuations in population size, the variance in the change in population size ΔN conditioned on population size N may be partitioned into two components, which are demographic and environmental variances (Engen et al., 1998). Assuming that these components are constant and denoting them by σ_d^2 and σ_e^2, respectively, $\mathrm{var}(\Delta N|N) = \sigma_d^2 N + \sigma_e^2 N^2$. Engen et al. (1998) presented general definitions of the demographic and environmental variances in terms of the lifetime reproductive contributions of individuals to the next generation, R_i. They showed that the demographic variance σ_d^2 is half of the variance in the difference of the R_i values for pairs of individuals (conditioned on current population size). Thus, if all individuals would make exactly the same contribution to the next generation, the demographic variance would be zero, that is, there would be no "demographic stochasticity." In reality, of course, this will not happen because of the intrinsic uncertainty involved in individual births and deaths. The environmental variance is a covariance of the R_i values (Engen et al., 1998). "Environmental stochasticity" is hence great when R_i values vary in parallel, as will happen if the performance of all individuals is influenced by the same common (environmental) factors. Note that positive covariance of the individual R_i values means that the population growth rate exhibits temporal variation. It is also noteworthy that environmental covariance may be negative, that is, an environmental effect may reduce the variance of the change in population size. Engen et al. (1998) gave the (hypothetical) example of space limitation and territoriality leading to a completely constant population size. In this case, the individual R_i values would be necessarily negatively correlated.

The approach developed by Engen et al. (1998) to characterize population fluctuations can be applied to real populations to estimate the demographic and environmental variances and to predict changes in population size, including the risk of population extinction. The drawback of this approach, however, is that one requires data on individual lifetime reproductive contributions, which data are not often available. Saether et al. (1998a) analyzed long-term data on the great tit population at Wytham Wood near Oxford. The environmental variance turned out to be large in this case, but the population was not expected to go extinct because the growth rate was also large. In contrast, in a brown bear population the environmental variance was very small and smaller than the demographic variance (Saether et al.,

1998b). This is consistent with the general expectation that large-bodied vertebrates (like the brown bear) are less influenced by environmental stochasticity than small-bodied vertebrates (like the great tit) and invertebrates. We have more to say about this in the next section.

As a more detailed example, we outline the analysis by Engen et al. (2001) of the stochastic population dynamics of the barn swallow population studied by A.P. Møller at Kraghede, Denmark, since 1970. At this site, the barn swallow population had declined from 184 pairs in 1984 to 58 pairs in 1999. Reasons for the decline appear to be changes in agricultural practices reducing the reproductive success of the birds.

The model fitted by Engen et al. (2001) to data on barn swallows assumes that the stochasticity in the population size is described by a Markov process and that the year-to-year change in the logarithm of population size $X(= \ln N)$ is normally distributed with the expectation

$$E(\Delta X|X = x) = r - {}^1\!/_2\sigma_e^2 - {}^1\!/_2\sigma_d^2/N \tag{14.1}$$

and variance

$$\mathrm{var}(\Delta X|X = x) = \sigma_e^2 + \sigma_d^2/N \tag{14.2}$$

The quantity $r_0 = r - {}^1\!/_2\sigma_e^2$ is defined as the stochastic growth rate and indicates the extent to which stochastic fluctuations in population size reduce the long-term ("long-run") growth rate (Tuljapurkar, 1982; Lande and Orzack, 1988; Lande, 1993). Demographic stochasticity also reduces the long-term growth rate, and the combined effects of demographic and environmental stochasticity lead to the expectation in Eq. (14.1).

Engen et al. (2001) obtained an estimate of the demographic variance σ_d^2 from data on the individual contributions of breeding females to the next generation, R_i (number of female offspring recorded in the next or following generations plus 1 if the female itself survived), calculated as

$$1/(k - 1) \sum (R_i - \overline{R})^2, \tag{14.3}$$

where $\overline{R}$ is the mean contribution of the individuals and k is the number of recorded contributions in 1 yr. If data are available for several years, σ_d^2 is estimated as the weighted average of the yearly estimates (Saether and Engen, 2002). In the case of the barn swallow, there were extensive data on individual reproduction and survival, and hence σ_d^2 was assumed to be accurately known as estimated from data for several years, $\sigma_d^2 = 0.180$. Next the values of r_0 and σ_e^2 were estimated from time series data on yearly population sizes by maximizing a likelihood function numerically (Engen et al., 2001). The maximum likelihood parameter estimates were $r_0{}^* = -0.076$ and $\sigma_e^{2*} = 0.024$. This barn swallow population has thus shown a mean decline of 7.6% per year.

Figure 14.1 shows the lower bound of the prediction interval, which includes the predicted population size with probability $1 - \alpha$. Comparison between Figs 14.1A and 14.1B demonstrates that ignoring uncertainty in parameter estimates (and using their maximum likelihood estimates) increases the predicted time to extinction. In other words, acknowledging the uncertainty in the

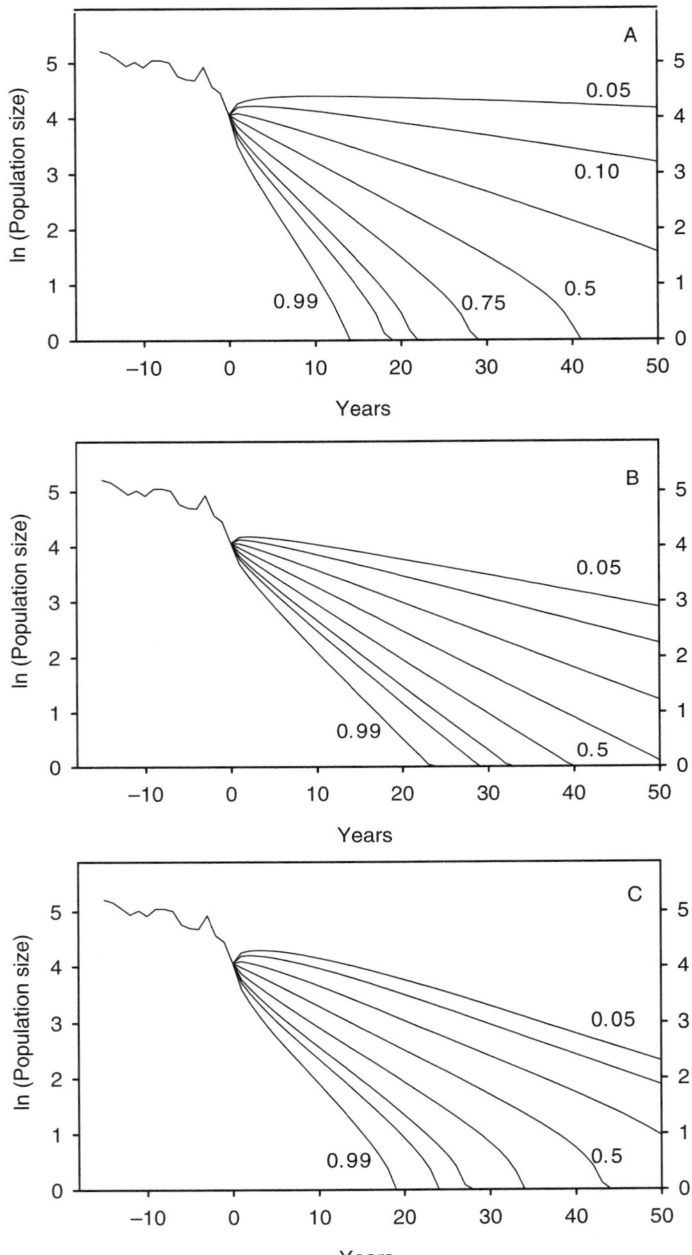

Fig. 14.1 Annual variation in the number of breeding pairs of the barn swallow at a study site in Denmark from 1984 until 1999 (the time period until zero on the *x* axis), followed by the lower bound of different prediction intervals for the future population size for different values of α. Results when (A) all available information is included, (B) uncertainty in parameter estimates is ignored, and (C) demographic variance is set to zero (from Engen et al., 2001).

parameter values leads to more cautious predictions: the population may go extinct sooner than the maximum likelihood estimates would suggest. In Fig. 14.1C, the demographic variance is assumed to equal zero. Ignoring this component increases the predicted time to extinction. Additionally, ignoring environmental variance reduces the range of variation of the prediction interval (Engen et al., 2001). In other words, the fate of the population would be much easier to predict without environmental stochasticity.

Scaling of Extinction Risk with Carrying Capacity

A useful framework for examining many ecological factors in population extinction is provided by the simple "ceiling" model of population dynamics (Lande, 1993; Foley, 1994, 1997; Middleton et al., 1995). Although this model does not incorporate any details of demography and life history of species, it is helpful in encapsulating in general terms the effects on extinction probability of those factors that should always be considered. This theory is also helpful in providing a submodel of local extinction that can be used in metapopulation models (Hanski, 1998a, 1999b; Chapter 4). The ceiling model is described in Box 14.1.

BOX 14.1 The Ceiling Model of Population Extinction

Population dynamics are assumed to obey the following equations:

$$n_{t+1} = n_t + r_t \quad \text{if } 0 \leq n_{t+1} \leq k$$
$$n_{t+1} = k \quad \text{if } n_{t+1} > k$$
$$n_{t+1} = 0 \quad \text{if } n_{t+1} < 0,$$

where n_t is the natural logarithm of population size (N) at time t, k is the logarithm of the population ceiling (K), and r_t is a normally distributed random variable with mean r and variance σ_e^2. The model assumes that the population size performs a random walk between the absorbing lower boundary of population extinction and the reflecting upper boundary of population ceiling. Population fluctuations are driven by environmental stochasticity. Using the diffusion approach to analyze this model (Foley, 1994; Lande, 1993; Middleton et al., 1995), the expected time to extinction of a population with $r > 0$ and starting at the ceiling K is given by

$$T(K) = K^s/sr\,[1 - (1 + sk)/\exp(sk)], \qquad \text{(B1)}$$

where $s = 2r/\sigma_e^2$. For reasonably large values of sk the term in square brackets is close to 1 and hence the result simplifies to

$$T(K) \approx K^s/sr. \qquad \text{(B2)}$$

These results were obtained for a model that ignores demographic stochasticity. Hanski (1998a) compared the scaling of time to extinction with population ceiling predicted by (B1) and by the comparable model (from Foley, 1997) with both demographic and environmental stochasticities. For values of σ_d^2/σ_e^2 less than 1, which is likely to be valid for most natural populations, the scaling result (B1) is little affected by the added demographic stochasticity (see Hanski, 1998a).

The most lucid and useful result is obtained by assuming that population fluctuations are driven solely by environmental stochasticity. In other words, we assume, for simplicity, that the demographic variance equals zero. The key parameters are then the population ceiling (absolute carrying capacity) K, which the population size cannot exceed (Box 14.1), and the stochastic population growth rate r_0 discussed in the previous section and given by $r_0 = r - {}^1\!/_2\sigma_e^2$. Note that if $r < {}^1\!/_2\sigma_e^2$, the population will go extinct with probability 1 even in the absence of any density dependence. For convenience, we denote the ratio $2r/\sigma_e^2$ by s. Assuming that $r > 0$ and that sk is reasonably large (where k is the logarithm of K), the time to extinction scales asymptotically as

$$T \approx K^s/sr. \tag{14.4}$$

Thus, if population fluctuations are caused solely by environmental stochasticity, the time to extinction scales as a power function of the population ceiling. In the other extreme, when there is no environmental stochasticity and population fluctuations are caused by demographic stochasticity alone, the scaling is nearly exponential (MacArthur and Wilson, 1967; Lande, 1993; Foley, 1994). Exponential scaling means that for reasonably large r, only very small populations have an appreciable risk of extinction. The extreme case of only demographic stochasticity is of academic interest only, as all real populations are more or less influenced by both environmental and demographic stochasticities. Adding demographic stochasticity to the model leading to Eq. (14.4) will shorten the time to extinction (see Fig. 14.1), but the scaling is little affected unless both the ceiling and s are very small (Foley, 1997; Hanski, 1998a; Box 14.1). Hence we focus on the simple result given by Eq. (14.4).

Taking now the interpretation of the power-function scaling further, let us observe that the value of $s = 2r/\sigma_e^2$ is an inverse measure of the strength of environmental stochasticity, scaled by r. The greater the impact of environmental stochasticity on the population growth rate (the smaller the value of s), the shorter the expected lifetime of the population and the smaller the increase in lifetime with increasing population ceiling [Eq. (14.4)]. A high growth rate (r) has the net effect of increasing population lifetime and the opposite effect to that of σ_e^2 on the scaling with population ceiling.

A useful feature of Eq. (14.4) is that the value of the scaling constant s can be estimated with empirical data. Recording actual extinction rates ($1/T$) for particular populations is impractical, but in the context of metapopulations with many local populations in a patch network, one may use the spatially realistic metapopulation theory (Chapter 4) to estimate s from data on the incidence of patch occupancy. Hanski (1998a) applied a mainland-island metapopulation model (Hanski 1993) to data on the occurrence of four species of *Sorex* shrews on small islands. The key assumptions were that island area multiplied by an estimate of population density is an adequate surrogate of the population ceiling and that the occurrence of the species on islands represents a balance between stochastic extinctions and recolonizations [as supported by the results of Hanski (1986) and Peltonen and Hanski (1991)]. Figure 14.2 shows the relationship between the expected lifetime of populations and their carrying capacity for the four species based on the parameter values estimated with the metapopulation model (Hanski, 1993). This result shows wide

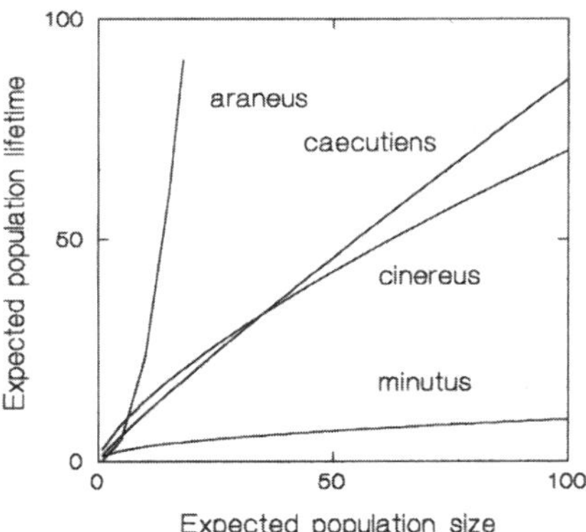

Fig. 14.2 Relationship between the expected population lifetime and the carrying capacity (island area times average density) in four species of *Sorex* shrews on islands. The result was calculated with the parameters of the incidence function metapopulation model fitted to data on island occupancy (from Hanski, 1993).

variation in the value of *s*, which can be interpreted as variation in the impact of environmental stochasticity among the species, as their *r* values are comparable. Furthermore, a positive correlation exists between the body size of the species and the value of *s*, suggesting that environmental stochasticity plays a greater role in the dynamics of small than large species of shrew [Cook and Hanski (1995) reported the same relationship for birds on oceanic islands]. This result makes biological sense because the smallest species of shrew, which weigh less than 3 g and starve in a few hours, are particularly vulnerable to temporal variation in food availability. Hanski (1998a) further estimated the values of r and σ_e^2 for the common shrew (*Sorex araneus*) from the parameter values of the metapopulation model, as $r = 0.75$ and $\sigma_e^2 = 0.42$. These values are consistent with the biology of shrews, which live for 1 yr only and produce one to three litters of seven young on average (Sheftel, 1989). The coefficient of variation calculated from these values of r and σ_e^2 is 0.86, which is consistent with the observed CV calculated from trapping data, 0.67 (average of four independent estimates; Hanski and Pankakoski, 1989). These results are encouraging in highlighting a clear connection between parameters of the extinction model for single populations and parameters of the metapopulation model. One general difficulty, however, is that the estimates of r and σ_e^2 thus obtained are sensitive to the estimate of population density (Hanski, 1998a). Luckily, the scaling constant *s* is not similarly affected.

Complex Population Dynamics and Extinction

Population dynamics may be called simple if the growth rate is a monotonically decreasing function of population density and if the density-dependent feedback itself does not suffice to generate population oscillations. In this case, exemplified among others by the continuous-time logistic model and the ceiling model in Box 14.1, population density would settle to a stable state with constant population size in the absence of environmental perturbations and demographic stochasticity. Population extinction is typically caused by a low growth rate (e.g., due to poor habitat quality), a high variance in growth rate (environmental stochasticity), or a small population size due to low carrying capacity or other factors, which increases extinction risk for many reasons (Section 14.4).

Not all populations exhibit such simple dynamics, however, and their extinction risk may be affected by the extra complexities of population dynamics. Most commonly, the population growth rate may be expected to be reduced at very low densities due to difficulty of locating a mate or performing other cooperative behaviors; this is called the Allee effect (Allee, 1938; Allee et al., 1949). If the reduction in growth rate is severe enough, a small population will go deterministically extinct. Demographic stochasticity also substantially increases the risk of extinction of very small populations, especially if their growth rate is low, and there can be a threshold population size below which the most likely population trajectory is a decreasing population size. In this sense, demographic stochasticity creates a sort of stochastic Allee effect (Lande, 1998; Dennis, 2002). In models with both conventional Allee effect and demographic stochasticity, there is an inflection point in the probability of reaching a small population size before reaching a large size. This inflection point, which corresponds to the unstable equilibrium in the underlying deterministic model, represents a threshold in the probabilistic prospects for the population (Dennis, 2002). The incidence and importance of the Allee effect have been reviewed most recently by Saether et al. (1996), Kuussaari et al. (1996), Wells et al. (1998), Courchamp et al. (1999), and Stephens and Sutherland (1999). It should be recognized that small populations have a high risk of extinction for many reasons, including both ecological and genetic factors (Section 14.4), and factors that reduce the expected growth rate as well as factors that increase the variance in growth rate (Stephens et al., 1999; Dennis, 2002). Therefore, it is generally difficult to conclusively isolate the operation of any particular mechanism, including the Allee effect. Many mechanisms are often likely to operate in concert.

A strong Allee effect creates an unstable equilibrium point below which the population goes extinct in a deterministic model. In this case there are two alternative stable equilibria, one corresponding to large population size (set by density dependence at high density) and the other one corresponding to population extinction. If the dynamics exhibit such alternative stable equilibria, a small population below the unstable equilibrium is unlikely to become large, although it may do so and cross the unstable equilibrium thanks to a favorable environmental perturbation. Likewise, a large population above the unstable equilibrium is expected to remain large, but a perturbation may take it below the treshold population size and send it toward extinction. This is a worrying possibility because it implies that currently large populations may have a much greater risk of extinction than one might expect and predict with models that fail to include the mechanism creating alternative equilibria. Unfortunately, it is difficult to assess how likely this scenario is for real populations.

Complex population dynamics in the sense of cyclic or chaotic fluctuations maintained by population dynamic processes (as opposed to environmental effects) have received much attention during the past decades (May, 1974; Schaffer, 1985; Turchin, 2003). Population variability generated by intraspecific and interspecific interactions is expected to increase the risk of extinction just like variability generated by environmental stochasticity. It has even been argued that extinctions caused by chaotic dynamics would exert a (group) selection pressure that would make chaotic dynamics less likely and that local extinctions due to chaotic dynamics would enhance metapopulation persistence because the extinctions would be asynchronous (Allen et al., 1993; Gonzalez-Andujar and Perry, 1993; Bascompte and Solé, 1994; Ruxton, 1996). Although these issues involve many challenges for further research, it seems unlikely that complex dynamics in this sense would be a major factor in population extinctions.

14.3 POPULATION EXTINCTION: GENETIC FACTORS

Natural populations are also subject to extinction due to genetic factors even in the absence of any human impact and the threat posed by ecological processes. Genetic threats are a function of the effective population size, N_e. Strictly speaking, N_e is defined as the number of individuals in an ideal population that would give the same rate of random genetic drift as observed in the actual population (Wright, 1931, 1938). The ideal population consists of N individuals with nonoverlapping generations that reproduce by a random union of gametes. More intuitively, N_e can be defined as the number of individuals in a population that contribute genes to the following generation. This number can be much lower than the observed population size because of unequal sex ratios, variance in family size, temporal fluctuations in population size, and so forth (for a review, see Frankham, 1995). Thus, apparently large populations may still be quite small in a genetic sense and hence face genetic problems. Small N_e can have multiple effects that include loss of genetic variability, inbreeding depression, and accumulation of deleterious mutations. The time scales at which these factors operate differ and, to a large extent, determine the risk of population extinction that they entail (Table 14.1).

Loss of Genetic Variability

Genetic variation comprises the essential material that allows natural populations to adapt to changes in the environment, to expand their ranges, and even to reestablish following local extinctions (e.g., Hedrick and Miller, 1992). The types of genetic variation considered most often are the heterozygosity of

TABLE 14.1 Time Scales at Which Genetic Factors Operate and Their Importance for Population Extinction[a]

Factor	Time scale	Extinction risk involved	Extinction vortex
Inbreeding depression	Short	High	F
Loss of genetic diversity	Long	Low	A
Mutational meltdown	Medium/long	Unknown	A

[a] The last column indicates the extinction vortex (as defined by Gilpin and Soulé, 1986) under which each genetic factor operates.

neutral markers, H, and the additive genetic variance, V_a, which underlies polygenic characters such as life history traits, morphology, and physiology.

In small populations, random genetic drift leads to stochastic changes in gene frequencies due to Mendelian segregation and variation in family size. In the absence of factors that would replenish genetic variance, such as mutation, migration, and selection favoring heterozygotes, populations lose genetic variance according to

$$V_a(t+1) = V_a(t)\left(1 - \frac{1}{2N_e}\right), \tag{14.5}$$

where $V_a(t)$ is the additive genetic variance in the tth generation. A similar equation is obtained for heterozygosity by replacing V_a with H. When a population is reduced to a small effective size N_e and maintained at that size for more than $2N_e$ generations, its genetic variability is reduced greatly (Wright, 1969). Genetic variability can be restored to its original level through mutation if the population grows back to its original size. The number of generations required to attain the original level is of the order of the reciprocal of the mutation rate, μ. Thus, for a nuclear marker with a mutation rate of 10^{-6}, genetic variation is restored after 10^6 generations, but genetic variation of quantitative characters can be restored after only 1000 generations because the relevant mutation rate is two orders of magnitude higher.

The maximum fraction of genetic variation lost during a bottleneck is a function of the population growth rate (Nei et al., 1975). Populations that recover quickly after the bottleneck lose little genetic variation even if the population was reduced to a few individuals only. For example, a growth rate of $r = 0.5$ ($\lambda = e^r = 1.65$) allows a population that is reduced to only two individuals to retain 50% of its genetic variability (Fig. 14.3). If the population is reduced to 10 individuals, then a growth rate of $r = 0.1$ ($\lambda = 1.10$) would allow it to retain 60% of its variability. Additionally, generation overlap can buffer the effect of environmental fluctuations on population sizes. In general,

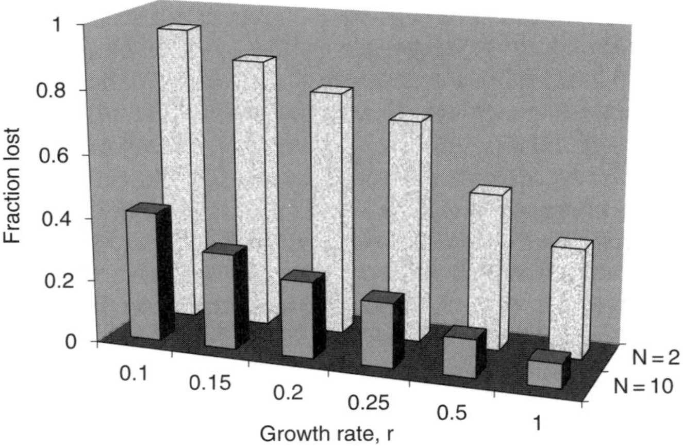

Fig. 14.3 Fraction of the genetic variation lost during a population bottleneck of $N = 2$ or 10 individuals. Calculated using Eq. (8) in Nei et al. (1975).

reductions in population size are brought about by environmental changes that cause fluctuations in vital rate parameters (environmental stochasticity; Section 14.2). The effect of these fluctuations on N_e depends on the life history of the species. The ratio of N_e to census size is directly proportional to the total reproductive value of a population, but the sensitivity of this ratio to environmental fluctuations is proportional to the generation overlap. The larger the generation overlap, the smaller the effect of environmental fluctuations on the level of genetic variability maintained by natural populations (Gaggiotti and Vetter, 1999). Thus, genetic variability is maintained through the "storage" of genotypes in long-lived stages. Adult individuals representing these stages reproduce many times throughout their lives and, therefore, the genetic variability present in a given cohort is more likely to be transferred to future generations than in the case of organisms with discrete generations.

These buffering mechanisms may explain why there are very few clear examples of populations that have lost a very large fraction of their genetic variability due to a bottleneck. One of the few cases is that of the Mauritius kestrel, which was reduced to a single pair in the 1950s. A comparison of microsatellite diversity present in museum specimens collected before the bottleneck and in extant individuals reveals that at least 50% of the heterozygosity was lost due to the bottleneck (Groombridge et al., 2000). Another example is the northern elephant seal, which was exploited heavily during the 19th century and reduced to a bottleneck population size estimated to be 10–30 individuals (Hoelzel et al., 2002). A comparison of genetic diversity in prebottleneck and postbottleneck samples shows a 50% reduction in *mt*DNA-haplotype diversity. The reduction in heterozygosity at microsatellite loci was less pronounced, however.

An important caveat concerning the effect of reductions in population size on genetic diversity is that although such reductions may not have a very large effect on H, they will have a large impact on allelic diversity because random genetic drift will eliminate low-frequency alleles very rapidly (Nei et al., 1975). This is of particular concern because the long-term response of a population to selection is determined by the allelic diversity that remains after the bottleneck or that is gained through mutations (James, 1971). A second caveat is that in the case of quantitative genetic characters, genetic variability may not always be beneficial. Using a model with overlapping generations and assuming weak stabilizing selection, Lande and Shannon (1996) showed that the effects of additive genetic variance on the average deviation of the mean phenotype from the optimum, and the corresponding "evolutionary" load, depend on the pattern of environmental change. In an unpredictable (random) environment, additive genetic variance contributes to the evolutionary load because any response to selection increases the expected deviation between the mean phenotype and the optimum. However, when environmental changes are unidirectional, cyclic, or positively correlated (predictable), additive genetic variance allows the mean phenotype to track the optimum more closely, reducing the evolutionary load.

Most empirical studies on the effects of population bottlenecks on genetic diversity focus on the heterozygosity of neutral markers. Although neutral genetic variation may become adaptive if the environment changes, the ability of a population to respond to novel selection pressures is proportional to the additive genetic variation underlying the traits that are the target of selection (Falconer and Mackay, 1996). Unfortunately, direct quantification of the genetic variation underlying polygenic traits is difficult to measure, and hence heterozygosity of nuclear markers is used as an indicator of additive genetic variation [see Pfrender et al. (2001) and references therein]. This practice is unwarranted, however, because of the different rates at which genetic variation is replenished in neutral and quantitative markers (Lande 1988; see earlier discussion). Indeed, Pfrender et al. (2001) detected no significant relationship between heritability for reproductive traits and heterozygosity in natural populations of *Daphnia pulex* and *D. pulicaria*. Thus, the absence of genetic diversity in nuclear markers does not necessarily indicate an immediate genetic threat.

In general, the loss of genetic variation is detrimental for the long-term survival of populations. However, as pointed out by Allendorf and Ryman (2002), there is one case where a reduction in genetic variability can represent an imminent extinction threat. This is the case for loci associated with disease resistance, such as the major histocompatibility complex (MHC), which is one of the most important genetic systems for infectious disease resistance in vertebrates (Hill, 1998; Hedrick and Kim, 2000). Allelic diversity at these loci is extremely high; for example, Parham and Otha (1996) documented 179 alleles at the MHC class I locus in humans. However, species that have been through known bottlenecks have very low amounts of MHC variation. A study of the Arabian oryx found only three alleles present at the MHC class II DRB locus in a sample of 57 individuals (Hedrick et al., 2000). Hunting pressure led to the extinction of this species in the wild in 1972. Captive populations have been susceptible to tuberculosis and foot-and-mouth disease, which is consistent with low genetic variability at MHC loci. Low genetic diversity at the MHC complex was also observed in the bison, which went through a bottleneck at the end of the 19th century (Mikko et al., 1997). In the Przewalski's horse, in which the entire species is descended from 13 founders, Hedrick et al. (1999) observed four alleles at one locus and two alleles at a second locus. The northern elephant seal is another example of low MHC diversity, as Hoelzel et al. (1999) found only two alleles at the MHC class II DQB gene in a sample of 69 individuals.

To summarize, we may conclude that loss of genetic variation as measured by heterozygosity and additive genetic variance represents a long-term extinction threat. In the short term, the loss of allelic diversity can have important consequences if it occurs at loci associated with disease resistance.

Inbreeding Depression

The decrease in fitness due to mating between related individuals is known as inbreeding depression and results from the segregation of partially recessive deleterious mutations maintained by the balance between selection and mutation. Deleterious mutations occur continuously in all populations and most mutations are at least partially recessive. In large populations, selection keeps these detrimental mutations at low equilibrium frequencies. Thus, under random mating, most copies of detrimental alleles are present in a heterozygous state and hence their detrimental effects are partially masked. Mating between relatives, however, increases homozygosity and, therefore, the deleterious effects become fully expressed, decreasing the fitness of inbred individuals.

Although it is generally agreed that increased expression of deleterious partially recessive alleles is the main cause of inbreeding depression, there is an additional mechanism that can contribute to inbreeding depression. If the fitness of a heterozygote is superior to that of both homozygotes (heterozygous advantage or overdominance), the reduced frequency of heterozygotes will reduce the opportunities to express heterozygous advantage. This mechanism may be important for certain traits (e.g., sperm precedence in *Drosophila melanogaster*) and may contribute to the very high inbreeding depression for net fitness observed in *Drosophila* and outcrossing plants (Charlesworth and Charlesworth, 1999).

The degree of inbreeding in a population is measured by the inbreeding coefficient F, which can be defined as the probability that the two alleles of a gene in an individual are identical by descent. The effect of inbreeding in a population with inbreeding coefficient F can be measured in terms of the logarithm of the ratio of the mean fitness values for the outbred, W_O, and the inbred, W_I, populations (Charlesworth and Charlesworth, 1999),

$$\ln\left(\frac{W_I}{W_O}\right) = BF. \tag{14.6}$$

The coefficient B can be interpreted as the reduction in log fitness associated with complete inbreeding ($F = 1$).

In small populations, the opportunities for mating are restricted, even under random mating. Thus, mating among relatives is common and the proportion of individuals that are homozygous at many loci increases, which results in inbreeding depression. The amount of inbreeding depression manifested by a population depends not only on F, but also on the opportunity for selection to purge recessive lethal and semilethal mutations. Gradual inbreeding by incremental reductions in population size over many generations allows selection to eliminate the lethal and sublethal mutations when they become homozygous (Falconer, 1989). However, the component of inbreeding depression due to more nearly additive mutations of small effect is difficult to purge by inbreeding (Lande, 1995). As to empirical results, recent reviews indicate that purging is inefficient in reducing inbreeding depression in small inbred populations [see Allendorf and Ryman (2002) and references therein].

Most of the evidence for inbreeding depression comes from domesticated or captive populations. This, together with the theoretical expectation that a large fraction of inbreeding depression can be purged in small populations and the numerous mechanisms of inbreeding avoidance observed in many species, has led many researchers to question the importance of inbreeding depression for the persistence of natural populations (Keller and Waller, 2002). However, in the last decade there has been a rapid accumulation of evidence showing that many populations do exhibit inbreeding depression. For example, the Soay sheep on the island of Hirta (Saint Kilda archipelago, UK) suffer of significant inbreeding depression in survival (Coltman et al., 1999). More homozygous sheep suffered higher rates of parasitism and, in turn, lower overwinter survival than heterozygous sheep. Another example comes from song sparrows living on Mandarte Island (western Canada). In this case, inbred birds died at a much higher rate during a severe storm than outbred birds (Keller et al., 1994). A more recent study (Keller, 1998) was able to quantify inbreeding depression in this population and estimated that inbreeding depression in progeny from a mating between first-degree relatives was 49%. The negative effect of inbreeding has also been documented in the red-cockaded woodpecker living in the southeastern United States. Inbreeding reduced egg hatching rates, fledgling survival, and recruitment to the breeding population (Daniels and Walters, 2000). Extensive long-term data sets can help uncover inbreeding depression in large populations with a low rate of inbreeding. An 18-yr study of a large population of the collared flycatcher revealed that inbreeding was rare, but when it did occur it caused a significant reduction in egg hatching rate, in fledgling skeletal size, and in postfledging juvenile survival (Kruuk et al., 2002). This study also found that the probability of mating between close relatives ($F = 0.25$) increased throughout the breeding season, possibly reflecting increased costs of inbreeding avoidance. Inbreeding depression is also evident in plants. Byers and Waller (1999) documented many examples of inbreeding depression in natural populations and indicated that purging does not appear to act consistently as a major force in natural plant populations.

Evidence shows that stressful environmental conditions can amplify inbreeding depression. Crnokrak and Roff (1999) gathered and analyzed a data set that included seven bird species, nine mammal species, four species of poikilotherms, and 15 plant species. They were able to show that conditions experienced in the wild increase the cost of inbreeding. A more recent study by Keller et al. (2002) showed that the magnitude of inbreeding depression in juvenile and adult survival of cactus finches living in Isla Daphne Major (Galápagos Archipelago) was strongly modified by two environmental conditions; food availability and number of competitors. In juveniles, inbreeding depression was present only in years with low food availability, whereas in adults, inbreeding depression was five times more severe in years with low food availability and large population size.

Demonstrating the importance of inbreeding depression in the wild does not necessarily imply that it will cause natural populations to decline (Caro and Laurenson, 1994). However, recent papers have demonstrated that this may happen. Saccheri et al. (1998) studied the effect of inbreeding on local extinction in a large metapopulation of the Glanville fritillary butterfly (*Melitaea cinxia*) and found that extinction risk increased significantly with decreasing heterozygosity due to inbreeding, even after accounting for the effects of ecological factors. Larval survival, adult longevity, and egg hatching rate were all affected adversely by inbreeding and seem to be the fitness component responsible for the relationship between inbreeding and extinction. An experiment by Nieminen et al. (2001) provided further support to the results of Saccheri et al.'s (1998) field study. Nieminen et al. (2001) established inbred and outbred local populations of the Glanville fritillary at previously unoccupied sites using the same numbers of individuals. The extinction rate was significantly higher in populations established with inbred individuals. Similar evidence for plants is provided by Newman and Pilson (1997). They established experimental populations of the annual plant *Clarkia pulchella* that differed in the relatedness of the founders. All populations were founded by the same number of individuals but persistence time was much lower in those populations whose founders were related. Additional evidence for inbreeding influencing population dynamics comes from the study of an isolated population of adders in Sweden (Madsen et al., 1999), which declined dramatically in the late 1960s and was on the brink of extinction due to severe inbreeding depression. The introduction of 20 adult male adders from a large and genetically variable population led to a rapid population recovery due to a dramatic increase in recruitment.

The evidence discussed here indicates that inbreeding depression is common in natural populations and can represent a short-term extinction threat to small populations, especially if populations are subject to stressful conditions or to sharp population declines.

Accumulation of Slightly Deleterious Mutations

Under more or less constant environmental conditions, mutations with phenotypic effects are usually deleterious because populations tend to be well adapted to the biotic and abiotic environmental conditions which they experience. A random mutation is likely to disrupt such adaptation. In populations with moderate or large effective sizes, selection is very efficient in eliminating detrimental mutations with large effects on fitness. However, mildly deleterious mutations with selection coefficient $s < 1/2N_e$ are difficult to remove because they behave almost as neutral mutations (Wright, 1931). Thus, small population size hampers selection and increases the role of genetic drift in determining allele frequencies and fates. This increases the chance fixation of some of the deleterious alleles supplied constantly by mutation and results in the reduction of population mean fitness, which eventually leads to population extinction (Muller, 1964). Initially, this process was assumed to represent a threat to asexual populations only because in the absence of recombination, their offspring carry all the mutations present in their parent as well as any newly arisen mutation (Muller, 1964). Mathematical models of this process (Lynch and Gabriel, 1990; Lynch et al., 1993, 1995a) show that the process of mutation accumulation can be divided into three phases. During the first two phases, deleterious mutations accumulate and fitness declines, but population size remains close to carrying capacity. During the third phase, fitness drops below 1 and population size declines. This population decline increases the effect of random genetic drift, which enhances the chance fixation of future deleterious mutations, leading to further fitness decline and reduction in population size. Due to this positive feedback, the final phase of population decline (when growth rate is negative) occurs at an accelerating rate, a process known as "mutational meltdown."

Although recombination can slow down the mutational meltdown to some extent, sexual populations are also at risk of extinction due to mutation accumulation (Lande, 1994; Lynch et al., 1995a). Lande (1994) modeled a randomly mating population with no demographic or environmental stochasticity and considered only unconditionally deleterious mutations of additive effects. He derived analytical approximations for the mean time to extinction for two cases: (a) when all mutations had the same selection coefficient s and (b) when there was variance in s. Lynch et al. (1995a) provided a more detailed analysis of scenario (a) and checked the analytical results using computer simulations. With constant s, the mean time to extinction, $\bar{t}_e$, is an approximately exponential function of the effective population size. Because the mean time to extinction increases very rapidly with increasing N_e, the fixation of new mutations poses little risk of extinction for populations with N_e of about 100 (Lande, 1994). However, with variance in s, the mean time to extinction increases as a power of N_e. For instance, if s is distributed exponentially, $\bar{t}_e$ is asymptotically proportional to N_e^2. As an increase in $\bar{t}_e$ with population size is now more gradual than for constant s, the risk of extinction is much elevated. For reasonable variance in s (coefficient of variation around 1), the mutational meltdown is predicted to pose a considerable risk of extinction for populations with N_e as large as a few thousand individuals (Lande, 1994). If, as is generally agreed, the ratio of N_e to census population size is around 0.1 to 0.5, moderately sized populations of several thousand individuals may face extinction due to genetic stochasticity.

Unfortunately, there is a paucity of empirical evidence for or against the mutational meltdown. What we have is experimental evidence for the accumulation of deleterious mutations due to genetic drift, but these studies do not directly address the risk of extinction (Zeyl et al., 2001). As of today, only Zeyl et al. (2001) explicitly explored the plausibility of the mutational meltdown. They established 12 replicate populations of the yeast *Saccharomyces cerevisiae* from two isogenic strains whose genome-wide mutation rates differed by approximately two orders of magnitude. They used a transfer protocol that resulted in an effective population size of around 250. After more than 100 daily bottlenecks, yeast populations with elevated mutation rates showed a tendency to decline in size, whereas populations with wild-type mutation rates remained constant. Moreover, there were two actual extinctions among the mutant populations. These results provide support for the mutational meltdown models.

Despite this preliminary empirical support, there are a number of issues that remain unresolved. The first one relates to a controversy about the estimates of per-genome mutation rates, U, and the average fitness cost per mutation, s, used in the meltdown models. The values that have been assumed were based on mutation accumulation experiments using *Drosophila melanogaster*, suggesting values of $U = 1$ and a reduction in fitness of about 1–2% (Lande, 1994; Lynch et al., 1995a). Studies reviewed by Garcia-Dorado et al. (1999) on *D. melanogaster*, as well as on *Caenorhabditis elegans* and *S. cerevisiae*, yielded values of U orders of magnitude less than 1. However, some mutation accumulation experiments (Caballero et al., 2002; Keightley and Caballero, 1997) reported average fitness effects one order of magnitude higher than those reported previously. The assumption of additive effects is also questioned by Garcia-Dorado et al. (1999), who reported estimates of 0.1 for the average coefficient of dominance. The new estimates of U and s would lead to much lower rates of fitness decline, making the mutational meltdown less likely. Caballero et al. (2002) used a combination of mutation accumulation experiments and computer simulations and concluded that a model based on few mutations of large effect was generally consistent with their empirical observations.

Finally, an additional criticism of the existing mutational meltdown models relates to the fact that the models ignore the effect of beneficial and back mutations. Models including these types of mutations suggest that only very small populations would face the risk of extinction due to genetic stochasticity (Poon and Otto, 2000; Whitlock, 2000). Estimates of mutational effects using mutation accumulation experiments with *Arabidopsis thaliana* indicate that roughly half of the mutations reduce reproductive fitness (Shaw et al., 2002). The genome-wide mutation rate was around 0.1–0.2. These new results suggest that the risk of extinction for small populations may be lower than initially thought. This issue is reviewed in greater detail in Chapter 7.

At the moment it is not possible to draw definite conclusions about the importance of the mutational meltdown process. This will only be possible once the existing controversy over the rate and nature of spontaneous mutations is resolved (Poon and Otto, 2000). The resolution of this question in turn requires knowledge of the distribution of mutational effects and the extent to which these effects are modified by environmental and genetic background. Additionally, it is necessary to better understand the contribution of basic biological features such as generation length and genome size to interspecific differences in the mutation rate (Lynch et al., 1999).

14.4 POPULATION SIZE, HABITAT PATCH SIZE, AND EXTINCTION RISK

The most robust generalization that we can make about population extinction is that small populations face a particularly high risk of extinction. Holsinger (2000) digged up statements to this effect from the writings of Darwin (1859), E.B. Ford (1945), and (not surprisingly) Andrewatha and Birch (1954). More recent empirical support for the extinction-proneness of small populations has been found practically whenever this issue has been examined; Diamond (1984), Newmark (1991, 1995), Ouborg (1993), Burkey (1995), and Fischer and Stöcklon (1997) represent a small sample of the literature covering different kinds of taxa and spatial scales. The high extinction risk of small populations is not surprising because this is the expectation based on several mechanisms of extinction: demographic and environmental stochasticity, Allee effect, inbreeding depression, mutational meltdown, and so forth. Furthermore, as the different mechanisms tend to make populations ever smaller, they reinforce the effect of each other and lead to what Gilpin and Soulé (1986) termed extinction vortices. Gilpin and Soulé (1986) identified four extinction vortices. Two of them, the R and D vortices, involve only demographic and ecological factors (demographic stochasticity and population fragmentation). The two other ones, F and A vortices, consider the feedback among demographic, ecological, and genetic factors. One way of gauging how much our understanding of the interactions among demographic, ecological, and genetic factors has improved in the last decade or so is to evaluate to what extent the current knowledge calls for a reformulation or refinement of the F and A vortices.

As originally formulated, the F vortex is the consequence of reduced fitness due to inbreeding depression and loss of heterozygosity in initially large populations that have been reduced to a small size. The decrease in fitness further reduces population size, which in turn further increases inbreeding depression and loss of heterozygosity, increasing the probability of extinction via this and all other vortices. Theoretical and empirical advances made in the last few years and reviewed earlier indicate that the enhanced vigor that is often associated with increased heterozygosity is most likely due to a reduced homozygosity of deleterious alleles rather than to heterozygosity per se (see Section 14.3). Furthermore, it is becoming increasingly clear that purging the genetic load leading to inbreeding depression is generally not that efficient in natural populations (Section 14.3). Therefore, the F vortex in the form of inbreeding depression remains a likely mechanism of population extinction.

The A vortex was also attributed to genetic drift and loss of genetic variance, but in this case, Gilpin and Soulé (1986) proposed that a reduction in population size and the increased genetic drift that ensues could reduce the efficiency of stabilizing and directional selections, in turn causing an increasing and accelerating "lack of fit" between the population phenotype and the environment it faces. This was hypothesized to reduce population size and growth rate even further until the population goes extinct. This mechanism was not formulated very precisely, but it is related to the mutational meltdown discussed in Section 14.3. The reduction in the efficiency of stabilizing and directional selections leads to an accumulation of slightly deleterious mutations, which will progressively reduce population growth rate until it becomes negative. Once this happens, the population size will decrease and the rate at which deleterious mutations accumulate will increase further. This feedback mechanism will eventually lead to population extinction. Another mechanism that was proposed for this vortex is loss of genetic variance, which will impair the capacity of populations to track environmental changes.

An additional short-term mechanism could be added to the A vortex. The loss of habitat reduces population sizes and may lead to a loss of variation at MHC loci, making individuals less able to resist infectious diseases. At the same time, habitat destruction might, in some cases, lead to an initial increase in local density, as individuals crowd in the remaining suitable habitat. High density following fragmentation might in turn increase the disease transmission rate (McCallum and Dobson, 2002). Additionally, land degradation increases the opportunity for contact among humans, domesticated animals, and wildlife, also possibly increasing the transmission of diseases (Deem et al., 2001). An increased transmission rate and a lowered disease resistance will further decrease population size and lead to a further decrease in genetic variability at MHC loci. This feedback loop will increase progressively the extinction probability via this and all other vortices.

Delayed Population Responses to Environmental Deterioration

Although it is abundantly clear that small populations exhibit a high rate of extinction, we cannot rest assured that large populations have a low risk of extinction. Consider the familiar deterministic continuous-time logistic model, with growth rate r and carrying capacity K. The equilibrium population size, without any consideration for stochasticity, is given by K. Now, many forms of deterioration in habitat quality affecting the birth and death rates may be reflected in a reduction in the value of r while K remains unchanged (or is only little affected). In this case, the deteriorating environmental conditions are not expected to be reflected in population size until r drops below zero and the population collapses rather abruptly to extinction or, in a metapopulation context, turns from a source population to a sink population. Incidentally, the genetic meltdown models discussed in the previous section envision a similar gradual decline in r, although now because of an accumulation of deleterious mutations. Although the deterministic logistic model can hardly be considered a realistic description of the dynamics of real populations, the phenomenon we have just outlined occurs in all population models. Things can be even worse, from the perspective of a manager who is trying to read the early signs of approaching trouble, in multispecies models, in which interspecific interactions can compensate for environmental deterioration (Abrams, 2002). The bottom line is that a large population size is not necessarily a reliable indicator of a small risk of extinction.

Even if the equilibrium population size would fairly reflect the environmental conditions, such that a large population would indicate favorable conditions and a low risk of extinction, there are still two other concerns that should not be ignored: (1) the possibility of alternative stable states, which was discussed in Section 14.2, and (2) the time it takes for the population to respond to changing environmental conditions. In other words, in a changing environment the current size of the population to some extent reflects the past rather than the present environmental conditions. If the environment has deteriorated rapidly, the population size is therefore larger than the long-term expected (equilibrium) population size, and evaluation of extinction risk based on population size only would lead to an overly optimistic assessment. Ovaskainen and Hanski's (2002; Hanski and Ovaskainen, 2002) analysis of transient dynamics in metapopulation models demonstrates that the time lag is especially long when the environment is close to the extinction threshold of the species following environmental change (see Section 4.4 in Chapter 4). Thus, whenever the changing environmental conditions lead to a relatively quick change in the parameters that set the extinction threshold, we may expect long transient times in exactly those species that we are most concerned about.

Effect of Habitat Patch Size on Extinction

Assuming constant population density, which implies uniform habitat quality, larger habitat patches have larger expected population sizes than smaller patches. Therefore, other things being equal, we could expect large habitat patches to have populations with a lower risk of extinction than populations in small patches. Although other things are usually by no means equal, and population density varies because of variation in habitat quality and for other reasons, a relationship between habitat patch size and extinction risk has typically been documented whenever this relationship has been examined (Hanski, 1994a,b, 1999b). This finding has been employed in the dynamic theory of island biogeography (MacArthur and Wilson, 1967) and, more recently, in the spatially realistic metapopulation theory (Chapter 4). More generally, the relationship between patch size and extinction risk provides a key rule of thumb for conservation: other things being equal, it is better to conserve a large than a small patch of habitat or to preserve as much of a particular patch as possible. One important caveat relates to the position of a habitat patch in a patch network (Section 4.4). Naturally, if empirical information exists on variation in patch quality, such information should be used in assessing the relative values of different patches (most simply by multiplying true patch area by the population density, estimated on the basis of habitat quality; for an example, see Chapter 20).

If habitat patches of very different sizes are compared, there are likely to be many complementary reasons why large patches have populations with a low risk of extinction. Hanski (1999b) discussed three different scenarios. In the small-population scenario, the reason for a low rate of population extinction in large patches is large population size itself, as discussed in Section 14.2 [Eq. (14.4)]. In the changing environment scenario, large patches support populations with a small extinction risk because the greater environmental heterogeneity in large than small patches reduces the risk of population extinction. Examples are discussed by Kindvall (1996) for a species of bush cricket and by several chapters in Ehrlich and Hanski (2004) for checkerspot butterflies. Finally, in the metapopulation scenario, large patches in fact consist of patch networks for the focal species, and metapopulation dynamics increase the lifetime of the population in the patch as a whole (Holt, 1993). Regardless of the actual reason why large patches of habitat support populations with a low risk of extinction, the conservation implications remain the same.

14.5 LOCAL EXTINCTION IN THE METAPOPULATION CONTEXT

The previous sections discussed the ecological and genetic processes that operate in the extinction of isolated populations. Although habitat fragmentation increases the isolation of populations, few populations are completely isolated. In contrast, innumerable local populations interact regularly via migration with other local populations in metapopulations. It is appropriate to ask what new processes influencing the extinction risk of local populations might operate in metapopulations. Not surprisingly, these new processes relate to migration and gene flow. Migration and gene flow can both increase and decrease local extinction risk.

Migration and Gene Flow Decreasing Extinction Risk

The beneficial effect of migration arises because immigrants from surrounding populations may prevent the extinction of small local populations, a process known as the rescue effect. In the literature on metapopulations, the rescue effect is occasionally extended to cover recolonization following extinction, but more properly the rescue effect refers to processes that reduce the extinction risk in the first place. A demographic rescue occurs because immigration increases the population size, thereby making extinction less likely (Brown and Kodric-Brown, 1977). An extreme case is presented by source–sink systems, where a (true) sink population has a negative growth rate (e.g., due to poor habitat quality) and may only survive with sufficient immigration from one or more source populations (Chapter 16). Immigration reducing extinction risk is also common in the case of small populations inhabiting small habitat patches located close to large populations, a common situation in many metapopulations. Table 14.2 gives an example on the Glanville fritillary butterfly (*M. cinxia*) in the Åland Islands, Southwest Finland, where the butterfly has a metapopulation consisting of several hundred local populations (Hanski, 1999b). Larvae live gregariously, and population sizes are often very small in terms of the number of larval groups, even though populations have tens of

TABLE 14.2 The Rescue Effect Reduces the Risk of Extinction in Small Local Populations of the Glanville Fritillary Butterfly (*Melitaea cinxia*)[a]

Number of larval groups	Extinct	*n*	Average *S*	The rescue effect *t*	*P*
1	Yes	150	2.55		
	No	76	2.84	−2.97	0.003
2	Yes	46	2.78		
	No	58	3.12	−2.24	0.025
3–5	Yes	46	2.88		
	No	202	2.75	−0.63	0.527
> 5	Yes	14	3.31		
	No	204	2.83	1.42	0.155

[a] Sizes of local populations are given in terms of the number of larval groups in autumn 1993, the numbers of these populations that went extinct and survived, a measure of connectivity (*S*) to nearby populations, and a *t* test of the rescue effect, which was measured by the effect of *S* on extinction (from a logistic regression, which also included the effects of patch area and regional trend in population sizes on extinction; Hanski, 1999b).

butterflies. Comparing the numbers of populations of given size that did or did not go extinct in 1 yr, it is apparent that populations that were well connected to other populations had a lower risk of extinction than more isolated populations (Table 14.2). It also makes sense that this effect was statistically significant in the case of the smallest populations only because the influence of a given amount of immigration in increasing population size is greatest in the case of the smallest populations. Note that large populations have a much smaller risk of extinction than small populations in Table 14.2.

Local populations may be rescued demographically, as we have just discussed, but they may also be rescued genetically. Gene flow may increase the mean population fitness due to heterosis and the arrival of immigrants with high fitness (outbred vigor). Heterosis refers to increased fitness among offspring from crosses among local populations; different populations tend to fix different random subsets of deleterious alleles, which mask each other when populations are crossed (Crow, 1948; Whitlock, 2000). Therefore, initially rare immigrant genomes are at a fitness advantage compared to resident genomes because their descendants are more likely to be heterozygous for deleterious recessive mutations that cause inbreeding depression in the homozygous state (Ingvarsson and Whitlock, 2000; Whitlock et al., 2000).

Several studies have provided fairly conclusive evidence supporting this expectation. Saccheri and Brakefield (2002) carried out an experimental study with the butterfly *Bicyclus anynana*. They focused on the consequences of a single immigration event between pairs of equally inbred local populations. The experiment involved transferring a single virgin female from an inbred (donor) population to another inbred (recipient) population. The spread of the immigrant's and all the residents' genomes was monitored during four consecutive generations by keeping track of the pedigree of all individuals in the treatment populations. They replicated this experimental design and observed a rapid increase in the share of the initially rare immigrant genomes in local populations. Ball et al. (2000) reported similar evidence for *D. melanogaster*, measuring the relative frequency of immigrant marker alleles in the first and second generations following a transfer to inbred populations. When immigrants were outbred, the mean frequency of the immigrant allele in the first and second generation after migration was significantly higher than its initial frequency. They attributed this result to the initial outbred vigor of immigrant males, but the possibility of heterosis having played a role was not excluded completely.

Ebert et al. (2002) carried out experiments using a natural *Daphnia* water flea metapopulation in which local extinctions and recolonizations, genetic bottlenecks, and local inbreeding are common events. Their results indicate that because of heterosis, gene flow was several times greater than would be predicted from the observed migration rate. Somewhat less conclusive evidence comes from Richards's (2000) experiments with the dioecious plant *Silene alba*, in which isolated populations suffer substantial inbreeding depression. Richards (2000) measured gene flow among experimental populations separated by 20 m and used paternity analysis to assign all seeds to either local males or to immigrants from other nearby experimental populations. When the recipient populations were inbred, unrelated males from the experimental population 20 m away sired more offspring than expected under random mating. This may be due to some form of pollen discrimination that may be influenced by early acting inbreeding depression (Richards, 2000) or to heterosis per se. Incidentally, the rescue effect in Table 14.2 for the Glanville fritillary butterfly could also involve a genetic component, as it is known that inbreeding depression increases the risk of extinction of small populations of this butterfly (Saccheri et al., 1998; Nieminen et al., 2001).

Migration can have a long-term beneficial effect on population persistence. The arrival of migrants from large populations can increase genetic variability in the recipient populations and, thereby, enhance the evolutionary potential of the species as a whole. The extent to which migration can replenish genetic variability depends on population dynamics and the pattern of migration among populations. Populations with positive growth rates can recover lost genetic variability rapidly, but sink populations will only be able to maintain genetic variability when the variance in the migration process is low (Gaggiotti, 1996; Gaggiotti and Smouse, 1996).

Migration and Gene Flow Increasing Extinction Risk

Migration may increase the extinction risk of local populations for several reasons. In the landscape ecological literature, the role of corridors in maintaining viable (meta)populations in fragmented landscapes has been discussed for a long time. Corridors enhance recolonization and the rescue effect (Bennett, 1990; Merriam, 1991; Haas, 1995; Andreassen et al., 1996b; Haddad, 1999a), but it has been pointed out that corridors may also facilitate the spread of disease agents and predators that might actually increase the extinction risk of the focal populations (Simberloff and Cox, 1987; Hess, 1994). More generally, it is well established both theoretically (Hassell et al., 1991; Comins et al., 1992; Nee et al., 1997) and empirically (Huffaker, 1958; Nachman, 1991; Eber and Brandl, 1994; Lei and Hanski, 1998; Schöps et al.,

1998) that specific natural enemies in prey–predator metapopulations may substantially increase the extinction risk of local prey populations.

Just like immigration into small populations may reduce their risk of extinction, emigration from small populations may increase extinction risk (Thomas and Hanski, 1997; Hanski, 1998b). Theoretical studies have elucidated the critical minimum size of habitat patches that would allow the persistence of viable populations (Okubo, 1980); populations in patches smaller than this critical size go extinct because they lose individuals too fast in comparison with the rate of reproduction. However, just like with the rescue effect in saving small populations, it is hard to prove conclusively that small populations go extinct because of emigration, as small populations are likely to go extinct for many other reasons as well. Nonetheless, emigration compromising the viability of local populations is a potentially important consideration in the conservation of some species. For instance, it has been suggested that small reserves for butterflies should not be surrounded by completely open landscape because this will increase the rate of emigration greatly (Kuussaari et al., 1996).

Migration can also have negative genetic effects on population persistence. In principle, gene flow may reintroduce genetic load fast enough to prevent the purging of inbreeding depression, although we are not aware of any clear evidence for this. More importantly, the long-term beneficial effects of migration may be offset by the introduction of maladapted genes, which may lead to a loss of local adaptation in some populations, the appearance of source–sink dynamics, and the evolution of narrow niches (Kirkpatrick and Barton, 1997; Ronce and Kirkpatrick, 2001). This process, called migrational meltdown (Ronce and Kirkpatrick, 2001) because small populations experience a downward spiral of maladaptation and shrinking size, is discussed in the next section.

The introduction of immigrant genomes from a highly divergent population can reduce mean population fitness, a phenomenon known as outbreeding depression (Fig. 14.4). Outbreeding depression will be expressed in the F_1

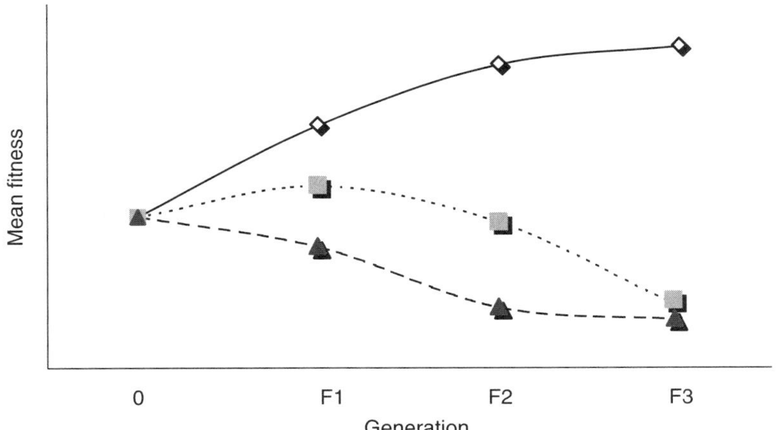

Fig. 14.4 Potential effects of migration on population fitness: (a) heterosis increases fitness (solid line and diamonds), (b) heterosis followed by outbreeding depression leads to a short-lived fitness increase followed by a decline, and (c) outbreeding depression leads to a steady decline in fitness.

generation if the favorable between-population dominance effects (masking the effect of deleterious recessive genes present in the homozygote state in parental lines but in the heterozygote state in the F_1) are outweighed by the loss in favorable additive x additive interactions within populations (Lynch and Walsh, 1998). However, even if this does not occur, outbreeding depression may still be expressed in the F_2 generation or later. The reason for this is that F_1s carry a haploid set of chromosomes from each parental line, and segregation and recombination begin to break apart coadapted genes from a single line in the F_2 generation (Dobzhansky, 1950, 1970). Thus, outbreeding depression is demonstrated when the performance of F_2s is less than the average of immigrants and residents (Lynch and Walsh, 1998). Unfortunately, only few studies of natural populations have tracked the contribution of immigrants beyond the F_1 generation (Marr et al., 2002). The few studies that go beyond F_1 indicate that outbreeding depression may be common in the wild. Marr et al. (2002) showed that the same population of song sparrows in the Mandarte Islands that manifested heterosis among immigrant offspring also displayed signs of outbreeding depression in the F_2 generation. Studies of the tidepool copepod *Tigriopus californicus* show that crosses between populations typically result in F_1 hybrid vigor and F_2 hybrid breakdown for a number of measures related to fitness (Burton 1987, 1990a,b; Edmands and Burton, 1998; Burton et al., 1999). Edmands (1999) showed that the detrimental effects of breaking up coadaptation are magnified by increasing genetic distance between populations. This same effect was shown for the shrub *Lotus scoparius*, but in this case outbreeding depression was already present in the F_1 generation (Montalvo and Ellstrand, 2001). Other plant species for which outbreeding depression has been demonstrated include *Ipomopsis aggregata* (Waser et al., 2000) and *Silene diclinis* (Waldmann, 1999).

14.6 METAPOPULATION EXTINCTION

Not only populations but also metapopulations consisting of many local populations possess a smaller or greater risk of extinction — the metapopulation is extinct when the last remaining local population is extinct. Chapter 4 presents a thorough account of the metapopulation theory, albeit largely from the perspective of one particular class of models, stochastic patch occupancy models. A primary focus of this theory is to dissect the conditions of long-term metapopulation persistence (in deterministic models) and the factors determining the expected lifetime of metapopulations (in stochastic models). Chapter 2 complements this analysis from the perspective of landscape ecology. The spatially realistic metapopulation theory in Chapter 4 is concerned primarily with just one factor in increasing the risk of metapopulation extinction, namely habitat loss and fragmentation, but as we all know, this is currently the main cause of population, metapopulation, and species extinctions. Rather than repeating what has already been written in Chapters 2 and 4 and discussed in the context of particular metapopulations in Chapters 20 and 21, we highlight here one ecological factor that is often critical in metapopulation extinction. We also discuss two genetic processes that have been proposed to increase the risk of metapopulation extinction, mutational and migrational meltdowns, both of which stem from an interaction between demographic and genetic processes in metapopulation dynamics.

Regional Stochasticity

The counterpart of environmental stochasticity in local populations is regional stochasticity in metapopulations — spatially correlated environmental stochasticity affecting local populations in metapopulations (Hanski, 1991). Just as environmental stochasticity amplifies population fluctuations in local populations and is the major cause of population extinction, regional stochasticity amplifies fluctuations in the size of metapopulations (Fig. 4.11 in Chapter 4 gives a theoretical example and Chapter 21 reviews regional stochasticity in small mammal metapopulations). There is a large literature on spatial synchrony in population dynamics (Ranta et al., 1998; Bjørnstad et al., 1999; Paradis et al., 1999; Engen et al., 2002a) with the same general message. The two mechanisms of spatial synchrony that have been most discussed are migration and regional stochasticity (typically spatially correlated weather conditions influencing birth and death rates). As shown by Lande et al. (1999), even low rates of short-distance migration may affect population synchrony greatly if population regulation is weak. Engen et al. (2002b) examined the probability of quasiextinction for a population distributed continuously in space and affected by regional stochasticity (quasiextinction was defined as the population size dropping below 10% of the carrying capacity). The expected time to quasiextinction decreases with increasing strength of environmental stochasticity, with decreasing rate of migration, and with increasing area within which changes in population size are recorded. The expected population density decreases, and hence the probability of quasiextinction increases, with increasing spatial scale of regional stochasticity.

Metapopulation Meltdown

The accumulation of slightly deleterious mutations can have detrimental effects at the metapopulation level. Higgins and Lynch (2001) extended the mutational meltdown theory described in Section 14.3 to metapopulations using an individual-based model that includes demographic and genetic mechanisms and environmental stochasticity. The metapopulation structure was modeled as a linear array of patches connected by nearest-neighbor (stepping-stone), global (island), or intermediate dispersal. The mutational effect was modeled in such a way that mutations of large effect are almost recessive, whereas those of small effect are almost additive. Results show that for metapopulations with more than a few patches, an accumulation of deleterious mutations accelerates extinction time by many orders of magnitude compared to a globally dispersing metapopulation without mutation accumulation. Moreover, extinction due to mutation accumulation can be quite rapid, on the order of tens of generations. In general, results indicate that the mutational meltdown may be a significant threat to large metapopulations and would exacerbate the effects of habitat loss or fragmentation on metapopulation viability. These conclusions were reached under the assumptions of an expected genome-wide mutation rate of 1 per generation and unconditionally deleterious mutational effects. As mentioned before, these two assumptions have been placed under close scrutiny, and preliminary evidence indicates that they may not be generally valid.

Migrational Meltdown

Another genetic mechanism for metapopulation extinction stems from the idea that peripheral populations receive gene flow from the center of the species' range. These immigrant genes will typically be adapted to the conditions at the range center and could inhibit adaptation in the periphery (Mayr, 1963). Kirpatrick and Barton (1997) used a quantitative genetic model to study the evolution of a species range in a linear habitat with local migration. The model tracks evolutionary and demographic changes across space and time and assumes that variation in the environment generates patterns of selection that change in space but are constant in time. Among other things, results show that a species' range may contract as the dispersal rate increases and extinction may follow if conditions change too rapidly as one moves across space, even if the species remains perfectly adapted to the habitat at the range center. Ronce and Kirpatrick (2001) also studied the maladaptive effect of migration but they considered a model with two discrete habitat types connected by migration. In this case, an increasing migration rate above a threshold value results in the collapse of the total population size and the complete loss of one of the populations. However, in contrast to Kirpatrick and Barton's (1997) analysis, there is no metapopulation extinction. Ronce and Kirpatrick (2001) attributed this disagreement between the two models to the assumption of infinite space made by Kirpatrick and Barton: the distance traveled by migrants and thus the maladaptation of such migrants to local conditions increase indefinitely with the migration rate. This assumption is unlikely to be valid for real situations and, therefore, complete metapopulation extinction due to migrational meltdown is unlikely to occur.

14.7 CONCLUDING REMARKS

The major causes of population and species extinctions worldwide are habitat loss and interactions among species. The models discussed in this chapter address the adverse effects of habitat loss in terms of the reduced sizes of populations and metapopulations that are the inevitable and direct result of habitat loss. With metapopulation models, we may additionally examine the consequences of habitat loss that occur in the surroundings of the focal population, and which consequences influence the focal population via metapopulation dynamics (Chapter 4).

Considering interactions with other species, it may at first appear surprising that this would be an important cause of population extinction — if this were the case, would such extinctions not have already happened a long time ago? This argument does not hold in two situations: in metapopulations with recurrent extinctions and colonizations (Section 14.5) and when species are spreading into areas where they did not use to occur and become hence engaged in novel interactions. We all know that such invasions, with often adverse consequences for native species, have become rampant in the modern world, where humans have helped, in one way or another, innumerable species to spread beyond their past geographical ranges. The actual mechanisms of extinction of native species include hybridization with the invasive species (Simberloff, 1994; Wolf et al., 2001; Levin, 2002; Perry et al., 2002). The spreading of *Homo sapiens* itself, in the far past, was the likely cause of extinction of a large fraction of the megafauna in North America, Australia, and many large islands (Martin and Klein, 1984; Caughley and Gunn, 1996) at a time when humans could be placed among other animals in their lack of concern for the survival of other species. No wonder, then, that modern humans are able to hunt and drive many species to extinction or near extinction. Harvesting of populations has been and continues to be a major threat to both terrestrial and marine populations. Models and ecological knowledge could and should be used to guide harvesting of economically valuable populations (Getz and Haight, 1989; Lande et al., 1995), but generally this is not what happens in reality.

Interactions with invasive species, persecution, and harvesting, along with habitat loss, are *the* major ultimate threats to populations and species, and the threats with which most practical conservation efforts have to be concerned. From this perspective, many of the population ecological and genetic mechanisms discussed in this chapter may appear insignificant. Nonetheless, the matter of fact is that increasing numbers of species are being reduced to a state in which the small-population issues (Caughley, 1994) covered here are relevant and interact with the primary causes of threat (Hedrick et al., 1996). Clearly, population biologists alone cannot solve the current extinction crisis, but we can provide improved knowledge of many specific biological issues. Finally, of course, just like the study of population regulation has been of great intrinsic interest to population ecologists for more than a century, so are the inevitable "failures" of regulation in finite populations.

One of the largely open scientific issues in the study of population extinction relates to the current controversy surrounding genome-wide mutation rates and the average effect of deleterious mutations (Section 14.3). Before these questions have been resolved, it is premature to draw definite conclusions about the importance of mutational meltdown in population and metapopulation extinctions. More research on the mutation process underlying the mutational meltdown and more extensive empirical research on the feasibility of this phenomenon are needed. Additionally, models such as that of Higgins and Lynch (2001) should be extended to include beneficial as well as deleterious mutations. Likewise, additional work has to be carried out to evaluate the importance of the genetic rescue effect due to heterosis and, in particular, to understand how outbreeding influences the mean fitness of natural populations. It is likely that the extent of outbreeding depression depends on how inbred the local populations that receive the migrants are. Highly inbred populations whose fitness is very low may react positively to the influx of migrants and show no signs of outbreeding depression at all. However, less inbred populations whose fitness has not been impaired dramatically may show heterosis in the F_1 generation but outbreeding depression in the F_2 and subsequent generations or outright outbreeding depression. Unraveling the effects of immigration on fitness will require carrying out experiments that follow the fate of the descendants of immigrants beyond the F_2 generation and control for the inbreeding level of the target populations.

We have commented in the introduction and in later sections of this chapter on the changing views about the relative roles of ecological and genetic factors in population and metapopulation extinction. The theoretical and empirical work done in the past decade makes it clear that genetic factors can contribute significantly to population extinction. In particular, there is a rapidly expanding body of literature demonstrating that inbreeding depression in natural populations is often sufficiently severe to have significant consequences for population dynamics and thereby for extinction. The most clear-cut demonstrations of inbreeding increasing the risk of population extinction, such as in the Glanville fritillary butterfly (Saccheri et al., 1998; Nieminen et al., 2001), relate to very small populations. For this reason, some might dismiss the new evidence as of little general importance. However, this is not so in the metapopulation context, where small populations are often frequent and matter for the dynamics of the metapopulation as a whole. This is also the context that shows very clearly how Caughley's (1994) declining-population paradigm and small-population paradigm interact. Very often, habitat loss and fragmentation are the root causes of metapopulation decline (declining-population paradigm), but the actual metapopulation response to environmental changes is largely determined by what happens in the often small local populations (small-population paradigm). The relative roles of genetic and ecological factors in extinction are also likely to vary among taxa with different biologies. For instance, environmental stochasticity is generally the overriding cause of extinction in insects and other invertebrates, whereas inbreeding might be expected to play a relatively greater role in vertebrate populations that are less influenced by random variation in environmental conditions.

15. MULTILOCUS GENOTYPE METHODS FOR THE STUDY OF METAPOPULATION PROCESSES

Oscar E. Gaggiotti

15.1 INTRODUCTION

Three fundamental processes are at the heart of metapopulation biology: local population extinction, (re)colonization, and migration (Hanski, 1999a). The problems being addressed in the metapopulation context are diverse and range from the effect of migration on local and global population dynamics (Chapters 4, 13, 16, and 20) to the effect of extinction, colonization, and migration on the evolutionary potential of metapopulations (Chapters 7, 10–12, and 16). In all these studies, a common interest is the estimation of the rates at which these three events take place. More detailed information is also required when studying colonization and migration processes. In these cases, it is also necessary to estimate additional parameters, such as the size of founding/migrant groups and their composition, and to identify the factors that force individuals to move away from their place of birth (Chapter 13). All these different problems can be studied using purely ecological approaches, such as mark–release–recapture methods (MRR), but only on a

0-12-323448-4

limited number of species. Moreover, these methods are time-consuming and cannot be applied to study large and/or spatially extended metapopulations. Population genetic approaches, however, are easier to implement in these situations, as, in general, they only require a carefully planned sampling program aimed at collecting tissue samples for DNA extraction and analysis.

The application of population genetic methods in the metapopulation context is not problem free. The population turnover that characterizes many metapopulations can decrease genetic variability greatly and, therefore, molecular markers such as allozymes and mtDNA may not be polymorphic enough. The more recently developed microsatellites markers (Goldstein and Schlötterer, 1999), however, are much more variable and useful in this context. Another type of problem that can be found concerns the power of the statistical methods that are available. Classical population genetic methods for the inference of demographic or ecological parameters have relied on measures that summarize the information contained in genetic data. Among these we have F_{ST}, sample heterozygosity, the distribution of pairwise differences between DNA sequences, and the number of segregating sites. A serious drawback of these approaches is that they assume constancy in demographic parameters and genetic equilibrium conditions. These assumptions are violated by all natural metapopulations. Additionally, most of these methods lack statistical power because they use only information provided by frequency distributions of alleles or haplotypes. More sophisticated methods that make better use of the various types of information contained in genetic data have been developed. These methods can be grouped into two types of approaches: (1) coalescent or genealogical approaches that use the genealogical information contained in DNA sequences and (2) multilocus genotype approaches that use gametic disequilibrium information (see later). It is important to realize that these two types of methods differ not only in the type of information they use, but also on the nature of the parameters they estimate. Coalescent methods (and those based on summary statistics) estimate long-term evolutionary parameters, whereas multilocus genotype methods estimate short-term ecological parameters. Chapter 8 discusses the application of the coalescent in metapopulations and explains how this approach can be used to make inferences about demographic processes. This chapter discusses the second type of approaches, those based on multilocus genotype data. It first provides some examples of applications of classical population genetic approaches and explains their limitations. Then the chapter introduces multilocus genotype approaches with a brief account of their short history and some details about their implementation followed by some examples. Finally, it discusses the need to integrate the information provided by genetic data with that coming from demographic and environmental data and provides examples of how to achieve this goal.

15.2 CLASSICAL POPULATION GENETIC APPROACHES

Until recently, the most widely used genetic approach in population biology was the estimation of migration rates from Wrights' F_{ST} statistic (Wright, 1931). This method is based on the island model of population structure (see Chapter 7), which leads to a simple relationship between F_{ST} and Nm, the effective number of migrants:

$$F_{ST} \approx 1/(4Nm + 1), \tag{15.1}$$

where N is the local population size and m is the migration rate. The power of this method is very limited, as a small amount of migration is enough to wipe out the genetic signal (Fig. 15.1). Additionally, the use of this method has been criticized repeatedly (for a review, see Whitlock and McCauley, 1999) because it is based on a large number of unrealistic assumptions, such as constant and equal local population sizes, symmetric migration, and probability of migration between populations independent of geographic distance. Most of these assumptions are violated in the case of metapopulations and therefore its use is unwarranted.

Some few studies have tried to make inferences about the composition of colonizing propagules (e.g., Giles and Goudet, 1997; Ingvarsson, 1998). This is an important problem because the composition of colonizing propagules has a substantial effect on the genetic structure of metapopulations (Whitlock and McCauley, 1990). Slatkin (1977) introduced two extreme models of colonizing group formation: the *propagule pool* model, in which colonizers are drawn from a single source population, and the *migrant pool* model, in which colonizers are drawn at random from the entire metapopulation. Intermediate cases can be considered if the formulation includes the probability, ϕ, that two alleles in a newly formed population come from the same parental population (Whitlock and McCauley, 1990). In the propagule pool model, $\phi = 1$, whereas in the migrant pool model, $\phi = 0$. Further extensions of Slatkin's

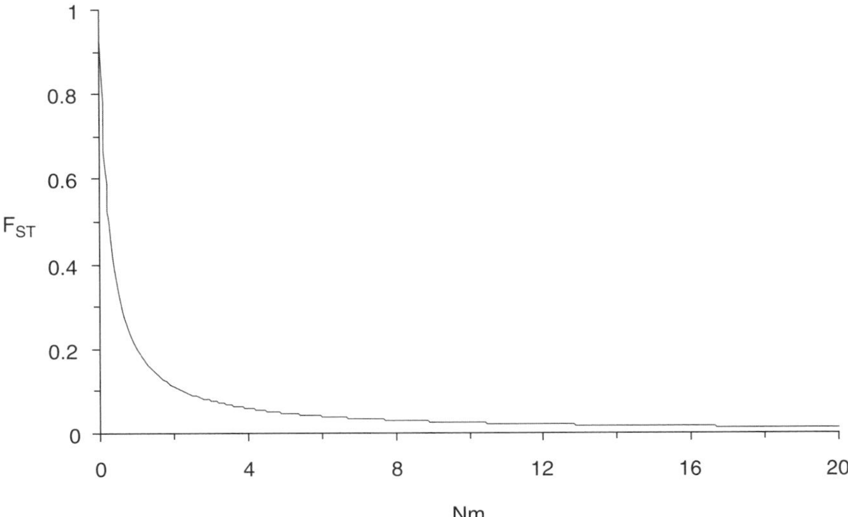

Fig. 15.1 Degree of population subdivision, F_{ST}, as a function of the effective number of migrants under the island model.

(1977) models show that the decrease in genetic diversity due to population turnover is more pronounced under the propagule pool model (Pannell and Charlesworth, 1999).

Giles and Goudet (1997) investigated the genetic structure of a metapopulation of *Silene dioica* inhabiting the Skeppsvik Archipelago in Sweden, where new islands are created due to rapid land uplift. They used information on the age of the local populations inhabiting the different islands to make inferences about the probability of common origin among migrants. They argued that isolation by distance among young populations or among both young and intermediate populations would indicate that colonizers of new islands were drawn from a limited number of source populations, in which case the propagule pool model would best describe the colonization process in the archipelago. However, isolation by distance among intermediate but not among young populations would indicate that colonizers represent a sample of the whole metapopulation, in which case the migrant pool model would be more appropriate. Because they only detected isolation by distance among populations of intermediate age, they concluded that the most appropriate colonization model for the *S. dioica* metapopulation was the migrant pool model.

Ingvarsson (1998) investigated how mating patterns of female *Phalacrus substriatus* beetles influenced both the effective size of newly colonized populations and the probability of common origin of individuals in the founding groups. He used Eq. (7) in Whitlock and McCauley (1990) to estimate the probability of common origin of two alleles in newly founded populations (ϕ). This equation requires estimates of the probability of common origin of diploid founders, which he estimated from mark–recapture experiments, and of the effective size of founding groups, which he estimated using F statistics. The estimate of $\phi = 0.8$ thus obtained indicated that the most appropriate colonization model for *P. substriatus* was the propagule pool model. The different approaches used in the studies of Giles and Goudet (1997) and Ingvarsson (1998) lead to rough estimates of colonization parameters and rely on equilibrium models that make unrealistic assumptions similar to those used by F_{ST} approaches.

Additionally, genetic methods based on summary measures have been developed for the estimation of the effective size of ancestral and descendant populations and divergence times between descendant populations (e.g., Gaggiotti and Excoffier, 2000; Wakeley and Hey, 1997) These methods have also been used to detect population declines or bottlenecks (for references, see Luikart et al., 1998) and for the estimation of the effective size of panmictic population (for a review, see Schwartz et al., 1998). These methods have low statistical power and, in general, they only provide point estimates of the parameters of interest, although approximate confidence intervals can be obtained using randomization techniques.

15.3 METHODS BASED ON MULTILOCUS GENOTYPES

As opposed to classic population genetics and coalescent approaches that are aimed at studying processes that take place on an evolutionary time scale, multilocus genotype methods can be used to study processes that occur on an ecological time scale. The realization that the use of multilocus genotype data was a powerful tool for the genetic study of populations dates back to the late 1970s and early 1980s (Smouse, 1978). These methods were applied in many different contexts, among which the most relevant to us is that of subdivided populations where they were used to allocate individuals to groups and to estimate the genetic contribution of different source populations to admixed populations [e.g., Smouse et al. (1982) and references therein]. The rationale underlying these latter applications is that a multilocus or gametic disequilibrium approach uses the information provided by the correlations between alleles at different loci (Waples and Smouse, 1990). In a population in gametic equilibrium, the expected frequency of gametes containing allele i at locus 1 and allele j at locus 2 is uncorrelated in the sense that an individual's genotype at locus 1 provides no information about its genotype at locus 2. In a mixture of gene pools with different frequencies of i and j, observed gametic frequencies will depart from the independence expectations, resulting in gametic disequilibrium. Thus, the presence of allele i at locus 1 may indicate that allele j is more or less likely to be present at locus 2. Note that the term linkage disequilibrium is also used to describe this phenomenon but is rather misleading because the nonrandom association of alleles across loci is not caused by physical linkage.

One of the most successful multilocus genotype methods for making inferences about demographic parameters was introduced two decades ago by fishery geneticists interested in estimating the contribution of different populations to the salmon mixed fishery operating off the northwest coast of North America. This genetic stock identification (GSI) method was described in a series of unpublished manuscripts dating back to 1979 and cited by Milner et al. (1985). Smouse et al. (1990) developed a statistically rigorous formulation of this method that allowed for the estimation of allele frequencies in the source populations. The underlying model assumes that a sample is taken from a mixed population, composed of unknown proportions $x_1, x_2, \ldots, x_s$ from a known or partially known set of panmictic source populations, $1, 2, \ldots, s$, respectively. The proportions $x_1, x_2, \ldots, x_s$ are treated as parameters that need to be estimated (see Fig. 15.2). The source populations are assumed to be at linkage and Hardy–Weinberg equilibrium. The gametic disequilibrium generated by the mixing of individuals coming from different source populations is used to estimate the proportionate contribution of each source population to the genetic mixture. As it was shown by Gaggiotti et al. (2002), this method is ideally suited to determine the composition of colonizing groups and, therefore, for the study of the colonization process in a metapopulation. This application is described in greater detail later.

Multilocus genotype methods have been applied recently to identify immigrants and assign them to a particular source population. The first assignment test was developed by Paetkau et al. (1995) and simply calculated the likelihood of drawing a single multilocus genotype from several potential sources based on the observed allele frequencies at each locus in each source. A more rigorous assignment test developed by Rannala and Mountain (1997) can identify individuals that are immigrants or have recent immigrant

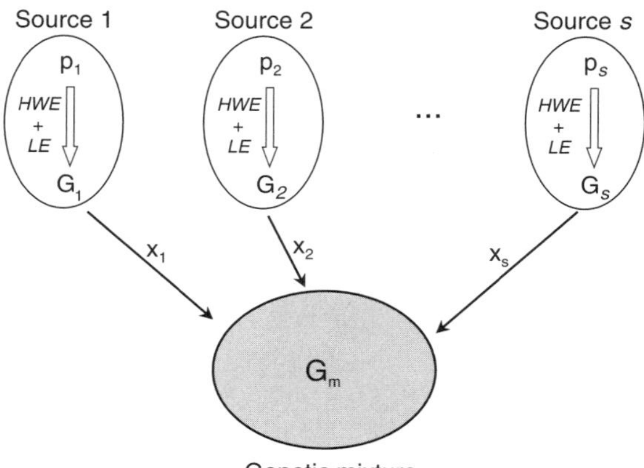

Fig. 15.2 Schematic description of the genetic stock identification method. Basic inputs needed by the GSI method are allele frequency distributions, **p**, in the source population and genotypes in the genetic mixture, G_m. The genotype frequencies in source populations, G_i, are obtained under the assumption of Hardy–Weinberg equilibrium (HWE) and linkage equilibrium (LE) across loci and are used as the probability of observing each mixture genotype in that population.

ancestry using likelihood ratio tests. A more general approach by Pritchard et al. (2000) can be used to identify migrants but assumes that the migration rate is known and small. The main objective of this latter method was to infer population structure and assign individuals to populations. This is an important problem, as in many cases the demarcation of local population based on the geographic location of sampled individuals is not possible. Other methods for the identification of panmictic populations and assignment of individuals are provided by Corander et al. (2003) and Dawson and Belkhir (2001).

Although these approaches may be able to identify immigrants, they are not appropriate to estimate migration rates. In principle, one could repeat the test for all the individuals in a sample and then simply count those individuals identified as migrants. However, this would be erroneous because making many pairwise comparisons between populations for each of a large number of individuals means that some individuals will appear to be immigrants purely by chance, which would lead to overestimation of the number of immigrants. Wilson and Rannala (2003) developed a multilocus genotype method that estimates rates of recent immigration among local populations. This method detects asymmetric migration between pairs of populations and estimates the total number of nonimmigrants, first-generation immigrants, and second-generation immigrants. This detailed information can be very useful for studies of metapopulation dynamics and for the design of management strategies for conservation.

All the methods just described use likelihood-based approaches that rely on either maximum likelihood or Bayesian estimation procedures. These approaches are described in Box 15.1.

BOX 15.1 Likelihood-Based Approaches

Likelihood-based methods proceed by assuming that observed data arose from some probabilistic model with unknown parameters. Their objective is to use data to estimate the parameters of the model and to assess the degree of uncertainty associated with these estimates. The core of the method consists in the calculation of the probability $P(G|\theta)$ of observing genetic data G if the parameters of the model take the value θ. This probability is the so-called likelihood function, $L(\theta|G)$, which, by definition, is a function of θ.

In the context of a GSI method, data G are the individual genotypes observed in the sample from the genetic mixture and the allele frequency distributions observed in the source populations. The aim of GSI is to estimate the vector $\mathbf{x} = \{x_i\}$, where x_i is the contribution of source population i. One possible formulation for the GSI likelihood function is

$$L(\mathbf{x}|G) = \prod_{k=1}^{m} \sum_{i} x_i \left(\prod_{l=1}^{L} \delta p_{il,a_{1lk}} p_{il,a_{2lk}} \right), \qquad \text{(B1)}$$

where, $p_{ila_{nlk}}$ is the frequency of the nth ($n = 1, 2$) allele at the lth locus of individual k in population i, m is the number of individuals in the sample from the genetic mixture, L is the number of loci scored, and

$$\delta = \begin{cases} 1 & \textit{if} \quad a_{1lk} = a_{2lk} \\ 2 & \textit{if} \quad \textit{otherwise} \end{cases}.$$

Maximum Likelihood Inference

Maximum likelihood (ML) inference consists of finding the value of θ that maximizes the likelihood function $L(\theta|G)$. One problem with this approach is that the uncertainty associated with the estimate $\hat{\theta}$ is expressed by a 95% confidence region that has a rather obscure interpretation. The precise interpretation is that the probability that the confidence region contains the true value of θ is 0.95. Note that this is not equivalent to saying that the probability that θ lies in the confidence region is 0.95. An important advantage of ML inference is that for a large sample size, the maximum likelihood estimate, $\hat{\theta}$, will have an approximate normal distribution centered on the true parameter value θ. Thus, an approximate 95% confidence interval can be calculated as the range of θ values that are within two log likelihood units of the maximum log likelihood. Additionally, we can test whether the maximum likelihood estimate is significantly different from another fixed value, θ_0, using the likelihood ratio test. This test uses the fact that the log-likelihood ratio statistic,

$$\Lambda = -2\log\frac{L(\theta_0)}{L(\hat{\theta})}, \qquad \text{(B2)}$$

has asymptotically a χ^2 distribution, if θ_0 is the "true" value of θ. Λ can then be assessed for statistical significance using standard χ^2 significance levels.

ML estimates can be obtained analytically for simple models, but the application of this method in population genetics leads to complex likelihood functions that need to be explored using computer approaches such as the expectation maximization algorithm. Equation (B1) is a good example of this situation.

Bayesian Inference

In order to make probability statements about the parameter θ given data G, we must begin with a model providing a joint probability distribution for θ and G. The joint probability mass, $P(\theta,G)$, can be written as a product of two probability distributions: the *prior* distribution $P(\theta)$ and the *sampling* distribution, given by the likelihood function $L(G|\theta)$: $P(\theta,G) = P(\theta)L(G|\theta)$. Using Bayes' theorem, we obtain postdata or *posterior* distribution:

$$P(\theta|G) = \frac{L(\theta|G)P(\theta)}{P(G)}. \tag{B3}$$

The posterior distribution represents our knowledge about the parameters, taking into account both our prior information (represented by the prior distribution) and observed data. The primary task of any specific application is to develop the model $P(\theta,G)$ and perform the necessary computations to summarize $P(\theta|G)$ in appropriate ways.

Visual inspection of the posterior distribution provides information that is unavailable when using ML estimation. Additionally, this distribution can be described by point estimates such as the mode or the mean. The uncertainty around the estimate is expressed by the 95% credible region for θ. The intuitive interpretation of this region is that the probability that θ lies in it is 0.95. Another advantage of Bayesian over ML estimation approaches is that the former does not rely on asymptotic arguments and, therefore, is valid in situations where the standard likelihood theory fails.

Simple problems in estimation lead to closed form solutions for the posterior distribution, but typical applications in population genetics require the use of numerical integration methods, such as Markov chain Monte Carlo (e.g., Brooks, 1998).

15.4 INTEGRATION OF GENETIC, DEMOGRAPHIC, AND ENVIRONMENTAL DATA

There is an increasing interest in finding ways of efficiently combining genetic, demographic, and other sources of information in order to make inferences about demographic, evolutionary, and ecological processes. Some examples are the works of Estoup et al. (2001), Burland et al. (2001), and Charbonnel et al. (2002). Estoup et al. (2001) use a Bayesian approach that combines microsatellite and enzyme data with information about demographic parameters describing the major phases of the introduction history of the cane toad *Bufo marinus* in various Caribbean and Pacific islands. The parameters for which limited prior information was available are the size of founding groups, effective size during population expansion, and population size at equilibrium. Burland et al. (2001) combined genetic data with information obtained from mark–release–recapture studies in order to identify the evolutionary determinants of social organization in brown long-eared bat *Pletocus auritus*. Charbonnel et al. (2002) combined demographic and genetic data in order to study evolutionary aspects of the metapopulation dynamics of *Biomphalaria pfeifferi*.

It is clear that understanding complex population histories requires the combination of all available information. Additionally, such a strategy can help overcome the problems generated by the lack of sufficient information in genetic data, a frequent problem when applying population genetic methods for estimating population parameters. This is typically the case, for example, when studying migration patterns in species with high dispersal capabilities. In this case, genetic differentiation is low and, therefore, the estimates of dispersal rates can have a very large variance. In order to solve this problem we need to complement genetic data with other types of data, such as demographic, geographic distance, or environmental data. The incorporation of these data should decrease the variance of the estimates without biasing the results of the analysis. Bayesian methods provide the framework needed for achieving these goals.

An important limitation of the statistical genetic methods developed is that they are simply aimed at estimating the parameters of population genetics models (Nm, N_e, etc.) Once the estimates of the parameters are obtained, the researcher proposes alternative hypotheses that focus on specific processes and that are consistent with these estimates. Under these circumstances it is difficult to obtain a clear idea of the relative importance of the different ecological or genetic processes responsible for the results. Hierarchical Bayesian methods (e.g., Gelman et al., 1995) are ideal for addressing this problem because they can be used to explicitly model the relationship between the likelihood function parameters and relevant ecological or genetic processes. This approach leads to alternative models that consider different processes/factors and whose significance can be evaluated using computationally intensive methods such as deviance information criterion (DIC) (Spiegelhalter et al., 2002) or reversible jump MCMC (Green, 1995).

This section exemplifies the implementation of these ideas using a study of colonization patterns in a metapopulation of gray seals (Gaggiotti et al., 2002).

The Gray Seal Metapopulation in the Orkney Isles

The gray seal (*Halicoerus grypus*) is a colonially breeding marine mammal that produces only a single offspring each year. During autumn, gray seals gather at breeding colonies for the females to give birth and suckle the pups. Toward the end of lactation, the females are mated by one or more males and then return to sea, leaving their weaned pup on land (Anderson et al., 1975). Pups remain on land for 1 or 2 more weeks and then go to sea where they spend the next 4–6 yr without returning to the breeding grounds. The average age of recruitment to the breeding population is 5 yr and new recruits usually arrive at the breeding grounds after the first pups have been born. One-third of the world population of gray seals breeds at 48 colonies around the British Isles, with the majority on offshore islands to the north and west of Scotland. Although the whole population is growing exponentially, individual colonies exhibit diverse dynamics. Some of them fluctuate around a long-term constant value, whereas others are increasing exponentially or logistically in size. Furthermore, some colonies have become extinct and others have been colonized recently.

The focus of the case study is the Orkney Isles, a group of 50 islands lying off northeast Scotland (see map in Fig. 15.3). In about 1992, three vacant islands (Stronsay, Copinsay, and Calf of Eday) were colonized. At more or less the same time, two large colonies reached their carrying capacity (Holm of Huip and Faray). Therefore, the question arises as to whether density-dependent effects in these populations may have played a role in the colonization of the three previously unoccupied islands, in which case most of the colonizers would come from the colonies that are at or close to their carrying capacity. Additionally, it is important to determine the composition of founding groups because of its substantial effect on the genetic structure of metapopulations (see earlier discussion). Finally, there is an interesting behavioral question that could be investigated: the possibility that individuals in the newly founded colonies mate assortatively depending on their origin. In other words, it is possible that individuals that come from the same source colony are more likely to mate among themselves than with individuals from other source populations.

The genetic stock identification method (see earlier discussion) is the logical approach to answer these questions because the vector of genetic mixture coefficients, $\mathbf{x} = \{x_i\}$, can be used to describe the composition of colonizing groups. Thus, genetic samples were obtained from the three newly founded colonies and from seven potential source populations that were deemed to be the most likely sources of founders. Some few smaller colonies

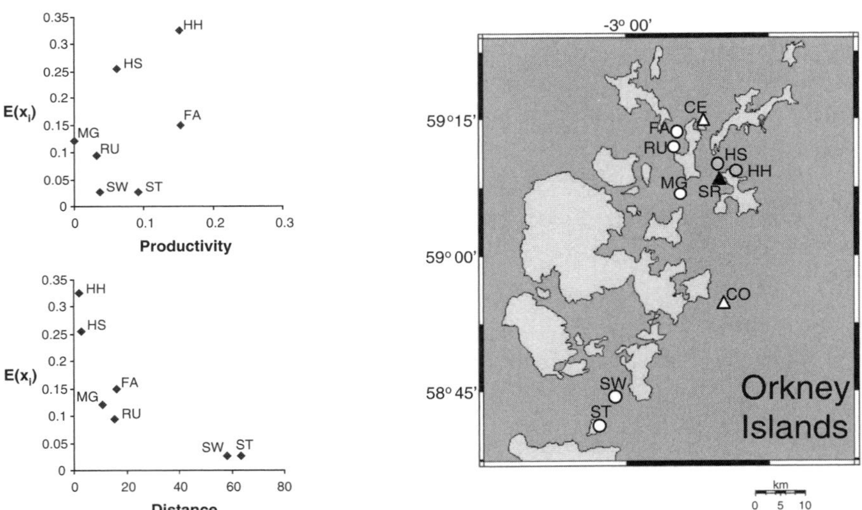

Fig. 15.3 Results of the hierarchical Bayesian analysis for the colonization event observed in Stronsay. Triangles identify new colonies, and circles identify potential source populations. The black triangle identifies Stronsay, and gray circles identify the two main contributors to its colonizing group. The source colonies are Faray (FA), Holm of Huip (HH), Holm of Spurness (HS), Muckle Greenholm (MG), Ruskholm (RU), Swona (SW), and Stroma (ST). The three newly founded colonies are Stronsay (SR), Calf of Eday (CE), and Copinsay (CO). The plot of $E(x_i)$ versus productivity shows a weak association, whereas that of $E(x_i)$ versus distance shows a strong association. (see Color Plate 1).

were not sampled because they were very difficult to reach. Due to the difficulties associated with the sampling of adult gray seals, and in order to obtain large samples (at least 150 individuals per colony), samples were obtained from pups and were scored for nine highly variable microsatellite loci. All samples from the recently founded colonies were collected sufficiently soon after founding to be sure that they were from the F_1 generation. The standard GSI method assumes that individuals in the genetic mixture are adults that came directly from the source populations. Our samples, however, consisted of descendants of the original colonizers and could include hybrid individuals. Thus, the GSI likelihood function [see Box 15.1, Eq. (B1)] needs to be modified.

The Likelihood Function

There are s source populations contributing to each newly founded population. Baseline data come from samples of n_i pups from each of the contributing populations and consist of the multilocus genotypes for each pup, denoted by y_{ij}, the genotype of the jth juvenile in the ith source population — where $i = 1, 2, \ldots, s$ and $j = 1, 2, \ldots, n_i$. Data for the newly founded populations (i.e., the genetic mixtures) also come from pups and are denoted by $y_{(m)j}$, with $j = 1, 2, \ldots, n_{(m)}$, where m denotes mixture. The complete data set is denoted by $y = \{y_{ij}: i = 1, 2, \ldots, s; j = 1, 2, \ldots, n_i\} \text{U} \{y_{(m)j}: j = 1, 2, \ldots, n_{(m)}\}$. The allele frequencies at each locus in each of the s source populations are parameters that need to be estimated. We denote these parameters $\mathbf{p}_\mathrm{i} = \{p_{hli}\}$, where p_{hli} is the frequency of allele h at locus l in population i and let $\mathbf{p} = \{\mathbf{p}_1, \ldots, \mathbf{p}_s\}$.

In order to include the possibility that some pups in the newly founded colonies have parents that came from two different source populations, we assume that there is a tendency w for individuals of the same colony to mate together. Thus, w can be interpreted as an assortative mating coefficient. The conditional probability of genotype k in the mixture given that both its parents came from source i is $P(K|ii) = \prod_l^L \delta p_{il,a_{1lk}} p_{il,a_{2lk}}$, where $p_{il,a_{nlk}}$ is the frequency of the nth ($n = 1, 2$) allele at the lth locus of individual k in population i, and

$$\delta = \begin{cases} 1 & if \quad a_{1lk} = a_{2lk} \\ 2 & if \quad a_{1lk} \neq a_{2lk} \end{cases}.$$

Conversely, the conditional probability of genotype k in the mixture given that its parents came from two different populations i and j, is

$$P(k|ij) = \prod_l^L p_{il,a_{1lk}} p_{jl,a_{2lk}} + \gamma p_{il,a_{2lk}} p_{jl,a_{1lk}},$$

where

$$\gamma = \begin{cases} 0 & if \quad a_{1lk} = a_{2lk} \\ 1 & if \quad a_{1lk} \neq a_{2lk} \end{cases}.$$

Given the aforementioned assumptions and notations, the probability of finding genotype of a given individual k in the mixture is

$$P(k|w,\mathbf{x},\mathbf{p}) = w \sum_i \left(x_i P(k|ii)\right) + (1-w) \left[\sum_i (x_i^2 P(k|ii)) + \sum_i \sum_{i \neq j} (x_i x_j P(k|ij)) \right]. \quad (15.2)$$

The first term on the right-hand side of Eq. (15.2) represents the probability of genotype k given that its parents mated assortatively, whereas the second term represents the same probability given that the parents mated at random. Using the aforementioned expression, we define the likelihood of w, $\mathbf{x}$, and $\mathbf{p}$ given data, y,

$$L(w,\mathbf{x},\mathbf{p}|y) = P(y|w,\mathbf{x},\mathbf{p}) = \prod_{k=1}^{n_{(m)}} P(k|w,\mathbf{x},\mathbf{p}), \quad (15.3)$$

where $n_{(m)}$ is the number of individuals sampled from the newly founded colony.

The Hierarchical Bayesian Approach

We want to develop a method for testing the hypothesis that density-dependent effects in the source populations are responsible for the new colonizations. If this is the case, we expect that the composition of colonizing groups, which is described by the vector $\mathbf{x} = \{x_i\}$, will be dominated by individuals from colonies at or near their carrying capacity. We note, however, that $\mathbf{x}$ may also be a function of the geographic distance between the newly founded colonies and the potential source populations; nearby colonies may send more colonizers than far away sources. Additionally, we want to reduce the variance of the estimate of $\mathbf{x}$ by combining genetic data with other sources of information. Both of these goals, testing of hypothesis and improving the precision of the estimates, can be achieved by incorporating demographic and geographic distance data in the context of a hierarchical Bayesian approach. Demographic data available consist of time series of pup production estimates for the different Orkney colonies. From these time series it is possible to calculate a colony-specific productivity index π_i, which describes both how strong the density-dependent effects within a source are and how large the source population is. Details of its calculation are presented in Gaggiotti et al. (2002). The geographic distance, δ_i, is obtained by measuring the distance along the path that a seal would use to move between colonies.

The integration of demographic and geographic data is achieved in a natural way if we use a prior distribution in which the x_i values are viewed as samples from a distribution whose parameters are functions of productivity and geographic distance. Because $\mathbf{x}$ is a vector of proportions, we can assume that it follows a Dirichlet distribution (e.g., Gelman et al., 1995) with parameters $\alpha_i (i = 1, 2, \ldots, s)$. There are two alternative ways of implementing this approach. The first one, employed by Gaggiotti et al. (2002), uses a model that makes specific assumptions about the dependence between x_i and productivity and distance. The second one, more recently developed (Gaggiotti et al., in press), uses a more general linear model that links the α_i with any pair of factors that could be but are not restricted to distance and productivity.

Figure 15.4 shows a graph describing the model used by the Gaggiotti et al. (2002) approach. Commonly used models of dispersal, such as the normal and Laplace dispersal kernels (e.g., Neubert et al., 1995), assume that the proportion of individuals that move from patch i to patch j, located i–j units away, decays exponentially with distance. Following these models, we assume that the proportion of individuals in a newly founded colony that came from a source population δ_i distance units away decays exponentially with distance. However, we assume that the effect of productivity is linear so that more productive source colonies contribute more individuals to the founding groups. Thus the expected contribution of a given source population to the founding groups, $E(x_i)$, is

$$E(x_i) = \frac{[(1-S) + S\pi_i]e^{-R\delta_i}}{\sum_i [(1-S) + S\pi_i]e^{-R\delta_i}}, \tag{15.4}$$

where R is the rate of decay with distance and S is the contribution of productivity. Because $\mathbf{x}$ is assumed to follow a Dirichlet distribution, the expectation of its elements can be written as $E(x_i) = \alpha_i/\alpha_0$, where $\alpha_0 = \Sigma_i \alpha_i$. Thus, $\alpha_i = \alpha_0 E(x_i)$ and α_0 is a nuisance parameter that needs to be estimated.

Equation (15.4) can be used to formulate four alternative models (Table 15.1). The first one can be considered as the null model because it assumes that the composition of founding groups is independent of the two factors. Thus, the prior distribution is the Dirichlet with $\alpha_i = 1$ for all i. The following two models include only one factor and the last one is the full model that includes both factors.

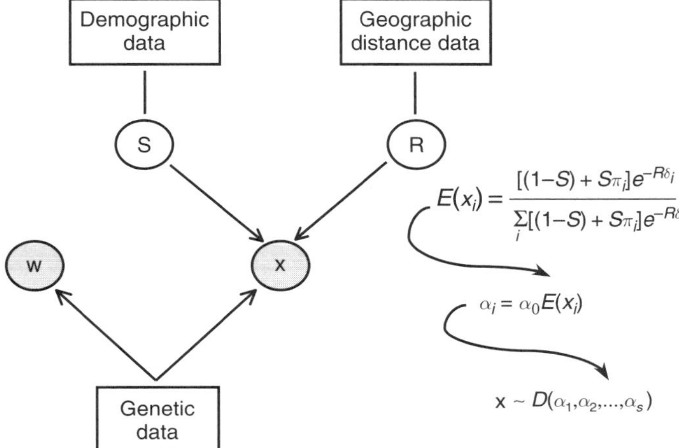

Fig. 15.4 Schematic description of Bayesian approach 1. Genetic data are combined with demographic and genotypic information by focusing on the expected contribution of each source population.

TABLE 15.1 Alternative Models Obtained from Eq. (4)

Model	$E(x_i)$
Founding group composition is independent of both density and distance	$E(x_i) = \frac{1}{s}$
Founding group composition depends only on density	$E(x_i) = \frac{[(1-S) + S\pi_i]}{\sum_i [(1-S) + S\pi_i]}$
Founding group composition depends only on distance	$E(x_i) = \frac{e^{-R\delta_i}}{\sum_i e^{-R\delta_i}}$
Founding group composition depends on both density and distance	$E(x_i) = \frac{[(1-S) + S\pi_i]e^{-R\delta_i}}{\sum_i [(1-S) + S\pi_i]e^{-R\delta_i}}$

The Bayesian formulation for the full model is

$$P(w,S,R,\alpha_0,\mathbf{x}|y) \propto P(w)P(\alpha_0)P(R)P(S)P(\mathbf{x}|S,R,\alpha_0)P(y|w,\mathbf{x}). \quad (15.5)$$

The likelihood $P(y|w,\mathbf{x})$ is given by Eq. (15.3). The prior for $P(\alpha_0)$ is uniform (noninformative) from s to 100, whereas those for $P(w)$ and $P(S)$ are uniform from zero to one. The prior for $P(R)$ is uniform from zero to five. Note that, for the sake of simplicity and given the large sample sizes used, this formulation assumes that allele frequencies are known [thus, the parameter $\mathbf{p}$ is not included in Eq. (15.5)]. Thus, the estimation is carried out using their maximum-likelihood estimates. A more general formulation allowing the use of smaller sample sizes would place a Dirichlet prior on the allele frequency distributions following the approach of Rannala and Mountain (1997) and would estimate them at the same time as all other parameters.

The Bayesian formulations for each of the three remaining models are obtained by eliminating from Eqs. (15.4) and (15.5) the factors that are not included in the respective model. The posterior distribution under each model is estimated separately using an MCMC approach (e.g., Brooks, 1998).

The simplest approach for comparing the fit of the different models to genetic data is the use of the DIC (Spiegelhalter et al., 2002), described in Box 15.2.

BOX 15.2 Bayesian Model Choice

We wish to compare alternative model formulations with the aim of identifying a model that appears to describe the information in data adequately. More precisely, we want to know if the incorporation of the effect of productivity and/or geographic distance leads to a better fit to data. Model choice is a relative measure: we choose the best-fitting model from those that are available. A model may be the best choice but it may still be inadequate by absolute standards. The likelihood ratio test used in maximum likelihood inference (see Box 15.1) is a model choice test: it measures relative merits of competing models but reveals little about their overall adequacy. Two alternative model choice approaches used in Bayesian statistics are described briefly.

Deviance Information Criterion

Spiegelhalter et al. (2002) developed Bayesian measures of model complexity (p_D, the "effective number of parameters") and fit ($\bar{D}$, the posterior mean deviance) and used them to obtain a deviance information criterion that can be used for model comparison.

The measure of model complexity is estimated as the difference between the sample mean of the simulated values of the deviance (mean deviance) minus an estimate of the deviance using the simulated values of the parameters θ (deviance of the means):

$$P_D = \overline{D(\theta)} - D(\bar{\theta}). \tag{B1}$$

The function $D(\theta)$ is the Bayesian deviance given by

$$D(\theta) = -2\log P(y|\theta) + 2\log f(y), \tag{B2}$$

where $P(y|\theta)$ is the likelihood function and $f(y)$ is a standardizing function of data, y, alone. In the gray seal example presented in the text, we used the null standardization obtained by assuming $f(y)$ is the perfect predictor that gave probability 1 to each observation.

The deviance information criterion is defined as the estimate of fit plus twice the effective number of parameters:

$$DIC = D(\bar{\theta}) + 2P_D. \tag{B3}$$

More complex models may be preferred if they give a sufficient improvement of fit or, equivalently, the preferred model will have the lower value of the DIC.

This approach requires running a separate MCMC for each model from which we calculate the quantities required to obtain the model-specific DIC.

Reversible Jump Markov Chain Monte Carlo (RJMCMC)

The Bayesian paradigm provides a very natural framework for considering several models simultaneously, assigning probabilities to each model. This involves moving between parameter spaces with different dimensions, as the alternative models may include different numbers of parameters. Green (1995) extended the basic Metropolis–Hastings algorithm to deal with jumps between states of different dimensions. RJMCMC allows for the estimation of the joint probability distribution of (M,Θ_m), where $M = (1,2, \ldots, K)$ is a "model indicator" and $\Theta_m = \{\theta_1, \theta_2, \ldots, \theta_{km}\}$ is a real stochastic vector whose dimension, k_m, depends on each model m. We assign a prior probability to each model, commonly assuming that all models are equally likely unless there is prior information that may suggest some models are more likely than others. Priors for the model and their corresponding parameters can be combined with the likelihood to obtain a full joint posterior distribution over both the model and the parameter space. RJMCMC allows us to sample from this joint posterior distribution, thereby providing estimates of model probabilities within the MCMC simulation itself by simply observing the number of times that the chain visits each distinct model.

Results of this analysis are presented by Gaggiotti et al. (2002), who show that the relative importance of distance and productivity in each of the colonization events appears to have been determined by the location of potential sources around the vacant site. The Isle of Copinsay is more or less equidistant from all potential sources, reducing the evidence of any association with distance and leading to a better fit for models that include productivity. Conversely, some potential source colonies are much closer than others to Stronsay and the main contributors are the closest leading to a better fit for the models that include distance. For Calf of Eday, results indicate that both population density and distance act concurrently. For the purpose of illustration, Fig. 15.3 shows the results for Stronsay.

For the sake of brevity, Gaggiotti et al. (2002) did not present results for the assortative mating coefficient. Figure 15.5 shows the posterior distribution for each new colonization event. The posterior distributions obtained are fairly similar to the uniform distribution used prior for the assortative mating coefficient. This indicates that information contained in the genetic samples is not enough to provide clear answers to the question of assortative mating. Note, however, that in the case of Stronsay, the posterior distribution of w is skewed toward $w = 1$, suggesting that there is a tendency for

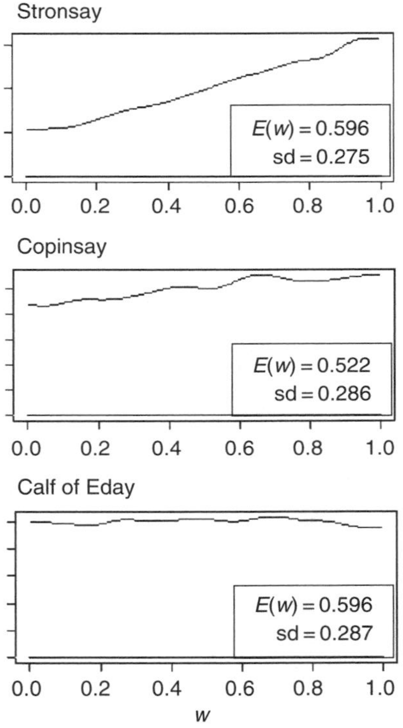

Fig. 15.5 Posterior distribution for the assortative mating coefficient. Also shown are the posterior means of w and the standard deviation.

individuals from the same source colony to mate more often among themselves than with individuals from another source colony. A larger number of highly polymorphic microsatellites might provide a more definitive answer to this question.

A more sophisticated approach to the study of colonization (Gaggiotti et al., in press) is to assume a more general relationship for the vector **x** and the factors that are hypothesized to determine the composition of colonizing groups. A full description of this approach is published elsewhere, but a brief description is presented in Box 15.3 as another possible way of combining different types of data and testing for alternative hypothesis.

BOX 15.3 A More Sophisticated Hierarchical Bayesian Approach

In the second hierarchical Bayesian approach, we no longer focus on the expected contribution of each source population to the founding groups, $E(x_i)$. Instead, we focus directly on the parameters, α_{xi}, of the Dirichlet distribution and assume that they are distributed lognormally (Fig. 15.6). In other words, the log of the ith element of the vector $\alpha_{\mathbf{x}} = \{\alpha_{xi}\}$ has a normal distribution with mean μ_i and variance σ^2. We further assume that means μ_i values are linear functions of the productivity of each source and the distance between the newly founded colony and the source populations,

$$\mu_i = a + b\delta_i + c\pi_i + d\delta_i\pi_i. \tag{B1}$$

Although we are using productivity and distance in this example, this approach could be used to study the effect of any other factors, such as inbreeding avoidance, frequency of environmental perturbations, and habitat quality. Additionally, this approach allows us to address possible interactions between different factors.

Equation (B1) can be used to generate nine alternative models (see Table B15.1). In order to discriminate among them, we derive probabilities associated with each model using reversible jump MCMC techniques (Green, 1995; for a simple derivation, see Waagepetersen and Sorensen, 2001).

TABLE B15.1 Nine Alternative Models obtained from Eq. (B1)

Model	μ_i	Dimension of Θ
Constant effect	$\mu_i = a$	6
Distance effect only	$\mu_i = c\delta_i$	6
Density effect only	$\mu_i = c\pi_i$	6
Constant and distance effects	$\mu_i = a + b\delta_i$	7
Constant and density effects	$\mu_i = a + c\pi_i$	7
Distance and density effects	$\mu_i = b\delta_i + c\pi_i$	7
Constant, distance, and density effects	$\mu_i = a + b\delta_i + c\pi_i$	8
Distance, density, and interaction effects	$\mu_i = b\delta_i + c\pi_i + d\delta_i\pi_i$	8
Full model	$\mu_i = a + b\delta_i + c\pi_i + d\delta_i\pi_i$	9

15.5 POTENTIAL PROBLEMS

This review would be incomplete without discussing the problems that may be found when applying Bayesian approaches. As already mentioned, the result of Bayesian approaches can be highly influenced by the prior distributions used in their formulation (see Box 15.1). This potential drawback has elicited harsh criticism from statisticians that use classical statistical inference or maximum likelihood methods. However, numerous tools can be used to investigate the potential biasing effect of prior distributions. Such knowledge allows the researcher to change priors so as to eliminate the bias. More importantly, the use of a modeling approach for the formulation of priors minimizes the subjectivity involved in their selection or, at the very least, it equalizes it with that involved in formulation of the likelihood function. Indeed, researchers commonly make subjective judgements about the parameters that should be included in the likelihood function. Moreover, the use of approaches such as DIC and RJMCMC provides measures of fit for the alternative models considered by the different prior distributions. The choice of the prior can therefore be based on an objective measure, namely the fit of the model to genetic data.

The fact that all these strategies are available invalidates the criticism concerning the subjectivity of Bayesian approaches, but it is necessary to acknowledge the fact that their implementation is less than straightforward. These complications may limit the use of Bayesian methods to scientists with substantial training in statistics. The development of sophisticated computer software that includes easy-to-use functions implementing diagnostic tests to detect potential biases should make Bayesian methods more accessible to a wider range of users.

Another problem associated with Bayesian methods is that they require substantial computing power. This may not be the case when they are applied to answer simple questions, but the type of problems that arise in metapopulation

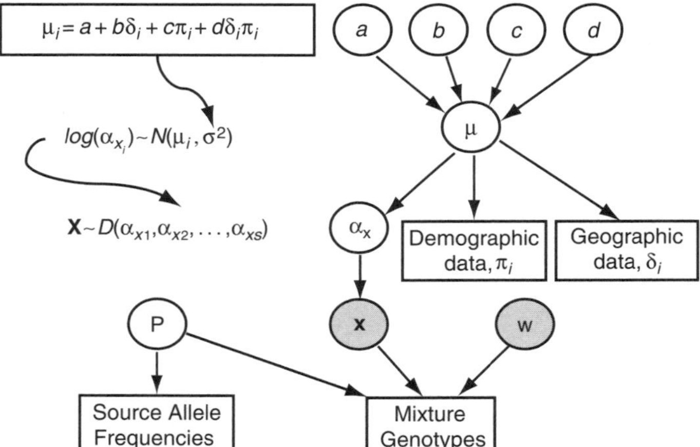

Fig. 15.6 Schematic description of the Bayesian approach 2. The genetic data is combined with demographic and genotypic information by focusing directly on the parameters of the Dirichlet distribution used as the prior of the vector of genetic mixture coefficients.

biology do lead to complex statistical models and their analysis requires substantial computing time. Luckily, the power of desktop workstations increases substantially every year and will soon be enough to allow the study of complex problems in metapopulation biology.

15.6 CONCLUDING REMARKS

Recent developments in the field of statistical genetics have paved the way for the development of new statistical approaches for studying the fundamental processes that characterize metapopulations, in particular those that involve dispersal of individuals. The possibility of combining genetic data with other sources of information in a single statistical framework is particularly promising in the context of metapopulation studies, as population turnover and dispersal tend to decrease the amount of information contained in genetic data. This is due to the predominantly negative effect that these processes have on the effective metapopulation size and the degree of differentiation among local populations (Chapter 7).

The potential for the development of new methods of inference depends on the particular process that we envision to study. In the case of colonization processes, the extension of the GSI method described in this chapter should prove very useful. The challenge here lies in the building up of the databases needed for its application. Nevertheless, as the example of the gray seal illustrates, demographic data could be simple estimates of abundance. Additionally, molecular ecologists routinely collect geographic distance information. Application of the GSI method to study the effect of other factors (e.g., kin competition and habitat quality) that can also influence the movement of individuals away from their patch of origin may prove more difficult. Still, as Chapter 13 highlights, there is a need for multifactorial studies of dispersal; the method discussed earlier is likely to be very useful for the analysis of data collected in such studies.

Studying migration in a metapopulation context is more complicated than studying colonization processes. There are available maximum likelihood methods based on the coalescent that can be used to estimate the effective number of migrants (Beerli and Felsenstein, 2001). However, this is a parameter that measures the long-term effect of migration under equilibrium conditions and, therefore, is of little relevance in a metapopulation context where nonequilibrium dynamics are pervasive. The Bayesian method described by Wilson and Rannala (2003) is more appropriate because it does not assume equilibrium conditions. One of the limitations of this method is that accurate estimates of migration rates are only possible when levels of genetic differentiation among local populations are large. This is rarely the case in metapopulations with high turnover rates, but it may be possible to extend the method in order to incorporate different sources of information much in the same way as the GSI method has been modified for the same purpose.

A related problem that could be addressed using some of the methods discussed earlier is the study of the effect of different factors such as geographic distance, environmental factors, and cultural affinities on the degree of genetic differentiation between pairs of local populations. The test generally used for these purposes is the partial Mantel test (Smouse et al., 1986), the generality of which has been recently questioned (Raufaste and Rousset, 2001). When more than one factor is included in the analysis, the Monte Carlo randomization procedure used to test for the significance of the correlation is inadequate. The use of hierarchical Bayesian approaches that relate the prior distribution for F_{ST}, or a related parameter, with the different factors would provide a proper way of testing for their effects.

Another related problem is the detection of sex-biased dispersal. Goudet et al. (2002) described methods based on biparental inherited genetic markers and summary statistics and concluded that they have a limited power, being able to detect biased dispersal only when the bias is extreme. Methods that make full use of the information contained in genetic data may prove more powerful. Additionally, it may be possible to devise statistical tests to identify the factors or attributes (e.g., size, colour, social status) that may influence the probability of dispersal.

In summary, the development of statistical methods that make full use of all available data is an area that will expand in the coming years and, as already shown by some of the existing studies, will be of great help in the study of metapopulation processes.

16. ECOLOGICAL AND EVOLUTIONARY CONSEQUENCES OF SOURCE–SINK POPULATION DYNAMICS

Tadeusz J. Kawecki

16.1 INTRODUCTION

The change of population density at a given area reflects the balance among local births, local deaths, immigration, and emigration. A local population may thus remain stable even though births do not equal deaths, the difference being compensated by net emigration to, or immigration from, neighboring populations. Trivial as this statement may be, its ecological and evolutionary consequences became appreciated only relatively recently. Although several earlier papers considered consequences of differences between emigration and immigration (e.g., Lidicker, 1975; Keddy, 1982; Holt, 1983, 1985), the concept of source–sink population structure was brought to the general attention of ecologists by Pulliam (1988). He saw it as a consequence of differences in

0-12-323448-4

habitat quality. Local births on average exceed deaths in some (usually high-quality) habitats, with surplus individuals dispersing to other (usually low-quality) habitats; the latter become net importers of individuals. Hence the definition of source and sink habitats based on the difference between emigration and immigration: in source habitats, emigration exceeds immigration; the reverse holds in sink habitats (Pulliam, 1988). This is the definition used in this chapter.

However, in the same paper Pulliam implied that a sink habitat cannot sustain a population in the absence of dispersal. This will generally not be the case except in models with no population regulation. A given habitat may well be able to sustain a population of a certain density, but immigration from a nearby higher quality habitat can lead to a state of permanent overcrowding (Holt, 1985), in which births will not compensate for deaths. When the distinction is necessary, I refer to such a habitat as a relative sink, and to a habitat unable to sustain a population as an absolute sink [Watkinson and Sutherland (1995) refer to them as "pseudosinks" and "true sinks," respectively]. Differentiating between relative and absolute sinks in natural heterogeneous environments will often be impossible without actually preventing immigration and emigration. Except where specified otherwise, the results discussed later are valid for both relative and absolute sinks.

This chapter reviews the consequences of imbalance between immigration and emigration for population dynamics and distribution and for adaptive evolution. Section 16.2 reviews models of source–sink dynamics. First a simple patch model is introduced and used to discuss the concept and meaning of habitat-specific reproductive value; then extensions of the basic model are discussed. Section 16.3 contains a discussion of main theoretical predictions concerning the effect of source–sink population structure on population dynamics, size, distribution, persistence, and stability, followed by a review of relevant empirical data. Section 16.5 focuses on the reasons for which individuals may disperse into sink habitats. Consequences of source–sink structure for adaptive evolution are discussed in Section 16.6. Integration of the concept of source–sink population structure with extinction–colonization dynamics and the metapopulation concept is the focus of Section 16.7. The final section includes some thoughts about future research directions and neglected applied aspects of source–sink population structure.

16.2 MODELS OF SOURCE–SINK POPULATION DYNAMICS

Throughout this chapter the word "habitat" is used to describe a certain set of environmental conditions (including abiotic conditions, available resources, and predator pressure), whereas "patch" refers to an actual physical space. Thus, numerous patches of two habitats may form a more or less fine-grained mosaic in the physical landscape. If variation in the environmental conditions is continuous (e.g., along a gradient), there would be no discrete patches, but still each point in the landscape can be defined as a certain habitat, characterized by given environmental parameters. A spatially explicit approach (considered in one of the following subsections) would be more appropriate than a patch model in such a case. This section first uses a simple patch model to introduce and define the concept of source–sink population dynamics; some of its assumptions are relaxed in the subsequent subsections.

A Patch Model of Source–Sink Population Structure

Consider a species with discrete generations, inhabiting an environment composed of patches of two types of habitats. Assume that habitat 1 is of better quality than habitat 2, i.e., at any particular population density the net reproductive rate (the expected lifetime reproductive success) is greater in habitat 1 (Fig. 16.1).

If the local populations are isolated from each other, the population dynamics in each habitat is fully determined by the respective density-dependent net reproductive rate. Assuming that a stable equilibrium exists, each population is expected to equilibrate at the local carrying capacity, i.e., the density at which births balance deaths, and thus the net reproductive rate equals 1 (Fig. 16.1a).

This will not any more be the case if local populations in the two habitat types are connected by dispersal. Because dispersal usually reduces variation in density, it tends to keep density in better habitats below the local carrying capacity, whereas poor habitats tend to be overcrowded relative to the density they would support in the absence of dispersal. Figure 16.1b illustrates the extreme case of complete mixing, whereby dispersing individuals (propagules) from both habitats form a common pool, which then becomes distributed between the two habitats in proportion to their relative area. In this case the population density, censused just after dispersal, will be the same in both habitats. Consequently, at equilibrium, the population in the better habitat will be below the local carrying capacity and its reproductive rate will be greater than 1, with the excess of births over deaths compensated by an excess of emigration over immigration. The reverse will be the case for the poor habitat. The equilibrium population will thus have a source–sink structure, with a net flow of dispersers from habitat 1 (source) to habitat 2 (sink).

An intermediate case is that of limited dispersal, where a certain fraction of individuals (smaller than 50%) exchange their habitats each generation. In this case, population densities will, in general, be different from each other and from the local carrying capacities. If the propensity to disperse is habitat independent, the equilibrium density will be greater in habitat 1 than in habitat 2,

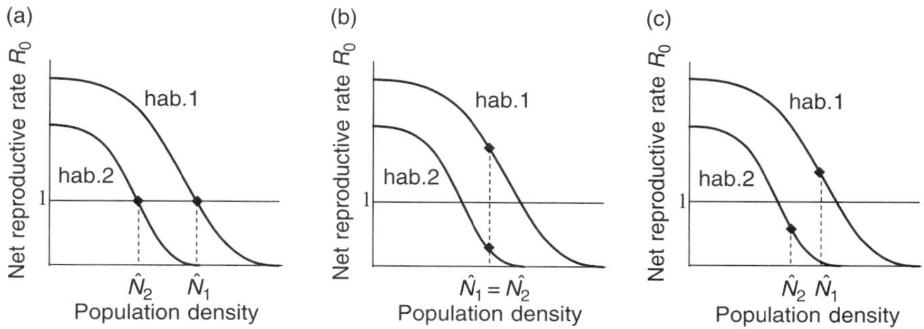

Fig. 16.1 A simple graphical model of the source–sink population structure; for explanations, see the text.

but it still will be below the local carrying capacity in habitat 1 and above the carrying capacity in habitat 2 (Fig. 16.1c). Thus, the population as a whole will still have a source–sink structure, although the net flow of dispersers from the source to the sink habitat will be smaller than under complete mixing.

This graphical model can be formalized and extended to an arbitrary number of habitat patches by application of the general matrix approach to population dynamics (Caswell, 1989):

$$\mathbf{n}(t + 1) = \mathbf{A}[\mathbf{n}(t)]\mathbf{n}(t), \tag{16.1}$$

where $\mathbf{n}(t)$ is the (column) vector of population sizes n_i at time (generation) t in the respective habitats, and $\mathbf{A}[\mathbf{n}(t)]$ is a density-dependent transition matrix. The element in the ith row and jth column of $\mathbf{A}[\mathbf{n}(t)]$ is given by

$$[a_{ij}(n_j)] = [f_j(n_j)m_{ji}], \tag{16.2}$$

where $f_j(n_j)$ is the net reproductive rate (expected lifetime reproductive success) of an individual living in habitat j, and m_{ji} is the dispersal rate from habitat j to habitat i. This model thus assumes that population regulation takes place within each habitat independently, followed by dispersal; census takes place after dispersal. The dispersal rate is defined as the probability that an individual present in habitat j before the dispersal phase will end up in habitat i after dispersal. I assume here that this probability is the same for all patches of a given habitat type, i.e., variation in patch connectivity is negligible. This model also assumes that parameters are constant, generations are discrete, and each individual spends most of its life in a single habitat and only this habitat affects its survival and reproduction (coarse-grained environment in the sense of Levins, 1968a). Note that this last assumption would be violated if there were nonnegligible habitat-related maternal effects on fitness, e.g., if the viability of seedlings was affected by the habitat from which the seeds originated.

Assume that the system has a stable nonzero equilibrium $\hat{\mathbf{n}}$. The dominant eigenvalue λ of matrix $\mathbf{A}(\hat{\mathbf{n}})$ equals 1, and the equilibrium population sizes are a corresponding right eigenvector. The normalized eigenvector $\mathbf{u} = \hat{\mathbf{n}}/(\Sigma n_i)$ describes the distribution of individuals among habitats. The corresponding left eigenvector $\mathbf{v}$, normalized so that the scalar product $<\mathbf{u}\cdot\mathbf{v}> = 1$, consists of the reproductive values of individuals in the respective habitats (Caswell, 1989; Rousset, 1999a). The importance of habitat-specific reproductive values is discussed in the following section.

The number of individuals that disperse from habitat h to other habitats at equilibrium is

$$E_h = \hat{n}_h f_h(\hat{n}_h) \sum_{i \neq h} m_{hi} \tag{16.3}$$

while the number of immigrants to habitat h from other habitats is

$$I_h = \sum_{i \neq h} \hat{n}_i f_i(\hat{n}_i) m_{ih}. \tag{16.4}$$

According to Pulliam's (1988) definition, habitat h is a sink if $E_h < I_h$. Noting that, from Eq. (16.2) $f_i(\hat{n}_i)m_{ih} = a_{hi}(\hat{n}_i)$, and that, from the definition of right eigenvector, $\Sigma_i a_{hi}(\hat{n}_i)\hat{n}_i = \hat{n}_h$, one can show that the difference between the numbers of emigrants and immigrant at equilibrium equals

$$E_h - I_h = \hat{n}_h\left(f_h(\hat{n}_h)\sum_i m_{hi} - 1\right) \tag{16.5}$$

Thus, habitat h is a sink (following Pulliam's definition) if $f_h(\hat{n}_h)\Sigma_i m_{hi} < 1$. If mortality during dispersal is negligible or is absorbed into $f_i(n_i)$, $\Sigma_i m_{hi} = 1$. In this case, Pulliam's definition of a sink implies that $f_h(\hat{n}_h) < 1$, i.e., that the local density in a sink habitat is above the local carrying capacity [defined as the density at which $f_h(n_h) = 1$].

Reproductive Value and the Definition of Sources and Sinks

Equation (16.5) formalizes the definition of source and sink habitats as net exporters and importers of dispersing individuals. However, as noted by Kawecki and Stearns (1993) and Rousset (1999a), habitat-specific reproductive values may be more closely related to the ecological and evolutionary consequences of environmental heterogeneity than the difference between emigration and immigration.

First, the reproductive value measures the expected long-term contribution of an individual to population growth and the future gene pool (Caswell, 1989). The asymptotic contribution of the local population in habitat h to a future gene pool is $u_h v_h$ (note the analogy between this quantity and the patch value as defined in Chapter 4; the two are, however, not identical as the latter focuses on colonization of empty patches). Local populations in habitats with $v_h > 1$ contribute more to future generations than would be expected based on their share of individuals; habitats with $v_h < 1$ contribute less. Similarly, the reproductive value quantifies the notion of "phylogenetic envelope" introduced by Holt and Gaines (1992) to describe tracing the ancestry of individuals alive at time t by looking Δt generations back. As Δt increases (the ancestry is traced into increasingly distant past), the contribution of habitat h to the ancestry of an individual converges to $u_h v_h$ (assuming constant conditions). In a sexual population, the contribution of habitat h to an individual's ancestry is interpreted as the proportion of genes descended from genes present in habitat h at time $t - \Delta t$; in a clonal lineage as the probability that the ancestor was present at time $t - \Delta t$ in habitat h. The effective population size of a subdivided population is also a function of the reproductive values of the local populations (Whitlock and Barton, 1997).

Second, the definition based on the difference between emigration and immigration is difficult to apply in a nonequilibrium situation, when the population sizes and the numbers of migrants fluctuate, whether due to inherent instability of population dynamics or due to fluctuations in the environment. If emigration from a given habitat exceeds immigration into that habitat in some generations, but the reverse is true in others, how should they be averaged over generations, especially that the population sizes fluctuate as well? In contrast, the concept of reproductive value can be generalized to transition matrices that vary in time (Kim, 1987; Tuljapurkar, 1990).

A logical consequence of the aforementioned arguments would be to base the definition of source and sink habitats on the reproductive value (Rousset,

1999a); habitats with $v_h > 1$ would be sources, whereas those with $v_h < 1$ would be sinks. To avoid contributing to terminological confusion, this redefinition is not advocated here. However, it is useful to see when the definition based on the reproductive value would classify habitats differently than the one based on net immigration. First, note that from definition of the left eigenvector $v_h = \Sigma_i v_i a_{ih}$. Hence, if $v_h = 1$ for all h, then $\Sigma_i a_{ih} = f_h \Sigma_i m_{hi} = 1$ for all h. In words, if immigration balances emigration in all habitats (i.e., the system does not have a source–sink structure), then the reproductive values in all habitats are 1. However, if immigration balances emigration in some habitats but not in others, the reproductive values in those habitats will generally be different from 1. Second, if there are only two habitats, then (1 and 2) $f_1(m_{11} + m_{12}) > 1 > f_2(m_{21} + m_{22})$ implies $v_1 > 1 > v_2$, i.e., the reproductive value is greater in the habitat classified as source according to Pulliam's definition (Rousset, 1999a). This is not any more the case when there are more than two habitats. Rousset (1999a) illustrated this with an example with one-way dispersal. However, a discrepancy between Pulliam's source–sink definition and the pattern of habitat-specific reproductive values may also occur when dispersal rates are symmetric (i.e., $m_{ij} = m_{ji}$). Such a case is illustrated in Fig. 16.2. In that example, habitat 1 and habitat 2 are both sources according to Pulliam's definition, with the excess of emigration over immigration being greater for habitat 2 than for habitat 1 (in absolute terms and relative to equilibrium population sizes). Consistent with this, the net reproductive rate at equilibrium, $f_h(\hat{n}_h)$, is largest in habitat 2. Yet, $v_2 < 1 < v_1$, reflecting the fact that most emigrants from habitat 2 end up in habitat 3, which is a strong sink.

To summarize, the application of the concept of source–sink population structure is most straightforward when there are only two habitat types. When there are more than two habitats, the fact that for a given habitat emigration exceeds immigration does not necessarily imply that habitat contributes more to a future gene pool than would be expected based on its share u_h of the total population. The reason for this discrepancy is that emigration and immigration refer to the movements of individuals within a single generation, whereas reproductive values take into account the consequences of chains of migration events among habitat types happening over many generations. The following subsections summarize some special cases of the model described by Eq. (16.1) and (16.2), as well as its extensions to include age structure and explicit spatial dimensions.

Habitat Area versus Habitat Quality

At the first approximation, a spatially heterogeneous environment can be characterized in terms of the area and quality of the habitats it consist of. A reasonable way to describe the quality of different habitats would be to compare the reproductive success that is expected in each of them at the same population density. Because the aforementioned model is formulated in terms of local population sizes rather than densities, $f_1(n) > f_2(n)$ does not imply that habitat 1 is of higher quality if the habitats cover different areas. If spatial variation in density within a habitat is negligible, it is straightforward to reformulate the model by setting $f_h(n_h) = F_h(n_h/b_h)$, where b_h is the area of habitat h and

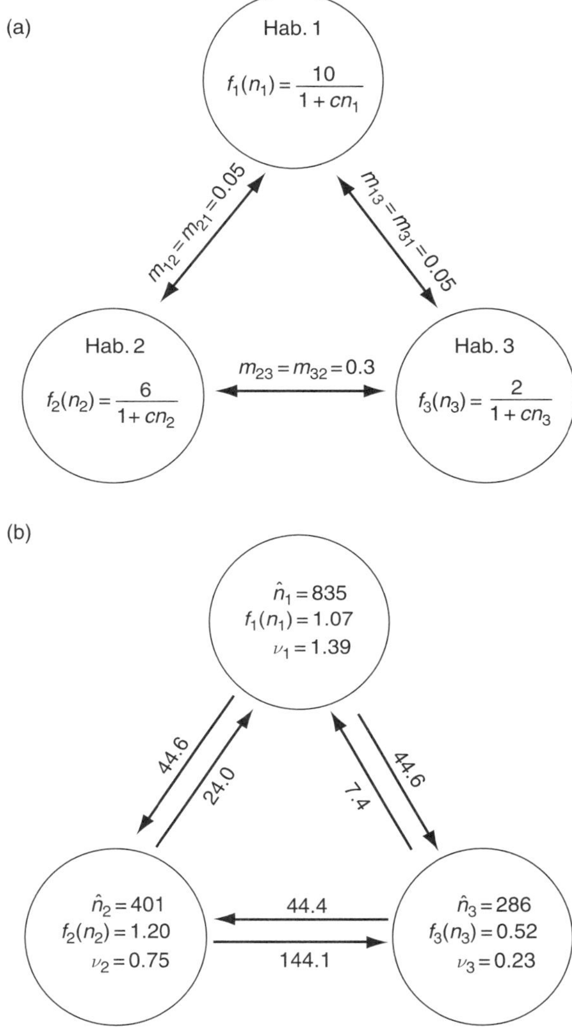

Fig. 16.2 A three-patch model illustrating the discrepancy between the classification of habitats as source or sink and the habitat-specific reproductive value. (a) Parameters of the model; $c = 0.01$. (b) Properties of the equilibrium; numbers next to arrows indicate the number of individuals dispersing from one habitat to the other each generation.

$F_h(n_h/b_h)$ is the net reproductive rate in habitat h as a function of the local population density. Other things being equal, habitats covering a larger area are likely to receive more immigrants, especially with passive dispersal. One way to implement such a relationship is to assume that a fraction $1 - \mu$ of potential dispersers remain in the habitat of origin while the rest end up in various habitats (including the habitat of origin) in proportion to their area, i.e.,

$$m_{ii} = 1 - \mu + \mu b_i/\Sigma_h b_h \tag{16.6a}$$

$$m_{ij} = \mu b_i/\Sigma_h b_h \tag{16.6b}$$

This model of dispersal was implemented to study the effects of habitat quality versus habitat area on adaptive evolution (Kawecki and Stearns, 1993; Kawecki, 1995). More realistically, the dispersal rates will be also affected by the size and arrangement of habitat patches and their connectivity (see later). For the sake of the argument, unless specified otherwise, most of the chapter assumes that all habitats have the same area.

Asymmetric Dispersal Rates, "Reverse" Source–Sink Structure, and Black Hole Sinks

In the model described in the preceding paragraph, individuals in all habitats show the same propensity to disperse, and the source–sink structure results from differences in habitat quality. However, individuals may change their propensity to disperse in response to their habitat. Environmental factors such as river or ocean current, wind, or gravity may also lead to an asymmetry of dispersal rates, increasing the probability of dispersing from an "upstream" habitat to a "downstream" habitat and reducing the probability of dispersing in the opposite direction. The equilibrium properties of a set of populations connected by dispersal depend on both habitat-specific net reproductive rates $f_i(n_i)$ and dispersal rates m_{ij}. Asymmetries of dispersal rates will thus have consequences for source–sink population dynamics.

In particular, asymmetric dispersal rates can create a source–sink structure in the absence of differences in habitat quality. In a system of two habitats of equal size, characterized by the same $f_i(n)$, habitat 1 will be a source and habitat 2 a sink if $m_{12} > m_{21}$, and vice versa. More generally, $m_{12} > m_{21}$ will reinforce the source–sink structure if $f_1(n) > f_2(n)$. Conversely, $m_{12} > m_{21}$ will make the source–sink structure less pronounced if $f_1(n) < f_2(n)$ — up to a point. If $f_1(n) < f_2(n)$, but m_{12} exceeds m_{21} by a sufficient margin, the source–sink structure will become reversed — habitat 1 will become a sink. In other words, an upstream habitat of lower quality (but still good enough to sustain a population despite the drain due to emigration) may become a source if the asymmetry of dispersal rates is sufficient, whereas the better downstream will act as a (relative) sink. For specific models of such populations, see Doebeli (1995) and Kawecki and Holt (2002).

An extreme case of asymmetric dispersal is one-way dispersal, resulting in what Holt and Gomulkiewicz (1997) termed a "black hole sink" — a habitat that receives immigrants but sends no emigrants back to the source. Within the framework of the model described by Eq. (16.1) and (16.2), the existence of black hole sinks implies that matrix $\mathbf{A}(\mathbf{n})$ is reducible (Caswell, 1989). This means that eliminating the rows and columns corresponding to the black hole sinks would have no effect on the equilibrium population sizes and reproductive values in the remaining habitats. In other words, population dynamics in the source habitat is unaffected by what happens in the sink; from the viewpoint of the source habitat, emigration to the sink; is not different from mortality. The reproductive value of black hole sink habitats is 0. For such a system to exist, the source habitat(s) must be good enough to sustain a population, despite the drain imposed by emigration. Note that a black hole sink may still send some migrants to another black hole sink, as in the example given by Rousset (1999a).

Balanced Dispersal

A special case worth considering in the context of asymmetric dispersal rates is the balanced dispersal scenario, whereby asymmetries in the dispersal rate exactly compensate for differences in habitat quality (Doebeli, 1995; Lebreton et al., 2000). Under the balanced dispersal scenario, $v_h = 1$ and $f_h(\hat{n}_h)\Sigma_i m_{hi} = 1$ for all habitats, i.e., there is no source–sink structure. Dispersal rates leading to a balanced dispersal situation are expected to be favored when dispersal is cost free (Doebeli, 1995; Lebreton et al., 2000). This is equivalent to the ideal free distribution (Fretwell and Lucas, 1970). Reasons why the evolution of balanced dispersal may be prevented, and thus the source–sink population structure may persist over evolutionary time, are discussed in Section 16.5.

Age- or Stage-Structured Populations

Generalization of the model described by Eq. (16.1) and (16.2) to multiple age classes (or stages) is, in principle, straightforward, provided that the vital rates (survival and fecundity) are assumed to be a function of age (stage) and the current habitat only (Lebreton, 1996). Nevertheless, the consequences of source–sink population structure in age-structured populations remain rather unexplored. The definition of sources versus sinks based on the number of emigrants versus immigrants can still be upheld if dispersal occurs at a well-defined prereproductive stage, as is the case, e.g., in perennial plants or corals. However, this definition does not seem appropriate if an individual can change its habitat at different ages or stages, and do it repeatedly, as is the case in birds and mammals. This can be illustrated by considering an equivalent of the balanced dispersal scenario discussed in the previous paragraph. Lebreton (1996) has shown that under cost-free dispersal, natural selection should favor a combination of age-specific dispersal rates that would equalize the vector of age-specific reproductive values across habitats. However, in contrast to the discrete generations case, this case does imply balanced dispersal (Lebreton et al., 2000). It is thus difficult to derive general predictions from this model and more work is needed. The problem will become even more complicated if, as is biologically realistic, survival and fecundity depend not only on the current habitat, but on the habitats an individual has experienced in the past. Nonetheless, incorporating both age structure and habitat heterogeneity will, in many cases, substantially improve the predictive power of management-oriented models of specific natural populations (e.g., Doak, 1995).

Spatially Explicit Models

The above discussion assumed environmental variation in the form of a set of discrete habitats, such that within a given habitat individuals become mixed thoroughly and density is the same everywhere. This may be a sufficient approximation for systems such as herbivorous insects that use two host plant species occurring in the same area or in other cases where well-defined discrete habitat patches form a relatively fine-grained mosaic (e.g., Blondel et al., 1992). However, the spatial location of individuals must be explicitly considered if variation in environmental factors is continuous. This can be done with a diffusion approximation (e.g., Kirkpatrick and Barton, 1997) or with an individual-based model. A spatially explicit approach will also be necessary if there are discrete habitat types, but the patches are large relative to the dispersal distance (e.g., Boughton, 2000). Such a case is illustrated in Fig. 16.3, where high-quality habitat 1 borders low-quality habitat 2 along a sharp ecotone (model details in the figure legend). As expected, at equilibrium, habitat 1 is a source and habitat 2 a sink, but the spatial model reveals that the source–sink nature of the two habitats is most pronounced close to the ecotone. That is, in habitat 1 the excess of births over deaths ($f_h(\hat{n}_h) > 1$), and of thus emigration over immigration, is greatest just left of the ecotone (light solid line in Fig. 16.3). The same holds for the excess of deaths over births ($f_h(\hat{n}_h) < 1$) on the other side of the ecotone. As one moves away from the ecotone, the population density (heavy line) converges to the local carrying capacity and $f_h(\hat{n}_h)$ converges to 1. Note, however, that the reproductive value (dotted line) does not follow the pattern of $f_h(\hat{n}_h)$ within the habitats. Instead, in the better habitat it declines somewhat as the ecotone is approached, indicating that the improved lifetime reproductive success due to lower density does not quite compensate for the fact that some of the offspring will end up in the poor habitat. This is thus another case where the pattern based on births versus deaths and emigration versus immigration does not agree with the pattern of reproductive values.

Even if the environment consists of discrete patches of different habitats, their size, shape, spatial arrangement, and connectivity will often cause different patches of the same habitat type to have different dispersal rates. Such a patch network may be modeled within the framework of the patch model described earlier [Eq. (16.1)]. However, the dispersal rates would now have to be defined on a patch-to-patch basis. Thus, in contrast to the simple model presented at the beginning of this section, patches of the same habitat type could not be lumped together. Instead, the vector of population sizes $\mathbf{n}(t)$ would have to have an entry for each patch, not only for each habitat type. Consequently, the definition of source versus sink could now be applied to individual patches; depending on their connectivity, size, and shape, some patches of a given

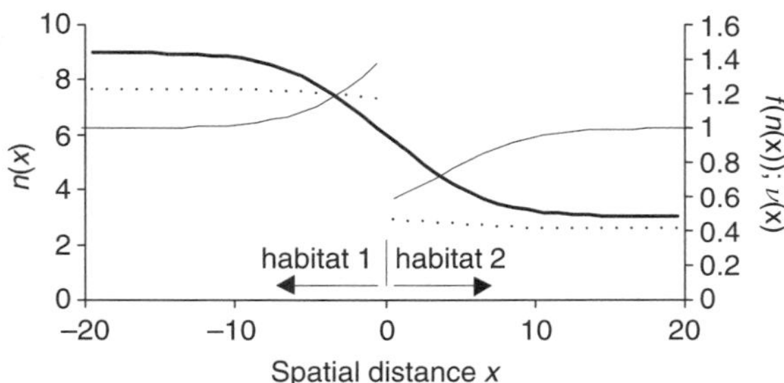

Fig. 16.3 A spatially explicit source–sink model with two habitat patches. The population density $n(x)$ (heavy line, left axis), net reproductive rate $f(n(x))$ (light line), and the reproductive value (dotted line) at equilibrium are plotted as a function of spatial location x. Discrete generations are assumed, with census after dispersal. The net reproductive rate $f(n(x)) = R_h/(1 + n(x))$, where $R_h = 10$ in habitat 1 and $R_h = 4$ in habitat 2. Dispersal distances follow a normal distribution with mean 0 and $\sigma = 5$.

habitat type may act as sources whereas others may act as sinks. Taking into account such a spatial effect is of particular importance in applied models developed for the management and conservation of particular species.

16.3 ECOLOGICAL CONSEQUENCES OF SOURCE–SINK DYNAMICS: THEORY

In addition to the defining feature of source–sink structure — net flow of dispersing individuals from source to sink habitats — a number of other ecological consequences of source–sink population structure have been predicted by mathematical models. These predictions are summarized in this section, whereas Section 16.4 reviews relevant empirical examples.

Species Range

Immigration can maintain a stable local population in a habitat, in which deaths exceed births even at low density (absolute sinks). Unless limited by barriers to dispersal, species ranges will therefore as a rule extend beyond the areas where habitat quality is sufficient to sustain a population without immigration (i.e., where the local conditions satisfy the species' niche requirements; Pulliam, 1988, 2000). This applies both to the geographical range of the species and to its distribution on the scale of local habitat variation (habitat occupancy). In practice, it will often be difficult to distinguish between an absolute sink and a habitat that is not quite optimal and acts as relative sink, but still satisfies the species' niche requirements. Successful reproduction may take place in absolute sinks and population density may be relatively high and stable; it may not be apparent that the population would deterministically go extinct without immigration.

Population Size and Distribution

What is the effect of source–sink population structure on the total (global) population size? An answer will depend on the precise formulation of this question.

First, one may compare a set of habitat patches connected by dispersal (and thus potentially having source–sink structure) with the same set of patches each inhabited by an isolated population. This perspective thus focuses on the effect of changing the dispersal rate(s) while keeping the landscape unchanged. In a two-patch model with symmetric passive dispersal, Holt (1985) showed that no simple general prediction about the effect of dispersal on the total population size can be made. Whether the total population size will increase or decrease as a result of dispersal will depend on the shape of the functions relating local density to the local birth and death rates. This applies even if the poorer habitat is an absolute sink. Doebeli (1995) considered two patches of the same habitat quality and showed that asymmetric dispersal, which resulted in a source–sink structure, led to an increase of the total population size. It is not clear how general this result is (only a numerical example is presented). A more general prediction concerns the effect of dispersal on the distribution of the population among the habitats: with increasing dispersal the fraction of the total population living in the source habitats tends to decrease (e.g., Holt, 1985; Pulliam, 2000; Kawecki and Holt, 2002).

Second, one may ask how adding a sink habitat or changing its size affects the total population size and the population in a high-quality source habitats, assuming that the dispersal pattern is unchanged. It is not surprising that replacing some good habitat with poor habitat will reduce the overall population size. It is more interesting to ask how the population size is affected if some sink habitat patches are eliminated (converted into hostile "nonhabitat" or "matrix") while keeping the amount of the source habitat constant. Under passive dispersal this will lead to a greater fraction of propagules perishing in the "nonhabitat," causing a reduction of the total population size. This is not necessarily the case if dispersal is active and thus the dispersing individuals avoid the "nonhabitat." In one model that made this assumption (Pulliam and Danielson, 1991), the number of individuals in the source habitats increased as the area of sink habitat was reduced. The effect on the total population size in Pulliam and Danielson's model depended on the degree of habitat selection. With poor habitat selection ability, the total population size increased as the area of sink habitat decreased. With a better habitat selection ability, the total population size peaked at an intermediate amount of sink habitat. In contrast, an individual-based model (Wiegand et al., 1999) predicted that eliminating sink habitat will lead to a reduction of the total population size. This discrepancy suggests that no simple general predictions can be made about the effect of eliminating patches of sink habitat on the overall population size.

Population Stability and Persistence

If too many dispersing individuals end up in a habitat that is an absolute sink, the entire population will go deterministically extinct (Pulliam, 1988; Donovan and Thompson, 2001). This is the most obvious effect of source–sink population structure on population persistence. More generally, as extinction risk, at least on the short term, tends to be correlated negatively with population size (Chapter 14), the effects of source–sink population structure on equilibrium population size are likely to have implications for population persistence.

However, the existence of sink habitats may affect population persistence by affecting the population dynamics independently of their effects on the equilibrium population size. Several models (Holt, 1984, 1985; McLaughlin and Roughgarden, 1991) predict that adding a habitat that is a sink for the prey can stabilize an otherwise unstable or neutrally stable predator–prey model. The source–sink structure also tends to have a stabilizing effect on the dynamics of a host–parasitoid model (Holt and Hassell, 1993). Finally, Doebeli (1995), generalizing results of Hastings (1993), showed that dispersal between two patches of the same quality tends to stabilize intrinsically chaotic population dynamics (see also Gyllenberg et al., 1996). The stabilizing effect is stronger if dispersal rates are asymmetric so that at equilibrium there is a source–sink population structure. One intuitive explanation of those results is that sink habitats act as a buffer, absorbing surplus individuals produced in source habitats. This prevents the population from greatly overshooting the equilibrium density, thus reducing or averting a population crash due to overcompensating density dependence. In contrast, dispersal to a sink habitat that is available only seasonally can destabilize population dynamics; this mechanism has been proposed to contribute to rodent cycles (Lomnicki, 1995).

Existence of a sink habitat may make the population less sensitive to environmental fluctuations affecting birth and death rates, provided that the sink habitat is less affected by the fluctuations (Holt, 1997). An extreme case of this type involves a source habitat subject to occasional catastrophes that wipe out the local population.

Age Structure

In organisms with overlapping generations, the age structure in source and sink habitats may differ as a consequence of differences in local survival. More interestingly, dispersal into sink habitats may be age dependent. In territorial species, young individuals may be more likely to be excluded from breeding in optimal, source habitats. Sink habitats will then contain a disproportionately large fraction of young adults.

16.4 ECOLOGICAL CONSEQUENCES OF SOURCE–SINK DYNAMICS: EMPIRICAL EVIDENCE

Basic Source–Sink Structure

There is increasing evidence of source–sink structure in natural populations, involving habitat variation at various spatial scales. At a continental scale, it has been reported in reindeer, in which low recruitment due to wolf predation causes boreal forests to act as sink habitats; the tundra is the source (Bergerud, 1988). Similarly, the reproductive success of pied flycatchers (*Ficedula hypoleuca*) at the northern range limit does not, on average, compensate for mortality (although it may do so in good years); these northernmost populations must thus be maintained by immigration (Järvinen and Väsäinen, 1984). In black-throated blue warbler (*Dendroica caerulescens*), population density and estimated habitat quality decline as one moves away in either direction from the Appalachian mountains (Graves, 1997).

The source–sink structure at a more local spatial scale has been well characterized in blue tits (*Parus caeruleus*) in southern France, where patches of good (deciduous) and poor (sclerophyllous) habitat form a mosaic landscape with a patch size on the order of 1 to 100 km^2 (Blondel et al., 1992). Even though the breeding density in the sclerophyllous habitat is less than half that in the deciduous habitat, birds in the sclerophyllous habitat have a smaller clutch size and a lower breeding success (Dias and Blondel, 1996). The breeding performance in the sink is impaired additionally by a locally maladaptive laying date (see Section 16.6) and possibly by a smaller size of individuals breeding there (Dias and Blondel, 1996). Genetic marker data are also consistent with an asymmetric gene flow from the deciduous to the sclerophyllous habitat patches (Dias et al., 1996). A number of North American migratory songbirds suffer extreme rates of nest parasitism and predation in fragmented forest patches of agricultural and suburban landscapes. These highly fragmented habitats constitute sinks supported by immigration from more extensive forests (Robinson et al., 1995). A similar source–sink structure related to habitat fragmentation has been observed in the reed warbler in The Netherlands (Foppen et al., 2000). A source–sink structure at a more local scale has been found in the checkerspot butterfly (*Euphydryas editha*), where forest clearings and rocky outcrops constitute two spatially separated habitats, each with a different host plant for the caterpillars (Boughton, 2000).

A source–sink structure at the scale of meters occurs in the snow buttercup (*Ranunculus adoneus*), a perennial alpine-zone plant confined to deep snow beds of the Rocky Mountains. The beginning of the vegetative season and flowering time are determined by the snowmelt (Stanton and Galen, 1997). As the pattern of snow accumulation is fairly constant from year to year and patches of old snow tend to melt from the edges, the relative timing of snowmelt at different localities is fairly constant from year to year. Localities separated by only tens of meters may become clear of snow several weeks apart. As seed size is correlated positively with season length, plants at later-melting sites produce smaller seeds (Galen and Stanton, 1993). These small seed have a low establishment rate, and most individuals at all localities come from large seeds produced in early melting sites (Stanton and Galen, 1997). The source–sink population structure in this species is thus at least partially mediated by maternal effects.

A source–sink structure dominated by asymmetry in dispersal rates imposed by wind has been described in the sand dune plant *Cakile edentula* (Keddy, 1981, 1982; Watkinson, 1985). In that system, the base of a dune on the seaward side is the source habitat where most seeds are produced. However, because most seeds are transported by wind to the sink habitat closer to the dune crests, plant density in the latter habitat is considerably higher than in the source habitat. At the same time, seed emigration from the source habitat reduces competition and boosts the reproductive output from that habitat (Watkinson, 1985). In this case the source and sink habitats are only separated by several meters.

The above review of examples of source–sink structure in natural populations is not meant to be exhaustive, and as the interest in this aspect of spatial ecology increases, more evidence will accumulate. Relatively unexplored remain cases of potential source–sink dynamics caused by biotic interactions, particularly the source–sink structure of parasite populations caused by variation in host susceptibility (e.g., Lively and Jokela, 1996).

Other Ecological Consequences

Despite accumulating evidence for the ubiquity of source–sink structure in natural populations, data directly addressing specific predictions concerning its ecological consequences are scarce. Addressing these predictions directly would involve experimental intervention, e.g., changing the amount of source or sink habitat or altering the dispersal pattern. Applying this approach to natural populations may not only be technically difficult, but also questionable on ethical or legal grounds. For example, it is likely that some of the examples mentioned earlier involve populations persisting in absolute sinks, unable to sustain a population without immigration (e.g., Bergerud, 1988; Robinson et al., 1995). However, definitive confirmation would require "closing" the population, i.e., preventing immigration and emigration.

Given the problems with a direct experimental approach, monitoring the consequences of "natural experiments," i.e., natural or anthropogenic changes in the environment, has a particular value. For example, the importance of a sink habitat for population persistence was demonstrated clearly in a source–sink population of the checkerspot butterfly *E. editha*. In 1992, an unusual summer frost killed all larvae in the source habitat (forest clearings). The population persisted only because larvae in the sink habitat (rocky outcrops) survived (Thomas et al., 1996; Boughton, 1999) (the outcrops were presumably the main habitat of this species before humans created forest clearings). A population structure with the source populations subject to repeated catastrophes has also been reported for a midge (Frouz and Kindlmann, 2001). Another study (Luttrell et al., 1999) suggested that extinction of numerous local populations of a cyprinid fish was due to disruption of dispersal between source and sink habitats by artificial reservoirs. The problem with such "natural experiments" is often the lack of replication and controls.

An alternative approach involves spatial analysis of landscape ecology (Chapter 2), whereby the properties of local populations can be correlated not only with the local habitat conditions, but with the composition of the regional habitat matrix (e.g., the presence and size of nearby source and or sink patches). Foppen et al. (2000) used this approach to show that the existence of sink habitat patches leads to a greater size and stability of reed warbler populations in source patches. Graves (1997) has shown that the proportion of yearlings among breeding males of black-throated blue warbler (*D. caerulescens*) is correlated negatively with habitat quality, indicating an effect of source–sink dynamics on the population age structure. The influence of source–sink dynamics on the population size structure has been demonstrated in blue tits in southern France, where males breeding in the source habitat are larger than those breeding in the sink habitat (size measured as tarsus length; Dias and Blondel, 1996). However, because male fledglings produced in the two habitat types do not differ in tarsus length (Dias and Blondel, 1996), the difference with respect to breeding males must reflect the displacement of smaller individuals from the source. As any approach based on correlations, this approach does not directly address causation and can be potentially confounded by factors not included in the analysis. This problem can be illustrated by results from the same study of blue tits. The population density in the sclerophyllous habitat in southern France, where it acts as a sink, is much lower than in the same habitat in Corsica, where it is a dominant habitat not affected by immigration (Dias and Blondel, 1996). These results seem to contradict the prediction that immigration from a source should boost the density in the sink (see earlier discussion). The discrepancy is explained by the fact that reproductive success in the sclerophyllous habitat in Corsica is higher than in the same habitat in southern France (Dias and Blondel, 1996).

A powerful but rarely used approach to study consequences of the source–sink population structure involves setting up controlled experimental source–sink systems in the laboratory or in outdoor enclosures or "mesocosms." Davis and collaborators (1998) used this approach to study the effect of dispersal on population size and distribution along an environmental gradient. Their system involved four *Drosophila* population cages, arranged along a series of temperatures (10, 15, 20, and 25°C), to simulate four habitat patches along a thermal gradient. In one treatment, adjacent cages were connected with plastic tubes, enabling dispersal (dispersal rate about 6% per day) and thus creating conditions under which the source–sink structure was expected. This could be contrasted with a no-dispersal treatment, which simulated isolated populations living at different temperatures. Three *Drosophila* species were tested separately. As predicted by source–sink models, in *D. melanogaster* and *D. simulans* permitting dispersal led to reduced density in patches that had high density under no dispersal and to increased density at marginal temperatures. The pattern was less clear in *D. subobscura*, in which a reduction of density at the optimal temperature was not accompanied by a marked increase of population size at suboptimal temperatures. In all three species, dispersal led to maintenance of local populations in absolute sinks, i.e., at temperatures at which local populations went extinct in the absence of dispersal (10°C for *D. melanogaster* and *D. simulans*, 25°C for *D. subobscura*). Although Davis and colleagues did not address this question statistically, in all three species the overall (global) population size tended to be larger in the absence of dispersal. This study points to the potential usefulness of experimental source–sink model systems to study ecological and evolutionary consequences of the source–sink structure. Because of scale issues, it can only be used with some, mostly invertebrate, model systems. However, use of such model laboratory systems enabled important advances in other areas of ecology and evolutionary biology, and their use to address source–sink-related questions should be promoted.

16.5 NATURAL SELECTION ON DISPERSAL AND EVOLUTIONARY STABILITY OF SOURCE–SINK POPULATION STRUCTURE

Given that the expected reproductive success is lower in a sink than in a source habitat, one would expect that dispersal from source to sink habitats should be countered by natural selection. As a result, the dispersal pattern should evolve toward retaining more individuals in the source, up to the point at which differences in local density compensate for differences in habitat quality and the source–sink structure disappears (balanced dispersal scenario, Section 16.2). This intuitive argument has been supported by formal analysis of a patch model assuming passive dispersal (Doebeli, 1995; Lebreton et al., 2000); it also underlies the ideal free distribution model for actively dispersing organisms (Fretwell and Lucas, 1970). To explain why the source–sink population structure should persist over evolutionary time, one must find reasons why the above prediction should not hold. These reasons are likely to be different for passively and actively dispersing organisms.

Passive Dispersal

By definition, passively dispersing individuals cannot choose their destination. Dispersal to sink habitats in such organisms can be understood easily as a consequence of a general propensity to disperse. The balanced dispersal scenario requires that the dispersal rate (m_{ij} as defined in Section 16.2) from a high to a low-quality habitat is lower than the dispersal rate in the opposite direction (Doebeli, 1995). Such an asymmetry of dispersal rates is possible if propagules produced in poorer habitats have a greater propensity to disperse, reflecting the plasticity of behavioral and morphological traits affecting dispersal. However, the evolution of such plasticity is likely to be constrained, in particular because the probability of dispersing from a source to a sink not only depends on the propensity to disperse, but also on the relative area of different habitats types within an individual's dispersal shadow. If plasticity of dispersal rates is constrained, simple source–sink models predict that natural selection should drive dispersal to minimum (e.g., Balkau and Feldman, 1973; Holt, 1985). However, this tendency will be counteracted by advantages of dispersing within a given habitat type, such as avoidance of inbreeding and sib competition or assurance against temporal unpredictability of the environment (see Chapter 10). The optimal dispersal propensity will reflect a balance between these two forces.

Active Dispersal with Habitat Choice

Three general reasons have been proposed to explain deviations from an ideal free distribution and dispersal into sink habitats in actively dispersing organisms capable of habitat choice (Holt, 1997). First, territoriality or other forms of contest competition may prevent some individuals from breeding in the source habitat. It will often pay for such individuals to attempt breeding in sink habitats rather than be nonbreeding "floaters" in source habitats (Pulliam, 1988; Pulliam and Danielson, 1991). Thus, in this scenario individuals breeding in a sink do the best of a bad job. Second, ideal free distribution requires that individuals can assess not only the quality of different habitats, but also the distribution of individuals among habitats. Gaining this information is likely to be constrained by the cognitive abilities of the species, particularly if the environment is changing in time (Remes, 2000). Even if the species is capable of evaluating habitats accurately, inspecting many habitat patches will be costly in terms of energy, time, and mortality. Thus, it may pay to settle in the first more or less suitable habitat patch (van Baalen and Sabelis, 1993). Third, if the environment is temporally variable in such a way that fitness in the sink habitat occasionally exceeds that in the source habitat and dispersal back from the sink to the source is possible, genotypes that choose sink habitat with a small but nonzero probability will have advantage over those that avoid sink completely (Holt, 1997; Wilson, 2001). In this scenario, dispersal into a sink habitat is thus a form of bet hedging (Seger and Brockmann, 1987).

16.6 EVOLUTIONARY CONSEQUENCES OF SOURCE–SINK STRUCTURE

From an evolutionary perspective, "habitat quality," which determines whether a habitat is a source or sink, reflects an interaction between the properties of the habitat and the characteristics of the species; the latter can evolve. It is thus of interest to know how the relative performance of a population in source and sink habitats should change over evolutionary time. Adaptation to initially marginal sink habitats has important implications for the evolutionary dynamics of species distributions.

Adaptation to marginal habitats may be constrained by a lack of genetic variation (Lewontin and Birch, 1966; Parsons, 1975; Blows and Hoffmann, 1993), which in turn may reflect biochemical, physiological, and developmental constraints resulting from the species' evolutionary history (Stearns, 1994). This factor is not specific to source–sink populations and is not discussed here. Instead this section focuses on predictions concerning the effect of source–sink population structure on adaptive evolution, assuming that genetic variation for fitness in both source and sink habitats exists. Two intuitive arguments have been made as to why source–sink dynamics make it difficult for a population to evolve improved performance in habitats that function as sinks. The first argument notices that sink habitats contribute relatively little to the reproduction of the entire population. Therefore, their contribution to the overall fitness, averaged over habitats, is relatively small, and natural selection on performance in sink habitats is relatively weak. The second argument stresses gene flow swamping locally adapted genotypes in sink habitats. These two arguments and the relationship between them are discussed in the following two subsections. The third subsection discusses the predictions of the theory, while the last subsection reviews the empirical evidence.

Reproductive Value and Sensitivity of Fitness

In the classic model of quantitative traits under natural selection (Price, 1970; Lande and Arnold, 1983), the expected direct response of a trait to selection is proportional to the strength of selection, measured as the derivative of fitness with respect to the trait value. For a source–sink population at a density equilibrium, the dominant eigenvalue λ of the transition matrix $\mathbf{A}(\hat{\mathbf{n}})$ is an appropriate measure of fitness (Caswell, 1989; Charlesworth, 1994). Thus the strength of selection on trait z can be partitioned according to its effect on the net reproductive rate $f_h(\hat{n}_h)$ in each habitat:

$$\frac{\partial \lambda}{\partial z} = \sum_h \frac{\partial \lambda}{\partial f_h} \frac{\partial f_h}{\partial z}, \tag{16.7}$$

all derivatives are evaluated at $\hat{\mathbf{n}}$; the arguments of f_h are left out for transparency of the formula. From the general equation for eigenvalue sensitivity (Caswell, 1989, Eq. 6.6), one gets

$$\frac{\partial \lambda}{\partial f_h} = \sum_i \frac{u_h v_i}{<\mathbf{u} \cdot \mathbf{v}>} \frac{\partial a_{ih}}{\partial f_h} = \frac{u_h}{<\mathbf{u} \cdot \mathbf{v}>} \sum_i v_i m_{hi}. \tag{16.8}$$

To proceed further, note that from Eq. (16.2) $m_{hi} = a_{ih}/f_h$, and that $\Sigma_i v_i a_{ih} = v_h$ (this follows from the definition of left eigenvector). Using these relationships in Eq. (16.8), noting that $<\mathbf{u}{\cdot}\mathbf{v}> = 1$ and substituting Eq. (16.8) into Eq. (16.7), one arrives at

$$\frac{\partial \lambda}{\partial z} = \sum_h u_h v_h \frac{\partial f_h}{f_h \partial z}. \tag{16.9}$$

Thus the relative effect of trait z on the reproductive rate (i.e., local fitness) in each habitat is weighed by the pooled reproductive value of individuals present in that habitat (Rousset, 1999a; see also Kawecki and Stearns, 1993; Holt, 1996b). The reproductive value tends to be smaller in sink habitats (Section 16.2), and sink habitats tend to harbor fewer individuals than sources. The evolution of trait z will thus be affected more strongly by its impact on performance in source habitats. If increasing z has a positive effect on performance in the sink, but a negative effect on performance in the source, the trait will evolve toward smaller values unless the positive effect in the source is considerably larger than the negative effect in the sink (Holt and Gaines, 1992; Kawecki, 1995; Holt, 1996b). Following this logic, one would predict that the optimal trait value will satisfy $\partial\lambda/\partial z = 0$ (Holt and Gaines, 1992).

Gene Flow versus Local Selection

The approach just given is simple and elegant and has been used to generate interesting predictions (e.g., Holt and Gaines, 1992; Houston and McNamara, 1992; Brown and Pavlovic, 1992; Kawecki and Stearns, 1993; Kawecki, 1995; Holt, 1996b). It is, however, problematic because it neglects genetic differentiation between populations, which may be substantial if the dispersal rate is low in relation to selection coefficients operating on individual genetic loci (Felsenstein, 1976; Chapter 7 of this volume).

The importance of accounting for genetic differentiation can be illustrated by considering adaptation to a black hole sink habitat (i.e., a habitat that receives immigrants but sends no dispersers back to the source). As the reproductive value in a black hole sink is 0 (see Section 16.2), the above approach would predict that an allele beneficial in the sink and deleterious in the source should never be maintained in the population. In contrast, explicit genetic models (Holt and Gomulkiewicz, 1997; Gomulkiewicz et al., 1999) demonstrate that, although eliminated deterministically from the source habitat, such an allele will be maintained in the sink if the local net reproductive rate of its carriers exceeds 1.

The effect of a passive dispersal rate on adaptive evolution in a source–sink system is another issue where qualitative discrepancies arise between the predictions of fitness sensitivity approach and explicit genetic models. In a two-patch model with a symmetric dispersal rate ($m_{12} = m_{21}$), the pooled reproductive value of the subpopulation in the sink ($u_h v_h$) typically increases monotonically with increasing dispersal rate (for a numerical example, see Fig. 16.4). This is largely because a greater dispersal rate shifts the spatial distribution of the population (Section 16.3), exposing a greater fraction of the total population to natural selection in the sink. The fitness sensitivity approach would thus suggest that high dispersal rates are most favorable and low dispersal rates least favorable for adaptation to a sink habitat (Holt and Gaines, 1992; Kawecki, 1995; Holt, 1996a).

However, the dispersal rate also affects the amount of gene flow between habitats, and thus the degree of genetic differentiation between source and sink habitats. The fitness sensitivity approach is likely to provide a reasonable approximation if gene flow is already strong enough to prevent any substantial

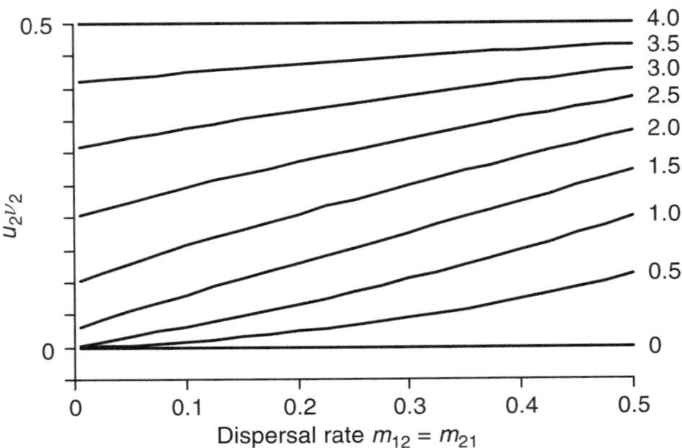

Fig. 16.4 The pooled reproductive value of individuals in a sink habitat, u_2v_2, as a function of dispersal rate and habitat quality. The model follows Eq. (16.1) and (16.2) with two patches and symmetric passive dispersal $m_{12} = m_{21}$; $f_h(n_h) = R_h/(1 + n_h)$. For all results $R_1 = 4$ is assumed; the different lines are for different values of R_2 indicated on the right. If $R_2 < 4$, habitat 2 is a sink.

genetic differentiation. In contrast, if dispersal is low, and thus gene flow restricted, some degree of local adaptation may be possible: alleles beneficial in the sink but deleterious in the source may increase in the sink while remaining rare in the source. In this case, increasing the dispersal rate will first of all result in greater swamping of the local gene pool in the sink by gene flow from the source. This negative effect of increased dispersal on adaptation to the sink is likely to outweigh any positive effect due to exposing a greater fraction of the population to the sink habitat. This argument predicts that, at least under some circumstances, the relationship between dispersal rate and the expected degree of adaptation to a sink habitat will be U-shaped rather than monotonic, with an intermediate dispersal rate being least favorable. Furthermore, for a given amount of gene flow, selection can maintain greater allele frequency differentiation between the habitats at loci with larger effects (Felsenstein, 1976). For that reason the range of dispersal rates over which the conditions for adaptation to a sink become more favorable with increasing dispersal should be greater when the adaptation involves loci with small effects (Kawecki, 2000).

These predictions are confirmed by the results of a polygenic model of evolution in a two-patch source–sink system described in Fig. 16.5. This model assumes a fitness trade-off between the habitats, mediated by a quantitative trait determined by up to eight additive loci each with two alleles. The total variability range of the trait is kept constant by adjusting the effects of single loci. The results of the genetic model (symbols) are compared to the predictions of an optimality model based on the fitness sensitivity approach (lines). The latter approach predicts that the mean fitness in the sink habitat should increase monotonically with the dispersal rate (lower line in each panel).

When the trade-off is mediated by eight loci, the optimality approach accurately predicts the outcome of the genetic model except for dispersal rates less than 0.05 (Fig. 16.5a). Only at such low dispersal rates can the local populations differentiate, which allows the local population in the sink to adapt locally. Genetic differentiation between the local populations causes the mean

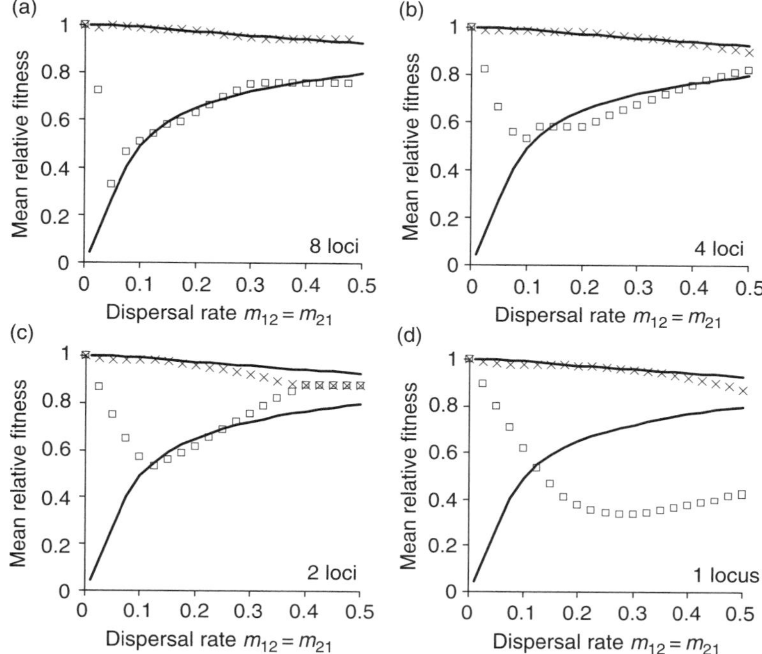

Fig. 16.5 Results of a genetic model of adaptive evolution in a source–sink system contrasted with predictions of an optimality approach. The model is based on Eq. (16.1) and (16.2) and assumes a trade-off in relative fitness across habitats, mediated by a quantitative trait z, which is under antagonistic directional selection in the two habitats. This is implemented by setting $f_h = R_h w_h(z)/(1 + n_h)$, where $R_1 = 4$ and $R_2 = 2$ and the relative fitness is $w_1 = 1 - z^3$ in the source and $w_2 = 1 - (1 - z)^3$ in the sink. Symmetric dispersal rates ($m_{12} = m_{21}$) are assumed. In the genetic model, trait z is determined by one to eight freely recombining loci with equal and additive effects, and codominance. The phenotypic effect of each locus is inversely proportional to the number of loci so that z always ranges from 0 (maximum possible adaptation in the source, zero fitness in the sink) to 1 (maximum adaptation to the sink, zero fitness in the source). Results were obtained using deterministic computer iterations (details in Kawecki and Holt, 2002) until an evolutionary equilibrium was reached; initial allele frequencies at all loci have been set to about 0.5 (slightly different among loci). Plots show mean relative fitnesses (w_h) in source (×) and sink (□) habitats as functions of the dispersal rate and the number of loci coding for trait z. Solid lines show relative fitnesses in the source $w_1(z^*)$ and sink $w_2(z^*)$ predicted with an optimality approach (the same for all panels). The optimal trait value z^* satisfies $\lambda = 1$, $\partial\lambda/\partial z = 0$, and $\partial^2\lambda/\partial z^2 < 0$, where the derivatives are evaluated at the equilibrium.

fitnesses in the two habitats to become less bound by the trade-off. As the number of loci that mediate the trade-off decreases, and thus the effect of each single locus increases, the range of dispersal rates permitting local adaptation in the sink increases. This causes the results of the genetic model to deviate increasingly from the predictions of the optimality approach; the minimum of the mean fitness in the sink habitat is shifted toward higher dispersal rates (Figs. 16.5b–16.5d). With only a single locus, the fit is very poor (Fig. 16.5d). An additional factor that reduces the mean fitness in the one-locus model is the segregational load — as the trade-off is convex, variance reduces mean fitness. It is also interesting to note that the two-locus version of the model predicts the same mean relative fitness in both habitats at high dispersal rates — at equilibrium the two loci are fixed for the alleles with opposite effects and no genetic variation remains. Analyzing the properties of equilibria in polygenic models goes beyond the scope of this chapter, but it should be kept in mind that details of the genetic system will affect the outcome of adaptive evolution in source–sink populations.

This example illustrates the importance of using explicit genetic models to study evolution in source–sink systems. The overall effect of dispersal on adaptive evolution in a sink habitat will depend on the relative importance of the demographic effect of dispersal and the homogenizing effect of gene flow.

Source–Sink Population Dynamics and Evolutionary Dynamics of Ecological Niches

In the model described above the mean relative fitness in the sink is typically lower than in the source, thus magnifying differences in habitat quality. Similar predictions have been reached by many published models. Alleles with a small positive effect on fitness in the source habitat will tend to be favored even if they have large negative effects in the sink (Holt and Gaines, 1992; Holt, 1996a; Kawecki, 2000). An allele beneficial in a black hole sink (no dispersal back to the source) may be eliminated deterministically even if neutral in the source (e.g., Haldane, 1948; Nagylaki, 1975; Slatkin, 1995; Holt and Gomulkiewicz, 1997). Source–sink populations are prone to accumulate mutations deleterious in the sink but neutral in the source (Kawecki et al., 1997). A quantitative trait affecting fitness may remain far from its local optimum in a sink habitat if the optimum in the source habitat is different (Garcia-Ramos and Kirkpatrick, 1997; Kirkpatrick and Barton, 1997). To summarize, natural selection is expected to maintain or improve adaptation in habitats, where the population is already well adapted, and be ineffective in improving adaptation to marginal habitats. This implies that ecological niches should usually be evolutionarily conserved (Holt and Gaines, 1992; Kawecki, 1995; Holt, 1996b).

This conclusion has also been reached in models in which habitat-specific parameters are symmetric so there are no a priori differences in habitat quality. A symmetric model will usually have a symmetric evolutionary equilibrium, at which the mean fitness in all habitats would be the same. However, such an equilibrium may be unstable, and even when it is stable, alternative asymmetric equilibria may exist; which equilibrium is reached will depend on the initial genetic composition of the population. Such alternative asymmetric and symmetric equilibria exist in a symmetric two-patch model by Ronce and Kirkpatrick (2001). If the population is initially well adapted to habitat 1 and poorly adapted to habitat 2, it will tend to remain so or may even evolve toward even greater adaptation in habitat 1 and reduced fitness in habitat 2. The reverse happens if the population is initially adapted to habitat 2. A symmetric equilibrium is only reached if the allele frequencies are initially intermediate so that the population is initially moderately well adapted to both habitats. Similarly, in a model of a species range evolving on an environmental gradient, source–sink population dynamics lead to evolution of a limited range, centered at the point along the gradient to which the population was initially best adapted (Kirkpatrick and Barton, 1997). This effect is augmented by character displacement caused by interspecific competition (Case and Taper, 2000). At the population genetic level it implies that the source–sink population structure generates epistasis among fitness effects of different loci, such that a positive effect of a particular allele on performance in a given habitat is augmented by a genetic background adapted to that habitat. Conversely, selection against an allele with a habitat-specific deleterious effect becomes weaker as alleles with similar effects increase in frequency, which may lead to a mutational erosion of fitness in a sink habitat (Kawecki et al., 1997).

Nonetheless, the prediction that ecological niches should be conserved evolutionarily is not absolute. A number of evolutionary changes of ecological niches have been directly observed, including host shifts in herbivorous insects or repeated evolution of tolerance of plants to high concentrations of heavy metals. This raises the question about environmental factors and properties of the organism, which make it more likely that a population will adapt to a novel habitat, which is initially a sink.

Dispersal rate and pattern are obviously of crucial importance. Given the tension between local adaptation and gene flow, one-time colonization of the novel habitat followed by complete isolation would seem most favorable. However, foundation of a persistent population by a few locally maladapted colonizers must be rare, spectacular exceptions like Darwin's finches notwithstanding. If the population initially performs poorly, it will likely become extinct before it has time to adapt (Gomulkiewicz and Holt, 1995), especially that a single colonization event will typically be associated with a bottleneck causing loss of heritable variation. If so, gene flow following the initial colonization may facilitate adaptation to the novel habitat by replenishing genetic variation (Caprio and Tabashnik, 1992; Gaggiotti, 1996; Gaggiotti and Smouse, 1996; Chapter 15). Finally, complete elimination of gene flow may be impossible. The above model suggests that high dispersal rates will often be more favorable for adaptation to a marginal habitat than intermediate dispersal rates, particularly if genes with small effects are involved (see also Kawecki and Holt, 2002). This conclusion is, however, contradicted by spatially explicit models of populations adapting to an environmental gradient (Kirkpatrick and Barton, 1997; Salathe and Kawecki, unpublished results), where high dispersal rates are most unfavorable for adaptation to sink habitats. Another model (Garcia-Ramos and Rodriguez, 2002) predicts a nonlinear relationship between dispersal and evolutionary invasions of novel habitats. It is not clear which of the differences in assumptions of these models were responsible for these different predictions.

Gene flow can occur through both sexes, but in species without paternal care, only female dispersal contributes to the maintenance of local populations in sink habitats. One would therefore expect that female-biased dispersal would be more favorable for adaptation to a sink habitat than sex-independent or male-biased dispersal. A genetic model assuming independent male and female dispersal rates confirms this intuition, although depending on the parameters, the conditions for adaptation to the sink may be least favorable under moderately rather than extremely male-biased dispersal (Kawecki, 2003).

Finally, Kawecki and Holt (2002) considered the evolutionary effect of the reverse source–sink structure, whereby an environment-imposed asymmetry of dispersal rates causes an "upstream" poorer habitat to act as an effective source and a "downstream" high-quality habitat as a sink (Section 16.2). In their model, selection tended to be more effective in the source habitat even if it was of lower quality than the sink habitat. They concluded that, assuming sufficient genetic variance, over evolutionary time the population should adapt to the upstream habitat at the expense of reduced fitness in the downstream habitat. In this case, source–sink population dynamics would thus promote an evolutionary shift of the ecological niche.

The effect of factors other than dispersal rates on adaptation to a sink habitat has not been investigated systematically. Fitness sensitivity analysis of a model described by Eq. (16.6) suggested that increasing the relative area of the sink habitat makes the conditions for adaptation to the sink more favorable, but only when the differences in habitat quality are not large (Kawecki, 1995). This conclusion still needs to be supported by a genetic model. Several models suggest that adaptation to a sink habitat should be more likely if few major loci are involved compared to many loci with small effects (e.g., Holt and Gomulkiewicz, 1997; Kawecki, 2000). Density dependence in the sink makes the conditions for adaptation to the sink habitat less favorable (Holt, 1996a; Gomulkiewicz et al., 1999). It is, however, not clear how general these predictions are. Most of them were derived from two-patch, spatially implicit models. In turn, spatially explicit models combining source–sink population dynamics and evolution have been based on the diffusion equation and infinitesimal quantitative genetic approximation (e.g., Kirkpatrick and Barton, 1997; Case and Taper, 2000). Future modeling of evolution in source–sink systems should combine spatially explicit and genetically explicit approaches.

Evidence for Maladaptation in Sink Habitats

The average reproductive success in a sink habitat is poor. The difficult part is to show that it is poor at least partially because of gene flow from source habitats. This has been demonstrated convincingly in only a few cases.

The best evidence for gene flow hampering adaptation in a sink habitat comes from the blue tit system described in Section 16.4. Populations in mainland southern France have a high breeding success in the deciduous habitat, whereas in the sclerophyllous habitat the breeding success and population density are low. However, on the island of Corsica, where the sclerophyllous forest is the dominant habitat the breeding success in that habitat type is higher than on the mainland, despite much higher local density (Blondel et al., 1992; Dias and Blondel, 1996). Furthermore, the breeding success of the Corsican population in small pockets of deciduous habitat on the island is poorer than in the sclerophyllous habitat; i.e., the deciduous habitat tends to act as a sink (Dias and Blondel, 1996).

It could still be argued that the difference in breeding success in the sclerophyllous habitat between Corsica and the mainland reflects different productivity of the sclerophyllous habitat on the island than on the mainland, rather than differential adaptation. However, the argument of maladaptation is also supported by data on breeding phenology. The breeding phenology is expected to be synchronized with the availability of caterpillars, which are the main food for the young, so that the peak demand of the brood for food coincides with the peak of caterpillar availability. This peak of food availability occurs about a month earlier in the deciduous than in the sclerophyllous habitat. Rather than showing a pattern of local adaptation, the breeding phenology of birds on the mainland does not differ between habitats and is synchronized with caterpillar availability in the source (deciduous) habitat. The birds in the sink (sclerophyllous) habitat lay their eggs almost a month too early and, as a consequence, suffer additional reduction of breeding success. The reverse holds in Corsica, where the breeding phenology is synchronized with caterpillar availability in the sclerophyllous habitat; birds breeding in small pockets of deciduous habitat lay their eggs much too late (Dias and Blondel, 1996). Thus in both regions the breeding phenology is well adapted to the source habitat and maladapted to the sink habitat. The difference in the laying date is genetic (Blondel et al., 1990), and it is unlikely that the lack of adaptation to the sink habitat is due to a lack of heritable variation for the laying date.

The conclusion about maladaptation of the blue tits in the sink habitats is thus supported both by an optimality analysis and by comparison of the performance of the local populations in patches of the same habitat located in different landscapes. A similar optimality approach has been used to demonstrate maladaptation of clutch size in great tits (Pettifor et al., 1988) and reproductive effort and offspring size of mosquitofish in a marginal population (Stearns and Sage, 1980). A spectacular counterexample is the repeated evolution of heavy metal tolerance by numerous plant species that colonized abandoned heavy metal mining sites and zinc-polluted areas around the bases of electricity pylons (e.g., Jain and Bradshaw, 1966; Coulaud and McNeilly, 1992; Alhiyaly et al., 1993; Nordal et al., 1999). Initially, these sites must have constituted small pockets of a sink habitat surrounded by a large source habitat. However, the colonizers were in a short time able to adapt to the toxic environment, despite continuous gene flow. Genetic studies reveal that in most cases, heavy metal tolerance in plants involves several major loci, although the contribution of minor loci is not excluded (e.g., MacNair, 1993; Schat et al., 1996). This finding is consistent with the prediction that adaptation to a sink habitat would be more likely if it involved few major genes rather than many genes with small effects (see earlier discussion).

Using reciprocal transplants of seeds, seedlings, and adults, Stanton and Galen (1997; see Section 16.4) have shown that snow buttercup populations living at early and late melting sites do not show a pattern of local adaptation to their respective sites. They do not seem to be differentiated genetically with respect to any fitness-related character. Instead, irrespective of the destination habitat, seeds originating from late melting sites are 25% less likely to germinate despite being only 8% smaller. One can speculate that in the absence of gene flow, local populations at late-melting sites would evolve toward producing fewer larger seeds, and that this change is prevented by the gene flow.

Research on the checkerspot butterfly (see Section 16.4) provides some evidence for alternative equilibria, similar to those predicted by Ronce and Kirkpatrick (2001). After the local populations in the original source habitat (forest clearings) had been wiped out by a frost, in several localities the original source–sink structure was not recreated. Instead, the population density became much higher in the former sink habitat (rocky outcrops), whereas individuals attempting to recolonize the former source habitat had poor reproductive success. Thus, the source–sink structure became reversed. This reversal was not due to an evolutionary change but was a consequence of phenological differences between the habitats: migrants from the outcrops arrived too late to reproduce successfully in the clearings. Once a resident population was established in a clearing, it expanded quickly (Boughton, 1999). Nonetheless, this example illustrates a potential for alternative source–sink equilibria.

A promising approach to study evolutionary consequences of a source–sink population structure would be to set up laboratory source–sink systems and let experimental populations evolve in them for generations. This "experimental evolution" approach has been applied successfully to other evolutionary questions, concerning, e.g., reproductive isolation (Rice and Salt, 1990), life history (Stearns et al., 2000), or learning ability (Mery and Kawecki, 2002). Although many studies involved experimental evolution in novel habitats, few included experimental populations evolving in heterogeneous environments, with different habitats connected by dispersal. Several of those studies focused on the role of environmental heterogeneity in the maintenance of genetic variation at allozyme loci (McDonald and Ayala, 1974; Powell and Wistrand, 1978; Haley and Birley, 1983) and quantitative traits (MacKay, 1981; Garcia-Dorado et al., 1991; Hawthorne, 1997). In those studies the habitats contributed equally to the total reproduction (soft selection). This design eliminated the relationship between mean performance in a habitat and this habitat's contribution to the total reproduction, which is an important characteristics of source–sink populations. Other studies were focused on the evolution of habitat choice (Bird and Semeonoff, 1986; Rice and Salt, 1990). Only a few compared adaptation to a novel habitat between lines exposed only to the novel habitat and lines exposed to both habitats (Wasserman and Futuyma, 1981; Mark, 1982; Verdonck, 1987; Taper, 1990). Because these studies were also concerned with habitat choice, the adults could choose the habitat for oviposition, and the amount of gene flow was not controlled. Verdonck (1987) let *D. melanogaster* populations evolve in cages containing two media: a standard medium and a medium supplemented with NaCl. The latter medium created a sink habitat, with low larval survival (although not an absolute sink). Despite the asymmetric gene flow, the experimental populations did evolve improved tolerance to NaCl, but to a lesser degree than control populations bred exclusively to the NaCl-supplemented medium. Thus, in this case, asymmetric gene flow slowed down, but did not completely prevent adaptation to a sink habitat. Taper (1990) maintained populations of the cowpea weevil (*Callosobruchus maculatus*) on a mixture of two host seed species, either on its own, or together with a competing species specializing on one of the hosts. In this latter treatment the competition pressure from the other species caused that host to become effectively a sink habitat. As predicted, competition with the specialist competitor caused the generalist species to become less well adapted to the host species used by the competitor and better adapted to the other host species (character displacement). These studies suggest that the "experimental evolution" approach has a great potential to provide insights into evolution in heterogeneous environments.

16.7 SOURCE–SINK METAPOPULATIONS

The concept of source–sink population structure emphasizes the effect of dispersal on the local population dynamics, whereas the metapopulation concept has originally been motivated by local extinctions and colonizations (Levins, 1968a). Source–sink population structure results from differences in habitat quality, whereas metapopulation structure reflects patchiness of the environment. Both concepts are concerned with the role of dispersal, but the source–sink structure requires much greater dispersal rates (i.e., greater connectivity of habitat patches), which would prevent habitat patches from remaining unoccupied. However, many real spatially structured populations are likely to be affected by both processes. First, in a "classic" metapopulation, immigration may significantly reduce the local extinction rate (rescue effect; Brown and Kodric-Brown, 1977; Chapters 4 and 14). It may also boost the local population size and the number of propagules it produces, thus potentially increasing the colonization rate. Second, some local populations (those in large habitat patches, or in the vicinity thereof) may show typical source–sink dynamics with negligible extinction probability, whereas the fate of others (those in small and more isolated patches) will be dominated by extiction–recolonization dynamics. This idea is explicit in metapopulation models of limits of species ranges (e.g., Lennon et al., 1997; Holt and Keitt, 2000).

The concept of source–sink dynamics can also be extended to extinction–recolonization dynamics by allowing the extinction rate or the contribution to the pool of colonizers to vary among patches (Chapter 4). Most empty patches would then be colonized by individuals originating from patches with more persistent and larger populations (sources). Colonizers from such source patches may maintain a significant level of patch occupancy in neighboring sink patch networks, in which otherwise extinction rate would exceed colonization. The mainland–island metapopulation model, in which all colonizing individuals originate in a permanent "mainland" population, is an extreme case, analogous to the black hole sink (Section 16.2). Such source–sink extinction–colonization dynamics is implicit in most structured or spatially explicit metapopulation models (e.g., Hanski and Gyllenberg, 1993; Hanski, 1994; Chapters 4 and 5). The distinction between source–sink dynamics at the level of extinction–colonization dynamics versus at the level of local population dynamics disappears in individual-based models (e.g., Wiegand et al., 1999).

16.8 CONCLUSIONS AND PROSPECTS

This necessarily incomplete review of ecological and evolutionary aspects of the source–sink population structure elucidates its importance for population dynamics, size, distribution, and persistence, as well as for the understanding of evolutionary dynamics of ecological niches and species ranges. The importance of source–sink dynamics for biodiversity conservation and pest management has been widely recognized.

As in many other areas of population biology, the development of theory has outpaced the accumulation of empirical data. In particular, direct experimental data addressing ecological and evolutionary consequences of source–sink population dynamics are scarce. One reason is the fact that most research motivated by the source–sink concept has concentrated on birds, mammals, and long-lived flowering plants. Experimental manipulations of spatial population structure (e.g., preventing dispersal, changing the amount of source or sink habitat) should be more feasible in insects or mites. Their shorter generation time would allow one to see the effects of those manipulations sooner. Some insects or mites are also ideal model organisms for laboratory source–sink systems. Such systems could be combined with the "experimental evolution" approach to study the evolutionary consequences of the source–sink population structure. This approach should be promoted.

Some applied aspects of the source–sink concept have also remained neglected. In particular, the concept has important implications for epidemiology and public health; the human population is a sink habitat for numerous parasites and pathogens (e.g., the rabies virus). The concept also applies to the dynamics of pathogens within the host's body, whereby some organs may be sources and others sinks for the pathogen. This has medical implications, as antipathogen drugs will be ineffective if they only target pathogens in sink organs. Some dangerous human diseases are caused by pathogens invading organs from which they cannot transmit; such organs are thus black hole sinks. Finally, our own population has a source–sink structure, with important economic and social consequences.

To summarize, although much progress has been made since Pulliam's (1988) seminal paper, much work remains to be done before we can fully understand the ecological and evolutionary consequences of the source–sink population structure.

17. METAPOPULATION DYNAMICS OF INFECTIOUS DISEASES

Matt J. Keeling, Ottar N. Bjørnstad, and **Bryan T. Grenfell**

17.1 INTRODUCTION

John Donne's famous line "No man is an island, entire of itself" has deep resonances for the dynamics of parasites. This is particularly true for microparasitic infections, such as viruses and bacteria, for which each susceptible host is a potential patch of favourable habitat. Propagules from infected "patches" can colonize others, followed by parasitic multiplication and "local" growth of the parasite population. Thus, at the scale of the host population, infectious dynamics bears strong analogies to metapopulation dynamics. Furthermore, host individuals are, more often than not, structured into local populations, within which contact among hosts may be very frequent and between which contacts may be less frequent. In this way, the spatiotemporal dynamics and persistence of parasites are determined at two scales: the *infrapopulation* scale (a local population scale; parasites within hosts) and the *metapopulation* scale (spatial and/or social aggregation of hosts). The spatiotemporal dynamics of infection in human and domestic systems are of particular academic interest because of the wealth of data combined with well-described natural histories.

As a result of the dual spatial scales of regulation, an extended metapopulation paradigm is central to infectious disease dynamics in two important

0-12-323448-4

ways. First, the metapopulation approach can help us understand disease dynamics at the different spatial scales. This topic is the main concern here, we use extensive data sets and realistic dynamic models to discuss the metapopulation dynamics of infectious disease. Second, there are important conceptual insights about the eradication by vaccination of infections to be gained from studies of the persistence of metapopulations (Nee, 1994; Grenfell and Harwood, 1997; Ovaskainen and Grenfell, 2003). This chapter therefore explores two main topics: (i) the analogies between the disciplines of ecology and epidemiology at the metapopulation-level and (ii) how metapopulation theory at a variety of scales can aid our understanding of epidemiological dynamics. We discuss these issues in the face of a set of detailed models and high-resolution space–time data of disease incidence.

Metapopulation-like disease dynamics occur whenever the environment, in this case the population of susceptibles, is sufficiently patchy that isolated clumps of suitable habitat exist. This is always the case at the microscale; each host is an island to be colonized and a resource patch to be depleted. At the macroscale, hosts are usually aggregated in local communities within which transmission is relatively frequent and between which infection spreads at a lower rate. Our dominant focus is on the metapopulation (macro)scale. To illustrate the key issues, we first introduce a simple epidemic model and then use this to illuminate the basic processes in the spatiotemporal dynamics of epidemics. Two distinct modeling scenarios are considered: a fully stochastic metapopulation where the individual level processes within each habitat (or community) are modeled explicitly and a spatially implicit (Levins-type) metapopulation where habitats are classified into a limited set of discrete classes. Both formulations have associated benefits and allow different insights into the dynamic processes in disease spread. We then revisit how metapopulation processes operate at various spatial scales (individual level, local, and regional epidemics). The resultant spatiotemporal dynamics are then illustrated through a series of case studies, which explore diseases metapopulation dynamics at the interface of models and data. We conclude with a section on fruitful areas for future work.

17.2 THE SIR MODEL FOR EPIDEMIC DYNAMICS

We focus here on microparasite infections (mainly viruses and bacteria), where direct reproduction of the pathogen in the host allows us to model disease dynamics by dividing the host population between compartments, classified by their infection status (Anderson and May, 1991). In contrast, macroparasitic helminth infections, where parasite burden matters, are much harder to model spatially (and not considered here), although strong analogies have been found between macroparasite and metapopulation dynamics (Cornell et al., 2000). The most studied microparasite system is the SIR model, where individuals are susceptible (S), infected (I), or recovered (R). This classification holds analogies to the "compartmental" Levins metapopulation models in which patches are classified as either occupied or empty (Chapter 4). As discussed in the next section, the "reversibility" of true metapopulations (such that local patch populations can become extinct, then reestablished by colonization) is a closer match to the SIS dynamics (susceptible–infectious–susceptible, such that recovered individuals do not possess immunity) associated with many sexually transmitted diseases (Anderson and May, 1991). In the SIR paradigm, susceptible individuals can catch the disease from contact with infected individuals; infected individuals then recover at a given rate, after which time they are assumed to be immune to further infection. This leads to the following set of differential equations:

$$\begin{aligned}\frac{dS}{dt} &= BN - \beta\frac{SI}{N} - dS \\ \frac{dI}{dt} &= \beta\frac{SI}{N} - gI - dI \\ \frac{dR}{dt} &= gI - dR \\ N &= S + I + R\end{aligned} \tag{17.1}$$

where B is the birth rate, d is the natural death rate, β is the transmission rate between infected and susceptible individuals, and g is the recovery rate. Many improvements and variations on this underlying framework have been developed successfully to describe the behavior of particular diseases and hosts (Anderson and May, 1991; Grenfell and Dobson, 1995; Hudson et al., 2002). In essence, Eq. (17.1) predicts a stable equilibrium level of susceptibles and infected, which is reached through a series of damped epidemics.

17.3 THE SPATIAL DIMENSION

Spatial structure and the aggregation of hosts into discrete patches can have dramatic effects on the dynamics of infectious diseases (May and Anderson, 1979; Grenfell and Bolker, 1998). We subdivide these effects into four main groups, which we consider with respect to the dynamics of one large, homogeneously mixed host population versus the dynamics of several smaller, more isolated ones.

Isolation and Coupling: A Simple Two-Patch Model

The most obvious aspect of spatial separation is the isolation of one or more local populations. The degree of isolation is controlled by the coupling between patches. In the absence of coupling, the dynamics in each patch are independent, and as the coupling increases, so does the correlation between them. We generally envisage coupling as the result of the movement of hosts; in such cases it is important to realize that the movement of both susceptibles and infecteds plays an equal role. We also note that two patches can be coupled directly due to the mixing of individuals in a third patch (e.g., people from two outlying towns might meet, and transmit infection, at a nearby large town). As we are concerned primarily with the spread of infection between human communities, we envisage coupling as the result of short duration commuter movements. For other host species, coupling could be generated by the permanent movement of hosts or simply the movement of pathogens between local populations (Keeling et al., 2001).

A key question for understanding the ensuing spatial dynamics is how to accurately allow for the movement of infection. Consider, first, a metapopulation of just two patches (Keeling and Rohani, 2002). In this model, individuals commute from their home population to the other patch, but return rapidly (Sattenspiel and Dietz, 1995). We label individuals by two subscripts such that S_{ij} is the number of susceptibles currently in patch j, whose home is patch i. We also assume that individuals from patch i commute at rate ρ_i and return at rate τ_i, independent of their infectious state. If we assume frequency-dependent transmission (de Jong et al., 1995; McCallum et al., 2001), then equations for the number of susceptibles and infecteds in each patch are given by

$$\begin{aligned}\frac{dS_{ii}}{dt} &= bN_{ii} - \beta S_{ii}\frac{I_{ii} + I_{ji}}{N_{ii} + N_{ji}} - dS_{ii} + \tau_i S_{ij} - \rho_i S_{ii} \\ \frac{dI_{ii}}{dt} &= \beta S_{ii}\frac{I_{ii} + I_{ji}}{N_{ii} + N_{ji}} - gI_{ii} - dI_{ii} + \tau_i I_{ij} - \rho_i I_{ii} \\ \frac{dS_{ij}}{dt} &= bN_{ij} - \beta S_{ij}\frac{I_{ij} + I_{jj}}{N_{ij} + N_{jj}} - dS_{ij} - \tau_i S_{ij} + \rho_i S_{ii} \\ \frac{dI_{ij}}{dt} &= \beta S_{ij}\frac{I_{ij} + I_{jj}}{N_{ij} + N_{jj}} - gI_{ij} - dI_{ij} - \tau_i I_{ij} + \rho_i I_{ii}\end{aligned} \tag{17.2}$$

where $i \neq j$. Here, equations for the recovered class (R_{ii} and R_{ij}) have not been given explicitly, as they can be calculated from the fact that $S + I + R = N$. If we allow the distribution of individuals to equilibrate, then $N_{ii}/N_{ij} = \tau_i/\rho_i$. Now, summing over all individuals whose home is patch i and assuming that time spent away from the home patch is relatively short compared to the disease dynamics, we get

$$\begin{aligned}\frac{dS_i}{dt} &= bN_i - \beta S_i\left[\sigma_{ii}I_i + \sigma_{ij}I_j\right] - dS_i \\ \frac{dI_i}{dt} &= \beta S_i\left[\sigma_{ii}I_i + \sigma_{ij}I_j\right] - gI_i - dI_i\end{aligned} \tag{17.3}$$

where σ are the conventional rates of coupling between populations. These are given by

$$\begin{aligned}\sigma_{ii} &= \frac{(1-\gamma_i)^2}{(1-\gamma_i)N_i + \gamma_j N_j} + \frac{\gamma_i^2}{\gamma_i N_i + (1-\gamma_j)N_j} \\ \sigma_{ij} = \sigma_{ji} &= \frac{(1-\gamma_i)\gamma_j}{(1-\gamma_i)N_i + \gamma_j N_j} + \frac{\gamma_i(1-\gamma_j)}{\gamma_i N_i + (1-\gamma_j)N_j}\end{aligned} \tag{17.4}$$

where $\gamma_i = \frac{\rho_i}{\tau_i}$ is the ratio of commuting to return rates and as such can be calculated from the expected amount of time an individual from patch i spends away from home $\left(= \frac{\gamma_i}{1+\gamma_i}\right)$. In the much simplified case where the population sizes and movement patterns are equal in both patches,

$$\sigma_{ii} = \frac{\gamma^2 + (1-\gamma)^2}{N} \qquad \sigma_{ij} = 2\gamma(1-\gamma) \tag{17.5}$$

The factor of two in σ_{ij} originates because coupling can come from either the movement of susceptibles or the movement of infecteds. Quadratic terms occur due to two individuals with the same home patch meeting in the away patch.

If we assume global coupling, such that commuter movement occurs equally to all other local populations irrespective of distance between them, then the n patch generalization is

$$\begin{aligned} \frac{dS_i}{dt} &= bN_i - \beta S_i\left[(1-n\sigma)I_i + \sigma\Sigma_{j=1}^{n} I_j\right]/N - dS_i \\ \frac{dI_i}{dt} &= \beta S_i\left[(1-n\sigma)I_i + \sigma\Sigma_{j=1}^{n} I_j\right]/N - gI_i - dI_i \\ \sigma &= 2\gamma(1-\gamma). \end{aligned} \tag{17.6}$$

where γ is again the ratio of the rate of commuting to a given patch to the rate of return. The proportion of time spent away from the home patch is now $\frac{n\gamma}{1+n\gamma}$.

These models [Eqs. (17.3) and (17.6)] illustrate that even the complex mechanistic movement of commuters can generally be expressed as a distributed force of infection from each infected individual across multiple local populations. Thus the complex patterns of human movements can be subsumed into a set of parameters σ, which specify the relative strengths of within-patch to between-patch transmission. These equations (17.3 and 17.6) are identical to those derived when the movement between local populations is permanent immigration rather than short-duration commuter travel (Kot et al., 1996; Smith et al., 2002) and to those formulated when the transmission of infection between different local populations is via wind-borne spread (Bolker, 1999; Park et al., 2001). Therefore, the simple and intuitive method of coupling local populations is applicable to a wide variety of diseases and interaction scenarios.

The aforementioned framework for studying the dynamics of a disease in a spatially structured population is founded on the premise of deterministic interactions and very rapid movement of commuters back to their home patch. Now we consider how this translates into a more realistic stochastic framework, where the population is individual based and events are assumed to occur at random; this is often termed demographic stochasticity. In such a framework, the coupled model [Eq. (17.3)], which has far fewer equations than the full mechanistic model [Eq. (17.2)], is a reliable approximation if the movement rate of individuals between the populations is rapid. However, as the movement rate slows (e.g., if commuters generally spend the entire day or longer away from home), the individual nature of the population plays an ever greater role. If just one individual is infected, then the level of coupling will be influenced greatly by whether that individual commutes. Figure 17.1 shows the distribution of cases caused by a single infectious case in their nonhome patch. Clearly the number of cases produced is highly dependent on whether

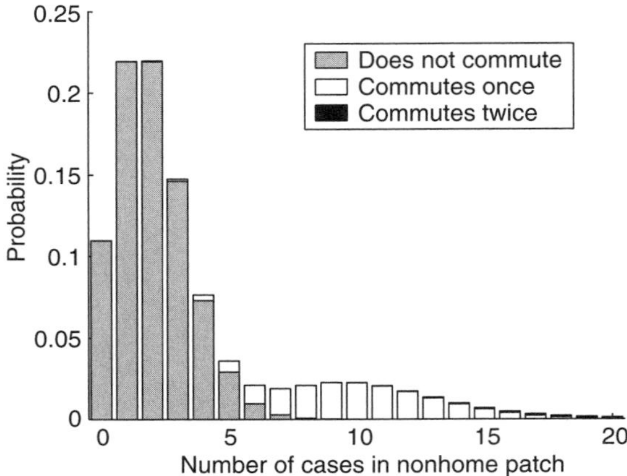

Fig. 17.1 Probability of a single infectious individual causing different numbers of secondary cases in their nonhome patch. If the infectious individual commutes, there is a dramatic shift in the expected number of cases. In this scenario, the infectious period is exactly 3 days, commuters always spend a full day away from home, and the basic reproduction ratio $R_0 = 30$. The susceptible population is considered to be very large.

the infectious person commutes, although even when they remain in their home patch the disease can still spread due to the movement of susceptibles.

In stochastic metapopulation models, therefore, when the level of infection is low and the commuter time is of the same order as the infectious period, we must be very cautious in our use of approximations to the true mechanistic dynamics. The occasional rare event, when the infected individual commutes, can have large repercussions and leads to a far wider range of outcomes than would be expected from a stochastic version of the simple coupling model [Eq. (17.3)].

Stochastic and Seasonal Forcing

The main manifestation of random fluctuations explored in epidemic theory is the impact of demographic stochasticity. As for conventional metapopulations, a major impact of demographic stochasticity is on the extinction rate, here of epidemics in small populations (see next section). However, due to the inherent oscillatory nature of epidemics, *stochastic forcing* of epidemics can give rise to regular or irregular cycles. This issue has strong parallels with the recurrent debate in ecology on the relative impact of noise and deterministic forces on dynamics (e.g., Bjørnstad and Grenfell, 2001). In epidemiology, the interaction between deterministic nonlinearity and forcing has been most studied in terms of the perturbing forces, which may maintain strong recurring epidemics of measles in the prevaccination era; these epidemics are predicted to dampen to an equilibrium by simple deterministic nonseasonal models (May and Anderson, 1991). The seminal work here is by Bartlett (1956, 1957), who showed that both stochastic forcing or the marked seasonality in transmission due to the aggregation of children in schools could excite the measles oscillator into sustained epidemics. In the case of childhood diseases, seasonality appears to play a major role in the maintenance of measles cycles (Schenzle, 1984; Bjørnstad et al., 2002; Grenfell et al., 2002).

In general, most observations and stochastic model results agree that the average number of cases in a population scales linearly with population size ($\bar{I} \propto N$). The variance in the number of cases, however, can be approximated by a power law, with an exponent between 1 and 2, [$var(I) \propto N^{2\alpha}, \frac{1}{2} \leq \alpha \leq 1$] (Keeling and Grenfell, 1999; Keeling, 2000a). This underlines how large populations have relatively lower standard deviations in the number of cases compared to the mean [$SD(I)/I \propto N^{\alpha-1}$] and thus behave more like deterministic systems. These subtleties lead to nontrivial consequences of spatial subdivision of hosts on epidemic dynamics.

Work on whooping cough illustrates the dramatic influence of demographic stochasticity on epidemic dynamics due to the intricate interaction between stochasticity and nonlinearity (Rohani et al., 2000; Keeling et al., 2001; Rohani et al., 2002; see also Rand and Wilson, 1994). As predicted by standard theory, demographic stochasticity becomes increasingly important in small populations because one individual in smaller populations is a comparatively larger fraction of the entire population and therefore each stochastic event represents a relative larger change to the susceptible and infected proportions — one infectious individual in a small village is likely to infect a greater proportion of the population than one infectious individual in a large city. Thus we can illustrate the complex roles of noise on epidemics by considering the stochastic dynamics of whooping cough across a range of host community sizes (Rohani et al., 2000). Small model populations are seen to display 4-yr cycles driven by stochastic resonance at, or close to, their natural frequency, whereas large populations possess more annual dynamics constrained by the deterministic attractor (Fig. 17.2).

We can extend the concept of power-law variances to a metapopulation with n local populations. If the level of coupling between the populations is weak, such that the dynamics are almost independent, then the average number of cases across all local populations is the same as for one large population. For independent populations, the variance of the sum is the sum of the variances; hence, an increase in the number of local populations causes a linear increase in the total variance. However, a similar increase in the population size of one large patch causes a faster than linear rise due to the scaling of the power law. Naively, then, one could be tempted to conjecture that by breaking the habitat into multiple (independent) patches, one would effect a decrease in the relative variability observed in the aggregate dynamics. This is analogous to the statistical averaging discussed as the "portfolio effect" in community ecology (Tilman, 1999). However, in practice, the significant levels of coupling and the complex interactions between nonlinear transmission dynamics and demographic stochasticity mean that no such general assertions are possible. Wilson and Hassell (1997) have shown that such complexities also take place in other host–enemy systems.

Extinctions

A major effect of demographic stochasticity in small populations is the tendency for chance extinctions. This behavior is highlighted in Bartlett's classical work on measles, where the number of fadeouts (or localized extinctions) decreases exponentially with population size (Bartlett, 1957). Bartlett (1957,

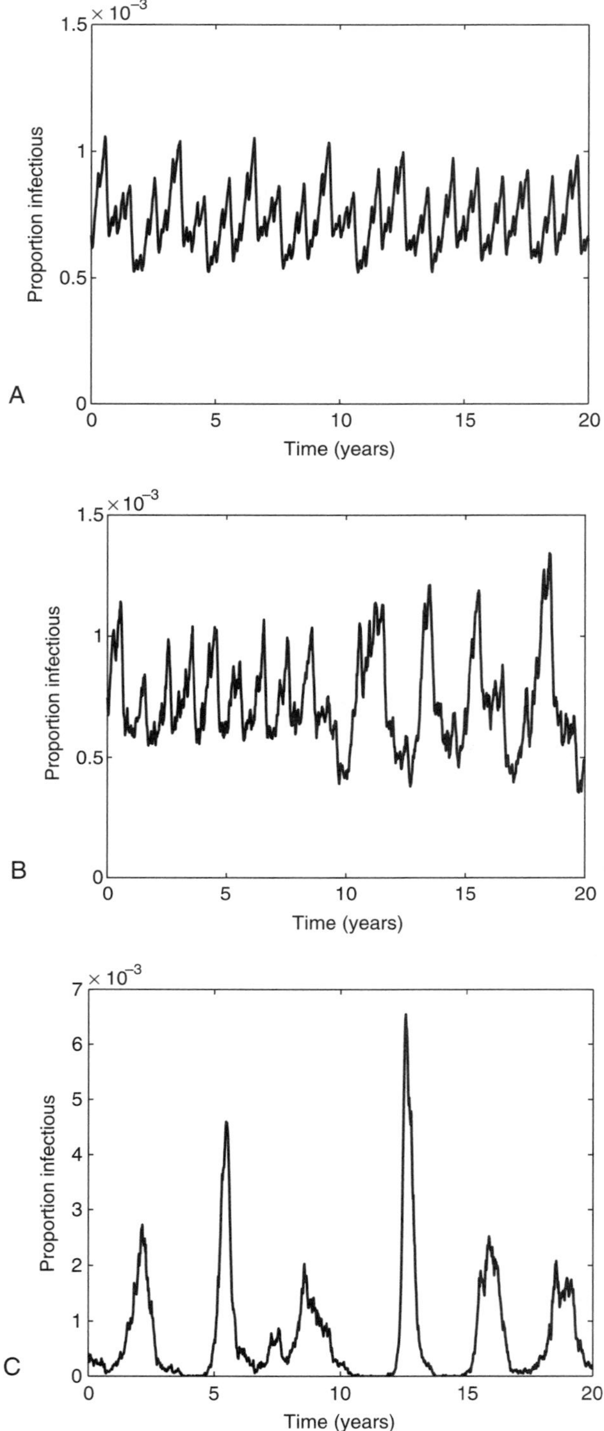

Fig. 17.2 The dynamics of a stochastic SEIR model (which contains a noninfectious incubation or exposed class) for whooping cough for three population sizes (graph A is with a population size of 10 million, graph B is 1 million and graph C is 100,000). At a population size of 10 million, the dynamics are very close to the deterministic attractor, whereas for populations of 100,000, stochastic forces dominate and 3- to 4-yr epidemic cycles are observed.

1960) identified a critical community size (CCS) for disease persistence, such that above this population size the disease is endemic and is rarely subject to stochastic extinctions. Interestingly, this emergent critical community size is remarkably robust, and similar values of around 300,000 for measles occur for communities in England, the United States, and isolated islands (Bartlett, 1957, 1960; Black, 1966). The CCS is arguably the best empirically documented local extinction threshold in metapopulation biology (Grenfell and Harwood, 1997; Keeling and Grenfell, 1997).

When considering the regional persistence of an infectious disease across several small populations versus one large population, there are conflicting elements. Small isolated populations exhibit more frequent local extinctions than large populations. However, in a metapopulation consisting of many small populations, extinction (or eradication) at national or regional scales requires the concerted collapse of all local epidemics; in contrast, regional eradication, where there is just one large population, only requires a single extinction event. In classic metapopulation models, coupling enhances persistence through local recolonization, but erodes persistence through synchronizing the local dynamics (Chapter 4). For epidemic metapopulations, the relationship between regional persistence and coupling is complex and depends critically on the disease parameters and the demography and movement of the hosts. Thus, it is far from obvious whether one large patch or several smaller patches have the greater extinction rate.

The effects of coupling between local populations on global disease eradication have received much attention due to the interesting trade-offs that arise (Keeling, 2000b). If the coupling is very small, then the local populations act independently and there is little or no chance of the disease being reintroduced from another local population; there is no rescue effect. Thus using the language associated with Levins metapopulations, the local populations have a large extinction rate and a very low probability of colonization. If the coupling is very large, then the local populations act like one large homogeneously mixed population and thus stochastic effects may lead the entire metapopulation to extinction. In disease models, heterogeneity plays an important role as low levels of infection allow the susceptible population to recover, which in turns promotes future cases. Persistence is therefore maximised at intermediate levels of coupling: there is sufficient coupling to allow recolonization and sufficient variability between patches for the metapopulation to absorb stochastic fluctuations.

The global eradication (extinction) of disease metapopulations is obviously a key aim in public health terms. This is generally investigated using computer simulations, as analytical techniques have difficulty dealing with the complexities of spatial heterogeneities and the stochastic dynamics that permeate the problem. Figure 17.3 illustrates the aforementioned principles using extinction probabilities for a spatial SIR epidemic model. When the coupling is very low such that recolonization is rare, local extinctions (at the local population level, Fig. 17.3A) are common, as are global extinctions (at the metapopulation level, Fig. 17.3B). When the coupling is high, local populations rarely go extinct. However, because of the synchrony induced by coupling, rescue effects are less effective as all epidemic declines are aligned and therefore prone to simultaneous local extinctions. Hence the global extinction

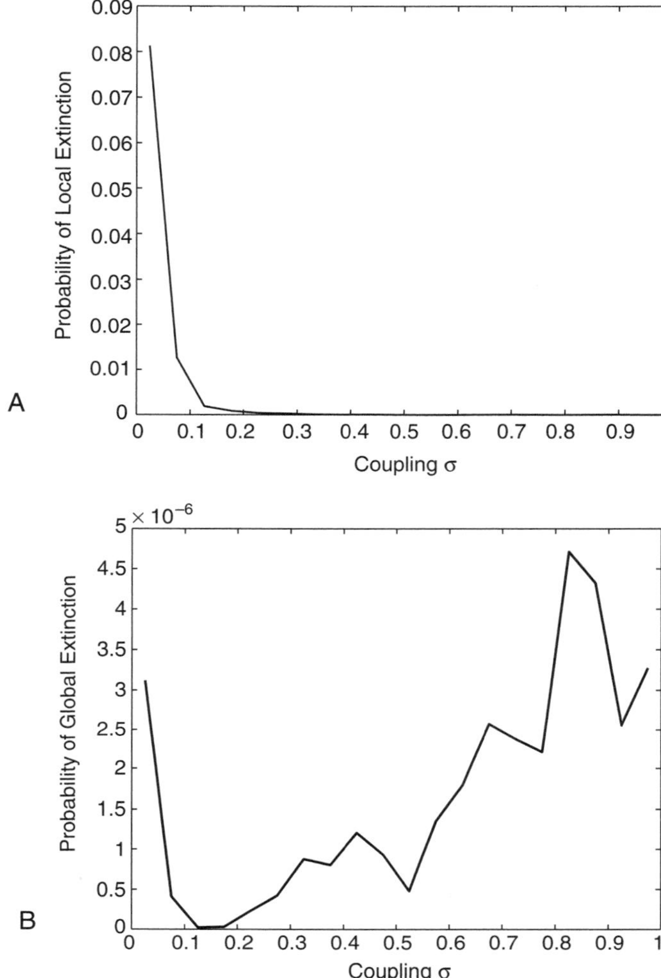

Fig. 17.3 For a stochastic SIR model with 20 coupled local populations, graph A gives the probability of local population level extinctions, whereas graph B is the probability of a global metapopulation level extinction across all 20 local populations. (Infectious period is 13 days, local population size is 30,000, basic reproductive ratio $R_0 = 14$, birth rate is 5.5×10^{-5} per person per day).

probability is again enhanced. There is a clear intermediate minimum for which disease persistence is the greatest. This compares with the diminished persistence of classic metapopulations when embedded in a correlated landscape (Harrison and Quinn, 1989). It is still an open problem to relate disease characteristics, host demography, and coupling to the extinction risk at the metapopulation scale for a wide range of microparasites (Keeling, 2000b). Changes in coupling between populations due to social changes and ease of long-distance travel have important implications for disease extinction and eradication — this is a major question for the theoretical epidemiology of the future.

Dynamic Heterogeneity

In this context, "heterogeneity" is taken to mean the total degree of variation (or asynchrony) between epidemic dynamics at different locations. This includes variation due to asynchronous timing of epidemics at different locations, as well as heterogeneities in local dynamics due to differences in local host demography. Heterogeneities are thus a fundamental difference between spatial and nonspatial processes. As outlined earlier, heterogeneity is promoted by stochasticity but is reduced by coupling. The level of heterogeneity further depends on the relative and absolute differences in host community size, movement, and demography, as well as subtle characteristics of the transmission dynamics. To better understand the causes and consequences of such heterogeneity, we contrast a range of metapopulation models. The simplest assumes identical demography within each local population, deterministic dynamics and global coupling, so that the interaction is the same between all patches. Under these simplifying assumptions, and even for very low levels of coupling, we generally observe phase locking where the interaction between patches leads to complete synchronization of each local epidemic and zero heterogeneity.

When the internal dynamics are stochastic, the spatial dynamics are more complex. Coupling still acts to synchronize the dynamics by homogenizing the level of infection in each local population. In contrast, stochasticity acts to separate the dynamics as different populations experience different random events. Figure 17.4 shows the correlation in disease incidence between two stochastic local populations for various levels of coupling. When coupling is low, the two populations are unsynchronized and the correlation is zero; however, as coupling increases, the stochastic oscillations are increasingly correlated. As seen in Fig. 17.4, coupling has a greater effect for larger populations (results on populations of more than 10,000 did not differ significantly), which is primarily due to the diminished effect of stochasticity (Grenfell et al., 2002).

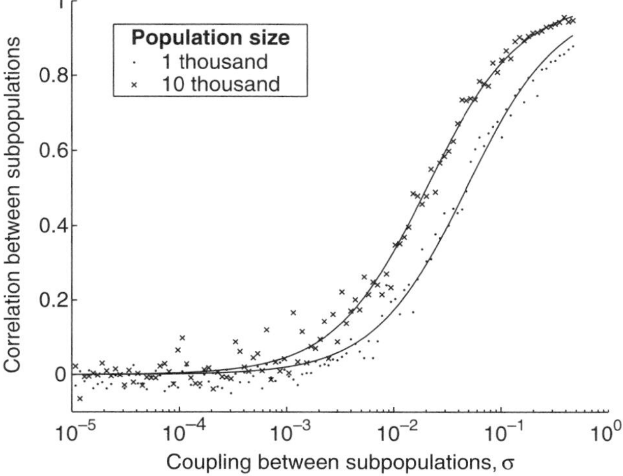

Fig. 17.4 Correlation between disease levels in two stochastic SEIR local populations. Two different population sizes are simulations; 1000 individuals and 10,000 individuals. Fitted curves are of the form $\sigma/(\xi + \sigma)$, which is based on theoretical predictions (Keeling and Rohani, 2002). Disease parameters match those of whooping cough, although seasonal forcing is ignored.

In the presence of seasonal forcing, space–time dynamics are more involved. In general, unforced stochastic epidemics can peak at any time of the year, whereas seasonal forcing usually constrains the epidemic cycle. Therefore, if seasonality tends to force the epidemics into a rigid annual cycle, populations appear partially or fully synchronized without the need for strong coupling — this echoes the operation of the Moran effect in ecology (Moran, 1953; Grenfell et al., 1998). If the epidemic period is multiannual, however, epidemics can become locked out of phase (Henson et al., 1998), as was the case for the 2-yr epidemics of measles in Norwich and London during the 1950s (Grenfell et al., 2001), for which high levels of coupling may be required to regain synchrony. In this latter case, greater levels of stochasticity and weaker attractiveness of the cyclic attractor can also help synchronization as there is a greater chance that a population will switch phases (as indeed happened for the Norwich measles epidemics during the 1960s).

Two other factors influence the synchrony and hence the level of heterogeneity. The first is the presence of inherent differences between the local populations, such as different host reproductive rates. In general, such heterogeneities will act to decorrelate the dynamics, as different populations will obey different underlying models. This was the case for the measles epidemics in Liverpool and Manchester during the prevaccination era when the higher birth rates in Liverpool led to annual epidemics whereas the rest of England and Wales was predominantly biennial (Finkenstädt et al., 1998; Grenfell et al., 2002). Heterogeneities in the size of local host populations may have contrasting effects. The presence of one large population may act to synchronize the behavior of many surrounding small populations; in such a scenario of mainland–island epidemic metapopulation, coupling to the large population is a main synchronizing force across the whole metapopulation (Grenfell et al., 2001; see also Fig. 17.10). Local, rather than global, coupling may furthermore lead to epidemic traveling waves, although strong seasonal forcing can again counteract this. Such epidemic waves are a common feature of many spatially explicit models of natural–enemy interactions (rabies, bubonic plague, parasitoid–host systems) and have been confirmed in both ecological and epidemiological systems (Nobel, 1974; Grenfell et al., 2001; Ranta et al., 1997; Smith et al., 2002; Bjørnstad et al., 2002).

The presence and absence of spatial synchrony can play important roles in the dynamics and persistence of disease. As discussed earlier, heterogeneity can vastly increase the long-term persistence of a disease through local recolonisation and repeated rescue events. This effect is heightened if there are demographic or size differences between the populations. Heterogeneities at a smaller scale can also alter the observed aggregate dynamics. Case reports are often aggregated at the community or regional level; however, such data may be composed of multiple smaller epidemic within wards or population cliques. As these subepidemics are likely to be somewhat out of phase, the aggregate picture is of a slower, longer duration epidemic. Figure 17.5 shows a simple example of this, while each localized epidemic (gray) is of short duration, the aggregate (black) is far longer with a much diminished epidemic peak.

Throughout the examples that follow, we refer continually to the aforementioned four basic elements of spatially structured disease dynamics: isolation, stochasticity, extinction, and heterogeneity. We discuss how

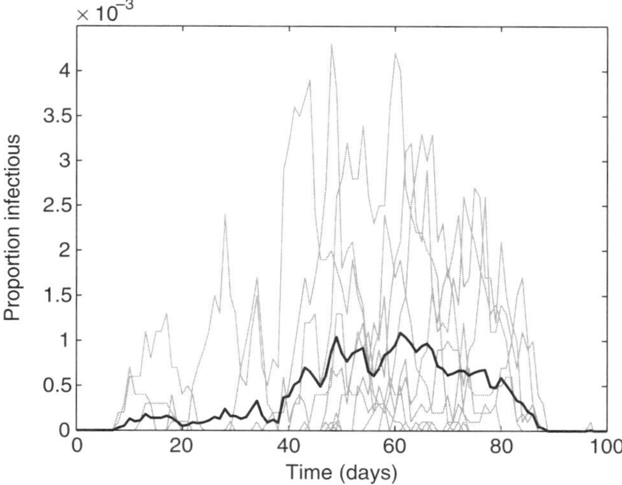

Fig. 17.5 The dynamics of 20 coupled SIR populations (gray) and the aggregate epidemic (black). (Population size is 10,000, $R_0 = 10$, infectious period is 1 day, coupling $\sigma = 0.0025$).

metapopulation models (both full stochastic metapopulations and the simpler Levins metapopulations) can be used to represent additional spatial structure and the insights that can be gained from such idealized models. We begin by studying the implications of spatial structure at a range of scales, starting with the individual and working up to the community or even country level. General conclusions of this exercise are illustrated in subsequent case studies.

17.4 DISEASE METAPOPULATIONS AT DIFFERENT SCALES

Individual Level

The standard SIR equations can be explored by considering each individual host as a patch using a modified form of the Levins metapopulation formulation. The mechanistic approach would model the interaction of the disease and the host's immune system, leading to models comparable to those used in the study of macroparasites, which classify hosts in terms of their burden of parasites (Anderson and May, 1991). However, the more classical models where hosts are described as unoccupied but suitable habitat (susceptible), occupied (infected), or unoccupied but exhausted habitat (recovered) have a much closer analogy to standard metapopulation models for successional habitats. From this perspective, birth and deaths correspond to the creation and destruction of habitat. Strictly speaking, SIR-type dynamics cannot correspond to "true" Levins metapopulation dynamics at the individual level because recovered patches cannot regain infection. However, most sexually transmitted diseases exhibit SIS dynamics; that is, once an infected individual recovers, it is once again susceptible to infection. In such cases there is a direct correspondence between the standard Levins model and the SIS equations (Ovaskainen and Grenfell, 2003).

The Levins metapopulation has an extensively developed theoretical armoury that can be applied to the description and understanding of disease dynamics (Chapter 4). Sexually transmitted diseases, which are characterized by a low transmission rate and long infectious period, can be thought of as poor colonizers, but a persistent species with low local extinction rates. In contrast, childhood diseases, which are short lived and transmitted rapidly, can be conceptualized as good colonizers that exploit the local resource rapidly, driving themselves extinct. This concept may be extended fruitfully to consider the competition between cross-resistant strains of disease (Gupta et al., 1998), in which case low transmission rates (poor colonizing ability) can be offset by good competitive ability within a patch.

The compartmental classification of "habitat" needs not necessarily operate at the level of the individual. Work on the 2001 foot-and-mouth epidemic in Great Britain considers each farm an epidemic unit ("patch") to be classified as susceptible, exposed, infectious, or removed (Keeling et al., 2001; Ferguson et al., 2001). Patch "removal" was in this case through massive culling of all potential host animals on a farm. In modeling the outbreak, the within-farm epidemic was ignored in favor of this Levins-type classification. Despite this great simplification, these models predicted the course of the epidemic with great accuracy at the regional level, justifying the approximation.

Within-Community Metapopulations

Many communities, especially large ones, can be subdivided into various weakly interacting components. This subdivision may take place along social, age, or simply spatial boundaries, but inevitably there are many factors that prevent the random mixing of the population and therefore break the assumptions underlying standard models. This necessitates the use of metapopulation-type equations to model the dynamics of these partially separated components.

Regional Metapopulations

The epidemic dynamics at regional or country level begs the use of metapopulation concepts. Here each local population represents a community, and hence there is a one-to-one correspondence between the scale at which the model operates and the available data. Exploring regional dynamics brings two main challenges: understanding the detailed consequences of demographic heterogeneities between the communities and analyzing the epidemic coupling between communities on real landscapes. The scientific development of the latter issue mimics the succession from the naive to more realistic models of metapopulation theory (see Chapters 4, 5, 20, and 22). The traditional models of identical local populations, with low levels of global coupling, have given way to models with distance-based coupling rates. Such models are slowly being replaced with models embracing heterogeneous patch sizes (with obvious parallels to the current generation of incidence function models and stochastic patch occupancy models as described throughout this book). However, for a complete understanding of epidemic metapopulations, it is becoming increasingly clear that a deeper knowledge of the complex geometries of the "transportation networks" for the infections is required (Cliff and Haggett,

1988). This is likely to provide an exciting area for future research with great theoretical, empirical, and statistical challenges.

17.5 CASE STUDIES

Prevaccination Measles in England and Wales

Of all infectious diseases, the dynamics of childhood microparasites, such as measles and whooping cough, are arguably among the best understood with respect to both local and regional dynamics. In particular, the rich data base and the comparatively simple natural history of measles have made this the prototypical system in the study of spatiotemporal dynamics of infectious disease (Anderson and May, 1991; Bartlett, 1957; Cliff et al., 1993; Grenfell and Harwood, 1997; Keeling and Grenfell, 1997; Grenfell et al., 2001, 2002; Keeling et al., 2001; Bjørnstad et al., 2002). Measles, along with other childhood infections, was made a notifiable disease in the United Kingdom in 1944. This resulted in the collection of weekly reports in 1400 communities in England and Wales through to the present. As such, this is likely to represent the longest and most detailed record of any epidemic metapopulation. Not surprisingly, these data have been studied extensively from epidemiological, mathematical modeling, and time-series analysis perspectives. Due to its very high basic reproductive ratio, $R_0 = 17$, most children were infected with measles (before the onset of mass vaccination campaigns in the late 1960s) with an average age of infection around 4–5 yr. Before mass vaccination was introduced in the United Kingdom, measles displayed predominantly biennial dynamics, with a major epidemic in odd years (Fig. 17.6A) (for further details see Bjørnstad et al., 2002; Grenfell et al., 2002).

Demographic stochasticity plays an important role in the dynamics of measles in small communities. This arises from the individual nature of populations (the fact that there must be whole numbers of cases) and the probabilistic nature of events, such that transmission of infection in particular occurs by chance. Stochasticity has two basic effects on the patterns of disease behavior: it can push trajectories away from the deterministic attractor such that transient dynamics play a more major role and it can lead to chance extinctions due to the random failure of chains of transmission (Figs. 17.6B and 17.6C). The role of patch size (host population size) on epidemic extinction rates is illustrated wonderfully in the public records. Extinction rates appear to decay exponentially with host population size so that above the critical community size of around 300,000 hosts, extinctions are rare (Bartlett, 1957). This pattern of size-based extinctions and recolonizations warrants interpretation from the metapopulation point of view.

In the prevaccination era childhood diseases, such as measles or whooping cough, were spread predominantly by school children mixing within the primary school environment. In this respect, the host populations can be thought of as subdivided into school catchment areas. Considering an average primary school has an intake of around 150 children in each year, then each school serves a population of around 10,000; this determines our basic unit of subdivision. A Levins-type metapopulation model (global dispersal, no local

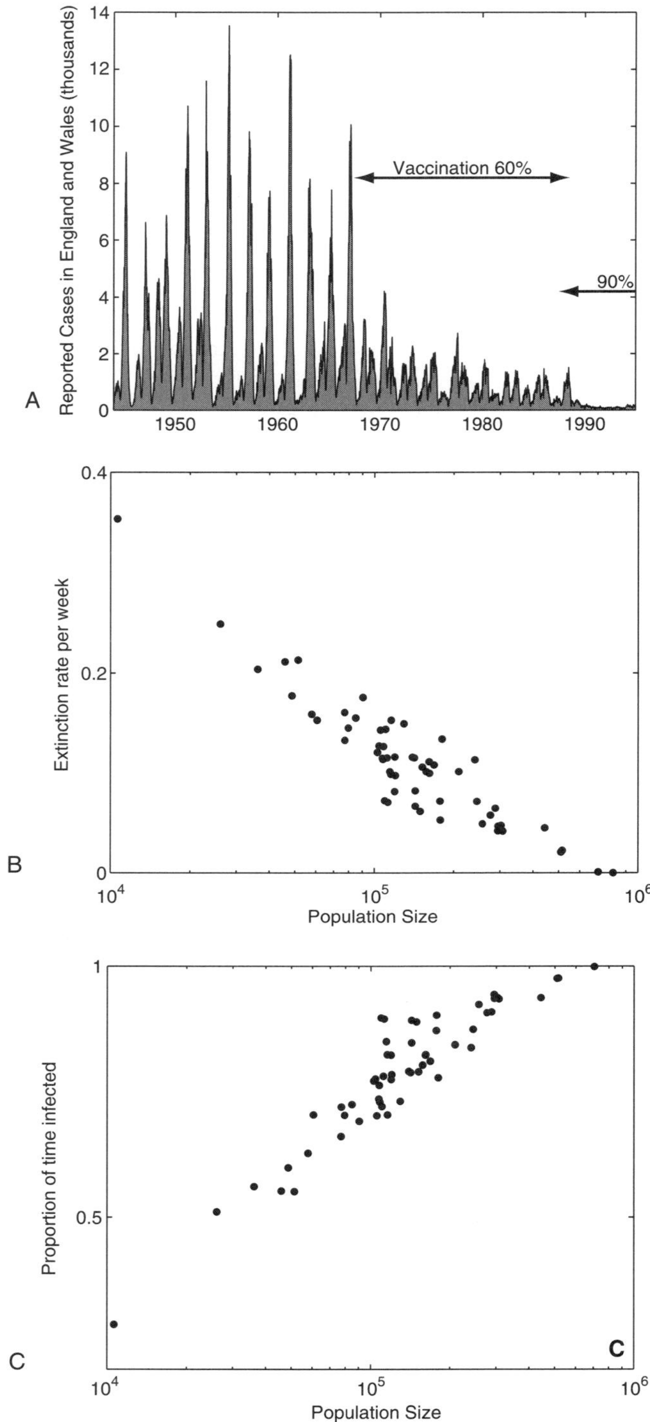

Fig. 17.6 Dynamics of measles in England and Wales. (A) Reported cases from 1944 to 1999 showing levels of vaccination. (B and C) Average rate of extinction and the proportion of time with recorded cases for measles in 60 towns and cities in England and Wales. Data taken from 1944 to 1968 before mass vaccination was begun. There are clear relationships with population size, such that large populations rarely suffer stochastic extinctions.

dynamics, etc.), which splits each community into school catchment areas of 10,000 people, motivates the following model for the proportion of "diseased" local populations (D) in the city:

$$\frac{dD}{dt} = -eD + cD(1 - D) + i(1 - D) \qquad (17.7)$$

where e is the extinction rate, c is the colonization rate from other catchment areas (local populations) within the city, and i is the rate at which external imports of infection arrive in a local population. This corresponds to an ecological model with migration both from a permanent mainland population and among local populations. If a metapopulation is composed of N such local populations, then the probability P_n that n catchment areas are infected is given by

$$\begin{aligned}\frac{dP_n}{dt} &= enP_{n+1} + c(n - 1)(N + 1 - n)P_{n-1} + i(N + 1 - n)P_{n-1} \\ &\quad -[en + cn(N - n) - i(N - n)]P_n\end{aligned} \qquad (17.8)$$

with the equilibrium solution,

$$P_{n+1} = \frac{(i + cn)(N - n)}{en}P_n \text{ such that } \sum_{n=0}^{N} P_n = 1 \qquad (17.9)$$

This provides easy calculation of the long-term distribution of disease within the metapopulation, the associated extinction rates ($= eP_1$), and the proportion of time the disease should be present ($= 1 - P_0$). We may compare these theoretical predictions with data on measles incidence. Figure 17.7 shows the resultant comparison as the total population size and hence the number of local populations increase. It thus appears that Levins-type local patch dynamics can reproduce the pattern in data.

The Levins approach, which completely ignores the local epidemic dynamics, works surprisingly well for predicting the presence or absence of the disease across a range of population sizes. However, it breaks down if we wish to predict the *level* of infection within the population. This is for three reasons. First the local dynamics of different cities are highly correlated, such that the level of infection is an increasing function of the colonization rate, which in turn is proportional to number of occupied patches (Fig. 17.8). Second, in the absence of infection, the level of susceptibles increases; any ensuing epidemic is therefore critically dependent on the local number of susceptibles, which in turn is dependent on the time since the last epidemic. This induces a level of memory to the local dynamics that breaks with the underlying Markovian assumption of the Levins model. Finally, the distribution of infection (Fig. 17.8) does not conform to the Levins-type metapopulation ideal, which assumes that local prevalence should be bimodal and dominated by a zero and a nonzero equilibrium level. For a detailed understanding of the population dynamics, we need to consider a metapopulation with detailed stochastic dynamics within each patch (Swinton et al., 1998).

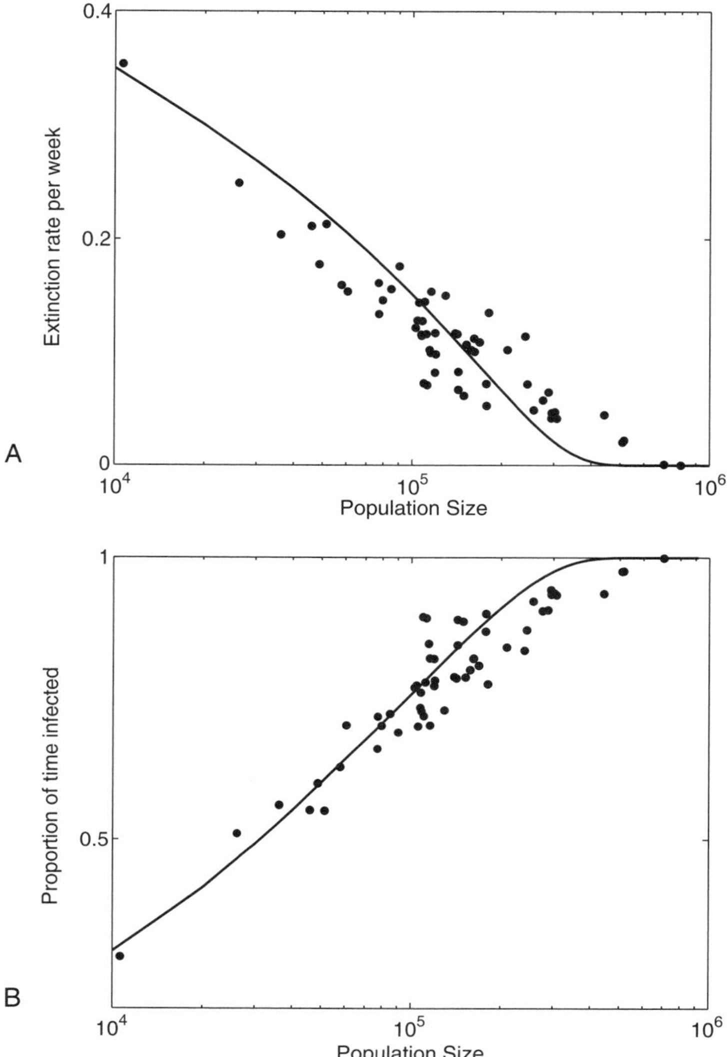

Fig. 17.7 Rate of extinction (A) and the proportion of time with infection (B) from a metapopulation model with local populations of 10,000 individuals. The extinction rate of each local population is $e = 0.35$, the colonization rate is $c = 0.01$, and the rate of external imports is taken as $i = 0.15/\sqrt{(N)}$. This local population level of imports means that the metapopulation level of imports scales with the square root of population size, as was conjectured originally by Bartlett (1957) and supported by measles report data from England and Wales. All timescales are measured in weeks to correspond to the aggregation of case reports. Theoretical results (line) compare well to measles data from England and Wales (dots).

We explore the breakdown of the Levins-type metapopulation model through a comparison with the full stochastic analogues across a variety of scenarios. A range of *single species* models, with a variety of forms of density dependence, have been explored elsewhere (Keeling, 2000b). These conform to Levins-type metapopulation behavior when (a) the distribution of population sizes falls into two distinct classes, extinct and close to carrying capacity, and

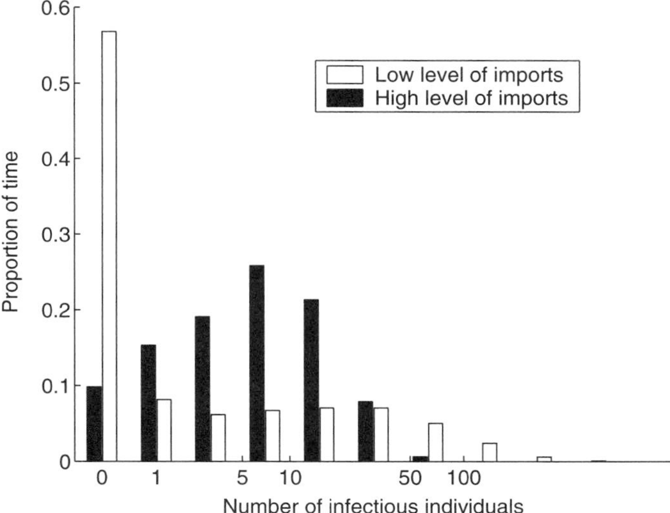

Fig. 17.8 Distribution of the number of infectious individuals in a stochastic SIR model of 10,000 with measles-like parameters ($R_0 = 17$, infectious period is 13 days) for two different levels of imports. The two levels of imports correspond to the effects of being part of a metapopulation where either all or few of the other patches are infected and close to the carrying capacity. Clearly the local population level dynamics are influenced strongly by the global metapopulation behavior.

(b) the carrying capacity and extinction rate are not significantly affected by the number of extinct patches. For such single-species systems the Levins framework is the ideal tool for describing the metapopulation dynamics (Figs. 17.9A and 17.9B). However, for disease models (and most likely many other enemy–host interactions), a different pattern of regional behavior arises due to the synchronization that occurs as an integral part of the space–time dynamics (Chapter 4). In particular, this synchrony of epidemics across the metapopulation will bias the extinction and colonization rates relative to the Levins assumption. Extinctions are, hence, far less common and colonizations more common than expected when the majority of patches are infected (Figs. 17.9C and 17.9D). In general, this leads to two distinct forms of global behavior: persistent endemic infections or irregular short-duration epidemics. Both of these states will be stable in the medium to long term. Which type of behavior is observed is critically dependent on the initial conditions.

Spatial coupling in epidemic metapopulations consisting of a geographic mosaic of cities and villages (May and Anderson, 1991, Grenfell and Bolker, 1998) has represented a thorny scientific question for more than half a century. No simple answer has as yet been found. As ever, Bartlett (1957), in his study of the scaling of epidemiological coupling, has been seminal in prompting detailed work in both spatial geography and epidemiology. Fifty years hence, the challenge of understanding epidemic coupling still stands. Progress is likely to lie in combining models for the nonlinear, seasonally forced local dynamics of measles with detailed transportation data. We see two strands of recent work that offer a way forward in the face of this daunting challenge.

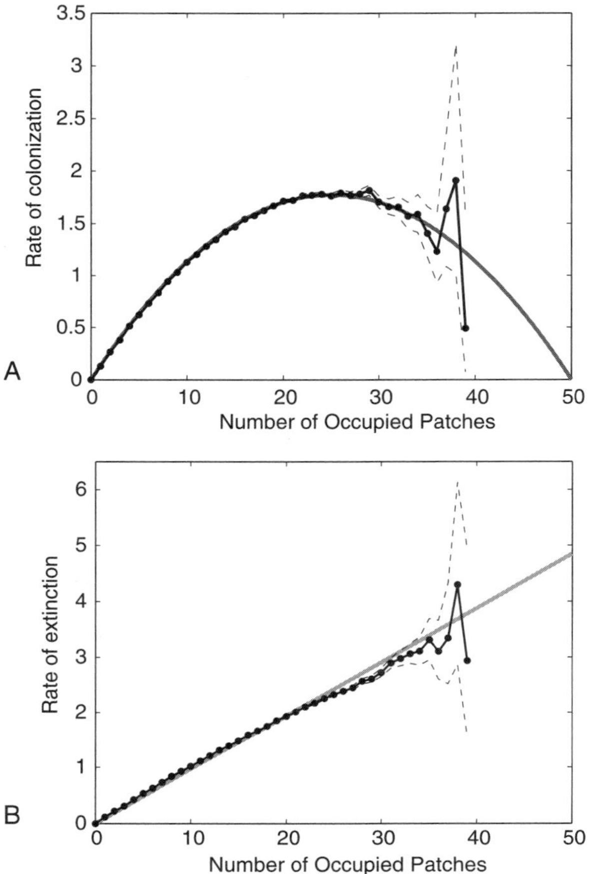

Fig. 17.9 Comparison between a full stochastic metapopulation and idealized functional forms of the Levins metapopulation. The average rates of extinction and colonization that occur whenever a simulation has a given number of occupied patches are shown. For simplicity, changes due to external imports are ignored. (A and B) Results for a single-species logistic model, which conforms well to the Levins ideal.

First, the development of discrete time versions of the SIR model (Finkenstädt and Grenfell, 2000; Bjørnstad et al., 2002; Grenfell et al., 2002) has allowed us to model the local scaling of dynamics and importation of infection over four orders of magnitude in host population size. Second, time series analysis of England and Wales data using wavelet phase analysis has testified to well-defined hierarchical traveling waves of infection, moving from large centers to the surrounding hinterland (Fig. 17.10; Grenfell et al., 2001).

These waves echo, on a larger spatiotemporal scale, the hierarchical waves detected in earlier geographical work (Cliff et al., 1981). Simplistic spatiotemporal models (Grenfell et al., 2001) show that the waves arise essentially from "forest fire"-like dynamics (Bak et al., 1990; Rand et al., 1995) in which epidemic "sparks" of infection from the large core populations ignite epidemics in smaller, locally extinct centers. These studies offers a glimpse of an ultimate understanding of the space–time dynamics of measles, but much is yet to be

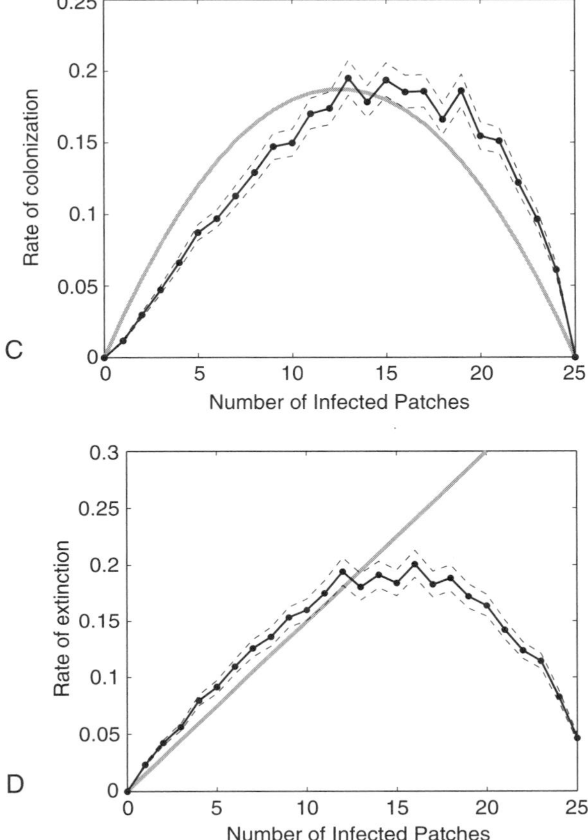

Fig. 17.9 *Continued.* (C and D) A SIR disease model with measles-like parameters and a between-patch coupling of $\mu = 0.001$. Clearly the latter contrasts with the Levins paradigm.

uncovered. In particular, developing models and theories that scale into the vaccination era appear to hold significant challenge.

Bubonic Plague in the Middle Ages

Bubonic plague invaded Europe in 1364, spreading rapidly north from the ports of southern Italy (Nobel, 1974). For the next 300 years or so the disease ravaged the towns and cities of Europe causing vast mortality (Shrewsbury, 1970). Bubonic plague is a disease of rodents that is generally transmitted by fleas; occasionally it spreads to humans, which is when cases are generally first noticed. Records show that although the disease was endemic in Europe as a whole through three centuries, each community displayed isolated epidemics in the human host population cases followed by "disease-free" periods. It has therefore long been thought that bubonic plague exhibited classic metapopulation behavior at the regional scale, with the infection continually going extinct and then recolonizing communities (Appleby, 1980). This conventional wisdom contradicts two pieces of

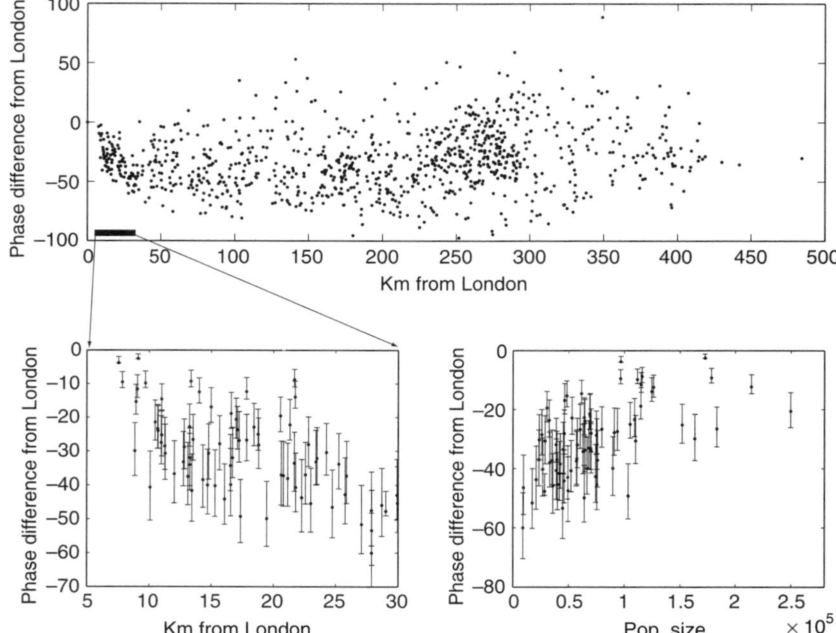

Fig. 17.10 Average phase differences for prevaccination urban and rural measles epidemic in the United Kingdom. (Top) Mean urban phase difference from London for major (biennial) epidemics during 1950–1966. Note that most places lag behind London. (Lower left) Mean urban phase difference from London for 1950–1966 in the London region (within 30 km) as a function of distance from the capital showing a significant correlation of phase difference and distance ($r = -0.59$, 99% bootstrap limits: -0.75 to -0.39); error bars are 99% bootstrapped confidence limits. (Lower right) Relationship between epidemic phase difference and local urban population size. The positive relationship between phase lag and population size is consistent with a "hierarchical forest fire" explanation for the waves (see Grenfell et al., 2001).

historical evidence. First, there is a fairly regular cyclic nature to human epidemics, which is unlikely to be caused by random imports of infection. Second, even in communities with tight quarantine controls, there is little change to the pattern of epidemics (Appleby, 1980).

Keeling and Gilligan (2000a,b) focused on the interaction among rats, fleas, and humans within a metapopulation setting. The life cycle of the plague can be partitioned into distinct stages and follows a general pattern for vector-borne diseases. Fleas that feed on an infected rat ingest the bubonic plague bacteria (*Yersinia pestis*) and become infectious. When an infected rat dies, its fleas, which are by now infectious, leave to search for a new host. Usually the fleas find other rats, infect them, and so spread the disease through the rodent community. Only when the density of rats is low are the fleas forced to feed on alternative hosts, such as humans, and spark off a human epidemic. Humans are considered a dead-end host, as transmission from humans to fleas is rare. Direct transmission between humans is possible if the pneumonic form of the disease develops, but due to the rapidity and virulence of such infection, pneumonic epidemics are small and short lived. These epidemiological observations can be translated into a mathematical model:

$$\frac{dS_R}{dt} = r_R S_R\left(1 - \frac{T_R}{K_R}\right) + R_R(1 - p) - d_R S_R - \beta_R \frac{S_R}{T_R} F\left[1 - \exp(-aT_R)\right],$$

$$\frac{dI_R}{dt} = \beta_R \frac{S_R}{T_R} F\left[1 - \exp(-aT_R)\right] - (d_R + m_R)I_R,$$

$$\frac{dR_R}{dt} = r_R R_R\left(p - \frac{T_R}{K_R}\right) + m_R g_R I_R - d_R R_R,$$

$$\text{where} \quad T_R = S_R + I_R + R_R \quad \text{(the total rat population size).}$$

$$\frac{dN}{dt} = r_F N\left(1 - \frac{N}{K_F}\right) + \frac{d_F}{T_R} F\left[1 - \exp(-aT_R)\right],$$

$$\frac{dF}{dt} = \left[d_R + m_R(1 - g_R)\right] I_R N - d_F F.$$

$$\frac{dS_H}{dt} = r_H(S_H + R_H) - d_H S_H - \beta_H S_H F \exp(-\mathrm{a}\mathrm{T}_\mathrm{R}),$$

$$\frac{dI_H}{dt} = \beta_H S_H F \exp(-aT_R) - (d_H + m_H)I_H,$$

$$\frac{dR_H}{dt} = m_H g_H I_H - d_H R_H. \qquad (17.10)$$

where S_R, I_R, and R_R refer to the number of susceptible, infectious, and resistant rats, respectively S_H, I_H, and R_H are similar quantities for the human population, N is the average number of fleas on a rat, and F is the number of free-living infected fleas that are searching for a host. Table 17.1 lists the meaning and value of all other parameters used in the model, which have been estimated from historical data or experiments (Keeling and Gilligan, 2000b).

The behavior of the theoretical model for this system is critically dependent on stochasticity and scaling. For large host populations, a deterministic solution gives rise to a constant level of infection in the rodents (as expected from most SIR-type models) and a negligible number of human cases. However, when stochastic effects play a major role, unusually large epidemics may drive the rat population to such low levels that the fleas are forced to feed on alternative hosts and a human epidemic occurs. This results in localized extinction of the disease. The subsequent local dynamics depends on the build-up of the susceptible rat population. Fairly rapid recolonizations of infection lead to an endemic persistence in the rat population and few, if any, human cases. In contrast, if recolonization is rare and hence the susceptible rat population has time to increase to high levels, major epidemic cycles with resultant spillover in human hosts occur. Thus, the epidemic behaviour is determined by the mixture of local transmission dynamics, stochasticity, and spatial coupling.

Good evidence suggests that in any large town or city, rats are unlikely to act as a homogeneously mixing host population, and therefore a spatially segregated metapopulation approach may be more appropriate. Studies performed by the Plague Commission in India (1906) showed that the spatial

TABLE 17.1 Parameters Used in the Bubonic Plague Model

Parameter	Value	Meaning
r_R	5	Reproductive rate of rat
p	0.975	Probability of inherited resistance
K_R	2500	Carrying capacity of rat
d_R	0.2	Death rate of rats
β_R	4.7	Transmission rate
m_R	20	(Infectious period)$^{-1}$
g_R	0.02	Probability of recovery
μ_R	0.03	Movement rate of rats
a	4×10^{-3}	Flea searching efficiency
r_F	20	Reproductive rate of flea
d_F	10	Death rate of fleas
K_F	$3.29 \rightarrow 11.17$ mean 6.57	Carrying capacity of flea per rat
μ_F	0.008	Movement rate of fleas
r_H	0.045	Reproductive rate of humans
d_H	0.04	Death rate of humans
β_H	0.01	Transmission rate to humans
m_H	26	(Infectious period)$^{-1}$
g_H	0.1	Probability of recovery

spread of the epidemic through the rodent population was extremely slow due to their largely territorial nature; this corresponds well with historical evidence of slow-moving waves of infection in the large medieval cities. Figure 17.11 shows the number of bubonic plague cases in rodents in a metapopulation model consisting of 25 local populations. Persistence of the metapopulation is due to the local populations that remain close to the endemic state (e.g., central site for the latter part of the simulation), whereas human cases (and thus historical reports) are due to the stochastically driven large epidemics. Due to the time necessary for the susceptible rat population to recover, these large epidemics have a period of around 10–12 yr, which corresponds remarkably well with the historical observations.

As observed earlier, the classic Levins metapopulation does not readily capture the dynamics of spatially structured epidemics due to the strong correlations that often exist between local and global levels of infection. However, for plague, such correlations are weak, and the local populations can be classified into three basic states: endemic (low level of infection and low risk of extinction), epidemic (high level of infection and high risk of extinction), and extinct (but susceptible). The extinct class is further subdivided so as to mimic the gradually increasing susceptible rat population. Figure 17.12 shows a caricature schematic of the Levins-type model for bubonic plague.

For this type of spatiotemporal dynamics, where the behavior is classified easily into a discrete set of states, the Levins approach provides great improvements in computational efficiency and clarity. The Levins formulation allows us to consider the dynamics at a far larger scale and hence observe the wave-like spread of the epidemics away from the endemic centers (Keeling and Gilligan, 2000b). From these models it is clear that the epidemic wave is often short lived and self-extinguishing, confirming the importance of endemic populations in allowing for long-term disease persistence.

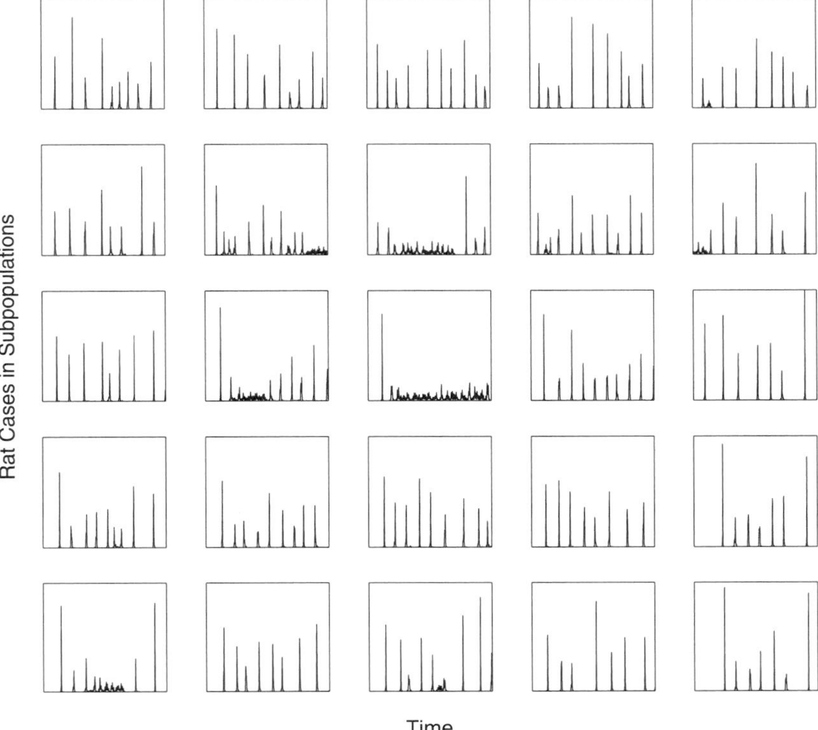

Fig. 17.11 Stochastic model dynamics of bubonic plague in a metapopulation. Each local population has a stochastic model for the behavior of rats, fleas, and humans, and the 25 local populations are coupled spatially with a low movement rate of rats and fleas between adjacent patches. For each local population, the number of cases in rodents over a 100-yr period is shown; during this time, the disease persists without the need for imports from outside the metapopulation.

Conservation or Contamination

An interesting extension to the classic metapopulation models for the population dynamics of endangered species is the inclusion of disease (Hess, 1996; Gog et al., 2002). In the absence of infection, increasing the spatial coupling between isolated habitats will increase the level of patch occupancy and decrease the risk of global extinction for one threatened species. Using the Levins metapopulation framework, with coupling SIGMA, the occupancy level x is given by:

$$\frac{dx}{dt} = \sigma c x(1 - x) - ex$$
$$x \longrightarrow 1 - \frac{e}{\sigma c} \tag{17.11}$$

where e is the extinction rate and c is the probability that invasion of an empty patch is successful. From this simple model it is clear that movement between largely isolated habitats improves the persistence of the endangered species.

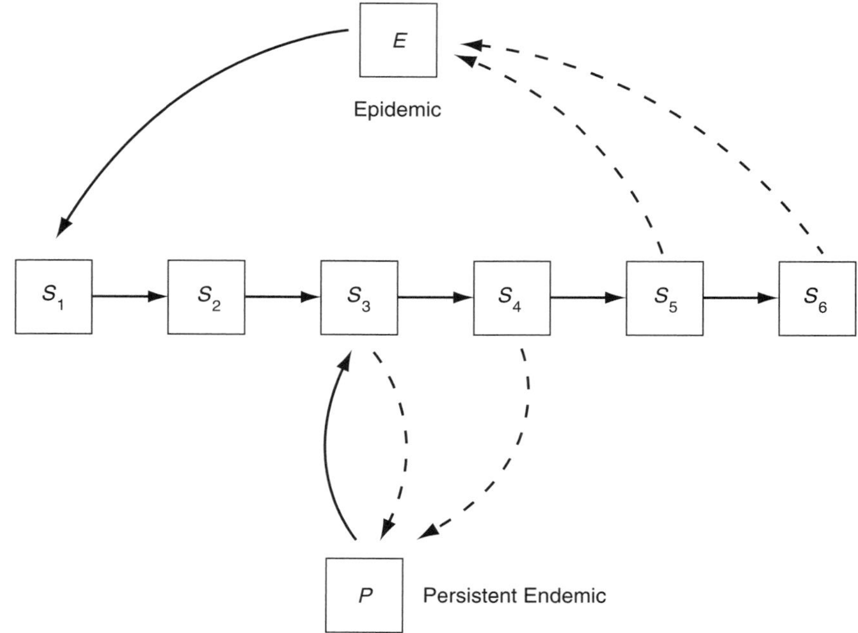

Fig. 17.12 Representation of transition states in a structured metapopulation model (Gyllenberg et al., 1997) of bubonic plague. Solid arrows represent probabilistic transitions, which occur independently of the surrounding environment. Dashed arrows show transitions that only occur due to the import of infection from a neighboring patch. Rates of transition can be measured from the full stochastic metapopulation model.

This effect occurs for two distinct reasons. Primarily, increased coupling σ leads to a higher colonization rate, which increases the likelihood of rescue events and the number of occupied local populations at equilibrium. Second, large amounts of movement between patches synchronize the dynamics, the local populations act effectively as one large habitat, and large populations suffer a relatively less extinction risk from stochasticity. The single-species model thus reveals no benefit of demographic heterogeneities between local populations as one large population (or several tightly coupled populations) shows the greatest persistence. This result is echoed by full stochastic metapopulation equations with explicit within-patch dynamics (Keeling, 2000b).

This conclusion can be altered radically in the presence of a virulent infectious disease, as coupling also facilitates the spread of infection (Hess, 1996). The resultant cost–benefit trade-off depends on the relative levels of host extinction with and without the disease, as well as the relative colonization rates. Gog et al. (2002) used the following model to explore the dynamics of infected (I) and uninfected (S) habitat:

$$\begin{aligned} \frac{dS}{dt} &= \sigma S(1 - I - S) - e_S S - \sigma\delta IS - gS \\ \frac{dI}{dt} &= \sigma I(1 - I - S) - e_I I + \sigma\delta IS + gS \end{aligned} \qquad (17.12)$$

where σ is the rate of movement to and colonization of empty habitat, e_S and e_I are the patch level extinction rates, δ is the chance that movement leads to the spread of infection, and g is the import rate of infection from outside the considered population. As this is a model of wildlife disease, the coupling between populations occurs as the random dispersal of organisms rather than the short-duration commuter movements associated with human disease transmission. The focus of this model is conservation of an endangered host, and therefore is the reverse of the scenarios discussed earlier where the eradication of infection was the main aim. In agreement with the earlier work of Hess (1996), this Levins-like model shows that under certain circumstances greater movement between patches (larger σ) can lead to a reduction in the number of occupied patches and an increased risk of global extinction to highlight an important conservation risk (Fig. 17.13).

It is informative to consider an extreme variation of this model. Suppose that the disease within an infected patch is severe and widespread so that animals from infected patches are unable to colonize a new habitat successfully. The model then can be rewritten as

$$\frac{dS}{dt} = \sigma S(1 - I - S) - (\sigma\delta)IS - e_S S - gS$$
$$\frac{dI}{dt} = (\sigma\delta)IS - (e_I - e_S)I - e_S I + gS \qquad (17.13)$$

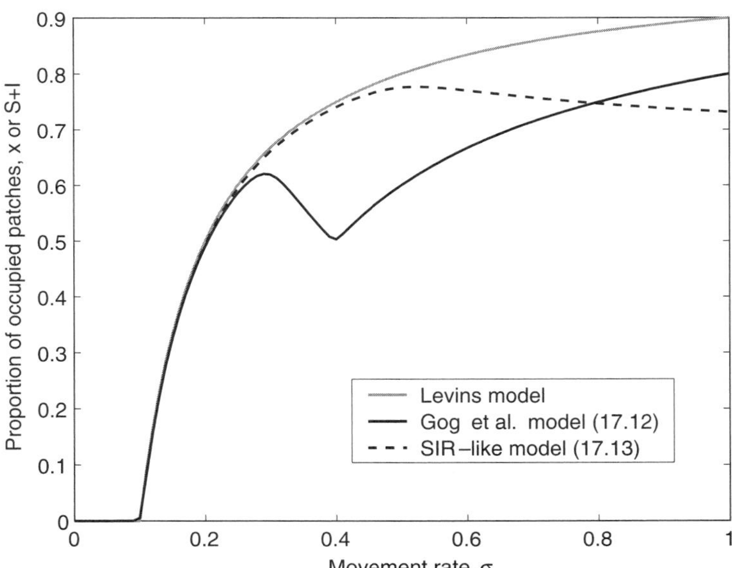

Fig. 17.13 Effects on patch occupancy of between-patch coupling σ for the three models are shown. Both the model of Gog et al. (2002) (17.12) and the simplified SIR-like version (17.13) show that increased levels of coupling can decrease patch occupancy in the presence of an infectious disease ($c = 1$, $e = e_S = 0.1$, $e_I = 0.2$, $\delta = 0.5$, $g = 0.001$).

This model then shares many elements with standard SIR disease models; $\sigma\delta$ plays the role of the transmission parameter β, e_S corresponds to the natural death rate, and $e_I - e_S$ equates to the recovery rate. The emergent parallel with classical disease models allows us to intuit about the resultant dynamics. For example, changes in the movement rate σ correspond to a trade-off between an increase in the birth rate and an increase in disease transmission.

Vaccination, Pulses, and Synchrony

A key issue of metapopulation modeling for infectious diseases is to compare different vaccination strategies to optimize the likelihood of disease eradication. As discussed earlier, the global persistence of a disease is determined by two main factors: the local extinction rate and the rate of recolonization, which in turn is related to the heterogeneity of the metapopulation. Figure 17.14 shows how these two facets change as the level of vaccination increases; we first consider the solid black line, which corresponds to continuous random vaccination. Below the critical vaccination level of 90% the local extinction rate shows only a moderate increase with the level of vaccination so that the expected length of an epidemic decreases slowly. In contrast, the correlation between two coupled local populations starts to decrease from the onset of vaccination. Therefore, in the Levins formulation, moderate levels of vaccination only cause a small increase in the extinction rate, which may be counteracted by the increase in asynchrony and therefore the increase in rescue effects when they are most needed.

The balance between vaccination increasing the stochastic extinction rate but reducing the synchrony between populations depends on the demographic and epidemiological parameters. Thus while moderate levels of vaccination will always act to reduce the total number of cases, they may surprisingly increase the global persistence of the disease if the loss of synchrony is dramatic enough. However, as the level of vaccination approaches the critical eradication threshold, the rapid rise in the rate of local extinctions will overwhelm any rescue effects and global extinction will inevitably follow.

Obviously, vaccination would be a much more effective tool if as well as reducing the number of cases it could also decrease the global persistence of the disease. In principle, this can be achieved by superimposing periodic "pulses" of vaccination on the overall background rate. Pulsed vaccination has been proposed to increase the efficiency of vaccination (Agur et al., 1993), but it could also have a spatial benefit by "lining up" epidemic troughs and therefore reducing rescue effects (Earn et al., 1998). The impact of a simple model for pulse vaccination (in the absence of background vaccination) is shown in gray in Fig. 17.14. The first observation is that pulse vaccination is associated with a slightly lower local extinction rate, and also more cases of the disease; this is because in the gaps between the vaccination pulses children that would have been immunized under continuous vaccination have a chance of catching the infection – in practice though, any pulsing would probably be superimposed on a constant 'background' rate, so that this effect is not realistic. The difference between pulsed and continuous vaccination is more dramatic in terms of the correlation between epidemics. The significant perturbation caused by a periodic vaccination campaign acts to synchronize the dynamics

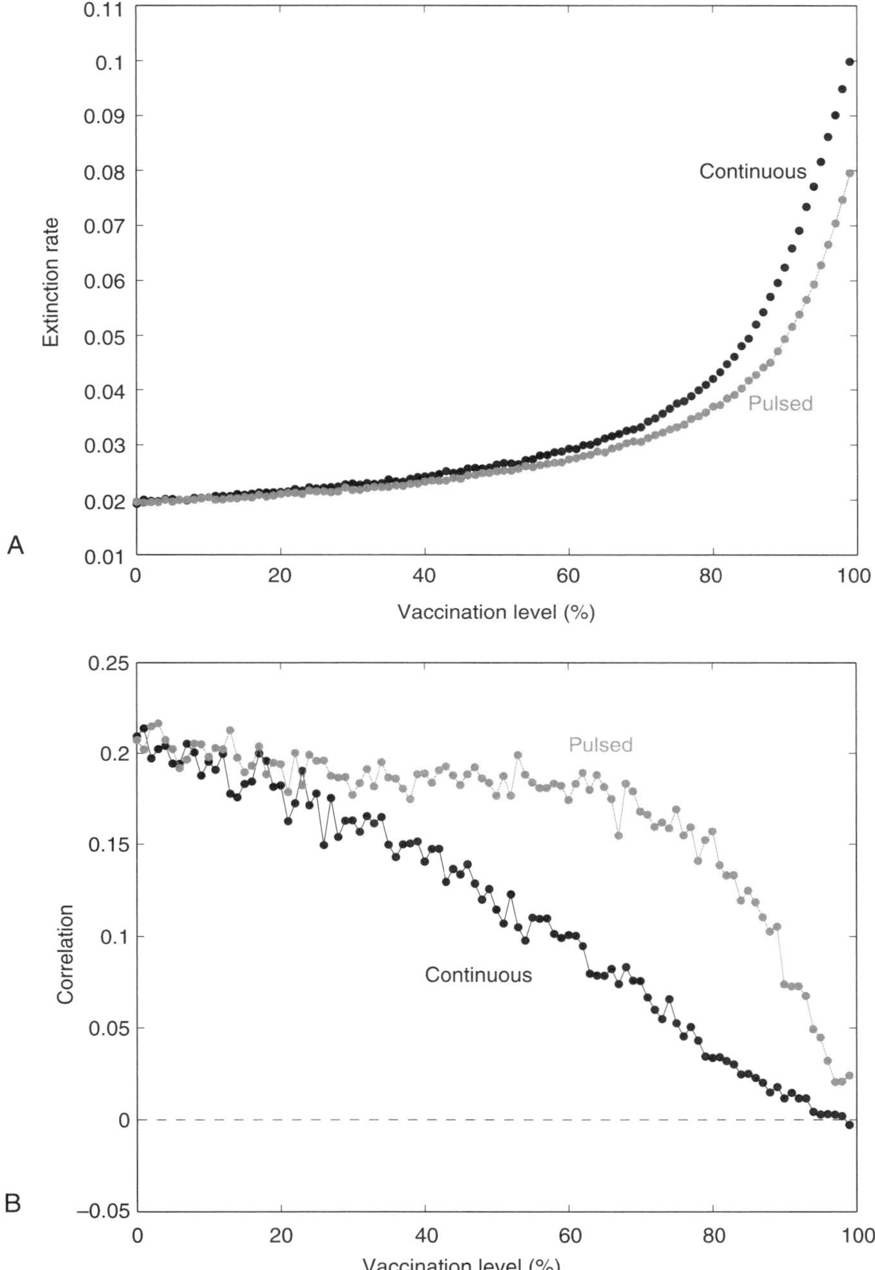

Fig. 17.14 Effects of vaccination on the characteristics of unforced SIR epidemics. Black symbols refer to constant random vaccination at birth, whereas gray symbols correspond to pulse vaccinating randomly at regular 4-yr intervals; similar results are achieved for more frequent yearly pulses. (A) Change in the local extinction rate (per day) of an isolated population. (B) Change in the correlation between two local populations coupled at a level $\sigma = 0.01$. (Population size is 10,000, $R_0 = 10$, $g = 10$ days, import rate is 5 per year.)

of the two populations, thus for pulsed vaccination the correlation remains approximately constant for vaccination levels below 60%.

Pulsed vaccination therefore provides a potentially important tool for increasing local extinctions, without increasing the effective rescue events, and therefore increasing the likelihood of global extinctions (compare to Levins, 1969). Results shown in Fig. 17.14 have ignored seasonal forcing, which naturally leads to greater synchronization of the dynamics. When seasonal forcing is important (such as for most childhood diseases), the interaction of background vaccination, vaccination pulses and seasonal effects may be very complex; the precise timing of vaccination could significantly increase the chance of regional eradication. However, this is very much an open problem for future research.

17.6 FUTURE DIRECTIONS

Metapopulation theory has a rich history in the ecological literature and has proved itself continually as both an applied tool and an insightful description of the complex world (Gilpin and Hanski, 1991; Hanski and Gilpin, 1997). The use of metapopulations has been somewhat more limited in epidemiology due to the more complex within-patch dynamics. However, in recent years this balance has begun to be redressed. Several key theoretical and practical issues still need to be dealt with successfully to allow the subject to develop further.

1. A better understanding of how the epidemiological and demographic parameters translate into the Levins-type metapopulation parameters of extinction and colonization rates. The ability to translate stochastic within-patch population dynamics into a simple set of population level states would lead to a vast increase in computational speed and provide powerful insights into the spatiotemporal dynamics of disease spread and extinction. Although moment-closure approximations and quasi-equilibrium solutions offer a likely approach, they have yet to be applied to realistic seasonally forced dynamics.
2. More detailed simulations of heterogeneous patches with complex connections (Chapter 4). So far the majority of studies have considered equally sized local populations and global coupling. While this is a natural starting point, the real world is far more complex, and developing models and intuition for such scenarios will be important if spatially targeted control measures are to be applied most effectively.
3. A range of more powerful statistical and mathematical techniques are also required to deal with coupling. First, there is the complex problem of how to estimate the coupling between communities from case reports. This estimation process is confounded by stochasticity, seasonality, and heterogeneities in demographic rates, although some progress has been made. Associated with this problem is developing mathematical rules for the coupling between populations as a function of their separation. In a metapopulation of N patches, there are $N(N-1)$ coupling terms, hence in large systems estimating or even storing all the coupling rates becomes

problematic so analytical approximations become necessary. Developing gravity (Cliff and Hagget, 1988) and other formulations of the relationship between human movement and disease spread is an important problem for both fundamental population biology and applied epidemiology.

Although these three problems present formidable challenges, metapopulations are likely to see far more use in the future as the degree of realism and resolution required from models increases.

18. TOWARD A METAPOPULATION CONCEPT FOR PLANTS

N.J. Ouborg and O. Eriksson

18.1 INTRODUCTION

The spatial structure of populations and communities has always been an important component of ecological and evolutionary theory (e.g., Wright, 1931; Andrewartha and Birch, 1954; den Boer, 1968). Introduction of the theories of island biogeography (MacArthur and Wilson 1967) and metapopulation dynamics (Levins, 1969) further enhanced the use of spatial considerations and arguments in explanations of ecological and evolutionary phenomena. Over the last several decades, the metapopulation concept has become the guideline for our understanding of issues as diverse as large-scale population dynamics, the spatial distribution of species, the dynamics of species interactions, and the effects of habitat fragmentation on biodiversity.

The development of metapopulation theory was inspired by the dynamics of animal populations, and many of the models implicitly have the features of a "model animal" as their basis. Husband and Barrett (1996) suggested that this may have led to a mismatch between the metapopulation concept and the features of the population biology of plants. Whether or not this is the case, a fact is that despite the patchy spatial structure of plant populations, and despite the establishment during the 1970s and 1980s of a population-oriented paradigm in plant ecology (Harper, 1977; Silvertown, 1987), metapopulation theory has received comparatively little attention in plant studies. Two reviews

0-12-323448-4

(Eriksson, 1996; Husband and Barrett, 1996) concluded that there is much indirect evidence for metapopulation dynamics. However, only a handful of studies have shown that regional plant populations exhibit features of typical metapopulations, for example, turnover of populations within a set of suitable sites with colonization rates dependent on distance between suitable sites and/or extinction rates dependent on local population size (Menges, 1990; van der Meijden et al., 1992; Ouborg, 1993; McCauley et al., 1995; Quintana-Ascencio and Menges, 1996; Barrett and Husband, 1997; Giles and Goudet, 1997; Bastin and Thomas, 1999; Harrison et al., 2000).

The applicability of the metapopulation concept to plants has even been questioned, both on empirical and on theoretical grounds. Scheiner and Rey-Benayas (1997) analyzed distribution–abundance patterns of a large sample of plant communities and concluded that metapopulation models were generally unable to predict patterns occurring at larger scales than 1 km^2. As explanations for the failure of the examined metapopulation models, Scheiner and Rey-Benayas (1997) suggested that assumptions concerning among-site variation in quality, site connectivity and equal migration and extinction rate, do not hold. A problem with this kind of test is that species are lumped together and implicitly assumed to behave similarly with regard to regional dynamics. As discussed later, plants as a group are heterogeneous with regard to their spatial dynamics, and therefore we may not expect that distribution patterns for all species occurring in a region will fit into a single metapopulation model.

The metapopulation concept has also been questioned as a useful tool for understanding regional dynamics of single species. Freckleton and Watkinson (2002) focused on the methodological difficulties to define suitable (but unoccupied) patches, to document turnover in sets of local populations, and to properly assess dispersal among patches and concluded that in many plant populations the assumptions of a metapopulation model do not hold and that metapopulations in plants may be restricted to relatively few cases. As alternatives to the metapopulation concept, they suggested other concepts, forming a new typology of plant regional dynamics, using “spatially extended populations” and “regional ensembles” as the main types. The key issue for this alternative typology is the fraction of suitable habitat in a region. If there is a lot of suitable habitat, Freckleton and Watkinson argued, populations will occur continuously throughout large areas. Thus, local processes will dominate the dynamics. If, however, there is very little suitable habitat (and patches occur more or less isolated), the populations will form regional ensembles. These consist of basically unconnected local populations, in turn leading to a dominance of local over regional processes. Of course, extinction is usually considered a local process also in metapopulation models, but strong isolation may imply that a regional turnover is more or less absent. In another review, Bullock et al. (2002) suggested that it is too early to determine whether “true” plant metapopulations exist because we lack the relevant data. Nevertheless, they cast doubt over the applicability of the concept in plants, mainly because local extinction and recolonization phenomena could well be explained by local rather than regional processes.

These critics of the metapopulation concept for plants illustrate two problems that any study of plant metapopulations must deal with. The first type of problem is habitat related. The distribution of plants, especially at larger spatial scales, tends to reflect abiotic conditions, climate, and soil. Why should we expect that metapopulation models based on colonization and extinction dynamics dependent on spatial configuration of habitats should add significantly to understanding regional dynamics at these scales? This chapter provides some reasons why the spatial configuration of habitats affects dynamics also when a set of local plant populations do not behave as typically envisaged in metapopulation models. Moreover, basic needs of plants (water, nutrients, light) are often quite unspecific, which means that indirect means of assessing habitat quality may prove useless. We will discuss experimental approaches to handle habitat suitability, and we stress the need for considering variation in habitat quality among sites.

A second problem relates to features of the plants themselves. Plants possess a number of life cycle features that make them different from the "model organism," short lived and mobile, typically envisaged in metapopulation models. Plants often have a long life span. Vegetative life cycle phases may persist for centuries or, in extreme cases, even millennia (e.g., Cook 1983). Commonly, the life span of established plants by far exceeds that of the human observer, making us perceive plants as being "static." Seeds may stay dormant in the soil for long periods, making it hard to document true extinctions and to distinguish colonization events from just reappearance of plants that have stayed on the site. Moreover, plants have mainly sessile life stages, but with two mobile phases: pollen dispersal and seed dispersal. For both these mobile phases, the common feature is that pollen and diaspores are not able to direct their dispersal, neither the distance traveled nor the target for deposition. Long-range dispersal is extremely difficult to assess, yet essential for understanding both gene flow (e.g., Young et al., 1996) and plant migration (e.g., Higgins and Richardson, 1999; Cain et al., 2000).

This chapter discusses how a metapopulation concept is useful for a range of issues related to studies of regional plant populations, despite the problems mentioned earlier. The basic tenets for our view are (1) that plant populations are characteristically patchy, at most (all) spatial scales, (2) that despite extensive life spans and an apparent "stasis" of many plant populations, viewed over a longer time period than are usually considered in research programs, most plants possess a turnover at a regional scale, (3) that there is convincing evidence that patterns of spatial habitat configuration do affect a number of processes that influence plants — pollination, herbivory, diseases, and seed dispersal — and (4) that the ongoing landscape transformation, including fragmentation of many natural and seminatural habitats, implies that the effects of landscape habitat configuration are likely to increase.

18.2 THE NEED FOR DEFINING PLANT METAPOPULATIONS

The discrepancy between the metapopulation concept as it is prevailing in current models and the specific features of plant populations is perhaps best exemplified by an overview of the discussion on the definition of a metapopulation. Hanski and Simberloff (1997) defined the metapopulation approach as taking into account that "populations are spatially structured into assemblages of local breeding populations and migration among local populations has some effect on local dynamics, including the possibility of local population reestablishment following extinction" (p. 6). The reviews that have been published on metapopulation dynamics in plants (Husband and Barrett, 1996; Eriksson, 1996; Freckleton and Watkinson, 2002; Bullock et al., 2002) all devote quite some text to the definition issue. While they vary in detail, there is general consensus about several components of the definition. Habitat should be distributed discontinuously with a mix of suitable occupied, suitable nonoccupied, and nonsuitable patches; some degree of migration should occur between local populations; and ultimately there should be extinction and (re)colonization at the local patch level.

Freckleton and Watkinson (2002) argued that distribution of a habitat is a key element determining whether there are metapopulations in plants. They claimed that the basic premise of metapopulation theory, that suitable habitat occurs as discrete patches within a matrix of unsuitable habitat, is frequently violated in plants, as regional populations of some plant species may exist on largely uninterrupted swathes of suitable habitat. In addition, Vandermeer and Carvajal (2001) showed that the dynamics in the supposedly nonsuitable habitat matrix surrounding suitable patches may be of great importance for the overall regional dynamics, thus reducing the importance of the suitable vs nonsuitable distinction. As this is likely to be a process relevant for many plant species, this further emphasizes the need for a revised plant metapopulation definition.

It is telling that Hanski and Simberloff (1997), in an attempt to defend the general applicability of the metapopulation approach, stress that ". . . empirical work has made good use of the metapopulation concept even when some tens of percents of individuals per generation leave their natal patch" (p. 9), whereas the hesitation of plant population biologists to embrace the metapopulation concept is (partly) based on the very limited dispersal distances that characterize almost all plant species. Husband and Barrett (1996) stated that this restricted dispersal makes plants particularly appropriate for metapopulation analyses. It is, however, clear that at least some migration should occur for the concept to be of value, and in several cases, authors place some doubt on whether dispersal in plants is not too restricted (e.g., Freckleton and Watkinson, 2002; Bullock et al., 2002).

Maybe the most prominent feature of metapopulation dynamics is the extinction and colonization dynamics at the regional population level. What is accepted as evidence for metapopulation dynamics follows a continuum from the animal-oriented side, where observations of extinctions and recolonizations are accepted as evidence (e.g., Harrison and Taylor, 1997) to various degrees of relaxation at the plant side. While all authors have some form of the original Levins metapopulation as a conceptual starting point, the extinction criterion is a matter of much debate when it comes to plant dynamics. Husband and Barrett (1996) stated that ". . . the metapopulation concept has been broadened to recognize that all species have local and regional dynamics" (p. 462). Eriksson (1996) argued that ". . . metapopulations in the strict sense . . . may not occur in all, perhaps not even in most, organisms" (p. 249). Freckleton and Watkinson (2002), following Hanski and Simberloff (1997), relaxed the extinction requirement from observed extinction to " . . . even the largest local population should have a measurable risk of extinction (unless the largest population is the source of a source–sink system)" (p. 420). This discussion around the extinction criterion is not so much motivated by theoretical objections, but rather has a methodological base: most plants have a much slower turnover than most animals, making observations of extinctions and colonizations impractical, if not impossible, in many cases. As a result, there is a clear need for reconsidering metapopulation definitions and concepts. This should be based on a thorough identification of plant-specific problems with the concept, which then may serve as the basis for redefining plant metapopulations.

18.3 PLANT-SPECIFIC PROBLEMS WITH THE METAPOPULATION CONCEPT

Suitable but Unoccupied Habitat

It is well known that the simplifying dichotomy suitable vs unsuitable habitat does not hold for most organisms. Habitats may have a fluctuating quality, for example, being open for colonization of seeds only certain years, and there may be a more or less continuous variation in abiotic and biotic properties determining the suitability of a site. For species inhabiting rather well-defined substrates, such as decomposing wood or small ponds, the concept of unoccupied suitable sites is relatively easy to use. In contrast, for most plants it is very difficult to exactly define the features of a suitable site without having the species under study present there. However, assuming that spatially delimited sites do vary with regard to quality, defined on the basis of their suitability to harbor populations of a focal species, the methodological problems of finding those sites are not an acceptable basis for refuting metapopulation theory.

One method used to overcome the problem of assessing site quality is to use seed sowing or transplantations of the focal species. Although experimental seed sowing has been much used to examine seed limitations in plant distributions (for a review, see Turnbull et al., 2000), this method has been employed only occasionally to analyze occupancy patterns. For example, Ehrlén and Eriksson (2000) estimated the occupancy of seven forest herb species among patches of deciduous and mixed coniferous–deciduous forests by sowing seed and transplanting juveniles. Results indicated that occupancy was related negatively to seed (diaspore) size; large-seeded species may thus be more restricted in their exploitation of available suitable sites. Interestingly, no relationship was found between the actual occurrence of the species, or the success of the sowing, and a number of measured abiotic soil factors. Thus, a study based only on measuring site factors would have yielded misleading results concerning the suitability of the sites and thus the actual occupancy of the species.

From a methodological viewpoint there are some important caveats with sowing/transplanting experiments that must be accounted for. First, plants may succeed in recruitment only in certain years. This means that failure to find recruitment may not mean that a site is altogether unsuitable.

Experiments may have to be repeated during several years. Second, a recorded recruitment may not mean that a population will be established. Plants may have differences in the requirements for recruits, juveniles, and adults (Schupp, 1995; Ohlson et al., 2001; Eriksson, 2002). Thus, in order to be reliable, recruitment experiments must be followed several years after the appearance of the seedlings. Using juvenile transplants in combination with seed sowing is, however, a means to reduce the time needed for the experiment. Ehrlén and Eriksson (2000) followed the recruits for 4 yr; in fact it took another 3 yr before the conclusion was reached that sowing really resulted in population establishment defined as sown plants development to maturity (Ehrlén and Eriksson, unpublished results). Assessing habitat quality by means of sowing and transplantation is therefore time-consuming, at least for perennial plants that have a long juvenile period. Yet we suggest that this method is the most appropriate for detecting suitable but unoccupied sites in plant metapopulation studies.

Long Life Spans

Plants are tremendously variable with regard to their life span. Annuals may live for a couple of weeks, whereas clonal plants and some trees may live for centuries. The oldest documented life span for trees is more than 4000 years (Currey, 1965), whereas the extreme age for some clonal plants is over 10,000 years (e.g., Vasek, 1980). Even if these ages are not representative for plants in general, life spans in the order of a century are common. Ehrlén and Lehtilä (2002) compiled demographic data for 71 perennial species and reanalyzed them with use of population matrix models. They found that over half of the species had a projected life span exceeding 35 years and a quarter of the species had a projected life span over 80 years. This means that local populations may be very persistent, even in cases where the population growth rate is negative, for example, due to lack of recruitment. In many changing habitats where the change is experienced as "deterioration" from the viewpoint of some of the inhabiting plants, such as abandoned grasslands and forests undergoing succession, reproduction and recruitment are likely to be the first population processes that are affected negatively. However, populations still persist, and they may do so for extended periods of time. Such remnant populations (Eriksson, 1996) may be a characteristic component of the landscape, especially in regions where land use has changed during the last century (i.e., in most parts of the world where humans live). Estimates of time to extinction following land use change in Scandinavia reveal that local populations of perennial plants may persist for periods of 50–100 years (Eriksson and Ehrlén, 2001). Thus, a fraction of the "occupied" sites in a regional perspective may reflect a historical habitat distribution rather than the actual one present in the landscape today (Fig. 18.1).

From a conservation perspective, such a time lag in the response of species to ongoing habitat changes represents a form of extinction debt (Tilman et al., 1994; Hanski and Ovaskainen, 2002). If no habitats are available where population growth is positive, these remnant populations are slowly moving toward extinction, although the species may be perceived as rather common based on a conventional survey. The expected time to extinction for the regional

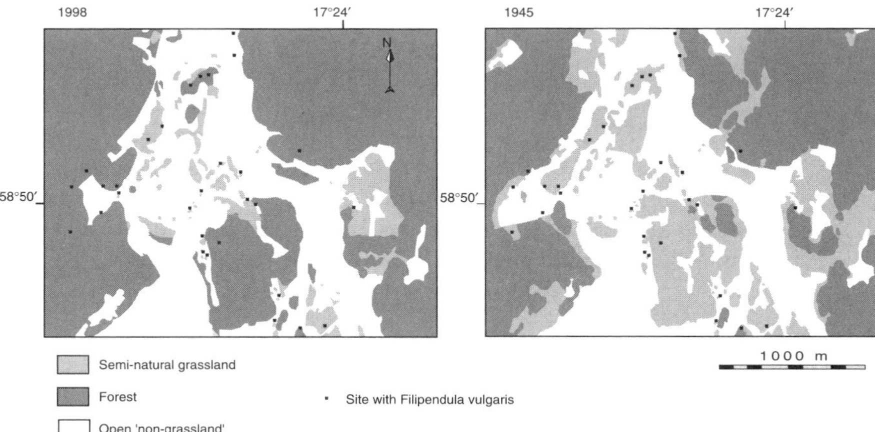

Fig. 18.1 Distribution of *Filipendula vulgaris*, a plant typical of seminatural grasslands, in a landscape in southern Sweden. Of the 30 sites recorded in a 1998 survey, 9 sites were found in what is presently forest. If the same sites are mapped on the 1945 landscape (as revealed by aerial photographs), all but 1 site is located in seminatural grasslands. Note that some sites that appear to be located in "open nongrasslands" actually occur on small grassland fragments. Most likely the forest sites in the 1998 map represent remnant populations (S. Cousins, unpublished result).

population equals the expected time to extinction for the most persistent local population. However, the likelihood of success after habitat restoration will increase if remnant populations are still present. Habitats dominated by species that develop remnant populations may therefore be the most suitable targets for restoration.

If remnant populations generally are common in plants, this has two important implications. First, occupancy patterns of long-lived perennial plants may be far from equilibrium with the present habitat configuration (Eriksson and Ehrlén, 2001). This means that estimates of colonization and extinction rates from incidence function models may yield very misleading results. To achieve better estimates of colonization and extinction, a combination of experimental studies and demographic studies of local populations, assessing the actual growth rate variation, is needed. The turnover of local populations will naturally be very slow in many perennial plants. However, the slow dynamics per se does not imply that there is no turnover also in regional populations of long-lived plants. The problem is a matter of time scale, but given sufficiently long observation series, long-lived plants probably have regional dynamics not fundamentally different from short-lived organisms (cf. Whittaker and Levin, 1977).

A second implication is that remnant population systems may function as a temporal source–sink population. Habitats may have a fluctuating quality (from a focal species viewpoint). If the persistence of local populations, despite a negative growth rate, is in the magnitude of the period of habitat quality fluctuation, time periods with a positive population growth rate compensate for those with population decline. Thus, using a dichotomized simplification of habitats into "suitable" and "unsuitable," colonization of suitable habitats may be considered through time as well as through space. This resembles the effects of a persistent seed bank, where the appearance of suitable habitat (e.g., after a disturbance) initiates development of a population of vegetative and reproductive plants.

The time lag of plant species response to habitat changes calls for analyses of population distributions that are based on historical data on the populations themselves and on the landscape. Estimates of colonization and extinction may be derived from combining knowledge of previous occupancy patterns, as found in old surveys, with reinventories (e.g., Ouborg, 1993; Lienert et al., 2002; Lindborg and Ehrlén, 2002). In most cases, such information does not exist, however, and considerations of historical effects can only be inferred indirectly by analyzing present occupancy patterns in relation to known habitat changes (Fig. 18.1). Such studies provide a direct link between metapopulation studies on the species level and landscape ecology (Wiens, 1997). This places a much stronger emphasis on the landscape matrix surrounding the target habitats. For example, habitats that are regarded as presently unsuitable should be examined for historical suitability. The historical time frame for landscape change must then be compared with the time lag of the local population response to the landscape changes.

Seed Banks

In addition to long life spans, many plant species have long-lived seed banks (Leck et al., 1989; Thompson et al., 1997), and for such species, estimates of colonization and extinction are complicated. Local populations (above ground) may vanish, but the species is nevertheless present in the seed bank. Documenting local population extinction is also therefore difficult if seed bank samples are gathered. Negative findings in seed bank samples are difficult to interpret, and it demands a large sampling effort to safely conclude that a species is in fact missing from the seed bank. Estimating colonization is also complicated by seed banks because of the difficulty of distinguishing between recruitment from seeds arriving at a site through ordinary dispersal and recruitment from the seed bank.

The extent of these problems for a plant metapopulation concept depends on the longevity of seeds in soil and on how often species occur in the seed bank but not in the vegetation (which is in turn likely to reflect the longevity of the seed bank). It is common knowledge that seed banks in many cases do not resemble the present vegetation, although this conclusion may partly rest on an insufficient sampling of the seed bank (Leck et al., 1989). Furthermore, the potential longevity of many seeds extends over several decades and, in some cases, centuries (Thompson et al., 1997). Although such extended longevity of seeds constitutes problems for assessments of colonization and extinction in metapopulation studies, the extent of the problem differs among vegetation types. Whereas highly disturbed habitats, (e.g., arable fields and spoil) contain long-lived seed banks, seeds in the seed bank of forests and permanent grasslands are generally not long lived (Thompson et al., 1998). For example, many species-rich grasslands do not generally have the majority of species represented in the seed banks (Bekker et al., 1997).

Only a few studies have examined the quantitative relationships between recruitment from seed rain vs the seed bank. Indirect studies (not estimating the resulting recruitment) indicate that contribution of the seed rain may be larger than contribution of the seed bank (e.g., Rabinowitz, 1981; Schott and Hamburg, 1997; Molau and Larsson, 2000). However, both seed rain and seed bank may contribute little to regeneration after disturbance (e.g., Owen et al., 2002), whereas other studies indicate that the seed bank is important. Kalamees and Zobel (2002) estimated that 36% of the regeneration in gaps in species-rich grasslands came from the seed bank. An important finding, however, was that rather few recruits (below 4%) from seed banks in permanent grassland belong to species not present in the vegetation (Eriksson and Eriksson, 1997; Kalamees and Zobel, 2002). Thus, at least for some vegetation types, the problems for assessing extinction and colonization based on aboveground vegetation seem to be rather limited. Although numerous studies have described the seed bank in different vegetation types, it is obvious that we still basically lack general quantitative knowledge on the direct importance of seed banks for regeneration. In particular, we need studies addressing how frequent recruitment of species from the seed bank that are not present in the vegetation occurs. In order to evaluate the potential problem with such "pseudo-colonizations," a crucial issue will be to estimate the time scale of the persistence of the seed bank after the aboveground population of conspecific plants has disappeared.

Time Lags and Population Turnover

There are several ways in which one can handle the problem of the time lag of population extinction resulting from the presence of a seed bank (after vegetative plants of the focal species have vanished) or the presence of long-lived vegetative plant individuals in a remnant population where, for example, reproduction and recruitment have ceased due to habitat change. If there are only seed banks or remnant populations left in a region, there will be no dynamics at all at the regional scale unless unsuitable but occupied sites improve. The regional population will go extinct when the last local seed bank or remnant population disappears. A more complex situation occurs when there are local populations at still suitable sites. Also, in this case, the actual pattern of occupancy is not in equilibrium with the present habitat configuration as it is lagging behind the habitat change. Thus, a key issue is to incorporate the time lag into metapopulation models. Assuming an equilibrium fraction of occupied sites, p_0, before habitat change and a new equilibrium (or close to equilibrium) fraction of occupied sites after habitat change, p_1, the time lag T will be equal to the time it takes to move from p_0 to p_1 (Hanski and Ovaskainen, 2002). T can be defined either as the expected time to local extinction, T_e, of the largest seed bank or remnant population or as the expected time to local extinction of the average (or median) sized seed bank or remnant population from the time where a local site starts to deteriorate (from a focal species point of view). For seed banks, T_e can be estimated from repeated seed samples from a selection of different sites. The decline of vegetative remnant populations may be estimated by analyses of transition matrix models based on a representative sample of populations (e.g., Eriksson and Ehrlén, 2001).

Estimates of colonization are also affected by seed banks and remnant populations because both can help bridge periods of suitable habitat conditions interrupted by unfavorable local conditions. If a population is able to persist through a period of unfavorable conditions, this implies that the population can be present at a site and start expanding again if conditions improve, without a new colonization through space. While this in principle may occur for all organisms, plants may, due to their unusual ability to withstand unfavorable conditions, be particularly prone to such a "dispersal in time" (Eriksson, 1996).

In metapopulation models, colonization is usually modeled as a function of some measure of habitat connectivity, for example, the total area of suitable habitat in a region weighted by the distance between the focal site and each surrounding site (Hanski, 1999). Assuming that a certain fraction of unsuitable (but formerly suitable) sites harbors seed banks or remnant populations and that there is a certain likelihood that such a site develops to become suitable, new suitable sites add to the ones already present, being "colonized" already from time zero. Again a key issue is the time it takes before local populations go extinct after a site has changed from suitable to unsuitable, T_e, which will set the temporal limit for colonization through time. Spatial relationships for this subset of colonizations are not expected to reflect the present-day habitat configuration. Rather, the occurrence of seed banks and remnant populations in the landscape will reflect the historical habitat connectivity within the time frame for the persistence of these populations.

Conceptually, this way of treating problems with time lags in extinction, habitat suitability, and colonization connects to models of patch dynamics (Watt, 1947; Whittaker and Levin, 1977) and "mosaic models" of communities (DeAngelis and Waterhouse 1987) in which there is a turnover of patches (sites) with regard to their quality for inhabitant species. Given that we use a strict delimitation of "suitable sites" (as sites where populations potentially have a positive population growth or, alternatively, as sites where reproduction is possible), sites where a focal species is present only in a seed bank or as a remnant population are, by definition, "unsuitable." Despite being unsuitable, these sites contribute to the overall dynamics of the focal species by (i) delaying regional extinction and by (ii) providing a means by which new suitable sites (re)appear already being colonized. Moreover, there is a link to community resilience (Eriksson, 2000); remnant populations (and seed banks) contribute to decrease the return time to community equilibrium (or any quasi-stable state) following disturbance.

Source Quality

Unlike animals, plants are sessile during most of their life cycle. In the mobile phase of the life cycle, seeds (or, for some aquatic plant species, vegetative propagules) disperse in a nondirected, random spatial process. Once established, individuals will have to cope with the local environment and the temporal variation in these conditions. To this goal, they only have a limited number of options. Seeds can arrive at sites where they will not germinate, or will not establish, despite the overall suitability of the patch. For instance, many plant species rely on the presence of gaps in the vegetation for successful establishment (e.g., Bullock et al., 1995; Moloney and Levin, 1996; Eriksson, 1997; Brokaw and Busing, 2000).

At the population level, plants can either adapt to local conditions or have enough plasticity to adapt their phenotype. Local adaptation has been demonstrated in many cases, including local adaptation to heavy metals (McNeilly and Bradshaw, 1968; Antonovics et al., 1971; Pollard, 1980), to competition with other species (Turkington and Harper, 1979; Burdon, 1980) and to grazing pressure (van Tienderen, 1989). Reciprocal transplant experiments have shown that local adaptation may occur at large spatial scales (Schmidt and Levin, 1988; Jordan, 1992), but also at small scales of 100 m or less (Antonovics, 1976; Schmitt and Gamble, 1990; Bell et al., 2002). Metapopulation models, when dealing with plants, should take this variance in source quality into account. The extent of genotypic adaptation will be the result of a selection process. The efficiency of this process is influenced strongly by the amount of migration. If migration among populations is common, a continuous mix of genotypes through the regional system occurs and local adaptation is unlikely (although not impossible; Barton (2001)) to evolve. However, if migration rates are low, as will often be the case in plants, gene flow does not disrupt results of the selective process and plants become adapted to their local environments. The adaptive differentiation will then be a function of the heterogeneity at the regional scale.

A prominent feature of plants is that genotypes often are able to change their phenotype in relation to environmental conditions. Although such phenotypic plasticity does occur in animals as well, it is certainly a prominent feature in plants. Whether plants will cope with environmental heterogeneity via genotypic adaptation or phenotypic plasticity, that is, whether they will be specialists or generalists, is a function of the amount of heterogeneity, the frequency of encountering various environments, the dispersal between various environments, and the fitness costs of establishing in each local environment (Gilchrist, 1995; van Tienderen, 1997; Reboud and Bell, 1998).

Thus, individuals throughout a regional system will not have the same value for the dynamics: regional systems will be characterized by a variance in source quality. This can be expressed either as a mean source quality value for each combination of source and target population or as a variance component, covering the variance in source quality throughout the regional system, that is added to the average colonization rates in models.

However, even at the within-population, genotypic level, a significant variance in source quality may exist. Ouborg et al. (2000) investigated the interaction between the host plant *Silene latifolia* and its specific pathogen *Microbotryum violaceum*. This host–pathogen system is characterized by metapopulation dynamics (Antonovics et al., 1994). After founding of a new local population, the development of that population will be characterized by a continuous increase in the inbreeding level. Ouborg et al. (2000) investigated the effect of inbreeding on the interaction between the host plant and the fungal pathogen and discovered that although inbreeding on average increased the resistance of the host, there was a strong and significant difference between genotypes in inbreeding effect, both in direction and in magnitude. Within the same population, in some genotypes, inbreeding increased resistance, whereas

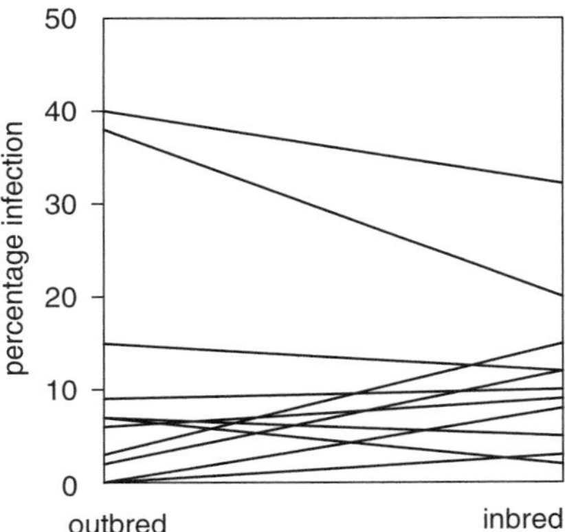

Fig. 18.2 Regression lines of inbreeding level versus percentage infection of *Silene alba*-inbred lines with *Microbotryum violaceum* fungus. Lines originate from the same field population. The variance between lines reflects the variance in inbreeding effects, both in magnitude (slope) and in direction (positive or negative slope). After Ouborg et al. (2000).

in other genotypes, inbreeding decreased the resistance (Fig. 18.2). This variance in inbreeding effects has been found in other plant species and in other traits (e.g., McCauley et al., 2001; Richards, 2000). As a consequence of this variation, the effects for the dynamics of local and regional populations will be very dependent on which genotype founds a new population. This illustrates the intricate, subtle effects of source quality variation.

Species Interactions

Many, if not all, plant species rely on interactions with other species for the completion of their life cycle. The most obvious example is the interaction many plant species have with specialist or generalist pollinating animal species. Plants may be pollinated by insects, birds, and mammals. Without the presence of these pollinators, the seed set will be low or even absent. Because pollinators tend to respond to the abundance of flowers, small plant populations are often subjected to a reduction in pollination. This leads to a positive density effect at low plant densities, (i.e., an Allee effect). Such effects may also be influenced by patch isolation. For example, Groom (1998) found that small patches of the annual *Clarkia concinna* suffered from reproductive failure due to a lack of pollinators when the patches were more than 26 m apart. For large patches, no isolation threshold occurred. Ample evidence exists that habitat fragmentation [which in this context can be perceived as the breaking up of previously (possibly) existing metapopulation structure] results in a decreased seed set (e.g., Oostermeijer et al., 1998; Fisher and Matthies, 1997; Luijten et al., 2000; Vergeer et al., 2003) or in reduced quality of the resulting seeds as a consequence of increased selfing in local, isolated populations (Rayman et al., 1994; van Treuren et al., 1993, 1994). Because seed limited recruitment, at both local and regional scales, is common (Turnbull et al., 2000), the regional dynamics of plant species cannot be understood and modeled completely without taking the local and regional dynamics of their pollinators into account. However, the effects on plant reproduction of habitat connectivity may be more complex than a simple patch size and isolation effect mediated by pollinators. Seed predators may also increase their patch incidence when connectivity increases, counteracting positive effects by pollinators. Steffan-Dewenter et al. (2001) found such counterbalancing effects of seed predators and pollinators, resulting in no net effect of habitat connectivity on seed production in *Centaurea jacea*.

Another example of a vital mutualistic relationship is the interaction many plant species have with arbuscular mycorrhizal fungi (AMF) (Smith and Read, 1997). Evidence has been presented (Gange et al., 1993; van der Heijden et al., 1998) that successful colonization and establishment of a plant species are only possible in the presence of the right species of mycorrhiza. Moreover, the presence or absence of AMF may alter the interaction between plants and herbivore insects (Brown and Gange, 2002). Thus, plant metapopulation models should take the local and regional dynamics of these fungal mutualists into account.

Plants constitute the habitat for herbivores and seed predators as well as pathogens, and plants are used as food for larger mobile herbivores, including, for example, frugivorous birds and mammals acting as seed dispersers. All these interactions may be potentially important for the plant. If the interactors respond to local population size and isolation (which seems reasonable), the plant will be affected by the spatial configuration of local populations, even if the plants themselves (by colonization and extinction processes) are not affected directly by the underlying habitat configuration. Evidence provided by Tscharntke and co-workers, demonstrated the relevance of multitrophic interactions in assessing the effects of habitat fragmentation. In manually established islands of red clover (*Trifolium pratense*), isolated patches were colonized by most of the available herbivores but only a few of the available parasitoid species. In isolated patches, herbivores experienced only 19 to 60% of the parasitoid attacks compared to nonisolated patches (Kruess and Tscharntke, 1994). In another experiment, 32 natural stinging nettle (*Urtica dioica*) patches of different size and degree of isolation were investigated. Habitat fragmentation reduced species richness, but not all species groups were affected to the same degree and in the same way. Monophagous insects were most affected by the area of the patch, whereas predatory insects were most affected by the degree of isolation (Zabel and Tscharntke, 1998). In experiments with bush vetch (*Vicia sepium*) plants in pots, the overall colonization success of insects decreased with increasing distance (Kruess and Tscharntke, 2000). Moreover, parasitism on the rape pollen beetle, a pest on *Brassica napus*, responded positively to increased habitat connectivity, thus enhancing seed production (Thies and Tscharntke, 1999).

In addition, insects that respond to landscape structure seem to perceive the landscape at different spatial scales (Roland and Taylor, 1997; Steffan-Dewenter et al., 2002). This implies that analyses of how landscape structure influences the whole range of processes ultimately determining plant fitness, pollination, seed production, seed predation, and seed dispersal should account for effects that appear in different spatial scales. Still rather few studies have addressed this complexity in plant–animal interactions in relation to landscape structure. From the studies that are at hand, however, we can conclude that effects of habitat connectivity do occur, but that these are not necessarily straightforward positive effects of patch size and negative effects of isolation. The regional dynamics of plants will be affected through variable responses of the next trophic levels. This exciting research field surely deserves more attention in the future.

18.4 ESTIMATING PLANT DISPERSAL

The amount of dispersal taking place as a function of distance between local populations is a core issue for any metapopulation concept. Thomas and Kunin (1999) argued that the type of dynamics that is inferred in studies of regional dynamics may be altered arbitrarily by chosing different spatial scales of study. It is therefore very important to define the appropriate scale for regional studies. Such an a priori definition should be based on estimates of dispersal rates and distances.

Dispersal may also be the most important issue underlying the discussion on the existence of plant metapopulations (e.g., Husband and Barrett, 1996; Eriksson, 1996; Bullock et al., 2002). Obviously, the most efficient way to solve this debate is to obtain reliable estimates of plant dispersal distances in a regional context. Unfortunately, quantifying dispersal, especially long distance dispersal, has always been one of the most difficult tasks in plant population biology.

An important topic for the discussion is interpretation of the plant dispersal distance curve. Typically, such curves are extremely leptokurtic, with the overwhelming majority of seeds dispersing over very short distances (in the order of a few meter) within local populations and only a very small proportion dispersing over longer distances, between local populations. Thus, long-distance dispersal events are rare, but have a great importance for plant migration, probabilities of colonization of suitable but unoccupied habitat, and for the metapopulation structure in general. Indeed, studies suggest that understanding these processes necessitates that the form of the tail of the dispersal curve, resulting from "chance events," is taken into account (Cain et al., 1998; Clark, 1998; Higgins and Richardson, 1999; Bullock et al., 2002).

Several approaches to studies of dispersal can be found in the literature (Table 18.1). Nathan (2001) mentioned three categories of approaches. In the first category (movement-redistribution methods; Nathan, 2001), the movement of individuals through space is measured directly by marking individuals. In the strict sense of traditional mark–recapture methods, which were designed for the study of animal dispersal, these methods are unsuitable for the study of dispersal in plants due to the impracticalities of marking and tracking large amounts of small seeds through potentially large amounts of space. Some studies have tried to measure the actual distance over which individual seeds disperse by trapping seeds at various distances from a source (e.g., Ruckelshaus, 1996; Thiede and Augspurger, 1996; Bullock and Clarke, 2000). However, because of the logistical problems with seed trapping at larger distances, these methods almost invariably rely on extrapolation to estimate the tails of the dispersal curve, and therefore the frequency and extent of long distance seed dispersal. Moreover, often these studies will be performed in situations where a single point source is placed in a habitat that is otherwise unoccupied by the focal species. Whether the resulting data can be translated without bias to more natural situations is unclear. Within this

TABLE 18.1 An Overview of Approaches to Studies of Plant Dispersal

Method	Type of estimate	Reliability of long-distance dispersal estimates	Sampling effort	Refs.
Mark–recapture studies of seeds	Instantaneous, dispersal	low	Very large	Platt and Weis (1977); Lee (1984); Nilsson et al. (1991); Johansson and Nilsson (1993)
Seed trap experiments	Instantaneous, extrapolation, dispersal	low	Large	Ruckelshaus (1996); Thiede and Augspurger (1996); Bullock and Clarke (2000)
Diffusion modeling	Prediction, dispersal	average	Minor	van Dorp et al. (1996); Greene and Johnson (1996); Cain et al. (1998); Jongejans and Schippers (1999); Soons and Heil (2002)
Demographic modeling	Prediction	Unknown	Average	Neubert and Caswell (2000); Bullock et al. (2002)
Indirect molecular techniques	Historical, dispersal + establishment	Case dependent	Average	cf. Ouborg et al. (1999)
Direct molecular techniques	Instantaneous, dispersal + establishment	Average to good	Very large	Meagher and Thompson (1987); Dow and Ashley (1996); Schnabel et al. (1998); Isagi et al. (2000)

same category of approaches, some studies use mark–recapture methods with natural seeds (e.g., Platt and Weis, 1977; Lee, 1984; Johansson and Nilsson, 1993) or with artificial seed mimics (Nilsson et al., 1991). In addition to the enormous effort that has to be put into these type of experiments, both in terms of labor and in the number of seeds to be used, the recovery rate of marked seeds at long distances is very low, making the estimations unreliable. In addition, the general value of dispersal estimates from these experiments is limited because they essentially (try to) measure one realization of a dispersal process, which may change from situation to situation and from time to time. Thus, it is very difficult to reliably estimate long-distance dispersal using this category of approaches (Silvertown, 1991; Bullock and Clarke, 2000).

A second category of approaches uses mathematical modeling to describe dispersal patterns and infer long-distance dispersal (Nathan, 2001). These methods predict rather than measure dispersal. Models that deal with wind dispersal are presumably the most advanced. The basic rationale in these models is that aerodynamic properties, which are measured in wind tunnel experiments (e.g., van Dorp et al., 1996; Jongejans en Schippers, 1999; Soons and Heil, 2002) and which result in terminal velocities of seeds, are fed into specific aerodynamic models (e.g., Greene and Johnson, 1996; Cain et al., 1998; Soons and Heil, 2002; Tackenberg et al., 2003; Tackenberg, 2003), which then transform wind and landscape profiles into distributions of dispersal distances. In general, these methods are reliable in the short dispersal range and less reliable in the biologically more relevant long dispersal range. It has been argued that the inaccurate description of long-distance dispersal in these models is the consequence of the relative inflexibility of the mathematical functions used (Bullock and Clarke,

2000; Nathan and Muller-Landau, 2000; Nathan, 2001). New functions have been proposed that allow "fatter" tails of the dispersal curve, thereby improving the fit of the model to actual dispersal (Nathan and Muller-Landau, 2000; Bullock et al., 2002; Fig. 18.3).

Another approach within this category has been proposed by Neubert and Caswell (2000) in analogy to the life table response analysis (Caswell, 2000). Their approach uses a combination of matrix projection modeling to describe and predict local demography, with integrodifference equations to capture dispersal. The method allows calculation of the sensitivity of spatial expansion speed to changes in local demographic parameters and regional dispersal parameters. Neubert and Caswell (2000) demonstrated that the rate of spatial expansion is governed by the long-distance component of dispersal, even when long-distance dispersal is rare. This result was confirmed by analyses of demographic and dispersal data for three heathland species (*Calluna vulgaris*, *Erica cinerea*, and *Rhinanthus minor*) (Bullock et al., 2002). Further development and application of this approach are needed to evaluate its usefulness and practicality in attempts to measure the implications of long-distance dispersal.

The third category of approaches is to use molecular markers and population genetic analyses (Ouborg et al., 1999; Cain et al., 2000; Ennos, 2001; Raybould et al., 2002). With the continuous invent of new marker techniques and the increasing automation of their application, these methods are promising. Two basic methods can be followed when applying molecular markers in the study of dispersal in plants (Ouborg et al., 1999). First, dispersal can be assessed from observed distributions of genetic variation in space. This indirect method is based on the quantification of genetic divergence between populations and the interpretation of this divergence in terms of the amount of past gene flow. Second, direct approaches try to establish the parent–offspring relationships between individuals in space. The various approaches using molecular markers to estimate dispersal rates are discussed in Chapter 15.

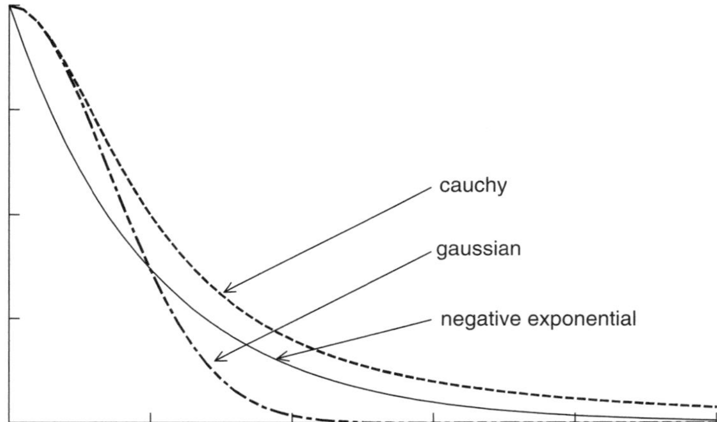

Fig. 18.3 Three dispersal functions demonstrating the difference in tail of the distribution. The negative exponential [$y = \exp(-x)$] is perhaps the most frequently used; the gaussian (normal) function [$y = \exp(-x^2)$] has a thinner tail, and the cauchy function [$y = 1/(1 + x^2)$] has a fatter tail, allowing more easy fitting of rare long-distance dispersal events.

It seems that long range dispersal, and its relevance for metapopulation dynamics, will never be quantified easily using a single method (Ouborg et al., 1999; Raybould et al., 2002). When studying plant dispersal, it seems advisable to apply several methods to the same study system for at least two reasons. First, one method may result in a hypothesis on dispersal and regional connection between local populations that can be tested independently with a second method. Second, various methods will provide different types of information about the dispersal process. Direct approaches will give estimates of instantaneous dispersal, whereas indirect methods will give estimates of historical gene flow. Mark–recapture estimates will address the seed dispersal component of dispersal only, whereas indirect (and some direct) approaches will estimate the combined results of seed dispersal and subsequent establishment.

This overview illustrates the perhaps somewhat depressing complexity of the study of dispersal; it is very likely that this complexity has resulted in the lack of data that is needed to build our arguments on the existence or absence of metapopulations in plants. However, there are some promising methods that reduce this complexity to manageable proportions by helping us to decide whether at least populations do exchange individuals at all. For instance, Rannala and Mountain (1997) presented a statistical method for detecting immigration using multilocus genotype data based on Bayesian statistical inference. In addition, other assignment methods have been developed where individuals can be assigned to a particular population; any individual assigned to another population than where it was sampled indicates a dispersal event (e.g., Pritchard et al., 2000). Some authors suggest that the maximum distance seeds disperse, rather than the frequency of dispersal events, determines the regional dynamics (Neubert and Caswell, 2000; Bullock et al., 2002). If this is true, it perhaps becomes more important to assess whether two populations ever exchange migrants rather than estimating the long-distance dispersal rate in detail.

18.5 EXAMPLES OF METAPOPULATION STUDIES IN PLANTS

An essential feature of regional population systems that exhibit metapopulation dynamics is that colonization and extinction processes are related to the configuration of habitats in the region. With configuration, we mean the size distribution of habitat patches, their shape, and the extent to which they are connected to each other by habitat corridors or by dispersal routes. Effects of habitat configuration on regional plant populations can result from different mechanisms. (1) Since seed dispersal from source populations is likely to be related to the distance the likelihood of colonization is expected to decline as habitat patches become more isolated. (2) If the same distance effects occur for pollen transport, isolated populations may suffer from reproductive limitations (e.g., in self-incompatible plants) and inbreeding. (3) Interactions other than related to pollination, e.g., herbivores and pathogens, may reflect landscape structure. (4) The size and shape of habitat patches may influence their capacity to harbor local populations. (5) The structure of the landscape surrounding habitat patches may influence the dispersal among patches, e.g., by "corridors" or by "stepping-stone dispersal." An issue related to these hypothesized mechanisms for configuration effects is whether there are threshold effects, (e.g., if there is a minimum amount of available habitat needed to sustain a persistent regional population or if there is a maximum distance between habitat patches above which no effective dispersal can be achieved). If such thresholds exist, the behavior of a local population may change drastically if the threshold value is passed.

Evidence supports all five mechanisms for habitat configuration effects on plants. Distance effects on colonization have been found in temperate woodlands (e.g., Grashof-Bokdam and Geertsema, 1998; Butaye et al., 2001; Jacquemyn et al., 2001; Verheyen and Hermy, 2001), open grasslands (e.g., Ouborg, 1993; Fig. 18.4), serpentine vegetation [Harrison et al., (2000), who also found evidence for rescue effects related to distance], scrub vegetation (Quintana-Ascencio and Menges, 1996), seasonal pools (Husband and Barrett, 1998), and urban vegetation (Bastin and Thomas, 1999). In practice, it may be very difficult to distinguish effects of size and isolation in studies of fragmented plant populations, as these two landscape features normally change along with each other. Moreover, factors other than isolation (and patch size) were also important for species occupancy patterns in some of these studies, for example, age of the sites (Bastin and Thomas, 1999; Jacquemyn et al., 2001). Small and isolated plant populations have been found to suffer from reduced reproduction (e.g., Jennersten, 1988; Groom, 1998; Jules, 1998; Morgan, 1999; Cunningham, 2000) and effects of inbreeding (e.g., Oostermeijer et al., 1994; Ouborg and van Treuren, 1994, 1995; Kéry et al., 2000; Richards, 2000; McCauley et al., 2001). The shape of habitat patches affects the edge-interior relationship, which in turn may influence colonization patterns (e.g., Restrepo et al., 1999; Kiviniemi and Eriksson, 2002). Thus, although it may well be that the general effects of habitat decline and deterioration have a dominating impact on the persistence of plants inhabiting fragmented landscapes (Harrison and Bruna, 1999; Fahrig, 2002), this evidence suggests that

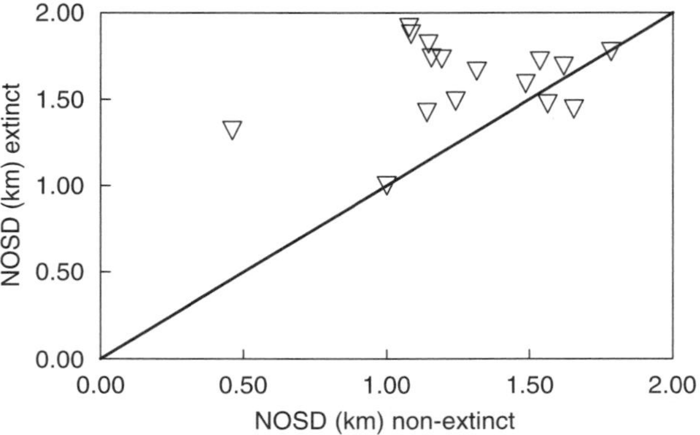

Fig. 18.4 In an analysis of 143 grassland sites along the Dutch Rhine, 16 plant species were examined in detail. For most species, populations that went extinct in a 32-yr period had a greater distance to nearest occupied sites (NOSD) than populations that remained extant. This illustrates the importance of regional processes for the local persistence of plant populations (after Ouborg, 1993).

changes in habitat configuration per se do influence the regional dynamics of plants in a range of different vegetation types.

In contrast, the effects of habitat corridors remain more controversial. Malanson and Cairns (1997) suggested that the effects of habitat isolation are more dependent on the capacity of source populations to produce diaspores than on the barrier effect of surrounding landscapes. This may, however, be treated conceptually by including the size of the source populations in the definition of habitat connectivity (Moilanen and Nieminen, 2002). Some authors have questioned whether habitat corridors are effective in enhancing dispersal (e.g., van Dorp et al., 1997; Cain et al., 1998). A rationale for this view is that long-range dispersal is mainly dependent on unlikely "chance events" (Clark, 1998; Higgins and Richardson, 1999; Cain et al., 2000; Bullock et al., 2002). Assuming this is the case, both range expansion and dispersal among relatively isolated habitat patches may be decoupled effectively from the structure of the landscape surrounding habitat patches, and even the distance among patches. Despite the plausibility of this argument, the studies listed earlier indicate that distance does play a role in dispersal also beyond the local neighborhood. Moreover, some evidence supports the hypothesized corridor effects: along rivers (Johansson et al., 1996; Burkart, 2001), hedgerows (Corbit et al., 1999), roads (Cousins and Eriksson, 2001), and railway verges (Tikka et al., 2001).

Theory suggests that an extinction threshold exists at the lowest amount of habitat in a region needed for sustaining a regional population (Lande, 1987). Under a set of simplifying assumptions, this amount equals the equilibrial unoccupied fraction of suitable habitat in a landscape (Lawton et al., 1994). A more realistic definition of a threshold value for the capacity of a certain landscape to harbor a metapopulation was suggested by Hanski and Ovaskainen (2000). This measure, the metapopulation capacity, incorporates both aspects of the configuration of habitat patches, i.e., their size and isolation, and dispersal features of the focal species. Even if improved realism makes studies of extinction thresholds more feasible, it is still very difficult to examine extinction thresholds empirically (Fahrig, 2002). One complicating factor is that extinction processes for most plants are subjected to time lags (Eriksson and Kiviniemi, 1999; Eriksson and Ehrlén, 2001; see also Hanski and Ovaskainen, 2002). Nevertheless, a few studies provide evidence suggesting that there are threshold effects manifested by a minimum amount of habitat or maximum allowed distance among patches. Husband and Barrett (1998) found that no populations of the water plant *Eichhornia paniculata* existed when the density of potential sites fell below a certain value. Butaye et al., (2001) found that isolation-sensitive woodland plants did not occur when the distance to source populations was above 200 m. In a study of an endemic herb, *Scutellaria montana*, Cruzan (2001) concluded that effects of habitat configuration appeared at different spatial scales. At a sampling scale of 2 km, small metapopulations had higher levels of selfing; at a sampling scale of 8 km, metapopulation size was related to levels of genetic diversity. Whether any of these genetic effects directly influence the persistence of the *S. montana* metapopulations were, however, not clear.

Even though there is much to be done before well-founded generalizations can be made on how habitat configuration influences regional plant populations, evidence suggests that there are configuration effects in many different kinds of vegetation. Skarpaas (2003; see Box 18.1) demonstrated that even though assumptions associated with a metapopulation approach may not strictly apply to (some) plant populations, the approach may still lead to valid results. A metapopulation concept that incorporates time lags (see Section 18.3) and allows for slow dynamics is likely to be a most productive approach for future studies.

BOX 18.1 Plant Population Dynamics in a Fragmented Landscape

In a comprehensive study on spatial dynamics of two plant species, Skarpaas (2003) tested various modeling approaches. He chose two contrasting species — the oyster plant *Mertensia maritima* (Boraginaceae) and the leafless hawk's beard *Crepis praemorsa* (Asteraceae) — to test whether predictions of metapopulation models would agree with results of a modeling effort to incorporate variance in local dynamics and more realistic dispersal-distance relationships. Both species are regionally rare and declining in northern Europe. *M. mertensa* is naturally fragmented with suitable sites surrounded by inhabitable unsuitable matrix (the sea), making the habitat structure of this species resemble an ideal metapopulation structure. In contrast, *C. praemorsa* is recently fragmented by landscape changes and is inhabiting seminatural grasslands, where the distinction between suitable and unsuitable habitat may be less strict. To test for spatial effects on occupancy, logistic regression models were fitted to either incidence (*M. maritima*) or extinction and colonization rates (*C. praemorsa*) using the model

$$\log\left(\frac{p}{1-p}\right) = \beta 0 + \beta 1 I + \beta 2 A$$

where p is incidence, extinction, or colonization, A is patch area, and I is isolation of occupied (when modeling extinction) or unoccupied (when modeling colonization) patches. (βj are regression coefficients). For *M. maritima*, there was no significant effect of area on incidence, but distance to the nearest occupied site or distance to all occupied sites significantly affected incidence. For *C. praemorsa*, extinction was not related to either area or isolation, whereas colonization was only influenced by isolation measures. The fitted regression models were used in a long-term simulation of metapopulation dynamics, leading to predictions of occupancy as a function of time.

Skarpaas next modeled both the local demography of both species, using matrix projection models, and the regional dispersal behavior with the use of various dispersal models (e.g., Greene and Johnson, 1989; Clarke et al., 1999; Nathan et al., 2001). In this way he explored the limits of a simple metapopulation approach when applied to these species. He presented evidence that the strict dichotomy between suitable and nonsuitable habitats did not apply to both species. In addition, he showed that, contrary to the assumption of simple metapopulation models that colonization is random, colonization for these species followed an isolation by distance relationship (Skarpaas and Stabbetorp, 2001; Skarpaas, 2003). Despite these deviations from the basic metapopulation models, he demonstrated that such models would still describe the regional dynamics of *M. maritima* fairly accurately. For *C. praemorsa*, the predictions of metapopulation models were not in agreement with the result of demographic and dispersal modeling. The study demonstrates therefore, that metapopulations in plants may exist, even though not all assumptions strictly apply.

18.6 REGIONAL DYNAMICS AND METAPOPULATION ARGUMENTS IN PLANT POPULATION BIOLOGY

Several areas in plant population biology have explicitly incorporated regional dynamics, or metapopulation dynamics, into their theories. This Section gives three examples. First, dispersal plays a central role in the metapopulation concept. Olivieri et al. (1995) and Olivieri and Gouyon (1997) developed a theory for how metapopulation dynamics may affect the evolution of dispersal ability. They suggested that two opposing selective forces, which they baptized as the metapopulation effect, affect dispersal traits, (i.e., fruit and seed traits such as plumes, bristles, showy fleshy fruits, and spines) that may be interpreted as adaptations to dispersal. At the local population level there will be selection against dispersal, as seeds may be dispersed to unsuitable sites. At the regional level, however, there may be selection promoting dispersal, especially in highly dynamic situations with a high extinction rate of local populations. In this last situation, dispersal will be a means of persistence at the regional scale.

Cody and Overton (1996) found that the evolution of dispersal traits may be very rapid in situations where suitable habitat is surrounded by large areas of unsuitable habitat. Any dispersal beyond the borders of the local population will lead to loss of individuals, making the selection pressure against dispersal very high. They measured the dispersal potential, defined as the ratio of pappus to achene volume, of *Lactuca muralis* in mainland and island populations around Vancouver, Canada. Newly colonized islands had populations with an increased dispersal potential, as only individuals with a good dispersal ability will be able to reach the islands. However, in the years following colonization, the dispersal potential dropped to levels below that of the mainland populations.

Roff and Fairbairn (2001) presented evidence that dispersal traits are often correlated genetically to other life history traits. Thus, evolution driven by the metapopulation effect on dispersal may result in correlated evolution of other traits, such as dormancy (Rees, 1996; Olivieri and Gouyon, 1997). Most evidence, however, comes from studies with animals (Roff and Fairbank, 2001), which is surprising given the suitability of plants to perform the large crossing studies necessary to estimate the genetic correlative structure. In conclusion, the balance between opposing forces imposed by local and regional processes may drive evolution of a range of life history traits. This theory is, however, awaiting experimental data to be tested.

A second area where metapopulation dynamics enters the theory is the evolutionary dynamics of reproductive systems. Various studies present evidence that attributes of the breeding system of plants may be affected by the metapopulation effect (Olivieri et al., 1995) of frequent extinction and colonization, for example, the frequency of females in local populations (Couvet et al., 1986), the frequency of various flower morphs (Husband and Barrett, 1995; Eckert et al., 1996), and the rate of self-fertilization in local populations (Husband and Barrett, 1992). It is probably not possible to understand the dynamics and evolution of these features without taking regional dynamics into account.

A third example is the evolutionary dynamics of plants and their herbivores or pathogens. Plants in natural populations are generally challenged by a wide variety of herbivores and pathogens and can be affected strongly by these natural enemies. Effects on individual hosts may extend to effects on population size, dynamics, and population structure (Burdon, 1987), as well as community structure (Dobson and Crawley, 1994; Peters and Shaw, 1996), for instance, by altering the relative competitive abilities of species (e.g., Paul and Ayres, 1990; Clay, 1990) that cause changes in their relative abundance or by affecting succession rates (Van der Putten et al., 1993) and local species diversity (Packer and Clay, 2000). Plants have evolved a variety of different mechanisms by which they defend themselves against natural enemies. Direct defenses of plants include at least three types of mechanisms: avoidance, resistance, and tolerance. Of these mechanisms, the molecular and genetic basis of gene for gene (GFG) resistance to pathogens is probably the best documented, and this mechanism has served as the basis for the majority of models of the evolution of host–parasite interactions (Thompson and Burdon, 1992). The GFG hypothesis states that for each gene determining resistance (R) in the host, there is a corresponding gene for avirulence (Avr) in the pathogen with which it interacts specifically.

An example of a GFG interaction in a natural system is the interaction between *Linum marginale* and the rust fungus *Melampsora lini*, endemic to Australia. Studies on the dynamics of host resistance types and pathogen virulence types in this system have yielded valuable knowledge about the spatial scale at which such GFG interactions occur. The frequency of different races within local pathogen populations in this system appears to be poorly correlated with the frequency of different resistance types within the corresponding local host populations (Jarosz and Burdon, 1991). Stochastic processes during population crashes rather than natural selection within local populations appear to be the main cause of large year-to-year variation in the frequencies of R and Avr alleles. For instance, after a severe epidemic, host genotypes resistant to the pathogen races that were present at high frequency during the epidemic surprisingly had not increased but decreased in frequency, opposing the view that natural selection in local populations is governing GFG coevolution (Burdon and Thompson, 1995). The authors suggested that individual populations are mainly influenced by genetic drift, extinction, and gene flow among populations within the same epidemiological region and that GFG coevolution is likely to take place at the metapopulation rather than the local population level. In other cases, authors have argued that the stability of host–pathogen interactions is only possible at regional scale levels, whereas at the local level, one or both of the interactors are bound to go extinct (e.g., Antonovics et al., 1994; Hess, 1996). These examples demonstrate that metapopulation theory forms an inextricable part of plant population biology.

18.7 CONCLUDING REMARKS

Considerable effort has been devoted to defining different types of regional populations (here used in the widest possible sense), for example, source–sink, mainland–island, patchy, and remnant; new suggestions include "regional ensembles" and "spatially extended populations" (Freckleton and Watkinson, 2002), in addition to metapopulations "in a strict sense" (Eriksson, 1996) or "true metapopulations" (Bullock et al., 2002). This question of classifying different types of regional populations would be important if it were shown that different concepts were productive in developing knowledge and insights on the pattern and process of plant regional populations. We agree that it would be erroneous to "force" a diversity of regional dynamics into a narrow set of concepts and definitions. For instance, studies on *Vulpia ciliata* (Carey et al., 1995; Watkinson et al., 2000) suggest that alternative regional models are more suitable than the metapopulation model. However, we are not aware of any alternative conceptual framework that has been even nearly as productive as the metapopulation concept in stimulating studies of regional populations. Thus, a broadening of the metapopulation concept may be more useful than developing new terms and concepts. Basically, metapopulation models focus on colonization and extinction processes and their relationship to the landscape structure. As discussed in this chapter, plant dynamics may be slow, nonequilibrial, determined by landscape history, and it includes plant features that present difficulties to any student working with regional dynamics: seed banks, long life spans, and elusive dispersal processes. Still, these difficulties can be incorporated into a conceptual framework based on colonization/extinction in relation to landscape structure. The choice of spatiotemporal scale is essential (Thomas and Kunin, 1999). Typologies of different forms of plant regional dynamics often make use of a hypothetical spectrum of landscapes with small to large amounts of suitable habitat (Freckleton and Watkinson, 2002) or, phrased differently, small to large amounts of migrants among sites (Bullock et al., 2002). Metapopulations ("true" or "strict") are placed at the intermediate portion of this spectrum. However, a broad-sense metapopulation concept can easily incorporate the whole spectrum (Ehrlén and Eriksson, 2003). When there is a small amount of suitable habitat and migration among sites is small, the temporal scale has to be extended; also, slowly fluctuating regional plant populations (where colonization appears to be almost nonexistent for a student working in the time scale of an ordinary research project) possess regional dynamics likely to obey the same colonization/extinction dynamics as more rapidly fluctuating regional populations (cf. Whittaker and Levin, 1977). When there is a large amount of suitable habitat and migration among sites is large, the spatial scale has to be extended; plants with large spatially extended populations are also likely to possess a patchiness at a larger spatial scale with colonization/extinction dynamics similar to more fine-grained populations. Thus, we believe that metapopulation studies will continue to be the most productive approach to advance the understanding of regional plant population systems.

19. LONG-TERM STUDY OF A PLANT–PATHOGEN METAPOPULATION

Janis Antonovics

19.1 INTRODUCTION

Although there has been a long-standing recognition that the numerical and gene frequency dynamics of natural populations may be affected by the interconnectedness of populations on a regional scale (Wright, 1931; Levins, 1969; MacArthur and Wilson, 1967), it is only since the early 1980s that attention has been given to the explicit study of interconnected sets of populations and to the theoretical exploration of the consequences of spatially explicit models for ecological and genetic processes (Silvertown and Antonovics, 2001). In the context of field studies, there is also increasing recognition that migration among interconnected populations and local extinction and recolonization are the rule rather than the exception in natural populations (Gilpin and Hanski, 1991; Hanski and Gilpin, 1996). Early metapopulation models assumed simplified within population dynamics driven largely by the effects of colonization, migration, and extinction (Levins, 1969; Caswell, 1978). More recently, with the advent of increased computational power, it has been possible to explore the consequences of within population dynamics on spatially extended systems and in multiple interconnected populations (Comins et al., 1992; Kareiva, 1994).

0-12-323448-4

The major feature to emerge from theoretical studies of spatially explicit systems is that conclusions regarding equilibrium states and dynamics derived from single populations are changed drastically when interactions among these populations are included. With regard to ecological dynamics, the best-known conclusion is that systems that show locally unstable population dynamics (such as would lead to extinction) can be stabilized readily when extended spatially (Comins et al., 1992; Antonovics et al., 1994; Molofsky et al., 2001). Conversely, it has been shown that changes in connectedness of populations can, in and of itself, drastically influence the overall prevalence of a species, without changes in the local dynamics (Carter and Prince, 1988; May and Anderson, 1990; Hanski, 1991). With regard to evolutionary dynamics, genetic change within populations can also be stabilized and allelic diversity can be maintained for extended periods in spatially explicit models (Frank, 1991). Many of the statistics of among population differentiation are also altered by explicit consideration of extinction and colonization processes (McCauley, 1993). Metapopulation structure can enhance the importance of kin selection or group selection by altering the local frequency of phenotypes (Gilpin, 1975; McCauley and Taylor, 1997; O'Keefe and Antonovics, 2002).

Short-term studies of a metapopulation can lead to estimates of population turnover and can be used to parameterize models that can be used as "surrogates" for experimental studies (Antonovics et al., 1998). Even single season studies of metapopulations can provide useful data for assessing distance dependence and size dependence of habitat occupancy and as a guide to conservation decisions (Hanski, 1991). However, whether a metapopulation is itself stable can only be determined if there are historical data on the state of the system at some point in the past or by long-term studies.

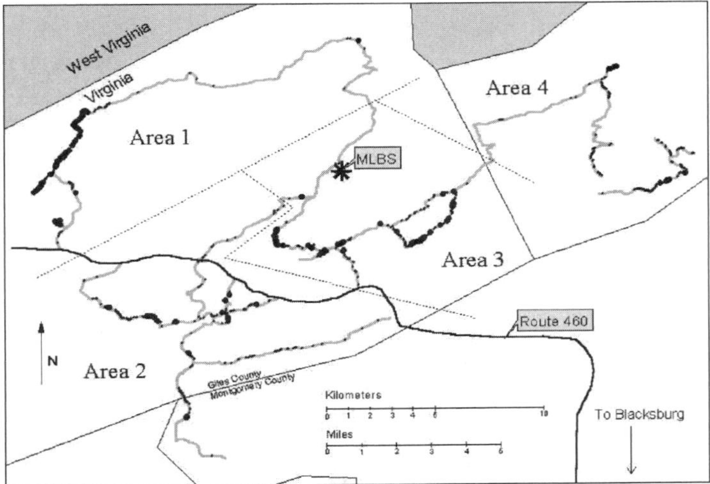

Fig. 19.1 Diagrammatic map of the census area showing the roads along which populations were censused (gray) and position of roadside segments that had healthy (small dots) or diseased (large dots) populations at some time during the census. Note that the scale results in an apparent overlap of populations that are often separated. Dotted lines separate the four "replicate" areas identified in the analyses. Populations were not censused along Route 460, which is a major highway; however, only rarely was the occasional plant seen along this highway.

We have been studying anther-smut disease caused by the fungal pathogen *Microbotryum violaceum* (= *Ustilago violacea*) in several hundred populations of the plant *Silene latifolia* (= *S. alba*) for 14 yr in the region of Mountain Lake Biological Station in western Virginia (Fig. 19.1). This chapter reports the results of these studies and discusses the factors that contribute to the overall stability of this metapopulation system.

19.2 THE STUDY SYSTEM

Life Cycles of Host and Pathogen

Silene latifolia, or white campion, is a short-lived perennial herb native to Europe commonly found in ruderal habitats throughout the northern regions of the United States and in upland areas farther south. Infection by *M. violaceum* results in the plant producing anthers that release fungal spores rather than pollen. In *S. latifolia*, which is dioecious, in addition to infecting the anthers in males, the pathogen induces the female flowers to produce stamens that bear diseased anthers. The ovary is aborted and sterile, although it is still visible as a rudimentary structure. The disease therefore has a large fitness effect by sterilizing the host, and diseased plants are identified easily in the field by their dark-smutted anthers. Anther smut is a relatively "slow" disease with a long latent period. The pathogen does not convert existing flowers into a smutted state, but grows into very young developing flower buds, which are then converted into smutted flowers. This process generally takes no less than 3 weeks, and sometimes more than 6 weeks from initial infection. Because *S. latifolia* in Virginia flowers from mid-May until early October, there are probably between one and three fungal generations per flowering season, depending when infection first takes place. The average life span of a plant that flowers is ca. 2 yr (see later). Initially, infected plants may be partially diseased, but the disease soon becomes systemic. The disease persists between seasons inside the overwintering rosette of the host plant.

The disease is transmitted when pollinators move from flower to flower. Because pollinators adjust flight distances to compensate for plant density, transmission at moderate plant densities depends on the frequency and not the density of infectious individuals, whereas at very high densities, when pollinators become limiting per capita, transmission rates decline (Alexander and Antonovics, 1992; Antonovics et al., 1995). Although the pathogen is actually vector transmitted, the frequency-dependent nature of the transmission and the expression of the disease in the sexual organs of the adult plants result in strong parallels between the biology of this host–pathogen system and other sexually transmitted diseases (Kaltz and Schmid, 1995; Lockhart et al., 1996).

There is substantial genetic variation in *S. latifolia* for disease resistance, and most populations are a mixture of genotypes that range from being almost completely resistant to completely susceptible (Alexander, 1989; Alexander et al., 1993; Biere and Antonovics, 1995). Although resistance has a high heritability (Alexander et al., 1993), the precise genetics underlying the resistance is not known. Additionally, large fitness costs are associated with resistance in the absence of the disease. More resistant plants flower later in the season and produce fewer flowers (Alexander, 1989; Biere and Antonovics, 1995). Unexpectedly, the fungus appears to be relatively uniform with regard to its pathogenicity, and therefore this host–pathogen system does not follow the classical gene-for-gene scenario (Jarosz and Burdon, 1991). Whether this is because the disease has been recently introduced into the United States from Europe and has gone through a bottleneck is not known (see Section 19.4).

All the evidence indicates that *M. violaceum* on *S. latifolia* is host specific in the United States, although a recent host shift to *S. vulgaris* has been observed (Antonovics et al., 2002). We have found anther smuts on two other native species in the southeast United States (*S. virginica* and *S. caroliniana*). However, these anther smuts are phylogenetically and chromosomally quite distinct from the one on *S. latifolia* (Perlin, 1996; Perlin et al., 1997). There is no evidence for any cross-species transmission with the native species. Moreover, Antonovics et al. (1995b) showed that the anther smut on *S. virginica* in this area is isolated reproductively from the anther smut on *S. latifolia*.

Study Area and Census Methods

In the study area, *S. latifolia* is a ruderal species that is largely confined to roadsides. Its roadside distribution allows us to gain rapid access to many populations over a large area (25 km from north to south and 30 km from east to west), while at the same time being confident that we are missing very few populations.

Because the plant is distributed in patches of differing sizes and spacing, which may coalesce or separate due to colonization and extinction events, we do not define a population in terms of the patches themselves but count numbers of diseased and healthy individuals within contiguous 40-m segments of roadsides (Antonovics et al., 1994). Therefore, in formal terms, we collect data on a one-dimensional grid system at a local scale, but at a larger scale the topology of this grid follows the pattern of the roads in the area. Distances on curves are estimated on the right-hand side of the road in the direction that the census is being made. Local landmarks (unusual trees, driveways, telephone poles, etc.) are used to demarcate each segment.

We have counted the number of diseased and healthy individuals within each roadside segment since 1988. The main census is done prior to seed dispersal in June, and a recensus in August is restricted to checking a much smaller subset of the populations that have been recorded as extinct or that have been recorded as having lost the disease. Although we make no attempt to map individuals within a segment to a precise location, we note the location (approximate distance from start of grid unit and distance from edge of road) of either healthy or diseased individuals when there are very few in a grid unit; this helps us relocate those individuals in subsequent censuses and/or confirm their absence. Our census is therefore simple and rapid; field work can be completed by three crews of two to three people in less than 1 week.

The one-dimensional grid units of 40 m include perhaps one or two, but not many, "genetic neighborhoods" (i.e., areas within which genetic exchange is essentially random) as estimated from spore, pollen, and seed dispersal distances (Alexander, 1990). They may include several distinct patches of *S. latifolia* and sometimes these patches are contiguous between grid units.

However, rarely is there a continuous "swath" of *S. latifolia* that spans more than two grid units. By analyzing field data based on pooling two, four, or eight adjacent segments, we have shown that the patterns of disease incidence are remarkably robust over a grid scale of 40–160 m (Antonovics et al., 1998).

Throughout the study we take pains not to disturb the system by our own activities during the census. Because the flowers usually close before midday, we census between 5:30 and 11:00 A.M., during which time we can determine the disease status visually without touching the plants or trampling on the sites. In this way we avoid becoming disease dispersal agents ourselves.

Not all populations of *S. latifolia* occur at roadsides. Some populations also occur on waste ground or along field edges away from the actual road. There are relatively few so-called "off-road sites" (on average 5.9 % of occupied grid units in any 1 yr) but the number of diseased and healthy plants in these sites was also recorded. Sketch maps were used to identify these sites. However, because the area of these sites is very variable, they were not included in the following analyses.

General Characteristics of the Silene–Microbotryum Metapopulation

It is clear that our system does not fit the simple conceptualization of metapopulations presented by Levins (1969) (uniform populations, no distance dependence, instantaneous within population dynamics). The populations are very different in size, dispersal is limited, and within population dynamics is important relative to the annual time scale of the study. Moreover, as in many plant metapopulations, it is not possible to define "suitable habitats" of *S. latifolia* by a clear environmental discontinuity (see Chapter 18 for a discussion of the metapopulation concept as applied to plants). Previous studies have shown that colonization and extinction of the host and pathogen populations are frequent (Antonovics et al., 1994, 1998; Thrall and Antonovics, 1995; see also results). These colonization events enhance the degree of genetic differentiation among populations (McCauley et al., 1995). The growth rate of healthy populations is density dependent, with the disease having a negative effect on population growth (Antonovics et al., 1998). In particular, high levels of disease shift population growth rates from positive to negative values (Antonovics et al., 1998). The impact of the disease on population extinction is gradual; the disease results in a declining population growth rate, and a small population size in turn presages an increased probability of extinction. Using simulations models, Antonovics (1999) showed that the presence of the pathogen can more than halve the number of occupied segments in the metapopulation as a whole.

19.3 LONG-TERM TRENDS

Analysis

In the first year, 1988, data were gathered on a 0.1-mile (ca. four grid units) scale, and no recensus was carried out. Therefore, although data from this first year were valuable in indicating the high rate of turnover in the populations and were a stimulus for embarking on the study, we did not use these data directly in the statistical analysis of population trends presented here.

To assess whether any long-term changes were general to the census as a whole, we divided the census region into four areas representing different valley systems and separated by high elevation areas (Fig. 19.1). Area 1 was the valley region of Big Stony Creek; Area 2 was the lowland area and foothills in the New River Valley; Area 3 was Clover Hollow and Route 700 up to the Biological Station; and Area 4 was Maggie Valley and the area east of Simmondsville, Route 42.

The scope of the census changed somewhat with circumstances. Thus access to one section of the census was denied by the land managers in 1994 (excluded nine grid units either side of the road). In Area 4, diseased sites present in the most northerly region were the result of artificial introduction of the disease from a spore dispersal experiment; this area was therefore not included in analyses of disease parameters. In 1998, we included a new section south of Area 4 when this became the focus of related demographic studies. Disease was present in this region. However, in order to avoid possible confounding effects, analyses of the long-term trends were based only on data from the roadside segments that were censused throughout the whole study period.

Host and Pathogen Occurrence

At a regional level, the abundance of a species can be assessed in terms of the number of populations as well as the average number of individuals within populations. Because the census was based on a grid system of roadside segments, we measured the number of host populations in terms of number of segments occupied and population size in terms of the number of individuals within each segment. The number of grid segments used in the analyses did not change, and therefore the former is a measure of regional abundance and the latter is a measure of local abundance (at the segment scale). We measured regional disease abundance as the fraction of segments occupied by *S. latifolia* that were diseased (we refer to this as "disease *incidence*") and the local abundance as the fraction of individuals that were diseased within each occupied segment (we refer to this as "disease *prevalence*").

The fraction of segments occupied by *S. latifolia* (Fig. 19.2A) did not change significantly overall ($P < 0.27$), but there was a significant area* year interaction ($P < 0.0001$). In Area 1 the occupancy declined significantly ($P < 0.0019$, $b = -0.0038$), while it increased in Areas 2 and 3 ($P < 0.015$, $b = 0.0029$ and $P < 0.0089$, $b = 0.0020$). There was no significant change in Area 4.

The average number of *S. latifolia* within each segment (Fig. 19.2B) declined markedly overall ($P < 0.0001$, $b = -0.0112$, $\log_{10}$ scale). The decline occurred in all four areas, significantly so in Areas 1–3 (P all < 0.0088), but not in Area 4 ($P < 0.13$).

The fraction of *S. latifolia* segments that were diseased, or "disease incidence," (Fig 19.3A) declined significantly overall ($P < 0.0001$; regression coefficient $b = -0.0153$, arcsin square root transformed data). Although disease incidence declined in Areas 1–3, the rate of decline differed among the areas (year*site interaction, $P < 0.0001$). Area 3 was particularly interesting in that it showed an initial increase in disease incidence, peaking in 1995

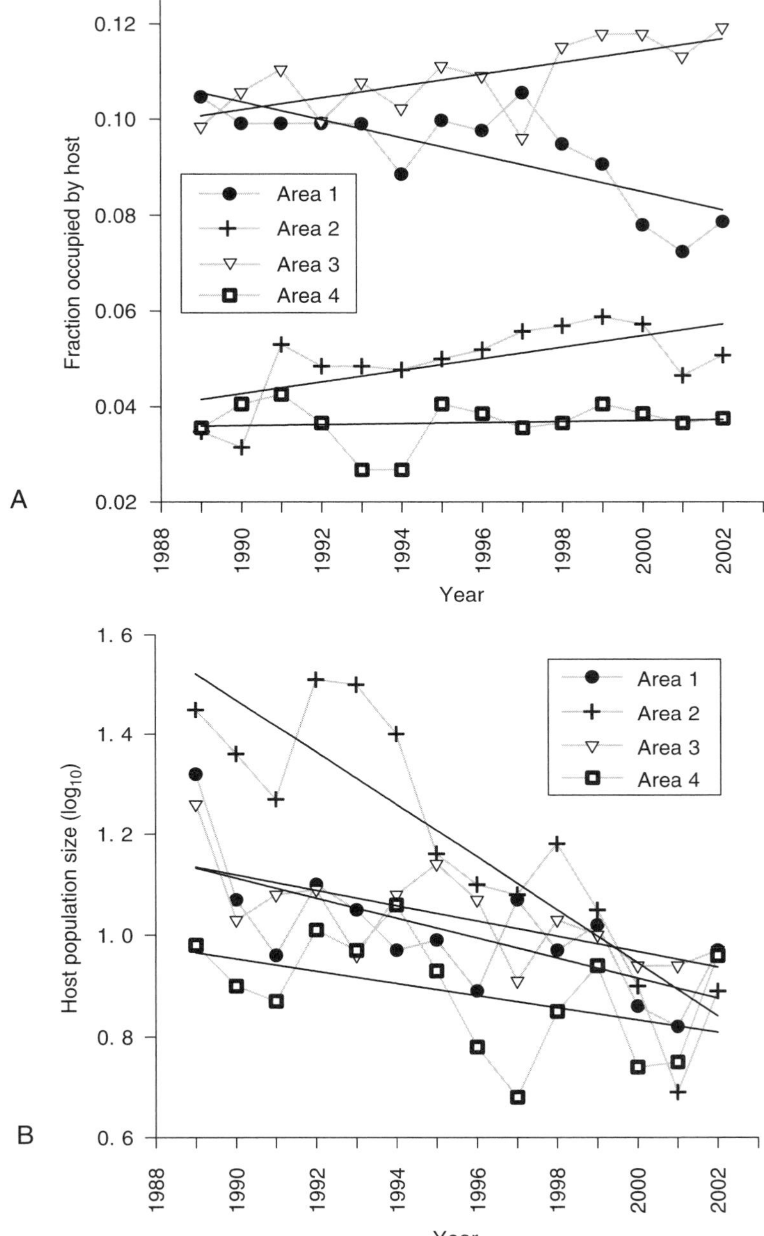

Fig. 19.2 (A) Fraction of roadside segments occupied by *S. latifolia* and (B) average number of *S. latifolia* individuals within each occupied segment in each year for the four areas of the metapopulation.

when nearly 35% of the populations were diseased, followed by a rapid decline. Three subareas were identified within this area on the basis of separation by long runs of unoccupied segments. All three subareas showed a similar pattern with disease incidence peaking in the mid-1990s and then declining (Fig. 19.4).

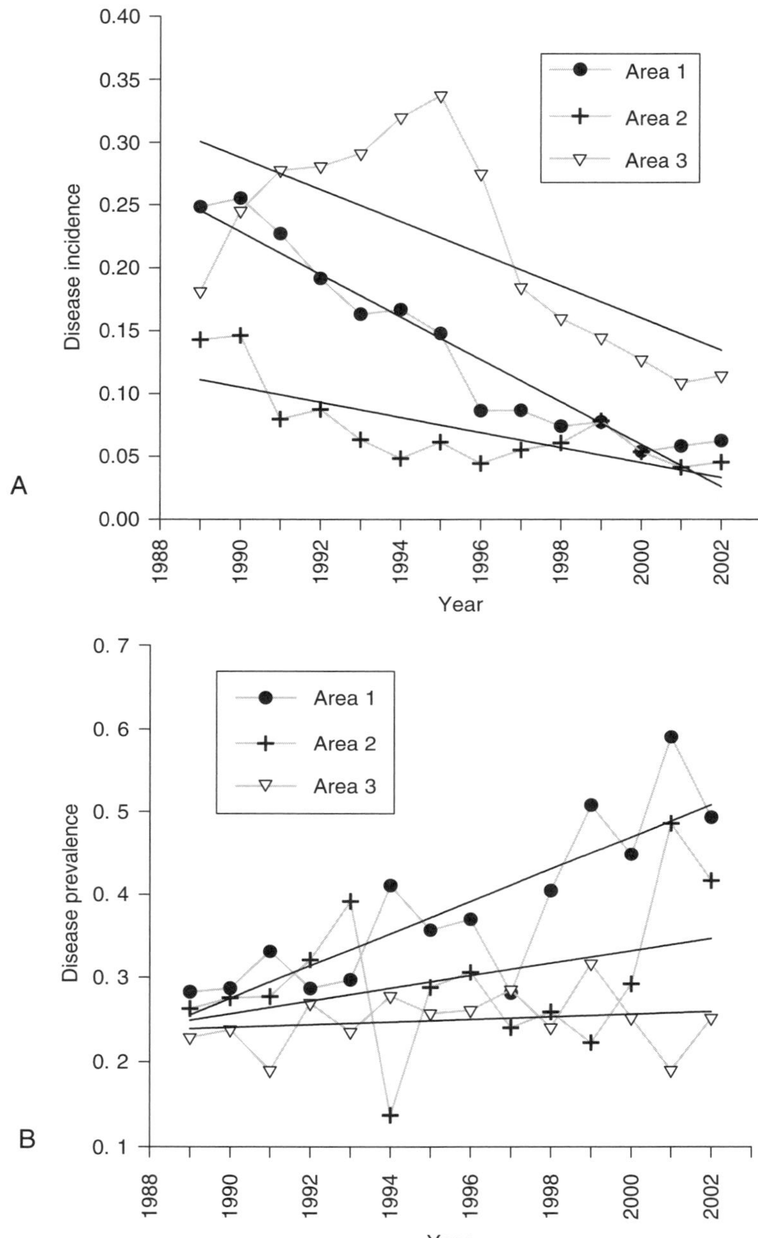

Fig. 19.3 (A) Fraction of *S. latifolia* populations that were diseased (disease incidence) and (B) fraction of individuals that were diseased (disease prevalence) within each diseased population for three areas of the metapopulation.

The fraction of individuals that were diseased, or "disease prevalence," within each diseased population (Fig. 19.3B) increased significantly overall ($P < 0.0042$, $b = 0.0051$, arcsin square root transformed data). All areas showed an increase in disease prevalence and the area*year interaction

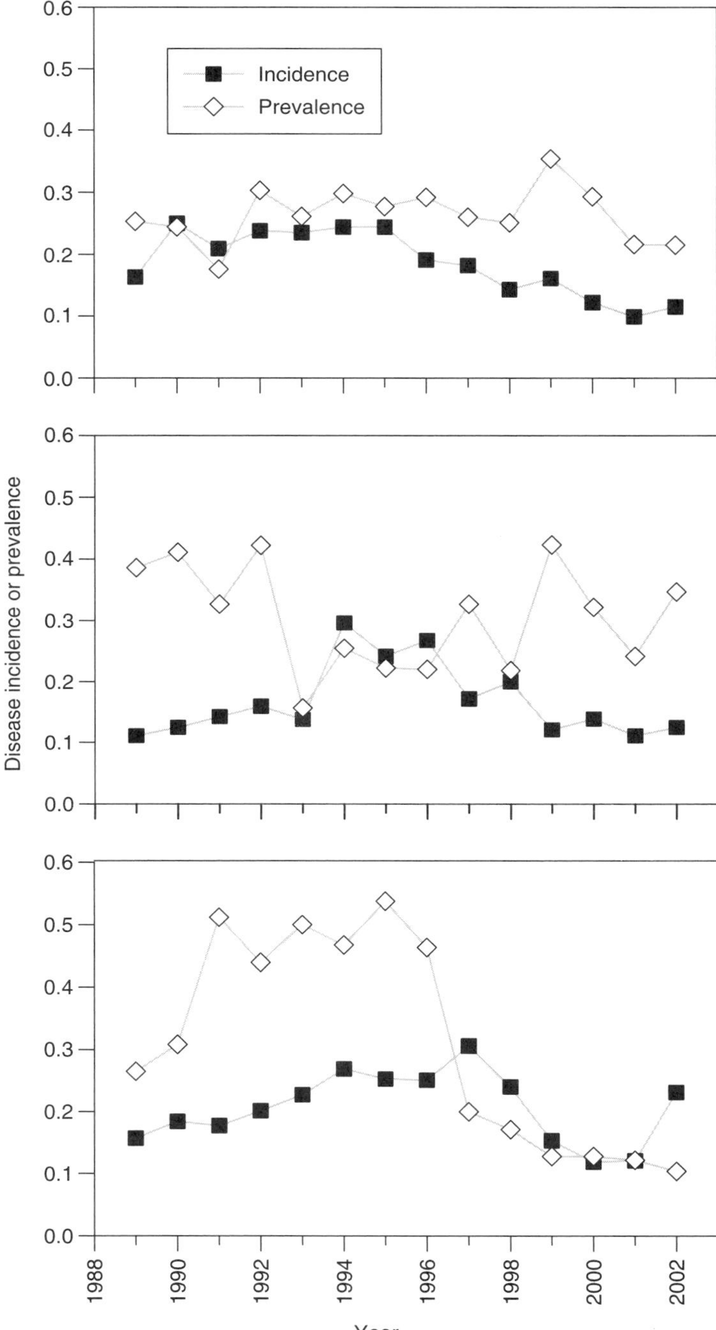

Fig. 19.4 Disease incidence and disease prevalence in three different sections of Area 3. Incidence is the fraction of roadside segments occupied by *S. latifolia* that contained at least one diseased plant, and prevalence describes the fraction of plants within each diseased segment that were diseased.

approached significance ($P < 0.061$). The increase was individually significant only in Area 1 ($P < 0.0005$, $b = 0.011$). The absolute number of diseased plants per segment decreased significantly ($P < 0.0037$, $b = -0.123$) and the area*year interaction was not significant ($P < 0.21$).

Within the subareas of Area 3 disease prevalence was positively but nonsignificantly correlated with incidence in two subareas ($r = 0.19, 0.38$; $P < 0.51, 0.17$), while in the other area they were negatively and nearly significantly correlated ($r = -0.50$, $P < 0.069$).

Host Colonization and Extinction

A host colonization was identified as the presence of a population in a roadside segment after a year when no plants were seen in that segment; the host colonization rate is therefore a compound measure that includes recruitments from the seed pool, recruitment of plants that had remained vegetative for a whole year, and immigration from other sites.

We calculated the colonization rates of the host *S. latifolia* as the number of new populations at time t per existing population at time $t - 1$. This "per capita" colonization rate does not take into account the number of empty segments available for colonization, as these were extremely numerous (1989: 6451, 1990–2002: 6616–6694) and did not vary appreciably with changes in host occupancy. Calculations on a "per unoccupied segment" basis (i.e., equivalent to Levins' "c" in the canonical metapopulation model, Levins, 1969) did not change the results appreciably. We included both healthy and diseased populations as sources because the latter also produced seed (except in the very rare case where there was 100% disease of females and/or males). Results (Fig. 19.5A) showed that the colonization rates of healthy populations declined over the time period of the study ($b = -0.0041$, $P < 0.040$) and that the rate of decline was not significantly different in the different areas (area*year interaction $P < 0.44$).

Host extinction was identified as the absence of a population in a roadside segment after a year when plants had been seen in that segment the previous year. Strictly speaking, it is an "apparent" host extinction rate because it refers to the absence of flowering individuals and does not preclude the persistence of the population as vegetative individuals or in the seed bank. Generally, most plants flower every year, except for very small individuals. When vegetative plants were occasionally seen, the population was not recorded as extinct; however, plants may have been missed because vegetative individuals are not very conspicuous. Results (Fig. 19.5B) showed that the extinction rates of the host tended to decline over the time period of the study, but this decline was not significant ($b = -0.017$, $P < 0.076$). The rate of decline was not significantly different in the different areas (area*year interaction $P < 0.21$).

Disease Colonization and Extinction

A disease colonization event was identified as the presence of the disease in a population of *S. latifolia* after a year when no disease had been seen in that population the previous year. Disease colonization is most probably by immigration,

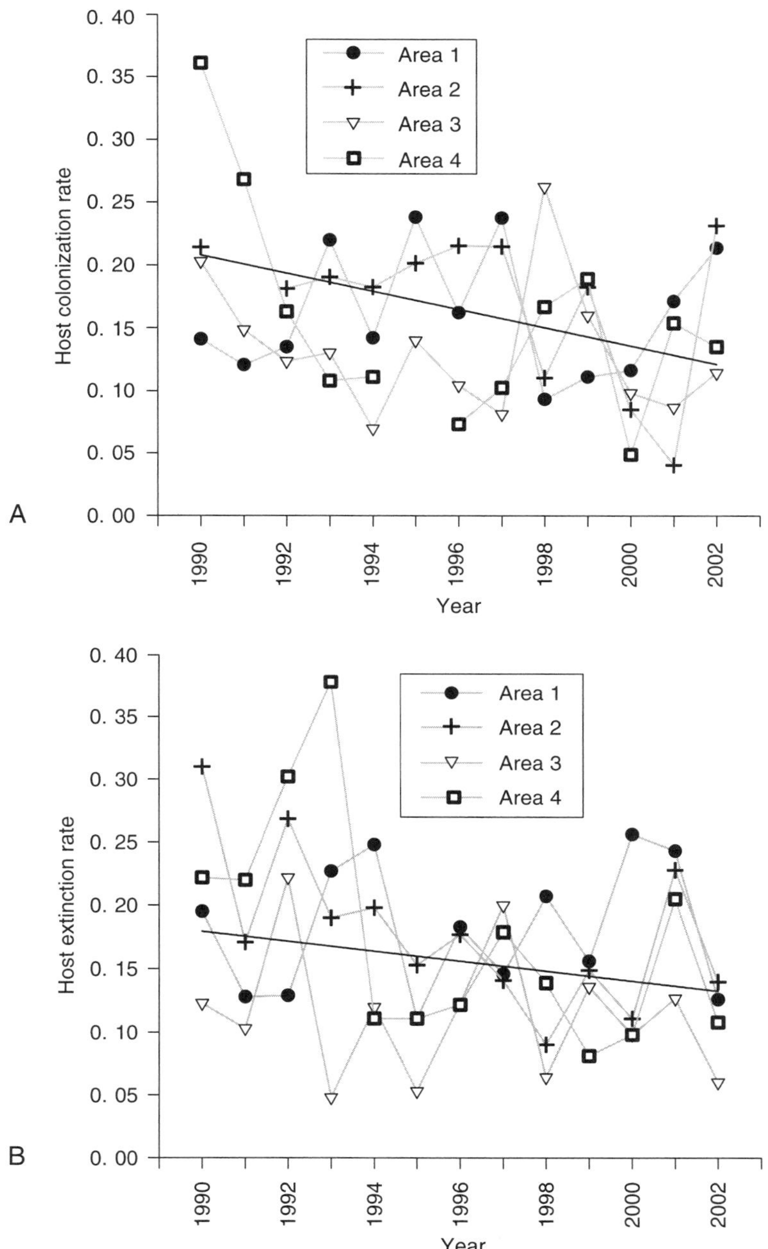

Fig. 19.5 (A) Colonization rate and (B) extinction rate of *S. latifolia* in each year for the four areas of the metapopulation. Colonization rate is measured as the number of new populations in a given year per existing population in the previous year. Extinction rate is measured as the number of populations that went extinct in a given year as a fraction of the number of populations in the previous year.

but the persistence of the disease in a vegetative plant (or a plant that was not flowering at the time of census) cannot be precluded. Across season soil-borne transmission and vertical transmission of the disease have never been observed. We calculated the colonization rate of the disease as the number of newly diseased populations at time t per existing population at time $t - 1$ divided by the number of healthy populations available for colonization in an area (i.e., Levins' "c"). Results (Fig. 19.6A) showed that the colonization rates of disease declined over the time period of the study ($b = -0.025$, $P < 0.029$) and that the rate of decline was not significantly different in the different areas (area*year interaction $P < 0.44$).

Disease extinction was identified as the absence of disease in a population that had been diseased in the previous year. Again this is an "apparent" extinction rate because the disease may have persisted in nonflowering individuals. Results (Fig. 19.6B) showed that the extinction rate of the disease did not change over the time period of the study ($b = 0.0094$, $P < 0.46$) and that the extinction rate was not significantly different in the different areas (area*year interaction $P < 0.20$). The correlation between disease extinction and colonization rate was not significant.

Disease Transmission Rates

Disease transmission rates were calculated using populations where disease had been present in two successive time intervals so as not to confound the estimates with disease colonization or extinction rates. Maximum likelihood methods were used to estimate the survival rate (S) and disease transmission rate (β) for each year by fitting the following model to the data (and minimizing the sum of squares of the log of predicted minus the log of observed):

$$Y_{t+1} = S(Y_t + X_t(1 - \exp(-\beta Y_t / N_t)) \qquad (19.1)$$

where X_t is the number of healthy plants in year t, Y_t, Y_{t+1} is the number of diseased plants in year t and $t + 1$, and $N_t = X_t + Y_t$. Note that the parameter β represents a within season transmission coefficient (assuming no summer mortality) and S represents overwinter survival. Equivalent analyses were also carried out using PROC NLIN in SAS (SAS Institute, 1999) and gave identical results.

The frequency-dependent transmission model always resulted in a better fit than the density-dependent model [where force of infection $= 1 - \exp(-\beta Y_t)$]; the latter also frequently produced unrealistic estimates of S (equal to or close to 1). A good fit was also obtained with a model where the force of infection was $= 1 - \exp(-\beta Y_t/N_t{}^*N_t)$, a model form appropriate for vector-based transmission, but because the relative values of S and β did not differ much between models, we present the results of the more familiar frequency-dependent model.

There was a strong colinearity in the estimates of S and β, such that high estimates of S were correlated with low estimates of β and vice versa. We therefore standardized the survival rate by taking the average over all years and including this average in the model to estimate β. Therefore, this estimate in effect represents an overall "cross-season" transmission coefficient that is a compound of the survival rate of diseased plants and the within season transmission.

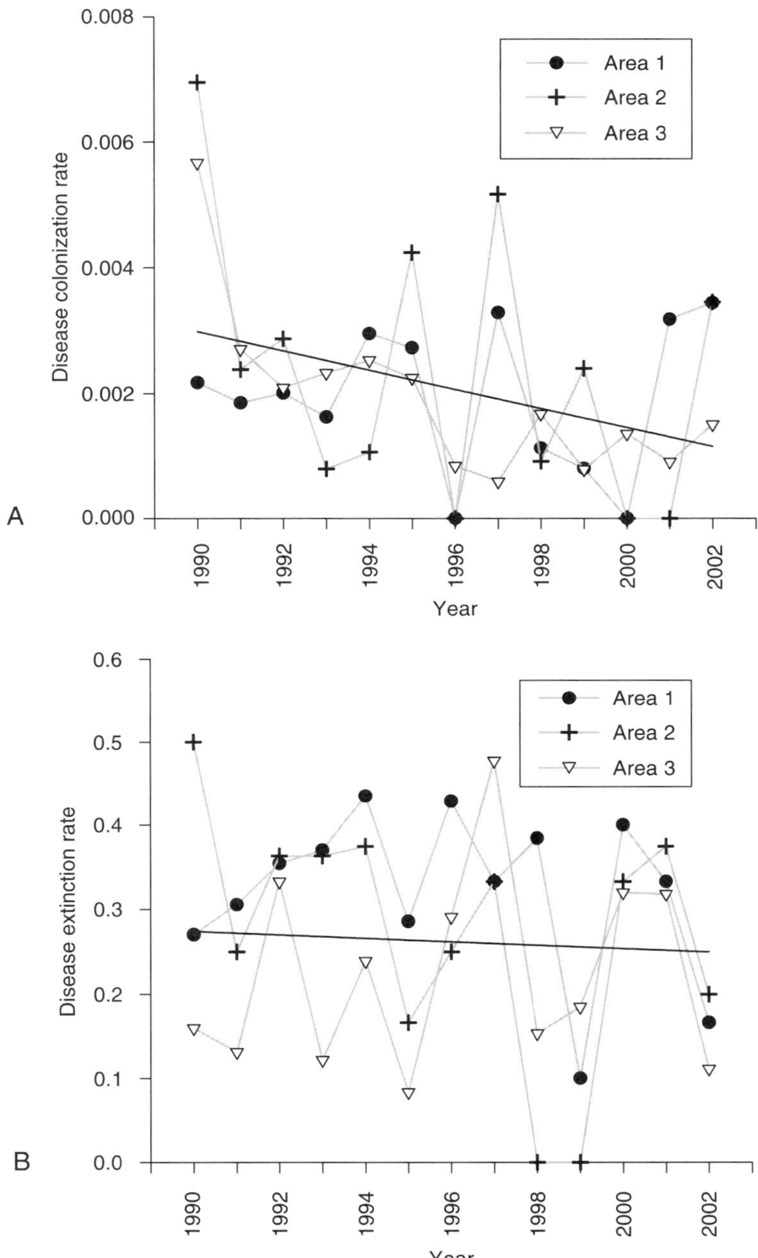

Fig. 19.6 (A) Colonization rate and (B) extinction rate of *M. violaceum* in each year for three areas of the metapopulation. Disease colonization rate is disease measured as the number of newly diseased populations in a given year per existing population in the previous year divided by the number of healthy populations the previous year. Disease extinction rate is measured as the number of populations that became healthy in a given year as a fraction of the number of diseased populations in the previous year.

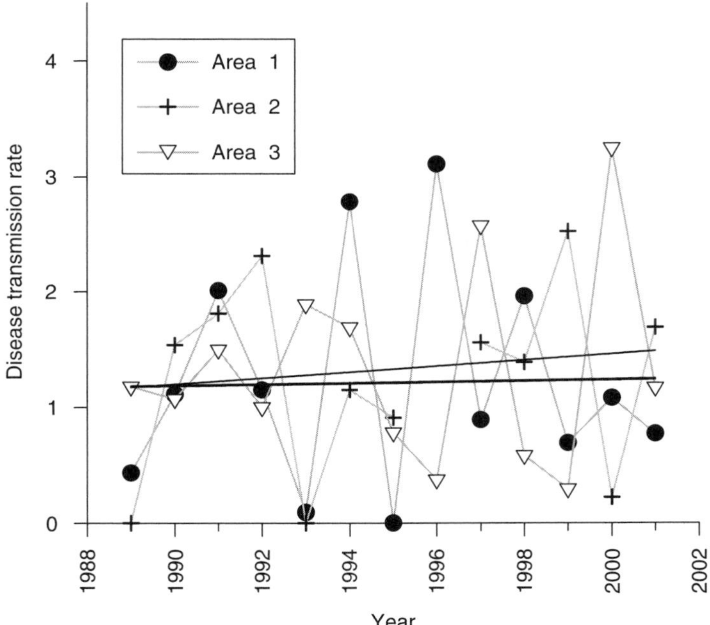

Fig. 19.7 Within population disease transmission rates per year for three areas of the metapopulation with diseased populations (see text for details of estimation).

Results showed that the transmission rates of the disease within populations (Fig. 19.7) did not change significantly over the time period of the study ($b = 0.015$; $P < 0.44$). Nor was there any evidence for a year*area interaction ($P < 0.93$); regressions for each area were slightly positive but did not individually approach significance, even when an outlier was removed ($P < 0.56$–0.82).

There was no significant relationship between disease transmission rates within populations and disease colonization rate (correlation coefficient, $r = -0.07$, $P < 0.83$). When an outlier was removed (1989 estimates), the relationship was positive ($r = 0.29$) but still not significant ($P < 0.36$). The relationship between disease transmission rates within populations and disease extinction rate was negative (-0.26) but not significant ($P < 0.40$) and was essentially unchanged when the 1989 estimate was removed. The same trends (greater colonization and lower extinction when the disease transmission rate was higher) were obtained when the analysis was carried out for each area individually, but these trends were not significant.

Weather Data

Prior to 1997, weather data at Mountain Lake Biological Station were gathered manually and were obtained from the National Climate Data Center. In 1994, a new weather station with automatic data acquisition was installed and run by the station. There was a close correspondence between weather data (monthly mean temperature, highest temperature, lowest temperature, and precipitation) during the 2 to 3 yr period when both types of data were being gathered. Therefore the two types of data were averaged during this overlap period and were used to span the period 1988 to the present.

We investigated a specific set of weather variables that we thought might be related to host and pathogen colonization and extinction, as well as to within population disease transmission. Based on our natural history observations of field experiments, we hypothesized that hot dry summers would decrease disease transmission and hence disease colonization. We also hypothesized that cold winters and/or unusually cold weather in early spring would increase host extinction rates. For each year of the census, for the summer (June, July, and August), we calculated precipitation and mean daily maximum and minimum temperatures; for the winter (December, January, and February) we calculated mean daily maximum and minimum temperatures. We also calculated the minimum temperature in March, as this represents the incidence of unusually cold weather in the early spring.

Over the period of the census, there was a significant decrease in summer daily maximum temperatures ($r = -0.57$, $P < 0.033$) and an increase in summer and winter minimum temperatures ($r = 0.76$, $P < 0.0015$; $r = 0.56$, $P < 0.045$). Analysis of weather data at Mountain Lake Biological Station from 1972 to 2001 showed a gradual but nonsignificant increase in mean, maximum, and minimum summer temperatures (0.021, 0.023, and 0.018°C per year, respectively); the decrease in summer maximum temperatures since 1988 was therefore contrary to the longer term trend. Summer precipitation did not change systematically with year, but was correlated negatively with maximum summer temperatures ($r = -0.60$, $P < 0.023$). No other weather relationships showed a significant change with year.

With a few exceptions, the population parameters were not correlated with weather data. Host extinction was negatively correlated with winter minimum temperatures ($r = -0.76$, $P < 0.0042$), and disease colonization rate (but not transmission rate) was significantly negatively correlated with summer mean minimum temperature ($r = -0.60$, $P < 0.031$). A Bonferroni correction of the $P < 0.05$ criterion for significance (given that 30 correlations were estimated) results in a value of $P < 0.0017$. Under this criterion none of the aforementioned relationships would be deemed significant.

Examination of the change in incidence and prevalence in Area 3 where incidence was initially low and then peaked in the mid-1990s (see Fig. 19.4) showed no obvious or even suggestive relationship with the weather variables.

19.4 DISCUSSION

This study provides clear evidence that the *Silene–Microbotryum* metapopulation that we have been studying since 1988 is not in a state of "global stability." This result came very much a surprise with regard to our ongoing impressions of the populations. Indeed, analysis of the 15 yr of data was stimulated by an assessment of whether it was "worthwhile" continuing with the census, given the resources and effort needed to carry it out every year (our attempts several years ago to get funding for the study were unsuccessful!). Year-by-year observations did not give us the sense that diseased populations were declining in frequency, as every year there were always reports of both disease extinctions and colonizations.

Several issues are raised by these data. First, what is the proximal cause of the decline? In particular, is it driven by changes in the external environment or is it intrinsic to the disease dynamics? Second, if it is the latter, is the instability related to the fact that both the host and the disease are relatively recent introductions into the United States? Finally, is the system moving toward some eventual equilibrium with host–pathogen coexistence or will the outcome be disease extinction?

It is well known from the crop literature that variation in weather can greatly influence the prevalence of disease. However, in the weather data we analyzed, only 2 out of a possible 30 correlations were significant. While the decrease in host–extinction rate with increasing winter minimum temperatures is hard to interpret causally, the increase in disease colonization rate with decreasing summer minimum temperatures is consistent with our own observations that disease transmission is highest at low temperatures and high humidity (Alexander et al., 1993). These low temperatures are most likely to occur during the night, which is also the period of greatest moth visitation (Altizer et al., 1998) and therefore the period most likely for the long distance transport of spores.

Other environmental changes unrelated to the weather may also have had an effect, although their relative importance is hard to judge. In Area 1, elimination of several heavily diseased off-road sites in 1995 by extensive relandscaping by a local lime-manufacturing company may have reduced the available disease sources. In one part of Area 2, road widening in 1990 eliminated five of six diseased sites and in 1995 it eliminated another two diseased sites nearby. However, it is doubtful that this had a cascading effect elsewhere in the area. The whole region of the census was also subject to early spring spraying to control gypsy moth (Sharov and Liebhold, 1998). However, because much of the spraying in this area has been with male mating pheromone whose effect is likely to be specific to gypsy moths (and which have not reached epidemic levels in the census area), the overall impact on moth pollinators (which are also disease vectors) has probably been small.

An alternative explanation for the disease decline is that it is intrinsic to the dynamics of the system as a whole. In a simulation of this host pathogen system (Antonovics et al., 1998), the disease could only be sustained in little over 50% of the runs. We have not reparameterized or reevaluated this model based on more recent data, but it is nonetheless interesting that our "best estimates" based on values from the earlier part of this census and from experimental studies often predicted that the disease would be lost from the metapopulation. Moreover, as the disease was lost, the prevalence of the disease within the remaining populations increased, as we have observed in this study. This is largely because newly founded populations with low levels of disease were no longer being produced. The decreasing disease incidence, the increasing prevalence within populations, and the declining of disease colonization rate observed here are all consistent with gradual disease extinction in the metapopulation.

In this region of Virginia there is extensive genetic variation in the host, yet no detectable variation in the infectiousness of the pathogen. Thus Antonovics et al. (1998) showed that if the simulation is carried out with a genetically uniform host population with a resistance that is intermediate between that of the most susceptible and most resistant genotypes, with an exponential β = 2.00, and a survival of 0.50, then the metapopulation would persist about 90% of the time. (Analysis of census data gave an average value of β over all years of 2.89 and an average survival of 0.55, remarkably close to values used in the earlier simulations.) However, when the simulation was carried out with a genetically variable host population, persistence was much less frequent (ca. 40%). Introduction of the disease into a population led to a rapid local spread of the resistance gene and the generation of resistant populations that were not colonized readily by the disease. Populations only become readily available for colonization by the disease when the gene for susceptibility increased because of the cost of resistance (estimated to be about 25%; Biere and Antonovics, 1995).

In experimental populations of *S. latifolia* where individuals are not replaced over successive years, disease transmission showed an extremely rapid decline to almost zero within 2 yr, due to the fact that the only individuals remaining healthy were from genetically resistant families (Alexander, 1989; Alexander et al., 1995). Disease prevalence also dropped rapidly in experimental populations started with progeny of resistant genotypes but not in populations started with progeny of susceptible genotypes (Thrall and Jarosz, 1994a,b). Moreover, detailed demographic studies of extant diseased populations have shown low transmission rates (Alexander, 1990). It is therefore possible that the decline in disease colonization rates may be due to an increased level of disease resistance in the metapopulation as a whole.

It is relevant to place our metapopulation in a broader geographical and historical context, as this may help with the interpretation of the local changes. In a survey of over a thousand herbarium specimens of *S. latifolia* in the eastern United States, there was no evidence that the plant had been collected south of the Pennsylvania line before 1914 (Antonovics et al., 2003), apart from a collection made in 1896 on the Biltmore estate in North Carolina. Biltmore House was opened in 1895, and it is likely that the estate imported seeds from New England for hay or for the meadows. The first record in Virginia was in 1924, and it was not until the 1930s that collections in Virginia became frequent. The first record we could find for Giles County, where the majority of the metapopulation is located, was 1938. Therefore, the weight of the evidence is that the host plant has only been in the Mountain Lake area for perhaps less than 80 years.

The history of the disease is completely unknown. Previously, *M. violaceum* had been noted on *S. caroliniana* in Virginia and New York State and on several species of *Silene* in the western United States (Farr et al., 1989), but there is no record of it on *S. latifolia*, even though other fungal diseases are recorded for this species in the United States (Farr et al., 1989). None of the herbarium specimens we examined were diseased so they did not help resolve the question of the disease origins. The current distribution of *S. latifolia* and *M. violaceum* in the eastern United States was studied by A. M. Jarosz and E. Lyons (personal communication). They found that the disease was largely confined to the ridge and valley system of western Virginia (where 16% of 102 populations were diseased). Further in the northeast, they only found 1 diseased population (in Pennsylvania) out of 169, except for 3 diseased populations on Nantucket Island, Massachusetts. Diseased plants have been known from Nantucket Island since the early 1980s (T. Meagher, personal communication). In the north central United States, a single diseased plant was found out of 387 populations sampled. The reason for the absence of the disease from more northern latitudes is unknown. In field experiments along a latitudinal gradient, A. M. Jarosz and E. Lyons (personal communication) showed that northern populations were susceptible to disease in their local areas, but that they were also somewhat more resistant than plants derived from seeds of a relatively susceptible parent from Mountain Lake that was used as a control. Artificial hand "pollination" with spores produced a higher incidence of disease than open visitation, suggesting a shortage of pollinators may limit disease transmission.

Given that the host has moved into this part of Virginia only recently and that the disease is near the southern edge of the current range of *S. latifolia*, yet is found sporadically in its former range, it is plausible that we may be seeing the movement of a disease "front" that is following the host as it colonizes new areas. The movement of this disease front may be driven by the evolution of more resistant populations in the wake of the disease. The spread of this disease in the United States may therefore be analogous to the spread of many other epidemics. In animal populations, "waves" of disease spread are often driven by the development of immunity in the wake of the epidemic, but a genetic component to this immunity has also been posited frequently. In the present metapopulation, this genetic component may be the major driving force. However, the issue of whether the changes we are observing are due to climatic and management changes or to intrinsic genetic and demographic factors cannot be determined by descriptive or simulation studies alone, but will require further experiments and more directed field studies of individual populations.

20. METAPOPULATION DYNAMICS IN CHANGING ENVIRONMENTS: BUTTERFLY RESPONSES TO HABITAT AND CLIMATE CHANGE

Chris D. Thomas and Ilkka Hanski

20.1 INTRODUCTION

A major criticism of the applications of metapopulation models in conservation has been that real metapopulations rarely conform to the assumptions of classic theory (Harrison, 1991; Harrison and Taylor, 1997). In metapopulation theory, it is usually assumed that the extinction of a particular local population generates one more patch of empty habitat that is subsequently available for colonization and that each new colonization removes a previously empty patch that is no longer available for colonization (Chapter 4). This assumption is the basis of the stochastic quasiequilibrium between colonizations and extinctions. An unusual number of extinctions in one generation

0-12-323448-4

would likely be followed by an excess of colonization events in subsequent generations, and an unusual number of colonization events would likely be followed by an excess of extinctions. However, if extinctions are generated by habitat deterioration and if colonizations follow an improvement in environmental conditions, then this feedback is broken and there is no logical reason why a metapopulation should exist in any kind of equilibrium (Thomas, 1994a,b). In this case, the metapopulation dynamics of an organism will be superimposed upon, and track (or fail to track), the dynamic distribution of suitable habitat.

At first glance, this criticism seems to be extremely serious because most of the practical applications of metapopulation models relate to habitat that is changing, in which context metapopulations at equilibrium might be expected to be particularly rare. However, the real issue is how fast species are tracking changing environments and whether metapopulation models can provide insight into these processes. If tracking is fast, the problem is in understanding and predicting how the environment is changing; if tracking is slow, there is additionally the problem of transient metapopulation dynamics responding to the changing environment. This chapter reviews the application of metapopulation concepts and models to situations where the environment is changing or the persistence of the metapopulation is precarious because the species occurs close to the extinction threshold (Chapter 4).

Butterfly metapopulations represent excellent systems with which to assess long-term and nonequilibrium dynamics because the quality of historical information on their distributions allows us to be confident whether populations are expanding or declining. Box 20.1 presents a brief history of butterfly metapopulation studies. In some cases, results of past mapping of distributions allow us to test model predictions against observed changes. Furthermore, knowledge of the often quite specific habitat requirements of many butterfly species allows us to define habitat networks, and changes in the structure of such networks, independently of the distribution of the species. We find that metapopulation models can have great predictive power in nonequilibrium systems and that they can be particularly useful in enhancing our understanding of the responses of species to landscape and climate change.

BOX 20.1 Brief History of Butterfly Metapopulation studies

Butterfly biologists developed the concept of "open" and "closed" population structures in the 1960s and 1970s (Ehrlich, 1961, 1965, 1984; Ehrlich et al., 1975; Thomas, 1984), following in the steps of E.B. Ford who, in the 1930s and 1940s, observed the sedentary behavior of many butterflies, confining most individuals to their natal habitat patch. The notion of fairly discrete and often small local populations paved the way to considerations of metapopulations, or assemblages of such local populations. The first full-fledged butterfly metapopulation study was due to Harrison et al. (1988), who demonstrated a mainland–island metapopulation structure in the checkerspot butterfly *Euphydryas editha* in California. The first study of the Glanville fritillary in Finland produced evidence for a classic metapopulation, and Hanski et al. (1994) concluded that "the *Melitaea cinxia* metapopulation . . . provides a contrasting example to the *Euphydryas editha* metapopulation reported by Harrison et al. (1988). Unlike the latter case, there is no large "mainland" population in the *M. cinxia* metapopulation, and its long-term persistence appears to depend on genuine extinction-colonization dynamics." Studies of several British butterflies found that almost all local breeding populations occur within a dispersal range of other local populations of the same species, which finding suggested, along with direct evidence of colonizations and extinctions, that metapopulation dynamics were likely to be commonplace (Thomas et al., 1992; Thomas and Harrison, 1992; Thomas and Jones, 1993; Thomas, 1994a,b).

From this point onward, metapopulation studies on butterflies have taken place mostly in Europe. Extinction–colonization dynamics have been researched intensively in several species, but most notably in *M. cinxia* (Hanski et al., 1994, 1995a,b, 1996; Kuussaari et al., 1998; Hanski, 1999b; Nieminen et al., 2004), *Plebejus argus* (Thomas, 1991; Jordano et al., 1992; Thomas and Harrison, 1992; Brookes et al., 1997; Lewis et al., 1997; Thomas et al., 1999a, 2002a), *Hesperia comma* (Thomas et al., 1986, 2001a; Thomas and Jones, 1993; Hill et al., 1996; Wilson and Thomas, 2002) and *Proclossiana eunomia* (Baguette and Nève, 1994; Nève et al., 1996a; Petit et al., 2001; Sawchik et al., 2002; Schtickzelle et al., 2002) in Europe and *E. editha* (Harrison et al., 1988; Harrison, 1989; Thomas et al., 1996; Boughton, 1999; McLaughlin et al., 2002) in North America. Studies of these and tens of other species (e.g., Warren, 1987, 1994; Settele et al., 1996; Gutiérrez et al., 1999, 2001; Knutson et al., 1999; Mousson et al., 1999; Shahabuddin and Terborgh, 1999; Baguette et al., 2002; Bergman and Landin, 2001; Bulman, 2001; Nekola and Kraft, 2002; Wahlberg et al., 2002a,b; Wilson et al., 2002) have shown great variation in metapopulation structure and that dynamics are nearly as variable within as among species. This latter conclusion underscores the pivotal role of landscape structure in influencing spatial dynamics. Nonetheless, these studies have confirmed that the general metapopulation notion provides valuable insight into the dynamics and distribution of many, although not all, butterfly species at the landscape level. The metapopulation approach can be applied to virtually all species that were formerly considered to have "closed" population structures (Thomas, 1984).

Since the mid-1990s, the emphasis on butterfly metapopulation studies has been in adding further details and evaluating how robust and useful the approach is under different circumstances. Studies have examined the validity of the major assumptions and processes of metapopulation dynamics, incorporated multispecies patterns and dynamics into the common framework (Lei and Hanski, 1997; van Nouhuys and Hanski, 1999, 2002), investigated the evolutionary and genetic dynamics of metapopulations (Nève et al., 1996b, 2000; Singer and Thomas, 1996; Brookes et al., 1997; Saccheri et al., 1998; Thomas et al., 1998; Barascud et al., 1999; Keyghobadi et al., 1999; Kuussaari et al., 2000; Nieminen et al., 2001; Saccheri and Brakefield, 2002), and applied metapopulation approaches at increasingly large scales in relation to conservation and climate change, as described in the main text.

Research on butterflies has played an important, and in some cases pivotal, role in the development of the science of metapopulation biology and in the application of the metapopulation approach to conservation. Many of the studies cited here and in this chapter were at least partially motivated by conservation concerns, and this pattern is likely to continue.

20.2 HABITAT FRAGMENTATION

Habitat loss typically results in fragmentation — smaller and more scattered fragments of habitat than existed formerly. In many cases, the habitat does not immediately all become unsuitable, and populations may be found, for some time at least, in patches of habitat that have recently become small, more isolated, or both. At equilibrium, many of these patches would be expected to be unoccupied most of the time (even though they may contain perfectly suitable habitat) because the rate of extinction of small populations is likely to be high and the recolonization rate of isolated patches will be low (Hanski, 1994, 1999b; Chapter 4). However, during and immediately following a period of fragmentation a species occupying the remaining fragments may show a period of decline, during which the rate of local extinction exceeds the rate of recolonization. In such situations, the potential contributions of metapopulation models are to help understand the timescale of decline, the spatial pattern of decline, and whether a species will decline to a reduced metapopulation size (restricted distribution) or become completely extinct from the areas where fragmentation has taken place. Such insights can be extremely important because they may provide an understanding of why some species continue to decline long after the damage to the environment has taken place. The following two examples illustrate how species may lag behind habitat loss.

Melitaea cinxia and the Speed of Metapopulation Decline

The distribution of the Glanville fritillary, *Melitaea cinxia*, in northern and western Europe has become greatly reduced over the past decades, and the species has gone regionally extinct in many areas (Hanski and Kuussaari, 1995; Maes and van Dyck, 1999; van Swaay and Warren, 1999). It is apparent that habitat loss is the primary or even the only significant cause of the decline. In Finland, *M. cinxia* went extinct in the mainland in the 1970s (Marttila et al., 1990), and it now occurs only in the Åland Islands in Southwest Finland (Hanski and Kuussaari, 1995). Luckily for this butterfly and many other species of insects and plants, land use practices have changed less drastically in the Åland Islands than in most other parts of northern Europe. Dry meadows with *Plantago lanceolata* and *Veronica spicata*, the two host plants of *M. cinxia*, still abound in Åland, partly because the general topography with numerous small granite outcrops prevents large-scale agricultural intensification. At present, the suitable habitat covers *ca.* 6 km^2, which is 0.6% of the total land area (Nieminen et al., 2004). Nonetheless, substantial habitat loss has occurred in parts of Åland in recent decades, as the following example shows, with adverse consequences for the occurrence of the butterfly.

Figure 20.1a shows one network of habitat patches in the Åland Islands, with 42 patches in 1992. Thanks to Hering's (1995) detailed analysis of old aerial photographs and interviews of local people, we know that 20 yr previously there had been 55 distinct patches in this network, and nearly three times more habitat for *M. cinxia*. In this case, the area of suitable habitat had declined largely because of reduced grazing pressure. Hanski et al. (1996) used the incidence function model (Hanski, 1994; Chapter 4), parameterized previously for *M. cinxia*, to assess the likely metapopulation dynamic consequences

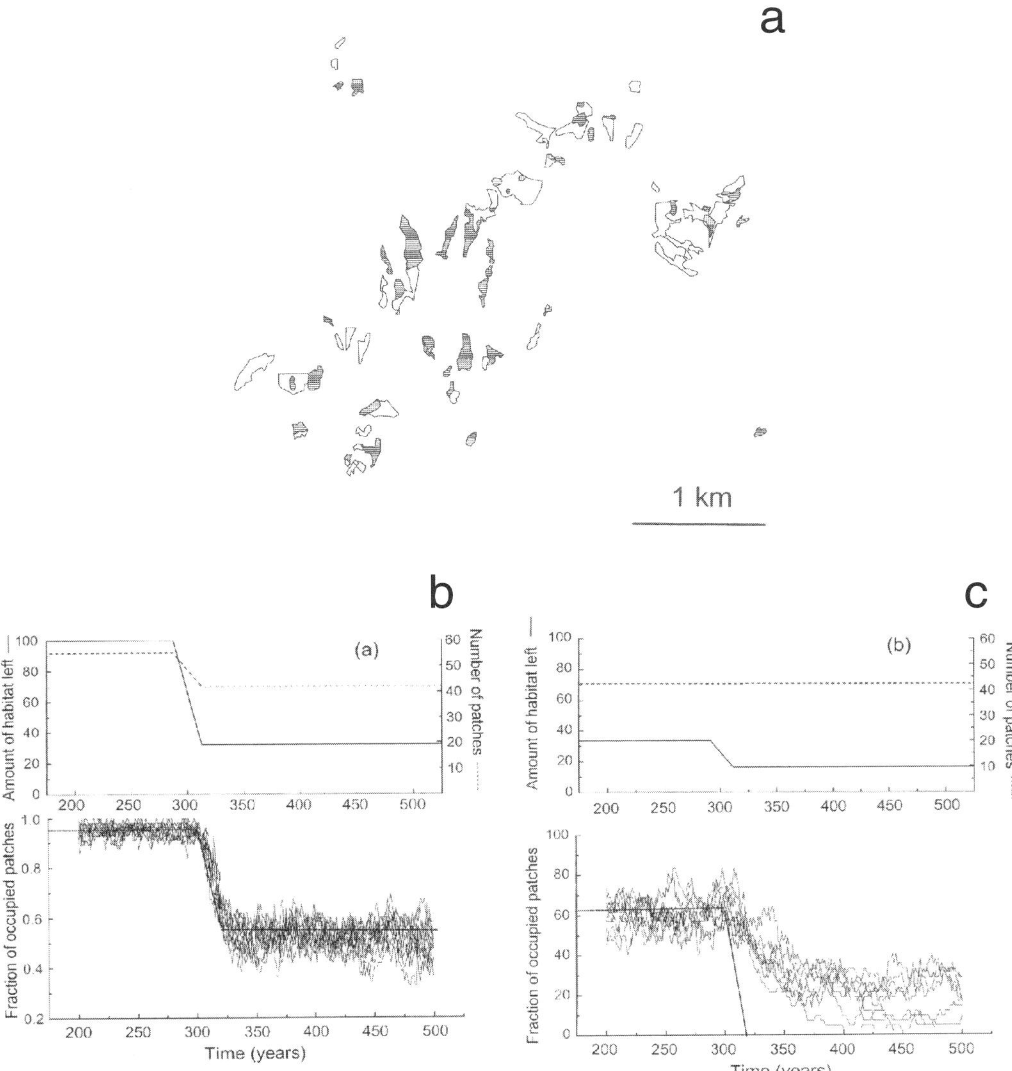

Fig. 20.1 (a) The extent of suitable habitat for *Melitaea cinxia* in one part of the Åland Islands in 1992 (shaded) and 20 yr previously (Hering, 1995). (b) Metapopulation dynamics of *M. cinxia* in the landscape shown in (a), modeled with the IFM. The size of the metapopulation is shown. (Top) The number of patches and the amount of habitat during the simulation, including the observed reduction in habitat over the 20-yr period documented in (a). (Bottom) The equilibrium metapopulation size (thick line) and 10 replicate predicted trajectories before, during, and after the period of habitat loss (thin lines). (c) Similar results for a hypothetical scenario of further loss of 50% of the area of each of the remaining patches in (b) (from Hanski et al., 1996; Hanski, 1999b).

of this particular scenario of habitat loss. Initially, the landscape had so many well-connected habitat patches (Fig. 20.1a) that the butterfly was able to occupy most of them most of the time. The modeling assumed that habitat was lost gradually over a period of 20 yr, and the equilibrium state of the metapopulation corresponding to the prevailing amount of habitat was calculated for each year. This equilibrium is shown by a thick line in Fig. 20.1b (Bottom). Next, 10 replicate simulations were run for 500 yr, with the 20-yr period of habitat loss starting in year 300. The result shows that butterfly dynamics tracked the changing environment very closly: the predicted metapopulation size is very close to the equilibrium during and following the period of environmental change, with a lag of only a few years (Fig. 20.1b).

In this particular case there was very little time lag in the dynamics of the butterfly in a changing environment, but it would be wrong to assume that the same result applies to all butterflies and to all environmental changes. As a matter of fact, and as Fig. 20.1c clearly shows, even the same species can respond very differently under different environmental conditions. Figure 20.1c is a continuation of the example in Fig. 20.1b: we now assume that each one of the remaining 42 patches that existed in 1992 loses 50% of its area in the next 20 yr. This amount of habitat loss is drastic enough in this particular case for the equilibrium metapopulation size to move all the way to zero (Fig. 20.1c). However, the model-predicted final extinction of the butterfly metapopulation took a very long time, with a median time to extinction of more than 200 yr (Fig. 20.1c). The striking contrast between the two examples in Fig. 20.1 can be understood in light of the general theory developed by Ovaskainen and Hanski (2002) and outlined in Chapter 4 (Section 4.4). The length of the time delay in metapopulation response to environmental change is expected to be especially long in cases where the new equilibrium following habitat loss (or any other perturbation) is close to the extinction threshold — as it evidently is in Fig. 20.1c. In reality, metapopulation extinction is hastened by temporal variation in environmental conditions (regional stochasticity; Chapter 4) even in the absence of further environmental changes, and the eventual extinction would probably take less time than in the example in Fig. 20.1c. Nonetheless, the qualitative result and message remain the same.

Euphydryas aurinia and Extinction Debt

The marsh fritillary, *Euphydryas aurinia*, has declined throughout Europe, especially in northern Europe (van Swaay and Warren, 1999). In northwest Europe, *E. aurinia* is restricted to rough and mainly moist pastures, where its host plant Devil's bit scabious, *Succisa pratensis*, grows among tall vegetation. The butterfly has been eliminated completely from eastern parts of Britain, where agriculture is most intensive, and it only survives as metapopulations of small local populations in the west. Even here, the original extent of suitable habitat has been reduced and fragmented greatly.

Euphydryas aurinia was studied in a 625-km^2 area in southwest England in the county of Dorset in 1981 (Warren, 1994) and again in 1998–2000 (Bulman, 2001). In this landscape, the butterfly shows typical "metapopulation patterns." It was found in only 14 out of the 123 habitat patches that were delimited in 1998–2000, occupying large patches of high quality (tall vegetation) that contained large quantities of larval food plants and patches that were closely connected to other patches also supporting local populations. Between the two time periods, from 1981 to 1998–2000, 10 populations became extinct, mostly in small patches, and four empty patches were colonized, three of which were very close to existing populations (Bulman, 2001).

Bulman (2001) fitted the incidence function model (IFM; Hanski, 1994; Chapter 4) to data from the part of the Dorset system that had the highest patch density and that was most heavily occupied by *E. aurinia*. As the butterfly had not declined in this area between 1981 and 1998–2000, it was reasonable to assume in parameter estimation that the metapopulation in this region was at quasiequilibrium (Bulman, 2001; see Chapter 5 for the importance of the equilibrium assumption). The model with parameter values thus estimated was then applied to 10 independent patch networks within areas of 16 km^2 scattered across Britain, as well as to two other networks in Dorset. Half of the 12 networks were centered on individual *E. aurinia* populations known to have become extinct in the last 15 yr and where it was known that the butterfly was extinct from the entire 16-km^2 square. The other half were centered on surviving *E. aurinia* populations, located nonrandomly in areas still considered to be strongly populated by *E. aurinia* (the best areas available within each region). Networks with surviving metapopulations contained more and larger habitat patches and were better connected to one another than patches in the networks in which the butterfly had gone extinct (Table 20.1; Bulman, 2001).

Metapopulation dynamics in each 16-km^2 square were predicted by the IFM with the initial condition of all patches occupied. All of the extinct networks were predicted to become extinct, with median times to extinction between 11 and 26 yr, depending on the network. This implies that not all such metapopulations would have become extinct within a 15-yr period and that some metapopulations might still be surviving for longer on borrowed time. The latter is illustrated by the analysis of the six surviving metapopulations in

TABLE 20.1 Habitat Availability and Metapopulation Capacity in 16-km^2 Areas Where *Euphydryas aurinia* Has either Survived or Become Extinct at the Landscape level[a]

	Network	No. of patches	Total patch area (ha)	Metapopulation capacity (λ_M)	Median time to extinction (yr)
Survived	North Wales	15	115	5.3	>200
	Mid Wales	8	41	3.8	116
	Southwest A	17	33	2.8	50
	Southwest B	15	116	4.7	>200
	Cumbria	6	14	1.5	24
	Dorset	18	80	4.0	130
Extinct	North Wales	7	10	0.6	15
	Mid Wales	14	20	1.0	21
	Southwest A	5	9	2.0	17
	Southwest B	3	8	1.2	22
	Cumbria	2	17	1.2	26
	Dorset	4	6	0.4	11

[a] Simulated times to extinction are given using the incidence function metapopulation model, starting simulations with all patches occupied (see text for further details). From Bulman (2001) and C. Bulman et al. (unpublished results).

Bulman's (2001) study, four of which had predicted median times to extinction between 24 and 130 yr (the remaining two metapopulations survived the entire duration of the simulations; Table 20.1). As the surviving metapopulations were chosen to be in the most favorable landscapes within each region, it is unlikely that these metapopulations would be rescued by immigrants from other landscapes still containing *E. aurinia*. There appears to be a substantial extinction debt in this species, which is likely to decline for many decades into the future even if there are no further habitat losses (Bulman, 2001). If the Dorset system, for which the model was parameterized, is actually in overall decline, even these projections are too optimistic.

The projected times to extinction should not be interpreted too literally, as the model was parameterized with a limited amount of information and no regional stochasticity was taken into account. However, the modeling exercise has helped reveal why this butterfly appears to be becoming extinct, region after region, even when the remaining populations fall within protected areas. Modeling results suggest that protection of all remaining populations and habitat patches in some regions may not be enough, whereas the butterfly is predicted to survive indefinitely in the two largest networks that contain over 100 ha of habitat. Despite the nonequilibrium nature of the system, a metapopulation approach has provided insight into the recent decline and has identified minimum viable network goals for the long-term conservation of the species.

Extinction Debt and Conservation

The concept of extinction debt is usually applied to communities and in the context of species diversity (Tilman et al., 1994; Hanski and Ovaskainen, 2002). Species–area calculations are performed to estimate the numbers of species that might eventually become extinct following habitat loss (Brooks and Balmford, 1996; Brooks et al., 1997; Cowlishaw, 1999). These calculations have provided insight into the extinction process — and into human impacts on biodiversity — but it is an approach without hope, as it does not provide a practical way forward for conservation action. However, a metapopulation approach provides a way forward even though the initial prognosis may be equally pessimistic.

Each species will have slightly different habitat requirements, which means that individual species have somewhat different habitat networks even in the same fragmented landscape (e.g., Gutiérrez et al., 2001; Thomas et al., 2001b). Species will also differ in local population densities and dispersal abilities, and hence local extinction and colonization rates will differ. Using a single-species approach, it is possible to identify which areas of the fragmented landscape are likely to be most important for particular species and potentially to assess whether a given species will eventually decline to extinction or become restricted to some limited area. The theory described in Chapter 4 has the potential to achieve this for species whose environments are highly fragmented.

If the prognosis is metapopulation extinction, the metapopulation approach can be used (1) to identify which landscapes are closest to the extinction threshold and (2) to identify how extinction and colonization rates could be altered, via management of landscape structure, to ensure that extinction does not actually take place. In other words, the theory provides means of targeting conservation action. In the case of *E. aurinia*, discussed earlier, this action might be ensuring that all grasslands within focal regions are grazed to have the right vegetation height, that management increases host plant densities, and that habitat areas are increased by the restoration of adjacent habitat — all measures to reduce extinction rates. Similarly, increasing habitat quality, restoring new habitats, and possibly providing stepping-stone habitats to connect semi-isolated patch networks could all increase colonization rates. By targeting these actions within the most favorable existing landscapes, rather than investing effort where the species is already doomed to extinction, real long-term success may be possible. Most conservation applications of metapopulation theory have stressed the need to increase habitat areas and minimize patch isolation, but often these particular suggestions are impractical. For example, changing the spatial locations of habitat patches is not usually an option. It is important to realize, however, that any actions that reduce extinction and increase colonization rates are equally valid applications of the "metapopulation approach" to conservation. For example, manipulation of habitat quality and the geometry of the landscape are equally legitimate means of altering extinction and colonization rates (Thomas, 1994a; Thomas et al., 2001b; Box 20.2).

BOX 20.2 Reconciling Habitat and Metapopulation Approaches in Butterfly Biology

Many simplifications were made during the early development of the metapopulation paradigm in butterfly biology. One of these relates to the emphasis on the spatial configuration (geometry) of suitable habitat in the landscape: what are the areas of habitat patches and how isolated they are from each other? Metapopulation studies appeared to pay less attention to the role of variation in habitat quality, which had previously been recognized as a major determinant of butterfly distributions (Thomas, 1984). However, this perception is somewhat misleading, as even the earliest butterfly metapopulation studies took account of habitat quality. For example, habitat quality thresholds were used to define habitat patches (Harrison et al., 1988), and variation in habitat quality was widely recognized as the driving force behind extinction and colonization dynamics within many metapopulations (Warren, 1987; Thomas, 1994a,b, 1996; Hanski, 1999b; Wahlberg et al., 2002a). Variation in habitat quality also underlies source–sink dynamics within butterfly metapopulations (Thomas et al., 1996) and may influence migration among patches (Box 20.3).

The perception that the "metapopulation approach" is somehow an alternative to the "habitat approach" has nonetheless persisted. Most recently, attempts have been made to tease apart the relative importance of variation in habitat quality and the spatial arrangement of habitats (Dennis and Eales, 1999; Thomas et al., 2001b; Fleishman et al., 2002). However, this is not very satisfactory because the metapopulation and habitat approaches operate at different levels of a hierarchy. At the metapopulation level, we are primarily interested in the probability of extinction of local populations. Habitat quality, habitat type and patch size all contribute to that probability. The term habitat quality is itself a "black box" simplifying complex interactions among species as well as responses to the physical environment (e.g., Hochberg et al., 1992; Jordano et al., 1992). A habitat quality approach is often a useful abstraction to summarize the consequences of multiple interactions within (usually) single landscape elements, just as a metapopulation approach is a useful abstraction to summarize the behavior of populations in many such elements.

Ultimately, the issue is not whether habitat quality matters, but how we deal with it in population biology and conservation. Most butterflies whose metapopulation biology has been studied are habitat specialists. In all cases where a metapopulation approach is deemed appropriate, the "first cut" is between habitat (patches) and nonhabitat (matrix). This is a given. After this first cut, the question is whether variation in habitat quality is great enough within patches, or within the matrix (Sutcliffe and Thomas, 1996; Ricketts, 2001; Keyghobadi et al., 2002), for it to be necessary to incorporate habitat variation within the metapopulation level of analysis. This can only be answered in specific terms; that is, whether and how variation in habitat quality should be treated in a particular landscape and for a particular species (Hanski et al., 2004). The example on *Hesperia comma* (Sections 20.5 and 20.6) illustrates one specific case.

The truly critical issue, and the one on which the relevance of the metapopulation approach rests, is whether population connectivity makes a difference to the distribution and spatial dynamics of the species. In this context, it is unfortunate that most empirical studies continue to use a simplistic measure of connectivity — distance to the nearest population (or even distance to the nearest habitat patch) — which measure lacks power. Instead, one should use a connectivity measure that takes into account the distances to and sizes of all neighboring populations (Hanski, 1999b; Moilanen and Nieminen, 2002). Such a measure of connectivity is also an integral part of the incidence function metapopulation model (Hanski, 1994, 1999b), which has been applied extensively in the research reviewed in this chapter.

20.3 PRECARIOUS METAPOPULATION PERSISTENCE

Some landscapes have ample habitat that ensure metapopulation persistence and high patch occupancy, whereas other landscapes have very little habitat and no chance for metapopulation persistence. Yet other landscapes contain intermediate amounts of habitat that may permit periodic, but not permanent occupancy, in which case metapopulations may flip back and forth between presence and absence (provided, of course, that recolonization from outside is possible following extinction). It is easy to misinterpret the dynamics of such systems, especially when studies cover too small a region and too short a time to encompass the long-term and large-scale dynamics of the system. The most serious implication is that researchers might study a patch network that is currently empty and conclude erroneously that it is of no consequence for conservation or the same network when it is well occupied and conclude that it is sufficient for long-term persistence. The following examples illustrate that such precarious metapopulation persistence may be commonplace.

Aricia agestis in North Wales

The brown argus butterfly, *A. agestis*, is a specialist on common rock rose *Helianthemum nummularium* plants in north Wales, where the plant is restricted to limestone grasslands and crags. Therefore, both the plant and the butterfly share a very patchy distribution in north Wales (Fig. 20.2). Within a 600-km^2 area, habitat patchiness at a coarse scale is determined by the distribution of limestone outcrops, and at a finer scale by the distribution of traditional flower-rich meadows and crags (Wilson et al., 2002).

The butterfly shows the usual metapopulation patterns. It is most likely to be present in *Helianthemum*-containing patches that are large and close together; some colonizations and extinctions have been observed, and individuals have been recorded moving between habitat patches (Wilson and Thomas, 2002; Wilson et al., 2002). Peripheral populations tend to contain only a subset of the genetic variation present within core areas, suggesting colonization by relatively small numbers of individuals (I. Wynne et al. unpublished result). Wilson et al. (2002) defined groups of meadows as semi-independent networks (SINs) of habitat if they were separated from other such groups by 3 km or more of unsuitable habitats. Because movements over distances greater than

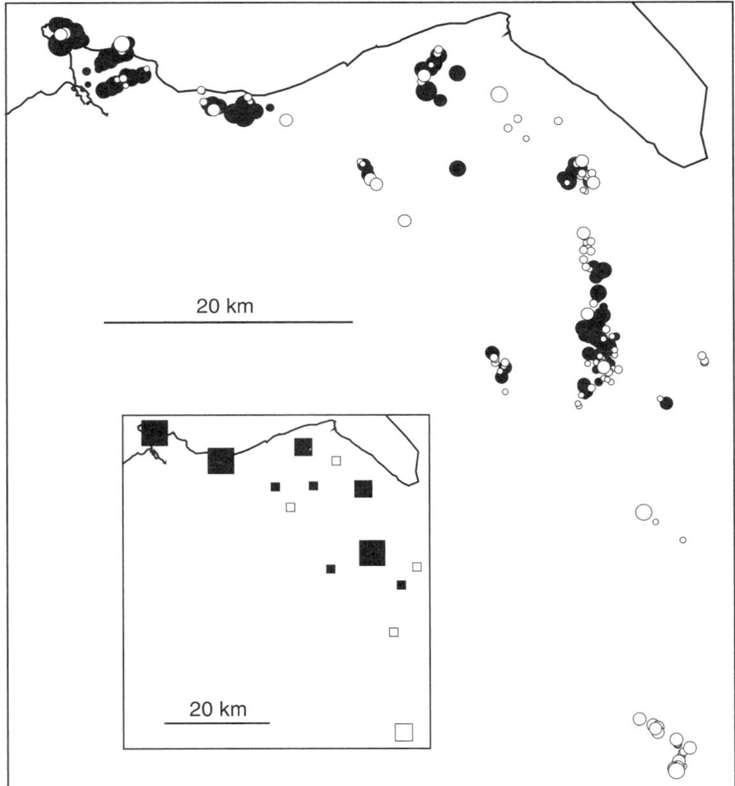

Fig. 20.2 Distribution of occupied (solid) and empty (open) limestone grassland habitat for *Aricia agestis* in north Wales in 1997 (northwest part) and 1999 (other areas). Circles denote individual habitat patches (all containing the host plant *Helianthemum nummularium*) in the main figure; squares represent semi-independent networks (SINs), separated by 3 km or more of unsuitable habitats from other networks (inset). Simulated median times to extinction of SINs are >200 yr (large squares), 25–125 yr (medium-sized squares), and <25 yr (small squares) (R. J. Wilson, unpublished result). Modified from Wilson et al. (2002).

3 km are rare, it is reasonable to interpret patch occupancy as a function of habitat availability within each network (extrapolating from mark–recapture study; 0.0004% of butterflies were estimated to move the 3.6-km distance separating the two habitat networks nearest to one another; Wilson and Thomas, 2002). Metapopulations in two of these networks have been observed to become extinct in the last 15 yr, even though there is no indication that the habitat has deteriorated recently in either network. Simulations using the IFM revealed that most of the semi-independent networks harbor extinction-prone metapopulations, with just the three largest networks exhibiting consistent long-term persistence (Wilson et al., 2002, unpublished result). The fourth most persistent network, as predicted by the IFM simulations, was one of the SINs that was observed to become extinct in reality (the southernmost network in Fig. 20.2). Furthermore, habitat patch networks were more likely to be occupied if they were large and close together. Crucially, extinction-prone networks were significantly more likely to be occupied if they were close to other, usually larger networks (Wilson et al., 2002).

The most likely interpretation of these results is that most of the habitat networks for *A. agestis* in north Wales are too small for consistent occupancy and that many of the smaller networks are occupied periodically, become extinct, and then again recolonized from the more persistent networks. By taking a metapopulation approach, it was possible to identify the key areas that were likely to be responsible for persistence at the regional scale. In county Durham in northern England, there may be no robustly surviving metapopulations of the related *Aricia artaxerxes*, and its regional persistence may depend on the dynamics of a "megapopulation" (Hanski, 1999b) — a "metapopulation of metapopulations." Here each semi-independent metapopulation is extinction prone, but metapopulations are recolonized by rare long-distance migrants. How stable such dynamics might be in this case is an open question, and the entire system may be slowly going extinct (Wilson et al., 2002).

Melitaea cinxia in Åland

Melitaea cinxia occupies a large network of ca. 4000 habitat patches in the Åland Islands in southwest Finland, including very small patches and patches of marginal quality (Nieminen et al., 2004). The patches have been divided into semi-independent patch networks (Hanski et al., 1996b). Within a SIN, patches are located so close to each other that butterfly movements are quite frequent, and a large fraction of butterflies, of the order of half of all the individuals, move from one patch to another during their lifetime (Hanski et al., 2000). In contrast, different SINs are separated by dispersal barriers or are so isolated from each other that butterfly movements are infrequent. Some migration among SINs still occurs, making it possible that a SIN from which a butterfly metapopulation has gone extinct will become recolonized.

Figure 20.3 shows the number of years in the period 1993–2001 that 53 SINs were occupied (the smallest SINs mostly consisting of just a single habitat patch have been omitted). The horizontal axis gives the metapopulation capacity of the network (Hanski and Ovaskainen, 2000), which is a proper measure of the "size" of the network, integrating the effects patch number, patch sizes, and their connectivities (Chapter 4). The vertical axis gives the

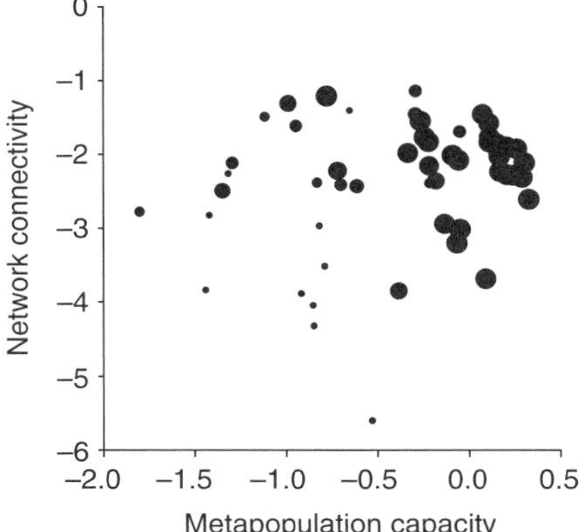

Fig. 20.3 Large-scale patterns of network occupancy for *Melitaea cinxia* in the Åland Islands. Each point represents a separate semi-independent network (SIN). The number of years that a particular SIN was occupied out of 9 yr is shown as a function of the size of the SIN (measured by metapopulation capacity) and its connectivity to other metapopulations in the surrounding SINs (calculation explained in the text). The size of the dot is proportional to the number of years that the network was occupied (smallest dot, network always unoccupied). Figure prepared by O. Ovaskainen, data from I. Hanski (unpublished).

connectivity of each network using an exponential dispersal kernel with the same parameter value as in the *M. cinxia* examples in Chapter 4, metapopulation capacity as a measure of network size, and p_λ calculated on the basis of the observed occupancy states of the patches as a measure of SIN occupancy (see Chapter 4 for the definition of p_λ). It is apparent that SINs with large metapopulation capacities tended to be occupied more regularly than SINs with small metapopulation capacities, but note that the connectivity of the SIN also makes a difference: SINs with small metapopulation capacities tended to be occupied more frequently if they were well connected to metapopulations in the neighboring SINs, exactly the same patterns as seen in *A. agestis*. The effects of metapopulation capacity and SIN connectivity were both significant in a logistic regression contrasting SINs that were never occupied in 1993–2001 versus those that were occupied during at least 1 yr (for both effects $p = 0.01$; I. Hanski, unpublished result).

Plebejus argus in North Wales

The silver-studded blue butterfly, *P. argus caernensis*, occurs within a restricted area of 35 km^2 in the Creuddyn Peninsula in north Wales. The subspecies (or "race") *caernensis*, endemic to the peninsula, has been described based on its morphology (unusually small, with especially blue females) and habitat requirements (medium-height turf with patches of bare ground, on southerly facing, species-rich limestone grasslands and crags), and on its interactions with host plants (especially *Helianthemum* species and *Lotus corniculatus*) and mutualistic ants (*Lasius alienus*) (Thomas, 1985a,b; Jordano and Thomas, 1992; Jordano et al., 1992; Thomas et al., 1999b). Such a level of local specialization implies that the butterfly has persisted in the peninsula for a long time, probably for several thousand to 11,000 or so years.

Sporadic information on the occurrence of the butterfly is available for the last 100 yr, and more detailed information on its distribution has been obtained in surveys carried out in 1970–1971, 1983, 1990, and 1996–1999 (Thomas et al., 2002a). The dynamics of the butterfly and its habitat are relatively slow, and hence only a few colonization and extinction events are observed in short study periods, potentially leading to the incorrect conclusion that the distribution is stable. However, since 1970–1971, 18 out of the 20 habitat patches that have been occupied at any time have shown population turnover, with 16 further patches remaining unoccupied; at the timescale of decades, the system has functioned as a metapopulation (Fig. 20.4). The long-term study has shown that the landscape actually consists of two parts for the butterfly, including a core metapopulation on Great Orme's Head to the northwest of the dashed lines in Fig. 20.4 and a more widely scattered patch network in the rest of the peninsula. The Great Orme's Head core area contains more habitat, and the patches are located much closer together, than those in the rest of the landscape. In the core area, a high percentage of

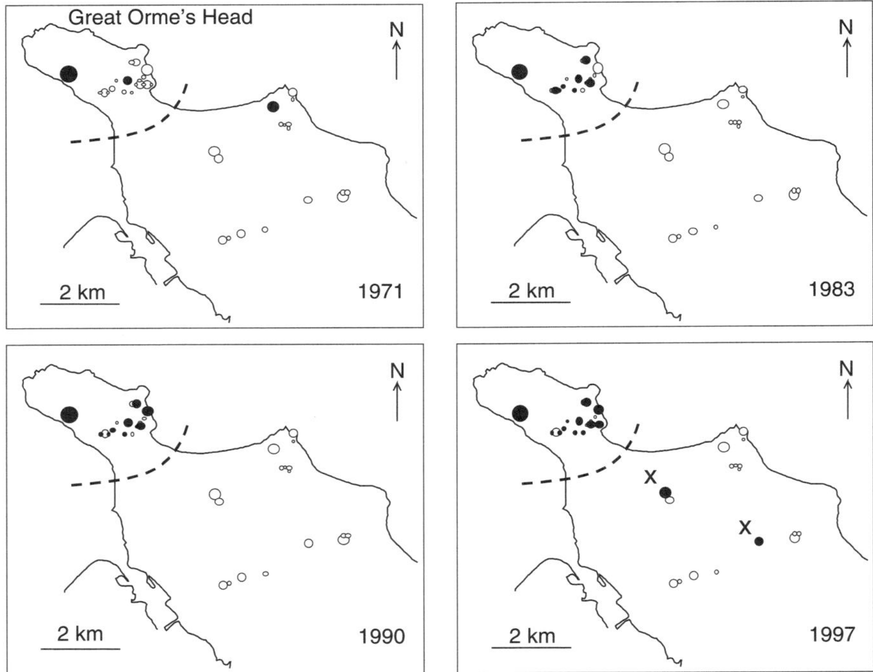

Fig. 20.4 Distribution of occupied (black) and empty (open) habitat patches for *Plebejus argus caernensis* on the Crueddyn Peninsula in north Wales since 1970. The solid line is the coast. Only two patches of habitat have remained constantly populated since 1970, both on Great Orme's Head, which forms the core of the metapopulation (to the northwest of the dashed line). The marginal landscape to the southeast of Great Orme's Head has been colonized sporadically, but the butterfly does not persist in this area: X shows 1997 populations that became extinct by 1999. Modified from Thomas et al. (2002a).

the suitable habitat is inhabited (Fig. 20.4). In the rest of the landscape, local populations have been established in the more scattered patches on at least three occasions during the last 100 yr: in the 1940s, after 1971, and again in the mid-late 1990s. Each time the populations in the peripheral areas became extinct again soon after coloniozation, apparently due to stochastic processes. A comparable metapopulation structure was found by Harrison et al. (1988) in a study of the Bay checkerspot butterfly, *Euphydryas editha*, in California, where a large core patch apparently maintained a persistent population, whereas small, marginal populations were ephemeral.

Thomas et al. (2002a) used the IFM to model the dynamics of *P. argus caernensis* in the peninsula as a whole and also separately in the marginal areas. Their first conclusion was that the longer the data series available to parameterize the model, the greater the amount of population turnover predicted. Short-term studies may give the impression of stability; relatively long-term empirical studies are required to obtain reliable insights into large-scale and long-term spatial dynamics. The large-scale and long-term study of *M. cinxia* in the Åland Islands provides further examples (Hanski, 1999b; Ovaskainen and Hanski, 2003b; Nieminen et al., 2004).

The second conclusion is that, in the case of *P. argus caernensis*, there is a core part of the distribution that is usually fully occupied and expected to be persistent, unless the habitat deteriorates, and that there is a marginal landscape where habitat patches are smaller and relatively isolated from one another and where habitat is occupied only sporadically. This marginal landscape was rarely occupied for more than a few years at a time during the 20th century because the populations always happened to become extinct before they had had time to grow to a large size. However, if populations would manage to grow to the local carrying capacity, there is no obvious reason why they could not survive for decades in the marginal landscape. Thomas et al. (2002a) inferred that the median time to extinction was 30 yr in the marginal landscape starting with the pattern of patch occupancy observed in 1997, with seven of 100 simulation runs surviving for more than 100 yr. Starting with all patches occupied, the median time to extinction was 59 yr in the marginal landscape.

Sporadic Metapopulation Occurrence and Conservation

In each of the cases just discussed, and also in the American pika at Bodie in California (Moilanen et al., 1998; Chapter 21), parts of the landscape were occupied only periodically, and long-term persistence was dependent on a combination of extinction-resistant core areas permitting recolonization of the more marginal landscapes, or reciprocal rescue and recolonization among semi-independent patch networks. These examples actually represent a substantial fraction of all metapopulation studies for which long-term and/or large-scale data are available, and they are probably not unusual. Therefore, we contend that a high proportion of habitat specialist species with restricted distributions will exhibit unstable dynamics in at least parts of their distributions, even if there is no further habitat degradation. In many cases, metapopulation dynamics without regional stochasticity are sufficient to explain these patterns, but in other cases they are likely to be driven by regional stochasticity causing the extinction of even relatively large metapopulations.

The implications for conservation are that planning may need to encompass even larger regions than currently considered, even when a metapopulation approach is being employed. There are often likely to be parts of species' distributions that are relatively safe from extinction, but other parts that are not. The former may occur in landscapes with high metapopulation capacities, with many large patches clustered close together, or they may be regions where semi-independent metapopulations are located sufficiently close together to allow some migration and gene flow. To allow the conservation of specialist species in such areas, it is clearly imperative first to identify them correctly and then target them for conservation. It would be tragic to lose the few remaining core areas to ongoing habitat loss while conservation measures were targeted at areas where eventual metapopulation decline was most likely, in marginal networks where long-term persistence is unlikely without migration from outside.

20.4 INTRODUCTIONS OF SPECIES

Successful introductions of species to empty habitat beyond their normal dispersal range provide experimental evidence for the presence of suitable but unoccupied habitat. Such introductions also allow us to test whether metapopulation models can be used to predict the invasion of empty habitat networks. Although some introductions fail instantly, usually for unknown reasons, many butterfly introductions have been successful (Holdren and Ehrlich, 1981; Oates and Warren, 1990; Warren, 1992; Nève et al., 1996b; Hanski et al., 2004). Even then, the timescale of success has often been limited, with populations surviving only for a few years before they die out again. In many cases, the area of habitat to which the population was introduced has been small, or the habitat has changed following introduction, and the population became extinct before it could spread to other potential habitats in the region (Oates and Warren, 1990). This is metapopulation failure: some suitable patches of habitat are present in the landscape, but the rate of colonization of new patches, starting from the site of introduction, was lower than the rate of extinction. A metapopulation approach can help understand which introductions will be successful and help predict how introduced metapopulations will spread. Fortunately, a large body of empirical information and mechanistic understanding already exists of migration and colonization of butterflies in metapopulations (Box 20.3).

Introductions of *Plebejus argus* in North Wales

As described previously, *P. argus caernensis* was restricted to limestone grassland on one peninsula in north Wales. Other outcrops of limestone in the same region were unoccupied by the butterfly, despite the fact that the habitat looked superficially similar and potentially suitable for the butterfly. In 1942, 90 females were introduced to one patch of grassland in this previously unoccupied area, from which they spread to occupy 17 out of 20 patches in 1990, establishing a metapopulation with roughly 90,000 adult butterflies (Thomas, 1985b; Thomas and Harrison, 1992). Using a metapopulation model parameterized for the original distribution of *P. argus caernensis*, Hanski and

BOX 20.3 Butterfly Studies on Migration and Colonization

One strong theme of research in butterfly metapopulation biology has been migration within metapopulations. Mark–release–recapture projects on butterflies started with Dowdeswell et al. (1940), followed by Ehrlich's (1961; Ehrlich and Davidson, 1961) important studies on the spatial structure of *Euphydryas editha* populations in California. Many multipopulation mark–release–recapture studies have been conducted on European butterflies in the 1990s (e.g., Baguette and Nève, 1994; Hanski et al., 1994; Hill et al., 1996; Nève et al., 1996; Thomas and Wilson, 2002), and new methods of data analyses have been developed (Hanski et al., 2000, Ovaskainen, 2003) and applied (Ricketts, 2001; Petit et al., 2001; Wahlberg et al., 2002b; Schtickzelle et al., 2002). Researchers have attempted to identify the factors that are responsible for variation in the observed rates of migration, such as habitat patch areas and distances between patches, the quality and quantity of larval and adult resources within patches, population density, the sex ratio, the type of boundary surrounding habitat patches, and the nature of the matrix between habitat patches (e.g., Thomas and Singer, 1987; Kuussaari et al., 1996; Hill et al., 1996; Sutcliffe and Thomas, 1996; Sutcliffe et al., 1997; Schultz, 1998; Haddad, 1999a; Haddad and Baum, 1999; Roland et al., 2000; Thomas et al., 2000; Ricketts, 2001; Fownes and Roland, 2002; Matter and Roland, 2002; Wilson and Thomas, 2002). Because almost all of these variables operate through the behavioral responses of individuals, this has led to the development of individual-based dispersal models, such as random walk simulations of the dispersal of individuals in habitat patches (Schultz and Crone, 2001; R. Setchfield, manuscript in preparation) and through ecological corridors (Haddad, 1999b). Ovaskainen (2003) has developed a diffusion model that allows quantitative analysis of individual movements in heterogeneous landscapes, including the cost of migration in terms of mortality in the matrix (see also Hanski et al., 2000).

Evolutionary studies have shown that migration rates, behaviors, and flight morphologies may evolve in response to the structure of the landscape (Thomas et al., 1998; Hanski and Singer, 2001; Heino and Hanski, 2001, Hanski et al., 2002), including factors such as the host plant composition of individual habitat patches (Hanski and Singer, 2001; Hanski et al., 2002; Hanski and Heino, 2003). In *Melitaea cinxia*, females in newly established populations are more dispersive than females in older populations (Hanski et al., 2002), suggesting that extinction–colonization dynamics select for an increased migration rate at the metapopulation level. However, the actual behavioral mechanisms that butterflies use to navigate and select habitat patches within metapopulations are largely unknown. Field experiments with individual butterflies removed from their natural habitat patches revealed apparently deliberate search behaviors and the potential of individuals to return preferentially to their patch of origin (Conradt et al., 2000, 2001). The behavioral means by which migrant individuals locate new patches of habitat requires further research.

Thomas (1994) predicted that the introduction should spread successfully and establish a persistent metapopulation in the Dulas Valley. The metapopulation still thrives in the Dulas Valley, 60 yr after the original introduction. Given the success of the metapopulation approach to predict the butterfly's eventual extinction from a landscape with limited habitat availability and its successful invasion of an empty habitat network, there is the basis to assess other areas of limestone for their potential for introductions.

Introductions of *Melitaea cinxia* to Unoccupied Patches and Patch Networks

Melitaea cinxia has been introduced since the early 1990s to some 10 previously unoccupied sites within the Åland Islands and in mainland Finland (M. Kuussaari and I. Hanski, unpublished results). All but two of these introductions occurred at sites that were either completely isolated and poorly connected to other (generally small) habitat patches; the introductions to single patches and sparse habitat networks all failed. Of the two successes, one site in mainland Finland (Fagervik) has persisted for at least 6 yr, but unfortunately the populations have not been surveyed systematically. In this case there are several suitable sites for the butterfly within its usual migration distance, up to 3–4 km (Hanski, 1999b; van Nouhuys and Hanski, 2002).

The other introduction that turned out to be successful took place in 1991, when 72 larval groups were translocated to 10 small meadows on the island of Sottunga, located 20 km east from the main Åland Island. Sottunga is 4 km long and 2 km wide; it had some 20 suitable meadows for *M. cinxia* when surveyed in June 1991, but no populations of *M. cinxia*. The metapopulation established in 1991 has persisted for the past 12 yr, although it went through a bottleneck of just three occupied meadows in 1999 (Fig. 20.5). It is noteworthy that none of the original local populations created in 1991 has survived, and hence this introduction provides an experimental demonstration of classic metapopulation dynamics in action. The metapopulation capacity (Chapter 4) of the Sottunga network is so large that, based on the occurrence

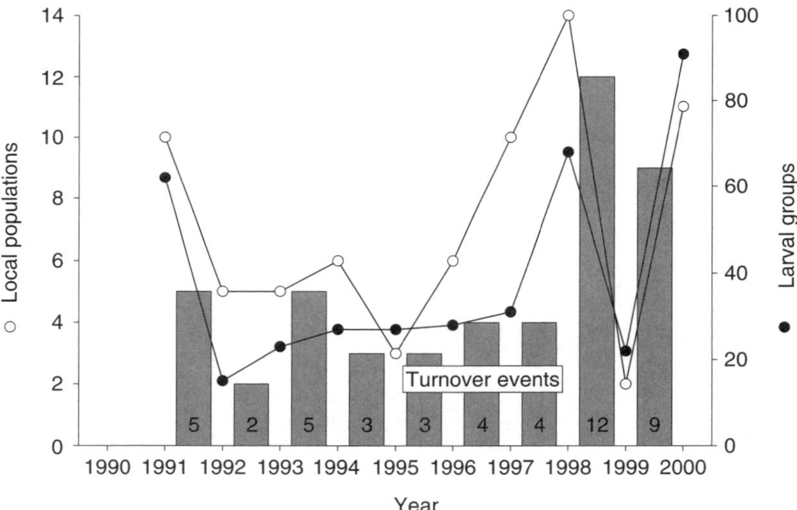

Fig. 20.5 Experimental metapopulation of *Melitaea cinxia* on the island of Sottunga in the Åland Islands. The metapopulation was established in a previously empty network of ca. 20 small patches in 1991 (72 larval groups placed in 10 meadows). Observed changes in metapopulation size in terms of the number of larval groups and the number of occupied meadows, as well as the number of turnover events (extinctions and colonizations) between consecutive years, are shown. None of the original populations that were established in 1991 has survived until present, and hence this metapopulation has survived in a stochastic balance between local extinctions and recolonizations of empty meadows as predicted by the theory (from Nieminen et al., 2004).

of the butterfly in patch networks on mainland Åland, persistence in Sottunga is expected (Hanski et al., 2004).

20.5 METAPOPULATIONS RESPONDING TO CLIMATE CHANGE

For thermally sensitive species such as butterflies, climate warming is predicted to alter the quantity, quality, and distribution of suitable habitats in a landscape. This section describes how climatic improvement can increase the density of suitable habitat patches in landscapes close to northern (poleward) range boundaries, permitting metapopulation invasion of previously empty habitat networks. We also evaluate whether metapopulation and invasion models accurately predict the expansion at range margins. At southern boundaries, deteriorating conditions may lead to reduced habitat availability, such that landscapes where species persisted in the past may no longer allow metapopulations to persist. Range contractions at southern margins have been demonstrated for *Euphydryas editha* in Mexico and California (Parmesan, 1996) and for a number of species in Europe (Parmesan et al., 1999), but no quantitative modeling of southern range contractions has yet been attempted.

Multispecies Patterns

A qualitative test of the role of habitat availability in range expansion comes from a cross-species comparison of range changes at the northern boundaries of butterfly distributions in Britain. Warren et al. (2001) evaluated the changes in the distribution of 46 southerly butterfly species, all of which were predicted to become more widespread and expand their distributions because of ameliorating climatic conditions. In practice, however, almost all of the habitat specialists and half of the habitat generalists declined in terms of their distribution between 1970–1982 and 1995–1999 due to habitat losses. The low habitat availability for habitat specialists means that most British landscapes fail to permit metapopulation expansion that would otherwise be expected to occur under a warmer climate.

Hesperia comma Expanding through Fragments of Chalk Grassland

The means by which climate warming can increase habitat availability is clearly seen in the silver-spotted skipper, *H. comma*. Like many other poikilothermic animals at their northern range margin (Thomas et al., 1999b), this butterfly was restricted in 1982 to unusually warm habitats: southerly facing hillsides, where it laid its eggs on the grass *Festuca ovina* in warm hollows close to bare ground (Thomas et al., 1986). However, by 2000 *H. comma* had shown a significant widening of the range of habitats used, spreading into north-facing habitats and laying eggs in taller and more continuous vegetation than had been used previously (Thomas et al., 2001a; Z. Davies et al., unpublished result). The species now occupies habitats that would have been regarded as too cool for the species in 1982. In the case of *H. comma*, availability of a suitable habitat has also increased due to increased grazing by expanding rabbit populations and livestock since 1982 (a deliberate conservation measure).

Habitat management and climate-induced habitat changes have approximately doubled the quantity of suitable habitat for this species in one part of its distribution in Britain, in the South Downs hills in southern England, a region where *H. comma* had been confined to a single large hillside in the late 1970s. The incidence function model was parameterized for *H. comma* using data from a different part of southern England and was then applied to predict the changes in distribution in the South Downs. This application represents a genuinely independent test of the predictive power of the model. When model simulations were run assuming the definition of suitable habitat that was valid in 1982, the model predicted an expansion of about 5 km in 18 yr (the distance from the source population to the furthest 10 populations in surviving metapopulations; the metapopulation went extinct in some simulation runs). However, assuming the relaxed definition of habitat as demonstrated empirically in 2000, expansion was much faster, with an average expansion of 14.3 km, and now none of the 100 simulation runs resulted in extinction (Fig. 20.6). The observed expansion after 18 yr (16.4 km) was not significantly different from this latter prediction, but was further than the expansion predicted in any simulation run using the 1982 definition of habitat. The reduced need for hot, south-facing hillsides means that there are now more and larger patches of habitat available in the landscape, and hence shorter distances between the patches. Shorter distances led to more numerous colonization events; more new populations increased the number of potential colonists, which led to a positive feedback between increased habitat availability and rate of expansion. An approximate doubling of the availability of habitat in this landscape resulted in a threefold increase in the predicted expansion rate.

These studies have been extended to other parts of the British distribution of *H. comma* with similar results (R. J. Wilson et al., unpublished result). Differences in expansion rates in different landscapes were predictable from differences in the level of habitat availability. Observed expansion distances were closely correlated with mean expansion distances predicted from 100 simulation runs for each landscape ($r^2 = 0.95$, $n = 5$ separate metapopulations, $P = 0.001$; R. J. Wilson et al., unpublished result). The prediction was accurate in four out of five landscapes studied. In the fifth, the model predicted faster expansion than was observed, with the observed expansion falling just outside the 95% confidence limits of IFM-predicted expansion. In this case, however, the average habitat quality (host plant density) was lower than in the other landscapes. When reduced habitat quality was taken into account by reducing effective habitat areas in the IFM (see later), the errant landscape fell back into line. These results give genuine hope that it may be possible to manage metapopulation recoveries and predict responses to climate change at a landscape level.

Scaling up from Metapopulations to Geographical Distributions

The expansion of *H. comma* can be regarded as an example of conventional metapopulation dynamics: increased regional habitat availability led to expansion rates up to 10 km per decade in habitat patch networks of up to a few

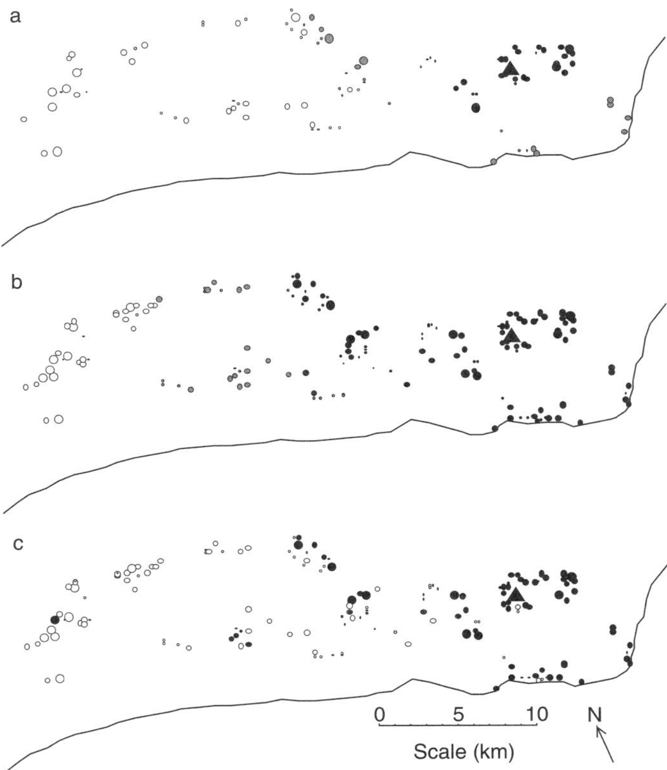

Fig 20.6 Simulated (a, b) and observed (c) expansion of *Hesperia comma* from a single large population (triangle) that existed in 1982 in the South Downs hills of southern England (solid line is the south coast of England). Circles are habitat patches that were unoccupied in 1982; those that had been colonized by 2000 are show in black and those that were still empty in 2000 are shown as open circles in (c). Simulated expansion using the incidence function model shows patches that were colonized in more than (black) and less than (gray) 50% of simulation runs: open circles were not colonized in any simulation run (out of 100). In (a), the 1982 definition of habitat was used, and expansion was predicted to be limited. By 2000 (b), the range of habitats available to *H. comma* had increased (principally as a result of climate warming), roughly doubling the availability of habitat in the landscape. Simulated expansion in the expanded habitat network (b) was much faster than in (a), and not significantly different from the observed expansion (c). Modified from Thomas et al. (2001a).

hundred patches. Other species that are expanding northward in response to climate change tend to have a broader range of habitats available to them, and in many cases this means that it is difficult to delimit distinct habitat patches. However, increased colonization potential and larger population sizes at newly colonized sites should increase with habitat availability even when the habitat is not fragmented into distinct patches. Similarly, the colonization rate is expected to depend on the distances to currently populated areas, whether they occur in discrete patches or not. Thus range dynamics of species that do not form recognizable metapopulations may nonetheless show metapopulation-like features. To this end, we adopted a 200-m resolution grid-based model (MIGRATE) to simulate range expansion in butterflies that do not always live in distinct habitat patches (Hill et al., 1999). The model focuses on colonizations only because

(a) it is generally difficult to identify extinctions when the landscape and hence population is not highly fragmented, (b) there have been very few observed extinctions, relative to the number of colonizations, in the expanding species since 1970, and (c) the computational time associated with modeling extinctions as well as colonizations would have been prohibitative at 200-m grid resolution across the whole of Britain.

The model was initially run on the speckled wood, *Pararge aegeria*, which occurs in woodland and scrub in the northern parts of its geographic distribution in Britain. Although habitat quality inevitably varies, almost all types of woodland and scrub contain suitable breeding areas and host plants (grasses), making it possible to identify potential habitat with satellite images. The model was run for two parts of Britain: for Yorkshire, where the woodland cover was estimated to be 2.72%, and for an area around Inverness in Scotland, where woodland cover was 3.58% (Hill et al., 2001). In both areas, the actual expansion of *P. aegeria* commenced around 1970, hence simulation and empirical results were compared for the period between 1970 and 2000.

The MIGRATE model was ran at 200 m resolution, but the fit was assessed at 5 km resolution, at which scale the field data had been collated (Hill et al., 2001). Most parameters of the MIGRATE model were obtained from independent field data (e.g., population density, population growth rate), but some parameters were estimated by maximizing the fit between the observed and the modeled expansion in Yorkshire. The latter involved adjusting the habitat threshold, permitting successful colonization of a grid square; the need to do this may stem from the lack of an extinction component in the MIGRATE model. In Yorkshire, the model predicted 53 occurrences in the 5-km grid squares, compared to 49 recorded occurrences. More interesting was the comparison between model prediction and empirical observation for the region around Inverness using the parameter values estimated in Yorkshire. The only extra element was the imposition of an upper elevational threshold for the occurrence of the butterfly, as it does not occur above 200 m. The model predicted 119 occurrences around Inverness, while 121 occurrences were observed (Fig. 20.7; Hill et al., 2001). This result strongly suggests that habitat availability is an important determinant of the rate of expansion in species responding to climate change and that metapopulation-like models can be used to predict distributional changes in regions where range expansions are taking place. The model has subsequently been applied to two other related species, *Pyronia tithonus* and *Aphantopus hyperantus*, as well as to *P. aegeria* across the whole of Britain. Results of these analyses show that differences in the rates of expansion of the three species can be predicted from differences in habitat availability and population parameters (especially population density; S. Willis et al., unpublished result).

20.6 CONCLUSIONS

Contrary to initial concerns about the applicability of the metapopulation approach to fragmented populations in changing environments, these situations actually provide some of the strongest evidence for the importance of

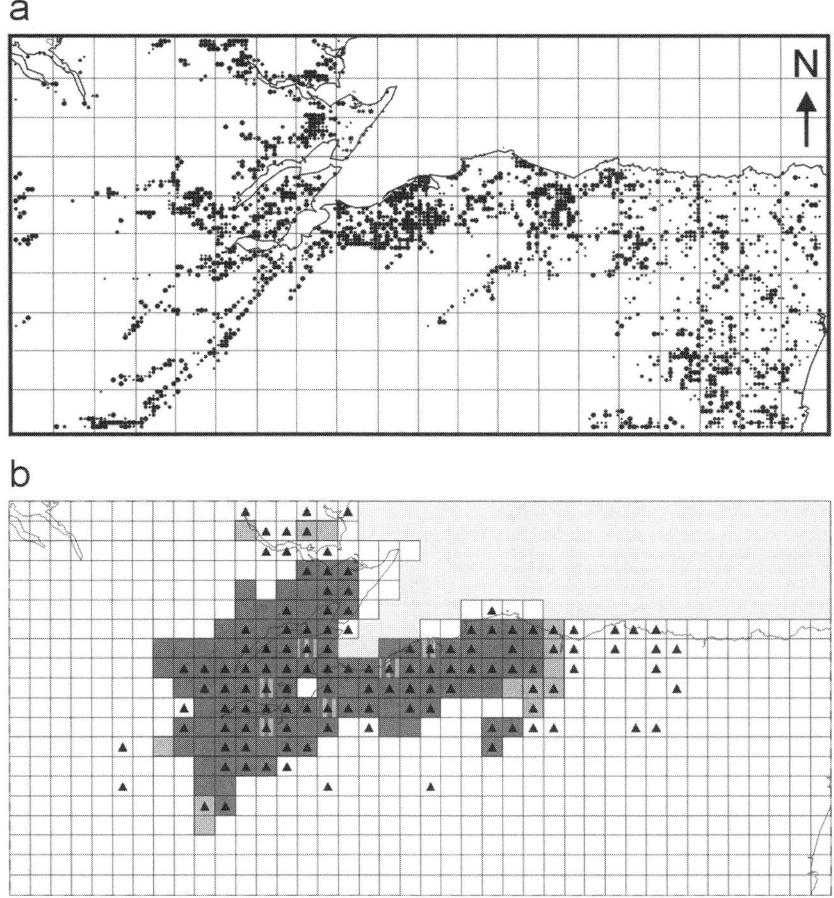

Fig. 20.7 (a) Distribution of woodland around Inverness in Scotland: small circle represents 5–10% woodland, medium-sized circle 11–25%, and large circle >25% woodland cover. Woodland above 200 m elevation (above sea level) asl was excluded. Lines at 10-km intervals. (b) Observed distribution in 1970 (striped squares) and 1995–1999 (triangles) at a 5-km grid resolution. Simulated distribution is based on five runs of MIGRATE showing areas normally (dark gray; colonized in ≥4 runs) and sometimes (pale gray; colonized <4 runs) colonized. Simulations were seeded in the six grid squares occupied in 1970.

metapopulation dynamics and for the predictive power of metapopulation models. Applied carefully, quantitative predictions are likely to be sufficiently accurate to use in practical conservation. Simple metapopulation models do not encompass all of the reality of natural systems, such as complex variation in habitat quality, behavior-driven dispersal, and so forth. These are important areas for further research. However, provided that one has an adequate knowledge of the habitat requirements (see later) and dispersal capacity of the focal species, simple pattern-based metapopulation models seem to capture enough of the dynamics to provide both insight and practical guidance for conservation and management. For example, we can manage for expansion in restoration ecology by targeting conservation management (increasing habitat quality and quantity) to increase metapopulation capacity in areas currently predicted to be only just below the threshold for invasion. Even if not all of the details of the dispersal process are known exactly, the recommendations are likely to be quite robust.

The same approach allows us to make the transition from landscape-level dynamics to geographic distributions and range dynamics. A metapopulation approach helps understand why some species are expanding at their northern range margins as the climate warms, but others are not, and helps us understand why some landscapes are permanently populated by a species while others are occupied only sporadically. More impressively, metapopulation models and spatially explicit colonization models make accurate predictions of differences in the rates of expansion in different landscapes and in different species (in the context of species introductions and climate change). The predictive power is sufficient for the current generation of models to be used in the development of environmental policy.

There are also important limits to the predictive power of the current models. Some of these relate to shortcomings in the simplest (and hence most practical) models. The most obvious concern has been the need to incorporate variation in habitat quality into metapopulation studies. The primary need is to be able to distinguish between suitable and unsuitable habitat, which often requires detailed studies of the ecology of the species in question. Because apparently suitable empty habitat may not actually be suitable, researchers have to be careful while mapping patch networks. Whenever possible, statistical habitat quality threshold models should be applied, and the suitability of empty habitat should be checked by experimental introductions and through long-term observations that permit the natural colonization of empty habitats to be observed (Harrison et al., 1988; Harrison, 1989). However, clear separation of suitable habitat from unsuitable areas is only possible in some cases. In the silver-studded blue butterfly (*Plebejus argus*), it is easy (after much research) to distinguish between suitable and unsuitable habitat in areas of limestone grassland, but much more difficult in heathland vegetation, where habitat quality appears to vary more continuously (Thomas, 1985b; Thomas and Harrison, 1992). In heathland, some unoccupied habitat is clearly suitable, as evidenced by successful introductions, but other patches of habitat seem to be of intermediate quality. Consequently, it has been possible to apply patch occupancy metapopulation models (Chapters 4 and 5) to *P. argus* where it inhabits limestone grasslands (Hanski and Thomas, 1994; Thomas et al., 2002a), but not in areas of heathland, even though extinction–colonization dynamics are even more important to the persistence of the species in successional heathland habitats than on limestone (Thomas and Harrison, 1992; Lewis et al., 1997). This is likely to be a widespread phenomenon, with only some species, and only some regions in many species, showing a clear distinction between habitat and nonhabitat. Identifying suitable habitat is an essential first step, and a failure to take account of variation in habitat quality is likely to result in inadequate or unrealistic metapopulation studies (Thomas, 1996; Dennis and Eales, 1999; Thomas et al., 2001b; Fleishman et al., 2002).

Once "suitable" habitat has been identified, habitat quality will still vary, and some occupied patches may represent population sinks (Thomas et al., 1996; Thomas and Kunin, 1999). One very simple and practical approach is to modify the true areas of habitat patches by their quality (Hanski, 1994;

Moilanen and Hanski, 1998). For example, in the study of *H. comma* we observed that range expansion was slower than predicted in one patch network (outside the 95% confidence limits of simulated expansion). In these networks, habitat patches vary greatly in host plant density, largely due to the management histories of the sites. Therefore, R. J. Wilson et al. (unpublished result) adjusted each patch area by the density of host plant so that patches with low densities were treated as if they were smaller patches than their actual physical dimensions would indicate, and patches with high plant densities were treated as larger than their physical size. Thereby low-quality patches would have higher rates of extinction than other patches of the same physical area and would contribute less to the colonization of empty habitat patches. The result of this adjustment was to predict the dynamics of *H. comma* accurately for the deviating patch network as well as for the other networks (Section 20.5). This example relates to host plant density, but it would be equally simple to apply the approach to any other measure of relative patch quality. Of course, this sort of approximation will not work in all situations, and it may conflict with the need to scale immigration and emigration rates with true patch area (Chapter 4). Nonetheless, we think that, in many or even most cases, relatively simple adjustments to existing models are likely to provide greater insight into the key dynamics of metapopulations than more complicated models that simultaneously take into account of a wide range of habitat quality variables.

A more fundamental problem is that distribution of a suitable habitat does not remain constant through time, particularly in species that track the distribution of successional habitats (Harrison, 1991; Thomas, 1994a,b). In modern landscapes, shrinking (due to habitat losses) and expanding (e.g., due to climate change) patch networks are particular cases in point. However, studies reviewed in this chapter demonstrate that metapopulation models can be applied successfully to these systems, provided that one *knows* how the patch network has changed (documented fragmentation) or *can predict* these changes (based on, e.g., changing climate or successional dynamics). In the situations examined, predictions were reasonably accurate over 20- to 30-yr periods (or generations, for these butterflies), allowing managers to predict species responses to given environmental changes.

Whether these predictions can be extended over longer periods of time is another question. Today, we know that various environmental changes have taken place and we can predict the responses of species to these changes with some degree of accuracy, including extinction debts and distributions lagging behind climate change. However, the longer the timescale of projection, the more important it will become to be able to predict changes to the environment itself, and this is of course outside the scope of metapopulation biology. Metapopulation projections can be applied to scenarios of hypothetical environmental changes, but they can never be better than the environmental projections on which they are based.

The second issue relating to long-term prediction is that metapopulation processes are stochastic, which will limit predictability. Events that are rare may nonetheless be critical determinants of large-scale metapopulation dynamics, and no models can ever predict particular realizations of such processes accurately. Models may be able to predict the possibility of such events, but never exactly when and where they will occur. Therefore, even "correct" predictions may not be of great practical use. This is best illustrated by two examples. Metapopulation models can include a wide variety of dispersal functions (see Box 4.3), some with more and some with less long-distance migration, but no model will ever be able to predict exactly when and where a new long-distance colonization will take place. *P. aegeria* colonizing Inverness (Section 20.5) is a case in point. Inverness was about 100 km from the nearest *P. aegeria* population when it was colonized in the late 1960s. Once Inverness had been colonized, however, it was possible to predict subsequent spread away from this population focus for another 50+ km, based on the frequency of more usual colonization events (Hill et al., 2001).

Second, rare climatic events may wipe out large and vigorous metapopulations in episodes of what are essentially extremes of regional stochasticity. In *E. editha*, an exceptional summer frost killed off all populations in forest clearings within one metapopulation (Thomas et al., 1996), and large-scale extinctions have also been caused by summer droughts (Ehrlich et al., 1980). If the distribution of stochastic events was known, the metapopulation consequences could be predicted statistically, although not of course in terms of exact timing. In practice, the situation is worse because we do not know the distribution of the relevant stochastic perturbations — the longer we observe any system, the more we know of ever rarer events, and these may be the ones that eventually cause the demise of the population. Rare events in space and time do not undermine the overall utility of the metapopulation approach, but they limit the spatial and temporal scales over which it is practical to make population predictions.

In summary, the application of metapopulation models to environments that are changing seems daunting. The environments occupied by virtually all rare and threatened species have changed over the past 100 yr, and in most cases are likely to continue to do so in the future as a result of continuing land use and climatic changes. Yet, the application of metapopulation models to butterfly distributions and dynamics has survived this test well: butterfly responses to environmental changes have been predicted accurately over periods of up to 30 yr (generations). Road testing with appropriate species has shown the metapopulation approach to be sufficiently robust not only to aid the theoretical understanding of large-scale dynamics, but also to guide the management of populations and metapopulations in practice over time periods of several decades. Rare stochastic events may make longer term predictions less certain, but high predictive power over several decades is likely to be useful for most management decisions. As a final caveat, we do not claim that all of the processes relevant to the regional persistence and dynamics of species are encompassed within simple metapopulation models such as IFM. Rather, metapopulation approaches capture a sufficient amount of the dynamics of many species to provide both useful insight and practical tools for conservation planning.

21. INFERRING PATTERN AND PROCESS IN SMALL MAMMAL METAPOPULATIONS: INSIGHTS FROM ECOLOGICAL AND GENETIC DATA

Xavier Lambin, Jon Aars, Stuart B. Piertney, and Sandra Telfer

21.1 INTRODUCTION

The metapopulation paradigm is increasingly being used to describe the structure and dynamics of mammalian populations, as it is with many other taxonomic groups. Previous reviews have, however, highlighted that metapopulation terminology in mammals is used to define a number of different population structures (Harrison, 1991; Harrison and Taylor, 1997) and observed that case studies of classical metapopulation structures are rare (Elmhagen and Angerbjorn, 2001). Most frequently, the observed pattern is of a "population of populations," or a suite of local populations, that is not always discrete but where limited dispersal gives rise to discontinuous spatial structure

0-12-323448-4

(e.g., Szacki, 1999; Entwistle et al., 2000; Gaggiotti et al., 2002). Metapopulation structure is also invoked frequently to describe those sets of populations where each population has become effectively isolated with no potential for the rescue effect, and so are ultimately destined for extinction (Brown, 1971). The aforementioned structures represent the extremes of a continuum of habitat fragmentation. Determining whether the processes of population turnover and dispersal between patches have a strong influence on overall persistence in a habitat with a given degree of fragmentation requires information on the dispersal behavior of individuals, as well as on the processes responsible for the extinction of local populations. Much progress has been made in understanding the process of dispersal with small mammals, but extinction processes remain poorly understood.

Relatively few studies to date have compellingly characterized metapopulation processes in small mammals. This begs the question whether metapopulation processes are not being observed because studies focus at inappropriate spatial or temporal scales or whether small mammal populations do not inherently show such processes. A short life span, specific habitat requirements (or at least preferences), and local organization into social units or habitat-defined aggregations may intuitively suggest that small mammals will form metapopulation structures. Conversely, however, mammals may fail to form metapopulations because (1) species with specific habitat requirements may have sufficiently high mobility relative to the grain of landscapes that the dynamics of local populations is typically synchronized; (2) the frequency of population turnover may be much reduced by effective and targeted immigration from adjacent local populations (Stacey et al., 1997); or (3) local population size in small mammals may normally be so large or local populations so well connected as to make extinction through demographic stochasticity and recolonization dynamics insignificant, except when larger scale dynamics processes bring about regional declines. Two separate reviews in Hanski and Gilpin (1997), of largely the same empirical studies of small mammals, failed to reach consensus on whether metapopulation structure should be considered the norm in small mammals. Stacey et al. (1997) wrote "Many diverse species of small mammals may be predisposed towards metapopulations because they show spatial population structure as a result of sociality or habitat fragmentation. In such rescue-effect metapopulations, local populations fluctuate fairly independently of one another, yet exchange low to moderate numbers of immigrants such that metapopulation structures have an important stabilizing effect at the regional level even without population turnover." However, Harrison and Taylor (1997) stated that they knew of no good example where metapopulation structure has an important stabilizing effect at the regional level without population turnover. This lack of consensus reflected the absence of any compelling empirical study of small mammal metapopulations available at that time that documented both the processes of dispersal and population turnover. Thus deciding whether metapopulation structure even existed among small mammals and deriving generalities as to their frequency was a matter of judgment. It is thus timely to ask whether new evidence has come to light to resolve the apparent contradiction between Stacey et al. (1997) and Harrison and Taylor (1997). This chapter first briefly reevaluates the evidence of metapopulation structures for three other small mammal species considered to conform most closely to the metapopulation paradigm — black-tailed prairie dogs (*Cynomys ludovicianus*), American pika (*Ochotona princeps*), and field voles (*Microtus agrestis*). Next, the chapter introduces new studies of water voles (*Arvicola terrestris*) that contain strong evidence of metapopulation processes and progresses our understanding of metapopulation processes. Throughout, ecological and genetic data are considered that can be used to identify metapopulation patterns in small mammals. We also ask whether common features of small mammal dispersal and of the processes responsible for extinction of small mammal local populations result in a distinct mammalian perspective of the metapopulation paradigm.

21.2 CLASSICAL SMALL MAMMAL METAPOPULATIONS

A small number of well-documented studies have proven highly influential in suggesting and subsequently understanding metapopulation processes in small mammals. With increasingly detailed study incorporating both ecological and genetic data, these investigations also highlight common problems associated with understanding mammalian metapopulation dynamics, namely that it is not necessarily straightforward to infer process from pattern.

American Pika at Bodie

Pikas are small lagomorphs closely associated with patchily distributed rocky outcrops. In their usual high-altitude habitat, patches are typically large and populations semicontinuous comprising up to 100 individuals. Smith (1974) reported that such populations rarely experience extinctions. At a single lower altitude site, the Bodie mining area (ca. 12 km^2) in California, population turnover does occur, and this set of populations has been presented as the best example of a classical metapopulation in small mammals (Moilanen et al., 1998). Evidence shows that migration maintains genetic diversity and rescues local populations from extinction (Smith and Gilpin, 1997; Stacey et al., 1997; Peacock and Ray, 2001). Pikas defend individual territories, such that upon weaning all juveniles must either inherit or disperse from their natal territory. There is an unusually inflexible link between the size of a patch and the number of individuals it can accommodate and acquiring an individual territory is a key transition in pika life history. Vagility is reduced by low tolerance to daytime high temperature at Bodie which lies at the lower altitudinal limit of the species (Smith, 1974). Unless a territory is vacant, residents aggressively inhibit immigration and the immigration rate is thus density dependent. In addition, immigrants settle preferentially next to opposite-sex conspecifics. Adjacent territories are typically occupied by opposite sex individuals and, in the study of Smith and Ivins (1984), territory owners that disappeared were always replaced by members of the same sex. Most young pikas are philopatric and settle in vacant territories in the natal patch. Dispersal between patches also occurs, although at low frequency, and tends to be directed toward the closest patch with vacancies (Peacock and Smith, 1997; Peacock and Ray, 2001).

Repeated surveys of patch occupancy by pikas at Bodie spanning 20 years reveal a sharp decline in the southern part of the patch network and a more constant rate of patch occupancy in its northern part (Smith, 1974; Smith and Gilpin, 1997). Moilanen Smith and Hanski (1998) used these data to parameterize Hanski's (1994) incidence function model (IFM) assuming a classical metapopulation structure and including the rescue effect (Chapter 4). The parameterized model successfully predicted observed spatial variation in the extent of the decline, suggesting that the dynamics is consistent with that of a metapopulation in a constant environment. When low levels of region-wide environmental stochasticity, implemented as a synchronized variation in the size of all patches, hence, indirectly, as variation in extinction probability, were included, model iterations suggested that the southern part of the metapopulation was extinction prone. Model simulations thus showed that the observed spatially correlated pattern in patch occupancy in pika at Bodie may have arisen from extinction–colonization dynamics without the need to assume that any spatially correlated processes affected the local dynamics and population extinction in a subset of patches (Moilanen et al., 1998).

In stark contrast to the interpretation that the Bodie pika population persists in a quasistable balance of extinction–recolonization (Moilanen et al., 1998), Clinchy et al. (2002) suggested that the observed dynamics of pika at Bodie could equally well be explained by spatially correlated extinction caused by predation in a slowly declining fragmented population. Predation by mustelids is the main cause of death and local population extinction (Smith and Gilpin, 1997). With such small and mobile predators, individuals in small local populations are expected to be exposed to a higher risk of predation mortality than individuals in large local populations. In addition, the fate of adjacent local populations potentially simultaneously exploited by a predator may become spatially correlated. In support of their argument, Clinchy et al. (2002) presented simulations that assumed spatially correlated extinctions caused by a mustelid in the Bodie patch network, but did not include dispersal or other demographic processes. The simulations suggested that more isolated patches became both less likely to be occupied and more likely to go extinct under the influence of spatially correlated extinction. Similar patterns of patch occupancy are observed at Bodie and, more generally, predicted by distance-dependent dispersal and the rescue effect. Clinchy et al. (2002) interpreted their simulations as showing that recolonization of vacant rock piles and the rescue effect played only minor roles in the overall pika dynamics at Bodie. They noted that recolonized patches were significantly more likely to go extinct than patches that were occupied in both of the preceding surveys. Clinchy et al. (2002) interpreted this pattern as evidence that extinctions are a direct result of recolonization. That colonizations often fail is not implausible for pika because of their low fecundity and the requirement for colonists to accumulate sufficient plant material in hay piles before winter.

The contrast between Moilanen et al. (1998) and Clinchy et al. (2002) interpretations of the same data shows that it is difficult to infer process from pattern. Neither model captures all patterns in the data and, in the real world, processes may mask each other. For instance, behavioral evidence of the rescue effect, such as dispersal directed at low-density patches where opposite sex relatives are present, may not translate into a detectable impact on extinction probability when predator-induced extinctions dominate. In addition, little is known about the dispersal behavior of pika and how it accords with the relationships used in the IFM. There are, however, more serious flaws with the Clinchy et al. (2002) argument. It implausibly assumes a mustelid capable of causing a 20-yr declining trend in pika but not overall extinction and causing correlated extinction in the southern half of the network without any noticeable influence on its northern part. It also fails to consider the likely relationship between probability of extinction and patch quality and/or size that may account for the observed relationship between extinction and recolonization. The assumed insignificance of dispersal is also inconsistent with the observation of Peacock and Ray (2001) that pika in Bodie had a level of heterozygosity as high as that observed in a more continuous habitat. This level of genetic diversity is consistent with a pattern of extinction–recolonization within Bodie combined with rare immigration from distant continuous populations (Peacock and Ray, 2001).

Black-Tailed Prairie Dogs

According to Stacey et al. (1997), studies of black-tailed prairie dogs provide detailed and convincing evidence that metapopulations with a strong rescue effect may be common among social mammals. Black-tailed prairie dog live socially in towns (local populations) of extended families. Early descriptions suggest that black-tailed prairie dog towns stretched over extensive areas of land and consisted of tens of thousands of individuals. As a result of human disturbance and the destruction of their habitat, prairie dogs over most of their range now live in small, scattered, and ephemeral local populations (Halpin, 1987; Lomolino and Smith, 2001; Antolin et al., 2002). The decline has been most pronounced in the western United States where an introduced pathogen, *Yersinia pestis*, the agent of sylvatic plague to which prairie dogs are highly susceptible, has become established starting in the 1930s (Antolin et al., 2002). Black-tailed prairie dogs have been studied most intensively in plague-free Wind Cave National Park, South Dakota, where colonies are still large (e.g., Garrett and Franklin, 1988; Hoogland, 1995). Behavioral studies of dispersal in this area show very strong conspecific attraction (Garrett and Franklin, 1988). Trajectories of radio-tracked dispersers were meandering when away from towns, but became directed when in sight of towns. All intertown dispersers in the study by Garrett and Franklin (1988) immigrated into existing town. This was despite being the target of outright hostility by residents that resulted directly or indirectly in the death of some newly established dispersers. Such behavior by dispersers would not be conducive to recolonization.

Studies of black-tailed prairie dogs in an area where plague is established show that towns have become fragmented and go extinct asynchronously, coinciding with plague epizootics decimating towns. Other causes of extinction include stochastic factors as well as persecution. Prairie dogs appear to now persist in a metapopulation-like state (Cully and Williams, 2001; Roach et al., 2001; Antolin et al., 2002). Despite their strong conspecific attraction, dispersing prairie dogs aggregated in and recolonized an empty habitat following a poisoning campaign (Cincotta et al., 1987). Colonization of towns was evident in plague-infected areas, and genetic data also demonstrated a substantial dispersal between existing towns (see also Halpin, 1987), consistent with the pattern of conspecific attraction documented in a more saturated environment (Garrett and Franklin, 1988).

Plague is epizootic not enzootic in prairie dogs due to their high susceptibility and is maintained within a community of more resistant rodent and carnivore species whose fleas reinfect prairie dogs (Biggins and Kosoy, 2001; Cully and Williams, 2001). During epizootics in prairie dogs, however, direct transmission between prairie dogs during their social encounters takes place (Antolin et al., 2002). Cully and Williams (2001) hypothesized that the transmission of plague is facilitated by the social cohesiveness of colonies such that, even though all local populations, regardless of size, are vulnerable to extinction from plague, large towns may be more susceptible, a pattern opposite that caused by demographic stochasticity. The pattern of extinction reported in Roach et al. (2001) is consistent with this.

Plague progressed only gradually over 3 years in three adjacent colonies spanning 30 km (Cully and Williams, 2001). Whether this pattern of slow spatial spread and moderately spatially correlated extinction was caused by dispersal by infective prairie dogs or a more complex process of amplification of the disease in the environment, infection and dispersal by resistant hosts and their fleas and reinfection of adjacent prairie dogs towns are not known (Cully and Williams, 2001; Biggins and Kosoy, 2001). Without a better understanding of the role of pathogens in prairie dogs metapopulations, it is not possible to establish whether long-term coexistence with plague is possible for highly social prairie dogs.

Field Voles in Baltic Archipelago

A 6-yr study of field voles (*M. agrestis*) living on islands in the Tvärminne Archipelago of the southeast of Finland's mainland (Pokki, 1981; Crone et al., 2001) represents a compelling example of a small mammal population closely approximating a metapopulation structure. Patterns of occupancy, changes in abundance, and within- and between-island dispersal by field voles were monitored on 71 islands ranging in size from treeless skerries (<1 ha) to larger islands (>10 ha) with heterogeneous vegetation including old fields, forested area, and less suitable heath. Levels of fragmentation have not been modified by anthropogenic influences, although the invasion by American mink (*Mustela vison*) over the last 50 years may have had some impact on vole dynamics (Banks et al., unpublished results). On treeless skerries, vole abundance mirrored the seasonal development of the vegetation. Vole populations grew during the early summer flush of vegetation growth, and voles on skerries had higher maturation rates, litter size, and mean densities than on larger islands during this period. However, skerry populations invariably declined in late summer when favored plant species were largely consumed or wilted. These declines resulted from both higher mortality and higher emigration movement to other islands. On larger islands, populations peaked in autumn and voles responded to seasonal crowding by intraisland movements from preferred old-field and meadows habitat to suboptimal heath and forests, with relatively low rates of interisland movements. Despite the proximity of the mainland (less than 3 km from the outermost islands), there was little evidence that it was the primary source of colonists for unoccupied islands and that the archipelago was part of an island–mainland metapopulation (Crone et al., 2001).

Extinctions and recolonizations of local island populations were common and affected all island size categories. Extinction rates were similar for large and medium islands, but populations on small islands turned over at a much higher rate than on larger islands. Despite the high extinction probability on skerries, seasonal extinction was not fully deterministic. The median period of occupancy for skerries was 2 yr. Only 1 of 13 large islands monitored was occupied for all 6 yr. It is suggested that two mechanisms were important for local population extinction: demographic stochasticity caused by low population sizes on all islands (mean population size on skerries in May was <9 voles, with mean peak size of 34.9 in July) and environmental variability, driven by annual variation in population density and rainfall, reducing carrying capacity on skerries (seasonal reduction in food availability). Interannual variation in rainfall must also have affected the development of vegetation on all islands simultaneously and is therefore a spatially correlated contributor to extinction (Banks et al. unpublished results). The relationship between island extinction probability and island size neatly matched the exponential function (Crone et al., 2001), as expected when demographic stochasticity is the main influence (Lande, 1993; Ray, 2001). A distinctive feature of the archipelago metapopulation was that a substantial fraction of extinctions were caused by deteriorating environmental conditions, leading to both high mortality and emigration (Pokki, 1981).

A much higher fraction of voles from skerries than from large islands dispersed between islands. Field voles on large islands dispersed within-island to suboptimal habitats in late summer. As suboptimal habitats were absent from small islands, field vole populations on small islands attained densities at least twice as high as typical peak densities in productive mainland habitat (Myllymäki, 1977; Agrell et al., 1992). Field voles escaped overcrowding and a deteriorating environment on small islands by swimming across open water. Immigration was disproportionately directed toward larger islands. Unlike mainland field voles that have strongly sex-biased dispersal (Sandell et al., 1990, 1991), there was no significant sex bias among dispersers in the Archipelago, nor was dispersal restricted to immature individuals. Dispersal was thus effective in causing colonization. Contrary to the standard assumption of metapopulation models that there is either no, or a negative, relationship between a local population extinction probability and its production of emigrants, field voles responded to deteriorating conditions by dispersing. *Per capita* emigration may have been positively related to extinction rate (Crone et al., 2001).

Overall, the Tvärminne metapopulation may be described as a combination of persistent, low-emigration local populations and ephemeral high-emigration sublocal populations. It meets all criteria to be considered a classical metapopulation. A high per capita dispersal of voles inhabiting small islands increased the relative importance of small islands in the metapopulation. The contrast in dispersal dynamics between large and small islands contributed to maintaining asynchrony in vole dynamics between islands of different sizes in the face of a synchronizing climatic influence. The important inference that *per capita* emigration was related positively to extinction rate stems from fitting a modified incidence function model to occupancy data (Crone et al., 2001). There are, however, strong correlations among parameters when fitting the IFM, and future analyses should verify this insight. In addition, despite evidence of life history evolution in a similar island archipelago in Sweden (Ebenhard, 1990), whether dispersal tendency has been subjected to different selection pressures than on the mainland is not known but would provide a useful test of model predictions (e.g., Gandon and Michalakis, 1999).

21.3 WATER VOLE METAPOPULATIONS

Given the lack of consensus on the very existence of metapopulation structures in small mammals, we first describe how water voles approximate, and sometimes deviate from, a classical metapopulation structure. Our studies of metapopulation dynamics and dispersal in water voles include both naturally and anthropogenically fragmented populations, broken up over a range of timescales, and encompass variation in patch size, levels of dispersal, and population isolation. By considering the dynamics of the same species in a range of ecological situations, we control for potentially confounding factors such as social structure and inherent dispersal ability. This section describes in detail how ecological and genetic approaches have helped characterize these metapopulations and shows that similar processes operate in contrasting habitats.

Water Vole Habitats

Water voles are exceptionally large microtines, with a mass reaching 300 g in Britain. British populations are confined to waterway edges, occupying a range of habitats from upland streams to agricultural ditches and wide rivers. Reproduction in the year of birth is much less frequent than in smaller microtines. Over the last 50 yr, water voles have suffered a catastrophic decline caused primarily by Britain's invasion by the American mink, an efficient generalist predator able to enter its burrow systems. Estimates of the magnitude of the decline vary geographically, but reach 98% of the 1950s population level in some regions (Strachan et al., 2000). The distribution of water voles along waterways is generally fragmented, irrespective of the effects of mink (Stoddart, 1970; Lawton and Woodroffe, 1991; Aars et al., 2001; Telfer et al., 2001), although large and more continuous populations can occur along productive lowland rivers (Macdonald and Strachan, 1999). Fragmented populations are thought to function as metapopulations (Lawton and Woodroffe, 1991; Aars et al., 2001; Telfer et al., 2001).

We have intensively studied several networks of populations from both low productivity upland moors and higher productivity lowland farmland areas within Scotland (Fig. 21.1): (1) four networks in the mountainous far northwest of Scotland (Upland Assynt hereafter), which lie to the north of the American mink invasion front. It is reasonable to assume that these populations fluctuate around their long-term equilibrium state: (2) Two networks in the mountains of northeast Scotland (Upland Grampians) where mink are present but distributed patchily (Lambin et al., 1998). Water voles were almost completely extirpated from one of these blocks in a localized mink advance in 1999, leaving the second

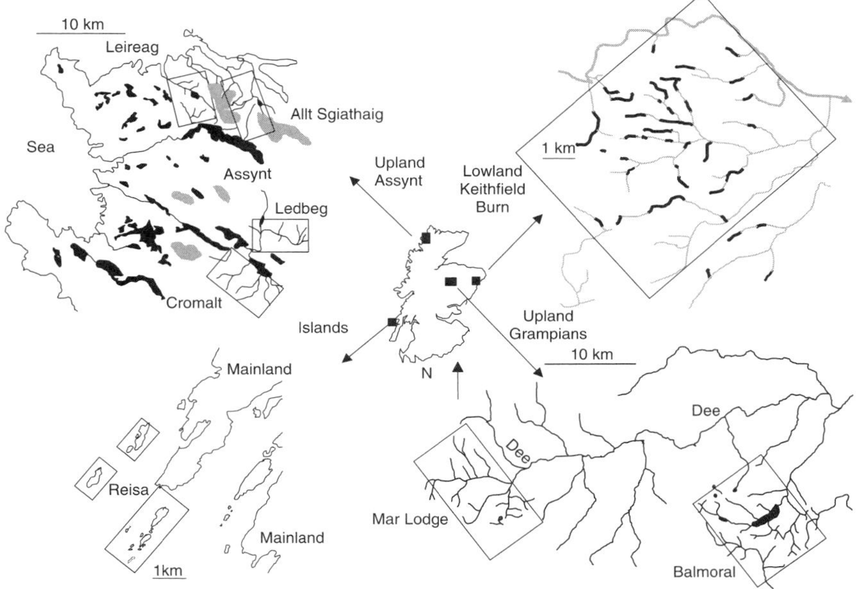

Figure 21.1 Maps showing water vole metapopulation study areas in Assynt and Grampians Mountains of upland Scotland and the lowland Keithfield Burn and Island study areas. Survey blocks are delineated by boxes. Areas shaded gray are mountains, those filled are lochs. In the lowland Keithfield Burn area, thick lines denote sections of waterways occupied in at least 1 yr by water voles. The arrow indicates the downstream flow of the River Ythan. Note that the four areas are not drawn to scale.

block 20 km to the East unaffected: (3) One network in a lowland farmland area of northeast Scotland (Lowlands). This area was first colonized by mink up to 30 yr ago, but clusters of fragmented populations of water voles still persist in upper tributaries that are reached only occasionally by mink from the main river (Telfer et al., 2001): (4) Island populations in the Sound of Jura, southwest coast of Scotland, where water voles are not associated with waterways but live fossorially as in continental Europe (Telfer et al., 2003a). The past history (e.g., degree of founder effect) of water voles on these islands is unknown, but population assignment and genetic analyses infer negligible migration between island groups so they may be distinct populations (Telfer et al., 2003a).

Whether in farmland or moorland, water voles select narrow watercourses with slow-flowing water and banks suitable for burrowing and with abundant vegetation (Aars et al., 2001; Telfer et al., 2001). These are often in headwaters of tributaries such that local populations in adjacent watersheds may be in close proximity (Fig. 21.2). The distribution of water vole habitat patches differs between areas. In upland areas, suitable habitat exists as distinct patches of grass surrounded by wholly unsuitable heather moorland (Aars et al., 2001). The boundaries of patches in the productive lowlands are less well defined and the probability of occupancy of a section of waterways is only predictable from a combination of habitat features (Lawton and Woodroffe, 1991; Macdonald and Strachan, 1999; Telfer et al., 2001). Changes in water vole distribution may consist of both expansions and contractions of populations centered on

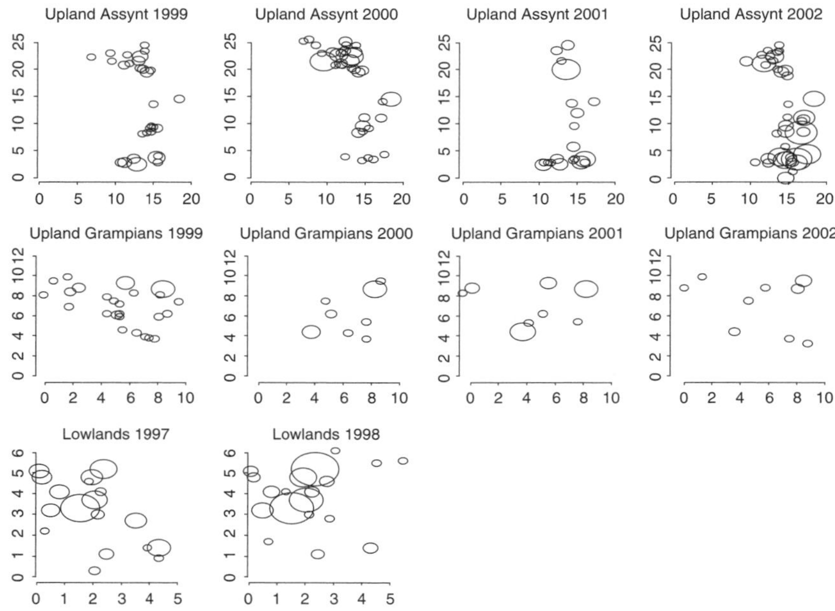

Figure 21.2 Changes in colony size for the four blocks in the Upland Assynt area are on the first line, those from the Upland Grampians are on the second line, and those for the lowland area are on the third line. Circle sizes reflect colony size, and values on the axis are distance in kilometers.

the most suitable habitat in addition to the extinction or recolonization of distinct sections of waterways.

Population Turnover in Water Vole Metapopulation

Water vole populations in both lowlands and uplands have a fragmented distribution with significant turnover between years (Figs. 21.2 and 21.3). The lowland network included 22 local populations between 1996 and 1999, occupying 21 km of waterways. Assuming this represents the length of suitable habitat, yearly occupancy rates ranged from 57 to 67% (Fig. 21.2; see also Telfer et al., 2001). Water voles were lost from between 15 and 17% of waterways each year with similar lengths being recolonized (Telfer et al., 2001). The proportion of suitable patches occupied in the uplands is more variable both spatially and between years. Observed occupancy per survey block in Upland Assynt ranged from 0 to 62% (Fig. 21.3). The Upland Grampians patch network had low occupancy and recolonization rates (17–53%, Fig. 21.3). Water voles were present on 5 of 15 surveyed islands in southwest Scotland (Table 21.1). No extinction has been observed since 1996, but two presently unoccupied islands held water voles 40 years ago (Corbet et al., 1970).

Local populations are larger in lowlands than uplands (Table 21.1). Only 13% of 76 upland local populations had more than 2 overwintered animals trapped, compared with 81% of 26 lowlands populations. The largest local population in the lowlands had only 18 animals in spring and was therefore

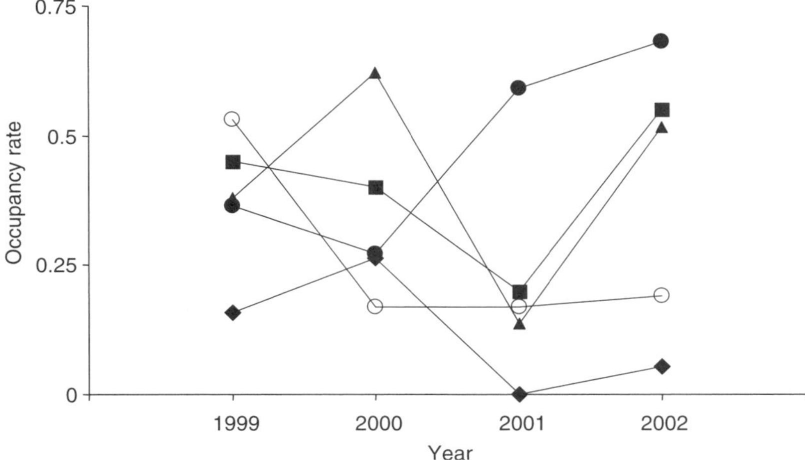

Figure 21.3 Occupancy rate of habitat patches in four upland survey blocks in Upland Assynt (filled symbols for the four survey blocks) and Upland Grampians (empty circle).

TABLE 21.1 Summary of Demographic Properties of Water Vole Populations[a]

Area		Median and range of population sizes (adults in spring) (*N*)	Number of patches	Mean distance to closest patch (km) (SE)	Range in annual median distance from population to closest other population (km)
Lowland area		4 (26) Range: 2–18	—	—	0.62 (0.08) to 0.70 (0.06)
Upland Assynt, four survey blocks	A	1 (30) Range: 1–5	20	0.58 (0.09)	0.46 to 0.85
	B		20	0.65 (0.08)	0.61 to 2.18
	C		19	0.60 (0.08)	1.41 to 2.33[b]
	D		29	0.55 (0.04)	0.60 to 1.76
Upland Grampians		1 (46) Range: 1–5	47	0.58 (0.04)	0.67 to 1.70
Islands		285 (5) Range: 50–621[c]	> 12	0.31 (0.02)[d]	0.2

[a] Although water voles populations in the lowlands are fragmented, discrete patches of suitable habitat are difficult to identify (see text) and therefore patch details are not presented. On the islands, water voles live fossorially. The prevalence of mink differs between areas. Mink have been resident in the lowland area for approximately 30 yr but are found predominantly along larger waterways, with some upper tributaries only being subjected to occasional incursions. In Upland Assynt, mink are absent. The Grampians uplands are at the edge of the invasion and are subjected to occasional invaders. Mink are long-term residents of the west coast and are caught occasionally on the islands. Values are reported for the four survey blocks in Upland Assynt.

[b] Distance to the closest population in 2002 for the single occupied patch in survey block C was to a local population in survey block D.

[c] Population sizes on the islands estimated from transects of signs, calibrated with trapping data (see Telfer et al., 2003).

[d] Mean distance between islands calculated for the 12 islands in the main group, excluding Jura and 2 islands surveyed 10 km to the northeast.

not immune to stochastic extinction. In contrast, each island population contained several hundred water voles (Telfer et al., 2003a). Thus water voles occur in fragmented populations in a range of habitats. On the mainland, water vole populations experience frequent turnover, especially in the uplands, indicating that metapopulation processes are important for regional dynamics. Similar patterns of distribution have been found in the closely related species, the southern water vole, *Arvicola sapidus,* with occupancy being related to both local attributes (abundant grass cover) and landscape attributes (distance to closest known pond holding water voles) (Fedriani et al., 2002).

True extinctions and recolonizations, as opposed to expansions and contractions of populations, were relatively rare in the lowlands with only 9 extinctions and 10 recolonizations observed over 4 yr. In contrast, extinction and recolonization processes were frequent in the uplands with 75 and 68 extinctions and recolonizations, respectively, between 1999 and 2002. Consequently, the life span of local populations was longer in lowland than upland populations. Fifty-nine percent of the 22 sections of lowland waterways occupied by water voles were at least partially occupied in all 4 yr. Only 5% of 98 upland patches inhabited between 1999 and 2002 held water voles in all 4 yr. Unlike the Bodie pika metapopulation (Clinchy et al., 2002), there was no evidence that most recolonizations were ephemeral and ultimately unsuccessful. Seven out of 10 recolonizations persisted into a second year in the lowlands and 84% (n = 36) in the uplands. The extinction rate of newly colonized patches in the uplands (16% year^{-1}) does not differ from the rate of extinction of longer lived patches (34% year^{-1}). The distance between local populations and their nearest neighbor, a measure of connectivity in metapopulations, was more variable and tended to be higher in upland than lowland metapopulations. In the lowlands the mean distance from each population to its nearest neighbor ranged from 0.62 to 0.70 km (Table 21.1). In contrast, changes in patch occupancy in the uplands resulted in much variation in the distance between local populations between blocks and years (Table 21.1) (range of annual block level means: 0.46–2.33 km). The highest values of isolation were observed in the absence of mink, suggesting that water voles, in their pristine environment, may persist with substantial turnover, despite being distributed very sparsely in some years.

Correlates of Extinction in Water Vole Metapopulations

Even though patterns of occupancy can be suggestive of mechanisms operating in fragmented populations, examining patterns of changes in occupancy is more revealing of the importance of metapopulation processes. Having documented similarities and differences in patterns of occupancy, we next ask whether similar processes influence the distribution of lowland, upland, and island water voles using logistic models. Analyses for the lowlands and islands have been published previously (Telfer et al., 2003, 2001) and we only summarize the main results here. Analyses for the lowland metapopulation consider the occupancy status of 200-m sections of waterway in relation to habitat characteristics and measures of population size and isolation in the preceding year (Telfer et al., 2001). In three analyses investigating patterns of patch occupancy, extinction, and recolonization in the upland metapopulations, the effects of survey block, year, and isolation in the preceding year were explored. The effect of local population size on extinction probability was also considered. Population size was taken as the number of voles trapped over a single 4-day period in June–September in the uplands (Aars et al., 2001) and as the length of a population in lowlands (Telfer et al., 2001). In order to quantify connectivity, we assumed that the number of immigrants a patch receives depends on both the sizes of the populations in the surrounding area and distance to those populations [e.g., Hanski, (1994) with $\alpha = 1/d$ in Telfer et al. (2001)]. Occupancy of islands was related to island size, distance to the nearest occupied island, and distance to the mainland.

The relationship between occupancy and population size changed in the lowlands, following the partial invasion of the network by American mink in the summer of 1998. Before mink invaded the area, large populations were more likely to persist than small ones, possibly reflecting demographic stochasticity (Telfer et al., 2001). The lack of any detectable influence of population size thereafter may have been a reflection of the ability of mink to decimate even the largest water vole population. Extinction events also became correlated spatially in 1999 with pairs of extinction events less than 500 m apart more frequent than expected by chance.

Occupancy of habitat patches in the uplands could not be predicted according to measures of connectivity within block. Instead, patch occupancy fluctuated asynchronously between blocks and years, with different blocks having high occupancy rates in different years, but block-wide synchrony. Our isolation measures reflect the number of voles within a relatively local area, as connectivity measures only considered populations within a distance of 6 km and values of d up to 2.5 km. No measure of isolation had a significant effect, suggesting that, unlike in the lowlands, local populations are not clustered within blocks. Extinction probability was also variable at the scale of block in each year but, as in the lowlands, small populations were more likely to go extinct than large populations, suggestive of a role of demographic stochasticity. There was no detectable influence of connectivity at any of the scales considered that would have reflected a rescue effect.

As with occupancy and extinction probabilities, recolonization rates varied with year and block. Patterns of patch occupancy are clearly dominated by factors that result in spatially correlated extinctions and recolonizations at about the scale of a block or higher. From an analysis of distribution patterns, there is no evidence of interdependency of populations over a smaller spatial scale. Upland water voles in an area little influenced by humans or mink thus persist in highly fragmented environments, with high rates of turnover, despite spatially correlated extinctions at the scale of 25–30 km^2. Local climate is unlikely to vary at such a small spatial scale and be responsible for the block level variation in occupancy, extinction, and recolonization rates. One possibility is that dispersal is so widespread that "block" statistically captures its influence better than the relatively small-scale connectivity measures considered (edge effects preclude consideration of larger scale dispersal coefficients). Alternatively, mobile native predators may be causing population turnover to be correlated spatially. Indeed, the very low biomass of alternative prey available to mustelids present in the area (*Mustela nivalis vulgaris*, *Mustela erminea*, and *Lutra lutra*) may magnify their impact on each local water vole population and cause them to exploit large areas.

No extinction was observed on the islands over 4 yr; however, large islands close to other occupied islands were more likely to support water vole populations. Occupancy was not related to distance from the mainland (Telfer et al., 2003a). Voles on occupied islands 50 to 100 m from each other had low genetic differentiation unlike those on islands separated by 1 km (Section 21.5). This suggests that dispersal between islands occurs over short distances, despite the fast tidal current in the area. Island archipelagoes may function as a metapopulation over evolutionary timescales, but, unlike field vole populations in the Baltic, metapopulation processes are unlikely to influence the dynamics over ecological timescales.

Dispersal in Water Vole Metapopulations

Dispersal is obviously paramount to the persistence of fragmented populations with turnover. Precise and unbiased estimates of dispersal rates and distances, especially for rare, long-distance events, are required to predict the effects of dispersal on population dynamics and genetic composition. Obtaining such estimates is extremely difficult (Ims and Yoccoz, 1997). If these estimates are to be used in predicting metapopulation persistence, characterizing density dependence in emigration and immigration is also essential.

Given the spatial scale involved in water vole metapopulations, even intensive standard capture–marking–recapture (CMR) techniques would underestimate the rates and scales of dispersal. We overcame this difficulty through combining CMR with microsatellite genotyping to identify parents and offspring in different populations (Telfer et al., 2003b). Voles sampled during monthly live trapping of 21 local populations over 2 yr in the lowlands area were genotyped and assigned to pools of putative parents. A similar approach was used in the uplands, although local populations were trapped only once each year (J. Aars et al., unpublished results). The parentage assignment approach does not require frequent sampling, and similar proportions of juveniles were assigned successfully to at least one parent in the two areas (lowlands: 68%, n = 675; uplands: 67%, n = 253). The percentages of assigned juveniles identified as dispersers were also similar in both the lowlands and the uplands [lowland females 13.6% (n = 214); upland females 12.8% (n = 78); lowland males 14.9% (n = 242); upland males 17.4% (n = 92)]. The proportion of overwintered individuals that have dispersed is the best reflection of the contribution of dispersal to regional dynamics as this cohort produces most recruits. In the lowland metapopulation, we estimated that 33% of males and 19% of females present in spring 1998 had dispersed between local populations. Comparable estimates based on CMR alone were 12 and 7%, respectively, reflecting that most water vole dispersal takes place early in life, before reproduction. A single instance of breeding dispersal by an adult female water vole in lowland populations is much cited (Stoddart, 1970). Despite the intensive trapping in our lowland study, no adult females (n = 211) and only 1% of adult males (n = 265) were trapped in different populations in the lowlands. Breeding dispersal was detected more frequently in the uplands, despite a less intensive sampling regime. Nine percent of adult males (n = 123) and 2% of adult females (n = 132) were caught in more than one population (n = 123). The average distance moved between populations by adult males was 0.61km (SE = 0.08).

The high rates of successful natal dispersal rates in water voles have the potential to influence regional dynamics in both the lowlands and the uplands. If immigration rates are negatively density dependent, as with other small mammal species (see later), dispersal could effectively reduce extinction rates through the "rescue effect." We used juveniles assigned to parents in the lowland study area to investigate how the density of breeding individuals and isolation influenced the proportion of immigrants in local populations in summer after members of the spring-born cohorts had dispersed. Increasing immigration in well-connected populations is the process assumed to underlie the rescue effect. We fitted logistic models with connectivity measures encompassing different scales and selected the best models following backward stepwise procedure. Immigration rates in males increased significantly with decreasing isolation, with $d' = 1.2$ km yielding the lowest deviance. The proportion of male immigrants in local populations also decreased significantly with increasing densities of breeding adults. Although immigration rates in females also decreased with increasing density, there was no detectable effect of isolation [odds ratio (95% CL) for density controlling for distance; males: 0.708 (0.658–0.762), females: 0.810 (0.761–0.861)]. However, underlying the expectation that connectivity influences immigration rates is the assumption that connectivity is a good measure of the number of dispersers reaching a site. This may not be the case if emigration or immigration rates are density dependent.

In metapopulations with small local populations, dispersal should contribute more to metapopulation growth if dispersers are attracted to opposite-sex conspecifics than if settlement were random with respect to the presence of mates. Single-sex local populations with no potential for reproduction would be most likely where overall occupancy is low and local populations are small, such as in pika at Bodie and water voles in the uplands. There was strong evidence of nonrandom settlement in the uplands with only 8 single-sex upland populations out of 54 with only two adults (Binomial test $P < 0.01$). Experimental translocation and radio tracking of juvenile water voles into good habitat sites, which were either occupied or vacant, show how well the settlement behavior of dispersing water voles is suited to locating conspecifics at low density (Fisher et al. unpublished results). Immigration by voles translocated to lowland-occupied sites was not inhibited. Voles translocated to sites that remained vacant for the subsequent week abandoned the sites and moved to new sites either overland or by water. In two cases out of seven, an opposite-sex natural immigrant arrived to join the translocated vole in a previously vacant site. Voles remained in the transient phase of dispersal for many days and often followed a "stepping stone" trajectory, stopping for several days at successive sites. Predation mortality during dispersal was, however, high (observed mortality of dispersers 4% day^{-1}) relative to daily disappearance rate estimates obtained by capture–recapture for voles from the same-age cohorts in the summer of two separate years [1997: 1.23% (95% CI: 0.41–1.28); 1998: 0.7% (95% CI: 0.48–2.68)] (Telfer and Lambin unpublished results).

If water voles disperse until they find a suitable mate and are not constrained by their inherent mobility, dispersal distances are predicted to increase with decreasing density of voles at the local scale (density of voles within populations) and at the regional scale (density of populations within an area). The mean dispersal distance from a fitted negative exponential was 1.39 km (95% CI = 1.19–1.64 km) in the lowlands (Telfer et al., 2003b) and 40% longer [1.96 km (95% CI = 1.77–2.18 km)] in the uplands. Similar rates of successful parentage assignment in both areas indicate that rates of successful dispersal are comparable and that longer dispersal was not accompanied by low immigration rates. Water vole dispersal thus appears relatively unconstrained by the distance between local populations and between habitat patches.

Regardless of habitat type, directed dispersal by water voles creates a degree of demographic interdependence between local populations over hitherto unsuspected large spatial scales. As dispersal is mostly an attribute of subadults, this interdependency is loose and does not entirely preclude extinctions caused by demographic stochasticity. It is sufficient to allow for metapopulation persistence, despite the synchronizing influence of predators acting over the scale of several local populations.

21.4 TOWARD A METAPOPULATION PARADIGM IN SMALL MAMMALS?

Metapopulation Processes in Nonclassical Metapopulation Structures in Small Mammals

The increasing use of metapopulation concepts in mammals (Elmhagen and Angerbjorn, 2001) must no doubt reflect the perception by authors that mammals show such structures because of their patchy distribution, habitat fragmentation, and social structure or that invoking the regional dynamics processes that underpin metapopulation adds to our understanding even when ecological processes do not operate on both local and regional scales. Indeed, in the following examples, regional dynamics akin to metapopulation processes operate, but their influence on long-term dynamics is limited either because of a dominant synchronizing influence or because of a mismatch between the dispersal behavior and the scale of habitat patchiness and disturbance.

Numerous small rodent populations fluctuate violently between years, sometimes with regular periodicity. In cyclic or irruptive populations, periods of high density may be separated by 1 or more years when numbers are so low as to be almost undetectable at a landscape scale. Because of their scarcity, little is known about the spatial structure of such populations, but it seems likely that they form temporary metapopulations in the low phase of population fluctuations, with small local populations in refuge habitats subjected to extinction–recolonization dynamics (e.g., Lima et al., 1996). The dynamics of tundra voles (*Microtus oeconomus*), inhabiting mires and bogs scattered in the taiga in southeast Norway, supports this conjecture (R.A. Ims, personal communication). Clearly, any metapopulation structure is a transient state occurring only for a limited fraction of the spatially correlated fluctuations of regional populations. As regional abundance increases, so would the contribution of dispersal, first linking extinction-prone local populations, then resulting in effective rescue effect, and finally binding local populations in a synchronized patchy population (see Harrison, 1991).

Populations of three species of shrews (*Sorex araneus*, *S. minutus*, and *S. caecutiens*) on islands in Finnish lakes experience turnovers. Colonization depends on distance from the mainland, and extinction probability relates to body size and competitive ability (Hanski, 1986; Peltonen and Hanski, 1991). These islands are excellent model systems for elucidating the processes of dispersal, colonization, and extinction also operating in classical metapopulations. As archetypal "mainland island metapopulations," a dominant influence on the occupancy of islands by shrews, however, is their dynamics on the mainland, which is loosely linked to that of microtine rodents (Henttonen, 1985).

Stacey et al. (1997) argued that spatial structures caused by sociality and group living would predispose mammals to metapopulation persistence. Indeed, the fates of social groups may be independent and influenced heavily by demographic and environmental stochasticity. New groups may be founded, usually by fission of existing groups (e.g., Rood, 1986). Group territory dynamics, however, differs from recolonization of habitat patches in a metapopulation as territories are not inherited as whole entities following the disappearance of a group, unless they are centered on a long-lived physical structure, such as traditional dens or burrows (e.g., Arctic foxes; Elmhagen and Angerbjorn, 2001). If new groups are formed by the gradual fission of otherwise socially cohesive groups, the requirement to enter a matrix of unsuitable habitat may hinder colonization, despite the obvious physical ability to do so, as appears the case with Samoango monkeys (*Cercopithecus mitis*) in recently fragmented habitats in South African forest (Lawes et al., 2000). In this scenario, once patches are disjoint, the spatial configuration of the habitat has no significant consequences for the dynamics of group territories.

Invoking metapopulation theory for managing previously continuous but now fragmented mammalian populations is increasingly common. Observing populations of the same species, such as water voles and field voles persisting both in highly fragmented and more continuous environment, suggests that ecological and behavioral traits are sufficiently flexible. Some degree of local adaptation of life history traits has taken place in natural field vole metapopulations (Ebenhard, 1990). Likewise, the increase of dispersal distance with degree of fragmentation in water voles may reflect adaptation or plasticity and lack of constraint on dispersal. Predicting whether species such as black-tailed prairie dogs in farmland or arboreal marsupials in recently fragmented forest patches (Lindenmayer and Possingham, 1996) possess the traits required for persistence in a metapopulation remains difficult. The predictors of extinction used within the metapopulation approach may be used to rank habitat degradation scenarios, but in the absence of empirical evidence of patch recolonization through an unsuitable habitat matrix (e.g., Lindenmayer and Possingham, 1996), the power of the metapopulation approach is limited. For instance, traits such as conspecific attraction by dispersers displayed in continuous colonies (Garrett and Franklin, 1988; Weddell, 1991) may contribute to a "rescue effect" but could also limit the colonization of empty patches regardless of the inherent mobility of the species (Ray et al., 1991). In fact, based on the aforementioned review, it appears less likely that mobility per se, rather than the details of dispersal behavior, would constrain persistence after fragmentation. Even where metapopulation processes operate, ill-adapted dispersal may cause a mismatch between the frequency of extinction and recolonization such that there may be no stable equilibrium, the so-called "extinction debt" (Hanski, 1999). Even then, managing such systems as metapopulations is a legitimate, if temporary, option, although long-term persistence obviously requires balancing of colonization and extinction rates. We are, however, not aware of any mammalian metapopulation created by human's activity having been shown to meet the conditions for persistence. In contrast, it may be easier to use metapopulation parameters to predict the persistence of natural metapopulations that are subject to further fragmentation. Examples of the latter include lowland water voles where the number of populations connected by dispersal is reduced by habitat destruction or mink invasion or Vancouver Island marmots (*Marmota vancouverensis*) following the creation of low-quality habitat patches through logging that may reduce the colonization of more productive distant patches (Bryant, 1996; Bryant and Janz, 1996).

Metapopulation Persistence despite Spatially Correlated Extinction

That spatial correlation in local population extinction is detrimental to metapopulation persistence is a long-established lesson of metapopulation biology (Hanski, 1999; Chapter 4). However, in all four instances given earlier, extinction processes were correlated spatially. The causes of such correlation were either abiotic (climatic) or biotic (predation and pathogens, and dispersal). These unexpected patterns include probabilities of extinction being independent of population size, as exemplified in the following paragraph.

Not surprisingly, given the small spatial scale involved, abiotic climatic factors have a synchronizing influence on disjoint small mammal populations. Eastern common voles (*Microtus rossiomeridionalis*) in high arctic Svalbard persist in a metapopulation-like state along a strip of cliff up to 10 km long. Vole populations have high growth rates and reach extremely high densities in some years during which numerous patches are colonized but this can be followed by catastrophic but irregular population crashes (Ims and Yoccoz, 1999). Population crashes occurred independent of prevailing population size and result from icing of soil and grass following rain events in early winter, which renders food inaccessible. Icing events are correlated spatially over altitudinal bands, but asynchrony in extinction and persistence of the metapopulation is made possible by the gradient in altitude of habitat patches such that precipitation falls as snow and allows persistence on some habitat patches (Ims and Yoccoz, 1999). Heterogeneity in aspect and associated variation in duration of snow cover also contributes to maintaining asynchrony within metapopulations of rock-dwelling pygmy possums (*Burramys parvus*) in Australia despite a region-scale variation in climate (Broome, 2001). In the two previous examples, local topography makes the scale of abiotic correlations small relative to the total population and allows for metapopulation persistence.

Predators of small mammals operate on larger spatial scale than their prey such that predation may result in a pattern of spatially correlated local population extinctions (Clinchy et al., 2002). There is no direct evidence that mustelid predation causes the observed spatially correlated patterns of extinctions either in the pika at Bodie (Smith and Gilpin, 1997) or in the water vole in Upland areas; however, in both instances, they are the most likely candidates. The change from a spatially uncorrelated pattern to a more aggregated pattern of extinction during a short-lived mink invasion in the lowland area water vole population amounts to more direct evidence of the synchronizing impact of predation. The scale of correlated extinction in the lowland (500 m) is much smaller than that observed in upland areas (25–50 km^2) where only native mustelids occur. Note that for both pika and water voles, there was also evidence of demographic stochasticity within colonies contributing to extinctions. The drastic decline of one survey block in Upland Assynt from 2000 to 2001 (Fig. 21.2) is not unlike the regional decline observed in the southern part of the Bodie mining area and, as suggested by Smith and Gilpin, (1997), it is quite possible that extinction–recolonization stochasticity slowed recovery. Although data are only fragmentary, the putative role of plague in causing colony extinction in black-tailed prairie dogs raises the interesting possibility that the pattern of density dependence in mortality and colony extinction may reveal the process involved. Predation mortality from small mobile predators such as mustelids is expected to be negatively density dependent, whereas the impact of a pathogen such as plague that requires large or dense host populations for transmission may cause positive or stepwise density-dependent mortality.

In all case studies considered here, small mammal populations persisted as metapopulations in naturally patchy habitats, despite being exposed to moderately spatially correlated extinctions. Thus, not only can spatially correlated extinction result in patterns resembling those caused by metapopulation processes (Clinchy et al., 2002), but empirical evidence suggests that small mammals adapted to metapopulation persistence may exist *despite* correlated extinctions. The scale of extinction due to predation is typically larger than that of a single patch but still sufficiently small to allow persistence as a metapopulation. In the extreme, exemplified by water voles in Upland Assynt (e.g., Fig. 21.3), regional asynchrony exists between population networks instead of within networks. Such persistence is only possible if dispersal is more effective than predicted if it were random.

Targeted Dispersal in Small Mammal Metapopulations

A general pattern emerges from recent studies with microtine rodents highlighting that they are flexible dispersers and that density-dependent immigration may ensure frequent rescue and colonization in a highly fragmented habitat. Mechanisms potentially important in metapopulations are well illustrated by experimental studies with root voles in patchy populations. Most dispersal by root voles takes place in spring or early summer (Aars et al., 1999), when densities are at their seasonal low. Female effective dispersal is very restricted when densities across local populations vary little, whereas most males leave their natal population (Aars and Ims, 1999). As with other vole and many mammal species, root voles appear to emigrate so as to avoid close contact with opposite-sex close relatives (Wolff, 1992; Lambin, 1994; Gundersen and Andreassen, 1998). At low density, animals are more likely to immigrate into patches with few members of the same sex and more members of the opposite sex (Andreassen and Ims, 2001). Negative density-dependent female immigration combined with a rigid male dispersal pattern can lead to local sex ratio bias variation, and thus to significant male variation in reproductive success between demes (Aars and Ims, 2000). As with water voles, root vole dispersal was costly in terms of survival (Aars et al., 1999; Ims and Andreassen, 2000). Over the relatively short spatial scale involved in the experimental enclosures, emigration from a patch was influenced heavily by the presence of a vacant habitat nearby, indicating that voles made exploratory movement followed by immigration into low-density patches (Gundersen et al., 2001). The patterns of opposite-sex attraction and the stepping-stone dispersal patterns by pikas and water voles suggest that similar patch sampling processes may also operate in metapopulations, despite larger distances between patches. Animals emigrating to escape unfavorable conditions (crowding, presence of relative) are likely to settle where conditions are more favorable. Despite the general pattern of negative density-dependent immigration, emigration was the main proximate cause of extinction of small experimental populations (Andreassen and Ims, 2001), as with field voles on small islands in the Baltic archipelago metapopulation and some upland water vole populations. Thus, contrary to the standard assumption of metapopulation models that there is either no or a negative relationship between local population extinction probability and production of emigrants, small mammals may respond to deteriorating conditions by dispersing rather than dying passively. As a result, per capita emigration may be related positively to the extinction rate (Crone et al., 2001). There was also evidence of the so-called "social fence effect" (Hestbeck, 1982) in root voles, whereby immigration almost completely ceased when densities were moderate or high, often in late summer (Aars et al., 1999). Patterns of low rates of migration at high densities seem to be a common trait in small mammals (see, e.g., Blackburn et al., 1998; Lambin et al., 2001; Lin and Batzli, 2001).

Overall, the aforementioned features of dispersal, including stepping-stone dispersal, active patch selection (sometimes preceded by sampling), conspecific attraction, and negative density-dependent immigration, should result in a strong rescue effect. These complications may also cause a mismatch between the effective impact of dispersal and that predicted from its frequency and scale, as was the case for both pikas and water voles (e.g., Moilanen et al., 1998). In addition, when patches were distant from each other, the distribution of dispersal distances by water vole was adjusted upward, suggesting that vagility does not necessarily limit dispersal in metapopulations.

Evidence of a rescue effect often involves no more than a positive relationship between the probability of local populations persisting in successive years and indices of proximity to other local populations, taken as reflecting the number of dispersers reaching a patch. By this measure, evidence of a rescue effect in small mammals was restricted to the Bodie pika metapopulation (but see Clinchy et al., 2002). This may be because the rescue effect is unimportant (Clinchy, 1997) or, more likely, because the assumption that connectivity is a good measure of the number of dispersers reaching a site is overly simplistic. Indeed, if the per capita emigration rate changes with density (or population size), there would not necessarily be a simple relationship between population size and the number of dispersers it produces. Considering more realistic relationships between the number of immigrants a patch receives and the regional metapopulation may be required to detect a rescue effect as pervasive as envisaged by Stacey et al. (1997).

Using genetic data to directly estimate immigration gives compelling evidence of the importance of immigration within local populations. A genetic-based assignment of individuals to populations or to putative parents indicated a high fraction of immigrants in colonies of both black-tailed prairie dogs (Roach et al., 2001) and water voles (Telfer et al., 2003b), including in the uplands where extinction rates were high, spatially correlated, and unrelated to levels of isolation. As the use of such genetic methods becomes more widespread to characterize metapopulations, a more quantitative evaluation of the role of the rescue effect will become possible. We expect this will help resolve the apparent contradiction between what can presently be inferred from patterns of occupancy and what is known about the process of dispersal and colonization.

21.5 THE GENETIC STRUCTURE OF SMALL MAMMAL METAPOPULATIONS

Considerable emphasis has been placed on trying to model the effects of metapopulation processes on patterns of population divergence between patches, and levels of diversity within individual patches and across entire metapopulations. Such efforts can provide information on the effects of extinction and recolonization on genetic diversity and concomitant viability, but also could potentially highlight metapopulation processes in the absence of ecological data on, for example, patch occupancy. Modeling approaches range from necessarily simplistic models derived from island or stepping-stone models to coalescent and simulation models that explicitly include ecologically realistic processes and exclude some of the simplifying assumptions added for logistical and mathematical expediency (see Chapter 8).

Models commonly predict that more fragmented populations will lose genetic variability faster than comparable unfragmented populations (Whitlock and Barton, 1997; Nunney, 1999; Chapter 7) and that population turnover can severely reduce neutral genetic diversity, particularly if combined with propagule-pool colonization (Pannell and Charlesworth, 1999; Chapter 7). However, Ray (2001) showed that the scale of correlated deaths or local extinction affects how diversity is retained in structured populations. In certain cases, subdivided populations with frequent local extinction may retain more diversity than populations of equivalent total size but fewer and larger units. Furthermore, higher diversity can be expected in a metapopulation with many small subpopulation local populations than in one with populations of variable size where larger source populations frequently provided immigrants to the smaller ones. This conclusion is thus comparable with that of the classic and more unrealistic island model, where the total amount of variation is high due to a lack of variance in reproductive success among local populations (see Chapter 7). Using a different approach, but reaching similar conclusions, Wakeley and Aliacar (2001) highlighted the contrasting effects of migration and population turnover on the site-frequency distribution at polymorphic sites. They warned that, from genetic data, it will be impossible to distinguish between changes in population number and changes in the rates and patterns of migration and turnover as explanations of variable effective size over time.

As such, it is not clear that any models can yet identify a signature of metapopulation dynamics in a species with little underlying ecological data nor determine the relative contributions of particular metapopulation processes to the overall patterns of diversity and divergence.

The problem of identifying a characteristic pattern of population genetic structure from a metapopulation is exacerbated for mammals. Those complex social structures and behaviors that are common if not inherent to mammals may either confound or negate the effects of metapopulation processes on patterns of genetic diversity or divergence. Indeed, parallel to the effort invested in trying to understand the effects of metapopulation dynamics on genetic parameters, considerable emphasis has been placed on developing a theoretical framework to model gene dynamics in socially structured populations (Chesser, 1991a,b; Sugg and Chesser, 1994; Sugg et al., 1996; Berg et al., 1998). Rarely have the two been combined, and certainly there remains a lack of replicated empirical genetic data from species in both fragmented and unfragmented landscapes where realistic estimates of turnover and dispersal are available.

A considerable number of studies have attempted to describe the effects of metapopulation processes on population genetic structure from empirical studies of allele frequencies at markers such as microsatellite DNA polymorphisms. While the choice of markers and subsequent analyses are generally sound, a majority of studies can still be criticized for attempting to examine metapopulation genetic structure when there is sparse evidence that a metapopulation framework is appropriate (e.g., Gerlach and Hoeck, 2001; Burland et al., 2001; Schulte-Hostedde et al., 2001). The patterns observed could thus be attributable to other contemporary or historical population processes such as ancestral bottlenecks or limited dispersal in a semicontinuous distribution of individuals.

Patterns of population genetic structure have been examined to a certain degree in populations of pika that are considered more consistent with the classical metapopulation paradigm. Low allozyme diversity has been detected, which has been attributed to repeated genetic bottlenecks and a concomitant reduction in N*e*, which suggests widespread extinction–recolonization dynamics (Hafner and Sullivan, 1995). In more recent studies, neutral markers (microsatellites) have been employed to compare fragmented and unfragmented populations. The fragmented Bodie metapopulation founded 40 generations ago had a level of heterozygosity as high as that observed in a more continuous habitat (Peacock and Ray, 2001). This was best explained by limited immigration from outside the metapopulation, combined with the small size limit on all local populations, which tends to equalize the reproductive success of colonists. Peacock and Ray (2001) suggest that this system retains high variability because variance in reproductive success due to local extinction–recolonization dynamics is minimal when local populations hold few individuals. In pikas, monogamy is the modal mating system in patchy populations where patches are too small to host more than one breeding female, and too dispersed for a male to defend several, but polygyny is more frequent in larger patches or continuous populations (Peacock and Ray 2001). This is thus also likely to be of importance for how genetic variance may be retained in the more fragmented populations (e.g., Sugg et al., 1996).

We have examined patterns of population genetic structure among our networks of water vole populations in Scotland with contrasting population size and turnover (Table 21.1). We have also surveyed three different continuous populations for comparison: (1) River Itchen in southern England, where water voles are present at high density and probably do not fluctuate much between years; (2) a linear population from Alajoki in southwestern Finland, where animals were sampled along a stream at trapping stations 1 km apart; and (3) a terrestrial (i.e., two-dimensional) population from Allkia, western Finland, sampled within a square km. Both the Finnish populations were sampled at very high density, but the populations are known to fluctuate (Aars et al. unpublished results).

Table 21.2 summarizes some of the genetic patterns observed in these populations (see also Aars et al. unpublished results). Further genetic characterization of the lowland populations can be found in Stewart et al. (1999). Comparing across all population networks, there was no pattern of genetic variability that correlated to the degree of habitat fragmentation. All populations were characterized by a high number of alleles per locus and high individual heterozygosity (typically 0.7 to 0.8). Thus the high level of fragmentation of water vole habitats was not associated with any significant loss of variability. Except for the lowland metapopulation, which had the lowest variability among all the surveyed populations, fragmented populations had variability comparable to the continuous populations (Table 21.2). This pattern was consistent over all six upland low-density survey blocks and over 2 to 5 yr of sampling. Thus even the extreme rates of local turnover observed do not reduce N*e* severely among fragmented water vole populations, consistent with Ray's (2001) prediction. The upland metapopulations are bound together by dispersal across large scales, with numerous small colonies thus experiencing frequent turnover due to stochastic demographic events. In all areas, environmental stochasticity also appeared influential with occupancy rates varying widely between blocks within years. According to Ray (2001),

TABLE 21.2 Summary of Genetic Properties of Water Vole Population Networks[a]

	Number of networks (sampling years)	Ho	A	F_{ST} spatial	F_{ST} temporal	(*f*)	*F*
Lowland area	1 (4)	0.65, 0.72	6.2, 7.4	0.10, 0.14	0.01, 0.03	−0.04, 0.01	0.06, 0.15
Upland (Assynt and Grampians)	6 (2–5)	0.73, 0.81	7.3, 10.0	0.02, 0.21	0.00, 0.12	−0.16, 0.00	−0.05, 0.12
Islands	5 (2–3)	0.46, 0.52[b]	2.6, 3.9	0.00, 0.38	0.00, 0.06	−0.02, 0.04	0.20, 0.32
Continuous	3 (1)	0.74, 0.79	6.0, 9.6	0.03 [0.03,0.04][c]	na	0.02 [−0.01,0.05][2]	0.03, 0.06

[a] Only colonies with more than 10 individuals are included in calculations of A, Ho. Continuous networks include two southwest Finland populations and one sample from the River Itchen in southern England.

[b] Island H_{exp} = 0.59 to 0.61 across three to five islands.

[c] Continuous n = 1 for F_{ST} and (*f*) (9-km transect, sampled each kilometer).

under such a pattern of extinction, highly fragmented populations will retain variability more effectively than less fragmented metapopulations with some persistent local populations. In metapopulations with very small local populations, founders of new colonized patches tend to be a mixture from different local populations rather than from one larger source population, which increases N*e*. The high dispersal ability observed in water voles implies that local populations often consist of animals from different patches, considerably increasing N*e* compared to what it would be if local populations were founded according to a propagule-pool pattern (Pannell and Charlesworth, 1999). Negative local inbreeding coefficients (f) are a common feature in fragmented water vole populations (Table 21.2) and the value is correlated with the local population size. This also reflects the fact that parents tend to be of different genotypes, partly because they will be so by chance when they are few (Wahlund, 1928), but more so the higher the dispersal rates. Within survey blocks, total inbreeding coefficients were frequently just slightly positive, consistent with the high dispersal rates between local populations revealed from CMR and assignment data. F_{ST} values for among-local populations differentiation within blocks were typically around 0.1 (Table 21.2). This reflects high kin similarity within local populations before juveniles have dispersed. Considering only adults, F_{ST} values in the upland populations are typically just above zero and frequently non-significantly positive (Aars et al. unpublished results). Temporally, genetic drift within survey areas is significant, and F_{ST} correlated positively with interval between sampling within survey blocks at least up to 5 yr, both on the islands and in the mainland metapopulations. Furthermore, it is correlated negatively to the harmonic mean of animals trapped within the blocks. The fact that local genetic drift can be quite profound, yet populations are still characterized by high genetic variability, indicates that these populations are part of much larger networks.

The high frequency of solitary breeding pairs or populations consisting of very few adults with about an even sex ratio may contribute to low variance in reproductive success between individuals, particularly among males, and thus contribute to a high N*e* in very fragmented populations. Variance of female and male reproductive success should be studied in detail across different fragmentation levels before conclusions are drawn about such effects, particularly as there also is a possibility that polyandry (a single litter sired by more than one father) could be more prevalent in denser less fragmented systems and have the opposite effect (i.e., increase N*e* more in continuous or less fragmented colonies).

Overall, these empirical studies of water vole population genetics confirm theoretical concerns that characteristic patterns of genetic divergence and diversity are difficult to detect. As such, it is apparent that (1) little inference can be made about the relative contributions of metapopulation dynamics, social structure, and contemporary and historical microevolutionary processes to the overall patterns observed; and (2) metapopulation processes may, in fact, retain comparable levels of genetic diversity within networks relative to populations with reduced population turnover and increased connectivity. Only the isolated island populations show any identifiable signature of reduction in genetic variability, and these populations do not experience extinction–recolonization dynamics.

21.6 CONCLUSION

Metapopulation persistence requires some degree of balance between extinctions and recolonizations through dispersal as well as sufficient independence between local populations so as to maintain asynchrony in extinctions. In nearly all instances of mammalian metapopulations reviewed in this chapter, there was a relatively high degree of spatial synchrony in extinction, and dispersal was nonrandom with respect to the presence and abundance of conspecifics. These may be general features of mammalian metapopulations and, taken singly, either of these features might call into question the adequacy of the metapopulation approach. Together they provide a strong rationale for using metapopulation theory to study these and other small mammal species.

Empirical studies show that dispersal by small mammal is highly targeted and effective in linking local populations over larger scales than dictated by dispersal distance alone. Migration among small populations is also a function of local demography with a higher immigration rate in smaller populations. These aspects of dispersal behaviors typical of small mammals are not addressed within most metapopulation models (but see Chapter 4). In most instances, however, seasonally restricted natal dispersal by juveniles and subadults and not movements by adults is linking populations. The dynamics of local populations thus remain sufficiently independent and subjected to local stochastic processes, extinctions, and recolonizations (e.g., water voles in the lowland area in the absence of mink). Our review also highlights the fact that spatially correlated extinctions are widespread among small mammal. In some instances, they dominate the overall dynamics to such an extent that the metapopulation paradigm is not useful for predicting their dynamics (e.g., cyclic microtines). In others, genuine metapopulations prevail where biotic (predation and pathogens) and abiotic (local variation in climate) factors external to the focal species impose independent local dynamics at the scale of several local populations. These agents of correlated extinction thus elevate the importance of metapopulation dynamics in systems that would otherwise be more influenced by demographic interdependence among small local populations. Two scales are thus relevant in small mammal metapopulations: the scale of habitat patchiness or modal dispersal movements and the scale of correlated extinctions. If dispersal is sufficiently directed and effective to overcome external synchronizing influence and allowing recolonization of areas that have suffered correlated extinctions (e.g., between blocks in upland water vole populations), small mammal metapopulations prevail, despite spatially correlated extinctions. If the scale of spatially correlated extinctions is too large relative to the tail of the distribution of dispersal distances, the metapopulation will be unable to persist.

In both the pika and in water voles, comparative studies on population genetics in continuous and in highly fragmented populations failed to reveal any loss of genetic variability in metapopulations. This questions whether subdivision commonly will reduce effective population sizes in most mammal species, and more important, the potential for a genetic signature associated with mammal metapopulations. Frequency-based descriptions of population genetic structures are unlikely to highlight the complex interplay between metapopulation and social processes that define the genetic structure among patches. This does not represent a major paradigm shift in thinking, but merely reiterates what has been said elsewhere, but is frequently overlooked (e.g., Pannell and Charlesworth, 1999). We advocate that genetic data can play an important role in defining a mammalian metapopulation, but more so in delimiting pedigrees within natural populations to identify relatives and determine social structure, and more directly in estimating migration. The challenge now lies in accurately characterizing the complex dynamics of dispersal and extinction and using these data to parameterize sufficiently realistic models that encompass the characteristics inherent yet unique to mammals.

22. METAPOPULATION DYNAMICS AND RESERVE NETWORK DESIGN

Mar Cabeza, Atte Moilanen, and **Hugh P. Possingham**

22.1 SYSTEMATIC METHODS FOR RESERVE NETWORK DESIGN

Choosing which sites to include in a reserve network is a fundamental problem in conservation biology. We will not be able to protect all places that contribute to biodiversity because that would be every place on earth. Given the realities of limited resources for conservation and other political and economical limitations, we need to identify priority areas to be set aside as reserves. Over the past two decades three philosophies have emerged in the reserve selection literature. First, there has been increasing emphasis on the design of entire reserve networks rather than the selection of individual sites. Second, there has been interest in reserve networks achieving goals such as adequacy, representativeness, and compactness. Third, there has been an interest in these goals being met efficiently. We now realize that ad hoc reserve selections, often a result of availability of sites, competition with alternative land uses, scenic value, and other factors, can lead to very inefficient steps toward constructing reserve networks that conserve the majority of species (Pressey, 1994, 1999; Cowling et al., 1999; Margules et al., 2002; Stewart et al., 2003).

0-12-323448-4

The ecological theories of island biogeography and metapopulation dynamics have provided some guidance for the selection of reserves, including rules such as "big reserves are better than small reserves," "aggregated reserves are better than dispersed reserves," "minimize patch loss/maximize the number of suitable habitats," and "protect source populations and ignore sink populations" (see, e.g., Hanski and Simberloff, 1997). While these general concepts provide broad guidance, they are not useful in helping us choose between different sites and they are not universally true (Possingham et al., 2001; Cabeza and Moilanen, 2003). Since the early 1980s, attention has focused on developing systematic quantitative methods that use empirical biodiversity data and economic considerations to select a set of sites for biodiversity conservation.

This chapter reviews quantitative methods for reserve selection and then concentrates on new ideas that incorporate concepts from spatial ecology, particularly metapopulation biology and landscape ecology. We compare a range of methods, which consider spatiotemporal dynamics in different ways, and show that persistence of species in reserve networks is enhanced when spatial considerations are taken into account in reserve network design. The chapter ends with a discussion of problems of dynamic reserve selection in an uncertain world where not only do we need to deal with dynamic populations but also a dynamic landscape.

Key Concepts in Reserve Network Design

Reserve network design methods seek to identify the adequate configuration of conservation areas for biodiversity, including issues of shape, area, connectedness, management, and scheduling (e.g., Pressey, 1999). Reserve selection algorithms are computational tools that are often used as part of the reserve network design procedure. Reserve selection algorithms (also known as area selection, or site selection algorithms) aim at meeting conservation goals efficiently in reserve networks.

In principle, the concept of biodiversity embraces the entire biological hierarchy, from molecules to ecosystems, including interactions and processes, although most often we consider our biodiversity elements to be simply species and/or habitat types. In marine conservation planning the features are more likely to be habitat types and biophysical domains due to a lack of comprehensive information on species distributions (e.g., Leslie et al., 2003). Likewise in terrestrial environments, planning is hindered by a lack of knowledge, for instance, about the existence and location of species. Consequently, selected areas will only protect a sample of biodiversity, hence the term "representation" is used to describe the goals for these reserve selection methods (Margules et al., 2002). "Represent" implicitly means "sample." Therefore, the *representativeness* of a system of reserves means the extent to which it adequately samples all the targeted natural features (e.g., species, vegetation types), of the region. The *efficiency* of the solution for a given problem is often measured as the cost of the solution (where cost may be measured as area, number of selection units, boundary length, acquisition cost, management cost, etc.; e.g., Pressey and Nicholls, 1989; Rodrigues et al., 1999).

Reserve selection algorithms have concentrated on the key concept of *complementarity*. Williams (2001) defined complementarity as a "property of sets of objects that exist when at least some of the objects in one set differ from the objects in another set." Complementarity-based methods were developed because they can achieve representation targets more efficiently (with less cost, area, or number of reserves) than, for instance, scoring approaches or approaches based on hot spots of richness or rarity (Williams, 2001; Possingham et al., 2001). Box 22.1 exemplifies the concept of complementarity by comparing scoring approaches to complementarity-based methods. In practice, the idea of complementarity refers to the amount of new features that a site would add to the already represented features in the set of selected sites in iterative heuristic algorithms (see section Reserve Selection Algorithms).

Complementarity-based methods were first described by Kirkpatrick (1983) and were developed simultaneously by several authors during the 1980s (Kirkpatrick, 1983; Ackery and Vane-Wright, 1984; Margules et al., 1988; Rebelo and Siegfried, 1990). The term complementarity was coined specifically in the context of reserve selection by Vane-Wright et al. (1991) (for a review, see Williams, 2001).

BOX 22.1 An Example of Advantages of Complementarity Methods

This sites × species matrix shows candidate sites and species occurring in them (indicated by a X) of a hypothetical system. Suppose that we could choose two sites. A scoring approach (Margules and Usher, 1981) would select sites in decreasing order of richness (number of endemisms, site quality) starting from the richest site (a hot spot of richness), which in this example is site 2 with six species. The next richest site would be site 3, with five species. Sites 2 and 3 would sample a total of seven species (c–i). However, another solution that represents more species with the same number of sites can be obtained when using the idea of complementarity. We can see that sites 3 and 4 would include all species and therefore complement each other better than sites 2 and 3. Scoring approaches ignore complementarity and tend to choose sets of sites that contain high levels of replication for some species while ignoring others and therefore are less efficient than complementarity methods (e.g., Williams, 2001).

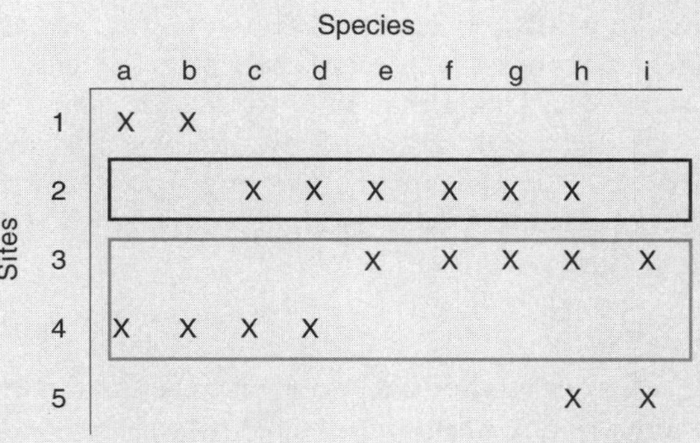

Reserve Selection Algorithms

Two common ways of defining the most basic reserve selection problem are (1) to choose the minimum set of reserves that contain all species (or other features) at least a given number of times, known as the minimum set covering problem (Underhill, 1994; Csuti et al., 1997; Pressey et al., 1997), and (2) to maximize the number of features represented when there is a limit on the number of reserves that may be chosen, known as the maximal coverage location problem (Camm et al., 1996; Church et al., 1996; Arthur et al., 1997).

These approaches share several desirable characteristics, such as being data driven, goal directed, efficient, explicit, repeatable, and flexible (Pressey, 1999). These features make these methods valuable for systematic reserve design in comparison to *ad hoc* decisions or rules of thumb derived from ecological theories.

Data Driven

The selection of a set of reserves is based on a "features × sites" matrix, usually indicating the occurrence (presence–absence or probability of occurrence) of a feature in an area. The features can be any chosen set of biodiversity surrogates and/or natural elements of interest, for instance, landscapes (Pressey and Nicholls, 1989; Pressey et al., 1996), plant communities (Cocks and Baird, 1989; Nicholls and Margules, 1993; McDonnell et al., 2002), habitat types (Olson and Dinerstein, 1998; Leslie et al., 2003), environmental variables (Araújo et al., 2001), and species (Kirkpatrick, 1983; Rebelo and Siegfried, 1992; Sætersdal et al., 1993; Kershaw et al., 1994; Church et al., 1996; Csuti et al., 1997). There are no rules for choosing the features to be represented, and often the choice is determined by the available information for the particular system of interest (see later for more discussion on biodiversity goals for conservation). The sites, or selection units, are any discrete part of the landscape to be evaluated for their contribution to the reserve system. They can be regular (e.g., grid cells, Freitag and Van Jaarsveld, 1998; Rodrigues et al., 2000; hexagons, Csuti et al., 1997) or irregular (e.g., woods, Sætersdal et al., 1993; pastoral holdings, Pressey et al., 1997) and can be continuous or discrete (e.g., forest fragments).

Goal Directed

Quantitative goals are set for all features. Most often the goals have been framed in terms of representation of the features by setting quantitative targets for each feature (e.g., at least one occurrence of all species). Increasingly, other goals that incorporate considerations of spatial distribution of reserves, or probabilities of persistence for the species, have been considered (Possingham et al., 2000; Noss et al., 2002). The targets may be different for each of the features.

Efficient

Reserve selection methods acknowledge the limited resources for conservation and therefore aim at minimum-cost solutions, given the goal. The cost is usually the number, or total area, of selected sites. Some authors consider more economically oriented costs, such as acquisition or opportunity costs (Ando et al., 1998; Polasky et al., 2001).

Biodiversity Conservation Goals and Data Requirements

Because all biodiversity cannot be measured, there is a need to base reserve selection on biodiversity surrogates. There are three major types of biodiversity surrogates: subsets of taxa, assemblages, and environmental variables. Different surrogates appear to be better than others under different circumstances (Williams et al., 2000; Araújo et al., 2001; Margules et al., 2002). In practice, because available data for setting priorities usually come from different sources, some combination of these surrogates should be used (Nix et al., 2000).

Distributional data for species can be constructed from records held by museums and herbariums, but this is presence-only data. Most datasets have spatial or taxonomical bias, as in the sampling process, certain kinds of sites and taxa are favored (i.e., easily accessible sites and vertebrates; Margules and Austin, 1994). Freitag and van Jaarsveld (1998) assessed the problem of using incomplete datasets for reserve selection and found a large variation in selected networks as well as decreases in efficiency. Andelman and Willig (2002) found that the source of data affects the location of reserves and the efficiency of the solutions. Therefore, systematic and intensive biodiversity surveys are an important element of efficient conservation planning. Reserve selection algorithms can also be applied to data on probabilities of occurrence (e.g., Araujo and Williams, 2000; Williams and Araújo, 2002). Analytical procedures can be used to address the problem of bias in data and provide probability values for all points in a map based, for instance, on environmental variables (Elith, 2000; Williams et al., 2002).

22.2 METAPOPULATION DYNAMICS, HABITAT LOSS, AND RESERVE SELECTION ALGORITHMS

Regional reserve networks (should) have two strategic goals: (i) to efficiently represent the full spectrum of biodiversity within a system of protected areas and (ii) to ensure the long-term persistence of that biodiversity (Frankel and Soulé, 1981). Until recently, reserve selection methods have generally focused on the goal of representation and the goal of persistence has been neglected. Moreover, few authors have paid attention to the role of spatiotemporal population dynamics and the effects of landscape changes. After formulating the basic reserve network design problem, most of this chapter focuses on these issues.

Simple approaches base the selection of reserves on static patterns of species presence/absence (often a single static snapshot of incidence patterns). Several studies have analyzed the persistence and turnover of species in reserves designed with these simple methods. For instance, Margules et al., (1994) evaluated a procedure for identifying a minimum set of sites that would capture all rare or uncommon plant species in the region. They compared the occurrences in the minimum set solution with the occurrences after 11 yr in the same locations and found that the original site selection was inadequate to preserve the initial species (the loss of species in the selected network was 36% even when there was no habitat loss in the region). Other studies have reached the same conclusion: the static representation approach does not account for species spatial turnover, and species may be naturally lost from the selected sites after a relatively short time (Nicholls, 1998; Rodrigues et al., 2000; Araújo et al., 2002).

Some reserve selection approaches have considered criteria for persistence, but mostly in an implicit way. These criteria affect (i) the selection units (e.g., by including sites of a minimum size), (ii) the features to be considered (ignoring representation of common species and concentrating on endangered species or on endemic species), or (iii) the goal. Criteria that affect the goal include, for example, approaches based on abundance data (that aim at selecting sites where the species is more abundant or where the total number conserved is above a viability threshold) or "bet-hedging" approaches (multiple representations for all species, i.e., include at least four populations of each species; see Cabeza and Moilanen, 2001) .

Spatiotemporal population dynamics and effects of landscape change (including habitat loss and degradation) have generally been ignored. Dynamics concerning landscape changes have been more commonly considered, but this has been basically from the perspective of cost effectiveness, e.g., scheduling and acquisition of sites (e.g., Pressey, 1999; Pressey and Taffs, 2001), not from the perspective of linking landscape dynamics, spatial population dynamics, and reserve network design.

We believe it is important to adopt a spatial population dynamics view in combination with a spatially structured landscape to determine the reserve network that will conserve biodiversity in the long term. For instance, Cabeza and Moilanen (2003) explored, by simulating species-specific spatial population dynamics, what happens in a reserve network selected by reserve selection algorithms with simple representation goals when all nonselected candidate reserves are lost. Simple reserve selection algorithms tended to select small and overdispersed reserve networks. Cabeza and Moilanen (2003) showed that when selected reserves are far apart and only one or a few representations of each species are considered, numerous extinctions can be expected. The chances of having species extinctions in such a reserve system are even larger if the habitat surrounding the reserve is lost. However, when the selected reserves were kept close together, the number of extinctions following habitat loss was much smaller (Cabeza and Moilanen, 2003). This is another way of saying that habitat loss around a reserve network will cause an extinction debt in the region and that extinctions in the region will follow (Tilman et al., 1994; Hanski and Ovaskainen, 2002; Ovaskainen and Hanski, 2002)

In summary, the ways in which persistence and spatiotemporal population dynamics have been considered in reserve selection algorithms can be classified into four broad problem categories (following Cabeza and Moilanen, 2003).

C0. In this category, no criteria for persistence or spatiotemporal dynamics are used. This is the "representation" problem, which implicitly assumes that a species will persist indefinitely in any site where it was observed. This might be appropriate when the probability of losing a species from a site is very low, which may be the case when sites are rather large.

C1. No spatial considerations are taken into account, although implicit or explicit criteria favoring local persistence are used, by acknowledging the effects of area or population size on persistence. This is often done by using thresholds to define the minimum size of selection units, by basing the selection on abundance data and species abundance targets, or by considering habitat quality on probability approaches (see Section 22.3 for a more extended description of probability approaches).

C2. Spatiotemporal population dynamics are considered implicitly by means of keeping reserves close together, which results in compact reserves/reserve networks that are less susceptible to negative external effects than scattered reserve networks. This class includes methods referred to as "spatial reserve design" (Possingham et al., 2000). A detailed description of these approaches is given later.

C3. These methods explicitly consider species-specific spatiotemporal population dynamics and regional persistence, and possibly the interaction between population dynamics and landscape change.

22.3 COMPARISON OF DIFFERENT ALGORITHMS

This section describes in increasing order of complexity the site selection algorithms used in the following comparison. We start with the minimum set covering formulation, with a simple (single) representation problem: "what is the smallest set of sites that contains at least one representation of each species":

$$\min \sum_{i=1}^{N} I_i$$

given that (22.1)

$$\sum_{i=1}^{N} I_i p_{ij} \geq 1 \quad \text{for all } j.$$

In this problem, and in later problem formulations, N is the number of candidate sites, M is the number of species, p_{ij} is an element of an $N \times M$ matrix giving the probability of presence of species j in site i, and I_i is an indicator variable indicating whether site i is selected or not.

The minimum set coverage formulation may be extended easily to cover a variety of situations. To do so, we introduce some new symbols: let Ω be the set of all sites, S be the index set of selected sites (for which $I_i = 1$), T_j the target level of conservation set for species j, c_i the cost of site i, A_i the area of site i, and $R_j(S)$ the representation of species j in S. A generalized form of Eq. (22.1) is now

$$\min \sum_{i \in S} c_i$$

given that (22.2)

$$R_j(S) = \sum_{i \in S} p_{ij} \geq T_j \quad \text{for all } j.$$

In this formulation $c_i = 1$ when minimizing the number of sites, $c_i = A_i$ if minimizing the area of the solution, or c_i may be equal to a real site cost if such information is available. The commonly used site selection algorithms operate on presence–absence data, in which case $p_{ij} \in \{0,1\}$, but p_{ij} can also be the probability of presence of the species if a statistical model for that exists. For the simple representation problem, $T_j = 1$, and for the multiple representation problem, $T_j > 1$ (often an integer). The proportional coverage problem is obtained by setting $T_j = \alpha \Sigma_\Omega \, p_{ij}$, where α is the proportion of populations that have to be protected for each species. Different weights can be given to different species by setting comparatively higher targets for species that are considered important in the region. Note that the problem definition applies equally to any biodiversity elements (habitat types, ecosystem type, etc.).

Variants of the minimum set coverage problem are solved commonly using a stepwise richness-based heuristic algorithm (e.g., Csuti et al., 1997; Pressey et al., 1997).

Algorithm 1: Forward Richness-Based Heuristic

1. set $S = \varnothing$
2. for all sites k not in S, calculate measure U_k, which gives the proportion of underrepresentation covered by the addition of site k; $U_k = \Sigma_j p_{kj} / (T_j - R_j(S))$, summing only over species j having $T_j > R_j(S)$
3. add site k with highest ratio U_k/c_k to S
4. if $R_j(S) < T_j$ for any j, go to 2 or else quit

For the simple representation problem, algorithm 1 adds the site iteratively, which adds the greatest number of new species to the solution until all species are represented at least once. At each step you add the patch with the highest value in complementarity richness per patch cost. At least for small and moderate problems, Eqs. (22.1) and (22.2) can be solved exactly using linear programming (Cocks and Baird, 1989; Possingham et al., 1993; Rodrigues and Gaston, 2002), although the simple stepwise heuristic runs orders of magnitude quicker (Pressey et al., 1996; Possingham et al., 2000). Note that in the final solution some species will have $R_j(S) > T_j$, i.e., the species are overrepresented. Algorithm 1 does not give any consideration to the distribution or value of overrepresentation among species.

Variants of the minimum set covering formulation that use presence–absence data belong to problem category C0 because they do not consider space and spatial population dynamics; effects of habitat loss outside the reserve network are not considered, and it can be interpreted that species are implicitly assumed to occur forever in the sites that they were observed. If p_{ij} is based on a probability model for the presence of species, the algorithm partly belongs to category C1 as it is acknowledged that probabilities of presence in sites partially reflect their capacity to support species (and presumably good sites are chosen), but the spatial configuration of the reserve network is still ignored.

The simplest site selection method, which attempts to consider the spatial configuration of the habitat, is based on an adjacency rule (Nicholls and Margules, 1993). This algorithm is similar to algorithm 1, but if there is a tie in step 3 (several sites give the same contribution to complementarity richness), then the site closest to any already selected site is chosen. In other words, proximity to sites in S is used to break ties. However, spatial configuration is only of secondary importance when using the adjacency rule (richness is the primary objective of optimization), and consequences of using the adjacency rule are entirely dependent on the occurrence of ties, which may be infrequent, especially if c_i is based on site area and A_i values vary. Consequently, we do not believe that the adjacency rule can be expected to be of much value in spatial reserve design in general.

The first method to properly consider the spatial configuration of a reserve is given by Possingham et al. (2000), which minimizes a linear combination of reserve area and boundary length. In this algorithm, it is possible to obtain reserves with different levels of aggregation by giving different relative weights to area and boundary length. Cabeza et al. (2004) described a method that combines the probability of occurrence approach (Araujo and Williams, 2000; Williams and Araujo, 2000, 2002) with the method of Possingham et al. (2000). In this method, the probabilities of presence of species j on patch i, p_{ij}, are obtained by fitting a habitat model (e.g., by logistic regression) to observations of species presence–absence. Optimization minimizes the cost of the reserve network penalized by the boundary length of the reserve, and targets are set as expected numbers of populations. By setting a high penalty for boundary length, a relatively compact reserve is obtained, whereas the spatial configuration of the reserve is of no consequence when the boundary length penalty approaches zero. More formally, the problem is

$$\min \sum_{i \in S} c_i + bL$$

subject to (22.3)

$$R_j(S) = \sum_{i \in S} p_{ij} \geq T_j \quad \text{for all } j$$

in which L is the boundary length and b is the penalty given to boundary length (relative to other costs). The objectives in Possingham et al. (2000) and Cabeza et al. (2004) are very similar; both essentially minimize a linear combination of cost (area) and boundary length. The difference between the methods is that in the latter the conservation targets T_j are framed in expected numbers of populations, as p_{ij} values are based on a probability model for the presence of the species. Cabeza et al. (2004) used the ratio of reserve boundary length to reserve area (L') instead of the boundary length of the reserve (L) directly. This is because L' is much less dependent on the absolute size of the system than L, which means that L' is more suitable to be used in the context of a stepwise heuristic optimization algorithm where the number of sites (and L) varies during the optimization process.

Note that the methods of Possingham et al. (2000) and Cabeza et al. (2004) both belong to problem category C2. This means that reserve aggregation is achieved in a qualitative manner, without any estimate of the species-specific effects of aggregation on spatial processes and persistence of populations per se. It is just assumed that aggregation is useful because it decreases edge effects and reserve maintenance costs (Possingham et al., 2000). It follows that an important question is how much aggregation in the reserve network can you get with little or no increase in reserve cost? In the example of Cabeza et al. (2004), it was typically possible to achieve a 50% decrease in L with a $<$2% increase in total reserve cost. Similar reductions in boundary length for minimal increases in area have been achieved for terrestrial and marine reserve network design problems (McDonnell et al., 2002, Leslie et al., 2003).

Cabeza (2004) presented an important development of Cabeza et al. (2004). In Cabeza et al. (2004), p_{ij} values are constant during optimization; they are not dependent on S. Connectivity and recolonization of empty habitat will, however, be important for some of the species. For such species it is possible to build a dependency between p_{ij} and S into the optimization. This means that habitat loss will decrease p_{ij} values, especially near the location where habitat has been lost, and p_{ij} will decrease possibly substantially when site i loses many of its neighbors. We start by defining a connectivity measure G_{ij} for species j in site i:

$$G_{ij} = \sum_{k \in N_j(S)} f(d_{ik})\, p_{kj}, \tag{22.4}$$

in which $N_j(S) \in S$ is the neighborhood of site j in S. $N_j(S)$ may be equal to S-$\{k\}$, but it can also be a buffer around patch j, which is what is used in the following example [see Moilanen and Nieminen (2002) for a discussion on connectivity measures used in metapopulation models]. d_{ik} is the distance between sites i and k, and $f(d_{ik})$ is a decreasing function of d_{ik}. Then

$$p_{ij}(S) = f[\mathbf{h}_i, G_{ij}(S)], \tag{22.5}$$

in which $\mathbf{h}_i$ is a vector of habitat variables. $p_j(S)$ may be fitted originally, for example, using logistic regression, the original presence–absence information, and $G_{ij}(\Omega)$ calculated assuming $p_{ij} = 1$ in sites i where species j was observed and 0 otherwise. If a species is not affected by connectivity, then simply put $p_{ij}(S) = f(h_i)$. The optimization problem is what Cabeza (2003) called the *dynamic probability problem*. "Dynamic" comes from the fact that the probabilities are reevaluated during optimization, meaning that the method takes into account that habitat outside the reserve will be lost:

$$\min\ F = \sum_{i \in S} c_i + bL'$$

subject to (22.6)

$$R_j(S) = \sum_{i \in S} p_{ij}(\mathbf{h}_i, S) \geq T_j \quad \text{for all } j.$$

To solve Eq. (22.6), all probabilities p_{ij} have to be recomputed by iterating Eqs. (22.4) and (22.5) after changing S until convergence. Any analogue of the forward stepwise heuristic (algorithm 1) will perform poorly with Eq. (22.6). Assume that the optimal solution consists of several essentially separate patch aggregates. The forward algorithm will start by selecting one site and thereafter it will overextend this first cluster because starting a new cluster from one site (and thus very low connectivity and low p_{ij}s) will have a lower marginal contribution than extending the existing cluster, which has high connectivity. Better solutions will be achieved by a backward algorithm (Cabeza, 2003; Cabeza et al., 2004); this algorithm starts by selecting all sites and then removes "bad" sites one by one until no site can be removed without the target being violated for at least one species. For the algorithm, we need to define the changes in F and R following the removal of site i:

$$\Delta F_i = -c_i + b\left(\frac{L + \Delta L}{A\left(S - \{i\}\right)} - \frac{L}{A(S)}\right)$$
$$\Delta R_i = \sum_{i \in S} \sum_{j=1}^{M} \frac{\Delta p_{ij}}{R_j(S) - T_j}, \qquad (22.7)$$

where $A(S)$ is the area of solution S and Δp_{ij} (≤ 0) is $p_{ij}' - p_{ij}$ in which p_{ij} and p_{ij}' correspond to the converged values of p_{ij} before and after the removal of site i, respectively. ΔF_i (<0) and $\Delta R_i \cdot (\leq 0)$ express changes in cost and boundary length and changes in species overrepresentation, respectively. During each optimization step, the site with the smallest ratio $\Delta R_i / \Delta F_i$ is chosen for removal. Note that a large $\Delta R_i / \Delta F_i$ means that either (i) ΔR_i is large and therefore removing site i would cause a large (negative) change in species overrepresentation (not desired) and/or (ii) ΔF_i is small, meaning that removing site i would increase the fragmentation of the reserve system (not desired) or the cost of site i is low, or both.

Westphal and Possingham (2003) considered a similar problem of optimal habitat reconstruction where the objective is to maximize the summed probabilities of occurrences and these probabilities are modeled as a function of the entire landscape. They use simulated annealing to find good solutions to their problem.

Algorithm 2: Backward Stepwise Heuristic

1. set $S = \Omega$
2. calculate ΔR_i (≤ 0) and ΔF_i for all sites $i \in S$
3. let $K \in S$ be the set of sites k, the removal of which does not violate the constraint T_j for any species j and for which $\Delta F_k < 0$
4. if $K = \emptyset$, optimal solution $= S$, quit
5. remove site $k \in K$ with smallest ratio $\Delta R_k / \Delta F_k$ from S, go to 2

Equations (22.1) and (22.2) assume implicitly that habitat outside the reserves will not change after the reserve has been established (or at least no effects of habitat loss are considered). In contrast, Eqs. (22.3) and (22.6) assume that changes will occur for unselected sites. Equation (22.6) can be extended to include an explicit model of landscape change (Cabeza and Moilanen, 2003).

Properties of different solutions that are of interest include cost of solution, total area, total boundary length, and the representation level calculated either from presence–absence observations or from the habitat model assuming (i) unchanged probabilities with habitat loss or (ii) dynamic probabilities [requires iteration of Eqs. (22.4) and (22.5)]. When comparing solutions obtained using two different algorithms, you should set one aspect (e.g., cost) of the solutions to be equal and compare the rest of the attributes (e.g., boundary length, persistence of species). The problems and algorithms are compared using the same data set.

22.4 A CASE STUDY

Data used here are from a survey of 26 butterfly and moth species, including two endemic races (*Plebejus argus caernensis* and *Hiparchia semele thyone*) living on the Creuddyn Peninsula, north Wales, United Kingdom (Cowley et al., 2000; Gutierrez et al., 2001; Menendez et al., 2002; Thomas et al., 2002; Wilson et al., 2003; Cabeza et al., 2004). The presence and density of these butterflies have been surveyed in a 35-km^2 region divided into 500 × 500-m squares (Cowley et al., 2000).

For the purpose of comparing reserve selection algorithms, two statistical models were fitted for each species (Cabeza, 2003; Cabeza et al., 2004). Both models were logistic regressions explaining the observed presence–absence of the species by a density index (a habitat quality-based index, see Cowley et al., 2000). The other model also included a connectivity measure as an explanatory variable. We used the proportion of occupied cells in the one cell neighborhood as connectivity, with neighboring cells that were on the ocean excluded from the computation on the assumption that the sea is a reflecting boundary for the butterflies. Even this simple connectivity measure was significant, and often highly significant, for 23 species out of 25 (Cabeza, 2003). For the species not having a significant effect of connectivity, the model was based only on the density index.

Reserve Networks Selected by Different Algorithms

We compare here the site selection methods described earlier using butterfly data from the Creuddyn Peninsula. The following denotes by BLM/PA and BLM/prob the boundary length minimization method [Eq. (22.3)] when using presence–absence data or probability data, respectively. The probability model is based on the grid density index of Cowley et al. (2000). By DP we denote the dynamic probability method [Eqs. (22.4–22.6); we used $b = 0$]. Targets in the following analyses are proportional coverage targets, with the same proportion used for all species (computed from full original data). Note that BLM/PA and BLM/prob methods with the penalty set to zero ($b = 0$) reduce to variants of the ordinary multiple representation problem [Eq. (22.2)].

Figure 22.1 shows sites with the highest species richness in the Creuddyn region. As shown later, however, the richest sites are not automatically the most valuable when considering complementarity richness. Some of the most species-rich sites are likely to be in the optimal selection, but the selection may also be influenced heavily by rare species, which do not always occur on species-rich sites. For this problem the minimum set coverage solution [Eq. (22.1)] consists of only three 500 × 500-m sites (not shown). It is obvious from studies of butterfly population dynamics that three scattered sites of this size cannot be expected to maintain viable populations of 26 butterfly and moth species (e.g., Thomas and Hanski, 1997; Hanski, 1999). Following habitat loss, such a "reserve network" could be expected to lose species quickly (see Chapter 20).

Results in Figs. 22.2–22.5 show summary information for all the 26 species. Optimal solutions for different site selection methods and increasing target levels are shown in Fig. 22.2. Rows one and four correspond to variants of the multiple representation problem. These rows show high scatter in the solution, which is expected because no spatial component is included in optimization. It is easy to imagine that the reserves systems selected in rows one and four would be difficult to start, expensive to maintain, and susceptible to habitat loss/degradation between reserve sites. Incidentally, use of the simplest method of spatial reserve design, the adjacency rule (Nicholls and Margules, 1993), does not differ from nonspatial reserve selection methods because very few ties occur in the optimization and thus the solutions are almost entirely unaffected by the adjacency rule (not shown).

Having a penalty for boundary length changes optimal solutions drastically. Reserves in Fig. 22.2 obtained using BLM/PA or BLM/prob with $b > 0$ are much more aggregated than the $b = 0$ solutions. Of these, BML/prob produces more compact reserves with less boundary — it will be difficult to find compact reserves using BLM/PA if the species is often missed in the field because many gaps will be left in the observed distribution of the species. Also, BLM/PA may be quite sensitive to errors and sampling artifacts in empirical data collection, which makes it more sensitive to data quality than BLM/prob, which relies on a habitat model. Nonetheless, BLM/prob is likely to be quite good for species for which the statistical habitat model explains a major proportion of the variance in the occurrences of the species, but

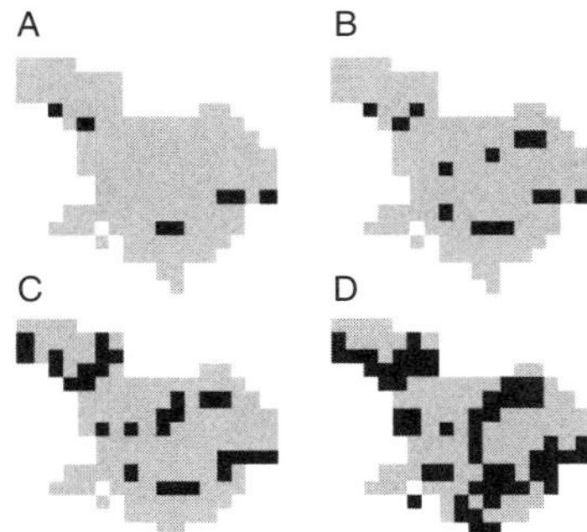

Fig. 22.1 Maps of the Creuddyn Peninsula showing sites having the highest species richness calculated from presence–absence data for the 25 butterflies: 5% (A), 10% (B), 20% (C), and 40% (D) of most species-rich sites.

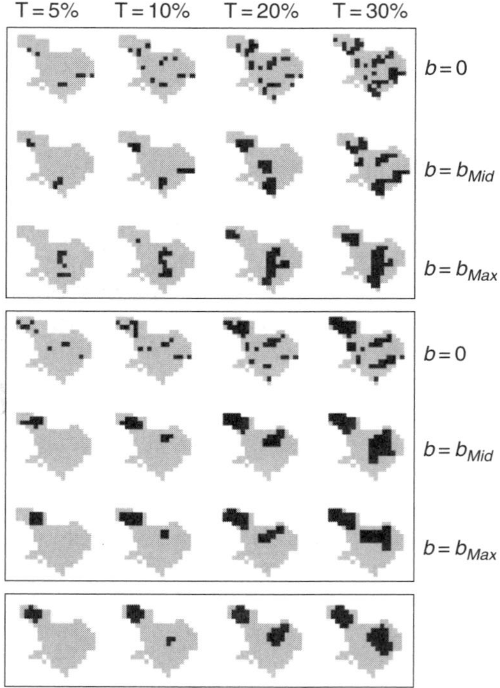

Fig. 22.2 A comparison of site selection methods. The *x* axis is the percentage level for the proportional coverage target (*T*). The three uppermost rows correspond to presence–absence data and the boundary length minimization method [Eq. (22.3)] with zero, intermediate, and high penalties for boundary length. Rows 4–6 are as rows 1–3, but using probabilities calculated from the habitat model instead of P/A observations. The bottom row is for the dynamic probability method [Eqs. (22.4–22.6), $b = 0$] with connectivity-dependent probabilities recalculated during each optimization step. As rows 1 and 4 have zero penalty for boundary length, they actually revert back to the multiple representation problem [Eq. (22.2)] with a proportional coverage target.

BLM/PA might be best for species for which a good habitat model cannot be obtained. Interestingly, the actual areas selected by BLM/PA and BLM/prob are quite different.

The dynamic probability method (Fig. 22.2, bottom row) consistently produces aggregated solutions, which correspond most closely to those obtained using BLM/prob ($b > 0$). This is not a coincidence as DP uses the habitat model augmented with connectivity for the 23 species for which connectivity was significant. No boundary length penalty was used for the DP method [$b = 0$ in Eq. (22.6)] and thus the clustering obtained with DP is a consequence of the use of connectivity. It is encouraging that including a component of spatial population dynamics into the site selection method consistently results in clear reserve aggregation.

The results of Fig. 22.2 are put into perspective when combined with those of Fig. 22.3, which shows the evolution of reserve network area and boundary length with increasing b when using BLM/PA and BLM/prob. The general trend is that a large reduction in reserve boundary length can be obtained with

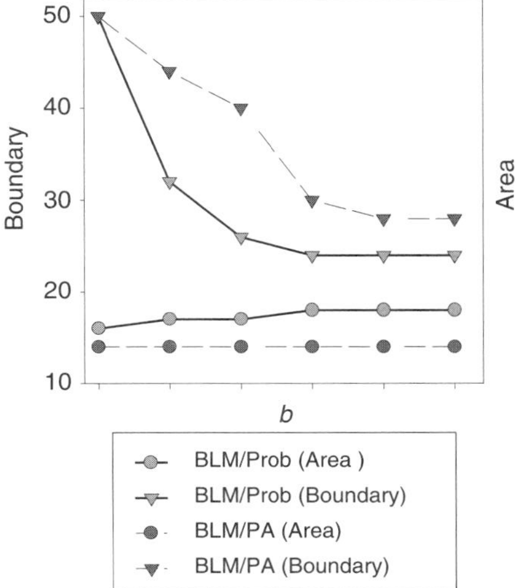

Fig. 22.3 Behavior of solution area and boundary length when using the boundary length minimization method [Eq. (22.3)]. Solid and dashed lines correspond to presence–absence and probability data, respectively. Lines with triangles and circles correspond to area and boundary length, respectively. With both problem variants, a significant reduction in reserve boundary length can be achieved with a minor increase in reserve network area. Using P/A observations produces more fragmented solutions than probability data. Results shown are for the 5% target level.

a minor (even zero) increase in reserve area. If reserve aggregation can be obtained for free in terms of cost (here cost = area), such aggregation should often be favored (but see Section 22.6). Very similar results were obtained also for targets other than 10% (Fig. 22.3).

Figure 22.4 compares different site selection methods in terms of the expected number of populations (per species) calculated using the effects of connectivity and assuming nonselected sites are lost. In this comparison, DP does best and averages about 30% higher in terms of populations than the simple multiple representation variants. Encouragingly, BLM/PA and BLM/prob with high b also do quite well, which indicates that the qualitative clustering achieved by BLM methods is a useful step in the direction of designing reserves that support long-term conservation of biodiversity. Another way of comparing the site selection methods looks at the difference between the realized (using DP) and target representation (Fig. 22.5). BLM/PA systematically fails to achieve the set target regardless of the choice of b. Best results are achieved with BLM/prob with high b or with DP. BLM/prob can produce an overall overrepresentation of the target even when evaluated using DP. This is because a high penalty for boundary length actually forces more area into the solution. When comparing solutions of the same size, DP still achieves highest expected numbers of populations (Fig. 22.4). Note that some overrepresentation in the solution does not mean that any site can be removed from the solution without the target failing for at least one species.

The effects of reserve aggregation are not equal for all the species; those species that show strongest effects of connectivity are likely to be affected most

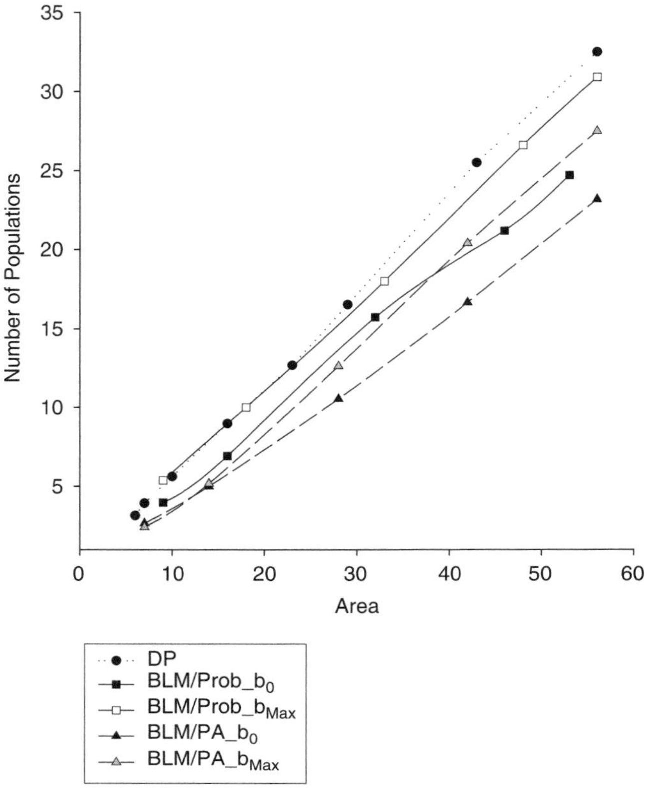

Fig. 22.4 Different site selection methods and the average expected number of populations as a function of solution area calculated over all species using the dynamic probability method (evaluating the effects of habitat loss). The dotted line is for the dynamic probability method [Eqs. (22.4–22.6)]. Solid lines are for the boundary length minimization problem [Eq. (22.3)] using the habitat model for probabilities and with zero (lower line) and high penalties (upper line) for boundary length. Dashed lines are as the solid lines but for presence–absence observations. The worst performers are problem variants with zero penalty for boundary length ($b = 0$). These solutions are fragmented (see Fig. 22.2), which shows a comparatively low expected number of populations when connectivity effects are accounted for.

adversely by fragmentation. In these particular data, *Plebeijus argus* is both an important endemic race and also a species showing strong effects of connectivity in statistical analysis. Figure 22.6 shows predicted effects of the site selection method for *P. argus*. When accounting for the effects of connectivity (right bars), the species is expected to be practically extinct from any solution with significant scatter (all solutions with P/A data or with $b = 0$). Thus the clustering of the reserve can be expected to be of primary importance for this species.

When applying site selection methods to real world problems, at least two factors that were ignored earlier should be considered: the weighting of the species and landscape dynamics. It makes sense to set different targets for different species according to their conservation status. The setting of species

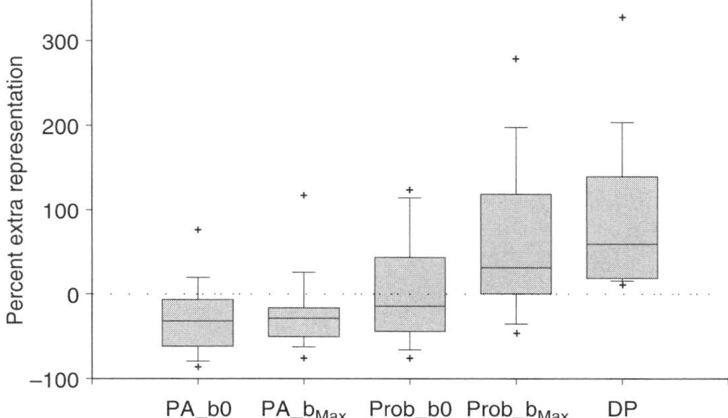

Fig. 22.5 Percentage of the optimization target realized for different site selection methods when calculated over all species and evaluated using dynamic probabilities. A positive value indicates that the realized representation is higher than the target, which is possible because some common species will have more than the minimum required number of populations. The methods producing scattered solutions perform worst.

weights is likely to be partially a political decision where local and global conservation needs are balanced. In this particular case, giving high weight to the two endemic races does not change the solution significantly from the solutions produced by the dynamic probability method (not shown). The reason is that the endemics have somewhat specialized habitat requirements that influence the solution disproportionately.

The BLM and DP methods applied to the case study assume a worst-case scenario in the sense that they explicitly assume that habitat outside the

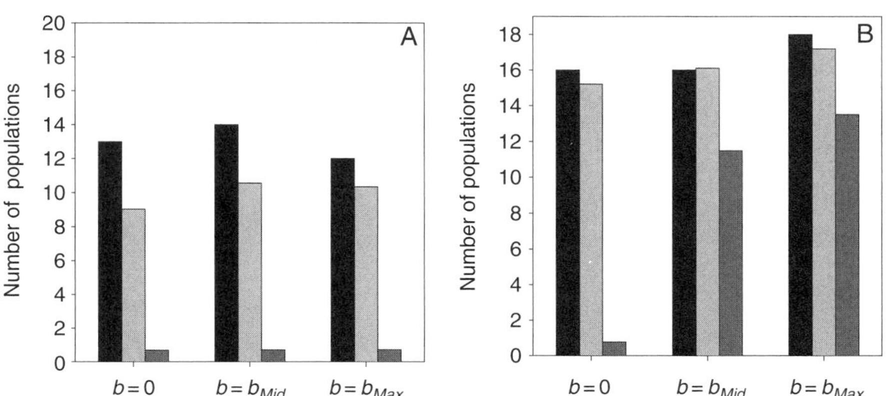

Fig. 22.6 Representation of an example species, *P. argus*, in the solution when increasing the boundary length penalty (b) in the boundary length minimization problem [Eq. (22.3)]. (A) Presence–absence data (BLM/PA) and (B) the habitat probability model (BLM/Prob). The three bars plotted for each solution are for the number of populations calculated from P/A observations (left bar), the expected number of populations calculated from the habitat model (middle bar), and the expected number of populations using dynamic probabilities (right bar).

selected reserve network is lost. In some cases, there may be more knowledge available about what can be expected to happen for nonselected habitat and this knowledge could be integrated into the reserve selection process.

22.5 USING A STOCHASTIC METAPOPULATION MODEL IN SITE SELECTION

Moilanen and Cabeza (2002) described how a stochastic metapopulation model can be used in the site selection process in order to explicitly incorporate spatiotemporal population dynamics into reserve network design (category C3, see Section 22.2). They ask the question: which subset of sites S do you select to maximize the long-term persistence of a metapopulation given that you have a parameterized metapopulation model, unselected habitat is lost, each site has a cost, and the amount of resource (e.g., money) available is limited? We show here what kind of results can be expected when applying this method. In our example, we use the incidence function model (IFM; see Chapter 4 and 5 for references and a description of the model).

The simplest way of integrating a stochastic metapopulation model in site selection is to use the metapopulation model to find the set of sites that gives the lowest metapopulation extinction rate for simulations of a specified length T. (Alternatively, one could find the set of sites that gives the metapopulation the longest average lifetime.) There are two significant problems with this appoach. First, if replicates go extinct, only rarely (or practically never), a very large number of simulation runs is needed to evaluate the extinction probability of the metapopulation reliably, which will slow down optimization considerably. Second, the simple measure is unable to distinguish between solutions that are always persistent and between solutions that always lead to extinction. Consequently, Moilanen and Cabeza (2002) used a measure of the persistence of the simulation, $F(S)$, which can distinguish the quality of solutions without actually observing extinctions. This is the average one-step global extinction probability of the metapopulation calculated over N simulation runs:

$$F(S) = \frac{1}{NT} \sum_{n=1}^{N} \sum_{t=1}^{T} \mu(X_{n,t}), \tag{22.8}$$

in which $X_{n,t}$ is the simulated patch occupancy pattern at time t in replicate simulation n and $\mu(X_{n,t})$ is calculated as the probability of simultaneous extinction of all local populations,

$$\mu(X_t) = P[\varnothing_{t+1} | X_t] = \prod_{i \in S} \begin{bmatrix} 1 - C_i(t), & if\ p_i(t) = 0 \\ [1 - C_i(t)]E_i, & if\ p_i(t) = 1 \end{bmatrix} \tag{22.9}$$

where $\varnothing_{t+1}$ denotes metapopulation extinction at time $t + 1$, $p_i(t)$ is the occupancy state of patch i at time t, $C_i(t)$ is the probability of patch i being colonized, E_i is the probability of patch i going extinct independently of colonization from other patches (the intrinsic extinction probability), and $[1 - C_i(t)]E_i$ is the extinction probability of patch i when considering the rescue effect. Functions $E()$ and $C()$ will naturally be determined by the structure of the metapopulation model.

Equation (22.8) is a function of the metapopulation model and its parameter distribution, the initial occupancy state of the metapopulation, S and T. It is related to, but not identical to, the extinction risk of the metapopulation and has the following properties: $F(S) \rightarrow 1$ when the metapopulation goes extinct almost immediately and $F(S) \rightarrow 0$ when the metapopulation is highly persistent. When the metapopulation goes extinct in a simulation at time t, $F(S)$ is increased by $(T - t)/(NT)$, and thus the value of $F(S)$ is always greater than the proportion of time the metapopulation is extinct in the simulations. Importantly, Eq. (22.8) is able to differentiate between two solutions that persist until the end of all simulation runs; $F(S)$ is smaller for the solution that is more persistent. Moilanen and Cabeza (2002) described an optimization technique that is able to efficiently solve the nontrivial optimization problem of finding the optimal set of sites S^* from the search space of size 2^m, where m is the number of patches in the metapopulation. Note that the difference between a spatial population viability analysis (PVA, see, e.g., Murphy et al., 1990; Coulson et al., 2001) and metapopulation site selection is that a PVA only compares a few alternatives, whereas metapopulation site selection actually searches for an optimal solution within the given constraints.

Important Factors Affecting the Selection of the Reserve Network

Here we apply the metapopulation site selection method to a metapopulation of the false heath fritillary butterfly, *Melitaea diamina*. *M. diamina* lives on moist meadows, which are nowadays being overgrown rapidly. This poses persistence problems to the butterfly if no restoration work is done for maintaining the quality of the meadows. A system of 125 habitat patches scattered in an area of 20×30 km in southern Finland (Fig. 22.7) was used to assess "which subset of sites should be maintained to maximize the long-term persistence of *M. diamina*, given the cost of the sites and the available amount of resources?" A brief overview of the effects of different factors on optimal selection is given: the value of the dispersal parameter α, the available amount of resources for setting the reserves, and the cost of the sites.

The dispersal ability of the species (average dispersal distance is given by $1/\alpha$) is possibly the most important factor in the metapopulation model affecting the configuration of the reserve network. When dispersal distances are short (large α), the best option is to protect sites that are close together (Fig. 22.8A). However, when the dispersal abilities of the species are not limiting and the individuals can reach any site in the system, the optimal solution does not consist of a compact cluster, but of a larger number of more scattered sites (Fig. 22.8B). In our example, to assess the effects of the dispersal parameter, we compared selections done with different values for the parameter: a small dispersal range ($\alpha = 1.5$) and a large dispersal range ($\alpha = 0.4$).

The configuration of the final reserve network might not be so intuitive as shown here when the real costs of the sites vary greatly. The real value of this algorithm comes to play when the costs of the sites are considered. An expert knowing the system and the dynamics of the species might be able to choose a good set of sites for species persistence. However, when the resources

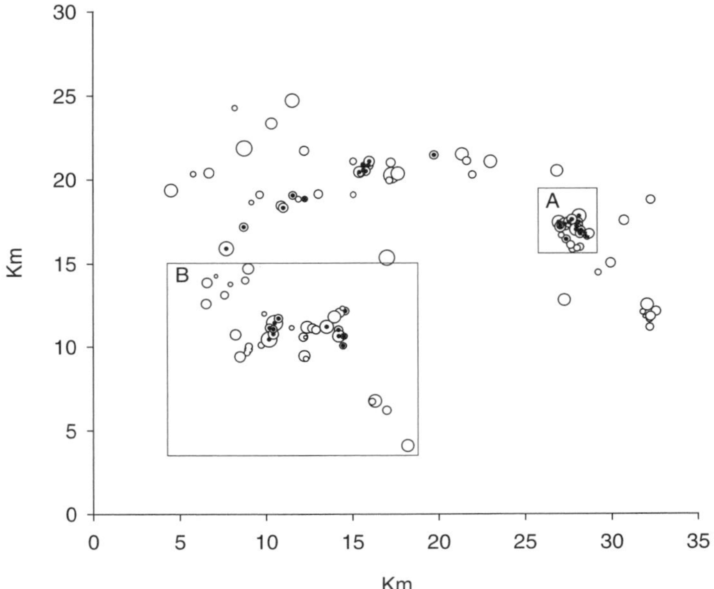

Fig. 22.7 The *M. diamina* patch system in southern Finland. Sizes of the circles are scaled according to the area of the patch. Black dots indicate patches occupied in the last survey year (1999). Rectangles A and B show areas presented in more detail in Figs 22.8 and 22.9, respectively.

are limiting and patch cost varies, it is very difficult to identify from a map, without computational aid, which sites would be the best ones given the amount of resources available for conservation. When some of the patches are considered to be proportionally more expensive than others (patches with commercial forest plantations were assumed to be 10 times more costly than

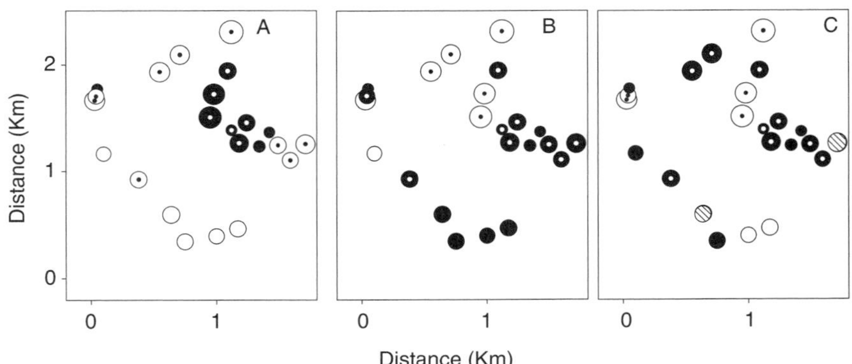

Fig. 22.8 Effects of dispersal ability and patch cost on the reserve network configuration when using the metapopulation approach. The area shown is a subregion of the complete patch system for *M. diamina* (see Fig. 22.7A). Sizes of the circles are scaled according to the area of the patch. Dark circles show the selected sites. (A) $\alpha = 1.5$, patch cost = patch area; (B) $\alpha = 0.4$, patch cost = patch area; and (C) $\alpha = 0.4$, patch cost (white circles) = patch area, patch cost (dashed circles) = 10x patch area. For the remaining IFM parameters, standard *M.diamina* parameter values were used (see Moilanen and Cabeza, 2002).

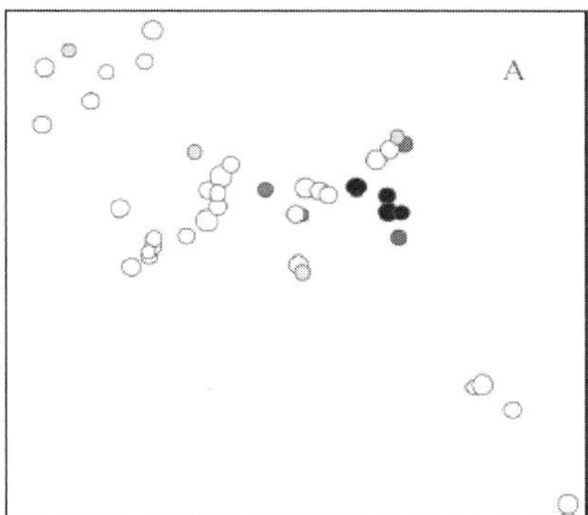

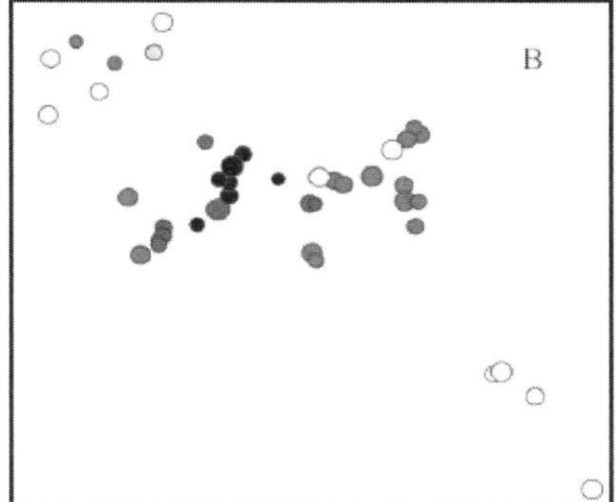

Fig. 22.9 Optimal reserve selection at different resource levels. The area shown is a subregion of the complete patch system for *M. diamina* (see Fig. 22.7B). Panels are based on 100 replicate optimizations with parameter values sampled from the joint four-parameter confidence limits for *M. diamina*. The color of the patch shows how often the patch was selected: White = never selected; Black = always selected. (A) Small resource (50,000). (B) Large resource (150,000).

those natural meadows in early successional stages; Moilanen and Cabeza, 2002), keeping all other factors equal, the optimal solution changes greatly (compare Figs. 22.8B and 22.8C).

Optimal solutions with a large amount of resources do not always build on solutions found with a smaller amount of resources. Figure 22.9 demonstrates the effect of increasing the amount of available resources. When the amount of resources is limited, it is optimal to select only a small cluster of sites (Fig. 22.9A). However, when the amount of resources is tripled, the solution consists of not only a larger amount of sites, but also a rather different set of sites (Fig. 22.9B). This result provides an important message for planners that often have thought that site selection algorithms only provide the core of the reserve network, which will be extended later on, by adding more sites to the core network when more resources are available. The optimal solution may strongly depend on the amount of resource that is available. An ordering in which patches should be conserved can only be given if the total amount of available resources is known.

22.6 DISCUSSION

In order to optimize *in situ* conservation of biodiversity, and given limited resources, major effort has been placed on the development of reserve network design problems and algorithms to solve those problems efficiently. Unfortunately, most of the existing problems have not been formulated in a way that is focused on persistence and hence solutions cannot guarantee the long-term persistence of biodiversity. Reserve selection problems have mostly been formulated so that the aim is to represent biodiversity, measured by a snapshot of species presence–absence information, in the most efficient way. More recent reserve selection problem formulations set reasonable targets for species viability (e.g., Noss et al., 2002) and allow for sensible spatial design (McDonnell et al., 2002; Leslie et al., 2003). However, the dynamics of populations and landscapes have been mostly overlooked, and much remains to be achieved in formulating and solving reserve network design problems in an ever-changing world.

This chapter addressed the effects of spatial (meta)population dynamics on the persistence of biodiversity in reserves designed using different site selection algorithms. These algorithms differed in the manner with which they consider spatial dynamics during the optimization process. We have shown how the selection of sites for the conservation of 26 butterfly species in the Creuddyn Peninsula may be very different depending on the reserve selection method. The methods used here range from the simplest "single representation heuristic" to more complex methods that explicitly consider spatiotemporal population dynamics during optimization.

For the example concerning the Creuddyn Peninsula, the single representation solution (i.e., a solution that includes the presence of the 26 butterfly species) requires only three sites. Studies of butterfly metapopulation dynamics support the view that we cannot expect all the 26 species to persist in three sites of 500 × 500 m if all the remaining suitable habitat would be lost. We can appreciate from subsequent results that the more realism included in the methods (e.g., larger number of representations per species, spatial considerations), the larger the amount of sites in the solution and the better the prospects for biodiversity persistence. Nonetheless, the different factors that need to be considered in a reserve selection procedure (including spatial population dynamics) depend on the spatial scale under consideration. Reserve selection algorithms have been applied at worldwide or continental scales. At these scales, and with sufficiently large selection units, a single representation for each species might be enough to ensure viability, especially if the aim is to demonstrate the most efficient way of concentrating conservation efforts. However, at smaller spatial scales and with smaller selection units, spatiotemporal dynamics should be considered when selecting reserve networks. Note that the scale where population dynamics need to be considered is species specific, and it depends mostly on the dispersal ability of the species — some bird species might show metapopulation dynamics at a continental scale, whereas metapopulation dynamics would be quite localized for snails.

The simplest site selection methods (problem categories C0 and C1, see Section 22.2) do not include any notion of the spatial configuration of the reserve, although populations may be chosen in a way that aims at local persistence (problem category C1). The simplest way of including spatial considerations to reserve selection is to use some computational technique to aggregate the reserve network (problem category C2), which implicitly improves biodiversity persistence by minimizing negative external effects. In the example of the Creuddyn Peninsula reserve, aggregation could actually be achieved with a very low cost in terms of increased area. In brief, we suggest that analysis of the cost of reserve aggregation should be done routinely as a part of the reserve selection process, and at least aggregation that can be obtained for free should, in most cases, be taken. Given that the maintenance cost of a compact reserve is likely to be smaller than that of a scattered reserve (Possingham et al., 2000), it is economically prudent to pay a little extra for a compact reserve. Nevertheless, from the perspective of species persistence, there might also be reasons to avoid reserve clustering. Where catastrophes can impact large areas causing extinctions, it may be less risky to conserve each species in at least two or three separate places rather than clustering those sites (Hess, 1996; Lei and Hanski, 1997). However, if reserves have to be selected far apart from each other, they should be large enough to allow species persistence independently.

Going one step further from qualitative reserve network clustering to the explicit consideration of spatial dynamics requires information on species-specific parameters of spatial population dynamics. In the example for the 26 butterflies, this was done by fitting for every species a statistical model for the probability of occurrence of the species. This model was made a function of habitat quality and connectivity (see also Westphal and Possingham, 2003). The inclusion of connectivity in the model enables us to consider the consequences of changes in habitat spatial pattern on species' occurrence probabilities. In practical terms, this means that habitat loss will have a negative effect in the probabilities of occurrence in the regions close to the site of habitat loss. This is a natural consequence of decreased immigration and increased edge effects. Estimates of population numbers in the selected reserve network differ quite significantly when effects of connectivity are excluded/included in optimization.

Another way of considering spatial dynamics explicitly is integrating stochastic metapopulation models into the reserve selection procedure. We have presented an approach for selecting the best reserve network that maximizes the persistence of a metapopulation for a given time frame. The extension of this approach for many species is challenging, but it is feasible technically (Moilanen and Cabeza, manuscript in preparation). One of the limitations of the approach is the availability of information to estimate all model parameters for all the species.

Reserve selection methods for problems C0 and C1 assume a best-case scenario in the sense that they implicitly assume that there will be no changes in the landscape outside the selected reserves. In contrast, methods for classes C2 and C3 assume a worst-case scenario in that all nonselected habitat is assumed to be lost, which of course will not always be the case. It is possible to improve the dynamic probability method by including information on threats and vulnerability of sites into the optimization model (Pressey et al., 1994; Pressey and Taffs, 2001, Cabeza and Moilanen, manuscript in preparation). At a general level, this means that a model of land-use change (Serneels and Lambin, 2001; Veldkamp and Lambin, 2001) would be integrated into the reserve selection algorithm and that the best-case/worst-case scenario would be relaxed and modeled more realistically.

The inclusion of landscape dynamics into reserve selection is in its infancy (Possingham et al., 1993; Costello and Polasky, 2003). Reserve networks are not generally constructed instantaneously (except perhaps in some marine areas). In many regions, sites can only be selected if they become available for acquisition. While sites are slowly being assembled into a network, some sites may be developed and lost to the system. To take this into account, we can formulate the problem as a dynamic programming problem and find optimal solutions using stochastic dynamic programming algorithms (Possingham et al., 1993; Costello and Polasky, 2003). These algorithms only work at present for small problems and we have yet to develop tools that can deliver adequate reserve networks when there are landscape dynamics as well as spatial population dynamics in large systems. Preliminary results suggest that some measure of irreplaceability (Pressey et al., 1994; Pressey and Taffs, 2001) may provide good solutions to the reserve network design problem when there is landscape change (E. Meir, personal communication).

In conclusion, the integration of spatial population dynamics, landscape modeling, and scheduling of conservation action into the reserve selection problem should lead to reserve network designs and acquisition strategies that are better at achieving the goal of long-term biodiversity persistence. Many challenges remain in the proper formulation of reserve network design problems and in the development of algorithms that deliver robust solutions in the face of several sources of uncertainty and change.

23. VIABILITY ANALYSIS FOR ENDANGERED METAPOPULATIONS: A DIFFUSION APPROXIMATION APPROACH

E.E. Holmes and B.X. Semmens

23.1 INTRODUCTION

Population viability analysis (PVA) assesses the rate of population decline and the risks of extinction or quasiextinction over a defined time horizon for a population of concern (Gilpin and Soule, 1986; Boyce, 1992; Morris and Doak, 2002). Although the techniques employed to conduct PVA are varied, they typically involve building quantitative models that are parameterized by demographic and environmental data. PVA was first used in the early 1980s (Shaffer, 1981), and in the past decade it has gained broad acceptance in the conservation community as a useful tool for assessing and managing "at-risk" species (Beissinger, 2002; Morris and Doak, 2002; Reed et al., 2002). This is particularly true for demogaphic PVAs, due in large part to the advancements in Monte Carlo techniques and desktop computers (Beissinger, 2002). The International Union for the Conservation of Nature (IUCN)'s Red List Criteria,

0-12-323448-4

probably the most widely applied set of decision rules for determining the status of at risk species, is partially defined by metrics that require some form of PVA (IUCN, 1994). For instance, under one of the Red List criteria, a taxon may be classified as endangered if a "reduction of at least 50%, projected or suspected to be met within the next ten years or three generations" is predicted.

Although many PVAs are focused on single populations in single sites, there are often needs for spatially explicit PVAs: many populations of conservation concern are distributed across multiple sites and additionally, the primary anthropogenic threats facing at-risk species are habitat destruction and alteration, which are fundamentally spatial processes (Wilcove et al., 1998). Several software packages have been written for spatially explicit PVA, including RAMAS Metapop (Akçakaya, 1997) and RAMAS GIS (Boyce, 1996), ALEX (Possingham and Davies, 1995), and VORTEX (Lacy, 1993). These models incorporate a diversity of demographic and spatial attributes such as distance-dependent migration, allee effects, social population structure, habitat quality and spatial arrangement, and genetic variability. The development of flexible sophisticated PVA software packages such as these has made the construction and simulation of spatially explicit PVA models feasible for those who are not highly skilled programmers and has greatly increased the number of managers and scientists capable of using spatially realistic PVA models.

As the use of PVA has grown in conservation science, so have concerns that PVAs are often overextended given limited data sets (Reed et al., 2002). Beissinger and Westpahl (1998) suggested that PVA should be limited to assessing relative risks over short time frames using the simplest models that can reasonably be justified. For single species with spatially simple structure, data needs can often be met when Beissinger and Westpahl's call for model moderation and simplicity are heeded. When one is faced with species with more complex spatial structure, a much larger amount of data is needed to parameterize the dynamics of individual local populations, the levels and patterns of dispersal, and the spatial pattern of temporal correlations among local populations (e.g., Ralls et al., 2002). Unfortunately, collection of data needed to parameterize a spatial model is rare for species of conservation concern, at least in the United States (Morris et al., 2002), and there is a disconnect between the parameter requirements for spatially explicit PVA models and the willingness and/or ability of management agencies to collect the types of data needed to appropriately apply such tools. Because it is usually impossible to retroactively fulfill data requirements for a spatial PVA and there will always be cases where collection of spatial data is infeasible, managers require PVA tools that can help guide conservation of metapopulations in the absence of large amounts of spatial data.

Diffusion Approximation for Metapopulations

One approach to the problem of limited population data is to find a diffusion approximation that correctly models the long-run statistical properties of a complex population process. This approach has been used successfully for single population models (Karlin and Taylor, 1981; Lande and Orzack, 1988; Lande, 1993; Dennis et al., 1991; Hill et al., 2002; see also Morris and Doak, 2002; Lande et al., 2003) and reduces the problem of parameterizing a large model with many parameters to the much simpler task of parameterizing a two-parameter diffusion model. One of the main practical implications of the diffusion approximation approach is that it is not necessary to know the multitude of parameters describing the local dynamics, dispersal levels, spatial patterns of dispersal, and spatial synchrony between local populations in order to make basic predictions about the statistical distribution of the long-term metapopulation or local population trajectories. The relevant two parameters for the diffusion approximation can be estimated from a simple time series of counts from the population process.

This chapter uses the diffusion approximation approach to model the long-run behavior of spatially structured populations. Our focus is on stochastic metapopulations characterized by structured population size, density-independent local dynamics, and, in keeping with the assumption of density independence, a metapopulation that is declining as a whole. Local populations are assumed to have patch-specific structured local dynamics and dispersal rates, with spatial structure among local populations in terms of both their local dynamics and dispersal patterns. Description of the long-run statistical behavior of the metapopulation trajectories using a diffusion approximation allows the estimation of PVA risk metrics such as the long-term rate of metapopulation decline and the probability of reaching different threshold declines over different time horizons (i.e., probabilities of extinction or quasiextinction). These methods for estimating metapopulation PVA metrics are illustrated using data from two chinook salmon metapopulations in the U.S. Pacific Northwest.

23.2 A STOCHASTIC METAPOPULATION MODEL

Our focus is on declining metapopulations, and thus what has been termed nonequilibrium metapopulations. We model a collection of local populations connected by dispersal where local populations have density-independent local dynamics, which may be "sources" or "sinks," but the metapopulation as a whole is declining. Dispersal levels could be very low, resulting in basically independent local populations, or extremely high, resulting in essentially one population. From a practical standpoint, this approach is most appropriate when dispersal is not insignificant (e.g., above 2% per year localized dispersal or 0.1% global dispersal), otherwise parameterization of the model requires inordinately long time series. Data from this type of metapopulation would be characterized by fluctuating local population trajectories, but actual extinctions would be unusual until the metapopulation has very few individuals. Our model assumes no density dependence nor carrying capacities within the individual local populations. Such a model is only appropriate in cases where the population is declining and all local populations are well below their carrying capacities. Our example using data on chinook salmon illustrates a situation that is likely to be well modeled as this type of metapopulation.

The following section gives a rather parameter-intensive mathematical description of a stochastic, declining metapopulation. However, the reader should keep in mind that this model will not be parameterized. Rather the asymptotic behavior of this model's trajectories will be derived and that information will be used to develop a diffusion approximation of the process. Time series data will then be used to parameterize the diffusion approximation.

The Model

Consider an individual local population i with stochastic yearly growth and stochastic dispersal to and from other local populations. Such a local population's numbers in year t, $N_i(t)$, could be described as follows:

$$\begin{aligned} N_i(t) &= \text{growth} - \text{dispersal out} + \text{dispersal in} \\ &= N_i(t-1)e^{z_i(t-1)} - d_i(t-1)N_i(t-1)e^{z_i(t-1)} \\ &\quad + \sum_{j \neq i} \alpha_{ji}(t-1)d_j(t-1)\,N_j(t-1)e^{z_j(t-1)} \end{aligned} \tag{23.1}$$

where $z_i(t)$ is the stochastic growth rate of local population i in year t and is a random variable with some unspecified statistical distribution with mean μ_i and variance σ_i^2. The μ_i term will be referred to as the local population's intrinsic growth rate; it will not be observed, as the local population is subject to immigration and emigration. Some fraction of individuals, $d_i(t)$, leaves local population i at year t and disperses to other local populations, and dispersal into local population i occurs from other local populations. The fraction of dispersers from local population j that go to local population i in year t is $\alpha_{ji}(t)$ and can vary depending on the destination, i, thus allowing for spatially structured dispersal. The dispersal parameters, $d_i(t)$ and $\alpha_{ji}(t)$, are assumed to be temporally random variables from some unspecified statistical distribution.

The Model in Matrix Form

The model for the entire metapopulation can be written using a random transition matrix, $\mathbf{A}(t)$, which encapsulates both dispersal and local growth:

$$\begin{bmatrix} N_1(t+1) \\ N_2(t+1) \\ N_3(t+1) \\ \cdots \\ N_k(t+1) \end{bmatrix} = \boldsymbol{A}(t) \times \begin{bmatrix} N_1(t) \\ N_2(t) \\ N_3(t) \\ \cdots \\ N_k(t) \end{bmatrix} \tag{23.2}$$

where

$$A(t) = \begin{bmatrix} (1-d_1)e^{z_1} & \alpha_{21}d_2e^{z_2} & \alpha_{31}d_3e^{z_3} & \cdots & \alpha_{k1}d_ke^{z_k} \\ \alpha_{12}d_1e^{z_1} & (1-d_2)e^{z_2} & \alpha_{32}d_3e^{z_3} & \cdots & \alpha_{k2}d_ke^{z_k} \\ \alpha_{13}d_1e^{z_1} & \alpha_{23}d_2e^{z_2} & (1-d_3)e^{z_3} & \cdots & \alpha_{k3}d_ke^{z_k} \\ \cdots & \cdots & \cdots & \cdots & \cdots \\ \alpha_{1k}d_1e^{z_1} & \alpha_{2k}d_2e^{z_2} & \alpha_{3k}d_3e^{z_3} & \cdots & (1-d_k)e^{z_k} \end{bmatrix} \tag{23.3}$$

The '(t)' on the d's, α's, and z's have been left off to remove clutter. There may be any level or spatial pattern of temporal correlation among the intrinsic local growth rates, z_i's, dispersal rates, d_i's, and dispersal patterns, α_{ji}'s.

In the matrix model, each row represents 1 unit of habitat. Local populations with multiple units of habitat appear as multiple rows with very high dispersal between the units of habitat in that local population. The habitat units within a local population could vary in quality (i.e., habitat within a local population need not be uniform) and different local populations certainly differ in the number of habitat units they contain. The d_i's and α_{ji}'s are assumed to be drawn from some distribution that can be different for each local population or local population pair. Although the d_i's, α_{ji} and z_i's are temporally random variables, they are assumed to be stationary, i.e., that there is no overall change in the mean values over time. For the purposes of this chapter, it will be assumed that the d_i's, α_{ji}'s, and z_i's are all strictly postitive, which means that all local populations are connected to each other to some (although possibly very low) degree and that mean yearly geometric growth rates, $\exp(\mu_i)$'s, while possibly very small are not zero. These assumptions imply that the $\mathbf{A}(t)$ describe an ergodic set of matrices (Caswell, 2001). The assumption of strict positivity is not strictly necessary. It is possible for $\mathbf{A}(t)$ to describe an ergodic set if some elements of $\mathbf{A}$ are zero; it depends on the pattern of zeros within $\mathbf{A}$ [cf. Caswell (2001) for a discussion of the conditions under which matrices are ergodic].

The model is very general, allowing some sites to be dispersal sources and others to be dispersal targets, allowing any spatial pattern of dispersal or spatially correlated local growth rates, allowing any pattern of temporal correlation amongst local growth rates, and allowing any combination or pattern of habitat sizes of local sites.

Using Random Theory to Understand the Model's Statistical Behavior

Together, Eqs. (2) and (3) describe a quite generic model of a declining metapopulation with density-independent local dynamics. From a viability analysis perspective, one might ask the question: "Can one predict the viability of the total metapopulation?" In more precise terms, this is asking what are the statistical properties of the metapopulation trajectories of this type of connected collection of local populations [of the form in Eqs. (2) and (3)]? Clearly, the matrix $\mathbf{A}(t)$ has a large number of parameters that would be difficult, if not impossible, to estimate for any given metapopulation of conservation concern. However, using random theory, it can be shown that the long-term dynamics can be described by only two parameters and that it is unnecessary to know the multitude of other parameters for the purpose of projecting long-run dynamics.

To use this random theory, we first need to recognize that this stochastic metapopulation model falls into the class of random processes that involve products of ergodic random matrices, in this case products of $\mathbf{A}(t)$, which can be seen by using Eq. (2) to project the vector of local population sizes forward:

$$\begin{aligned}
\mathbf{N}(1) &= \mathbf{A}(0)\mathbf{N}(0) \\
\mathbf{N}(2) &= \mathbf{A}(0)\mathbf{A}(1)\mathbf{N}(0) \qquad (23.4) \\
&\dots \\
\mathbf{N}(t) &= \mathbf{A}(0)\mathbf{A}(1)\mathbf{A}(2)\dots\mathbf{A}(t-1)\mathbf{N}(0)
\end{aligned}$$

where $\mathbf{N}(t)$ is the column vector of N_i values at time t in Eq. (2). Products of random ergodic matrices have a well-established theoretical foundation and have certain well-studied asymptotic statistical properties. A brief review of two of the key results from this theory is provided in Box 23.1 and a simulated

BOX 23.1 Key Results from Random Theory

Two of the fundamental results from the theory of products of random matrices are reviewed and interpreted in the context of our metapopulation model. The reader is referred to chapter 14.3 in Caswell (2001) and Tuljapurkar (1990) for other reviews interpreted in the context of demographic, single population models.

The Metapopulation and Local Populations Decline at the Same Rate

One of the basic results from Furstenberg and Kesten's "Products of Random Matrices" (1960) is that the product of ergodic random matrices asymptotically goes to an equilibrium. Say that $\mathbf{X}_t$ is an ergodic random "$k \times k$" matrix and that $\mathbf{Y}$ (also a $k \times k$ matrix) denotes the product of n of the $\mathbf{X}$ matrices: $\mathbf{X}_1, \mathbf{X}_2, \mathbf{X}_3, \ldots \mathbf{X}_n$. Then Furstenberg and Kersten's results say that $\mathbf{Y}$ goes to an equilibrium state such that

$$\lim_{t\to\infty} \frac{1}{t} \log \sum_{i \in a} \sum_{j} \mathbf{Y}_{ij} = \text{a constant which is the same for all } a \tag{B1}$$

We can use this result to show that the long-run exponential growth rate of the metapopulation and the local populations will be the same.

$$\mathbf{N}(t) = \mathbf{A}(0)\,\mathbf{A}(1)\,\mathbf{A}(2)\ldots\mathbf{A}(t-1)\mathbf{N}(0) \quad \text{our metapopulation model}$$

$$\text{Let } \mathbf{Y} = \mathbf{A}(0)\,\mathbf{A}(1)\,\mathbf{A}(2)\ldots\mathbf{A}(t-1)$$

$$\text{Then } \log N_i(t) = \log \sum_j \mathbf{Y}_{ij} + \log N_i(0)$$

$$\text{and } \log M(t) = \log \sum_i \sum_j \mathbf{Y}_{ij} + \log M(0)$$

Thus from Eq. (B1),

$$\lim_{t\to\infty} \frac{1}{t} \log \sum_j \mathbf{Y}_{ij} = \lim_{t\to\infty} \frac{1}{t} \log \sum_i \sum_j \mathbf{Y}_{ij} = \text{a constant} = \mu_m$$

The Distribution of Local Population and Metapopulation Sizes is Distributed Lognormally

One of the most powerful results, for our purposes at least, concerns the statistical distribution of the metapopulation and local trajectories. This tells us what distribution of sizes we would see if we ran our model over and over again and allows us to make population viability analyses for metapopulations since we have a prediction about the likelihood of different metapopulation futures. Random theory (Furstenberg and Kersten, 1960; Tuljapurkar and Orzack, 1980) shows that any sum of the $\mathbf{N}_i(t)$'s, such as the total metapopulation (all i's), a single local population (one i), or any other subset, goes to the same distribution:

$$\log \frac{\mathbf{c}^T \mathbf{N}(t)}{\mathbf{c}^T \mathbf{N}(0)} \xrightarrow[t\to\infty]{} \text{Normal}(t\mu_m,\, t\sigma_m^2) \tag{B2}$$

where the sum of local populations is denoted in matrix terms as $\mathbf{c}^{0}\mathbf{N}(t)$ and $\mathbf{c}$ is a column vector with 0's and 1's to show which local populations to sum together.

Example

These results are simple to see with simulations. An example of a linear chain of 10 local populations connected via 2% yearly dispersal to their nearest neighbors and 0.2% to nonnearest neighbors is shown. The local dynamics were e^{z_i} where z_i is a normally distributed random variable, Normal(μ_i, σ_i^2). The local growth rates, μ_i's, for local populations 1 to 10 were, respectively, 0.97, 1.00, 0.96, 0.83, 0.88, 1.00, 1.00, 0.89, 0.99, and 0.81. Figure 23.1A shows that the long-run growth rate of the local population and metapopulations is equal to the same constant. Figure 23.1B shows that the distribution of metapopulation size after 100 yr is Normal($100\mu_m$, $100\sigma_m^2$). The expected distribution was specified using the maximum likelihood (ML) estimates for μ_m and σ_m^2 [Eq. (9)] from a single 1000-yr time series of metapopulation counts. The ML estimate for σ_m^2 relies on an assumption of normality for $t = 1$, although strictly speaking normality only holds for t large. However, it does quite well as can be seen in Fig. 23.1B.

example is shown to illustrate these results. As described in Box 23.1, the theory demonstrates that this stochastic, density-independent metapopulation will have an asymptotic growth rate and that the metapopulation, $M(t) = \Sigma N_i(\mathrm{t})$, the individual $N_i(t)$'s, and sets of $N_i(t)$'s representing the units of habitat comprising a semi-independent local population will be distributed lognormally with the same parameters:

$$\begin{aligned}
&\log M(t)/M(0) \xrightarrow[t\to\infty]{} \mathrm{Normal}(t\mu_m, t\sigma_m^2)\\
&\log N_i(t)/N_i(0) \xrightarrow[t\to\infty]{} \mathrm{Normal}(t\mu_m, t\sigma_m^2) \qquad (23.5)\\
&\log \sum_{i\in a} N_i(t)/\sum_{i\in a} N_i(0) \xrightarrow[t\to\infty]{} \mathrm{Normal}(t\mu_m, t\sigma_m^2)\\
&\text{where } a = \{a_1, a_2, \ldots, a_m\}
\end{aligned}$$

Figure 23.1 shows an example of this behavior. A metapopulation is simulated (described in Box 23.1) and, over time, the metapopulation declines at a constant rate and all $N_i(t)$'s have the same long-term fate. When viewed over short time frames, t small in Fig. 23.1, the local sites show different growth rates with some declining more or less than the long-term rate, but over the long-term their rate of decline is the same.

The model studied here approximates the local dynamics by a simple exponential growth (or decline) model. However, it has been shown that results from random theory (presented in Box 23.1) also apply to a more complicated metapopulation model where local dynamics are described by stochastic age-structured Leslie matrices (Sanz and Bravo de la Parra, 1998). Essentially, this occurs because even when the local dynamics are described by a local matrix model, the system can still be described by products of random matrices.

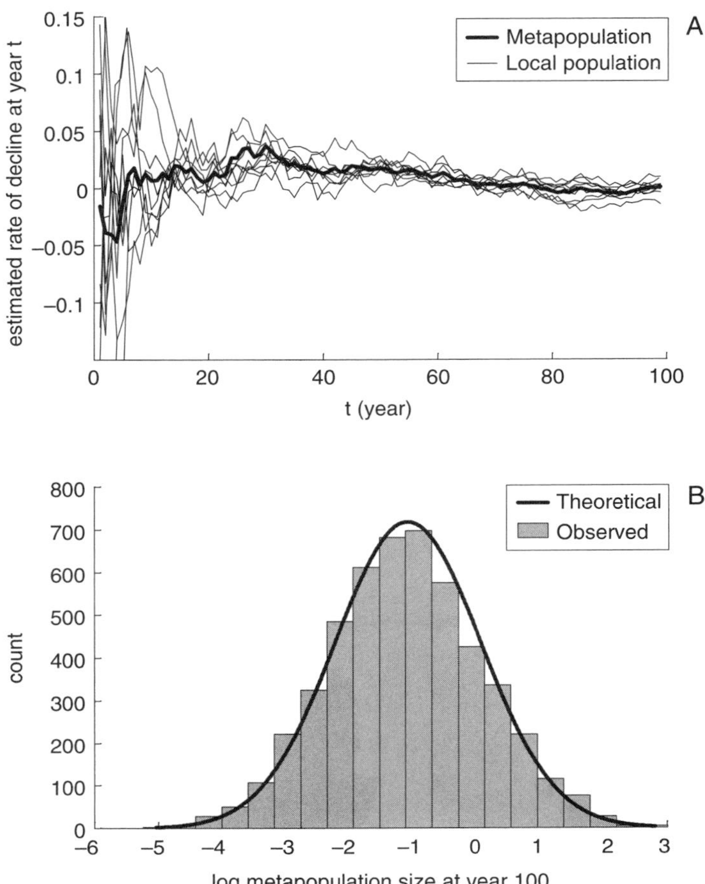

Fig. 23.1 Illustration of two of the main results from random theory. (A) All local populations go toward the same long-term rate of population growth (or decline) as t gets large. (B) The distribution of log $M(t)$ is a normal distribution with mean given by the long-term rate of growth (or decline) multiplied by t and the variance given by t multiplied by the rate that variance increases in an individual trajectory, i.e., $t \times (1/\tau)\log M(t + \tau)/M(t)$ for τ not overly small. Here the variance was estimated from one time series using $\tau = 10$ and this is used to predict the distribution at $t = 100$.

23.3 DIFFUSION APPROXIMATION

The asymptotic distribution of log $M(t)$ in Eq. (5) has the same properties as the distribution of a diffusion process with drift; it is normal and the mean and variance of the distribution of log $M(t)$ increase linearly with time, t. This observation in the context of age-structured matrix population models (Lande and Orzack, 1988; Dennis et al., 1991) led to the use of a diffusion approximation to enable parameterization using simple time series and to enable calculation of extinction probabilities. Diffusion approximation methods for single population populations are an important and established method for approximating stochastic trajectories (Lande and Orzack, 1988; Dennis et al.,

1991; Chapter 3 in Morris and Doak, 2002; Chapter 2 in Lande et al., 2003). Models for single populations are mathematically analogous to the models used here for metapopulations with a stochastic process involving products of random matrices. However, in single population models, the matrix represents a life history matrix rather than a growth and dispersal matrix, and the **N**(t) vector [in Eq. (2)] represents different age or stage classes, whereas in the metapopulation matrix, it represents different local sites and populations.

A diffusion approximation with drift is a stochastic process with the following properties (cf. Karlin and Taylor 1981):

$$\left.\begin{array}{l} X(t) - X(0) = \mu_m t + \varepsilon \\ \varepsilon \sim \text{normal}(0, \sigma_m^2 t) \end{array}\right\} \text{for } t = 1,2,3,\ldots \tag{23.6}$$

For any nonoverlapping pair of time periods, $t_1 < t_2$ and $t_3 < t_4$, $X(t_2) - X(t_1)$, and $X(t_4) - X(t_3)$ are independent random variables. $X(t + \tau)$ is a random variable with distribution Normal $(X(t) + \mu_m\tau,\ \sigma_m^2\tau)$. Correspondingly, the probability density function for $X(t + \tau)$ given log $X(t)$ is

$$p(X(t + \tau) \mid X(t)) = \frac{1}{\sqrt{2\pi\sigma_m^2\tau}} \exp\left[\frac{-(X(t + \tau) - X(t) - \mu_m\tau)^2}{2\sigma_m^2\tau}\right] \tag{23.7}$$

Behavior of Metapopulation Trajectories Versus Diffusion Trajectories

Diffusion approximation is based on the behavior of log $M(t)$ as t goes to infinity; however, in PVA settings the time frame of interest is substantially less than infinity and is typically in the range of 25 to 100 yr. How well does the diffusion approximation do over these finite time periods? To explore this, a collection of 50 local populations were simulated that were connected by global dispersal ranging from 0.1 to 5% per year and that had correlated local dynamics, $z_i(t)$, drawn from a Normal(mean = −0.05, variance = 0.09) and a temporal covariance of 0.2 * 0.09 between the $z_i(t)$'s of local populations in any given year.

If the log metapopulation trajectories behave like a diffusion process, and if we repeatedly generate a large sample of replicate metapopulation trajectories, the mean and variance of $(1/t)\log M(t)/M(0)$ from those trajectories should be a constants over the time period of interest. Additionally, $(1/t)\log M(t)/M(0)$ should be normally distributed. To examine whether the metapopulation trajectories had these properties, the simulations were started from a distribution of local population sizes selected from the equilibrium set of local population distributions and then run forward for 200 yr. This was repeated (using the same initial distribution of local populations) 1000 times to estimate the distribution of $(1/t)\log M(t)/M(0)$. This process was repeated for four randomly chosen initial distributions of local population sizes. The mean and variance of $(1/t)\log M(t)/M(0)$ are denoted as $\mu_m(t)$ and $\sigma_m^2(t)$, respectively, in Fig. 23.2 and in the discussion given later.

Figure 23.2 illustrates the results. For dispersal levels 1% or higher, the trajectories behaved like a diffusion process with $\mu_m(t)$ and $\sigma_m^2(t)$ roughly constant and the distributions approximately normal according to a Kolmogorov–Smirnov test at $P = 0.05$. For low dispersal, 0.1%, the trajectories did not behave like a diffusion process for t less than 200 at least. The variance $\sigma_m^2(t)$ was not constant, except for $t > 150$, and the normality assumption was generally violated except again at large t. This means that when dispersal is very low, diffusion approximations for this metapopulation would be more approximate than for metapopulations with higher dispersal.

Figure 23.2 illustrates results from one particular model. Repeating this process for a number of different models indicated some general behaviors. The higher the dispersal levels, the more trajectories behaved like a diffusion process. Global dispersal levels of at least 2 to 5% were generally high enough to result in diffusion-like behavior within a short time frame. Note that localized dispersal has the effect of lowering the effective dispersal rates. The higher the amount of temporal covariance between local populations in terms of their yearly growth rates, the more the trajectories behaved like a diffusion process. The simulations were done with the local population sizes within the equilibrium set of local population distributions — indeed the theory is predicated on the local populations being near equilibrium. For metapopulations with 2 to 5% dispersal, the local populations equilibrated fairly quickly starting from all local populations with equal numbers. However, at very low dispersal, equilibration took thousands of time steps. This suggests that the assumption of equilibrium should be viewed cautiously for metapopulations that have very low dispersal rates between local populations.

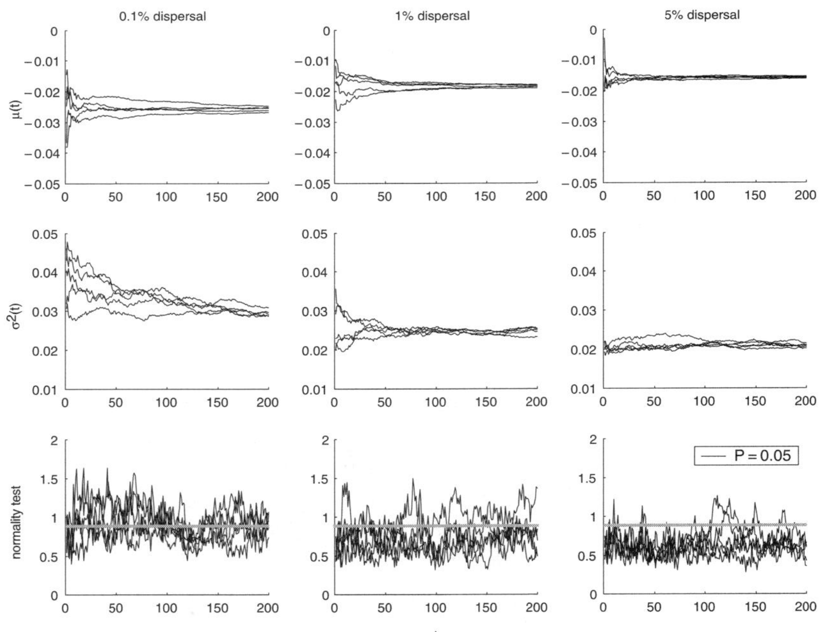

Fig. 23.2 Illustration of the performance of a diffusion approximation for modeling the behavior of a metapopulation with 50 local populations and uniform 0.1, 1, or 5% yearly dispersal. The diffusion approximation performs well for a given time frame when $\mu_m(t) = (1/t)\log M(t)/M(0)$ and $\sigma_m^2(t) = (1/t)$ var $[\log M(t)/M(0)]$ are constants over that time frame and when $\log M(t)/M(0)$ is normal.

23.4 ESTIMATING THE PARAMETERS

Maximum likelihood estimates of μ_m and σ_m^2 can be calculated using the diffusion approximation for log $M(t)$. Denote the observed time series as $M = M(0)$, $M(1)$, $M(2)$, . . . , $M(n)$. If we approximate log $M(t)$ as a diffusion process, the likelihood function $L(\mu_m, \sigma_m^2|M)$ is given by the product of the probability function distributions for the transitions from log $M(t)$ to log $M(t + 1)$, which is Eq. (7) with $\tau = 1$, over $t = 0, 1, 2, \ldots, n$. Thus the log likelihood function is

$$\log L(\mu_m, \sigma_m^2 \mid M) = -(n/2)\log(2\pi\sigma_m^2) - \frac{1}{2\sigma_m^2}\sum_{i=1}^{n}[\log(M(i)/M(i-1)) - \mu_m]^2 \tag{23.8}$$

Maximum likelihood estimates are obtained by solving for μ_m and σ_m^2, which maximize Eqn. (8),

$$\hat{\mu}_m = \frac{1}{n}\log\left(\frac{M(n)}{M(0)}\right)$$
$$\hat{\sigma}_m^2 = \frac{1}{n}\sum_{i=1}^{n}\left[\log\left(\frac{M(i)}{M(i-1)}\right) - \hat{\mu}_m\right]^2 \tag{23.9}$$

Note that the unbiased estimator for σ_m^2 divide by $(n - 1)$ rather than n. The $\hat{\mu}_m$ and $\hat{\sigma}_m^2$ are analogous to the estimates of mean and variance from n samples from a normal distribution, and confidence intervals on μ_m and σ_m^2 are analogous:

$$\left(\hat{\mu}_m - t_{\alpha/2,n-1}\sqrt{\hat{\sigma}_m^2/n},\ \hat{\mu}_m + t_{\alpha/2,n-1}\sqrt{\hat{\sigma}_m^2/n}\right)$$
$$(n\hat{\sigma}_m^2/\chi^2_{\alpha,n-1},\ n\hat{\sigma}_m^2/\chi^2_{1-\alpha,n-1}) \tag{23.10}$$

where t_{α}, q is the critical value of a t distribution at $P = \alpha$ and q degrees of freedom and $\chi^2_{\alpha,q}$ is the critical value of a χ^2 distribution at $P = \alpha$ and q degrees of freedom. See Dennis et al. (1991) for a more in-depth discussion of maximum likelihood estimates for diffusion processes. Following Dennis et al.'s monograph, parameter estimation based on the diffusion approximation has been widely used for the analysis of single population trajectories. For a discussion of parameter estimation that is not based on the diffusion approximation, the reader is referred to Heyde and Cohen (1985).

Maximum likelihood estimates assume that the metapopulation has reached a stochastic equilibrium and thus that the diffusion approximation is reasonable. When exploring these methods using simulations, it is important to allow the system to equilibrate, after starting the simulation with something peculiar like all local populations at the same size. Equilibruim can be monitored by waiting for the variance of $(\log(N_i(t)) - \log[\text{mean}(N_i(t))])$ to stabilize. In simulations done for this chapter, the distribution stabilized relatively quickly when dispersal was nonzero. If dispersal is zero, however, the distribution never stabilizes and the variance of $(\log(N_i(t)) - \log[\text{mean}(N_i(t))])$ increases continually. For an actual metapopulation, for which one wants to conduct a PVA, it is also critical to test the appropriateness of the diffusion approximation for one's time series data. Dennis et al. (1991) and Morris and Doak (2002) reviewed how to do this, which is based on diagnostic procedures for evaluating the appropriateness of linear models.

Parameter Bias

The estimators are unbiased maximum likelihood estimators for the diffusion approximation, $X(t)$. It is important to understand whether and how these estimates are biased when working with short time series of metapopulation trajectories, $M(t)$, as opposed to an actual diffusion process. In particular, $\hat{\sigma}_m^2$ is certain to be biased to some degree, as it relies on the diffusion approximation holding for $\tau = 1$ in log $M(t + \tau)/M(t)$, regardless of the length of the time series used for estimation. This is not the case for $\hat{\mu}_m$, which is also an unbiased predictor for $M(t)$ given a long time series (Heyde and Cohen, 1985).

To numerically explore parameter bias from short time series, simulations were used to look at the difference between $\hat{\mu}_m$ and $\hat{\sigma}_m^2$ from a 20-yr time series versus their true values μ_m and σ_m^2. An example metapopulation of 50 local sites was simulated with global dispersal and correlated local growth rates, $z_i(t)$, drawn yearly from a normal distribution with mean = μ_i, variance = ϕ, and covariance of $0.2^*\phi$ between any two local growth rates Two versions of the simulation were run: one to model uniform site quality (spatially uniform $\mu_i = -0.05$) and one to model highly variable site quality (spatially variable μ_i's). To explore biases over a range of different dispersal and variability levels, models were run with dispersal between 0.1 and 5% per year and local variability, ϕ, between 0.1 and 0.5. These parameters translated to metapopulation level rates, μ_m, in the range of 0.01 to -0.05 and metapopulation level variability, σ_m^2, in the range of 0.001 to 0.08. For each dispersal and local variability pair, 1000 replicate metapopulation trajectories were simulated, each with an initial distribution of local population sizes selected randomly from the equilibrium set.

The mean difference between $\hat{\mu}_m$ and μ_m over the dispersal and local variability parameter space was very low, <0.0015, for both uniform and variable μ_i simulations. Overall the lack of bias in $\hat{\mu}_m$ supports metrics that rely primarily on this parameter, such as the metapopulation λ (next section). For most of the parameter space explored, $0 < |\hat{\sigma}_m^2 - \sigma_m^2| < 0.01$, representing a 0 to 20% under- or overestimation of σ_m^2. Larger biases, $|\hat{\sigma}_m^2 - \sigma_m^2| > 0.01$, representing a >20% under- or overestimation, were seen for some parameter combinations. The impact of this bias depended on where $\hat{\sigma}_m^2$ was used. For instance, the effect on estimated confidence intervals on μ_m [Eq. (10)] was minimal with the width of the interval changing by a median 0.002. The effect on estimated passage probabilities was higher, although not dramatic. For example, the estimated probability that the metapopulation will be 10% of current levels at the end of 50 yr was decreased by 0 to 0.04 (on a scale from 0 to 1) for the uniform μ_i simulation and increased by 0 to 0.04 for the variable μ_i simulation. The estimated probability that the metapopulation will pass below 10% of current levels at any point during the next 50 yr was changed by 0 to 0.09. Overall, the effect of $\hat{\sigma}_m^2$ bias was low in these simulations, but this will depend on the particular metapopulation and will need to be investigated for individual cases of interest.

In practical applications, one must contend with other factors that can lead to parameter bias, but which are outside the scope of this chapter. In particular, observation error, nonequilibrium local population distributions, and temporal autocorrelation can lead to parameter bias. Such problems are being studied in the context single population PVA. Much of this work is likely to be relevant for metapopulation PVA. See Morris and Doak (2003) and Holmes (2004) for a review and discussion of current work in this area.

23.5 METAPOPULATION VIABILITY METRICS

One of the most basic viability metrics is the long-term geometric rate of decline (or growth) of a population, termed generally λ in the PVA literature. If λ is less than 1.0, the population ultimately declines to extinction and $100^*(1 - \lambda)$ is roughly the average yearly percent decline. The metapopulation λ is $\exp(\mu_m)$ and its estimate is then

$$\hat{\lambda} = \exp(\hat{\mu}_m). \tag{23.11}$$

This definition of λ follows Caswell's use of the symbol λ_s as the long-term average stochastic growth rate: $\lambda_s = [N(t)/N(0)]^{1/t}$ as $t \to \infty$ (Caswell, 2001). This is the long-run geometric growth rate that would be observed in almost every trajectory. Defined this way, if $\lambda < 1$, the population goes extinct with certainty, eventually. This differs from Dennis et al.'s use of the symbol λ where λ is used for $\exp(\mu + \sigma^2/2)$ and the long-term average geometric growth rate is instead denoted by $\alpha = \exp(\mu)$. The maximum likelihood estimate of λ is a biased estimator; because $\hat{\mu}_m$ is normally distributed, the median value of $\exp(\hat{\mu}_m)$ is $\exp(\mu_m)$ but the mean value is not. Dennis et al. (1991) gave an unbiased estimator [mean($\hat{\lambda}$) = λ] based on Shimizu and Iwase (1981), although Dennis and colleagues found negligible differences between biased and unbiased estimators in their examples.

From the asymptotic distribution of log $M(t)$, Eq. (5), the probability that the metapopulation is below a threshold b at the end of y years can be calculated as

$$P[M(t) \leq b \mid M(0)] = \Phi\left(\frac{\log(b/M(0)) - \mu_m t}{\sqrt{\sigma_m^2 t}}\right) \tag{23.12}$$

Although this uses the asymptotic distribution, this is mitigated by the fact that it is used for the distribution at the end of y years but not at any time before that. The estimate of $P[M(t) \leq b \mid M(0)]$ replaces μ_m and σ_m^2 by their estimates $\hat{\mu}_m$ and $\hat{\sigma}_m^2$. Like the estimate of λ, the median estimate of $P[M(t) \leq b \mid M(0)]$ is equal to the true value, but not the mean.

Some metapopulations can have a low long-term risk of being below a threshold due to a λ near 1.0, but high short-term risks of hitting that threshold due to high variability. Such quasiextinction or extinction probabilities are commonly used and very important PVA metrics. The diffusion approximation for log $M(t)$ can be used to estimate these probabilities for the metapopulation. The probability of that the diffusion process, $X(t)$, experiences a decline below a threshold log b at some time T less than y years is calculated by integrating over the probability density function for first passage times for a diffusion process with drift (Karlin and Taylor, 1981). Lande and Orzack (1988) go through the calculation, which leads to

$$P(T \leq y) = \Phi\left(\frac{-(X(0) - \log b) - \mu_m y}{\sqrt{\sigma_m^2 y}}\right)$$
$$+ \exp(-2(X(0) - \log b)\mu_m/\sigma_m^2) \times \Phi\left(\frac{-(X(0) - \log b) + \mu_m y}{\sqrt{\sigma_m^2 y}}\right) \quad (23.13)$$

$\Phi()$ is the cumulative distribution function for a standard Normal(mean = 0, variance = 1). The estimate of $P(T \leq y)$ for the metapopulation uses $\hat{\mu}_m$ and $\hat{\sigma}_m^2$ with $X(0) = \log M(0)$. The estimated probability of extinction (to 1 individual) is calculated using Eq. (13) and setting b equal to 1. The reader is cautioned that estimates of extinction are problematic and that estimates of quasiextinction (e.g., some threshold greater than 1 individual) are more robust (cf. Morris and Doak, 2002). Also, Eq. (13) uses diffusion approximation over short time scales, as it calculates the probability of hitting a threshold at any time, including short times, before y years. This makes Eq. (13) more approximate than other metrics.

Other viability metrics based on diffusion approximation, such as the mean time to extinction, and median time to extinction, are discussed in Lande and Orzack (1988) and Dennis et al. (1991).

Risk Metric Uncertainty

The $100(1 - \alpha)\%$ confidence intervals are often used as characterizations of uncertainty. These can be calculated for risk metrics using the estimated distributions of $\hat{\mu}_m$ and $\hat{\sigma}_m^2$. The confidence intervals for $\hat{\lambda}$ are

$$(\exp(\hat{\mu}_m - t_{\alpha/2,n-1}\sqrt{\hat{\sigma}_m^2/n}), \exp(\hat{\mu}_m + t_{\alpha/2,n-1}\sqrt{\hat{\sigma}_m^2/n})). \quad (23.14)$$

where t_{α},q is the critical value of a t distribution at $P = \alpha$ and q degrees of freedom. The corresponding significance level, α, for a hypothesis test, such as "Is $\lambda < b$" is the α such that

$$\frac{\hat{\mu}_m - \log b}{\sqrt{\hat{\sigma}_m^2/n}} = t_{\alpha,n-1}. \quad (23.15)$$

Confidence intervals on $P(T \leq y)$ and $P[M(y) \leq b \mid M(0)]$ can be calculated by parametric bootstrapping from the estimated distributions of $\hat{\mu}_m$ and $\hat{\sigma}_m^2$: Normal($\hat{\mu}_m$, $\hat{\sigma}_m^2/n$) and Gamma(shape = $(n - 1)/2$, scale = $2\hat{\sigma}_m^2/(n - 1)$). A large number of ($\hat{\mu}_b$, $\hat{\sigma}_b^2$) pairs are generated randomly by sampling from these distributions and the risk metric Ψ is calculated [Eqs. (13) or (12)] for each pair. The range of Ψ over the ($\hat{\mu}_b$, $\hat{\sigma}_b^2$) bootstrapped pairs, for which both parameters are within their respective $100(1-\alpha)\%$ confidence intervals, defines the $100(1 - \alpha)\%$ confidence interval for Ψ. This and other methods for calculating confidence intervals for diffusion approximation risk metrics are discussed in Dennis et al. (1991).

An alternate way to present the level of uncertainty is to estimate the data support for different values of a risk metric. There are both frequentist and Bayesian approaches for this [see Wade (2001) for a review geared toward conservation applications]. Holmes (2004) presented a Bayesian approach, which uses posterior probability distributions to illustrate data support. That method is adapted here for estimating the level of data support for the metapopulation risk metrics. Let Ψ be a risk metric. The probability that Ψ is greater than some threshold φ given the data is

$$P(\Psi > \varphi \mid \hat{\mu}_m, \hat{\sigma}_m^2) = \int_{\substack{\text{all } (\mu_m, \sigma_m^2) \\ \text{for which } \Psi > \varphi}} \frac{L(\mu, \sigma_m^2 | \hat{\mu}_m) L(\sigma_m^2 | \hat{\sigma}_m^2) \pi(\mu_m) \pi(\sigma_m^2)}{\eta(\hat{\mu}_m, \hat{\sigma}_m^2)} \tag{23.16}$$

where $L(\mu, \sigma_m^2 \mid \hat{\mu}_m)$ is the likelihood function given $\hat{\mu}_m \sim \text{Normal}(\mu_m, \sigma_m^2/n)$, $L(\sigma_m^2 \mid \hat{\sigma}_m^2)$ is the likelihood function given $\hat{\sigma}_m^2 \sim \text{Gamma}((n-1)/2, 2\sigma_m^2/(n-1))$, $\pi(\mu_m)$ and $\pi(\sigma_m^2)$ are the priors on μ_m and σ_m^2, and the normalizing constant is

$$\eta(\hat{\mu}_m, \sigma_m^2) = \int_{\mu_m=-\infty}^{\mu_m=+\infty} \int_{\sigma_m^2=0}^{\sigma_m^2=\infty} L(\mu_m, \sigma_m^2 | \hat{\mu}_m) L(\sigma_m^2 | \hat{\sigma}_m^2) \pi(\mu_m) \pi(\sigma_m^2) d\mu_m d\sigma_m^2 \tag{23.17}$$

The posterior distribution of Ψ is $[P(\Psi < \varphi + d\varphi \mid \hat{\mu}_m, \hat{\sigma}_m^2) - P(\Psi < \varphi \mid \hat{\mu}_m, \hat{\sigma}_m^2)]/d\varphi$ over all φ. Examples of this calculation for λ and the probability of being below thresholds at the end of 25 yr are shown in the salmon examples. Holmes (2004) supplied Splus code for these calculations.

23.6 A SIMULATED EXAMPLE

In this example, a collection of 49 local populations in a 7×7 grid was simulated with neighborhood dispersal. Local populations were specified with variable mean local growth rates; thus, some μ_i values were much larger than others. The local growth rates in any given year were slightly correlated between sites. Thus all sites were more likely than random to have good and bad years together. Dispersal was variable between 5 and 10% from year to year and was mainly to the four nearest neighbors (or two and three for corner or edge sites). In specific terms, $\mathbf{A}(t)$ was specified with $z_i(t)$'s drawn from a normal distribution with mean $= \mu_i$ and a variance of 0.0625. The μ_i were different for each local population and were chosen randomly between -0.22 and -0.01. Each year, new $z_i(t)$'s were selected from the normal distribution for that local population. The $z_i(t)$'s were correlated among the local populations such that the covariance of $z_i(t)$ and $z_j(t)$ was (0.1)(0.0625). The $d_i(t)$ varied from year to year. Each year and separately for each local population, $d_i(t)$ was selected from a uniform random distribution between 0.05 and 0.1; thus the dispersal varied from year to year and between local populations in any given year. Most of this dispersal, 80%, was to nearest neighbors. Thus for nearest neighbors, $\alpha_{ji} = 0.80\ d_j(t)/nn$, where nn is the number of nearest neighbors, and for nonnearest neighbors, $\alpha_{ji} = 0.2\ d_j(t)/nnn$; where nnn is the number of nonnearest neighbors.

The simulation was started from a set of local population sizes drawn randomly from the stochastic equilibrium, and starting sizes were drawn anew from this distribution for each replicate of the simulation. For each replicate, a 25-yr time series was generated, and from this time series, μ_m and σ_m^2 were estimated using Eq. (9). From the estimates, the probability that the metapopulation would be below different thresholds (50 or 75% of starting levels) at the end of 25-yr was predicted and compared to the actual probabilities obtained by repeatedly (1000 times) running the simulation for 25-yr starting from the point where the initial 25-yr time series stopped. This simulation was replicated 500 times to generate the distribution of estimated probabilities of 50 and 75% decline in 25-yr versus the true probability. Also, from each 25-yr simulation, the metapopulation λ was estimated and compared to the actual value calculated by running a 10000-yr simulation. For each estimated risk metric, confidence intervals were estimated via the methods in Section 23.5.

Figure 23.3 shows the distribution of λ estimates and the estimated probabilities of 50 and 75% decline versus true values. As expected, the median estimate of λ was equal to the true value ($\hat{\mu}_m$ is an unbiased estimator of μ_m).

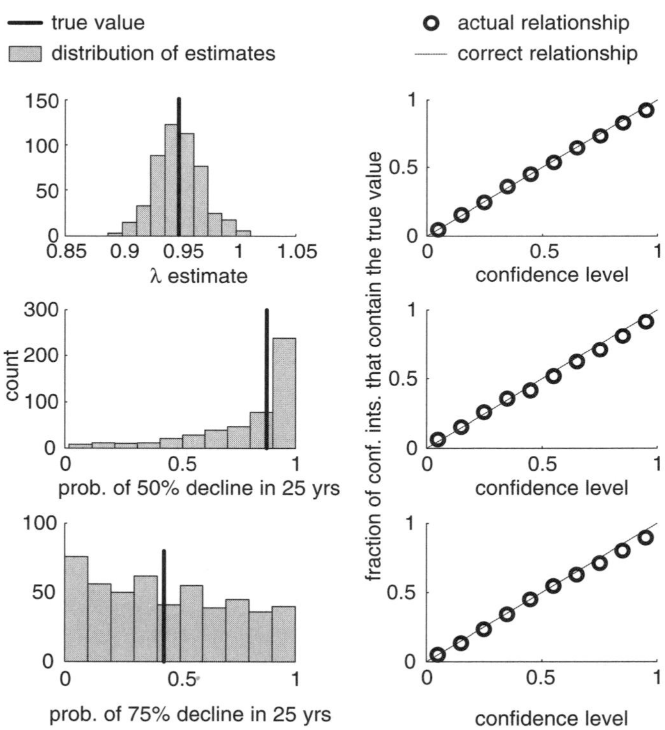

Fig. 23.3 Estimated viability metrics and their estimated confidence intervals versus the true values for a 49 site metapopulation in a 7 × 7 grid with 5–10% dispersal to the closest neighboring sites. (Left) True metrics compared to the distribution of estimated metrics from 500 simulations starting from the same initial conditions. (Right) Performance of the estimated confidence intervals by looking at the fraction of estimated 100(1 − α)% confidence intervals that contain the true values.

The median estimate of λ was 0.97 compared to the true value of 0.97. The median estimates of 50 and 75% decline were 0.63 and 0.14 compared to the true values of 0.62 and 0.12, respectively. Although the median estimates were very close to the true values, the estimates were variable. The estimates of λ ranged between 0.9 and 1.0. The estimates of declines to thresholds were also variable. The variability depended on the threshold and the time frame. In this example, there was low variability around the estimate of 50% decline in 25-yr, but high variability in the estimate of 75% decline. The true values for each of the metrics are shown by the solid lines in the middle of the distributions.

The variability of the estimates is due to the stochastic nature of the process and is not a fault of the estimation methods per se; by chance, short trajectories will appear to have underlying parameters that are different than the true underlying parameters, which leads to variability in the estimated viability metrics. When estimates are inherently variable, it is critical that the confidence intervals for the estimates be estimated correctly. Figure 23.3 (right) confirms that the estimated confidence intervals properly characterize the uncertainty for the estimate risk metrics: e.g., $100(1-\alpha)\%$ of the time the $100(1-\alpha)\%$ confidence intervals contain the true values.

23.7 SALMON AS METAPOPULATIONS

Salmonid populations (*Oncorhynchus* spp.) show strong spatial structuring and they have often been referred to as metapopulations (Reiman and McIntyre, 1995; Policansky and Magnuson, 1998; Cooper and Mangel, 1999; Hill et al., 2002). Spawning and rearing habitats of different salmon stocks occur on discrete and physically separated river or stream sections. Salmon have a well-known and strong tendency to return to their natal streams with a low (1 to 20%) dispersal to other stocks (Fulton and Pearson, 1981; Mathews and Waples, 1991; Quinn, 1993). Within the U.S. Pacific Northwest, collections of anadromous salmon stocks have been divided into "evolutionary significant units" (ESUs) (Waples, 1991), which represent substantially reproductively isolated conspecific groups that can be distinguished based on their coherence on a genetic level and known dispersal between the stocks. Salmon within a stock spawn on individual streams or river sections and the majority of offspring return to spawn in their natal stream or river. Straying of returning adults to nonnatal streams is spatially structured and occurs more frequently within subbasins. Stocks within an ESU have some level of synchrony due to exposure to common migratory corridors between the ocean and the natal stream and also due to exposure to similar large-scale ocean dynamics (Pearcy, 1992; Ware, 1995; Mantua et al., 1997). However, stocks also show a great deal of asynchrony due to exposure to their independent spawning and juvenile rearing habitats and variability in migration timing between stocks (e.g., PSTRT, 2001). Throughout the Pacific Northwest, most salmonid populations show regional decline with the majority of individual stocks showing steady declines with densities well below historical levels (Rieman and Dunham, 2000; McClure et al., 2003).

23.8 SNAKE RIVER SPRING/SUMMER CHINOOK ESU

The Snake river spring/summer chinook ESU (Fig. 23.4) includes all spring and summer chinook spawning within the subbasins of the Tucannon river, Grande Ronde river, and the south, middle, and east fork Salmon rivers, which flow into the Snake river below the Hells Canyon dam (Mathews and Waples, 1991). Juvenile fish rear in the mountain streams and then migrate down the Snake and Columbia rivers to the ocean. After maturing in the ocean, adult fish return to spawn at variable ages between 3 and 5 yr (mean = 4.5 yr). Tagging experiments in the Columbia river basin (which the Snake river basin is a part of) have found that the proportion of individuals that disperse and spawn away from their natal sites is on the order of 1–3% for wild-born individuals (Quinn, 1993).

The Snake river spring/summer chinook ESU was listed as threatened under the U.S. Endangered Species Act in 1992. Stocks within this large and complex basin, like salmon stocks throughout the Pacific Northwest, are impacted negatively by a variety of factors (Wissman et al., 1994) and many have experienced substantial declines (Myers et al., 1998; McClure et al., 2003). There is habitat degradation in many areas related to forestry, grazing, mining, and irrigation practices, resulting in lack of pools, high temperatures, low flows, poor overwintering conditions, and high sediment loads in many areas. At the same time, a substantial portion of the ESU is protected as part of federally designated wilderness (Mathews and Waples,

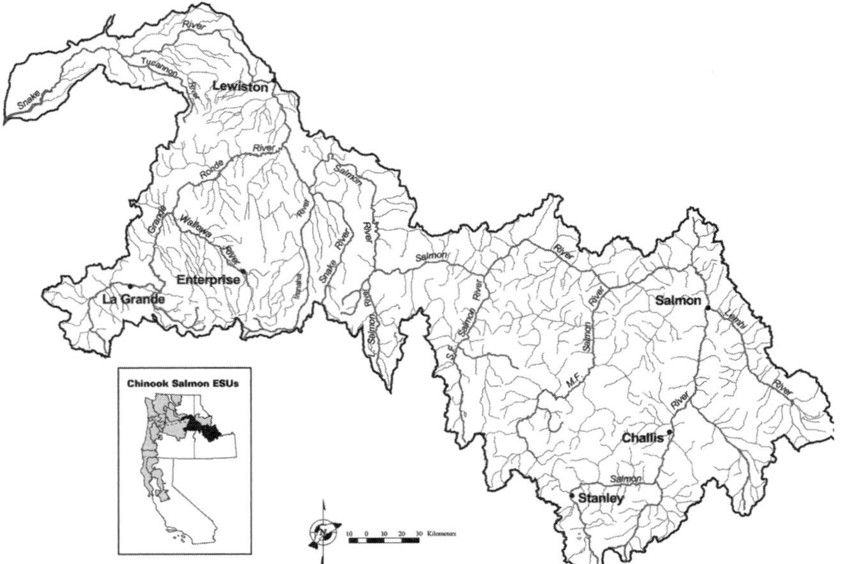

Fig. 23.4 Map of the Snake river spring/summer chinook ESU. The ESU includes stocks from the Snake river and its tributaries between Ice Harbor and Hells Canyon dams. The Hells Canyon hydropower dam has no passage facilities and blocks the migration of salmon into their historical habitat in the upper Snake river basin.

1991). The official ESU designation does not include salmon in the Clearwater basin, as chinook in this subbasin originate from hatchery fish that were stocked in the subbasin after the original natural fish were extirpated in the 1940s. However, from a metapopulation dynamic perspective, current stocks in the Clearwater river basin interact with stocks within other subbasins. Thus, in this analysis, all stocks in the entire Snake river basin were analyzed together.

A total metapopulation level time series was available for this ESU from counts of the total number of wild-born spawners returning through the Ice Harbor dam at the downstream end of the ESU (Fig. 23.4). Returning spawners can be either wild born or hatchery born as hatcheries have been operating in the basin since the early 1970s. McClure et al. (2003) discussed the effects of hatchery production on viability analyses. By focusing on the wild-born spawner time series and not incorporating a correction for hatchery production, the in-stream viability metrics assume that hatchery-born fish all return to the hatchery and do not spawn in stream (which would produce wild-born offspring). As discussed by McClure et al. (2003), this means our viability metrics are optimistic upper bounds, as some unknown fraction of hatchery fish do stray to the wild spawning grounds and potentially reproduce.

In addition to the metapopulation level dam count, time series of redds per mile (rpm), which are indices of the density of gravel egg nests made by spawning females, were available for the majority of stocks within the Snake river basin. Redds per mile are an index of the redds (and consequently returning spawners) trend within a stock, but the total redds are unknown, as the total spawning habitat is not surveyed. The majority of rpm and dam data are available in the digital appendices of McClure et al. (2003).

Parameter Estimation

Our Ice Harbor dam time series starts in 1962 and ends in 1999. The wild-born component of the dam count is denoted $M(0)$, $M(1)$, $M(2)$, . . . $M(37)$, where $M(0)$ is the 1962 count and $M(37)$ is the 1999 count. The maximum likelihood estimates presented in Eq. (9) assume that data do not contain sampling error or other nonprocess error; however, salmon data typically have high levels of sampling error and boom–bust cycles that confound estimation of μ_m and especially σ_m^2 (Holmes, 2001). An alternate approach uses data filtering and examination of the rate at which variance increases within the time series to improve parameter estimation and separate out sampling error variance from the time series (Holmes, 2001; Holmes and Fagan, 2002; Holmes, 2004 cf also Morris and Doak, 2002). These methods have been cross-validated extensively with salmon data (Holmes and Fagan, 2002; Fagan et al., 2003) and are used here to estimate parameters. First, data are transformed using a running sum:

$$\widetilde{M}(t) = \frac{1}{4}\sum_{j=0}^{3} M(t+j) \text{ for } t = 0 \text{ to } 34 \qquad (23.18)$$

The estimates of μ_m and σ_m^2 are then

$$\hat{\sigma}_m^2 = \frac{1}{3}\left(\mathrm{var}\left(\log\widetilde{M}(t+4)/\widetilde{M}(t)\right) - \mathrm{var}\left(\log\widetilde{M}(t+1)/\widetilde{M}(t)\right)\right) = 0.0353$$

$$\hat{\mu}_m = \frac{1}{34}\sum_{t=0}^{34}\log\widetilde{M}(t+1)/\widetilde{M}(t) = -0.0561 \tag{23.19}$$

The estimate of σ_m^2 uses the property that the variance of the underlying stochastic process should increase linearly with time: $E[\mathrm{var}(\log M(t)/M(0))] = \sigma_m^2 t$. The confidence intervals for $\hat{\mu}_m$ and $\hat{\sigma}_m^2$ using $\widetilde{M}(t)$ are slightly different than Eq. (10) (Holmes and Fagan, 2002):

$$\left(\hat{\mu}_m - t_{\alpha/2,df}\sqrt{\hat{\sigma}_m^2/(t-L+1)},\ \hat{\mu}_m + t_{\alpha/2,df}\sqrt{\hat{\sigma}_m^2/(t-L+1)}\right)$$
$$\left(df\,\hat{\sigma}_m^2/\chi^2_{\alpha,df},\ df\,\hat{\sigma}_m^2/\chi^2_{1-\alpha,df}\right) \tag{23.20}$$

where L is the number of counts summed together for the running sum and $df = 0.333 + 0.212\,(n+1) - 0.387\,L = 6.84$ ($L = 4$ and $n = 38$ here). The estimated 95% confidence intervals on μ_m and σ_m^2 are (−0.133, 0.020) and (0.017, 0.111), respectively.

Metapopulation Viability Metrics

The estimate of λ for the Snake river spring/summer chinook ESU is $\hat{\lambda} = \exp(\hat{\mu}_m) = 0.94$. To the extent that long-term trends continue, the expected population size in 25 yr is 21% of current levels ($= \hat{\lambda}^{25}$). The point estimate of the probability of that the ESU drops below 10% of current levels at any time over the next 25 yr is

$$\Pr(T \le 25) = \Phi\left(\frac{-\log(10/1) - \hat{\mu}_m 25}{\sqrt{\hat{\sigma}_m^2 25}}\right) + \exp(-2\log(10/1)\hat{\mu}_m/\hat{\sigma}_m^2)$$
$$\times\, \Phi\left(\frac{-\log(10/1) + \hat{\mu}_m 25}{\sqrt{\hat{\sigma}_m^2 25}}\right) \tag{23.21}$$
$$= 0.23$$

The corresponding estimate of 90% decline over the next 50 yr is 0.74. The probability of extinction was not estimated, as this requires an estimate of the total population size. The number of returning spawners is not the total population size, as nonmature fish remain in the ocean. However, if the true λ of the metapopulation is less than 1.0, the population will eventually go extinct.

The posterior probability density functions [Eq. (16)] for the estimated metrics are shown in Fig. 23.5. The posterior probability distributions give an indication of the degree to which data support different risk levels. The distribution for λ shows considerable data support for a $\lambda < 1$, indicating a declining metapopulation. There is also strong data support for a high risk of 90% decline over the next 50 yrs; however, the estimate of 90% decline over 25 yr is very uncertain. The mean value is 0.23, but the probability distribution is

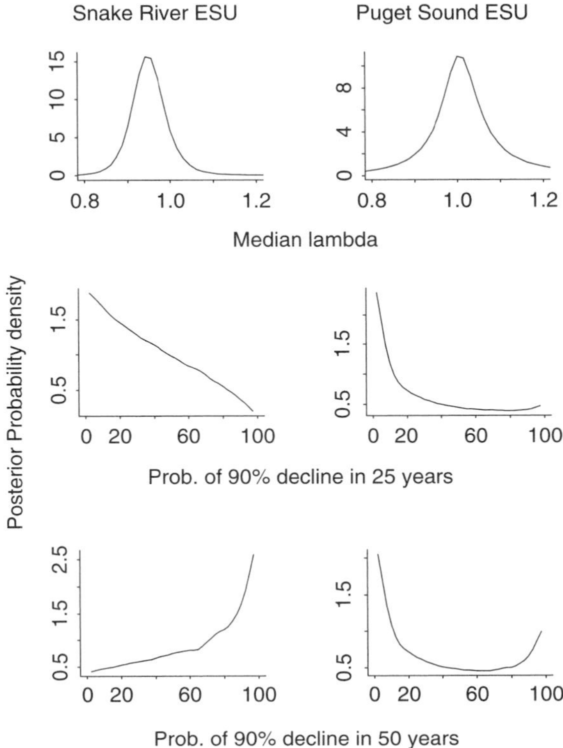

Fig. 23.5 Estimated posterior probability distributions for λ and the probability of 90% decline in 25 and 50 yr. Posterior probability distributions, which were calculated using uniform priors on μ_m and σ^2_m, indicate the relative levels of data support for different risk metric values. Distributions for Snake river spring/summer chinook (left) and Puget Sound ESUs (right).

very broad over the 0 to 1 range. This illustrates that uncertainty in estimates of probabilities of quasiextinction can vary widely depending on the time frame over which one is interested.

23.9 PUGET SOUND CHINOOK ESU

The Puget Sound ESU is a subset of the major chinook salmon group in Washington's northern coastal basins and Puget Sound. The ESU (Fig. 23.6) includes all spring, summer, and fall runs in the Puget Sound region from the north fork Nooksack river to the Elwha river on the Olympic peninsula (Myers et al., 1998). The Elwha and Dungeness coastal basins of the Strait of Juan de Fuca, Hood Canal, and the Puget Sound area north to the northern Nooksack river basin and the U.S. Canadian border are all a part of the Puget Sound ESU. Basin-to-basin dispersal rates have been observed at between 0.1 and 6% based on recoveries of tagged juveniles returning as adults (PSTRT, 2001). Fish in this ESU typically mature at ages 3 and 4 and are coastally oriented during the ocean phase of their life history. The Puget Sound ESU does not include Canadian or

Fig. 23.6 Map of the Puget Sound chinook ESU.

coastal Washington populations. The Puget Sound ESU was listed as threatened under the Endangered Species Act in March of 1999. Trends in abundance throughout the ESU are predominantly downward, with several populations exhibiting severe short-term declines. Degraded spawning and rearing habitats, as well as access restrictions to spawning grounds and migration routes, have all likely contributed to population declines. Salmon in this ESU do not migrate through a hydropower system as the Columbia river ESUs do.

Data for this ESU consist of yearly estimates of the total returning spawners (wild-plus hatchery-born) to the 44 separate river and creek systems feeding into the Puget Sound (Fig. 23.6). These time series were compiled by the National Marine Fisheries Service (Seattle, WA) based on a variety of data: redd counts, carcass counts, in-stream harvest records, weir counts, and hatchery return counts. An independent metapopulation level count was not available; unlike spawners returning to the Columbia river basin, spawners here do not pass through a hydropower system where they can be enumerated. Instead, a 1979–1997 index of the metapopulation was constructed by adding together the 29 time series for the local populations with data over the

1979–1997 period. As for Snake river analyses, our viability metrics implicitly assumes that hatchery fish have not been reproducing and will be optimistic if some hatchery fish do not return to the hatchery and instead spawn successfully in the wild.

Metapopulation Viability Metrics

Parameters were estimated as for the Snake river. The parameter estimates are $\hat{\mu}_m = 0.0036$, and $\hat{\sigma}_m^2 = 0.012$. The estimate of λ for the Puget Sound chinook ESU is $\hat{\lambda} = \exp(\hat{\mu}_m) = 1.003$. The point estimate of the probability that the ESU drops below 10% of current levels at any time over the next 25 yr is 0.000 and over the next 50 yr is 0.001.

The posterior probability distributions (Fig. 23.5, right) illustrate the high uncertainty, given the data, as to whether this ESU is declining, stable, or increasing. The most that can be said from these data is that there is low data support for a severely declining ($\lambda < 0.9$) or increasing ($\lambda > 1.1$) metapopulation. Interestingly, the low support for small λ values translates into high data support for a low risk of 90% decline in the short term (over 25 yr). Over the longer term, however, the uncertainty as to whether the metapopulation is declining or increasing gives rise to a U-shaped distribution, meaning that data give the most support to a probability of 0 or 1, reflecting that λ could be either less than or greater than 1.0. This example illustrates that while data may be equivocal on some questions of conservation concern, such as "is $\lambda < 1$?", data may still give information on other questions, such as "is the short-term risk of severe decline high?"

23.10 USING THE STOCHASTIC METAPOPULATION MODEL TO INVESTIGATE EFFECTS OF MANAGEMENT

Determining how to distribute effort in order to recover an at-risk species is a routine, and challenging, task of conservation managers. For salmon, management actions tend as a generality to affect an entire ESU or multiple ESUs or to affect individual stocks. Management actions such as harvest reductions or increases to survival during migration (between spawning areas and the ocean) or improvements to estuarine environments are examples of actions that will tend to improve conditions for all stocks within an ESU or multiple ESUs. Habitat improvements or protections that affect spawning areas and management of in-stream water levels are examples of actions that tend to affect individual stocks. Without knowing the local stock dynamics or dispersal rates, one can still give certain types of guidance about how much effort is required for recovery of a declining metapopulation and about how effort should be distributed across all local populations.

Metapopulation Level Actions

When management actions affect all local populations roughly equally, it can be estimated how change would change the metapopulation λ. Mathematically, this means that all μ_i values increase by some $d\mu$.

An absolute $d\mu$ change in all μ_i values is equivalent to multiplying all elements in $\mathbf{A}(t)$ by a constant $= \exp(d\mu)$. The mean of the distribution of $\log M(t)/M(0)$ becomes $(\mu_m + d\mu)t = (\log\lambda_{new})t$. Thus $\log(\lambda_{new}/\lambda_{old}) = d\mu$. The change, $d\mu$, can be translated into currency that is more meaningful from a management standpoint by using the relationship $\lambda = R_0^{1/T}$, between λ, the net reproductive rate, R_0, and the mean generation time, T (Caswell, 2001). This is illustrated here for harvest and hydropower effects on salmon in the Snake river spring/summer chinook ESU (cf. McClure et al., 2003).

Harvest

In the Pacific Northwest, harvest rates for salmon are generally expressed in terms of the fraction of spawners that did not return to the spawning grounds but that would have without harvest, e.g., a harvest rate of 0.8 indicates that the actual number of returning spawners is 20% of what it would have been if there had been no harvest. Harvest rates are expressed in this way so that harvest that occurs in the stream versus in the ocean can be compared via a common currency. We can write the net reproductive rate using fecundity and age-specific survival (cf. Caswell 2001) as

$$R_0 = s_1F_1(1-h)f + s_1(1-F_1)s_2F_2(1-h)f + s_1(1-F_1)s_2(1-F_2)s_3F_3(1-h)f\ldots \quad (23.22)$$

where h is the harvest rate, s_i is the survival from age $i-1$ to i, F_i is the fraction of spawners that return at age i, and f is the mean offspring per spawner. Using Eq. (22), the change in λ from a change in h alone is

$$\frac{\lambda_{new}}{\lambda_{old}} = \left(\frac{R_{0,new}}{R_{0,old}}\right)^{1/T} = \left(\frac{1-h_{new}}{1-h_{old}}\right)^{1/T} \quad (23.23)$$

Hydropower

Juvenile salmon from the Snake river basin must migrate through the mainstem of the Snake river, enter the Columbia river, and descend down the Columbia river on their journey to the ocean. This migration, and the return migration of spawning adults, involves passage through four large hydropower dams on the Columbia river and four Snake river hydropower dams. Improving the survival of both juvenile and adult fish migrating through the Columbia and Snake river hydropower systems has been the focus of much effort and is one of the human impacts that has been relatively well quantified.

Following a strategy similar to that used for harvest, the effect of changes in survival through the hydropower system on the rate of decline at the ESU level can be estimated. Denoting by c_d and c_u the proportional increase in down- and upstream passage survival due to improvement in the hydropower system, the improved net reproductive rate is

$$R_{0,new} = c_dc_u(s_1F_1f + s_1(1-F_1)s_2F_2f + s_1(1-F_1)s_2(1-F_2)s_3F_3f\ldots). \quad (23.24)$$

Thus, for assessing the impacts of increased survival through the hydropower system:

$$\frac{\lambda_{new}}{\lambda_{old}} = \left(\frac{R_{0,new}}{R_{0,old}}\right)^{1/T} = (c_d c_u)^{1/T} \tag{23.25}$$

Estimates of the Impacts of Harvest and Hydropower Changes to the Snake River ESU

The mean ocean and in-river 1980–1999 harvest rate for the Snake river spring/summer chinook ESU was $h = 0.08$ (McClure et al., 2003). By setting $h_{new} = 0$, we can examine the effect of successful selective harvest management that would substantially eliminate harvest impacts on salmon in this ESU. Using Eq. (23) and a mean generation time of 4.5 yr, the estimated increase in λ with $h_{new} = 0$ is roughly 2%. NMFS (2000) has required that agencies operating the Federal Columbia river power system implement a variety of activities, including increased spill, improved passage facilities, and increased barging of salmon around the dams as a means of improving survival through the system. The estimated improvement in passage survival from the improvements proposed by NMFS are on the order of 5–6% (i.e., $c_d c_u$ = 1.05–1.06) for the Snake river spring/summer chinook (McClure et al., 2003). This translates into a 1% improvement in λ for this particular ESU using Eq. (25). Thus if the combined effects of substantially reduced harvest and the proposed passage improvements are additive, then roughly a 3% increase in λ is estimated for these actions. If the true λ is less than 0.97, a 3% increase would not be sufficient to achieve $\lambda > 1$. Figure 23.5 indicates that data cannot rule out that the λ in this is ESU is greater than 0.97, but data certainly give more support to a lower λ. This suggests that other recovery actions, such as improvements at the stock level, will also be necessary.

Local Population Level Actions

The effects of changes to individual units of habitat are harder to quantify than the effects of metapopulaion level changes. The change in λ achieved by a change at the level of a specific unit of habitat depends on the level of dispersal, the spatial pattern of dispersal, whether that habitat is connected to source or sink habitat, the level and pattern of synchrony between sites, and so on. In other words, it depends on the type of detailed information that has traditionally been difficult to obtain for metapopulations of conservation concern. Interestingly, although it is difficult to determine how much change in λ can be achieved, it appears possible to estimate where the largest $d\lambda$ from a given $d\mu$ change (per unit of habitat) in the local growth rate is achieved, even though the size of the resultant $d\lambda$ cannot be determined.

Recall that each row of **A** represents a unit of habitat and that a local population is composed of some set of units of habitat with high connectivity. When the intrinsic growth rate, μ_j, in a unit of habitat j is changed by $d\mu$, to

$\exp(\mu_j + d\mu)$, all the a_{ij} elements of column j in matrix $\mathbf{A}(t)$ are multiplied by $\exp(d\mu)$. The goal is to calculate the total change in λ from this $d\mu$ change to all elements in column j by summing over rows i:

$$d\lambda = \sum_i \frac{\partial \lambda}{\partial \mu_{ij}} d\mu = \sum_i \frac{1}{\lambda} \frac{\partial \log \lambda}{\partial \log a_{ij}} d\mu \qquad (23.26)$$

where a_{ij} is the value in row i column j in matrix $\mathbf{A}$. The term $\partial\log\lambda/\partial\log a_{ij}$ is the elasticity of λ. Chapter 14 in Caswell (2001) presented the calculation for the elasticity of λ for products of stochastic matrices:

$$\frac{\partial \log \lambda}{\partial \log a_{ij}} = \lim_{t\to\infty} \frac{1}{n} \sum_{t=0}^{n-1} \frac{a_{ij} v_i(t+1) w_j(t)}{R(t)\mathbf{v}^{\mathsf{T}}(t+1)\mathbf{w}(t+1)} \qquad (23.27)$$

where $R(t)$ is the relationship between the right eigenvector and $\mathbf{A}(t)$, $R(t)\mathbf{w}(t+1) = \mathbf{A}(t)\mathbf{w}(t)$. Thus, the $d\lambda$ from a $d\mu$ change in a unit of habitat j can be solved for by summing Eq. (27) over i:

$$d\lambda_j = \frac{d\mu}{\lambda} \lim_{n\to\infty} \frac{1}{n} \sum_{t=0}^{n-1} \sum_i \frac{a_{ij} v_i(t+1) w_j(t)}{R(t)\mathbf{v}^{\mathsf{T}}(t+1)\mathbf{w}(t+1)} \qquad (23.28)$$

This can be translated into matrix form for all units of habitat 1 to K:

$$[d\lambda_1 \quad d\lambda_2 \ldots d\lambda_k] = \frac{d\mu}{\lambda} \lim_{n\to\infty} \frac{1}{n} \sum_{t=0}^{n-1} \frac{\mathbf{w}^{\mathsf{T}}(t) \circ \mathbf{v}^{\mathsf{T}}(t+1)\mathbf{A}(t)}{R(t)\mathbf{v}^{\mathsf{T}}(t+1)\mathbf{w}(t+1)} \qquad (23.29)$$

where "$\circ$" denotes the scalar product. Using the relationship between the left eigenvector and $\mathbf{A}(t)$, $Q(t+1)\mathbf{v}^{\mathsf{T}}(t) = \mathbf{v}^{\mathsf{T}}(t+1)\mathbf{A}(t)$,

$$[d\lambda_1 \quad d\lambda_2 \ldots d\lambda_k] = \frac{d\mu}{\lambda} \lim_{n\to\infty} \frac{1}{n} \sum_{t=0}^{n-1} \frac{Q(t+1)\mathbf{w}^{\mathsf{T}}(t) \circ \mathbf{v}^{\mathsf{T}}(t)}{R(t)\mathbf{v}^{\mathsf{T}}(t+1)\mathbf{w}(t+1)} \qquad (23.30)$$

The denominator reduces to a constant that depends on t but not j. Thus $d\lambda$ from a change in a unit of habitat j is a weighted temporal average of the reproductive value of local population j times its density:

$$d\lambda_j = \frac{1}{n} \sum_{t=0}^{n-1} c(t) w_j(t) v_j(t) \qquad (23.31)$$

where $c(t)$ is a constant that depends on t but not j. A local population a is composed of units of habitat in the set $a = \{a_1, a_2, a_3, \ldots, a_m\}$, where $\{a_1, a_2, a_3, \ldots, a_m\}$ denotes which rows of $\mathbf{A}$ corresponding to the units of habitat in local population a. The $d\lambda$ per $d\mu$ per unit habitat for a particular local population a is $d\lambda_a = (1/m) \sum_{j\in a} d\lambda_j$, where m is the number of units of habitat in local population a. In words, this means the change in λ is proportional to the product of the "average" density of individuals in a particular local population times the "average" reproductive value of its units of habitat.

Although reproductive values are unknown, there are many cases where the product $v_j w_j$ is a positive function of v_j as long as dispersal is not too unidirectional (meaning, dispersal from A to B but not B to A). This can be shown analytically in three extreme cases: (a) 100% uniform and equal dispersal, (b) all μ_j values equal, or (c) dispersal extremely low. In cases (a) and (b), the reproductive values are all equal and $v_j w_j$ = (a constant) $\times$ v_j. In case (c), $w_j \approx v_j$ and $v_j w_j \approx (v_j)^2$. However, this positive relationship was also found in simulations with variable local growth rates, neighborhood dispersal, and dispersal sources and targets. An obvious exception to this positive relationship is if dispersal is unidirectional, for example, a linear chain of local populations with dispersal via a steady directional wind or ocean current. However, as the following simulations illustrate, the general relationship can still hold even when dispersal is strongly, although not strictly, directional.

Density and λ Sensitivity

Three different types of metapopulation models were used to look at the relationship between average densities in units of habitat versus the $d\lambda$ from a small increase in the local growth rate in each unit of habitat. In each model, dispersal was nonuniform among the local populations so that some sites were dispersal sources (more dispersal out than in) and others dispersal targets (more dispersal in than out). In the first model, local growth rates were equal among all sites and dispersers were distributed globally among all sites. In the second model, local growth rates were variable so that some sites had much higher local growth rates than others and dispersal was mainly to nearest neighbors. In the third model, local growth rates were again variable and dispersal mainly to the two south and east neighbors; however, a small proportion of dispersers were distributed globally. Thus the three examples illustrate global, local, and directional dispersal.

A hundred randomly generated matrix models in each of these three categories were made and $d\lambda_j$ calculated via Eq. (30). Figure 23.7 shows the relationship between the average density of a local population and the $d\lambda$ from increasing the local growth rate in that unit of habitat. The x axis ranks the $d\lambda_j$, thus "1" indicates the local population with the highest $d\lambda_j$ in any simulation and "49" the lowest. The y axis shows the corresponding mean density rank of that local population; "1" indicates the population had the highest density among the 49 sites and "49" the lowest. Results from the 100 randomly generated models are summarized by showing a box plot, which shows the median and range of all density ranks for the sites with a given $d\lambda_j/d\mu$ rank. Thus, the box plot at the x axis position "1" shows the range of density ranks for the units of habitat with the highest $d\lambda/d\mu$ in each model. Model results show a strong positive relationship between the relative density rank within a unit of habitat and which unit of habitat produced the largest increases in the metapopulation λ for a given $d\mu$. The two to three units of habitat with the highest average densities were consistently the units that produced the largest $d\lambda$ for a given $d\mu$. This suggests that plotting the distribution of the relative densities within local populations in a metapopulation could give a rapid indication of the sensitivity of the metapopulation to changes to individual local populations.

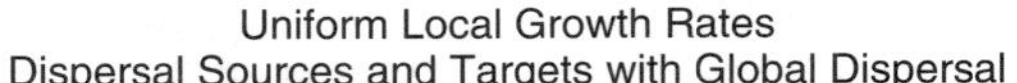

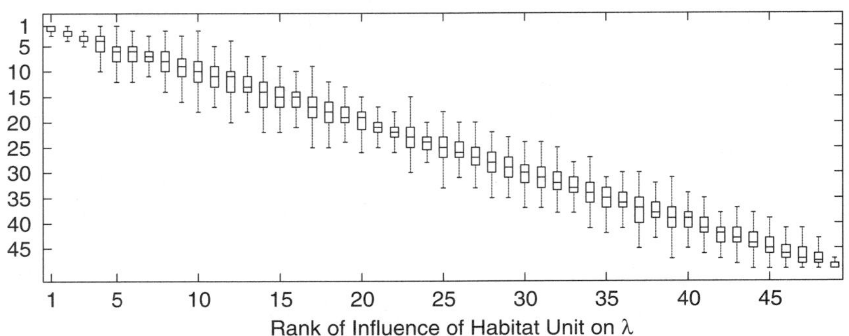

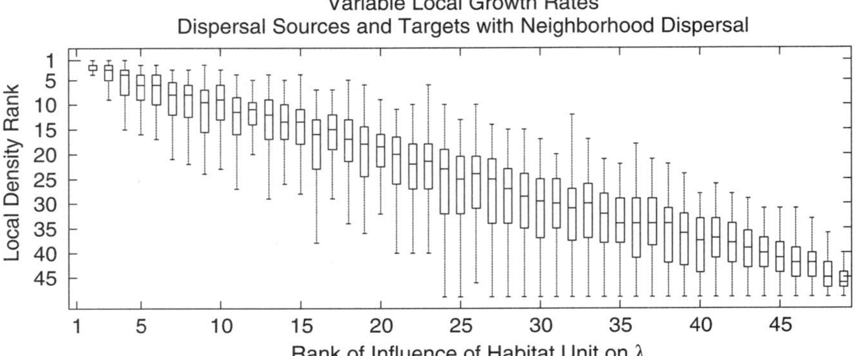

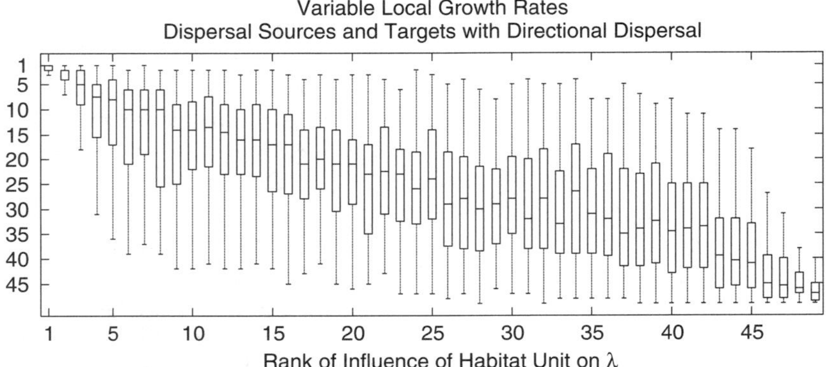

Fig. 23.7 Relationship between the influence of a given habitat unit on the metapopulation λ and the average density in that habitat unit. One hundred 7×7 metapopulations with spatially variable dispersal rates (some sites dispersal sinks and others targets) were generated randomly in each of three classes: (1) spatially uniform growth rates and global dispersal, (2) spatially variable growth rates with neighborhood dispersal, and (3) spatially variable growth rates with directional neighborhood dispersal to the S and E two neighbors only. The x axis shows the rank in terms of $d\lambda/d\mu$, and the y axis shows a box plot of the distribution of density ranks for sites with a given $d\lambda/d\mu$ rank across all 100 models in each class. Thus the box plot at $x = 1$ shows the distribution of ranks for the sites with the highest $d\lambda/d\mu$ in each model. The line in each box shows the median density rank for the sites with a given $d\lambda/d\mu$ rank, the box encloses 50% of the ranks, and the whiskers show the range from all 100 randomly generated models.

One application of this would be to estimate where negative impacts would lead to the greatest decrease in λ, thus suggesting where protection in most critical. It would also suggest where local improvements would be most effective for a given increase in the local growth rate. However, in actual management situations where improvements are being sited, one is generally trying to maximize the "bang per buck," $d\lambda_j/d\$ = d\lambda_j/d\mu_j \times d\mu_j/d\$$. The cost, $d\$$, is the actual monetary cost or some combination of monetary, logistical, and political costs and $d\mu_j/d\$$ is the cost of a unit improvement to a unit of habitat j. Thus $d\lambda_j/d\mu$ is one part of the equation, and the other part, the cost of a unit improvement in different habitats, would have to come from a specific analysis of the costs and estimated effects of management actions on different local populations.

Example Using the Snake River ESU

The overall level of salmon dispersal between and among stocks within this ESU is known to be fairly low and spatially localized (Mathews and Waples, 1991; Quinn, 1993). In addition, there is high variability in the habitat quality between stocks, with some stocks relatively pristine and protected within wilderness areas, whereas others are exposed to high and multiple impacts (such as stream degradation and disturbance, pollution, in-stream harvest, and irrigation impacts). Figure 23.8 (top) shows the distribution of average normed redds per mile for 50 Snake river spring/summer chinook stocks. For each year between 1980 and 1995, the redds-per-mile count for each stock was divided by the maximum count among the 50 stocks in that year. The average over the 16 yr was then used as an estimate of the average normed redds per mile. The long-tailed distribution is the expectation from theory given low dispersal and high variability in stock habitat quality.

Estimation of the average normed redds per mile was repeated using a variety of different time periods. Regardless of the time period or number of years used for averaging, six stocks consistently appeared among the top five stocks with the highest density of redds: Johnson Cr., Poverty Cr., and Secesh R. in the south fork of the Salmon R., the Lostine R. in the Grande Ronde subbasin, Marsh Cr. in the middle fork of the Salmon R., and the Imnaha R. Perhaps not surprisingly, all of these are in relatively low impacted regions of the ESU. At a subbasin level, the overall highest redd density was in the south fork of the Salmon river where summer-run chinook primarily occur (Fig. 23.8, bottom). The other regions are primarily spring-run chinook. The south fork of the Salmon river is relatively pristine and few hatchery fish have been released into this subbasin; the stocks presumably have experienced relatively low interbreeding with hatchery-reared stocks. In addition, the later run timing may somehow be associated with less straying, lower harvest, or lower hydropower impacts.

This analysis predicts that the λ of the Snake river spring/summer chinook ESU would be most sensitive to changes to the summer-run stocks in the south fork of the Salmon river and to the spring-run stocks, the Lostine R., Imnaha River, and Marsh Creek and should be protected preferentially from impacts. This can be counterintuitive in some situations. For example, imagine making choices about where to allow a limited catch-and-release sport

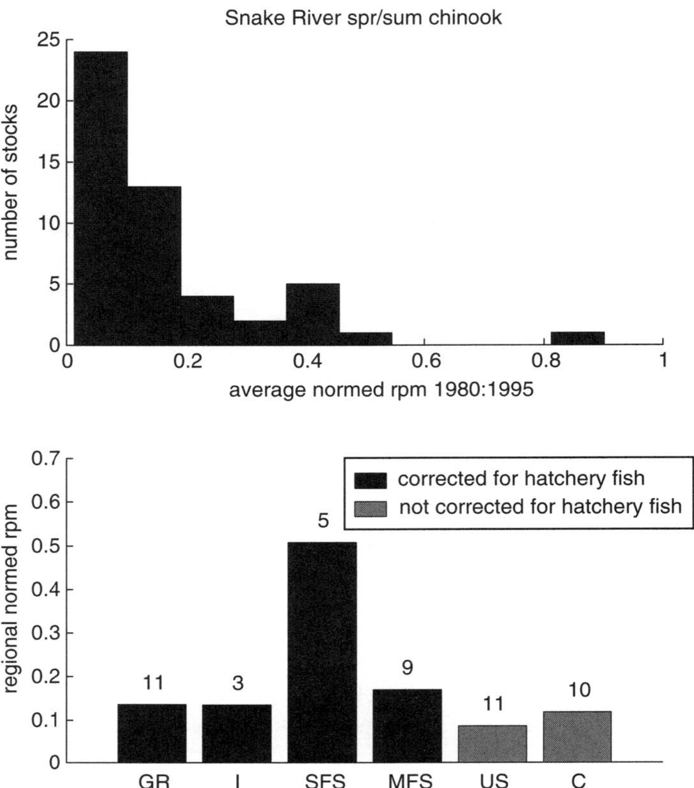

Fig. 23.8 Distribution of densities of redds in the Snake river spring/summer chinook ESU at a stock and subbasin level. The average normed redds densities (top) are shown for the 50 stocks with 1980–1995 data (the years were chosen to maximum the number of stocks with data). For each stock the normed redd density was averaged over the 16 yr to get an estimate of the normed average density. In the lower plot, relative average densities over all stocks within different basins are shown (with the number of stocks in each basin shown above the bars). The basin designations are GR, Grande Ronde; I, Imnaha; SFS, south fork salmon; MFS, middle fork salmon; US, upper salmon; C, Clearwater. Redds due to hatchery fish released into stocks were removed before doing these analyses, as the density will be artificially high simply due to hatchery fish releases. This correction could not be done for the upper salmon or Clearwater regions because the fraction of spawners that are hatchery strays were unknown; however, the hatchery releases are very high in these basins and thus the corrected relative densities would be much lower than shown.

fishery. Sites with the highest density would seem to be the prime candidates, whereas the analysis of $d\lambda/d\mu$ indicates just the opposite. In terms of determining where to direct improvements, the $d\lambda/d\mu$ suggests that these pristine sites are where a given $d\mu$ would produce the greatest metapopulation λ; however, the regions where $d\lambda/d\mu$ is the highest are not necessarily the regions where μ is improved most easily. Indeed a given unit of improvement may be more difficult in pristine sites. Choosing where to direct stock improvements requires consideration of the cost and difficulty of a given $d\mu$ for different stocks in combination with the estimate of the sensitivity of λ to local changes.

23.11 POPULATION VIABILITY ANALYSIS IN PRACTICE

The purpose of this chapter is to present a theoretical framework for metapopulation PVA using time series data and diffusion approximations. These methods are then illustrated using data from two salmon metapopulations. The salmon analyses are intended as an example of how to calculate the diffusion parameters and metrics. An actual PVA must grapple with other important issues that are outside the scope of this chapter, but which anyone contemplating an actual PVA must be aware. Morris and Doak (2002) gave a review of the criticisms and caveats surrounding the use of PVA and outlined general recommendations and cautions when conducting a PVA. In the context of diffusion approximation methods in particular, Holmes (2004) outlined an approach using matrix models to conduct sensitivity analyses in order to choose among different parameterization methods and metrics for a specific PVA application.

One of the issues that is especially pertinent for our chapter is the issue of variability in estimated risk metrics. A number of recent PVA cross-validations using actual data on a large number of different populations have shown that careful PVA analyses give unbiased risk estimates (Brook et al., 2000; Holmes and Fagan, 2002; Fagan et al., 2003). Although this is very encouraging, a difficult issue is the high inherent variability associated with estimated probabilities (such as probability of extinction), even though they may be unbiased (Ludwig, 1996, 1999; Fieberg and Ellner, 2000; Holmes, 2001; Ellner et al., 2002). How to properly use risk metrics that have high variability is currently being debated within the field with arguments ranging from "don't use them" (Ludwig, 1996, 1999; Fieberg and Ellner, 2000), to "use to estimate risks within collections of populations" (Fagan et al., 2001; Holmes and Fagan, 2002), to "use where data are extensive and high quality" (Coulson et al., 2001), to "PVA metrics based on data, even if variable, are better than the alternatives" (Brook et al., 2002). An encouraging aspect of diffusion approximation methods is that cross-validations using real time series data have indicated that the uncertainty in the estimated metrics appears to be characterized properly (Holmes and Fagan, 2002). Nonetheless, how to use and present metrics with high variability, albeit well characterized, is not an easy question to answer. Presentation of $100(1-\alpha)\%$ is an oft-used approach, but experience in the forum of salmon recovery planning in the Pacific Northwest has shown that it is easy to misinterpret confidence intervals. For example, it is easy to interpret 95% confidence intervals for λ that overlap 1.0 as an indication that data are equivocal as to whether the population is declining or increasing, whereas there may be considerable data support for a declining population. Graphic presentations of data support for different risk levels have been more compelling and informative, although translating levels of data support into numbers that policy makers can use to take uncertainty into account in policy decisions has been challenging.

23.12 DISCUSSION

This chapter focused on the calculation of metapopulation PVA metrics; however, there are other more general PVA insights from an examination of stochastic metapopulations and of this specific class of declining density-independent metapopulations. First, by definition the trajectory of a stochastic metapopulation is subject to random processes and thus the metapopulation trajectory observed in any one snippet of time is unlikely to capture the long-term dynamics. The shorter the time frame, the farther the observed trend is likely to be from the long-term trend. Thus the trends in any two adjacent time periods are unlikely to be identical, and the difference indicates not necessarily a change in the underlying rate of decline, but can be due simply to chance. The variability of observed rates of decline can be estimated from the level of the variability driving the long-term dynamics, and thus statistical tests performed to determine the likelihood that an apparent change in trend occurred due to the stochastic nature of the process rather than an underlying change in conditions.

Second, the local populations within a metapopulation are linked and experience the same long-term growth rates, regardless of the underlying difference in local population conditions (i.e., whether they are "sources" or "sinks"). However, measured over a short time period, there will be differences in the observed local population trends due to chance and local conditions. This means that over a given time period, local populations will appear to be declining at different rates, but this is not an indication the long-term trends and not necessarily related to local conditions being better or worse than other areas. That the long-term trends of the individual local populations are the same as the metapopulation has a direct impact on PVA for local populations within a metapopulation. The rate of decline observed among the different local populations will differ, as will the apparent level of variability in the local time series. Thus if an individual viability analysis is done using parameters estimated from local population time series alone, it will appear that there is tremendous variability among the local populations risk levels when in fact their long-term risks are similar. When looking at the long-term risks, use of metapopulation level parameters leads to better estimates of the long-term local population risks. Short-term risks, however, are still strongly influenced by local conditions. Clearly estimates of both short-term and long-term risks are needed to capture the whole viability picture for a metapopulation. Although local populations within the type of metapopulations modeled here will be eventually repopulated by dispersal if they undergo extreme declines, the resulting loss of genetic diversity leads to a gradual erosion of the genetic health of the metapopulation. Indeed this has happened for salmon species throughout the Pacific Northwest.

Recovery planning for endangered and threatened species typically requires determining where to put the most effort. Rarely is it the case that maximal effort can be applied everywhere. Using the stochastic metapopulation model, a sensitivity analysis was used to look for local characteristics that predict where local changes would produce the biggest change in the metapopulation growth rate. Interestingly, local density (not absolute numbers) was a strong predictor of where a unit change in local growth rates led to the largest metapopulation growth rate. This relationship was observed even in simulations with dispersal sources and targets and strongly directional dispersal, although it will break down when dispersal is strictly unidirectional. Determining which local populations are best suited for restoration efforts also requires assessing the feasibility, cost, and acceptance of restoration efforts. Indeed when it comes to actually implementing recovery actions, optimizing the efficiency of effort in terms affecting recovery requires solving a complex function of biological, economic, and political information. However, understanding the population dynamics of the species of concern and gaining insight regarding how the demography of the species will respond to alternative management actions are fundamental and primary components of this conservation equation.

REFERENCES

Aars, J., Dallas, J., Piertney, S., Marshall, F., Gow, J., Telfer, S., and Lambin, X. (2003b). Excess heterozygosity reveals widespread gene flow in highly fragmented populations of water voles *Arvicola terrestris*. Submitted for publication.

Aars, J., and Ims, R. A. (1999). The effect of habitat corridors on rates of transfer and interbreeding between voles demes. *Ecology* **80**, 1648–1655.

Aars, J., and Ims, R.A. (2000). Spatial density dependent transfer and gene frequency changes in patchy vole populations. *Am. Nat.*, **155**, 252–265.

Aars, J., Johannesen, E., and Ims, R.A. (1999). Demographic consequences of movements in subdivided root vole populations. *Oikos* **85**, 204–216.

Aars, J., Lambin, X., Denny, R., and Griffin, A. (2001). Water vole in the Scottish uplands: Distribution patterns of disturbed and pristine populations ahead and behind the American mink invasion front. *Anim. Conserv.*, **4**, 187–194.

Aars, J., Norrdahl, K., Marshall, F., Henttonen, H., and Lambin, X. (2003a). No evidence of genetic bottlenecks in strongly fluctuating populations of water voles (*Arvicola terrestris*). Submitted for publication.

Abrams, P.A. (2002). Will small population sizes warn us of impending extinctions? *Am. Nat.* **160**, 293–305.

Ackery, P.R., and Vane-Wright, R.I. (1984). "Milkweed butterflies". British Museum (Natural History), London.

Agrawal, A.F., Brodie, E.D., III, and Wade, M.J. (2001). On indirect genetic effects in structured populations. *Am. Nat.* **158**, 308–323.

Agrell, J., Erlinge, S., Nelson, J., and Sandell, M. (1992). Body weight and population dynamics: Cyclic demography in a noncyclic population of the field vole (*Microtus agrestis*). *Can. J. Zool.*, **70**, 494–501.

Agur, Z., Cojocaru, L., Mazor, G., Anderson, R.M., and Danon, Y.L. (1993). Pulse mass measles vaccination across age cohorts. *Proc. Natl. Acad. Sci. USA* **90**, 11698–11702.

Akçakaya, H.R. (1992). Population cycles of mammals: Evidence for a ratio-dependent predation hypothesis. *Ecol. Monogr.* **62**, 119–142.

Akçakaya, H.R. (1997). RAMAS Metapop: Viability analysis for stage-structured metapopulations. Version 2. Applied Biomathematics, Setauket, New York.

Akçakaya, H.R. (2000a). Population viability analyses with demographically and spatially structured models. *Ecol. Bull.* **48**, 23–38.

Akçakaya, H.R. (2000b). Viability analyses with habitat-based metapopulation models. *Popul. Ecol.* **42**, 45–53.

Akçakaya, H.R., and Ferson, S. (1992). RAMAS/Space user manual: Spatially structured population models for conservation biology. Applied Biomathematics, New York.

Akçakaya, H.R., and Sjögren-Gulve, P. (2000b). Population viability analysis in conservation planning: An overview. *Ecol. Bull.* **48**, 9–21.

Akimoto, J., Fukuhara, T., and Kikuzawa, K. (1999). Sex ratios and genetic variation in a functionally androdioecious species, *Schizopepon bryoniaefolius* (Cucurbitaceae). *Am. J. Bot.* **86**, 880–886.

Alexander, H.M. (1989). An experimental field study of anther-smut disease of *Silene alba* caused by *Ustilago violacea*: Genotypic variation and disease incidence. *Evolution* **43**, 835–847.

Alexander, H.M. (1990). Epidemiology of anther-smut infection of *Silene alba* caused by *Ustilago violacea*: Patterns of spore deposition and disease incidence. *J. Ecol.* **78**, 166–179.

Alexander, H.M., and Antonovics, J. (1995). Spread of anther-smut disease (*Ustilago violacea*) and character correlations in a genetically variable experimental population of *Silene alba*. *J. Ecol.* **83**, 783–794.

Alexander, H.M., Antonovics, J., and Kelly, A. (1993). Genotypic variation in disease incidence: Integration of greenhouse and field studies on *Silene alba* and *Ustilago violacea*. *J. Ecol.* **81**, 325–334.

Alhiyaly, S.A.K., McNeilly, T., Bradshaw, A.D., and Mortimer, A.M. (1993). The effect of zinc contamination from electricity pylons: Genetic constraints on selection for zinc tolerance. *Heredity* **70**, 22–32.

Allee, W.C. (1938). "The Social Life of Animals." W.W. Norton.

Allee, W.C., Park, O., Emerson, A.E., Park, T., and Schmidt, K.P. (1949). "Principles of Animal Ecology." Saunders, Philadelphia.

Allen, J.C., Schaffer, W.M., and Rosko, D. (1993). Chaos reduces species extinction by amplifying local population noise. *Nature* **364**, 229–232.

Allendorf, F.W., and Ryman, N. (2002). The role of genetics in population viability analysis. *In* "Population Viability Analysis." (S.R. Beissinger and D.R. McCullough, eds.), pp. 50–85. University of Chicago Press, Chicago.

Allmon, W. (1992). A causal analysis of stages in allopatric speciation. *Oxf. Surv. Evol. Biol.* **8**, 219–257.

Allmon, W.D., Morris, P.J., and McKinney, M.L. (1998). An intermediate disturbance hypothesis of maximal speciation. *In* "Biodiversity Dynamics" (M.L. McKinney and J.A. Drake, eds.), pp. 349–376. Columbia Univ. Press, New York.

Altizer, S., Thrall, P.H., and Antonovics, J. (1998). Pollinator behavior and disease transmission of the anther-smut disease of *Silene alba*. *Am. Midl. Nat.* **139**, 147–163.

Amarasekare P. (2000). Coexistence of competing parasitoids on a patchily distributed host: Local vs. spatial mechanisms. *Ecology* **81**, 1286–1296.

Amarasekare P., and Nisbet, R.M. (2001). Spatial heterogeneity, source-sink dynamics, and the local coexistence of competing species. *Am. Nat.* **158**, 572–584.

Andelman, S.J., and Willig, M.R. (2002). Alternative configurations of conservation reserves for Paraguayan bats: considerations of spatial scale. *Conserv. Biol.* **16**, 1352–1363.

Anderson, R.M., and May, R.M. (1985). Helminth infections of humans: Mathematical-models, population-dynamics, and control. *Ad. Parasito.* **24**, 1–101.

Anderson, R.M., and May, R.M. (1991). "Infectious Diseases of Humans." Oxford Univ. Press.

Anderson, S.S., Burton, R.W., and Summers, C.F. (1975). Behaviour of grey seals (*Halichoerus grypus*) during the breeding season at North Rona. *J. Zool. Lond.* **177**, 179–195.

Anderson, W.W. (1971). Genetic equilibrium and population growth under density-regulated selection. *Am. Nat.* **105**, 489–498.

Andersson, H., and Djehiche, B. (1998). A threshold limit theorem for the stochastic logistic epidemic. *J. Appl. Prob.* **35**, 662–670.

Andersson, M. (1994) "Sexual Selection". Princeton Univ. Press, Princeton, NJ.

Ando, A., Camm, J., Polasky, S., and Solow, A. (1998). Species distributions, land values, and efficient conservation. *Science* **279**, 2126–2128.

Andreassen, H.P., Halle, S., and Ims, R.A. (1996a). Optimal design of movement corridors in root voles — not too wide and not too narrow. *J. Appl. Ecol.* **33**, 63–70.

Andreassen, H.P. and Ims, R.A. (2001). Dispersal in patchy vole populations: Role of patch configuration, density dependence, and demography. *Ecology* **82**, 2911–2926.

Andreassen, H.P., Ims, R.A., and Steinse, O.K. (1996b). Discontinuous habitat corridors: Effects on male root vole movements. *J. Appl. Ecol.* **33**, 555–560.

Andreassen, H.P., Stenseth, N.C., and Ims, R.A. (2002). Dispersal behaviour and population dynamics of vertebrates. *In* "Dispersal Ecology" (J. Bullock, R. Kenward, and R. Hails, eds.), pp. 237–256. Blackwell Science, Oxford.

Andrewartha, H.G. (1957). The use of conceptual models in population ecology. *Cold Spring Harb. Symp. Quant. Biol.*

Andrewartha, H.G., and Birch, L.C. (1954). "The Distribution and Abundance of Animals." The University of Chicago Press, Chicago.

Andrewartha, H.G., and Birch, L.C. (1984). "The Ecological Web." The University of Chicago Press, Chicago.

Antolin, M.F., Gober, P., Luce, B., Biggins, D.E., Van Pelt, W.E., Seery, D.B., Lockhart, M., and Ball, M. (2002). The influence of sylvatic plague on North American wildlife at the landscape level, with special emphasis on black-footed ferret and prairie dog conservation. *In* "Transactions of the 67th North American Wildlife and Natural Resources Conference", pp.104–127. Wildlife Management Institute, Washington, DC.

Antonovics, J. (1976). The nature of limits to selection. *Ann. Miss. Bot. Gard.* **63**, 224–247.

Antonovics, J. (1999). Pathogens and plant population dynamics: The effects of resistance genes on numbers and distribution. *In* "Ecological Effects of Pest Resistance Genes in Managed Ecosystems" (P.M. Traynor and H.H. Westwood, eds.) pp. 49–55. Information Systems for Biotechnology, Blacksburg ,VA.

Antonovics, J., and Alexander, H.M. (1992). Epidemiology of anther-smut infection of *Silene alba* caused by *Ustilago violacea*: Patterns of spore deposition in experimental populations. *Proc. R. Soc. Lond. Ser. B* **250**, 157–163.

Antonovics, J., Bradshaw, A.D., and Turner, R.G. (1971). Heavy metal tolerance in plants. *Adv. Ecol. Res.* **7**, 1–85.

Antonovics, J., Hood, M.E., and Partain, J. (2002). The ecology and genetics of a host-shift: *Microbotryum* as a model system. *Am. Nat.* **160**, S40–S53.

Antonovics, J., Hood, M.E., Thrall, P.H., Abrams, J.Y., and Duthie, M. (2003). Distribution of anther-smut disease (*Microbotryum violaceum*) and *Silene* species in the eastern United States. *Am. J. Bot.* **90**, 1522-1531.

Antonovics, J., Iwasa, Y., and Hassell, M.P. (1995). A generalized model of parasitoid, venereal, and vector-based transmission processes. *Am. Nat.* **145**, 661–675.

Antonovics, J., Stratton, D., Thrall, P.H., and Jarosz, A.M. (1995). An anther smut disease (*Ustilago violacea*) of fire-pink (*Silene virginica*): Its biology and relationship to the anther smut disease of white campion (*Silene alba*). *Am. Midl. Nat.* **135**, 130–143.

Antonovics, J., Thrall, P.H., and Jarosz, A.M. (1998). Genetics and the spatial ecology of species interactions: The *Silene-Ustilago* system. *In* "Spatial Ecology: The Role of Space in Population Dynamics and Interspecific Interactions" (D., Tilman, and P. Kareiva, eds.) pp. 158–180. Princeton Univ. Press.

Antonovics, J., Thrall, P.H., Jarosz, A.M., and Stratton, D. (1994). Ecological genetics of metapopulations: The *Silene-Ustilago* plant-pathogen system. *In* "Ecological Genetics" (L.A. Real, ed.), pp. 146–170. Princeton Univ. Press, Princeton, NJ.

Appleby, A.B. (1980). The disappearance of plague: A continuing puzzle. *Econ. Hist. Rev.* **33**, 161–173.

Araújo, M.B., Densham, P.J., Lampinen, R., Hagemeijer, W.J.M., Mitchell-Jones, J.P., Gasc, J.P., and Humphries, C.J. (2001). Would environmental diversity be a good surrogate for species diversity? *Ecography* **24**, 103–110.

Araújo, M.B., and Williams, P.H. (2000) Selecting areas for species persistence using occurrence data. *Biol. Conserv.* **96**, 331–345.

Araújo, M.B., Williams, P.H., and Fuller, R.J. (2002). Dynamics of extinction and the selection of nature reserves. *Proc. R. Soc. Lond. B* **269**, 1970–1980.

Arthur, J.F., Hachey, M., Sahr, K., Huso, H., and Kiester, A.R. (1997). Finding all optimal solutions to the reserve site selection problem: Formulation and computational analysis. *Environ. Ecol. Stat.* **4**, 153–165.

Ås, S., Bengtsson, J., and Ebenhard, T. (1997). Archipelagoes and theories of insularity. *Ecol. Bull.* **46**, 88–116.

Asmussen, M.A. (1979). Regular and chaotic cycling in models of ecological genetics. *Theor. Popul. Biol.* **16**, 172–190.

Asmussen, M.A. (1983a). Density dependent selection incorporating intraspecific competition. 1. A haploid model. *J. Theor. Biol.* **101**, 113–127.

Asmussen, M.A. (1983b). Density-dependent selection incorporating intraspecific competition. 2. A diploid model. *Genetics* **103**, 335–350.

Avilés, L., Abbot, P., and Cutter, A. (2002). Population ecology, nonlinear dynamics, and social evolution. I. Associations among non relatives. *Am. Nat.* **159**, 115–127.

Baars, M.A., and Van Dijk, T.H. (1984). Population dynamics of two carabid beetles at a Dutch heathland. I. Subpopulation fluctuations in relation to weather and dispersal. *J. Anim. Ecol.* **53**, 375–388.

Baguette, M., and Nève G. (1994). Adult movements between populations in the specialist butterfly *Proclossiana eunomia* (Lepidoptera, Nymphalidae). *Ecol. Entomo.* **19**, 1–5.

Baguette, M., Petit, S., and Queva F. (2000). Population spatial structure and migration of three butterfly species within the same habitat network: Consequences for conservation. *J. Appl. Ecol.* **37**, 100–108.

Bailey, N.T.J. (1964). "The Elements of Stochastic Processes with Applications to the Natural Sciences". Wiley, New York.

Bak, P., Chen, K., and Tang, C. (1990). A forest fire model and some thoughts on turbulence. *Phys. Lett. A* **147**, 297–300.

Baker, H.G., and Stebbins, G.L. (1965). "The Genetics of Colonizing Species." Academic Press, New York.

Balkau, B.J., and Feldman, M.W. (1973). Selection for migration modification. *Genetics* **74**, 171–174.

Ball, S.J., Adams, M., Possingham, H.P., and Keller, M.A. (2000). The genetic contribution of single male immigrants to small, inbred populations: A laboratory study using *Drosophila melanogaster. Heredity* **84**, 677–684.

Barascud, B., Martin, J.F., Baguette, M., and Descimon, H. (1999). Genetic consequences of an introduction-colonization process in an endangered butterfly species. *J. Evol. Biol.* **12**, 697–709.

Barraclough, T.G., and Vogler, A.P. (2000). Detecting the geographic pattern of speciation from species-level phylogenies. *Am. Nat.* **43**, 419–434.

Barrett, S.C.H., and Husband, B.C. (1997). Ecology and genetics of ephemeral plant populations: *Eichhornia paniculata* (pontederiaceae) in northeast Brazil. *J. Hered.* **88**, 277–284.

Bartlett, M.S. (1956). Deterministic and stochastic models for recurrent epidemics. *Proc. Third Berkley Symp. Math. Stats. Prob.* **4**, 81–108.

Bartlett, M.S. (1957). Measles periodicity and community size. *J. R. Stat. Soc. A* **120**, 48–70.

Bartlett, M.S. (1960). The critical community size for measles in the U.S. *J. R. Stat. Soc. A* **123**, 37–44.

Barton, N.H. (1987). The probability of establishment of an advantageous mutation a subdivided population. *Genet. Res. (Camb.)* **50**, 35–40.

Barton, N.H. (1993). The probability of fixation of a favoured allele in a subdivided population. *Genet. Res. (Camb.)* **62**, 149–158.

Barton, N.H. (2001). The evolutionairy consequences of gene flow and local adaptation: Future approaches. *In* "Dispersal" (J. Clobert, E. Danchin, A.A. Dhondt, and J.D. Nichols, eds.), pp. 329–340. Oxford Univ. Press, Oxford.

Barton, N.H., Depaulis, F., and Etheridge, A.M. (2002). Neutral evolution in spatially continuous populations. *Theor. Popul. Biol.* **61**, 31–48.

Barton, N.H., and Whitlock, M.C. (1997). The evolution of metapopulations. *In* "Metapopulation Biology: Ecology, Genetics, and Evolution" (I. Hanski and M. Gilpin, eds.), pp. 183–210. Academic Press, San Diego.

Bascompte, J., and Solé, R.V. (1994). Spatially induced bifurcations in single-species population dynamics. *J. Anim. Ecol.* **63**, 256–264.

Bascompte, J., and Solé, R. (1996). Habitat fragmentation and extinction thresholds in spatially explicit models. *J. Anim. Ecol.* **65**, 465–473.

Bascompte, J. and Solé, R.V. (1997). Models of habitat fragmentation. *In* "Modeling Spatiotemporal Dynamics in Ecology" (J. Bascompte and R.V. Solé, eds.), pp. 127–150. Springer, New York.

Bastin, L., and Thomas, C.D. (1999). The distribution of plant species in urban vegetation fragments. *Landsc. Ecol.* **14**, 493–507.

Bateson, W. (1909). Heredity and variation in modern lights. *In* "Darwin and Modern Science" (A. C. Seward, ed.), pp. 85–101. Cambridge Univ. Press, Cambridge.

Beadle, G.W. (1980). The ancestry of corn. *Sci. Am.* **242**, 112–119.

Beaumont, M.A., Zhang, W., and Balding, D.J. (2003). Approximate bayesian computation in population genetics. *Genetics.*

Beerli, P., and Felsenstein, J. (1999). Maximum-likelihood estimation of migration rates and effective population numbers in two populations using a coalescent approach. *Genetics* **152**, 763–773.

Beerli, P., and Felsenstein, J. (2001). Maximum likelihood estimation of a migration matrix and effective population sizes in n subpopulations by using a coalescent approach. *Proc. Nat. Acad. Sci. USA* **98**, 4563–4568.

Beissinger, S.R. (2002). Population viability analysis: Past, present, future. *In* Population Viability Analysis (S.R. Beissinger and D.R. McCullough, eds.). University of Chicago Press, Chicago.

Beissinger, S.R., and McCullough, D.R. (2001). "Population Viability Analysis." University of Chicago Press, Chicago.

Beissinger, S.R., and Westphal, M.I. (1998). On the use of demographic models of population viability in endangered species management. *J. Wildl. Manage.* **62**, 821–841.

Bekker, R.M., Verweij, G.L., Smith, R.E.N., Reine, R., Bakker, J.P., and Schneider, S. (1997). Soil seed banks in European grasslands: Does land use affect regeneration perspectives? *J. Appl. Ecol.* **34**, 1293–1310.

Belhassen, E., Beltran, M., Couvet, D., Dommée, B., Gouyon, P.-H., and Olivieri, I. (1990). *Evolution des taux de femelles dans les populations naturelles de thym, Thymus vulgaris L. Deux hypothèses alternatives confirmées. C. R. Acad. Sci. Ser. III Sci. Vie* **310**, 371–375.

Bélichon, S., Clobert, J., and Massot, M. (1996). Are there differences in fitness components between philopatric and dispersing individuals? *Acta Oecol.* **17**, 503–517.

Bell, G. (2000). The distribution of abundance in neutral communities. *Am. Nat.* **155**, 606–617.

Bell, G. (2001). Neutral macroecology. *Science* **293**, 2413–2418.

Bell, G., Lechowicz, M.J., and Waterway, M.J. (2002). The scale of local adaptation in forest plants. *In* "Integrating Ecology and Evolution in a Spatial Context" (J. Silvertown and J. Antonovics, eds.), pp. 117–135. Blackwell Science, Oxford.

Bengtsson, G., Hedlund, K., and Rundgren, S. (1994). Food- and density-dependent dispersal: Evidence from a soil collembolan. *J. Anim. Ecol.* **63**, 513–520.

Bennett, A.F. (1990). Habitat corridors and the conservation of small mammals in a fragmented forest environment. *Landsc. Ecol.* **4**, 109–122.

Berg, L.M., Lascoux, M., and Pamilo, P. (1998). The infinite island model with sex-differentiated gene flow. *Heredity* **81**, 63–68.

Bergerud, A.T. (1988). Caribou, wolves and man. *Trends Ecol. Evol.* **3**, 68–72.

Bergman, K.O., and Landin, J. (2001). Distribution of occupied and vacant sites and migration of *Lopinga achine* (Nymphalidae: Satyrinae) in a fragmented landscape. *Biol. Conserv.* **102**, 183–190.

Bernstein, C., Auger, P., and Poggiale, J.C. (1999). Predator migration decisions, the ideal free distribution, and predator-prey dynamics. *Am. Nat.* **153**, 267–281.

Biere, A., and Antonovics, J. (1995). Sex-specific costs of resistance to the fungal pathogen *Ustilago violacea* (*Microbotryum violaceum*) in *Silene alba*. *Evolution* **50**, 1098–1110.

Biggins, D.E., and Kosoy, M.Y. (2001). Influences of introduced plague on North American mammals: Implications from ecology of plague in Asia. *J. Mammal.* **82**, 906–916.

Birch, L.C. (1960). The genetic factor in population ecology. *Am. Nat.* **94**, 5–24.

Bird, S.R., and Semeonoff, R. (1986). Selection for oviposition preference in *Drosophila melanogaster. Genet. Res. Camb.* **48**, 151–160.

Bjørnstad, O. and Bolker, B. (2000). Canonical functions for dispersal-induced synchrony. *Proc. R. Soci. Lond. Ser. B Biol. Sci.* **267**, 1787–1794.

Bjørnstad, O.N., Finkenstädt, B.F., and Grenfell, B.T. (2002a). Dynamics of measles epidemics. I. Estimating scaling of transmission rates using a time series SIR model. *Ecol. Monogr.* **72**, 169–184.

Bjørnstad, O.N., Ims, R.A., and Lambin, X. (1999). Spatial population dynamics: Analyzing patterns and processes of population synchrony. *Trends Ecol. Evol.* **14**, 427–432.

Bjørnstad, O.N., Peltonen, M., Liebhold, A.M., and Baltensweiler, W. (2002). Waves of larch budmoth outbreaks in the European Alps. *Science* **298**, 1020–1023.

Black, F.L. (1966). Measles endemicity in insular populations: Critical community size and its evolutionary implications. *J. Theor. Biol.* **11**, 207–211.

Blackburn, G.S., Wilson, D.J., and Krebs, C.J. (1998). Dispersal of juvenile collared lemmings (*Dicrostonys groenlandicus*) in a high-density population. *Can. J. Zool.* **76**, 2255–2261.

Blondel, J., Perret, P., and Maistre, M. (1990). On the genetical basis of the laying-date in an island population of blue tits. *J. Evol. Biol.* **3**, 469–475.

Blondel, J., Perret, P., Maistre, M., and Dias, P. C. (1992). Do harlequin Mediterranean environments function as a source sink for Tits (*Parus caeruleus* L.). *Lands. Ecol.* **7**, 213–219.

Blows, M.W., and Hoffmann, A.A. (1993). The genetics of central and marginal populations of *Drosophila serrata*. I. Genetic variation for stress resistance and species borders. *Evolution* **47**, 1255–1270.

Bolker, B.M. (1999). Analytic models for the patchy spread of plant disease. *Bull. Math. Biol.* **61**, 849–874.

Bolker, B.M. (2003). Combining endogenous and exogenous spatial variability in analytical population models. *Theor. Popul. Biol.* **64**, 255–270.

Bolker, B.M. and Pacala, S.W. (1997). Using moment equations to understand stochastically driven spatial pattern formation in ecological systems. *Theor. Popul. Biol.* **52**, 179–197.

Bolker, B.M. and Pacala, S.W. (1999). Spatial moment equations for plant competition: Understanding spatial strategies and the advantages of short dispersal. *Am. Nat.* **153**, 575–602.

Bolker, B.M., Pacala, S.W., and Levin, S.A. (2000). Moment methods for stochastic processes in continuous space and time. *In* "The Geometry of Ecological Interactions: Simplifying Spatial Complexity" (U. Dieckmann, R. Law, and J.A.J. Metz, eds.), 388–411. Cambridge Univ. Press, Cambridge.

Bolker, B.M., Pacala, S.W., and Neuhauser, C. (2003). Spatial dynamics in model plant communities: What do we really know? *Am. Nat.* **162**, 135–148.

Boonstra, R., and Krebs, C.J. (1977). A fencing experiment on a population of *Microtus townsendii. Can. J. Zool.* **55**, 1166–1175.

Boorman, S.A., and Levitt, P.R. (1973). Group selection on the boundary of a stable population. *Theor. Popul. Biol.* **4**, 85–128.

Boots, M., and Sasaki, A. (2000). The evolutionary dynamics of local infection and global reproduction in host-parasite interactions. *Ecol. Lett.* **3**, 181–185.

Boots, M., and Sasaki, A. (2002). Parasite-driven extinction in spatially explicit host-parasite systems. *Am. Nat.* **159**, 706–713.

Borcard, D., Legendre, P., and Drapeau, P. (1992). Partialling out the spatial component of ecological variation. *Ecology* **73**, 1045–1055.

Boudjemadi, K., Lecomte, J., and Clobert, J. (1999). Influence of connectivity on demography and dispersal in two contrasted habitats: An experimental approach. *J. Anim. Ecol.* **68**, 1–19.

Boughton, D.A. (1999). Empirical evidence for complex source-sink dynamics with alternative states in a butterfly metapopulation. *Ecology* **80**, 2727–2739.

Boughton, D.A. (2000). The dispersal system of a butterfly: A test of source-sink theory suggests the intermediate-scale hypothesis. *Am. Nat.* **156**, 131–144.

Boutin, V., Jean, R., Valero, M., and Vernet, P. (1988). *Gynodioecy in Beta maritima. Oecol. Plant.* **1**, 37–45.

Boyce, M.S. (1992). Population viability analysis. *Annu. Rev. Ecol. Syst.* **23**, 481–506.

Boyce, M.S. (1996). Review of RAMAS/GIS. *Q. Revi. Biol.* **71**, 167–168.

Braad, J. (2000). Nature restoration for the conservation of the tree frog in the Zuid-Eschmarke. *De Levende Natuur* **101**, 117–121. [In Dutch].

Braak, C.J.F. ter and Etienne, R.S. (2003). Improved Bayesian analysis of metapopulation data with an application to a tree frog metapopulation. *Ecology* **84**, 231–241.

Bramson, M., Cox, J.T., and Durrett, R. (1996). Spatial models for species area curves. *Ann. Prob.* **24**, 1727–1751.

Bramson, M., and Griffeath, D. (1980). Asymptotics for interacting particle systems on Z^d. *Z. Wahrsch. Verw. Gebiete* **53**, 183–196.

Bramson, M., Durrett, R., and Schonmann, R.H. (1991). The contact process in a random environment. *Ann. Prob.* **19**, 960–983.

Bray, J.R., and Curtis, J.T. (1957). An ordination of the upland forest communities of southern Wisconsin. *Ecol. Monogr.* **27**, 325–349.

Briscoe, B.K., Lewis, M.A., and Parrish, S.E. (2002). Home range formation in wolves due to scent marking. *Bull. Math. Biol.* **64**, 261–284.

Brodie, E.D., Jr., Ridenhour, B.J., and Brodie, E.D., III (2002). The evolutionary response of predators to dangerous prey: Hotspots and coldspots in the geographic mosaic of coevolution between garter snakes and newts. *Evolution* **56**, 22067–22082.

Brokaw, N., and Busing, N.T. (2000). Niche versus chance and tree diversity in forest gaps. *Trends Ecol. Evol.* **15**, 183–188.

Brook, B.W., Burgman, M.A., Akçakaya, H.R., O'Grady, J.J., and Frankham, R. (2002). Critiques of PVA ask the wrong questions: Throwing the heuristic baby out with the numerical bath water. *Conserv. Biol.* **16**, 262–263.

Brook, B.W., O'Grady, J.J. Chapman, A.P. Burgman, M.A. Akçakaya, H.R., and Frankham, R. (2000). Predictive accuracy of population viability analysis in conservation biology. *Nature* **404**, 385–387.

Brookes, M.I., Graneau, Y.A., King, P., Rose, O.C., Thomas, C.D., and Mallet, J.L.B. (1997). Genetic analysis of founder bottlenecks in the rare British butterfly *Plebejus argus*. *Conserv. Biol.* **11**, 648–661.

Brooks, S.P. (1998). Markov chain Monte Carlo method and its application. *J.R. Stat. Soc. Ser. D Stat.* **47**, 69–100.

Brooks, T., and Balmford, A. (1996). Atlantic forest extinctions. *Nature* **380**, 115.

Brooks, T.M., Pimm, S.L., and Collar, N.J. (1997) Deforestation predicts the number of threatened birds in insular southeast Asia. *Conserv. Biol.* **11**, 382–394.

Broome, L.S. (2001). Intersite differences in population demography of Mountain Pygmy-possums *Burramys parvus* Broom (1986–1998): Implications for metapopulation conservation and ski resorts in Koskiuszko National Park, Australia. *Biol. Conserv.* **102**, 309–323.

Brown, D.H. (2001). "Stochastic Spatial Models of Plant Diseases." Ph.D. thesis, UC Davis.

Brown, J. (1971). Mammals on mountaintops: Non-equilibrium insular biogeography. *Am. Nat.* **105**, 467–478.

Brown, J.H., and Kodric-Brown, A. (1977). Turnover rates in insular biogeography: Effect of immigration on extinction. *Ecology* **58**, 445–449.

Brown, J.S., and Pavlovic, N.B. (1992). Evolution in heterogeneous environments: Effects of migration on habitat specialization. *Evol. Ecol.* **6**, 360–382.

Brown, V.K., and Gange, A.C. (2002). Tritrophic below- and above-ground interactions in succesion. *In* "Multitrophic Level Interactions" (T. Tscharntke and B.A. Hawkins, eds.), pp. 197–222. Cambridge Univ. Press, Cambridge.

Bryant, A.A. (1996). Reproduction and persistence of Vancouver Island marmots (*Marmota vancouverensis*) in natural and logged habitats. *Can. J. Zool.* **74**, 678–687.

Bryant, A.A., and Janz, D.W. (1996). Distribution and abundance of Vancouver Island marmots (*Marmota vancouverensis*). *Can. J. Zool.* **74**, 667–677.

Bryant, E.H., McCommas, S.A., and Combs, L.M. (1986). The effect of an experimental bottleneck upon the quantitative genetic variation in the housefly. *Genetics* **114**, 1191–1211.

Buckling, A., Kassen, R., Bell, G., and Rainey, P.B. (2000). Disturbance and diversity in experimental microcosms. *Nature* **408**, 961–964.

Bullock, J.M., and Clarke, R.T. (2000). Long distance seed dispersal by wind: Measuring and modelling the tail of the curve. *Oecologia* **124**, 506–521.

Bullock, J.M., Clear Hill, B., Silvertown, J., and Sutton, M. (1995). Gap colonization as a source of grassland community change: Effects of gap size and grazing on the rate and mode of colonization by different species. *Okios* **72**, 273–282.

Bullock, J.M., Kenward, R.E., and Hails, R.S. (eds.) (2002). "Dispersal Ecology." The British Ecological Society, Blackwell, Oxford.

Bullock, J.M., Moy, I.L., Pywell, R.F., Coulson, S.J., Nolan, A.M., and Caswell, H. (2002). Plant dispersal and colonisation processes at local and landscape scales. *In* "Dispersal Ecology" (J.M. Bullock, R. Kenward, and R. Hails eds.), pp. 279–302. Blackwell Science, Oxford.

Bulman, C. (2001). "Conservation Biology of the Marsh Fritillary Butterfly *Euphydryas aurinia*". Ph.D. thesis, University of Leeds, UK.

Bulmer, M.G., and Taylor, P.D. (1981). Worker-queen conflict and sex-ratio theory in social hymenoptera. *Heredity* **47**, 197–207.

Bunn, A.G., Urban, D.L., and Keitt, T.H. (2000). Landscape connectivity: A conservation application of graph theory. *J. Environ. Manage.* **59**, 265–278.

Burdon, J.J. (1980). Intraspecific diversity in a natural population of *Trifolium repens. J. Ecol.* **68**, 717–736.

Burdon, J.J. (1987). "Diseases and Plant Population Biology." Cambridge Univ. Press, Cambridge

Burdon, J.J., and Thompson, J.N. (1995). Changed patterns of resistance in a population of *Linum marginale* attacked by the rust pathogen *Melampsora lini. J. Ecol.* **83**, 199–206.

Burkart, M. (2001). River corridor plants (Stromtalpflanzen) in central European lowland: A review of poorly understood plant distribution pattern. *Glob. Ecol. Biogeogr. Lett.* **10**, 449–468.

Burkey, T.V. (1995). Extinction rates in archipelagoes: Implications for populations in fragmented habitats. *Conserv. Biol.* **9**, 527–541.

Burkey, V.T. (1997). Metapopulation extinction in fragmented landscapes: Using bacteria and protozoa communities as model ecosystems. *Am. Nat.* **150**, 568–591.

Burland, T.M., Barratt, E.M., Nichols, R.A., and Racey, P.A. (2001). Mating patterns, relatedness and the basis of natal philopatry in the brown long-eared bat, *Plecotus auritus. Mol. Ecol.* **10**, 1309–1321.

Burland, T.M., Barratt, E.M., Beaumont, M.A., and Racey, P.A. (1999). Population genetic structure and gene flow in a gleaning bat, *Plecotus auritus. Proc. R. Soc. Lond. Ser. B Biol. Sci.*, **266**, 975–980.

Burton, R.S. (1987). Differentiation and integration in the genome of the marine copepod, *Tigriopus californicus. Evolution* **41**, 504–513.

Burton, R.S. (1990a). Hybrid breakdown in developmental time in the copepod *Tigriopus californicus. Evolution* **44**, 1814–1822.

Burton, R.S. (1990b). Hybrid breakdown in physiological-response: A mechanistic approach. *Evolution* **44**, 1806–1813.

Burton, R.S., Rawson, P.D., and Edmands, S. (1999). Genetic architecture of physiological phenotypes: Empirical evidence for coadapted gene complexes. *Am. Zool.* **39**, 451–462.

Butaye, J., Jacquemyn, H., and Hermy, M. (2001) Differential colonization causing non-random forest plant community structure in a fragmented agricultural landscape. *Ecography* **24**, 369–380.

Byers, D.L., and Waller, D.M. (1999). Do plant populations purge their genetic load? Effects of population size and mating history on inbreeding depression. *Annu. Rev. Ecol. Syst.* **30**, 479–513.

Byers, J.A. (2001). Correlated random walk equations of animal dispersal resolved by simulation. *Ecology* **82**, 1680–1690.

Byrom, A.E. (2002). Dispersal and survival of juvenile feral ferrets Mustela furo in New Zealand. *J. Appl. Ecol.* **39**, 67–78.

Caballero, A., Cusi, E., Garcia, C., and Garcia-Dorado, A. (2002). Accumulation of deleterious mutations: Additional *Drosophila melanogaster* estimates and a simulation of the effects of selection. *Evolution* **56**, 1150–1159.

Cabeza, M. (2003). Habitat loss and connectivity of reserve networks in probability approaches to reserve design. *Ecol. Lett.* **6**, 665–672.

Cabeza, M., and Moilanen, A. (2001). Design of reserve networks and the persistence of biodiversity. *Trends Ecol. Evol.* **16**, 242–248.

Cabeza, M., and Moilanen, A. (2003). Site-selection algorithms and habitat loss. *Conserv. Biol.* **17**, 1402–1413.

Cabeza, M., Araújo M.B., Wilson R.J., Thomas C.D., Cowley M.J.R. & Moilanen A. (2004). Combining probabilities of occurrence with spatial reserve design. *Journal of Applied Ecology*, in press.

Cadet, C., Ferrière, R., and Metz, J.A.J. (2003). The evolution of dispersal with density dependence and demographic stochasticity. *Am. Nat.*

Cain, M.L., Damman, H., and Muir, A. (1998). Seed dispersal and the Holocene migration of woodland herbs. *Ecol. Monogr.* **68**, 325–348.

Cain, M.L., Milligan, B.G., and Strand, A.E. (2000). Long-distance seed dispersal in plant populations. *Am. J. Bot.* **87**, 1217–1227.

Calow, P. (eds.) (1998). "The Encyclopedia of Ecology and Environmental Management." Blackwell Science, New York.

Camm, J.D., Polaski, S., Solow, A., and Csuti, B. (1996) A note on optimal algorithms for reserve site selection. *Biol. Conserv.* **78**, 353–355.

Cannings, C. (1974). The latent roots of certain Markov chains arising in genetics: A new approach. I Haploid models. *Adv. Appl. Prob.* **6**, 260–290.

Caprio, M.A., and Tabashnik, B.E. (1992). Gene flow accelerates local adaptation among finite populations: Simulating the evolution of insecticide resistance. *J. Econ. Entomol.* **85**, 611–620.

Carey, P.D., Watkinson, A.R., and Gerard, F.F.O. (1995). The determinants of the distribution and abundance of the winter annual grass *Vulpia ciliata* subsp. *ambigua. J. Ecol.* **83**, 177–188.

Carlson, A. (2000). The effect of habitat loss on a deciduous forest specialist species: The white-backed woodpecker (*Dendrocopos leucotos*). *For. Ecol. Manage.* **131**, 215–221.

Carlsson, A., and Kindvall, O. (2001). Spatial dynamics in a metapopulation network: Recovery of a rare grasshopper *Stauroderus scalaris* from population refuges. *Ecography* **24**, 452–460.

Caro, T.M., and Laurenson, M.K. (1994). Ecological and genetic factors in conservation: A cautionary tale. *Science* **263**, 485–486.

Carrière, Y., and Roitberg, B.D. (1995). Evolution of host-selection behaviour in insect herbivores: Genetic variation and covariation in host acceptance within and between populations of the oblique banded leaf-roller, *Choristoneura rosaceana* (family: Tortricidae). *Heredity* **74**, 357–368.

Carter, R.N., and Prince, S.D. (1988). Distribution limits from a demographic viewpoint. *In* "Plant Population Ecology" (A.J., Davy, M.J., Hutchins, and A.R. Watkinson, eds.), pp. 165–184. Blackwell Scientific, New York.

Casagrandi, R., and Gatto, M. (1999). A mesoscale approach to extinction risk in fragmented landscapes. *Nature* **400**, 560–562.

Casagrandi, R., and Gatto, M. (2002a). A persistence criterion for metapopulations. *Theor. Popul. Biol.* **61**, 115–125.

Casagrandi, R., and Gatto, M. (2002b). Habitat destruction, environmental catastrophes, and metapopulation extinction. *Theor. Popul. Biol.* **61**, 127–140.

Case, T.J., and Taper, M.L. (2000). Interspecific competition, environmental gradients, gene flow, and the coevolution of species' borders. *Am. Nat.* **155**, 583–605.

Caspersen, J.P., Silander, J.A., Canham, C.D., and Pacala, S.W. (1999). Modeling the competitive dynamics and distribution of tree species along soil moisture gradients, *In* "Spatial Modeling of Forest Landscape Change" (D. Mladenoff and W. Baher, eds.). Cambridge Univ. Press, Cambridge.

Caswell, H. (1978). Predator-mediated coexistence: A non-equilibrium model. *Am. Nat.* **112**, 127–154.

Caswell, H. (1989). "Matrix Population Models." Sinauer, Sunderland, MA.

Caswell, H. (2001). "Matrix Population Models: Construction, Analysis, and Interpretation." Sinauer Associates, Sunderland, MA.

Caswell, H., and Cohen, J.E. (1991). Disturbance, interspecific interaction, and diversity in metapopulations. *Biol. J. Linn. Soc.* **42**, 193–218.

Caughley, G. (1994). Directions in conservation biology. *J. Anim. Ecol.* **63**, 215–244.

Caughley, G.A., and Gunn, A. (1996). "Conservation Biology in Theory and Practice." Blackwell Science, Cambridge, MA.

Charbonnel, N., Angers, B., Rasatavonjizay, R., Bremond, P., Debain, C., and Jarne, P. (2002). The influence of mating system, demography, parasites and colonization on the population structure of *Biomphalaria pfeifferi* in Madagascar. *Mol. Ecol.* **11**, 2213–2228.

Charlesworth, B. (1994). "Evolution in Age-Structured Populations". Cambridge Univ. Press, Cambridge.

Charlesworth, B., and Charlesworth, D. (1978). A model for the evolution of dioecy and gynodioecy *Am. Nat.* **112**, 975–997.

Charlesworth, B., and Charlesworth, D. (1999). The genetic basis of inbreeding depression. *Genet. Res.* **74**, 329–340.

Charlesworth, D. (1984). Androdioecy and the evolution of dioecy. *Biol. J. Lin. Soc.* **22**, 333–348.

Charlesworth, D., and Laporte, V. (1998). The male-sterility polymorphism of *Silene vulgaris*: Analysis of genetic data from two populations and comparison with *Thymus vulgaris*. *Genetics* **150**, 1267–1282.

Charnov, E. (2002). Reproductive effort, offspring size and benefit-cost ratios in the classification of life histories. *Evol. Ecol. Res.* **4**, 749–758.

Charnov, E.L., and Shaffer, W.M. (1973). Life history consequences of natural selection: Cole's result revisited. *Am. Nat.* **107**, 791–793.

Chase, J. and Leibold, M.A. (2003). "Ecological Niches". University of Chicago Press, Chicago.

Chave, J. (1999). Study of structural, successional and spatial patterns in tropical rain forests using TROLL, a spatially explicit forest model. *Ecol. Model.* **124**, 233–254.

Chave, J., and Leigh, E.G. (2002). A spatially explicit neutral model of beta-diversity in tropical forests. *Theor. Popul. Biol.* **62**, 153–168.

Chave J., Muller-Landau, H.C., and Levin, S.A. (2002). Comparing classical community models: Theoretical consequences for patterns of diversity. *Am. Nat.* **159**, 1–23.

Cheptou, P.O., Lepart, J., and Escarré, J. (2001). Differential outcrossing rates in dispersing and non-dispersing achenes with heterocarpic plant *Crepis sancta* (Asteraceae). *Evol. Ecol.* **15**, 1–13.

Chesser, R. (1991a). Gene diversity and female philopatry. *Genetics* **127**, 437–447.

Chesser, R. (1991b). Influence of gene flow and breeding tactics on gene diversity within populations. *Genetics* **129**, 573–583.

Chesser, R.K., Smith, M.H., and Brisbin, I.L. (1980). Management and maintenance of genetic variability in endangered species. *Int. Zool. Yearb.* **20**, 146–154.

Chesson, P.L. (2000a). General theory of competitive coexistence in spatially varying environments. *Theor. Popul. Biol.* **58**, 211–237.

Chesson, P.L. (2000b). Mechanisms of maintenance of species diversity. *Annu. Revi. Ecol. Syst.* **31**, 343–366.

Cheverud, J.M., and Routman, E.J. (1995). Epistasis and its contribution to genetic variance components. *Genetics* **139**, 1455–1461.

Christiansen, F.B. (1975). Hard and soft selection in a subdivided population. *Am. Nat.* **109**, 11–16.

Church, R.L., Stoms, D.M., and Davis, F.W. (1996). Reserve selection as maximal covering location problem. *Biol. Conserv.* **76**, 105–112.

Cincotta, R.P., Uresk, D.W., and Hansen, R.M., (1987). Demography of black-tailed prairie dog populations reoccupying sites treated with rodenticide. *Great Bas. Nat.* **47**, 339–343.

Clark, A.G. (1990). Genetic components of variation in energy storage in *Drosophila melanogaster. Evolution* **44**, 637–650.

Clark, J.S. (1998). Why trees migrate so fast: Confronting theory with dispersal biology and the paleorecord. *Am. Nat.* **152**, 204–224.

Clark, J.S., Fastie, C., Hurtt, G., Jackson, S.T., Johnson, C., King, G.A., Lewis, M., Lynch, J., Pacala, S., Prentice, C., Schupp, E.W., Webb, T., and Wyckoff, P. (1998).

Reid's paradox of rapid plant migration: Dispersal theory and interpretation of paleoecological records. *Bioscience* **48**, 12.

Clark, J.S., Silman, M., Kern, R., Macklin, E., and HillRisLambers, J. (1999). Seed dispersal near and far: Patterns across temperate and tropical forests. *Ecology* **80**, 1475–1494.

Clay, K. (1990). The impact of parasitic and mutualistic fungi on competitive interactions among plants. *In* "Perspectives on Plant Competition" (J.B. Grace and D. Tilman, eds.), pp. 391–412. Academic Press, San Diego.

Clements, F.E. (1916). "Plant Succession." Carnegie Institute of Washington Publication 242.

Cliff, A.D., and Haggett, P. (1988). "Atlas of Disease Distributions: Analytic Approaches to Epidemiologic Data." Basil Blackwell, Oxford.

Cliff, A.P., Haggett, P., Ord, J.K., and Versey, G.R. (1981). "Spatial Diffusion: An Historical Geography of Epidemics in an Island Community." Cambridge Univ. Press, Cambridge.

Cliff, A.D., Haggett, P., and Smallman-Raynor M. (1993). "Measles: An Historical Geography of a Major Human Viral Disease from Global Expansion to Local Retreat, 1840–1990." Blackwell, Oxford.

Clinchy, M. (1997). Does immigration "rescue" populations from extinction? Implications regarding movement corridors and the conservation of mammals. *Oikos* **80**, 618–622.

Clinchy, M., Haydon, D.T., and Smith, A.T. (2002). Pattern does not equal process: What does patch occupancy really tell us about metapopulation dynamics? *Am. Nat.* **159**, 351–362.

Clobert, J., Danchin, E., Dhondt, A.A., and Nichols, J.D. eds. (2001). "Dispersal". Oxford Univ. Press, Oxford.

Clobert, et al (1994): Determinant of dispersal behavior: The common lizard as a case study. *In* Lizard Ecology: Historical and Experimental Perspectives (L.J. Vitt and E.R. Pianka, eds.), pp. 182–206. Princeton Univ. Press, Princeton, NJ.

Clobert, J., Wolff, J.O., Nichols, J.D., Danchin, E., and Dhondt, A.A. (2001). Introduction. *In* "Dispersal" (J. Clobert, E. Danchin, A.A. Dhondt, and J.D. Nichols, eds.), pp XVII–XXI. Oxford Univ. Press, Oxford.

Cocks, K.D., and Baird, I.A. (1989). Using mathematical programming to address the multiple reserve selection problem: An example from the Eyre Peninsula, South Australia. *Biol. Conserv.* **49**, 113–130.

Cody, M.L., and Overton, J.M. (1996). Short-term evolution of reduced dispersal in island plant populations. *J. Ecol.* **84**, 53–61.

Cohen, D., and Eshel, I. (1976). On the founder effect and the evolution of altruistic traits. *Theor. Popul. Biol.* **10**, 276–302.

Coltman, D.W., Pilkington, J.G., Smith, J.A., and Pemberton, J.M. (1999). Parasite-mediated selection against inbred Soay sheep in a free-living, island population. *Evolution* **53**, 1259–1267.

Comins, H.N., Hamilton, W.D., and May, R.M. (1980). Evolutionary stable dispersal strategies. *J. Theor. Biol.* **82**, 205–230.

Comins, H.N., Hassell, M.P., and May, R.M. (1992). The spatial dynamics of host-parasitoid systems. *J. Anim. Ecol.* **61**, 735–748.

Connell, J.H. (1978). Diversity in tropical rainforests and coral reefs. *Science* **199**, 1302–1310.

Conrad, K.F., Willson, K.H., Harvey, I.F., Thomas, C.J., and Sherratt, T.N. (1999). Dispersal characteristics of seven odonate species in an agricultural landscape. *Ecography* **22**, 524–531.

Conradt, L., Bodsworth, E.J., Roper, T., and Thomas, C.D. (2000). Non-random dispersal in the butterfly *Maniola jurtina*: Implications for metapopulation models. *Proc. R. Soci. Seri. B* **267**, 1505–1510.

Conradt, L., Roper, T.J., and Thomas, C.D. (2001). Dispersal behaviour of individuals in metapopulations of two British butterflies. *Oikos* **95**, 416–424.

Conradt, L., Zollner, P.A., Roper, T.J., Frank, K., and Thomas, C.D. (2003). Foray search: An effective systematic dispersal strategy in fragmented landscapes. *Am. Nat.* **161**, 905–915.

Cook, R.E. (1983). Clonal plant populations. *Am. Sci.* **71**, 244–253.

Cook, R.R., and Hanski, I. (1995). On expected lifetimes of small-bodied and large-bodied species of birds on islands. *Am. Nat.* **145**, 307–315.

Cooper, A.B., and Mangel, M. (1999). The dangers of ignoring metapopulation structure for the conservation of salmonids. *Fish. Bull.* **97**, 213–226.

Corander, J., Waldmann, P., and Sillanpää, M. (2003). Bayesian analysis of genetic differentiation between populations. *Genetics* **163**, 367–374.

Corbet, G., Cummins, J., Hedges, S., and Krzanowski, W. (1970). The taxonomic status of British water voles, genus *Arvicola*. *J. Zool.* **161**, 301–316.

Corbit, M., Marks, P.L., and Gardescu, S. (1999). Hedgerows as habitat corridors for forest herbs in central New York, USA. *J. Ecol.* **87**, 220–232.

Cornell, S.J., Isham, V.S., and Grenfell, B.T. (2000). Drug-resistant parasites and aggregated infection: Early-season dynamics. *J. Math. Biol.* **41**, 341–360.

Costello, C., and Polasky, S. (2003). Dynamic reserve site selection. *Res. Energy Econ.* in press.

Cottenie, K., Nuytten, N., Michels, E., and De Meester, L. (2001). Zooplankton community structure and environmental conditions in a set of interconnected ponds. *Hydrobiologia* **442**, 339–350.

Coulaud, J., and McNeilly, T. (1992). Zinc tolerance in populations of *Deschampsia cespitosa* (Gramineae) beneath electricity pylons. *Plant Syst. Evol.* **179**, 175–185.

Coulson, J.C., Mace, G.M., Hudson, E., and Possingham, H. (2001). The use and abuse of population viability analysis. *Trends Ecol. Evol.* **16**, 219–221.

Courchamp, F., Cluttonb-Brock, T., and Grenfell, B. (1999). Inverse density dependence and the Allee effect. *Trends Ecol. Evol.* **14**, 405–410.

Cousins, S.A.O., and Eriksson, O. (2001). Plant species occurrences in a rural hemiboreal landscape: Effects of remnant habitats, site history, topography and soil. *Ecography* **24**, 461–469.

Couvet, D. (2002). Deleterious effects of restricted gene flow in fragmented populations. *Conserv. Biol.* **16**, 369–376.

Couvet, D., Atlan, A., Belhassen, E., Gliddon, C., Gouyon, P.-H., and Kjellberg, F. (1990). Co-evolution between two symbionts: The case of cytoplasmic male-sterility in higher plants. *In* "Oxford Surveys in Evolutionary Biology" (D. Futuyma and J. Antonovics, eds.), pp. 225–247. Oxford Univ. Press, Oxford.

Couvet, D., Bonnemaison, F., and Gouyon, P.H. (1986). The maintenance of females among hermaphrodites : The importance of nuclear-cytoplasmic interactions. *Heredity* **57**, 325–330.

Couvet, D., Gouyon, P., Kjellberg, F., Olivieri, I., Pomente, D., and Valdeyron, G. (1985). From metapopulation to neighbourhood: Genetics of unbalanced populations. De la metapopulation au voisinage: La genetique des populations en desequilibre. *Genet. Selec. Evol.* **17**, 407–414.

Couvet, D., Ronce, O., and Gliddon, C. (1998). The maintenance of nucleocytoplasmic polymorphism in a metapopulation: The case of gynodioecy. *Am. Nat.* **152**, 59–70.

Cowley, M.J.R., Wilson, R.J., León-Cortés, J.L., Gutiérrez, D., Bulman, C.R., and Thomas, C.D. (2000). Habitat-based statistical models for predicting the spatial distribution of butterflies and day-flying moths in a fragmented landscape. *J. Appl. Ecol.* **37**, 60–72.

Cowling, R.M., Pressey, R., Lombard, A.T., Desmet, P.G., and Ellis, A.G. (1999). From representation to persistence: Requirements for a sustainable reserve system in the species-rich mediterranean-climate deserts of southern Africa. *Div. Distrib.* **5**, 51–71.

Cowlishaw, G. (1999). Predicting the pattern of decline of African primate diversity: An extinction debt from historical deforestation. *Conserv. Biol.* **13**, 1183–1193.

Coyne, J.A., Barton, N.H., and Turelli, M. (1997). Perspective: A critique of Sewall Wright's shifting balance theory of evolution. *Evolution* **51**, 643–671.

Coyne, J.A., Barton, N.H., and Turelli, M. (2000). Is Wright's shifting balance process important in evolution? *Evolution* **54**, 306–317.

Craig, D.M. (1982). Group selection versus individual selection: An experimental analysis. *Evolution* **36**, 271–282.

Crnokrak, P., and Roff, D.A. (1999). Inbreeding depression in the wild. *Heredity* **83**, 260–270.

Crombaghs, B.H.J.M., and Lenders, H.J.R. (2001) Protection Plan Tree Frog 2001–2005. Ministry of Agriculture, Nature Conservation and Fisheries, The Hague. [In Dutch]

Crone, E.E., Doak, D., and Pokki, J. (2001). Ecological influences on the dynamics of a field vole metapopulation. *Ecology* **82**, 831–843.

Crow, J.F. (1948). Alternative hypotheses of hybrid vigor. *Genetics* **33**, 477–487.

Crow, J.F. (1958). Some possibilities for measuring selection intensities in man. *Hum. Biol.* **30**, 1–13.

Crow, J.F. (1993). Mutation, mean fitness and genetic load. *Oxf. Surve. Evol. Biol.* **9**, 3–42.

Crow, J.F., and Kimura, M. (1970). "An Introduction to Population Genetics Theory". Harper and Row, New York.

Crowley, P.H., and McLetchie, D.N. (2002). Trade-offs and spatial life-history strategies in classical metapopulations. *Am. Nat.* **159**, 190–208.

Cruzan, M.B. (2001) Population size and fragmentation thresholds for the maintenance of genetic diversity in the herbaceous endemic *Scutellaria montana* (Lamiaceae). *Evolution* **55**, 1569–1580.

Csuti, B., Polaski, S., Williams, P.H., Pressey, R., Camm, J.D., Kershaw, M., Kiester, R., Downs, B., et al. (1997). A comparison of reserve selection algorithms using data on terrestrial vertebrates in Oregon. *Biol. Conserv.* **80**, 83–97.

Cuddington, K.M., and Yodzis, P. (2000). Diffusion-limited predator-prey dynamics in euclidean environments: An allometric individual-based model. *Theor. Popul. Biol.* **58**, 259–278.

Cuddington, K., and Yodzis, P. (2002). Predator-prey dynamics and movement in fractal environments. *Am. Nat.* **160**, 119–134.

Cuguen, J., Wattier, R., Saumitou-Laprade, P., Forcioli, D., Mörchen, M., Van Dijk, H., and Vernet, P. (1994). Gynodioecy and mitochondrial DNA polymorphism in natural populations of *Beta vulgaris* ssp. *maritima*. *Genet. Sel. Evol.* **26**, 87–101.

Cully, J.F., and Williams, E.S. (2001). Interspecific comparisons of sylvatic plague in prairie dogs. *J. Mammal.* **82**, 894–905.

Cumming, G.S. (2002). Habitat shape, species invasions, and reserve design: insights from simple models. Conservation Ecology 6. URL http://www.consecol.org/vol6/iss1/art3.

Cunningham, S.A. (2000). Effects of habitat fragmentation on the reproductive ecology of four plant species in Mallee woodland. *Conserv. Biol.* **14**, 758–768.

Currey, D.R. (1965). An ancient bristlecone pine stand in eastern Nevada. *Ecology* **46**, 564–566.

Dale, V.H., Pearson, S.M., Offerman, H.L., and O'Neill, R.V. (1994). Relating patterns of land-use change to faunal biodiversity in the central Amazon. *Conserv. Biol.* **8**, 1027–1036.

Danchin, E., Boulinier, T., and Massot, M. (1998). Conspecific reproductive success and breeding habitat selection: Implications for the study of coloniality. *Ecology* **79**, 2415–2428.

Danchin, E., Heg, D., and Doligez, B. (2001). Public information and breeding habitat selection. *In* "Dispersal" (J. Clobert, E. Danchin, A.A. Dhondt, and J.D. Nichols, eds.), pp. 243–258. Oxford Univ. Press, Oxford.

Danchin, E., and Wagner, R.H. (1997). The evolution of coloniality: The emergence of new perspectives. *Trends Ecol. Evol.* **12**, 342–347.

Daniels, S.J., and Walters, J.R. (2000). Between year breeding dispersal in red cockaded woodpeckers: Multiple causes and estimated cost. Ecology 81, 2473–2484.

Danielson, B.J., and Gaines, M.S. (1987). The influences of conspecific and heterospecific residents on colonization. *Ecology* **68**, 1778–1784.

Darroch, J.N., and Seneta, E. (1965). On quasi-stationary distributions in absorbing discrete-time finite Markov chains. *J. Appl. Prob.* **2**, 88–100.

Darroch, J.N., and Seneta, E. (1967). On quasi-stationary distributions in absorbing continuous-time finite Markov chains. *J. Appl. Prob.* **4**, 192–196.

Darwin, C. (1859). "On the Origin of Species by Means of Natural Selection." John Murray, London.

Darwin, C. 1876 (1993). "The Autobiography of Charles Darwin" (N. Barlow, ed.), Norton, New York.

Davies, C.M., Webster, J.P., Kruger, O., Munatsi, A., Ndamba, J., and Woolhouse, M.E.J. (1999). Host-parasite population genetics: A cross-sectional comparison of *Bulinus globosus* and *Schistosoma haematobium*. *Parasitology* **119**, 295–302.

Davis, A.J., Jenkinson, L.S., Lawton, J.H., Shorrocks, B., and Wood, S. (1998). Making mistakes when predicting shifts in species range in response to global warming. *Nature* **391**, 783–786.

Dawson, K.J., and Belkhir, K. (2001). A Bayesian approach to the identification of panmictic populations and the assignment of individuals. *Genet. Res.* **78**, 59–77.

DeAngelis, D. and Gross, L. (eds.) (1992). "Populations and Communities: An Individual-Based Perspective". Chapman and Hall, New York.

DeAngelis, D.L., Gross, L.J., Huston, M.A., Wolff, W.F., Fleming, D.M. Comiskey, E.J., and Sylvester, S.M. (1998). Landscape modeling for Everglades ecosystem restoration. *Ecosystems* **1**, 64–75.

Deem, S.L., Karesh, W.B., and Weisman, W. (2001). Putting theory into practice: Wildlife health in conservation. *Conserv. Biol.* **15**, 1224–1233.

de Fraipont, M., Clobert, J., John-Alder, H., and Meylan, S. (2000). Increased prenatal maternal corticosterone promotes philopatry of offspring in common lizards *Lacerta vivipara*. *J. Anim. Ecol.* **69**, 404–413.

de Jong, G. (1979). The influence of the distribution of juveniles over patches of food on the dynamics of a population. *Netherlands J. Zool.* **29**, 33–51.

de Jong, T.J., Klinkhamer, P.G.L., and de Heiden, J.L.H. (2000). The evolution of generation time in metapopulations of monocarpic perennial plants: Some theoretical considerations and the example of the rare thistle *Carlina vulgaris*. *Evol. Ecol.* **14**, 213–231.

de Jong, T.J., van Batenburg, F.H.D., and van Dijk, J. (2002). Seed sex ratio in dioecious plants depends on relative dispersal of pollen and seeds: An example using a chessboard simulation model. *J. Evol. Biol.* **15**, 373–379.

Delattre, M., and Felix, M.A. (2001). Microevolutionary studies in nematodes: A beginning. *Bioessays* **23**, 807–819.

Delmotte, F., Bucheli, E., and Shykoff, J.A. (1999). Host and parasite population structure in a natural plant-pathogen system. *Heredity* **82**, 300–308.

Dempster, E.R. (1955). Maintenance of genetic heterogeneity. *Cold Spring Harb. Symp. Quant. Biol.* **20**, 25–32.

Dempster, E.R. (1963). Concepts and definitions in relation to selection schemes. *In* "Statistical Genetics and Plant Breeding" (W.D. Hanson, and H.F. Robinson, eds.), pp. 34–44. National Academy of Sciences/National Research Council, Washington, DC.

den Boer, P.J. (1968). Spreading of risk and stabilization of animal numbers. *Acta Biotheor.* **18**, 165–194.

Dennis, B. (2002). Allee effects in stochastic populations. *Oikos* **96**, 389–401.

Dennis, B., Munholland, P.L., and Scott, J.M. (1991). Estimation of growth and extinction parameters for endangered species. *Ecol. Monogr.* **64**, 205–224.

Dennis R.L.H., and Eales, H.T. (1999). Probability of site occupancy in the large heath butterfly *Coenonympha tullia* determined from geographical and ecological data. *Biol. Conserv.* **87**, 295–301.

Denno, R.F. (1994). The evolution of dispersal polymorphism in insects: The influence of habitats, host plants and mates. *Res. Pop. Ecol.* **36**, 127–135.

Denno, R.F., and Peterson, M.A. (1995). Density-dependent dispersal and its consequences for population dynamics. *In* "Population Dynamics: New Approaches and Synthesis" (N. Cappucino and P.W. Price, eds.), pp. 113–130. Academic Press, San Diego.

Denno, R.F., Roderick, G.K., Olmstead, K.L., and Dobel, H.G. (1991). Density-related migration in planthoppers (Homoptera, Delphacidae). The role of habitat persistence. *Am. Nat.* **138**, 1513–1541.

Denno, R.F., Roderick, G.K., Peterson, M.A., Huberty, A.F., Döbel, H.G., Eubanks, M.D., Losey, J.E., and Langellotto, G.A. (1996). Habitat persistence underlies intraspecific variation in the dispersal strategies of planthoppers. *Ecol. Monogr.* **66**, 389–408.

Denslow, J. (1980). Gap partitioning among tropical rainforest trees. *Biotrop. Supp.* **12**, 47–55.

Diamond, J.M. (1984). "Normal" extinctions of isolated populations. *In* "Extinctions" (M.H. Nitecki, ed.), pp. 191–246. University of Chicago Press, Chicago.

Dias, P.C. (1996). Sources and sinks in population biology. *Trends Ecol. Evol.* **11**, 326–330.

Dias, P.C., and Blondel, J. (1996). Local specialization and maladaptation in the Mediterranean blue tit (*Parus caeruleus*). *Oecologia* **107**, 79–86.

Dias, P.C., Verheyen, G.R., and Raymond, M. (1996). Source-sink populations in Mediterranean blue tits: Evidence using single-locus minisatellite probes. *J. Evol. Biol.* **9**, 965–978.

Dieckmann, U., and Law, R. (2000). Relaxation projections and the method of moments. *In* The Geometry of Ecological Interactions: Simplifying Spatial Complexity (U. Dieckmann, R. Law, and J.A.J. Metz, eds.), pp. 412–455. Cambridge Univ. Press, Cambridge.

Dieckmann, U., and Law, R., and Metz, J.A.J. (eds.), (2000). The geometry of ecological interactions: Simplifying spatial complexity. Cambridge Univ. Press, Cambridge.

Diffendorfer, J.E. (1998). Testing models of source-sink dynamics and balanced dispersal. *Oikos* **81**, 417–433.

Dingle, H. (1996). "Migration: The Biology of Life on the Move". Oxford Univ. Press, Oxford.

Dixon, A.F.G. (1985). "Aphid Ecology." Chapman & Hall, London.

Doak, D.F. (1995). Source-sink models and the problem of habitat degradation: General models and applications to the Yellowstone grizzly. *Conserv. Biol.* **9**, 1370–1379.

Dobson, A., and Crawley, W. (1994). Pathogens and the structure of plant-communities. *Trends Ecol. Evol.* **9**, 393–398.

Dobson, A.P., and Hudson, P.J. (1992). Regulation and stability of a free-living host-parasite system—*Trichostrongylus tenuis* in red grouse. 2. Population models. *J. Anim. Ecol.* **61**, 487–498.

Dobzhansky, T.G. (1937). "Genetics and the Origin of Species." Columbia Univ. Press, New York.

Dobzhansky, T. (1950). Genetics of natural populations. XIX. Origin of heterosis through natural selection in populations of *Drosophila pseudoobscura*. *Genetics* **35**, 288–302.

Dobzhansky, T. (1970). "Genetics of the Evolutionary Process." Columbia Univ. Press, New York.

Doebeli, M. (1995). Dispersal and dynamics. *Theor. Popul. Biol.* **47**, 82–106.

Doebeli, M., and Ruxton, G.D. (1997). Evolution of dispersal rates in metapopulation models: Branching and cyclic dynamics in phenotype space. *Evolution* **51**, 1730–1741.

Doebley, J., and Stec, A. (1993). Inheritance of the morphological differences between maize and teosinte: Comparison of results for two F2 populations. *Genetics* **134**, 559–570.

Doebley, J., Stec, A., and Gustus, C. (1995). Teosinte branched1 and the origin of maize: Evidence for epistasis and the evolution of dominance. *Genetics* **141**, 333–346.

Doligez, B., Danchin, E., and Clobert, J., (2002). Public information and breeding habitat selection in a wild bird population. *Science* **297**, 1168–1170.

Doligez, B., Danchin, E., Clobert, J., and Gustafsson, L. (1999). The use of conspecifics reproductive success for breeding habitat selection in a non colonial, hole nesting species, the collared flycatcher *Ficedula albicollis*. *J. Anim. Ecol.* **68**, 1–15.

Dommée, B., Assouad, M.W., and Valdeyron, G. (1978). Natural selection and gynodioecy in *Thymus vulgaris* L. *Bot. J. Linn. Soc.* **77**, 17–28.

Donalson, D.D., and Nisbet, R.M. (1999). Population dynamics and spatial scale: effects of system size on population persistence. *Ecology* **80**, 2492–2507.

Doncaster, C.P. (2000). Extension of ideal free resource use to breeding populations and metapopulations. *Oikos* **89**, 24–36.

Doncaster, C.P., Clobert, J., Doligez, B., Gustaffson, L., and Danchin, E. (1997). Balanced dispersal between spatially varying local populations: An alternative to the source-sink model. *Am. Nat.* **150**, 425–445.

Doncaster, C.P., and Gustafsson, L. (1999). Density dependence in resource exploitation: Empirical test of Levins' metapopulation model. *Ecol. Lett.* **2**, 44–51.

Donnelly, P., and Tavaré, S. (1995). Coalescents and genealogical structure under neutrality. *Annu. Rev. Genet.* **29**, 401–421.

Donovan, T.M., Jones, P.W., Annand, E.M., and Thompson, F.R., III (1997). Variation in local-scale edge effects: Mechanisms and landscape context. *Ecology* **78**, 2064–2075.

Donovan, T.M., Lamberson, R.H., Kimber, A., Thompson, F.R., III, and Faaborg, J. (1995a). Modeling the effects of habitat fragmentation on source-sink demography of Neotropical migrant birds. *Conserv. Biol.* **9**, 1396–1407.

Donovan, T.M., and Thompson, F.R. (2001). Modeling the ecological trap hypothesis: A habitat and demographic analysis for migrant songbirds. *Ecol. Appl.* **11**, 871–882.

Donovan, T.M., Thompson, F.R., III, Faaborg, J., and Probst, J.R. (1995b). Reproductive success of migratory birds in habitat sources and sinks. *Conserv. Biol.* **9**, 1380–1395.

Dooley, J.L., Jr., and Bowers, M.A. (1998). Demographic responses to habitat fragmentation: Experimental tests at the landscape and patch scale. *Ecology* **78**, 2064–2075.

Dow, B.D., and Ashley, M.V. (1996). Microsatellite analysis of seed dispersal and parentage of saplings on bur oak, *Quercus macrocarpa*. *Mol. Ecol.* **5**, 615–627.

Dowdeswell, W.H., Fisher, R.A., and Ford, E.B. (1940). The quantitative study of populations in the Lepidoptera. *Ann. Eugen.* **10**, 123–136.

Drechsler, M., Frank, K., Hanski, I., O'Hara, R.B., and Wissel, C. (2003). Ranking metapopulation extinction risk for conservation: From patterns in data to management decisions. *Ecol. Appl.* **13**, 990–998.

Drechsler, M., and Wissel, C. (1997). Separability of local and regional dynamics in metapopulations. *Theor. Popul. Biol.* **51**, 9–21.

Dufty, A.M., Jr., and Belthoff, J.R. (2001). Proximate mechanisms of natal dispersal: The role of body condition and hormones. *In* "Dispersal" (J. Clobert, E.Danchin, A.A. Dhondt, and J.D. Nichols, eds.), pp. 217–229. Oxford Univ. Press, Oxford.

Dufty, A.M., Jr., Clobert, J., and Møller, A.P. (2002). Hormones, developmental plasticity, and adaptation. *Trends Ecol. Evol.* **17**, 190–196.

Dunning, J.B., Stewart, D.J., Danielson, B.J., Noon, B.R., Root, T.L., Lamberson, R.H., and Stevens, E.E. (1995). Spatially explicit population models: Current forms and future uses. *Ecol. Appl.* **5**, 3–11.

Durrett, R. (1988). Crabgrass, measles, and gypsy moths: An introduction to modern probability. *Bull. Am. Math. Soc.* **18**, 117–143.

Durrett, R. (1992). Multicolor particle systems with large threshold and range. *J. Theor. Prob.* **5**, 127–152.

Durrett, R., and Levin, S. (1994). The importance of being discrete (and spatial). *Theor. Popul. Biol.* **46**, 363–394.

Durrett, R., and Levin, S. (1996). Spatial models for species-area curves. *J. Theor. Biol.* **179**, 119–127.

Durrett, R., and Levin, S. (2000). Lessons on pattern formation from planet WATOR. *J. Theor. Biol.* **205**, 201–214.

Durrett, R., and Neuhauser, C. (1994). Particle systems and reaction-diffusion equations. *Ann. Prob.* **22**, 289–333.

Duryea, M., Caraco, T., Gardner, G., Maniatty, W., and Szymanski, B.K. (1999). Population dispersion and equilibrium infection frequency in a spatial epidemic. *Phys. D* **132**, 511–519.

Dushoff, J. (1999). Host heterogeneity and disease endemicity: A moment-based approach. *Theor. Popul. Biol.* **56**, 325–335.

Dybdahl, M.F., and Lively, C.M. (1996). The geography of coevolution: Comparative population structures for a snail and its trematode parasite. *Evolution* **50**, 2264–2275.

Earn, D., Rohani, P., and Grenfell, B.T. (1998). Spatial dynamics and persistence in ecology and epidemiology. *Proc. R. Soc. Lond. B* **265**, 7–10.

Ebenhard, T. (1990). A colonization strategy in field voles (*Microtus agrestis*): Reproductive traits and body size. *Ecology*, **71**, 1833–1848.

Eber, S., and Brandl, R. (1994). Ecological and genetic spatial patterns of *Urophora cardui* (Diptera: Tephritidae) as evidence for population structure and biogeographical processes. *J. Anim. Ecol.* **63**, 187–199.

Eber, S., and Brandl, R. (1996). Metapopulation dynamics of the tephritid fly *Urophora cardui*: An evaluation of incidence-function model assumptions with field data. *J. Anim. Ecol.* **65**, 621–630.

Ebert, D., Haag, C., Kirkpatrick, M., Riek, M., Hottinger, J.W., and Pajunen, V.I. (2002). A selective advantage to immigrant genes in a *Daphnia* metapopulation. *Science* **295**, 485–488.

Eckert, C.G., Manicacci, D., and Barrett, S.C.H. (1996). Genetic drift and founder effects in native and introduced populations of an invading plant, *Lythrum salicaria* (Lythraceae). *Evolution* **50**, 1512–1519.

Edelstein-Keshet, L. (1988). "Mathematical Models in Biology." Random House, New York.

Edmands, S. (1999). Heterosis and outbreeding depression in interpopulation crosses spanning a wide range of divergence. *Evolution* **53**, 1757–1768.

Edmands, S. (2002). Does parental divergence predict reproductive compatibility? *Trends Ecol. Evol.* **17**, 520–527.

Edmands, S., and Burton, R. (1998). Variation in cytochrome-C oxidase activity is not maternally inherited in the copepod *Tigriopus californicus. Heredity* **80**, 668–674.

Ehrlén, J., and Eriksson, O. (2000). Dispersal limitation and patch occupancy in forest herbs. *Ecology* **81**, 1667–1674.

Ehrlén, J., and Eriksson, O. (2003). Large-scale spatial dynamics of plants: A response to Freckleton & Watkinson. *J. Ecol.* **91**, 316–320.

Ehrlén, J., and Lehtilä, K. (2002). How perennial are perennial plants? *Oikos* **98**, 308–322.

Ehrlich, P.R. (1961). Intrinsic barriers to dispersal in checkerspot butterfly. *Science* **134**, 108–109.

Ehrlich, P.R. (1965). The population biology of the butterfly, *Euphydryas editha.* II. The structure of the Jasper Ridge colony. *Evolution* **19**, 327–336.

Ehrlich, P.R. (1984). The structure and dynamics of butterfly populations. *Symp. R. Entomol. Soc. Lond.* **11**, 25–40.

Ehrlich, P., and Hanski, I. (2004). "On the Wings of Checkerspots." Oxford Univ. Press, Oxford.

Ehrlich, P.R., and Daily, G.C. (1993). Population extinction and saving biodiversity. *Ambio* **22**, 64–68.

Ehrlich, P.R., and Davidson, S.E. (1961). Techniques for capture-recapture studies of Lepidoptera populations. *J. Lepidopterists' Soc.* **14**, 227–229.

Ehrlich, P.R., Murphy, D.D., Singer, M.C., Sherwood, C.B., White, R.R., and Brown, I.L. (1980). Extinction, reduction, stability and increase: The responses of checkerspot butterfly (*Euphydryas editha*) populations to California drought. *Oecologia* **46**, 101–105.

Ehrlich, P.R., White, R.R., Singer, M.C., McKechnie, S.W., and Gilbert, L.E. (1975). Checkerspot butterflies: A historical perspective. *Science* **188**, 221–228.

Eldredge, N. (2003). The sloshing bucket: How the physical realm controls evolution. *In* "Towards a comprehensive dynamics of evolution: Exploring the interplay of selection, neutrality, accident, and function" (J. Crutchfield and P. Schuster, eds.), pp. 3–32. Oxford Univ. Press, New York.

Elith, J. (2000). Quantitative methods for modeling species habitat: Comparative performance and an application to Australian plants. *In* "Quantitative Methods in Conservation Biology" (S. Ferson and M.A. Burgman, eds.). Springer, New York.

Ellner, S., and Shmida, A. (1981). Why are adaptations for long-range seed dispersal rare in desert plants? *Oecologia* **51**, 133–144.

Ellner, S.P. (2001). Pair approximation for lattice models with multiple interaction scales. *J. Theor. Biol.* **210**, 435–447.

Ellner, S.P., Fieberg, J., Ludwig, D., and Wilcox, C. (2002). Precision of population viability analysis. *Conserv. Biol.* **16**, 258–261.

Ellner, S.P., Sasaki, A., Haraguchi, Y., and Matsuda, H. (1998). Speed of invasion in lattice population models: Pair-edge approximation. *J. Math. Biol.* **36**, 469–484.

Ellstrand, N.C., Lord, E.M., and Eckard, K.J. (1984). The inflorescence as a metapopulation of flowers: Position-dependent differences in function and form in the cleistogamous species *Collomia grandiflora* (Polemoniaceae). *Botan. Gaz.* **145**, 329–333.

Elmhagen, B., and Angerbjorn, A. (2001). The applicability of metapopulation theory to large mammals. *Oikos* **94**, 89–100.

Elton, C. (1949). Population interspersion: An essay on animal community patterns. *J. Ecol.* **37**, 1–23.

Engen, S., Lande, R., and Saether, B.E. (2002a). Migration and spatiotemporal variation in population dynamics in a heterogeneous environment. *Ecology* **83**, 570–579.

Engen, S., Lande, R., and Saether, B.-E. (2002b). The spatial scale of population fluctuations and quasi-extinction risk. *Am. Nat.* **160**, 439–451.

Engen, S., Bakke, Ø., and Islam, A. (1998). Demographic and environmental stochasticity: Concepts and definitions. *Biometrics* **54**, 840–846.

Engen, S., Saether, B.-E., and Møller, A.P. (2001). Stochastic population dynamics and time to extinction of a declining population of barn swallows. *J. Anim. Ecol.* **70**, 789–797.

Ennos, R.A. (2001). Inferences about spatial processes in plant populations from the analysis of molecular markers. *In* "Integrating Ecology and Evolution in a Spatial Context" (J. Silvertown and J. Antonovics, eds.), pp. 45–71. Blackwell, Oxford.

Entwistle, A.C., Racey, P.A., and Speakman, J.R. (2000). Social and population structure of a gleaning bat, *Plecotus auritus*. *J. Zool.* **252**, 11–17.

Epperson, B.K. (1993a). Recent advances in correlation studies of spatial patterns of genetic variation. *Evol. Biol.* **27**, 95–155.

Epperson, B.K. (1993b). Spatial and space-time correlations in systems of subpopulations with genetic drift and migration. *Genetics* **133**, 711–727.

Epperson, B.K. (1995). Spatial distributions of genotypes under isolation by distance. *Genetics* **140**, 1431–1440.

Epperson, B.K., and Allard, R.W. (1989). Spatial auto-correlation analysis of the distribution of genotypes within populations of lodgepole pine. *Genetics* **121**, 369–377.

Eriksson, Å., and Eriksson, O. (1997). Seedling recruitment in semi-natural pastures: The effects of disturbance, seed size, phenology and seed bank. *Nord. J. Bot.* **17**, 469–482.

Eriksson, O. (1996). Regional dynamics of plants: A review of evidence for remnant, source-sink and metapopulations. *Oikos* **77**, 248–258.

Eriksson, O. (1997). Colonization dynamics and relative abundance of three plant species in dry semi-natural grasslands. *Ecography* **20**, 559–568.

Eriksson, O. (2000). Functional roles of remnant plant populations in communities and ecosystems. *Glob. Ecol. Biogeogr.* **9**, 443–449.

Eriksson, O. (2002). Ontogenetic niche shifts and their implications for recruitment in three clonal Vaccinium shrubs: *Vaccinium myrtillus, Vaccinium vitis-idaea*, and *Vaccinium oxycoccos*. *Can. J. Bot.* **80**, 635–641.

Eriksson, O., and Ehrlén, J. (2001). Landscape fragmentation and the viability of plant populations. *In* "Integrating Ecology and Evolution in a Spatial Contex." (J. Silvertown and J. Antonovics, eds.), pp. 157–175. Blackwell, Oxford.

Eriksson, O., and Kiviniemi, K. (1999). Site occupancy, recruitment and extinction thresholds in grassland plants: An experimental study. *Biol. Conserv.* **87**, 319–325.

Erwin, D.H. (1994). Early introduction of major morphological innovations. *Acta Palaentol. Polonica* **38**, 281–294.

Estoup, A., Wilson, I.J., Sullivan, C., Cornuet, J.M., and Moritz, C. (2001). Inferring population history from microsatellite and enzyme data in serially introduced cane toads, *Bufo marinus*. *Genetics* **159**, 1671–1687.

Etienne, R.S. (2000). Local populations of different sizes, mechanistic rescue effect and patch preference in the Levins metapopulation model. *Bull. Math. Biol.* **62**, 943–958.

Etienne, R.S. (2002a). "Striking the Metapopulation Balance. Mathamatical Models & Methods Meet Metapopulation Management" Ph.D. thesis, Wageningen University, the Netherlands.

Etienne, R.S. (2002b). A scrutiny of the Levins metapopulation model. *Comments Theor. Biol.* **7**, 257 – 281.

Etienne, R.S., and Nagelkerke, C.J. (2002). Non-equilibria in small metapopulations: Comparing the deterministic Levins model with its stochastic counterpart. *J. Theor. Biol.* **219**, 463–478.

Everitt, B.S. (1993). "Cluster Analysis." Arnold, London.

Ewens, W.J. (1972). The sampling theory of selectively neutral alleles. *Theor. Popul. Biol.* **3**, 87–112.

Ewens, W.J. (1982). On the concept of the effective population size. *Theor. Popul. Biol.* **21**, 373–378.

Ewens, W.J. (1990). Population genetics theory: The past and the future. *In* "Mathematical and Statistical Developments of Evolutionary Theory" (S. Lessard, ed.), pp. 177–227. Kluwer Academic, Amsterdam.

Ezoe, H. (1998). Optimal dispersal range and seed size in a stable environment. *J. Theor. Biol* **190**, 287–293.

Fagan, W.F., Holmes, E.E., Rango, J.J., Folarin, A., Sorensen, J.A., Lippe, J.E., and McIntyre, N.E. (2003). Cross-validation of quasi-extinction risks from real time series: An examination of diffusion approximation methods. Submitted for publication.

Fagan, W.F., Meir, E., Prendergast, J., Folarin, A., and Kareiva, P.M. (2001). Characterizing vulnerability to extinction for 758 species. *Ecol. Lett.* **4**, 132–138.

Fahrig, L. (1992). Relative importance of spatial and temporal scales in a patchy environment. *Theor. Popul. Biol.* **41**, 300–314.

Fahrig, L. (1997). Relative effects of habitat loss and fragmentation on population extinction. *J. Wild. Manage.* **61**, 603–610.

Fahrig, L. (1998). When does fragmentation of breeding habitat affect population survival? *Ecol. Model.* **105**, 273–292.

Fahrig, L. (2001). How much habitat is enough? *Biol. Conserv.* **100**, 65–74.

Fahrig, L. (2002). Effect of habitat fragmentation on the extinction threshold: A synthesis. *Ecol. Appl.* **12**, 346–353.

Fahrig, L., and Merriam, G. (1985). Habitat patch connectivity and population survival. *Ecology* **66**, 1762–1768.

Falconer, D.S. (1985). A note on Fisher's "average effect" and average excess." *Genet. Res. Camb.* **46**, 337–347.

Falconer, D.S. (1989). "Introduction to Quantitative Genetics." Longman Scientific & Technical, Essex, UK.

Falconer, D.S., and Mackay, T.F.C. (1996). "Introduction to Quantitative Genetics." Longman, Harlow.

Farr, D.F., Bills, G.F., Chamuris, G.P., and Rossman, A.Y. (1989). "Fungi on Plants and Plant Products in the United States." The American Phytopathological Society, St. Paul, MN.

Fay, J.C., and Wu, C.-I. (2000). Hitchhiking under positive Darwinian selection. *Genetics* **155**, 1405–1413.

Fearnhead, P., and Donnelly, P. (2001). Estimating recombination rates from population genetic data. *Genetics* **159**, 1299–1318.

Fedriani, J.M., Delibes, M., Ferreras, P., and Roman, J. (2002). Local and landscape habitat determinants of water vole distribution in a patchy Mediterranean environment. *Ecoscience* **9**, 12–19.

Felsenstein, J. (1975). A pain in the torus: Some difficulties with models of isolation by distance. *Am. Nat.* **109**, 359–368.

Felsenstein, J. (1976). The theoretical population genetics of variable selection and migration. *Annu. Rev. Genet.* **10**, 253–280.

Ferguson, N., Donnelly, C., and Anderson, R. (2001). The foot-and-mouth epidemic in Great Britain: Pattern of spread and impact of interventions. *Science* **292**, 1155–1160.

Ferreras, P. (2001). Landscape structure and asymmetrical inter-patch connectivity in a metapopulation of the endangered Iberian lynx. *Biol. Conserv.* **100**, 125–136.

Ferreras, P., Gaona, P., Palomares, F., and Delibes, M. (2001). Restore habitat or reduce mortality? Implications from a population viability analysis of the Iberian lynx. *Anim. Conserv.* **4**, 265–274.

Fieberg, J., and Ellner, S.P. (2000). When is it meaningful to estimate an extinction probability? *Ecology* **81**, 2040–2047.

Filipe, J.A.N. (1999). Hybrid closure-approximation to epidemic models. *Phys. A* **266**, 238–241.

Filipe, J.A.N., and Gibson, G.J. (2001). Comparing approximations to spatiotemporal models for epidemics with local spread. *Bull. Math. Biol.* **63**, 603–624.

Filipe, J.A.N., and Maule, M.M. (2003). Analytical methods for predicting the behaviour of population models with general spatial interactions. *Math. Biosci.* **183**, 15–35.

Finkenstädt, B., and Grenfell, B. (2000). Time series modelling of childhood diseases: A dynamical systems approach. *J. R. Stat. Soc. C* **49**, 187–205.

Finkenstädt, B., Keeling, M.J., and Grenfell, B.T. (1998). Patterns of density dependence in measles dynamics. *Proc. R. Soc. Lond. B.* **265**, 753–762

Fischer, M., and Matthies, D. (1997). Mating structure and inbreeding and outbreeding depression in the rare plant *Gentianella germanica* (Gentianaceae). *Am. J. Bot.* **84**, 1685–1692.

Fischer, M., and Stöcklon, J. (1997). Local extinctions of plants in remnants of extensively used calcareous grasslands 1950–85. *Conserv. Biol.* **11**, 727–737.

Fisher, D., Lambin, X., and Yletyinen, S. (2003). Dispersal in water voles *Arvicola terrestris*: Transience and immigration. Submitted for publication.

Fisher, R.A. (1930). "The Genetical Theory of Natural Selection." Clarendon, Oxford.

Fix, A.G. (1985). Evolution of altruism in kin-structured and random subdivided populations. *Evolution* **39**, 928–939.

Flather, C.H., and Bevers, M. (2002). Patchy reaction-diffusion and population abundance: The relative importance of habitat amount and arrangement. *Am. Nat.* **159**, 40–56.

Fleishman, E., Ray, C., Sjogren-Gulve, P., Boggs, C.L., and Murphy, D.D. (2002). Assessing the roles of patch quality, area, and isolation in predicting metapopulation dynamics. *Conserv. Biol.* 16: 706–716.

Foley, P. (1994). Predicting extinction times from environmental stochasticity and carrying capacity. *Conserv. Biol.* **8**, 124–137.

Foley, P. (1997). Extinction models for local populations. *In* "Metapopulation Biology" (I. Hanski and M.E. Gilpin, eds.), pp. 215–246. Academic Press, San Diego.

Foote, M. (1992). Paleozoic record of morphological diversity in blastozoan echinoderms. *Proc. Nat. Acad. Sci. USA* **89**, 7325–7329.

Foote, M. (1999). Morphological diversity in the evolutionary radiation of Paleozoic and post-Paleozoic crinoids. *Paleobiology* **25**, 1–115 Suppl.

Foose, T.J. (1977). Demographic models for management of captive populations. *Int. Zool. Yearb.* **17**, 70–76.

Foppen, R.P.B., Chardon, J.P., and Liefveld, W. (2000). Understanding the role of sink patches in source-sink metapopulations: Reed Warbler in an agricultural landscape. *Conserv. Biol.* **14**, 1881–1892.

Forbes, A.E., and Chase, J.M. (2002). The role of habitat connectivity and landscape geometry in experimental zooplankton metacommunities. *Oikos* **96**, 433–440.

Ford, E.B. (1945). "Butterflies." Collins, London.

Fowler, K., and Whitlock, M.C. (1999). The variance in inbreeding depression and the recovery of fitness in bottlenecked populations. *Proc. R. Soc. Lond. B Biol. Sci.* **266**, 2061–2066.

Fownes, S. and Roland, J. (2002). Effects of meadow suitability on female behaviour in the alpine butterfly *Parnassius smintheus. Ecol. Entomol.* **27**, 457–466.

Frank, K., and Wissel, C. (1998). Spatial aspects of metapopulation survival: From model results to rules of thumb for landscape management. *Landsc. Ecol.* **13**, 363–379.

Frank, K., and Wissel, C. (2002). A formula for the mean lifetime of metapopulations in heterogeneous landscapes. *Am. Nat.* **159**, 530–552.

Frank, S.A. (1986). Dispersal polymorphism in subdivided populations. *J. Theor. Biol* **122**, 303–309.

Frank, S.A. (1989). The evolutionary dynamics of cytoplasmic male sterility. *Am. Nat.* **133**, 345–76.

Frank, S.A. (1991). Spatial variation in coevolutionary dynamics. *Evol. Ecol.* **5**, 193–217.

Frankel, O.H., and Soulé, M.E. (1981). "Conservation and Evolution." Cambridge Univ. Press, Cambridge.

Frankham, R. (1995). Effective population-size adult-population size ratios in wildlife: A review. *Genet. Res.* **66**, 95–107.

Freckleton, R.P., and Watkinson, A.R. (2002). Large-scale spatial dynamics of plants: Metapopulations, regional ensembles and patchy populations. *J. Ecol.* **90**, 419–434.

Freitag, S., and Van Jaarsveld, A.S. (1998). Sensitivity of selection procedures for priority conservation areas to survey extent, survey intensity and taxonomic knowledge. *Proc. R. Soc. Lond. B* **265**, 1475–1482.

Fretwell, S.D., and Lucas, H.L. (1970). On territorial behaviour and other factors influencing habitat distribution in birds. I. Theoretical development. *Acta Biotheoret.* **19**, 16–36.

Friedenberg, N.A. (2003). Experimental evolution of dispersal in spatiotemporally variable microcosms. *Ecol. Let.* **6**, 953–959.

Frouz, J., and Kindlmann, P. (2001). The role of sink to source re-colonisation in the population dynamics of insects living in unstable habitats: An example of terrestrial chironomids. *Oikos* **93**, 50–58.

Fu, X.-Y. (1994). Estimating effective population size or mutation rate using the frequencies of mutations in various classes in a sample of DNA sequences. *Genetics* **138**, 1375–1386.

Fu, X.-Y. (1995). Statistical properties of segregating sites. *Theor. Popul. Biol.* **48**, 172–197.

Fu, X.-Y., and Li, W.-H. (1993). Statistical tests of neutrality of mutations. *Genetics* **133**, 693–709.

Fu, X.-Y., and Li, W.-H. (1997). Estimating the age of the common ancestor of a sample of DNA sequences. *Mol. Biol. Evol.* **14**, 195–199.

Fulton, L.A., and Pearson, R.E. (1981). Transplantation and homing experiments on salmon, *Oncorhynchus* spp., and steelhead trout, *Salmo gairdneri*, in the Columbia River system: Fish of the 1939–44 broods. U.S. Dept. Commer., NOAA Tech. Memo. NMFS F/NWC-12.

Furstenberg, H., and Kesten, H. (1960). Products of random matrices. *Ann. Math. Stat.* **31**, 457–469.

Futuyma, D.J. (1986). "Evolutionary Biology." Sinauer, Sunderland, MA.

Gadgil, M. (1971). Dispersal: Population consequences and evolution. *Ecology* **52**, 253–261.

Gaggiotti, O.E. (1996). Population genetic models of source-sink metapopulations. *Theor. Popul. Biol.* **50**, 178–208.

Gaggiotti, O.E., and Excoffier, L. (2000). A simple method of removing the effect of a bottleneck and unequal population sizes on pairwise genetic distances. *Proc. R. Soc. Lond. Ser. B Biol. Sci.* **267**, 81–87.

Gaggiotti, O.E., Jones, F., Lee, W.M., Amos, W., Harwood, J., and Nichols, R.A. (2002). Patterns of colonization in a metapopulation of grey seals. *Nature* **416**, 424–427.

Gaggiotti, O.E., Lee, C.E., and Wardle, G.M. (1997). The effect of overlapping generations and population structure on gene-frequency clines. *In* "Structured Population Models in Marine, Terrestrial, and Freshwater Systems" (S. Tuljapurkar and H. Caswell, eds.), pp. 355–369. Chapman and Hall.

Gaggiotti, O.E., and Smouse, P.E. (1996). Stochastic migration and maintenance of genetic variation in sink populations. *Am. Nat.* **147**, 919–945.

Gaggiotti, O.E., and Vetter, R.D. (1999). Effect of life history strategy, environmental variability, and overexploitation on the genetic diversity of pelagic fish populations. *Can. J. Fish. Aqua. Sci.* **56**, 1376–1388.

Gaines, M.S., and McClenaghan, L.R., Jr. (1980). Dispersal in small mammals. *Annu. Rev. Ecol. Syst.* **11**, 163–196.

Galen, C., and Stanton, M.L. (1993). Short-term responses of alpine buttercups to experimental manipulations of growing season length. *Ecology* **74**, 1052–1058.

Gandon, S. (1999). Kin competition, the cost of inbreeding and the evolution of dispersal. *J. Theor. Biol* **200**, 345–364.

Gandon, S. (2002). Local adaptation and the geometry of host-parasite coevolution. *Ecol. Lett.* **5**, 246–256.

Gandon, S. and Michalakis, Y. (1999). Evolutionarily stable dispersal rate in a metapopulation with extinctions and kin competition. *J. Theor. Biol.* **199**, 275–290.

Gandon, S., and Michalakis, Y. (2001). Multiple causes of the evolution of dispersal. *In* "Dispersal" (J. Clobert, E.Danchin, A.A. Dhondt, and J.D. Nichols, eds.), pp. 155–167. Oxford Univ. Press, Oxford.

Gandon, S., and Rousset, F. (1999). Evolution of stepping-stone dispersal rates. *Proc. R. Soc. Lond. B Biol. Sci.* **266**, 2507–2513.

Gange, A.C., Brown, V.K., and Sinclair, G.S. (1993). VA mycorrhizal fungi: A determinant of plant community structure in early succession. *Funct. Ecol.* **7**, 616–622.

García-Dorado, A., and Caballero, A. (2000). On the average coefficient of dominance of deleterious spontaneous mutations. *Genetics* **155**, 1991–2001.

Garcia-Dorado, A., Lopez-Fanjul, C., and Caballero, A. (1999). Properties of spontaneous mutations affecting quantitative traits. *Genet. Res.* **74**, 341–350.

Garcia-Dorado, A., Martin, P., and Garcia, N. (1991). Soft selection and quantitative genetic variation: a laboratory experiment. *Heredity* **66**, 313–323.

Garcia-Ramos, G., and Kirkpatrick, M. (1997). Genetic models of adaptation and gene flow in peripheral populations. *Evolution* **51**, 21–28.

Garcia-Ramos, G., and Rodriguez, D. (2002). Evolutionary speed of species invasions. *Evolution* **56**, 661–668.

Gardner, R.H., Milne, B.T., Turner, M.G., and O'Neill, R.V. (1987). Neutral models for the analysis of broad-scale landscape pattern. *Landsc. Ecol.* **1**, 19–28.

Garrett, M. and Franklin, W. (1988). Behavioral ecology of dispersal in the black-tailed prairie dog. *J. Mammal.* **69**, 236–250.

Gaston, K.J. (1994). "Rarity." Chapman & Hall, London.

Gaston, K.J. (1996). Species-range-size distributions: Patterns, mechanisms and implications. *Trends Ecol. Evol.* **11**, 197–201.

Gaston, K.J. (1998). Species-range size distributions: Products of speciation, extinction and transformation. *Phil. Trans. R. Soc. Lond. B* **353**, 219–230.

Gavrilets, S. (1997). Evolution and speciation on holey adaptive landscapes. *Trends Ecol. Evol.* **12**, 307–312.

Gavrilets, S. (1999a). A dynamical theory of speciation on holey adaptive landscapes. *Am. Nat.* **154**, 1–22.

Gavrilets, S. (1999b). Dynamics of clade diversification on the morphological hypercube. *Proc. R. Soc. Lond. B* **266**, 817–824.

Gavrilets, S. (2000). Waiting time to parapatic speciation. *Proc. R. Soc. Lond. B* **267**, 2483–2492.

Gavrilets, S. (2003). Evolution and speciation in a hyperspace: The roles of neutrality, selection, mutation and random drift. *In* "Towards a Comprehensive Dynamics of Evolution: Exploring the Interplay of Selection, Neutrality, Accident, and Function" (J. Crutchfield and P. Schuster, eds.), pp. 135–162. Oxford Univ. Press, New York.

Gavrilets, S., Acton, R., and Gravner, J. (2000a). Dynamics of speciation and diversification in a metapopulation. *Evolution* **54**, 1493–1501.

Gavrilets, S., and Gibson, N. (2002). Fixation probabilities in a spatially heterogeneous environment. *Popul. Ecol.* **44**, 51–58.

Gavrilets, S., and Gravner, J. (1997). Percolation on the fitness hypercube and the evolution of reproductive isolation. *J. Theor. Biol.* **184**, 51–64.

Gavrilets, S., Li, H., and Vose, M.D. (2000b). Patterns of parapatric speciation. *Evolution* **54**, 1126–1134.

Geffeney, S., Brodie, E.D., Jr., Ruben, P.C., and Brodie, E.D., III (2002). Mechanisms of adaptation in a predator-prey arms race: TTX-resistant sodium channels. *Science* **297**, 1336–1339.

Gelman, A., Carlin, J.B., Stern, H.S., and Rubin, D.B. (1995). "Bayesian Data Analysis." Chapman & Hall, London.

Gerlach, G., and Hoeck, H.N. (2001). Islands on the plains: Metapopulation dynamics and female biased dispersal in hyraxes (Hyracoidea) in the Serengeti National Park. *Mol. Ecol.* **10**, 2307–2317.

Getz, W.M., and Haight, R.G. (1989). "Population Harvesting." Princeton Univ. Press, Princeton, NJ.

Gilchrist, G.W. (1995). Specialists and generalists in changing environments. I. Fitness landscapes of thermal sensitivity. *Am. Nat.* **146**, 252–270.

Giles, B.E., and Goudet, J. (1997). Genetic differentiation in *Silene dioica* metapopulations: Estimation of spatiotemporal effects in a successional plant species. *Am. Nat.* **149**, 507–526

Gill, D.E. (1978a). Effective population size and interdemic migration rates in a metapopulation of the red-spotted newt, *Notophthalmus viridescens* (Rafinesque). *Evolution* **32**, 839–849.

Gill, D.E. (1978b). Meta-population ecology of red-spotted newt, *Notophthalmus viridescens* (Rafinesque). *Ecol. Monogr.* **48**, 145–166.

Gilmour, J.S.L., and Gregor, J.W. (1939). Demes: A suggested new terminology. *Nature* **144**, 333.

Gilpin, M.E. (1975). "Group Selection in Predator – Prey Communities." Princeton Univ. Press, Princeton, NJ.

Gilpin, M.E., and Hanski, I.A. ((eds.)1991). "Metapopulation Dynamics: Empirical and Theoretical Investigations." Academic Press, New York.

Gilpin, M.E., and Soulé, M.E. (1986). Minimum viable populations: Processes of species extinction. *In* "Conservation Biology: The Science of Scarcity and Diversity." (M. E. Soulé, ed.), pp. 19–34. Sinauer, Sunderland, MA.

Gleason, H.L. (1917). The structure and development of the plant association. *Bull. Torrey Bot. Club* **53**, 7–26.

Gobeil, J.F., and Villard, M.A. (2002). Permeability of three boreal forest landscape types to bird movements as determined from experimental translocations. *Oikos* **98**, 447–458.

Gog, J.R., Woodroffe, R., and Swinton, J. (2002). Disease in endangered metapopulations: The importance of alternative hosts. *Proc. R. Soc. Lond. A* **269**, 671–676.

Goldberg, D.E., and Gross, K.L. (1988). Disturbance regimes of midsuccessional old fields. *Ecology* **69**, 1677–1688.

Goldstein, D.B., and Schlötterer, C. (1999). "Microsatellites Evolution and Applications." Oxford Univ. Press, Oxford.

Gomulkiewicz, R., and Holt, R.D. (1995). When does evolution by natural selection prevent extinction? *Evolution* **49**, 201–207.

Gomulkiewicz, R., Holt, R.D., and Barfield, M. (1999). The effects of density-dependence and immigration on local adaptation in a "black-hole" sink environment. *Theor. Popul. Biol.* **55**, 283–296.

Gomulkiewicz, R., Thompson, J.N., Holt, R.D., Nuismer, S.L., and Hochberg, M.E. (2000). Hot spots, cold spots, and the geographic mosaic theory of coevolution. *Am. Nat.* **156**, 156–174.

Gonzalez, A., and Chaneton, E.J. (2002). Heterotroph species extinction, abundance and biomass dynamics in an experimentally fragmented microecosystem. *J. Anim. Ecol.* **71**, 594–602.

Gonzalez, A., Lawton, J.H., Gilbert, F.S., Blackburn, T.M., and Evans-Freke, I. (1998). Metapopulation dynamics, abundance, and distribution in a microecosystem. *Science* **281**, 2045–2047.

Gonzalez-Andujar, J.L., and Perry, J.N. (1993). Chaos, metapopulations and dispersal. *Ecol. Model.* **65**, 255–263.

Goodnight, C.J. (1985). The influence of environmental variation on group and individual selection in a cress. *Evolution* **39**, 545–558.

Goodnight, C.J. (1987). On the effect of founder events on epistatic genetic variance. *Evolution* **41**, 80–91.

Goodnight, C.J. (1988). Epistatic genetic variance and the effect of founder events on the additive genetic variance. *Evolution* **42**, 441–454.

Goodnight, C.J. (1990). On the relativity of quantitative genetic variance components. *Behav. Brain Sci.* 13:134–135.

Goodnight, C.J. (1991). Intermixing ability in two-species communities of flour beetles. *Am. Nat.* **138**, 342–354.

Goodnight, C.J. (1995). Epistasis and the increase in additive genetic variance: Implications for phase 1 of Wright's shifting balance process. *Evolution* **49**, 502–511.

Goodnight, C.J. (1999). Epistasis and heterosis. *In* "The Genetics and Exploitation of Heterosis in Crops." (J. Coors and S. Pandey, eds.), pp. 59–68. American Society of Agronomy, Inc/Crop Science Society of America, Inc., Madison, WI.

Goodnight, C.J. (2000a). Quantitative trait loci and gene interaction: The quantitative genetics of metapopulations. *Heredity* **84**, 587–598.

Goodnight, C.J. (2000b). Modeling gene interaction in structured populations. *In* "Epistasis and the Evolutionary Process" (J.B. Wolf, E.D. Brodie, III, and M.J. Wade, eds.), pp. 213–231. Oxford Univ. Press, Oxford.

Goodnight, C.J. (2003). Gene interaction and selection. *In* "Long Term Selection: A Celebration of 100 Years of Selection for Oil and Protein in Maize" (K. Lamkey, ed.),

Goodnight, C.J., and Craig, D.M. (1996). The effect of coexistence on competitive outcome in *Tribolium castaneum* and *T. confusum*. *Evolution* **50**, 1241–1250.

Goodnight, C.J., and Stevens, L. (1997). Experimental studies of group selection: What do they tell us about group selection in nature? *Am. Nat.* **150**, S59–S79.

Goodnight, C.J., and Wade, M.J. (2000). The ongoing synthesis: A reply to Coyne et al. (1999). *Evolution* **54**, 317–324.

Gopasalmy, K. (1977a). Competition and coexistence in spatially heterogeneous environments. *Math. Biosci.* **36**, 229–242.

Gopasalmy, K. (1977b). Competition, dispersion and coexistence. *Math. Biosci.* **33**, 25–33.

Gosselin, F. (1998). Reconciling theoretical approaches to stochastic patch-occupancy metapopulation models. *Bul. Math. Biol.* **60**, 955–971.

Gotelli, N.J. (1991). Metapopulation models: The propagule rain, the rescue effect, and the core-satellite hypothesis. *Am. Nat.* **138**, 768–776.

Gotelli, N.J., and Graves, G.R. (1996). "Null Models in Ecology." Smithsonian Institution Press, Washington, DC.

Goudet, J., N. Perrin, and P. Waser. 2002. Tests for sex-biased dispersal using biparentally inherited genetic markers. Molecular Ecology 11:1103–1114.

Gradshteyn, I.S., and Ryzhik, I.M. (1994). "Tables of Integrals, Series, and Products." Academic Press, San Diego.

Grashof-Bokdam, C.J., and Geertsema, W. (1998). The effect of isolation and history on colonization patterns of plant species in secondary woodland. *J. Biogeogr.* **25**, 837–846.

Graves, G.R. (1997). Geographic clines of age ratios of black-throated blue warblers (*Dendroica caerulescens*). *Ecology* **78**, 2524–2531.

Green, P. (1995). Reversible jump MCMC computation and Bayesian model determination. *Biometrika* **82**, 711–732.

Greene, D.F., and Johnson, E.A. (1989). A model for wind dispersal of winged and plumed seeds. *Ecology* **70**, 339–347

Greene, D.F., and Johnson, E.A. (1996). Wind dispersal of seeds from a forest into a clearing. *Ecology* **77**, 595–609.

Greenwood, P. (1980). Mating systems, philopatry and dispersal in birds and mammals. *Anim. Behav.* **28**, 1140–1162.

Greenwood, P.J. (1983). Mating systems and evolutionary consequences of dispersal. *In* "The Ecology of Animal Movement," (I.R. Swingland and P.J. Greenwood, eds.), pp. 116–131. Clarendon Press, Oxford.

Greenwood, P.J., and Harvey, P.H. (1982). The natal and breeding dispersal of birds. *Annu. Rev. Ecol. Syst.* **13**, 1–21.

Grenfell, B.T., Bjørnstad, O.N., and Finkenstädt, B. (2002). Dynamics of measles epidemics: Scaling noise, determinism and predictability with the TSIR model. *Ecol. Monogr.* **72**, 185–202.

Grenfell, B.T., Bjørnstad, O.N., and Kappey J. (2001). Travelling waves and spatial hierarchies in measles epidemics. *Nature* **414**, 716–723.

Grenfell, B.T., and Bolker, B.M. (1998). Cities and villages: Infection hierarchies in a measles metapopulation. *Ecol. Lett.* **1**, 63–70.

Grenfell, B.T., and Dobson, A. (1995). "Ecology of Infectious Diseases in Natural Populations." Cambridge Univ. Press, Cambridge.

Grenfell, B.T., and Harwood, J. (1997). (Meta)population dynamics of infectious diseases. *Trends Ecol. Evol.* **12**, 395–399.

Grenfell, B.T., Wilson, K., Finkenstdt, B.F., Coulson, T.N., Murray, S., Albon, S.D., Pemberton, J.M., Clutton-Brock, T.H., and Crawley, M.J. (1998). Noise and determinism in synchronized sheep dynamics. *Nature* **394**, 674–677.

Grenfell, B.T., Wilson, K., Isham, V.S., Boyd, H.E.G., and Dietz, K. (1995). Modeling patterns of parasite aggregation in natural populations: Trichostrongylid nematoderuminant interactions as a case study. *Parasitology* **111**, S135–S151.

Griffing, B. (1967). Selection in reference to biological groups. I. Individual and group selection applied to populations of unordered groups. *Aust. J. Biol. Sci.* **10**, 127–139.

Griffing, B. (1977). Selection for populations of interacting phenotypes. *In* "Proceedings of the International Conference on Quantitative Genetics." (E. Pollak, O. Kempthorne, and T.B. Bailey, eds.), pp. 413–434. Iowa State Univ. Press, Ames, IA.

Griffing, B. (1981). A theory of natural selection incorporating interactions among individuals. I. The modeling process. *J. Theor. Biol.* **89**, 636–658.

Griffing, B. (1989). Genetic analysis of plant mixtures. *Genetics* **122**, 943–956.

Griffith, A.B., and Forseth, I.N. (2002). Primary and secondary seed dispersal of a rare, tidal wetland annual, *Aeschynomene virginica. Wetlands* **22**, 696–704.

Griffiths, R.C. (1980). Lines of descent in the diffusion approximation of neutral Wright – Fisher models. *Theor. Popul. Biol.* **17**, 37–50.

Griffiths, R.C., and Marjoram, P. (1996). Ancestral inference from samples of DNA sequences with recombination. *J. Comp. Biol.* **3**, 479–502.

Griffiths, R.C., and Tavaré, S. (1994a). Simulating probability distributions in the coalescent. *Theor. Popul. Biol.* **46**, 131–159.

Griffiths, R.C., and Tavaré, S. (1994b). Ancestral inference in population genetics, *Stat. Sci.* **9**, 307–319.

Griffiths, R.C., and Tavaré, S. (1995). Unrooted genealogical tree probabilities in the infinitely-many-sites model. *Math. Biosci.* **127**, 77–98.

Grime, J.P. (1977). Evidence for the existence of three primary strategies in plants and its relevance to ecological and evolutionary theory. *Am. Nat.* **111**, 1169–1194.

Grimm, V. (1999). Ten years of individual-based modelling in ecology: What have we learned and what could we learn in the future? *Ecol. Model.* **115**, 129–148.

Grimm, V., and Wissel, C. (2004). The intrinsic mean time to extinction: A unifying approach to analyzing persistence and viability of populations. Oikos, in press.

Grimm, V., Wyszomirski, T., Aikman, D., and Uchmanski, J. (1999). Individual-based modelling and ecological theory: Synthesis of a workshop. *Ecol. Model.* **115**, 275–282.

Grimmet, G., and Stirzaker, D. (2001). "Probability and Random Processes," 3rd Ed. Oxford Univ. Press, Oxford.

Groom, M.J. (1998). Allee effects limit population viability of an annual plant. *Am. Nat.* **151**, 487–496.

Groom, M.J., and Preuninger, T.E. (2000). Population type can influence the magnitude of inbreeding depression in *Clarkia concinna* (Onagraceae). *Evol. Ecol.* **14**, 155–180.

Groombridge, B. (1992). "Global Biodiversity." Chapmann & Hall, London.

Groombridge, J.J., Jones, C.G., Bruford, M.W., and Nichols, R.A. (2000). 'Ghost' alleles of the Mauritius kestrel. *Nature* **403**, 616.

Grosberg, R.K., and Quinn, J.F. (1986). The genetic control and consequences of kin recognition by the larvae of a colonial marine invertebrate. *Nature* **322**, 456–459.

Gross, M.R. (1996). Alternative reproductive strategies and tactics: Diversity within sexes. *Trends Ecol. Evol.* **11**, 92–98.

Grünbaum, D. (1994). Translating stochastic density-dependent individual behavior with sensory constraints to an Eulerian model of animal swarming. *J. Math. Biol.* **33**, 139–161.

Grünbaum, D. (1998). Schooling as a strategy for taxis in a noisy environment. *Evol. Ecol.* **12**, 503–522.

Gu, W., Heikkilä, R., and Hanski, I. (2002). Estimating the consequences of habitat fragmentation on extinction risk in dynamic landscapes. *Landsc. Ecol.* **17**, 699–710.

Gundersen, G., and Andreassen, H.P. (1998). Causes and consequences of natal dispersal in root voles, *Microtus oeconomus. Anim. Behav.* **56**, 1355–1366.

Gundersen, G., Andreassen, H.P., and Ims, R.A. (2002). Individual and population level determinants of immigration success on local habitat patches: An experimental approach. *Ecol. Lett.* **5**, 294–301.

Gundersen, G., Johannesen, E., Andreassen, H.P., and Ims, R.A. (2001). Source-sink dynamics affect demography of sources. *Ecol. Lett.* **4**, 14–21.

Gupta, S., Ferguson, N., and Anderson, R. (1998). Chaos, persistence, and evolution of strain structure in antigenically diverse infectious agents. *Science* **280**, 912–915.

Gustafson, E.J., and Gardner, R.H. (1996). The effect of landscape heterogeneity on the probability of patch colonization. *Ecology* **77**, 94–107.

Gutiérrez, D., León-Cortés, J.L., Menéndez, R., Wilson, R.J., Cowley, M.J.R., and Thomas, C.D. (2001). Metapopulations of four lepidopteran herbivores on a single host plant, *Lotus corniculatus. Ecology* **82**, 1371–1386.

Gutiérrez, D., Thomas, C.D., and León-Cortés, J.L. (1999). Dispersal, distribution, patch network, and metapopulation dynamics of the dingy skipper butterfly (*Erynnis tages*). *Oecologia* **121**, 506–517.

Gutiérrez, R.J., and Harrison, S. (1996). Applying metapopulation theory to spotted owl management: A history and critique. *In* "Metapopulations and Wildlife Conservation" (D.R. McCullough, ed.), pp. 167–185. Island Press, Washington, DC.

Gutzwiller, K.J. (ed.) (2002). "Applying Landscape Ecology in Biological Conservation." Springer, New York.

Gyllenberg, M., and Hanski, I. (1992). Single-species metapopulation dynamics: A structured model. *Theor. Popul. Biol.* **42**, 35–61.

Gyllenberg, M., and Hanski, I. (1997). Habitat deterioration, habitat destruction, and metapopulation persistence in a heterogeneous landscape. *Theor. Popul. Biol.* **52**, 198–215.

Gyllenberg, M., Hanski, I., and Hastings, A. (1997). Structured metapopulation models. *In* "Metapopulation Biology." (I. Hanski and M. Gilpin, eds.), pp. 93–122. Academic Press, San Diego.

Gyllenberg, M., Osipov, A.V., and Soderbacka, G. (1996). Bifurcation analysis of a metapopulation model with sources and sinks. *J. Nonlinear Sci.* **6**, 329–366.

Gyllenberg, M., Parvinen, K., and Dieckman, U. (2002). Evolutionary suicide and evolution of dispersal in structured metapopulations. *J. Math. Biol.* **45**, 79–105.

Gyllenberg, M., Söderbacka, G., and Ericsson, S. (1993). Does migration stabilize local-population dynamics: Analysis of a discrete metapopulation model. *Math. Biosci.* **118**, 25–49.

Gyllenberg, M., and Silvestrov, D.S. (1994). Quasi-stationary distributions of a stochastic metapopulation model. *J. Math. Biol.* **33**, 35–70.

Haag, C.R., Hottinger, J.W., Riek, M., and Ebert, D. (2002). Strong inbreeding depression in a Daphnia metapopulation. *Evolution* **56**, 518–526.

Haas, C.A. (1995). Dispersal and use of corridors by birds in wooded patches of agricultural landscape. *Conserv. Biol.* **9**, 845–854.

Haddad, N.M. (1999a). Corridor and distance effects on interpatch movements: A landscape experiment with butterflies. *Ecol. Appl.* **9**, 612–622.

Haddad, N.M. (1999b). Corridor use predicted from behaviors at habitat boundaries. *Am. Nat.* **153**, 215–227.

Haddad, N.M., and Baum, K.A. (1999). An experimental test of corridor effects on butterfly densities. *Ecol. Appl.* **9**, 623–633.

Hafner, D.J., and Sullivan, R.M. (1995). Historical and ecological biogeography of nearctic pikas (Lagomorpha, Ochotonidae). *J. Mammal.* **76**, 302–321.

Haldane, J.B.S. (1927). A mathematical theory of natural and artificial selection. V. Selection and mutation. *Proc. Camb. Phil. Soc.* **23**, 838–844.

Haldane, J.B.S. (1948). The theory of a cline. *J. Genet.* **48**, 277–284.

Haley, C.S., and Birley, A.J. (1983). The genetical response to natural selection by varied environments. II. Observations on replicate populations in spatially varied laboratory environments. *Heredity* **51**, 581–606.

Halpin, Z. (1987). Natal dispersal and the formation of new social groups in a newly established social group in a newly established town of Black-tailed prairie dog (*Cynomys ludovicianus*). *In* "Mammalian Dispersal Patterns: The effects of Social Structure on Population Genetics" (B. Chepko-Sade and Z. Tang Halpin eds.), pp. 104–118. University of Chicago Press, Chicago.

Hamilton, W.D. (1964a). The genetical evolution of social behavior. I. *Theor. Biol.* **7**, 1–16.

Hamilton, W.D. (1964b). The genetical evolution of social behavior. II. *Theor. Biol.* **7**, 17–52.

Hamilton, W.D. (1987). Discriminating nepotism: expectable, common, overlooked. *In* "Kin Recognition in Animals" (D.J.C. Fletcher and C.D. Michener, eds.), pp. 417–437. Wiley, New York.

Hamilton, W.D., and May, R.M. (1977). Dispersal in stable habitats. *Nature* **269**, 578–581.

Hanski, I. (1981). Coexistence of competitors in patchy environment with and without predation. *Oikos* **37**, 306–312.

Hanski, I. (1982). Dynamics of regional distribution: The core and satellite species hypothesis. *Oikos* **38**, 210–221.

Hanski, I. (1983). Coexistence of competitors in patchy environment. *Ecology* **64**, 493–500.

Hanski, I. (1985). Single-species spatial dynamics may contribute to long-term rarity and commonness. *Ecology* **66**, 335–343.

Hanski, I. (1986). Population dynamics of shrews on small islands accord with the equilibrium theory. *Biol. J. Linn. Soc.* **28**, 23–36.

Hanski, I. (1990a). Dung and carrion insects. *In* "Living in a Patchy Environment" (B. Shorrocks and I. Swingland, eds.), pp. 127–145. Oxford Univ. Press, Oxford.

Hanski, I. (1990b). Density dependence, regulation and variability in animal populations. *Phil. Trans. R. Soc. Lond. B Biol. Sci.* **330**, 141–150.

Hanski, I. (1991). Single-species metapopulation dynamics: Concepts, models and observations. *Biol. J. Linn. Soc.* **42**, 17–38.

Hanski, I. (1992). Inferences from ecological incidence functions. *Am. Nat.* **139**, 657–662.

Hanski, I. (1993). Dynamics of small mammals on islands. *Ecography* **16**, 372–375.

Hanski, I. (1994a). A practical model of metapopulation dynamics. *J. Anim. Ecol.* **63**, 151–162.

Hanski, I. (1994b). Patch-occupancy dynamics in fragmented landscapes. *Trends Ecol. Evol.* **9**, 131–135.

Hanski, I. (1996). Metapopulation ecology. *In* "Population Dynamics in Ecological Space and Time" (O.E. Rhodes, Jr., R.K. Chesser, and M.H. Smith, eds.), pp. 13–43. University of Chicago Press, Chicago.

Hanski, I. (1997a). Habitat destruction and metapopulation dynamics. *In* "Enhancing the Ecological Basis of Conservation: Heterogeneity, Ecosystem Function and Biodiversity" (S.T.A. Pickett, R.S. Ostfeld, M. Shachak, and G.E. Likens, eds.), pp. 217–227. Chapmann & Hall, New York.

Hanski, I. (1997b). Predictive and practical metapopulation models: The incidence function approach. *In* "Spatial Ecology: The Role of Space in Population Dynamics and Interspecific Interactions" (D. Tilman and P. Kareiva, eds.), pp. 21–25. Princeton Univ. Press, Princeton, NJ.

Hanski, I. (1998a). Connecting the parameters of local extinction and metapopulation dynamics. *Oikos* **83**, 390–396.

Hanski, I. (1998b). Metapopulation dynamics. *Nature* **396**, 41–49.

Hanski, I. (1999a). Habitat connectivity, habitat continuity, and metapopulations in dynamic landscapes. *Oikos* **87**, 209–219.

Hanski, I. (1999b). "Metapopulation Ecology." Oxford Univ. Press, New York.

Hanski, I. (2001a). Spatially realistic theory of metapopulation ecology. *Naturwissenschaften* **88**, 372–381.

Hanski, I. (2001b). Population dynamic consequences of dispersal in local populations and metapopulations. *In* "Dispersal" (J. Clobert, E. Danchin, A.A. Dhondt, and J.D. Nichols, eds.), pp. 282–298. Oxford Univ. Press, Oxford.

Hanski, I., Alho, J., and Moilanen, A. (2000). Estimating the parameters of migration and survival for individuals in metapopulations. *Ecology* **81**, 239–251.

Hanski, I., Breuker, C.J., Schöps, K., Setchfield, R., and Nieminen, M. (2002). Population history and life history influence the migration rate of female Glanville fritillary butterflies. *Oikos* **98**, 87–97.

Hanski, I., Ehrlich, P.R., Nieminen, M., Murphy, D.D., Hellmann, J.J., Boggs, C.L., and McLaughlin, J.F. (2004). Checkerspots and conservation biology. *In* "On the Wings of Checkerspots" (P.R. Ehrlich and I. Hanski, eds.), Oxford Univ. Press, Oxford.

Hanski, I., and Gilpin, M.E. (1991). Metapopulation dynamics: Brief history and conceptual domain. *Biol. J. Linn. Soc.* **42**, 3–16.

Hanski, I.A., and Gilpin, M.E. ((eds.)1997). "Metapopulation Biology." Academic Press, San Diego.

Hanski, I., and Gyllenberg, M. (1993). Two general metapopulation models and the core-satellite species hypothesis. *Am. Nat.* **142**, 17–41.

Hanski, I., and Heino, M. (2003). Metapopulation-level adaptation of insect host plant preference and extinction-colonization dynamics in heterogeneous landscapes. *Theor. Popul. Biol.* **64**, 281–290.

Hanski, I., and Kuussaari, M. (1995). Butterfly metapopulation dynamics. *In* "Population Dynamics: New Approaches and Synthesis" (N. Capuccino and P. Price eds.), pp. 149–171. Academic Press, London.

Hanski, I., Kuussaari, M., and Nieminen, M. (1994). Metapopulation structure and migration in the butterfly *Melitaea cinxia. Ecology* **75**, 747–762.

Hanski, I., Moilanen, A., and Gyllenberg, M. (1996a). Minimum viable metapopulation size. *Am. Nat.* **147**, 527–541.

Hanski, I., Moilanen, A., Pakkala, T., and Kuussaari, M. (1996b). The quantitative incidence function model and persistence of an endangered butterfly metapopulation. *Conserv. Biol.* **10**, 578–590.

Hanski, I., and Ovaskainen, O. (2000). The metapopulation capacity of a fragmented landscape. *Nature* **404**, 755–758.

Hanski, I., and Ovaskainen, O. (2002). Extinction debt at extinction threshold. *Conserv. Biol.* **16**, 666–673.

Hanski, I., and Ovaskainen, O. (2003). Metapopulation theory for fragmented landscapes. *Theor. Pop. Biol.* **64**, 119–127.

Hanski, I., Pakkala, T., Kuussaari, M., and Lei, G. (1995a). Metapopulation persistence of an endangered butterfly in a fragmented landscape. *Oikos* **72**, 21–28.

Hanski, I., and Pankakoski, E. (1989). Population biology of Eurasian shrews: Introduction. *Ann. Zool. Fennici* **26**, 335–338.

Hanski, I., Pöyry, J., Kuussaari, M., and Pakkala, T. (1995b). Multiple equilibria in metapopulation dynamics. *Nature* **377**, 618–621.

Hanski, I., and Ranta, E. (1983). Coexistence in a patchy environment: Three species of *Daphnia* in rock pools. *J. Anim. Ecol.* **52**, 263–279.

Hanski, I., and Simberloff, D. (1997). The metapopulation approach, its history, conceptual domain and application to conservation. *In* "Metapopulation Biology: Ecology, Genetics and Evolution," (I.A. Hanski and M.E. Gilpin, eds.), pp. 5–26. Academic Press, London.

Hanski, I., and Singer, M.C. (2001). Extinction-colonization dynamics and host-plant choice in butterfly metapopulations. *Am. Nat.* **158**, 341–353.

Hanski, I., and Thomas, C.D. (1994). Metapopulation dynamics and conservation: A spatially explicit model applied to butterflies. *Biol. Conserv.* **68**, 167–180.

Hanski, I., and Woiwod, I.P. (1993). Spatial synchrony in the dynamics of moth and aphid populations. *J. Anim. Ecol.* **62**, 656–668.

Hanski, I., and Zhang, D.-Y. (1993). Migration, metapopulation dynamics and fugitive co-existence. *J. Theor. Biol.* **163**, 491–504.

Hansson, L. (1991). Dispersal and connectivity in metapopulations. *Biol. J. Linn. Soc.* **42**, 89–103.

Hanzelova, V., and Spakulova, M. (1992). Biometric variability of *Proteocephalus neglectus* (Cestoda: Proteocephalidae) in two different age groups of the rainbow trout from the Dobsina water reservoir (East Slovakia). *Folia Parasitol. Ceske Budejovice* **39**, 307–316.

Harada, Y., Ezoe, H., Iwasa, Y., Matsuda, H., and Satō, K. (1995). Population persistence and spatially limited social-interaction. *Theor. Popul. Biol.* **48**, 65–91.

Harada, Y., and Iwasa, Y. (1994). Lattice population dynamics for plants with dispersing seeds and vegetative propagation. *Res. Popul. Ecol.* **36**, 237–249.

Harding, K.C., and McNamara, J.M. (2002). A unifying framework for metapopulation dynamics. *Am. Nat.* **160**, 173–185.

Harper, J.L. (1977). "Population Biology of Plants." Academic Press, London.

Harrison, S. (1989). Long-distance dispersal and colonization in the bay checkerspot butterfly. *Ecology* **70**, 1236–1243.

Harrison, S. (1991). Local extinction in a metapopulation context: An empirical evaluation. *Biol. J. Linn. Soc.* **42**, 73–88.

Harrison, S. (1994). Metapopulations and conservation. *In* "Large-Scale Ecology and Conservation Biology." (P.J. Edwards, R.M. May, and N.R. Webb, eds.), pp. 111–128. Blackwell Scientific Press, Oxford.

Harrison, S., and Bruna, E. (1999). Habitat fragmentation and large-scale conservation: What do we know for sure? *Ecography* **22**, 225–232.

Harrison, S., and Hastings, A. (1996). Genetic and evolutionary consequences of metapopulation structure. *Trends Ecol. Evol.* **11**, 180–183.

Harrison, S., Maron, J., and Huxel, G. (2000). Regional turnover and fluctuation in populations of five plants confined to serpentine seeps. *Conserv. Biol.* **14**, 769–779.

Harrison, S., Murphy, D.D., and Ehrlich, P.R. (1988). Distribution of the Bay checkerspot butterfly, *Euphydryas editha bayensis*: Evidence for a metapopulation model. *Am. Nat.* **132**, 360–382.

Harrison, S., and Quinn, J.F. (1989). Correlated environments and the persistence of metapopulations. *Oikos* **56**, 293–298.

Harrison, S., and Taylor, A. (1997). Migration within metapopulations: The impact upon local population dynamics. *In* "Metapopulation Biology: Ecology, Genetics and Evolution" (I. Hanski and M.E. Gilpin, eds.), pp. 27–42. Academic Press, San Diego.

Harrison, S., Thomas, C.D., and Lewinsohn, T.M. (1995). Testing a metapopulation model of coexistence in the insect community on ragwort (*Senecio jacobaea*). *Am. Nat.* **145**, 563–593.

Hartl, D.L., and Clark, A.G. (1997). "Principles of Population Genetics." Sinauer Associates, Sunderland, MA.

Hassell, M.P. (1978). "The Dynamics of Arthropod Predator-Prey Systems." Princeton Univ. Press, Princeton.

Hassell, M.P. (2000). Host-parasitoid dynamics. *J. Anim. Ecol.* **69**, 543–566.

Hassell, M.P., Comins, H.N., and May, R.M. (1991). Spatial structure and chaos in insect population dynamics. *Nature* **353**, 255–258.

Hastings, A. (1980). Disturbance, coexistence, history, and competition for space. *Theor. Popul. Biol.* **18**, 363–373.

Hastings, A. (1983). Can spatial variation alone lead to selection for dispersal? *Theor. Popul. Biol.* **24**, 244–251.

Hastings, A. (1991). Structured models of metapopulation dynamics. *In* "Metapopulation Dynamics" (M. Gilpin and I. Hanski, eds.), pp. 57–71. Academic Press, London.

Hastings, A. (1993). Complex interactions between dispersal and dynamics: Lessons from coupled logistic equations. *Ecology* **74**, 1362–1372.

Hastings, A., and Harrison, S. (1994). Metapopulation dynamics and genetics. *Ann. Rev. Ecol. Syst.* **25**, 167–188.

Hastings, A., and Wolin, C.L. (1989). Within-patch dynamics in a metapopulation. *Ecology* **70**, 1261–1266.

Hayman, G.I., and Mather, K. (1955). The description of genetic interactions in continuous variation. *Biometrics* **11**, 69–82.

Hatcher, M.J., Dunn, A.M., and Tofts, C. (2000). Co-existence of hosts and sex ratio distorters in structured populations. *Evol. Ecol. Res.* **2**, 185–205.

Hawthorne, D.J. (1997). Ecological history and evolution in a novel environment: Habitat heterogeneity and insect adaptation to a new host plant. *Evolution* **51**, 153–162.

Hedrick, P.N., and Kim, T.J. (2000). Genetics of complex polymorphisms: Parasites and maintenance of the major histocompatibility complex variation. *In* "Evolutionary Genetics: From Molecules to Morphology." (R.S. Singh and C.B. Krimbas, eds.), pp. 204–234. Cambridge Univ. Press, Cambridge.

Hedrick, P.W. (1986). Genetic polymorphism in heterogeneous environments: A decade later. *Annu. Rev. Ecol. Syst.* **17**, 535–566.

Hedrick, P.W. (2000). "Genetics of Populations," 2nd Ed. Jones and Bartlett, London.
Hedrick, P.W., Ginevan, M.E., and Ewing, E.P. (1976). Genetic polymorphism in heterogeneous environments. *Annu. Rev. Ecol. Syst.* 7, 1–32.
Hedrick, P.W., Lacy, R.C., Allendorf, F.W., and Soulé, M.E. (1996). Directions in conservation biology: Comments on Caughley. *Conserv. Biol.* **10**, 1312–1320.
Hedrick, P.W., Lee, R.N., and Parker, K.M. (2000). Major histocompatibility complex (MHC) variation in the endangered Mexican wolf and related canids. *Heredity* **85**, 617–624.
Hedrick, P.W., and Miller, P.S. (1992). Conservation genetics: Techniques and fundamentals. *Ecol. Appl.* **2**, 30–46.
Hedrick, P.W., Parker, K.M., Miller, E.L. and Miller, P.S. (1999). Major histocompatibility complex variation in the endangered Przewalski's horse. *Genetics* 152: 1701–1710.
Heino, M., and Hanski, I. (2001). Evolution of migration rate in a spatially realistic metapopulation model. *Am. Nat.* **157**, 495–511.
Heino, M., Kaitala, V., Ranta, E., and Lindström, J. (1997). Synchronous dynamics and rates of extinction in spatially structured populations. *Proc. R. Soc. Lond. B* **264**, 481–486.
Hels, T. (2002). Population dynamics in a Danish metapopulation of spadefoot toads *Pelobates fuscus*. *Ecography* **25**, 303–313.
Henson, S.M., Cushing, J.M., Costantino, R.F., Dennis, B., and Desharnais, R.A. (1998). Phase switching in population cycles. *Proc. R. Soc. Lond. B* **265**, 2229–2234.
Henttonen, H. (1985). Predation causing extended low densities in microtine cycles: Further evidence from shrew dynamics. *Oikos* **45**, 156–157.
Hering, F. (1995). "Habitat Patches of the Threatened Butterfly Species *Melitaea cinxia* (L.) on the Åland Islands, Finland: Vegetation Characteristics and Caterpillar – Host Plant Interactions." M.Sc. Dissertation, Münster, Germany.
Herkert, J.R. (1994). The effects of habitat fragmentation on Midwestern grassland bird communities. *Ecol. Appl.* **4**, 461–471.
Hess, G.R. (1994). Conservation corridors and contagious disease: A cautionary note. *Conserv. Biol.* **8**, 256–262.
Hess, G. (1996). Disease in metapopulation models: Implications for conservation. *Ecology* **77**, 1617–1632.
Hestbeck, J. (1982). Population regulation of cyclic mammals: The social fence hypothesis. *Oikos* **39**, 157–163.
Heyde, C.C. and Cohen, J.E. (1985). Confidence intervals for demographic projections based on products of random matrices. *Theor. Popul. Biol.* **27**, 120–153.
Hiebeler, D. (2000). Populations on fragmented landscapes with spatially structured heterogeneities: Landscape generation and local dispersal. *Ecology* **81**, 1629–1641.
Higgins, K., and Lynch, M. (2001). Metapopulation extinction caused by mutation accumulation. *Proc. Natl. Acad. Sci. USA* **98**, 2928–2933.
Higgins, S.I., and Richardson, D.M. (1999). Predicting plant migration rates in a changing world: The role of long-distance dispersal. *Am. Nat.* **153**, 464–475.
Hill, A.V.S. (1998). The immunogenetics of human infectious diseases. *Annu. Rev. Immunol.* **16**, 593–617.
Hill, J.K., Collingham, Y.C., Thomas, C.D., Blakeley, D.S., Fox, R., Moss, D., and Huntley, B. (2001) Impacts of landscape structure on butterfly range expansion. *Ecol. Lett.* **4**, 313–321.
Hill, J.K., Thomas, C.D., Fox, R., Telfer, M.G., Willis, S.G., Asher, J., and Huntley, B. (2002). Responses of butterflies to twentieth century climate warming: Implications for future ranges. *Proc. R. Soc. Lond. Ser. B Biol. Sci.* **269**, 2163–2171.
Hill, J.K., Thomas, C.D., and Huntley, B. (1999) Climate and habitat availability determine 20th century changes in a butterfly's range margin. *Proc. R. Soc. Ser. B* **266**, 1197–1206.

Hill, J.K., Thomas, C.D., and Lewis, O.T. (1996). Effects of habitat patch size and isolation on dispersal by *Hesperia comma* butterflies: Implications for metapopulation structure. *J. Anim. Ecol.* **65**, 725–735.

Hill, J.K., Thomas, C.D., and Lewis, O.T. (1999). Flight morphology in fragmented populations of a rare British butterfly, *Hesperia comma. Biol. Cons.* **87**, 277–283.

Hill, M.F., and Caswell, H. (1999). Habitat fragmentation and extinction thresholds on fractal landscapes. *Ecol. Lett.* **2**, 121–127.

Hill, M.F., Hastings, A., and Botsford, L.W. (2002). The effects of small dispersal rates on extinction times in structured metapopulation models. *Am. Nat.* **160**, 389–402.

Hochberg, M.E., Michalakis, Y., and de Meeus, T. (1992). Parasites as constriants on the rate of life-history evolution. *J. Evol. Biol.* **5**, 491–504.

Hoelzel, A.R., Fleischer, R.C., Campagna, C., Le Boeuf, B.J., and Alvord, G. (2002). Impact of a population bottleneck on symmetry and genetic diversity in the northern elephant seal. *J. Evol. Biol.* **15**, 567–575.

Hoelzel, A.R., Stephens, J.C., and O'Brien, S.J. (1999). Molecular genetic diversity and evolution at the MHC DQB locus in four species of pinnipeds. *Mol. Biol. Evol.* **16**, 611–618.

Hoelzer, G. (2001). Self-organization of population structure in biological systems. Inter-Journal of Genetics p. Article 345.

Holdren, C.E., and Ehrlich, P.R. (1981). Long-range dispersal in checkerspot butterflies: Transplant experiments with *Euphydryas gillettii. Oecologia* **50**, 125–129.

Holmes, E.E. (2001). Estimating risks in declining populations with poor data. *Proc. Nat. Acad. Sci. USA* **98**, 5072–5077.

Holmes, E.E. (2004). Beyond theory to application and evaluation: Diffusion approximations for population viability analysis. Ecological Applications, in press.

Holmes, E.E., and Fagan, W.F. (2002). Validating population viability analysis for corrupted data sets. *Ecology* **83**, 2379–2386.

Holmes, E.E., Lewis, M.A., Banks, J.E., and Veit, R.R. (1994). Partial differential equations in ecology: Spatial interactions and population dynamics. *Ecology* **75**, 17–29.

Holmes, E.E, and Wilson, H.B. (1998). Running from trouble: Long-distance dispersal and the competitive coexistence of inferior species. *Am. Nat.* **151**, 578–586.

Holsinger, K.E. (2000). Demography and extinction in small populations. *In* "Genetics, Demography, and the Viability of Fragmented Populations." (A. Young and G. Clark, eds.), pp. 55–74. Cambridge Univ. Press, Cambridge.

Holt, R.D. (1983). Immigration and the dynamics of peripheral populations. *In* "Advances in Herpetology and Evolutionary Biology" (A.G.J. Rhodin and K. Miyata, eds.), pp. 680–694. Harvard Univ., Cambridge.

Holt, R.D. (1984). Spatial heterogeneity, indirect interactions, and the coexistence of prey species. *Am. Nat.* **124**, 377–406.

Holt, R.D. (1985). Population dynamics in two-patch environments: Some anomalous consequences of an optimal habitat distribution. *Theor. Popul. Biol.* **28**, 181–208.

Holt, R.D. (1993). Ecology at the mesoscale: The influence of regional processes on local communities. *In* "Species Diversity in Ecological Communities: Historical and Geographical Perspectives." (R.E. Ricklefs and D. Schluter, eds.), pp. 77–88. University of Chicago Press, Chicago.

Holt, R.D. (1996a). Adaptive evolution in source-sink environments: Direct and indirect effects of density-dependence on niche evolution. *Oikos* **75**, 182–192.

Holt, R.D. (1996b). Demographic constraints in evolution: Towards unifying the evolutionary theories of senescence and niche conservatism. *Evol. Ecol.* **10**, 1–11.

Holt, R.D. (1997). From metapopulation dynamics to community structure: Some consequences of spatial heterogeneity. *In* "Metapopulation Biology: Ecology, Genetics, and Evolution." (I.A. Hanski and M.E. Gilpin, eds.), pp. 149–165.

Holt, R.D. (2002). Food webs in space: On the interplay of dynamics instability and spatial processes. *Ecol. Res.* **17**, 261–273.

Holt, R.D., and Barfield, M. (2001). On the relationship between the ideal free distribution and the evolution of dispersal. *In* "Dispersal," (J. Clobert, E. Danchin, A.A. Dhondt, and J.D. Nichols, eds.), pp. 83–95. Oxford Univ. Press, Oxford.

Holt, R.D., and Gaines, M.S. (1992). Analysis of adaptation in heterogeneous landscapes: Implications for the evolution of fundamental niches. *Evol. Ecol.* **6**, 433–447.

Holt, R.D., and Gomulkiewicz, R. (1997). How does immigration influence local adaptation? A reexamination of a familiar paradigm. *Am. Nat.* **149**, 563–572.

Holt, R.D., and Hassell, M.P. (1993). Environmental heterogeneity and the stability of host – parasitoid interactions. *J. Anim. Ecol.* **62**, 89–100.

Holt, R.D., and Keitt, T.H. (2000). Alternative causes for range limits: A metapopulation perspective. *Ecol. Lett.* **3**, 41–47.

Holt, R.D., and McPeek, M.A. (1996). Chaotic population dynamics favors the evolution of dispersal. *Am. Nat.* **148**, 709–718.

Holyoak, M. (2000). Habitat subdivision causes changes in food web structure. *Ecol. Lett.* **3**, 509–515.

Holyoak, M., Leibold, M.A., and Holt, R.D. (eds.). (2003). Metacommunities.

Hood, M.E., Antonovics, J., and Heishman, H. (2003). Karyotypic similarity identifies multiple host-shifts of a pathogenic fungus in natural poulations. *Infect. Genet. Evol.*

Hoogland, J.L. (1995). "The Black-Tailed Prairie Dog: Social Life of a Burrowing Mammal." University of Chicago Press, Chicago.

Horn, H.S., and MacArthur, R.H. (1972). Competition among fugitive species in a harlequin environment. *Ecology* **53**, 749–752.

Hosseini, P.R. (2003). How localized consumption stabilizes predator-prey systems with finite frequency of mixing. *Am. Nat.* **161**, 567–585. in press.

Houle, D., Hughes, K.A., Assimicopoulos, S., and Charlesworth, B. (1997). The effects of spontaneous mutation on quantitative traits. II. Dominance of mutations with effects on life-history traits. *Genet. Res. Camb.* **70**, 27–34.

Houssard, C., and Escarré, J. (1995). Variation and covariation among life-history traits in *Rumex acetosella* from a successional old-field gradient. *Oecologia* **102**, 70–80.

Houston, A.I., and McNamara, J.M. (1992). State-dependent life-history theory, and its implications for phenotypic plasticity and clutch size. *Evol. Ecol* **6**, 243–253.

Hovestadt, T., Messner, S. and Poethke, H.J. (2001). Evolution of reduced dispersal mortality and 'fat-tailed' dispersal kernels in autocorrelated landscapes. *Proc. R. Soc. Lond. B Biol. Sci.* **268**, 385–391.

Hubbell, S.P. (1997). A unified theory of biogeography and relative species abundances and its application to tropical rain forests and coral reefs. *Coral Reefs* **16**, S9–S21.

Hubbell, S.P. (2001). "The Unified Neutral Theory of Biodiversity and Biogeography." Princeton Univ. Press, Princeton.

Hudson, P., Annapaola, R., Grenfell, B., Heesterbeek, H., and Dobson, A. (2002). "The Ecology of Wildlife Diseases." Oxford Univ. Press.

Hudson, R.R. (1983a). Testing the constant-rate neutral allele model with protein sequence data. *Evolution* **37**, 203–217.

Hudson, R.R. (1983b). Properties of a neutral allele model with intragenic recombination. *Theor. Popul. Biol.* **23**, 183–201.

Hudson, R.R. (1987). Estimating the recombination parameter of a finite population model without selection. *Genet. Res. Camb.* **50**, 245–250.

Hudson, R.R. (1990). Gene genealogies and the coalescent process. *In* "Oxford Surveys in Evolutionary Biology" (D.J. Futuyma and J. Antonovics, eds.), Vol. 7, pp. 1–44, Oxford Univ. Press, Oxford.

Hudson, R.R., and Kaplan, N.L. (1985). Statistical properties of the number of recombination events in the history of a sample of DNA sequences. *Genetics* **111**, 147–164.

Huffaker, C.B. (1958). Experimental studies on predation: Dispersion factors and predator-prey oscillations. *Hilgardia* **27**, 343–383.

Huntly, N. (1991). Herbivores and the dynamics of communities and ecosystems. *Annu. Revi. Ecol. Syst.* **22**, 477–503.

Hurst, L.D., Atlan, A., and Bengtsson, B.O. (1996). Genetic conflicts. *Q. Rev. Biol.* **71**, 317–364.

Husband, B.C., and Barrett, S.C.H. (1992). Genetic drift and the maintenance of the style length polymorphism in tristylous populations of *Eichhornea paniculata* (Pontederiaceae). *Heredity* **69**, 440–449.

Husband, B.C., and Barrett, S.C.H. (1995). Estimates of gene flow in *Eichhornea paniculata*: Effects of range substructure. *Heredity* **75**, 549–560.

Husband, B.C., and Barrett, S.C.H. (1996). A metapopulation perspective in plant population biology. *J. Ecol.* **84**, 461–469.

Husband, B.C., and Barrett, S.C.H. (1998). Spatial and temporal variation in population size of *Eichhornia paniculata* in ephemeral habitats: Implications for metapopulation dynamics. *J. Ecol.* **86**, 1021–1031.

Hutchinson, G.E. (1951). Copepodology for the ornithologist. *Ecology* **32**, 571–577.

Imbert, E. (2001). Capitulum characters in the seed heteromorphic species *Crepis sancta* (Asteraceae): Variance partioning and inference for the evolution of dispersal rate. *Heredity* **86**, 78–86.

Ims, R.A. (1990). Determinants of natal dispersal and space use in grey-sided voles, *Clethrionomys rufocanus*: A combined field and laboratory experiment. *Oikos* **57**, 106–113.

Ims, R.A. (1995). Movements patterns related to spatial structures. *In* "Mosaic Landscape and Ecological Processes" (L. Hansson, L. Fahrig, and G. Merriam, eds.), pp. 85–109. Chapman & Hall, London.

Ims, R.A., and Andreassen, H.P. (2000). Spatial synchronization of vole population dynamics by predatory birds. *Nature* **408**, 194–196.

Ims, R.A., and Steen, H. (1990). Geographical synchrony in microtine population cycles: A theoretical evaluation of the role of nomadic avian predators. *Oikos* **57**, 381–387.

Ims, R.A., and Yoccoz, N.G. (1997). Studying transfer processes in metapopulations: Emigration, migration, and colonization. *In* "Metapopulation Biology: Ecology, Genetics and Evolution" (I. Hanski and M.E. Gilpin eds.), pp. 247–265. Academic Press, San Diego.

Ims, R.A., and Yoccoz, N.G. (1999). Østmarkmus: Den russiske invasjonen. *In* "Svalbardtundraens økologi" (S.-A. Bengtson, F. Mehlum, and T. Severinsen eds.), pp. 149–156. Norsk Polar Institutt, Tromsø.

Ingvarsson, P.K. (1997). The effect of delayd population growth on the genetic differentiation of local populations subject to frequent extinctions and recolonisations. *Evolution* **51**, 29–35.

Ingvarsson, P.K. (1998). Kin-structured colonization in *Phalacrus substriatus. Heredity* **80**, 456–463.

Ingvarsson, P.K. (1999). Group selection in density-regulated populations revisited. *Evol. Ecol. Res.* **1**, 527–536.

Ingvarsson, P.K., and Giles, B.E. (1999). Kin-structured colonization and small-scale genetic differentiation in *Silene dioica. Evolution* **53**, 605–611.

Ingvarsson, P.K., Olsson, K., and Ericson, L. (1997). Extinction-recolonization dynamics in the mycophagous beetle *Phalacrus substriatus. Evolution* **51**, 187–195.

Ingvarsson, P.K., and Whitlock, M.C. (2000). Heterosis increases the effective migration rate. *Proc. R. Soci.* **267**, 1321–1326.

Irwin, A.J., and Taylor, P.D. (2000). Evolution of dispersal in a stepping-stone population with overlapping generations. *Theor. Popul. Biol.* **58**, 321–328.

Irwin, D.E., Irwin, J.H., and Price, T. (2001). Ring species as bridges between microevolution and speciation. *Genetica* **112**, 223–243.

Isagi, Y., Kanazashi, T., Suzuki, W., Tanake, H., and Abe, T. (2000). Microsatellite analysis of the regeneration process of *Magnolia obovata*. *Heredity* **84**, 143–151.

IUCN (1994). IUCN Red List Categories, Prepared by the IUCN Species Survival Commission As approved by the 40th Meeting of the IUCN Council, Gland, Switzerland.

Ives, A.R., and Whitlock, M.C. (2002). Perspective: On inbreeding and metapopulations. *Science* **295**, 454–455.

Iwasa, Y. (2000). Lattice models and pair approximation in ecology. *In* The Geometry of Ecological Interactions: Simplifying Spatial Complexity (U. Dieckmann, R. Law, and J.A.J. Metz, eds.), pp. 227–251. Cambridge Univ. Press, Cambridge.

Jacquemyn, H., Butaye, J., Dumortier, M., Hermy, M., and Lust, N. (2001). Effects of age and distance on the composition of mixed deciduous forest fragments in an agricultural landscape. *J. Veget. Sci.* **12**, 635–642.

Jacquez, J.A., and Simon, C.P. (1993). The stochastic SI model with recruitment and deaths I. Comparison with the closed SIS model. Math. Biosci. **117**, 77–125.

Jain, S.K., and Bradshaw, A.D. (1966). Evolution in closely adjacent plant populations. I. The evidence and its theoretical analysis. *Heredity* **20**, 407–441.

James, J.W. (1971). The founder effect and response to artificial selection. *Gene. Res.* **12**, 249–266.

Jánosi, I.M., and Scheuring, I. (1997). On the evolution of density dependent dispersal in a spatially structured population model. *J. Theor. Biol.* **187**, 379–408.

Jansen, V.A.A., and Mulder, G. (1999). Evolving biodiversity. *Ecol. Lett.* **2**, 379–386.

Jarosz, A.M., and Burdon, J.J. (1991). Host-pathogen interactions in natural populations of *Linum marginale* and *Melampsora lini*. II. Local and regional variation in patterns of resistance and racial structure. *Evolution* **45**, 1618–1627.

Jarosz, A.M., and Lyons, E. (2003). Distribution of anther-smut *Microbotryum violaceum* on *S. latifolia* (=*S. alba*) in the eastern US. "pers.comm."

Järvinen, A., and Väsäinen, R.A. (1984). Reproduction of pied flycatchers (*Ficedula hypoleuca*) in good and bad breeding seasons in a northern marginal area. *Auk* **101**, 439–450.

Jennersten, O. (1988). Pollination in *Dianthus deltoides* (Caryophyllaceae): Effects of habitat fragmentation on visitation and seed set. *Conserv. Biol.* **2**, 359–366.

Johannesen, E., and Andreassen, H. (1998). Survival and reproduction of resident and immigrant vole (*Microtus oeconomus*). *Can. J. Zool.* **76**, 763–766.

Johansson, M.E., and Nilsson, C. (1993). Hydrochory, population dynamics and the distribution of the clonal aquatic plant *Ranunculus lingua*. *J. Ecol.* **81**, 81–91.

Johansson, M.E., Nilsson, C., and Nilsson, E. (1996). Do rivers function as corridors for plant dispersal? *J. Veget. Sci.* **7**, 593–598.

Johnson, M.P. (2000). The influence of patch demographics on metapopulations, with particular reference to successional landscapes. *Oikos* **88**, 67–74.

Johnson, M.P., Burrows, M.T., and Hawkins, S.J. (1998). Individual based simulations of the direct and indirect effects of limpets on a rocky shore *Fucus* mosaic. *Mar. Ecol. Progr. Series* **169**, 179–188.

Johst, K., and Brandl, R. (1997). Evolution of dispersal: The importance of the temporal order of reproduction and dispersal. *Proc. R. Soc. Lond. B Biol. Sci.* **264**, 23–30.

Joly, P., and Grolet, O. (1996). Colonization dynamics of new ponds, and the age structure of colonizing Alpine newts, *Triturus alpestris*. *Acta Oecol.* **17**, 599–608.

Jongejans, E., and Schippers, P. (1999). Modeling seed dispersal by wind in herbaceous species. *Oikos* **87**, 362–372.

Jonsen, I., and Taylor, P.D. (2000). *Calopteryx* damselfly dispersions arising from multiscale responses to landscape structure. Conservation Ecology 4. URL http://www.consecol.org/vol4/iss2/art4/.

Jordano, D., Rodríguez, J., Thomas, C.D., and Fernández Haeger, J. (1992) The distribution and density of a lycaenid butterfly in relation to *Lasius* ants. *Oecologia* **91**, 439–446.

Jordano, D., and Thomas, C.D. (1992). Specificity of an ant-lycaenid interaction. *Oecologia* **91**, 431–438.

Jordan, N. (1992). Path analysis of local adaptation in two ecotypes of the annual plant *Diodia teres* Walt (Rubiaceae). *Am. Nat.* **140**, 149–165.

Jules, E.S. (1998). Habitat fragmentation and demographic change for a common plant: Trillium in old-growth forest. *Ecology* **79**, 1645–1656.

Kalamees, R., and Zobel, M. (2002). The role of the seed bank in gap regeneration in a calcareous grassland community. *Ecology* **83**, 1017–1025.

Kaltz, O., Gandon, S., Michalakis, Y., and Shykoff, J.A. (1999). Local maladaptation in the anther-smut fungus *Microbotryum violaceum* to its host plant *Silene latifolia*: Evidence from a cross-inoculation experiment. *Evolution* **53**, 395–407.

Kaltz, O., and Schmid, B. (1995). Plant venereal disease: A model for integrating genetics, ecology and epidemiology. *Trends Ecol. Evol.* **10**, 221–222.

Kaplan, N.L., Darden, T., and Hudson, R.R. (1988). Coalescent process in models with selection. *Genetics* **120**, 819–829.

Kaplan, N.L., and Hudson, R.R. (1985). The use of sample genealogies for studying a selectively neutral *m*-loci model with recombination. *Theor. Popul. Biol.* **28**, 382–396.

Kaplan, N.L., Hudson, R.R., and Iizuka, M. (1991). Coalescent processes in models with selection, recombination and geographic subdivision. *Genet. Res. Camb.* **57**, 83–91.

Kareiva, P. (1994). Space: The final frontier for ecological theory. *Ecology* **75**, 1–47.

Kareiva, P., and Odell, G.M. (1987). Swarms of predators exhibit "preytaxis" if individual predators use area restricted search. *Am. Nat.* **130**, 223–270.

Karlin, S., and McGregor, J. (1972). Addendum to a paper of W. Ewens. *Theor. Popul. Biol.* **3**, 113–116.

Karlin, S., and Taylor, H. (1981). "A Second Course in Stochastic Processes." Academic Press, New York.

Kaul, M.H.L. (1988). "Male-Sterility in Natural Populations of Hermaphrodite Plants." Springer-Verlag, Berlin.

Kawecki, T.J. (1993). Age and size at maturity in a patchy environment: Fitness maximization versus evolutionary stability. *Oikos* **66**, 309–317.

Kawecki, T.J. (1995). Demography of source-sink populations and the evolution of ecological niches. *Evol. Ecol.* **9**, 38–44.

Kawecki, T.J. (2000). Adaptation to marginal habitats: Contrasting influence of dispersal on the fate of rare alleles with small and large effects. *Proc. R. Soc. Lond. B* **267**, 1315–1320.

Kawecki, T.J. (2003). Sex-biased dispersal and adaptation to marginal habitats. *Am. Nat.* **162**, 415–426.

Kawecki, T.J., Barton, N.H., and Fry, J.D. (1997). Mutational collapse of fitness in marginal habitats and the evolution of ecological specialisation. *J. Evol. Biol.* **10**, 407–429.

Kawecki, T.J., and Holt, R.D. (2002). Evolutionary consequences of asymmetric dispersal rates. *Am. Nat.* **160**, 333–347.

Kawecki, T.J., and Stearns, S.C. (1993). The evolution of life histories in spatially heterogeneous environments: Optimal reaction norms revisited. *Evol. Ecol.* **7**, 155–174.

Keddy, P.A. (1981). Experimental demography of the sand-dune annual, *Cakile edentula*, growing along an environmental gradient in Nova Scotia. *J. Ecol.* **69**, 615–630.

Keddy, P.A. (1982). Population ecology on an environmental gradient: *Cakile edentula* on a sand dune. *Oecologia* **52**, 345–355.

Keeling, M.J. (1999a). Correlation equations for endemic diseases: Externally imposed and internally generated heterogeneity. *Proc. R. Soc. Lond. Ser. B Biol. Sci.* **266**, 953–960.

Keeling, M.J. (1999b). The effects of local spatial structure on epidemiological invasions. *Proc. R. Soci. Lond. Ser. B Biol. Sci.* **266**, 859–867.

Keeling, M.J. (2000a). Simple stochastic models and their power-law type behaviour. *Theor. Popul. Biol.* **58**, 21–31.

Keeling, M.J. (2000b). Metapopulation moments: Coupling, stochasticity and persistence. *J. Anim. Ecol.* **69**, 725–736.

Keeling, M.J. (2002). Using individual-based simulations to test the Levins metapopulation paradigme. *J. Anim. Ecol.* **71**, 270–279.

Keeling, M.J., and Gilligan, C.A. (2000a). Metapopulation dynamics of bubonic plague. *Nature* **407**, 903–906.

Keeling, M.J., and Gilligan, C.A. (2000b). Bubonic plague: A metapopulation model of a zoonosis. *Proc. R. Soc. Lond. B* **267**, 2219–2230.

Keeling, M.J., and Grenfell, B.T. (1997). Disease extinction and community size: Modeling the persistence of measles. *Science* **275**, 65–67.

Keeling, M.J., and Grenfell, B.T. (1998). Effect of variability in infection period on the persistence and spatial spread of infectious diseases. *Math. Biosci.* **147**, 207–226.

Keeling, M.J., and Grenfell, B.T. (1999). Stochastic dynamics and a power law for measles variability. *Phil. Trans. R. Soc. Lond. B* **354**, 768–776.

Keeling, M.J., Rand, D.A., and Morris, A.J. (1997). Correlation models for childhood epidemics. *Proc. R. Soc. Lond. Ser. B Biol. Sci.* **264**, 1149–1156.

Keeling, M.J., and Rohani, P. (2002). Estimating spatial coupling in epidemiological systems:

Keeling, M.J., Rohani, P., and Grenfell, B.T. (2001). Seasonally-forced disease dynamics explored as switching between attractors. *Physica D* **148**, 317–335.

Keeling, M.J., Wilson, H.B., and Pacala, S.W. (2000). Reinterpreting space, time lags, and functional responses in ecological models. *Science* **290**, 1758–1761.

Keeling, M.J., Wilson, H.B., and Pacala, S.W. (2002). Deterministic limits to stochastic spatial models of natural enemies. *Am. Nat.* **159**, 57–80.

Keeling, M.J., Woolhouse, M.E.J., Shaw, D.J., Matthews, L., Chase-Topping, M., Haydon, D.T., Cornell, S.J., Kappey, J., Wilesmith, J., and Grenfell, B.T. (2001). Dynamics of the 2001 UK foot and mouth epidemic: Stochastic dispersal in a heterogeneous landscape. *Science* **294**, 813–817.

Keightley, P.D., and Bataillon, T.A. (2000). Multi-generation maximum likelihood analysis applied to mutation accumulation experiments in *Caenorhabditis elegans*. *Genetics* **154**, 1193–1201.

Keightley, P.D., and Caballero, A. (1997). Genomic mutation rates for lifetime reproductive output and lifespan in *Caenorhabditis elegans*. *Proc. Nal. Acad. Sci. USA* **94**, 3823–3827.

Keightley, P.D., and Eyre-Walker, A. (2000). Deleterious mutations and the evolution of sex. *Science* **290**, 331–333.

Keightley, P.D., and Lynch, M. (2003). Towards a realistic model of mutations affecting fitness. *Evolution* **57**, 683–685.

Keister, A.R., Lande, R., and Schemske, D.W. (1984). Models of coevolution and speciation in plants and their pollinators. *Am. Nat.* **124**, 220–243.

Keitt, T.H. (1997). Stability and complexity on a lattice: Coexistence of species in an individual-based food web model. *Ecol. Model.* **102**, 243–258.

Keitt, T.H., Urban, D.L., and Milne, B.T. (1997). Detecting critical scales in fragmented landscapes. *Conserv. Ecol.* **1**, 4.

Keller, L.F. (1998). Inbreeding and its fitness effects in an insular population of song sparrows (*Melospiza melodia*). *Evolution* **52**, 240–250.

Keller, L.F., Arcese, P., Smith, J.N.M., Hochachka, W.M., and Stearns, S.C. (1994). Selection against Inbred song sparrows during a natural population bottleneck. *Nature* **372**, 356–357.

Keller, L.F., Grant, P.R., Grant, B.R., and Petren, K. (2002). Environmental conditions affect the magnitude of inbreeding depression in survival of Darwin's finches. *Evolution* **56**, 1229–1239.

Keller, L.F., and Waller, D.M. (2002). Inbreeding effects in wild populations. *Trends Ecol. Evol.* **17**, 230–241.

Kelly, F.P. (1977). The asymptotic behaviour of an invasion process. *J. Appl. Prob.* **14**, 584–590.

Kelly, J.K. (1996). Kin selection in the annual plant Impatiens capensis. *Am. Nat.* **147**, 899–918.

Kennedy, M., and Gray, R.D. (1993). Can ecological theory predict the distribution of foraging animals? A critical analysis of experiments on the Ideal Free Distribution. *Oikos* **68**, 158–166.

Kermack, W.O., and McKendrick, A.G. (1927). A contribution to the mathematical theory of epidemics. *Proc. R. Soc. Lond. Ser. A* **115**, 700–721.

Kershaw, M., Williams, P.H., and Mace, M. (1994). Conservation of Afrotropical antelopes: Consequences and efficiency of using different site selection methods and diversity criteria. *Biodiv. Conserv.* **3**, 354–372.

Kéry, M., Matthies, D., and Spillman, H.-H. (2000). Reduced fecundity and offspring performance in small populations of the declining grassland plants *Primula veris* and *Gentiana lutea*. *J. Ecol.* **88**, 17–30.

Keyghobadi, N., Roland, J., Fownes, S., and Strobeck, C. (2002). Ink marks and molecular markers: Examining effects of landscape on dispersal using both mark-recapture and molecular methods. *In* "Butterflies as Model Systems: Evolution and Ecology Taking Flight" (C.L. Boggs, W.B. Watt, and P.R. Ehrlich, eds.). University of Chicago Press, Chicago, IL.

Keyghobadi., N., Roland, J., and Strobeck, C. (1999). Influence of landscape on the population genetic structure of the alpine butterfly *Parnassius smintheus* (Papilionidae). *Mol. Ecol.* **8**, 1481–1495.

Keymer, J.E., Marquet, P.A., Velasco-Hernández, J.X., and Levin, S.A. (2000). Extinction thresholds and metapopulation persistence in dynamic landscapes. *Am. Nat.* **156**, 478–494.

Khaladi, M., Grosbois, V., and Lebreton, J.D. (2000). An explicit approach to evolutionarystable dispersal strategies with a cost of dispersal. *Nonlinear Anal. real World Appl.* **1**, 137–144.

Kierstead, H., and Slobodkin, L.B. (1953). The size of water masses containing plankton blooms. *J. Mar. Res.* **12**, 141–147.

Kim, Y.J. (1987). Dynamics of populations with changing vital rates: Generalizations of the stable population theory. *Theor. Popul. Biol.* **31**, 306–322.

Kim, Y., and Stephan, W. (2000). Joint effects of genetic hitchhiking and background selection on neutral variation, *Genetics* **155**, 1415–1427.

Kimura, M. (1964). Diffusion models in population genetics. *J. Appl. Prob.* **1**, 177–232.

Kimura, M. (1969). The number of heterozygous nucleotide sites maintained in a finite population due to the steady flux of mutations. *Genetics* **61**, 893–903.

Kimura, M. (1983). "The Neutral Theory of Molecular Evolution." Cambridge Univ. Press, New York.

Kimura, M., and Ohta, T. (1969). The average number of generations until fixation of a mutant gene in a finite population. *Genetics* **61**, 763–771.

Kimura, M., and Weiss, G.H. (1964). The stepping stone model of population structure and the decrease of genetic correlation with distance. *Genetics* **49**, 561–576.

Kindvall, I. (1996). En naturvårdsbiologisk betraktelse över de österlenska gröngrodornas undergång. *In* (U. Gärdenfors, and A. Carlson, eds.), "Med huvudet före—Festskrift till Ingemar Ahléns 60-årsdag." pp. 69–80. Rapport 33, Department of Wildlife Ecology, Swedish Univ. of Agricult. Sci., Uppsala. [In Swedish].

Kindvall, O. (1996). Habitat heterogeneity and survival in a bush cricket metapopulation. *Ecology* **77**, 207–214.

Kindvall, O. (1999). Dispersal in a metapopulation of a bush cricket, *Metrioptera bicolor* (Orthoptera: Tettigoniidae). *J. Anim. Ecol.* **68**, 172–185.

King, A.W., Mann, L.K., Hargrove, W.W., Ashwood, T.L., and Dale, V.H. (2000). Assessing the persistence of an avian population in a managed landscape: A case study with Henslow's sparrow at Fort Knox, Kentucky. ORNL/TM-13734. Oak Ridge National Laboratory, Oak Ridge, TN.

King, A.W., and With, K.A. (2002). Dispersal success on spatially structured landscapes: When do spatial pattern and dispersal behavior really matter? *Ecol. Model.* **147**, 23–39.

Kingman, J.F.C. (1982a). The coalescent. *Stochast. Process. Appl.* **13**, 235–248.

Kingman, J.F.C. (1982b). On the genealogy of large populations. *J. Appl. Prob.* **19A**, 27–43.

Kingman, J.F.C. (1982c). Exchangeability and the evolution of large populations. *In* "Exchangeability in Probability and Statistics" (G. Koch and F. Spizzichino, eds.), pp. 97–112. North-Holland, Amsterdam.

Kinzig, A.P., Levin, S.A., Dushoff, J., and Pacala, S. (1999). Limiting similarity, species packing, and system stability for hierarchical competition-colonization models. *Am. Nat.* **153**, 371–383.

Kirchner, J.W., and Roy, B.A. (1999). The evolutionary advantages of dying young: Epidemiological implications of longevity in metapopulations. *Am. Nat.* **154**, 140–159.

Kirkpatrick, J.B. (1983). An iterative method for establishing priorities for the selection of nature reserves: An example from Tasmania. *Biol. Conserv.* **25**, 127–134.

Kirkpatrick, M., and Barton, N.H. (1997). Evolution of a species' range. *Am. Nat.* **150**, 1–23.

Kisdi, E. (2002). Dispersal: Risk spreading versus local adaptation. *Am. Nat.* **159**, 579–596.

Kisdi, E., and Meszena, G. (1995). Life histories with lottery competition in a stochastic environment: ESSs which do not prevail. *Theor. Popul. Biol.* **47**, 191–211.

Kiviniemi, K., and Eriksson, O. (2002). Size-related deterioration of semi-natural grassland fragments in Sweden. *Diversity Distrib.* **8**, 21–29.

Klausmeier, C.A. (1999). Regular and irregular patterns in semiarid vegetation. *Science* **284**, 1826–1828.

Klausmeier, C.A. (2001). Habitat destruction and extinction in competitive and mutualistic metacommunities. *Ecol. Lett.* **4**, 57–63.

Klinkhamer, P.G.L., de Jong, T.J., and de Heiden, L.H. (1996). An eight-year study of population dynamics and life-history variation of the "biennial" *Carlina vulgaris*. *Oikos* **75**, 259–268.

Kneitel, J.M. (2003). Scale-dependent trade-offs among coexisting species found in pitcher-plants (*Sarracenia purpurea*). Submitted for publication.

Kneitel, J.M., and Miller, T.E. (2002). Resource and top-predator regulation in the pitcher plant (*Sarracenia purpurea*) inquiline community. *Ecology* **83**, 680–688.

Kneitel, J.M., and Miller, T.E. (2003). Dispersal rates affect species composition in metacommunities of *Sarracenia purpurea* inquilines. *Am. Nat.*

Knopf, F.L., and Samson, F.B. (1994). Conservation of grassland vertebrates. *In* "Ecology and Conservation of Great Plains Vertebrates" (F.L. Knopf and F.B. Samson, eds.), pp. 273–289. Springer, New York.

Knutson R.L., Kwilosz, J.R., and Grundel, R. (1999). Movement patterns and population characteristics of the Karner blue butterfly (*Lycaeides melissa samuelis*) at Indiana Dunes National Lakeshore. *Nat. Areas J.* **19**, 109–120.

Koella, J.C., and Agnew, P. (1999). A correlated response of a parasite's virulence and life cycle to selection on its host's life history. *J. Evol. Biol.* **12**, 70–79.

Kokko, H., Sutherland, W.J., and Johnstone, R.A. (2001). The logic of territory choice: Implications for conservation and source-sink dynamics. *Am. Nat.* **157**, 459–463.

Komdeur, J., Daan, S., Tinbergen, J., and Mateman, C. (1997). Extreme adaptive modification in sex ratio of the Seychelles warbler' eggs. *Nature* **385**, 522–525.

Kondrashov, A.S. (2001). Speciation: Darwin revisited. *Trends Ecol. Evol.* **16**, 412.

Kornfield, I., and Smith, P.F. (2000). African cichlid fishes: Model systems for evolutionary biology. *Annu. Rev. Ecol. Syst.* **31**, 163–196.

Kot, M., Lewis, M.A., and van den Driessche, P. (1996). Dispersal data and the spread of invading organisms. *Ecology* **77**, 2027–2042.

Krebs, C.J. (1992). The role of dispersal in cyclic rodent populations. *In* "Animal Dispersal: Small Mammals as a Model." (N.C. Stenseth and W.Z. Lidicker, eds.), pp. 160–176. Chapman & Hall, London.

Krebs, C.J., Keller, B.L., and Tamarin, R.H. (1969). *Microtus* population biology: Demographic changes in fluctuating populations of *Microtus ochrogaster* and *M. pennsylvanicus* in southern Indiana. *Ecology* **50**, 587–607.

Kreitman, M. (1983). Nucleotide polymorphism at the alcohol dehydrogenase locus of *Drosophila melanogaster. Nature* **304**, 412–417.

Krone, S.M., and Neuhauser, C. (1997). Ancestral processes with selection. *Theor. Popul. Biol.* **51**, 210–237.

Kruess, A., and Tscharntke, T. (1994). Habitat fragmentation, species loss, and biological control. *Science* **264**, 1581–1584.

Kruess, A., and Tscharntke, T. (2000). Species richness and parasitism in a fragmented landscape: Experiments and field studies with insects on *Vicia sepium. Oecologia* **122**, 129–137.

Kruuk, L., Sheldon, B., and Merilä, J. (2002). Severe inbreeding depression in collared flycatchers (*Ficedula albicollis*). *Proc. R. Soc. Lond. Ser. B Biol. Sci.* **269**, 1581–1589.

Kryscio, R.J., and Lefèvre, C. (1989). On the extinction of the S-I-S stochastic logistic epidemic. *J. Appl. Prob.* **27**, 685–694.

Kuhner, M.K., Yamato, J., and Felsenstein, J. (1995). Estimating effective population size and mutation rate from sequence data using Metropolois-Hastings sampling. *Genetics* **140**, 1421–1430.

Kuhner, M.K., Yamato, J., and Felsenstein, J. (2000). Maximum likelihood estimation of recombination rates from population data. *Genetics* **156**, 1393–1401.

Kuussaari, M., Nieminen, M., and Hanski, I. (1996). An experimental study of migration in the Glanville fritillary butterfly *Melitaea cinxia. J. Anim. Ecol.* **65**, 791–801.

Kuussaari, M., Saccheri, I., Camara, M., and Hanski, I. (1998). Allee effect and population dynamics in the Glanville fritillary butterfly. *Oikos* **82**, 384–392.

Kuussaari, M., Singer, M., and Hanski, I. (2000) Local specialization and landscape-level influence on host use in an herbivorous insect. *Ecology* **81**, 2177–2187.

Lacy, R.C. (1993). VORTEX: A computer simulation model for population viability analysis. *Wild. Res.* **20**, 45–65.

Lacy, R.C. (2000). Structure of the VORTEX simulation model for population viability analysis. *Ecol. Bull.* **48**, 191–203.

Lambin, X. (1994). Natal philopatry, competition for ressources, and inbreeding avoidance in Townsend's voles (*Microtus townsendii*). *Ecology* **75**, 224–235.

Lambin, X., Aars, J., and Piertney, S.B. (2001). Dispersal, intraspecific competition, kin competition and kin facilitation. *In* "Dispersal" (J. Clobert, E. Danchin, A.A. Dhondt, and J.D. Nichols, eds.), pp. 110–122. Oxford Univ. Press, Oxford.

Lambin, X., Fazey, I., Sansom, J., Dallas, J., Stewart, W., Piertney, S., Palmer, S., Bacon, P., and Webb, A. (1998). Aberdeenshire Water vole survey: The distribution of isolated water vole populations in the upper catchments of the rivers Dee and Don., Rep. No. C/LF1/BAT/97/2. Scottish Natural Heritage.

Lambin, X., and Yoccoz, N.G. (1998). The impact of population kin-structure on nestling survival in Townsend's voles, *Microtus townsendii*. *J. Anim. Ecol.* **67**, 1–16.

Lande, R. (1979). Effective deme size during long-term evolution estimated from rates of chromosomal rearrangement. *Evolution* **33**, 234–251.

Lande, R. (1984). The expected fixation rate of chromosomal inversions. *Evolution* **38**, 743–752.

Lande, R. (1985). The fixation of chromosomal rearrangements in a subdivided population with local extinction and colonization. *Heredity* **54**, 323–332.

Lande, R. (1987). Extinction thresholds in demographic models of territorial populations. *Am. Nat.* **130**, 624–635.

Lande, R. (1988). Genetics and demography in biological conservation. *Science* **241**, 1455–1460.

Lande, R. (1988b). Demographic models of the northern spotted owl (*Strix occidentalis caurina*). *Oecologia* **75**, 601–607.

Lande, R. (1992). Neutral theory of quantitative genetic variance in an island model with local extinction and colonization. *Evolution* **46**, 381–389.

Lande, R. (1993). Risks of population extinction from demographic and environmental stochasticity and random catastrophes. *Am. Nat.* **142**, 911–927.

Lande, R. (1994). Risk of population extinction from fixation of new deleterious mutations. *Evolution* **48**, 1460–1469.

Lande, R. (1995). Mutation and conservation. *Conserv. Biol.* **9**, 782–791.

Lande, R. (1998). Demographic stochasticity and Allee effect on a scale with isotropic noise. *Oikos* **83**, 353–358.

Lande, R., and Arnold, S.J. (1983). The measurement of selection on correlated characters. *Evolution* **37**, 1210–1226.

Lande, R., Engen, S., and Saether, B.-E. (1995). Optimal harvesting of fluctuating populations with a risk of extinction. *Am. Nat.* **145**, 728–745.

Lande, R., Engen, S., and Saether, B.-E. (1998). Extinction times in finite metapopulations with stochastic local dynamics. *Oikos* **83**, 383–389.

Lande, R., Engen, S., and Saether, B.E. (1999). Spatial scale of population synchrony: Environmental correlation versus dispersal and density regulation. *Am. Nat.* **154**, 271–281.

Lande, R., Engen, S., and Saether, B. (2003). "Stochastic Population Models in Ecology and Conservation: An Introduction." Oxford Univ. Press, Oxford.

Lande, R. and Orzack, S.H. (1988). Extinction dynamics of age-structured populations in a fluctuating environment. *Proc. Natl. Acad. Sci. USA* **85**, 7418–7421.

Lande, R., and Shannon, S. (1996). The role of genetic-variation in adaptation and population persistence in a changing environment. *Evolution* **50**, 434–437.

Lavigne, C., Reboud, X., Lefranc, M., Porcher, E., Roux, F., Olivieri, I., and Godelle, B. (2001). Evolution of genetic diversity in metapopulations: *Arabidopsis thaliana* as an experimental model. *Genet. Sel. Evol.* **33**, S399–S423.

Lavorel, S., and Chesson, P. (1995). How species with different regeneration niches coexist in patchy habitats with local disturbances. *Oikos* **74**, 103–114.

Law, R., and Dieckmann, U. (2000). Moment approximations of individual-based models. *In* "Cambridge Studies in Adaptive Dynamics" (U. Dieckman, R. Law, and J.A.J. Metz, eds.), pp. 252–270. Cambridge Univ. Press, Cambridge.

Law, R., and Leibold, M.A. (in press). Assembly dynamics in metacommunities. *In* "Metacommunities" (M. Holyoak, M.A. Leibold, and R.D.Holt, eds.). University of Chicago Press, Chicago, IL.

Law, R., Murrell, D.J., and Dieckmann, U. (2003). Population growth in space and time: Spatial logistic equations. *Ecology* **84**, 252–262.

Law, R., Purves., D.W., Murrell, D.J., and Dieckmann, U. (2001). *In* Integrating Ecology and Evolution in a Spatial Context (J. Silvertown and J. Antonovics, eds.), pp. 21–44. Blackwell Science, Oxford.

Lawes, M.J., Mealin, P.E., and Piper, S.E. (2000). Patch occupancy and potential metapopulation dynamics of three forest mammals in fragmented afromontane forest in South Africa. *Conserv. Biol.* **14**, 1088–1098.

Lawton, J.H., Nee, S., Letcher, A.J., and Harvey, P.H. (1994). Animal distribution: patterns and process. *In* "Large-Scale Ecology and Conservation Biology" (P.J. Edwards, R.M. May, and N.R. Webb, eds.), pp. 41–58. Blackwell, Oxford.

Lawton, J.H., and Woodroffe, G.L. (1991). Habitat and the distribution of water Voles: Why are there gaps in a species range. *J. Anim. Ecol.* **60**, 79–91.

Lebreton, J.D. (1996). Demographic models for subdivided populations: The renewal equation approach. *Theor. Popul. Biol.* **49**, 291–313.

Lebreton, J.D., Khaladi, M., and Grosbois, V. (2000). An explicit approach to evolutionarily stable dispersal strategies: No cost of dispersal. *Math. Biosci.* **165**, 163–176.

Leck, M.A., Parker, V.T., and Simpson, R.L. (1989). "Ecology of Soil Seed Banks." Academic Press, San Diego.

Lecomte, J., Boudjemadi, K., Sarrazin, F., Cally, K., and Clobert, J. (2003). Connectivity and homogenisation of population sizes: an experimental approach in *Lacerta vivipara*. *Journal of Animal Ecology* (in press).

Le Corre, V., and Kremer, A. (1998). Cumulative effects of founder events during colonization on genetic diversity and differentiation in an island and stepping-stone model. *J. Evol. Biol.* **11**, 495–512.

Lee, T.D. (1984). Effects of seed number per fruit on seed dispersal in *Cassia fasciculata* (Caesalpiniaceae). *Bot. Gaz.* **145**, 136–139.

Lefranc, A. (2001). "Dispersal Behaviour in *Drosophila melanogaster*." Ph.D. Thesis, University of Pierre and Marie Curie, Paris, France.

LeGalliard, J.-F., Ferrière, R., and Clobert, J. (2003a). Kinship-dependent dispersal: Evidence from an experimental metapopulation of the common lizard. *Proc. R. Soc. Lond. B.*

LeGalliard, J.-F., Ferrière, R., and Dieckmann, U. (2003b). The adaptive dynamics of altruism in spatially heterogeneous populations. *Evolution*

Legendre, P., and Legendre, L. (1998). "Numerical Ecology." 2nd English Ed. Elsevier, Amsterdam.

Lehman, C.L., and Tilman, D. (1997). Competition in spatial habitats. *In* "Spatial Ecology: The Role of Space in Population Dynamics and Interspecific Interactions" (D. Tilman and P. Kareiva, eds.), pp. 185–203. Princeton Univ. Press, Princeton, NJ.

Lei, G.C., and Hanski, I. (1997). Metapopulation structure of *Cotesia melitaearum*, a specialist parasitoid of the butterfly *Melitaea cinxia*. *Oikos* **78**, 91–100.

Lei, G., and Hanski, I. (1998). Spatial dynamics of two competing specialist parasitoids in a host metapopulation. *J. Anim. Ecol.* **67**, 422–433.

Leibold, M.A. (1996). A graphical model of keystone predation: Effects of productivity on abundance, incidence and ecological diversity in communities. *Am. Nat.* **147**, 784–812.

Leibold, M.A. (1998). Similarity and local coexistence of species from regional biotas. *Evol.* **12**, 95–110.

Leibold, M.A., Chase, J.M., Shurin, J.B., and Downing, A.L. (1997). Species turnover and the regulation of trophic structure. *Annu. Rev. Ecol. Syst.* **28**, 467–494.

Leibold, M.A., and Mikkelson, G.M. (2002). Coherence, species turnover, and boundary clumping: Elements of meta-community structure. *Oikos* **97**, 237–250.

Leibold, M.A., and Norberg, J. (2003). Plankton biodiversity in metacommunities, plankton as complex adaptive systems? *Limnol. Oceanog.*

Leimar, O., and Norberg, U. (1997). Metapopulation extinction and genetic variation in dispersal- related traits. Oikos **80**, 448–458.

Lemel, J.-Y., Bélichon, S., Clobert, J., and Hochberg, M.E. (1997). The evolution of dispersal in in a two-patch system: Some consequences of differences between migrants and residents. *Evol. Ecol.* **11**, 613–629.

Léna, J.-P., Clobert, J., de Fraipont, M., Lecomte, J., and Guyot, G. (1998). The relative influence of density and kinship on dispersal in the common lizard. *Behav. Ecol.* **9**, 500–507.

Lennon, J.J., Turner, J.R.G., and Connell, D. (1997). A metapopulation model of species boundaries. *Oikos* **78**, 486–502.

Leslie, H., Ruckelshaus, M., Ball, I.R., Andelman, S., and Possingham, H.P. (2003). Using siting algorithms in the design of marine reserve networks. *Ecol. Appl.* **13**, 185–198.

Lessard, S., and Wakeley, J. (2003). The two-locus ancestral graph in a subdivided population: convergence as the number of demes grows in the island model. "in press" *J. Math. Biol.* DOI: 10.1007/s00285-003-0230-x.

Leturque, H., and Rousset, F. (2002). Dispersal, kin competition and the ideal free distribution in a spatially heterogeneous population. *Theor. Popul. Biol.* **62**, 169–180.

Levene, H. (1953). Genetic equilibrium when more than one ecological niche is available. *Am. Nat.* **87**, 331–333.

Levin, D.A. (2002). Hybridization and extinction: In protecting rare species, conservationists should consider the dangers of interbreeding, which compound the more well-known threats to wildlife. *Am. Sci.* **90**, 254–261.

Levin, S.A., Cohen, D., and Hastings, A. (1984). Dispersal strategies in patchy environments. *Theor. Popul. Biol.* **26**, 165–191.

Levine, J.M., and Rees, M. (2002). Coexistence and relative abundance in annual plant assemblages: The roles of competition and colonization. *Am. Nat.* **160**, 452–467.

Levins, R. (1968). "Evolution in Changing Evironments." Princeton Univ. Press, New Jersey.

Levins, R. (1969). Some demographic and genetic consequences of environmental heterogeneity for biological control. *Bull. Entomol. Soc. Am.* **15**, 237–240.

Levins, R. (1970). Extinction. *Lect. Notes Math.* **2**, 75–107.

Levins, R., and Culver, D. (1971). Regional coexistence of species and competition between rare species. *Proc. Nat. Acad. Sci. USA* **68**, 1246–1248.

Levitt, P.R. (1978). Mathematical theory of group selection. 1. Full solution of a nonlinear Levins E=E(X) model. *Theor. Popul. Biol.* **13**, 382–396.

Lewis, D. (1941). *Male sterility in natural populations of hermaphrodites plants. The equilibrium between females and hermaphrodites to be expected with different types of inheritance. New. Phytol.* **40**, 56–63.

Lewis, M.A. (2000). Spread rate for a nonlinear stochastic invasion. *J. Math. Biol.* **41**, 430–454.

Lewis, M.A., and Pacala, S. (2000). Modeling and analysis of stochastic invasion processes. *J. Math. Biol.* **41**, 387–429.

Lewis, O.T., Thomas, C.D., Hill, J.K., Brookes, M.I., Crane, T.P.R., Graneau, Y.A., Mallet, J.L.B., and Rose, O.C. (1997). Three ways of assessing metapopulation structure in the butterfly *Plebejus argus*. *Ecol. Entomo.* **22**, 283–293.

Lewontin, R.C., and Birch, L.C. (1966). Hybridization as a source of variation for adaptation to new environments. *Evolution* **20**, 315–336.

Lidicker, W.Z., Jr. (1962). Emigration as a possible mechanism permitting the regulation of population density below carrying capacity. *Am. Nat.* **96**, 29–33.

Lidicker, W.Z. (1975). The role if dispersal in the demography of small mammals. *In* "Small Mammals: Their Productivity and Population Dynamics" (F.B. Golley, K. Petrusewicz, and L. Ryszkowski, eds.) pp. 103–128. Cambridge Univ. Press, Cambridge.

Lidicker, W.Z., Jr., and Koenig, W.D. (1996). Responses of terrestrial vertebrates to habitat edges and corridors. *In* "Metapopulations and Wildlife Conservation" (D.R. McCullough, ed.), pp. 85–109. Island Press, Washington, DC.

Lidicker, W.Z., and Stenseth, N.S. (eds.) (1992). "Animal Dispersal: Small Mammals as a Model." Chapman & Hall, London.

Lienert, J., Fischer, M., and Diemer, M. (2002). Local extinctions of the wetland specialist *Swertia perennis* L. (Gentianaceae) in Switzerland: A revisitation study based on herbarium records. *Biol. Conserv.* **103**, 65–76.

Lima, M., Marquet, P.A., and Jaksic, F.M. (1996). Extinction and colonization processes in subpopulations of five neotropical small mammal species. *Oecologia* **107**, 197–203.

Lin, Y.K., and Batzli, G.O. (2001). The effect of interspecific competition on habitat selection by voles: An experimental approach. *Can. J. Zool.* **79**, 110–120.

Lindborg, R., and Ehrlén, J. (2002). Evaluating the extinction risk of a perennial herb: Demographic data versus historical records. *Conserv. Biol.* **16**, 683–690.

Lindenmayer, D.B., Ball, I., Possingham, H.P., McCarthy, M.A. and Pope, M.L. (2001). A landscape-scale test of the predictive ability of a spatially explicit model for population viability analysis. *J. Appl. Ecol.* **38**, 36–48.

Lindenmayer, D.B., McCarthy, M.A., and Pope, M.L. (1999). Arboreal marsupial incidence in eucalypt patches in south-eastern Australia: A test of Hanski's incidence function metapopulation model for patch occupancy. *Oikos* **84**, 99–109.

Lindenmayer, D.B., and Possingham, H.P. (1996a). Modelling the inter-relationships between habitat patchiness, dispersal capability and metapopulation persistence of the endangered species, Leadbeater's possum, in south-eastern Australia. *Landsc. Ecol.* **11**, 79–105.

Lindenmayer, D.B., and Possingham, H.P. (1996b). Ranking conservation and timber management options for Leadbeater's possum in southeastern Australia using population viability analysis. *Conserv. Biol.* **10**, 235–251.

Liston, A., Rieseberg, L.H., and Elias, T.S. (1990). *Functional androdioecy in the flowering plant Datisca glomerata. Nature* **343**, 641–642.

Lively, C.M., and Jokela, J. (1996). Clinical variation for local adaptation in a host-parasite interaction. *Proc. R. Soc. Lond. B* **263**, 891–897.

Lloyd, M.B. (1968). Self-regulation of adult numbers by cannibalism in two laboratory strains of flour beetles (*Tribolium castaneum*). *Ecology* **49**, 245–259.

Lockhart, A., Thrall, P.H., and Antonovics, J. (1996). Sexually-transmitted diseases in animals: Ecological and evolutionary implications. *Biol. Rev. Cambr. Phil. Soc.* **71**, 415–471.

Lomnicki, A. (1955). Why do populations of small rodents cycle-a new hypothesis with numerical model. *Evol. Ecol.* **9**, 64–81.

Lomolino, M.V., and Smith, G.A. (2001). Dynamic biogeography of prairie dog (*Cynomys ludovicianus*) towns near the edge of their range. *J. Mammal.* **82**, 937–945.

Loreau, M., and Mouquet, N. (1999). Immigration and the maintenance of local species diversity. *Am. Nat.* **154**, 427–440.

Losos, J.B. (1998). Ecological and evolutionary determinants of the species-area relationship in Carribean anoline lizards. *In* "Evolution on Islands" (P. Grant, ed.), pp. 210–224. Oxford Univ. Press, Oxford.

Loveless, M.D., and Hammrick, J.L. (1984). Ecological determinants of genetic structure in plant populations. *Annu. Rev. Ecol. Syst.* **15**, 65–95.

Ludwig, D. (1996). Uncertainty and the assessment of extinction probabilities. *Ecol. Appl.* **6**, 1067–1076.

Ludwig, D. (1999). Is it meaningful to estimate a probability of extinction? *Ecology* **80**, 298–310.

Luijten, S.H., Dierick, A., Oostermeijer, J.G.B., Raijmann, L.E.L., and den Nijs, J.C.M. (2000). Population size, genetic variation and reproductive success in the rapidly declining, self-incompatible *Arnica montana* in The Netherlands. *Conserv. Biol.* **14**, 1776–1787.

Luikart, G., Sherwin, W.B., Steele, B.M., and Allendorf, F.W. (1998). Usefulness of molecular markers for detecting population bottlenecks via monitoring genetic change. *Mol. Ecol.* 7, 963–974.

Lundberg, P., Ranta, E., Ripa, J., and Kaitala, V. (2000). Population variability in space and time. *Trends Ecol. Evol.* **15**, 460–464.

Lupia, R. (1999). Discordant morphological disparity and taxonomic diversity during the Cretaceous angiosperm radiation: North American pollen record. *Paleobiology* **25**, 1–28.

Luttrell, G.R., Echelle, A.A., Fisher, W.L., and Eisenhour, D.J. (1999). Declining status of two species of the *Macrhybopsis aestivalis* complex (Teleostei: Cyprinidae) in the Arkansas River Basin and related effects of reservoirs as barriers to dispersal. *Copeia* **1999**, 981–989.

Lynch, M., Blanchard, J., Houle, D., Kibota, T., Schultz, S., Vassilieva, L., and Willis, J. (1999). Perspective: Spontaneous deleterious mutation. *Evolution* **53**, 645–663.

Lynch, M., Burger, R., Butcher, D., and Gabriel, W. (1993). The mutational meltdown in asexual populations. *J. Hered.* **84**, 339–344.

Lynch, M., Conery, J., and Bürger, R. (1995a). Mutational meltdown in sexual populations. *Evolution* **49**, 1067–1080.

Lynch, M., Conery, J., and Bürger, R. (1995b). Mutation accumulation and the extinction of small populations. *Am. Nat.* **146**, 489–518.

Lynch, M., and Gabriel, W. (1990). Mutation load and the survival of small populations. *Evolution* **44**, 1725–1737.

Lynch, M., and Walsh, B. (1998). "Genetics and Analysis of Quantitative Traits." Sinauer, Sunderland, MA.

Lytle, D.A. (2001). *Disturbance regimes and life-history evolution. Am. Nat.* **157**, 525–536.

MacArthur, R.H. (1962). Some generalized theorems of natural selection. *Proc. Nat. Acad. Sci. USA* **48**, 1893–1897.

MacArthur, R.H., and Wilson, E. O. (1963). An equilibrium theory of insular zoogeography. *Evolution* **17**, 373–387.

MacArthur, R.H., and Wilson, E.O. (1967). "The Theory of Island Biogeography." Princeton University Press.

Macdonald, D.W., and Johnson, D.D.P. (2001). Dispersal in theory and practice: Consequences for conservation biology. *In* "Dispersal" (J. Clobert, E. Danchin, A.A. Dhondt, and J.D. Nichols, eds.), pp. 358–372. Oxford Univ. Press, Oxford.

Macdonald, D.W., and Strachan, R. (1999). "The Mink and the Water Vole: Analyses for Conservation." The Environment Agency and Wildlife Conservation Research Unit, Oxford.

MacKay, P.A., and Wellington, W.G. (1977). Maternal age as a source of variation in the ability of an aphid to produce dispersing forms. *Res. Popul. Ecol.* **18**, 195–209.

MacKay, T.F.C. (1981). Genetic variation in varying environments. *Genet. Res.* **37**, 79–93.

MacNair, M.R. (1993). The genetics of metal tolerance in vascular plants. *New Phytol.* **124**, 541–559.

Madsen, T., Shine, R., Olsson, M., and Wittzell, H. (1999). Conservation biology: Restoration of an inbred adder population. *Nature* **402**, 34–35.

Maes, D., and Van Dyck, H. (1999). Dagvlinders in Vlaanderen: Ecologie, verspreiding en behoud. Stichting Leefmilieu vzw/KBC, Antwerp.

Maes, D., and Van Dyck, H. (2001). Butterfly diversity loss in Flanders (north Belgium): Europe's worst case scenario? *Biol. Conserv.* **99**, 263–276.

Malanson, G.P., and Cairns, D. (1997). Effects of dispersal, population delays and forest fragmentation on tree migration rates. *Plant Ecol.* **131**, 67–79.

Malecot, G. (1948). "Les mathématiques de l'hérédité." Masson, Paris.

Manicacci, D., Atlan, A., and Couvet, D. (1997). Spatial structure of nuclear factors involved in sex determination in the gynodioecious *Thymus vulgaris* L.J. *Evol. Biol.* **10**, 889–907.

Manicacci, D., Couvet, D., Belhassen, E., Gouyon, P.H., and Atlan, A. (1996). Founder effects and sex ratio in the gynodioecous *Thymus vulgaris* L. *Mol. Ecol.* 5, 63–72.

Mantua, N.J., Hare, S.R., Zhang, Y., Wallace, J.M., and Francis, R.C. (1997). A Pacific interdecadal climate oscillation with impacts on salmon production. *Bull. Am. Meteorol. Soc.* **78**, 1069–1079.

Margules, C.R., and Austin, M.P. (1994). Biological models for monitoring species decline: The construction and use of data bases. *Phil. Trans. R. Soc. Lond. B* **344**, 69–75.

Margules, C.R., Nicholls, A.O., and Pressey, R.L. (1988). Selecting networks of reserves to maximize biological diversity. *Biol. Conserv.* **43**, 63–76.

Margules, C.R., Nicholls, A.O., and User M.B. (1994). Apparent species turnover, probability of extinction and the selection of nature reserves: a case study of Ingleborough limestoe pavements. Conserv. Biol., 8, 398-409.

Margules, C.R., Pressey, R.L., and Williams, P.H. (2002) Representing biodiversity: Data and procedures for identifying priority areas for conservation. *J. Biosci.* **4**, 309–326.

Margules, C.R., and Usher, M.B. (1981) Criteria used in assessing wildlife conservation potential: A review. *Biol. Conserv.* **21**, 79–109.

Mark, G.A. (1982). An experimental study of evolution in heterogeneous environments: Phenological adaptation by a bruchid beetle. *Evolution* **36**, 984–997.

Marr, A.B., Keller, L.F., and Arcese, P. (2002). Heterosis and outbreeding depression in descendants of natural immigrants to an inbred population of song sparrows (*Melospiza melodia*). *Evolution* **56**, 131–142.

Martin, K., Stacey, P.B., and Braun, C.E. (1997). Demographic rescue and maintenance of population stability in grouse – beyond metapopulations. *Wildlife Biol.* **3**, 295–296.

Martin, P.S., and Klein, R.G. (1984). "Quaternary Extinctions: A Prehistoric Revolution." University of Arizona Press, Tucson.

Marttila, O., Saarinen, K., and Jantunen, J. (1997). Habitat restoration and a successful reintroduction of the endangered Baton Blue butterfly (*Pseudophilotes baton schiffermuelleri*) in SE Finland. *Ann. Zool. Fenn.* **34**, 177–185.

Maruyama, T. (1970). On the fixation probability of mutant genes in a subdivided population. *Genet. Res.* **15**, 221–225.

Maruyama, T. (1974). A simple proof that certain quantities are independent of the geographical structure of population. *Theor. Popul. Biol.* **5**, 148–154.

Maruyama, T., and Kimura, M. (1980). Genetic variability and effective population size when local extinction and recolonization of subpopulations are frequent. *Proc. Nat. Acad. Sci. USA* **77**, 6710–6714.

Massot, M., and Clobert, J. (1995). Influence of maternal food availability on offspring dispersal. *Behav. Ecol. Sociobiol.* **37**, 413–418.

Massot, M., and Clobert, J. (2000). Processes at the origin of similarities in dispersal behaviour among siblings. *J. Evol. Biol.* **13**, 707–719.

Massot, M., Clobert, J., Lorenzon, P., and Rossi, J.M. (2002). Condition dependent dispersal and ontogeny of the dispersal behavior: An experimental approach. *J. Anim. Ecol.* **71**, 235–261.

Mathews, G.M., and Waples, R.S. (1991). Status review of Snake River spring and summer chinook salmon. U.S. Dept. Commer., NOAA Tech. Memo. NMFS F/NWC-200.

Mathias, A., Kisdi, E., and Olivieri, I. (2001). Divergent evolution of dispersal in a heterogeneous landscape. *Evolution* **55**, 246–259.

Matter, S.F., and Roland, J. (2002). An experimental examination of the effects of habitat quality on the dispersal and local abundance of the butterfly *Parnassius smintheus*. *Ecol. Entomol.* **27**, 308–316.

May, R.M. (1974). Biological populations with non-overlapping generations: Stable points, stable cycles and chaos. *Science* **186**, 645–647.

May, R.M. (1975). Patterns of species abundance and diversity. *In* "Ecology and Evolution of Communities." (M.L. Cody and J.M. Diamond, eds.), pp. 81–120. Belknap, Cambridge, MA.

May, R.M., and Anderson, R.M. (1979). Population biology of infectious diseases part II. *Nature* **280**, 455–461.

May, R.M., and Anderson, R.M. (1990). Parasite-host coevolution. *Parasitology* **100**, S89–S101.

Maynard Smith, J. (1964). Group selection and kin selection. *Nature* **201**, 1145–1147.

Maynard Smith, J. (1966). Sympatric speciation. *Am. Nat.* **100**, 637–650.

Maynard Smith, J. (1974). "Models in Ecology." Cambridge Univ. Press, Cambridge.

Maynard Smith, J., and Hoekstra, R. (1980). Polymorphism in a varied environment: How robust are the models? *Genet. Res. Camb.* **35**, 45–57.

Mayr, E. (1942). "Systematics and the Origin of Species." Columbia Univ. Press, New York.

Mayr, E. (1963). "Animal Species and Evolution." Belknap Press, Harvard.

McCallum, H., Barlow, N., and Hone, J. (2001). How should pathogen transmission be modelled? *Trends Ecol. Evol.* **16**, 295–300.

McCallum, H., and Dobson, A. (2002). Disease, habitat fragmentation and conservation. *Proc. R. Soc. Lond. Ser. B Biol. Sci.* **269**, 2041–2049.

McCauley, D.E. (1993). Evolution in metapopulations with frequent local extinction and recolonization. *Oxf. Surv. Evol. Biol.* **9**, 109–134.

McCauley, D.E., and Brock, M.T. (1998). *Frequency-dependent fitness in Silene vulgaris, a gynodioecious plant. Evolution* **52**, 30–36.

McCauley, D.E., Olson, M.S., Emery, S.N., and Taylor, D.R. (2000). Population structure influences sex ratio evolution in a gynodioecious plant. *Am. Nat.* **155**, 814–819.

McCauley, D.E., Raveill, J., and Antonovics, J. (1995). Local founding events as determinants of genetic structure in a plant metapopulation. *Heredity* **75**, 630–636

McCauley, D.E., Richards, C.M., Emery, S.N., Smith, R.A., and McGlothlin, J.W. (2001). The interaction of genetic and demographic processes in plant metapopulations: A case study of *Silene alba*. *In* "Integrating Ecology and Evolution in a Spatial Context." (J. Silvertown and J. Antonovics, eds.), pp. 177–196. Blackwell, Oxford.

McCauley, D.E., and Taylor, D.R. (1997). Local population structure and sex ratio: Evolution in gynodioecious plants. *Am. Nat.* **150**, 406–419.

McCauley, D.E., and Wade, M.J. (1980). Group selection: The genetic and demographic basis for the phenotypic differentiation of small populations of *Tribolium castaneum*. *Evolution* **34**, 813–821.

McCauley, E., Wilson, W.G., and De Roos, A.M. (1993). Dynamics of age-structured and spatially structured predator-prey interactions: Individual-based models and population-level formulations. *Am. Nat.* **142**, 412–442.

McClure, M., Holmes, E., Sanderson, B., and Jordan, C. (2003). A large-scale, multispecies risk assessment: anadromous salmonids in the Columbia River Basin. Submitted for publication.

McCullagh, P., and Nelder, J.A. (1989). "Generalized Linear Models," (2nd edn.). Chapman and Hall: London, U.K.

McDonald, J.F., and Ayala, F.J. (1974). Genetic response to environmental heterogeneity. *Nature* **250**, 572–574.

McDonnell, M.D., Possingham, H.P., Ball, I.R., and Cousins, E.A. (2002). Mathematical methods for spatially cohesive reserve design. *Environ. Model. Assess.* **7**, 107–114.

McGarigal, K., and McComb, W.C. (1995). Relationships between landscape structure and breeding birds in the Oregon coast range. *Ecol. Monogr.* **65**, 235–260.

McIntyre, N.E., and Wiens, J.A. (1999). Interactions between habitat abundance and configuration: Experimental validation of some predictions from percolation theory. *Oikos* **86**, 129–137.

McKay, J.K., Bishop, J.G., Lin, J.-Z., Richards, J.H., Sala, A., Stranger, B., and Mitchell-Olds, T. (2001). Local adaptation across a climatic gradient despite small effective size in the rare sapphire rockcress. *Proc. R. Soc. Lond. B* **268**, 1715–1721.

McLaughlin J.F., Hellmann J.J., Boggs C.L., and Ehrlich P.R. (2002). Climate change hastens population extinctions. *Proce. Nat. Acad. Sci. USA* **99**, 6070–6074.

McLaughlin, J.F., and Roughgarden, J.D. (1991). Patterns and stability in predator-prey communities: How diffusion in spatially variable environments affects the Lotka-Volterra model. *Theor. Popul. Biol.* **40**, 148–172.

McNeilly, T., and Bradshaw, A.D. (1968). Evolutionary processes in populations of copper tolerant *Agrostis tenuis* Sibth. *Evolution* **22**, 108–118.

McPeek, M.A., and Holt, R.D. (1992). The evolution of dispersal in spatially and temporally varying environments. *Am. Nat.* **140**, 1010–1027.

McVean, G. (2002). A genealogical interpretation of linkage disequilibrium. *Genetics* **162**, 987–991.

Meagher, T.R., and Thompson, E. (1987). Analysis of parentage for naturally established seedlings of *Chamaelerium luteum* (Liliaceae). *Ecology* **68**, 803–812.

Menéndez, R., Gutiérrez, D., and Thomas, C.D. (2002). Migration and Allee effects in the sixspot burnet moth, *Zygaena filipendulae*. *Ecological Entomology* **27**, 317–325.

Menges, E.S. (1990). Population viability analysis for an endangered plant. *Conserv. Biol.* **4**, 52–62.

Menges, E.S., and Dolan, R.W. (1998). Demographic viability of populations of *Silene regia* in midwestern prairies: Relationships with fire management, genetic variation, geographic location, populations size and isolation. *J. Ecol.* **86**, 63–78.

Merriam, G. (1988). Landscape dynamics in farmland. *Trends Ecol. Evol.* **3**, 16–20.

Merriam, G. (1991). Corridors and connectivity: animal populations in hetergenous environments. pp. 133–142 in D.A. Sanders and R.J. Hobbs, eds. Nature conservation 2: The role of corridors. Surrey Beatty & Sons, Chipping Norton, N.S.W.

Mery, F., and Kawecki, T.J. (2002). Experimental evolution of learning ability in fruit flies. *Proc. Natl. Acad. Sci. USA* **99**, 14274–14279.

Metz, J.A.J., and Gyllenberg, M. (2001). How should we define fitness in structured metapopulation models? Including an application to the calculation of evolutionarily stable dispersal strategies. *Proc. R. Soc. Lond. Ser. B Biol. Sci.* **268**, 499–508.

Metz, J.A.J., and van den Bosch, F. (1995). Velocities of epidemic spread. *In* "Epidemic Models: Their Structure and Relation to Data" (D. Mollison, ed.). pp. 150–186. Cambridge Univ. Press, Cambridge.

Meylan, S., Belliure, J., Clobert, J., and de Fraipont, M. (2002). Stress and body condition as prenatal and postnatal determinants of dispersal in the common lizard (*Lacerta vivipara*). *Horm. Behav.* **42**, 319–326.

Michalakis, I., and Olivieri, I. (1993). The influence of local extinctions on the probability of fixation of chromosomal rearrangements. *J. Evol. Biol.* **6**, 153–170.

Middleton, D.A.J., Veitch, A.R., and Nisbet, R.M. (1995). The effect of an upper limit to population size on persistence time. *Theor. Popul. Biol.* **48**, 277–305.

Mikko, S., Spencer, M., Morris, B., Stabile, S., Basu, T., Stormont, C., and Andersson, L. (1997). A comparative analysis of Mhc DRB3 polymorphism in the American bison (*Bison bison*). *J. Hered.* **88**, 499–503.

Miller, T., Kneitel, J.M., and Burns, J. (2002). Effects of community structure on invasion success and rate. *Ecology* **83**, 898–905.

Miller, T.E., Kneitel, J.M., and Mouquet, N. (2003). The effects of patch heterogeneity on the role of dispersal among experimental microcosms. Submitted for publication.

Mills, L.S., and Smouse, P.E. (1994). Demographic consequences of inbreeding in remnant populations. *Am. Nat.* **144**, 412–431.

Milne, A. (1957). The natural control of insect populations. *Can. Entomol.* **89**, 193–213.

Milne, A. (1962). On the theory of natural control of insect populations. *J. Theor. Biol.* **3**, 19–50.

Milner, G., Teel, D., Utter, F., and Winans, G. (1985). A genetic method of stock identification in mixed populations of pacific salmon, *Oncorhynchus spp. Mar. Fish. Rev.* **47**, 1–8.

Mitteldorf, J., and Wilson, D.S. (2000). Population viscosity and the evolution of altruism. *J. Theor. Biol.* **204**, 481–496.

Möhle, M. (1998a). A convergence theorem for Markov chains arising in population genetics and the coalescent with partial selfing. *Adv. Appl. Prob.* **30**, 493–512.

Möhle, M. (1998b). Coalescent results for two-sex population models. *Adv. Appl. Prob.* **30**, 513–520.

Möhle, M. (1998c). Robustness results for the coalescent. *J. Appl. Prob.* **35**, 438–447.

Moilanen, A. (1999). Patch occupancy models of metapopulation dynamics: Efficient parameter estimation using implicit statistical inference. *Ecology* **80**, 1031–1043.

Moilanen, A. (2000). The equilibrium assumption in estimating the parameters of metapopulation models. *J. Anim. Ecol.* **69**, 143–153.

Moilanen, A. (2002). Implications of empirical data quality to metapopulation model parameter estimation and application. *Oikos* **96**, 516–530.

Moilanen, A., and Cabeza, M. (2002). Single-species dynamic site selection. *Ecol. Appl.* **12**, 913–926.

Moilanen, A., and Hanski, I. (1998). Metapopulation dynamics: Effects of habitat quality and landscape structure. *Ecology* **79**, 2503–2515.

Moilanen, A., and Hanski, I. (2001). On the use of connectivity measures in spatial ecology. *Oikos* **95**, 147–151.

Moilanen, A., and Nieminen, M. (2002). Simple connectivity measures in spatial ecology. *Ecology* **84**, 1131–1145.

Moilanen, A., Smith A.T., and Hanski, I. (1998). Long-term dynamics in a metapopulation of the American pika. *Am. Nat.* **152**, 530–542.

Mollison, D. (1991). Dependence of epidemic and population velocities on basic parameters. *Math. Biosci.* **107**, 255–287.

Molofsky, J., Bever, J.D., and Antonovics, J. (2001). Coexistence under positive frequency dependence. *Proc. R. Soc. Ser. B* **268**, 273–277.

Moloney, K.A., and Levin, S.A. (1996). The effects of disturbance architecture on landscape-level population dynamics. *Ecology* **77**, 375–394

Montalvo, A.M., and Ellstrand, N.C. (2001). Nonlocal transplantation and outbreeding depression in the subshrub *Lotus scoparius* (Fabaceae). *Am. J. Bot.* **88**, 258–269.

Mooij, W.M., Bennetts, R.E., Kitchens, W.M., and DeAngelis, D.L. (2002). Exploring the effect of drought extent and interval on the FLORIDA snail kite: Interplay between spatial and temporal scales. *Ecol. Model.* **149**, 25–39.

Mooij, W.M., and DeAngelis, D.L. (1999). Error propagation in spatially explicit population models: A reassessment. *Conserv. Biol.* **13**, 930–933.

Moorcroft, P.R., Lewis, M.A., and Crabtree, R.L. (1999). Home range analysis using a mechanistic home range model. *Ecology* **80**, 1656–1665.

Moore, A.J., Brodie, E.D., III, and Wolf, J.B. (1997). Interacting phenotypes and the evolutionary process. I. Direct and indirect genetic effects of social interactions. *Evolution* **51**, 1352–1362.

Moore, A.J., Wolf, J.B., and Brodie, E.D., III. (1998). The influence of direct and indirect genetic effects on the evolution of social behavior: Sexual and social selection meet maternal effects. "Maternal Effects as Adaptations". pp. 22–41 *in* T.A. Mousseau, and C.W. Fox, eds., Oxford Univ Press, New York.

Moran P.A.P. (1953). The statistical analysis of the Canadian lynx cycle. II. Synchronization and meteorology. *Aust. J. Zool.* **1**, 291–298.

Morand, S. (2002). Life history evolution of nematodes: Linking epidemiological modelling and comparative tests. *Nematology* **4**, 593–599.

Morgan, J.W. (1999). Effects of population size on seed production and germinability in an endangered, fragmented grassland plant. *Conserv. Biol.* **13**, 266–273.

Morgan, M.T. (2002). Genome-wide dele mutation favors dispersal and species integrity. *Heredity* **89**, 253–257.

Morris, W.F., Bloch, P.L., Hudgens, B.R., Moyle, L.C., and Stinchcombe, J.R. (2002). Population viability analysis in endangered species recovery plans: Past use and future improvements. *Ecol. Appl.* **12**, 708–712.

Morris, W.F., and Doak, D.F. (2002). "Quantitative Conservation Biology: Theory and Practice of Population Viability Analysis." Sinauer, Sunderland, MA.

Morton, R.D., and Law, R. (1997). Regional species pools and the assembly of local ecological communities. *J. Theor. Biol.* **187**, 321–331.

Motro, U. (1991). Avoiding inbreeding and sibling competition: The evolution of sexual dimorphism for dispersal. *Am. Nat.* **137**, 108–115.

Mouquet, N., and Loreau, M. (2002). Coexistence in metacommunities: The regional similarity hypothesis. *Am. Nat.* **159**, 420–426.

Mouquet, N., Moore, J.L., and Loreau, M. (2002). Plant species richness and community productivity: Why the mechanism that promotes coexistence matters. *Ecol. Lett.* **5**, 56–65.

Mouquet, N., Munguia, P., Kneitel, J.M., and Miller, T.E. (2003). Community assembly time and the relationship between local and regional species richness. Submitted for publication.

Molau, U., and Larsson, E.-L. (2000). Seed rain and seed bank along an alpine altitudinal gradient in Swedish Lapland. *Can. J. Bot.* **78**, 728–747.

Mousseau, T.A., and Fox, C.W. (1998). "Maternal Effects as Adaptations." Oxford Univ. Press, Oxford.

Mousson, L., Nève, G., and Baguette, M. (1999). Metapopulation structure and conservation of the cranberry fritillary *Boloria aquilonaris* (Lepidoptera, Nymphalidae) in Belgium. *Biol. Conserv.* **87**, 285–293.

Muir, W.M. (1996). Group selection for adaptation to multiple-hen cages: Selection program and direct responses. *Poult. Sci.* **75**, 447–458.

Muirhead, C.A. (2001). Consequences of population structure on genes under balancing selection. *Evolution* **55**, 1532–1541.

Muller, H.J. (1939). Reversibility in evolution considered from the standpoint of genetics. *Biol. Rev. Camb. Phil. Soc.* **14**, 261–280.

Muller, H.J. (1942). Isolating mechanisms, evolution and temperature. *Biol. Symp.* **6**, 71–125.

Muller, H.J. (1964). The relation of recombination to mutational advance. *Mutat. Res.* **1**, 2–9.

Mullon, C., Cury, P., and Penven, P. (2002). Evolutionary individual-based model for the recruitment of anchovy (*Engraulis capensis*) in the southern Benguela. *Can. J. Fish. Aquat. Sci.* **59**, 910–922.

Murphy, D.D., Freas, K.S., and Weiss, S.B. (1990). An environment-metapopulation approach to the conservation of an endangered invertebrate. *Conserv. Biol.* **4**, 41–51.

Murray, J. (1990). Mathematical biology, 2nd edn. Springer, New York.

Murrell, D.J., and Law, R. (2000). Beetles in fragmented woodlands: A formal framework for dynamics of movement in ecological landscapes. *J. Anim. Ecol.* **69**, 471–483.

Murren, C.J., Julliard, R., Schlichting, C.D., and Clobert, J. (2001). Dispersal, individual phenotype, and phenotypic plasticity. *In* "Dispersal" (J. Clobert, E. Danchin, A.A. Dhondt, and J.D. Nichols, eds.), pp. 261–272. Oxford Univ. Press, Oxford.

Myers, J.M., Kope, R.G., Bryant, G.J., Teel, D., Lierheimer, L.J., Wainwright, T.C., Grand, W.S., Waknitz, F.W., Neely, K., Lindley, S.T., and Waples, R.S. (1998). Status review of chinook salmon from Washington, Idaho, Oregon, and California. U.S. Dept. Commer., NOAA Tech. Memo. NMFS-NWFSC-35.

Myllymäki, A. (1977). Demographic mechanisms in the fluctuating populations of the field vole *Microtus agrestis*. *Oikos* **29**, 468–493.

Nachman, G. (1991). An acarine predator-prey metapopulation system inhabiting greenhouse cucumbers. *Biol. J. Linn. Soc.* **42**, 285–303.

Nachman, G. (2000). Effects of demographic parameters on metapopulation size and persistence: An analytical stochastic model. *Oikos* **91**, 51–65.

Nagelkerke, K.C.J., Verboom, J., van den Bosch, F., and van de Wolfshaar, K. (2002). Time lags in metapopulation responses to landscape change. *In* "Applying Landscape Ecology in Biological Conservation" (K.J. Gutzwiller, ed.), pp. 330–354. Springer, New York.

Nagylaki, T. (1975). Conditions for the existence of clines. *Genetics* **80**, 595–615.

Nagylaki, T. (1980). The strong-migration limit in geographically structured populations. *J. Math. Biol.* **9**, 101–114.

Nagylaki, T. (1982). Geographical invariance in population genetics. *J. Theor. Biol.* **99**, 159–172.

Nakamaru, M., Matsuda, H., and Iwasa, Y. (1997). The evolution of cooperation in a lattice-structured population. *J. Theor. Biol.* **184**, 65–81.

Nath, H.B., and Griffiths, R.C. (1996). Estimation in an island model using simulation. *Theor. Popul. Biol.* **50**, 227–253.

Nathan, R. (2001). The challenges of studying dispersal *Trends Ecol. Evol.* **16**, 481–483.

Nathan, R., and Muller-Landau, H.C. (2000). Spatial patterns of seed dispersal, their determinants and consequences for recruitment. *Trends Ecol. Evol.* **7**, 278–285.

Nathan, R., Safriel, U.N., and Noy-Meir, I. (2001). Field validation and sensitivity analysis of a mechanistic model for tree seed dispersal by wind. *Ecology* **82**, 374–388.

National Marine Fisheries Service (NMFS (2000). Operation of the federal Columbia River power system including the juvenile fish transportation program and the Bureau of Reclamation's 31 projects, including the entire Columbia Basin Project. U.S. Department of Commerce, NOAA, National Marine Fisheries Service, Northwest Region, NW Region, Seattle.

Naveira, H.F., and Masida, X.R. (1998). The genetic of hybrid male sterility in *Drosophila*. *In* "Endless Forms: Species and Speciation" (D.J. Howard and S.H. Berlocher, eds.), pp. 330–338. Oxford Univ. Press, New York.

Nee, S. (1994). How populations persist. *Nature* **367**, 123–124.

Nee, S., and May, R.M. (1992). Dynamics of metapopulations: Habitat destruction and competitive coexistence. *J. Anim. Ecol.* **61**, 37–40.

Nee, S., May, R.M., and Hassell, M.P. (1997). Two-species metapopulation models. *In* "Metapopulation Biology: Ecology, Genetics, and Evolution" (I. Hanski and M. Gilpin, eds.), pp. 123–148. Academic Press, San Diego.

Nei, M., Maruyama, T., and Chakraborty, R. (1975). The bottleneck effect and genetic variability in populations. *Evolution* **29**, 1–10.

Nekola, J.C., and Kraft, C.E. (2002). Spatial constraint of peatland butterfly occurrences within a heterogeneous landscape. *Oecologia* **130**, 53–61.

Nelson, E.M., and DeBault, L.E. (1978). Transformation in *Tetrahymena pyriformis*: Description of an inducible phenotype. *J. Protozool.* **25**, 113–119.

Neubert, M.G., and Caswell (2000). Demography and dispersal: Calculation and sensitivity analysis of invasion speed for structured populations. *Ecology* **81**, 1613–1628.

Neubert, M., Kot, M., and Lewis, M.A. (1995). Dispersal and pattern-formation in a discrete-time predator-prey model. *Theor. Popul. Biol.* **48**, 7–43.

Neuhauser, C., and Krone, S.M. (1997). The genealogy of samples in models with selection. *Genetics* **145**, 519–534.

Nève, G., Barascud, B., Descimon, H., and Baguette, M. (2000). Genetic structure of *Proclossiana eunomia* populations at the regional scale (Lepidoptera, Nymphalidae). *Heredity* **84**, 657–666.

Nève, G., Barascud, B., Hughes, R., Aubert, J., Descimon, H., Lebrun, P., and Baguette, M. (1996a). Dispersal, colonization power and metapopulation structure in the vulnerable butterfly *Proclossiana eunomia* (Lepidoptera: Nymphalidae). *J. Appl. Ecol.* **33**, 14–22.

Nève, G., Mousson, L., and Baguette, M. (1996b). Adult dispersal and genetic structure of butterfly populations in a fragmented landscape. *Acta Oecol.* **17**, 621–626.

Newman, D., and Pilson, D. (1997). Increased probability of extinction due to decreased genetic effective population size: Experimental populations of *Clarkia pulchella*. *Evolution* **51**, 354–362.

Newmark, W.D. (1991). Tropical forest fragmentation and the local extinction of understory birds in the eastern Usambara mountains, Tanzania. *Conserv. Biol.* **5**, 67–78.

Newmark, W.D. (1995). Extinction of mammal populations in western North American national parks. *Conserv. Biol.* **9**, 512–526.

Nicholls, A.O. (1998). Integrating population abundance, dynamics and distribution into broad-scale priority setting. *In* "Conservation in a Changing World" Cambridge Univ. Press, Cambridge.

Nicholls, A.O., and Margules, C.R. (1993). An upgraded reserve selection algorithm. *Biol. Conserv.* **64**, 165–169.

Nicholson, A.J. (1933). The balance of animal populations. *J. Anim. Ecol.* **2**, 132–178.

Nicholson, A.J. (1954). An outline of the dynamics of animal populations. *Australian Journal of Zoology* 2: 9–65.

Nicholson, A.J. (1957). The self-adjustment of populations to change. *Cold Spring. Harb. Symp. Quan. Biol.* **22**, 153–173.

Nielsen, R. (1998). Maximum likelihood estimation of population divergence times and population phylogenies under the infinite sites model. *Theor. Popul. Biol.* **53**, 143–151.

Nielsen, R., and Wakeley, J. (2001). Distinguishing migration from isolation: A Markov Chain Monte Carlo approach. *Genetics* **158**, 885–896.

Nieminen, M. (1996). Risk of population extinction in moths: Effect of host plant characteristics. *Oikos* **76**, 475–484.

Nieminen, M., Siljander, M., and Hanski, I. (2004). Structure and dynamics of *Melitaea cinxia* metapopulations. "On the Wings of Checkerspots" (P.R. Ehrlich and I. Hanski eds.), Oxford Univ. Press, Oxford.

Nieminen, M., Singer, M.C., Fortelius, W., Schöps, K., and Hanski, I. (2001). Experimental confirmation that inbreeding depression increases extinction risk in butterfly populations. *Am. Nat.* **157**, 237–244.

Nilsson, C., Gardfjell, M., and Grelsson, G. (1991). Importance of hydrochory in structuring plant communities along rivers. *Can. J. Bot.* **69**, 2631–2633.

Nisbet, R., and Gurney, W. (1982). "Modelling Fluctuating Populations." Wiley, New York.

Nix, H.A. (ed.) (1986). "A Biogeographic Analysis of Australian Elapid Snakes." Australian Government Publishing Service, Canberra.

Nix, H.A., Faith, D.P., Hutchinson, M.F., Margules, C.R., West, J., Allison, A., Kesteven, J.L., Natera, G., et al. (2000). The BioRap Toolbox: A national study of biodiversity assessment and planning for Papua New Guinea. Centre for Resource and Environmental Studies, Australian National University, Canberra.

Noble, J.V. (1974). Geographic and temporal development of plagues. *Nature* **250**, 726–728.

Noss, R.F., C. Carroll, K. Vance-Borland, and G. Wuerthner (2002). A multicriteria assessment of the irreplaceability and vulnerability of sites in the Greater Yellowstone Ecosysem, Conserv. Biol., **16**, 895-908.

Nordal, I., Haraldsen, K.B., Ergon, A., and Eriksen, A.B. (1999). Copper resistance and genetic diversity in *Lychnis alpina* (Caryophyllaceae) populations on mining sites. *Folia Geobot*. **34**, 471–481.

Nordborg, M. (1997). Structured coalescent processes on different time scales. *Genetics* **146**, 1501–1514.

Nordborg, M. (2001). Coalescent theory. *In* "Handbook of Statistical Genetics" (D.J. Balding, M.J. Bishop, and C. Cannings, eds.). Wiley, Chichester, England.

Nordborg, M., and Krone, S.M. (2002). Separation of time scales and convergence to the coalescent in structured populations. *In* "Modern Developments in Theoretical Population Genetics" (M. Slatkin and M. Veuille, eds.). Oxford Univ. Press, Oxford.

Noss, R.F., C. Carroll, K. Vance-Borland, and G. Wuerthner. 2002. A multicriteria assessment of the irreplaceability and vulnerability of sites in the Greater Yellowstone Ecosystem. Conservation Biology 16: 895–908.

Notohara, M. (1990). The coalescent and the genealogical process in geographically structured population. *J. Math. Biol.* **29**, 59–75.

Notohara, M. (1993). The strong migration limit for the genealogical process in geographically structured populations. *J. Math. Biol.* **31**, 115–122.

Nunes, S., Co-Diem, T.H., Garrett, P.J., Mueke, E.-M., Smale, L., and Holekamp, K.E. (1998). Body fat and time of year interact to mediate dispersal behaviour in ground squirrels. *Anim. Behav*. **55**, 605–614.

Nunes, S., Zugger, P.A., Engh, A.L., Reinhart, A.L., and Holekamp, K.E. (1997). Why do female Belding's ground squirrel disperse away from food ressources? *Behav. Ecol. Sociobio*. **40**, 199–207.

Nunney, L. (1999). The effective size of a hierarchically structured population. *Evolution* **53**, 1–10.

Oates, M.R., and Warren, M.S. (1990). A review of butterfly introductions in Britain and Ireland. Joint Committee for the Conservation of British Insects/World Wildlife Fund, Godalming.

O'Brien, S.J., Wildt, D.E., Goldman, D., Merril, C., and Bush, M. (1983). The cheetah is depauperate in genetic variation. *Science* **221**, 459–462.

O'Hara, R.B., Arjas, E., Toivonen, H., and Hanski, I. (2002). Bayesian analysis of metapopulation data. *Ecology* **83**, 2408–2415.

Ohlson, M., Økland, R.H., Nordbakken, J.-F., and Dahlberg, B. (2001). Fatal interactions between Scots pine and Sphagnum mosses in bog ecosystems. *Oikos* **94**, 425–432.

Ohta, T. (1972). Population size and rate of evolution. *Mol. Evol.* **1**, 305–314.

Ohta, T., and Cockerham, C.C. (1974). Detrimental genes with partial selfing and effects on a neutral locus. *Genet. Res. Camb*. **23**, 191–200.

O'Keefe, K.J., and Antonovics, J. (2002). Playing by different rules: The evolution of virulence in sterilizing pathogens. *Am. Nat.* **159**, 597–605.

Okubo, A. (1980). "Diffusion and Ecological Problems: Mathematical Models." Springler-Verlag, Berlin.

Okubo, A., and Levin, S.A. (eds.) (2002). "Diffusion and Ecological Problems: New Perspectives." Springer, New York.

Olivieri, I. (2001). The evolution of seed heteromorphism in a metapopulation: Interactions between dispersal and dormancy. *In* "Integrating Ecology and Evolution in a Spatial Context" (J. Silvertown and J. Antonovics, eds.), pp. 245–268. The British Ecological Society and Blackwell Science, Oxford.

Olivieri, I., and Gouyon, P.H. (1997). Evolution of migration rate and other traits: The metapopulation effect. *In* "Metapopulation Biology: Ecology, Genetics and Evolution," (I. Hanski and M.E. Gilpin, eds.), pp. 293–324. Academic Press, San Diego.

Olivieri, I., Michalakis, Y., and Gouyon, P.H. (1995). Metapopulation genetics and the evolution of dispersal. *Am. Nat.* **146**, 202–228.

Olson, D.M., and Dinerstein, E. (1998). The global 200: A representation approach to conserving the Earth's most biologically valuable ecoregions. *Conserv. Biol.* **12**, 502–515.

Oostermeijer, J.G.B., Luijten, S.H., Krenová, Z.V., and den Nijs, J.C.M. (1998). Relations between population and habitat characteristics and reproduction of the rare *Gentiana pneumonanthe* L. *Conserv. Biol.* **12**, 1042–1053.

Oostermeijer, J.G.B., van Eijck, M.W., and den Nijs, J.C.M. (1994). Offspring fitness in relation to population size and genetic variation in the rare perennial plant species *Gentiana pneumonanthe* (Gentianaceae). *Oecologia* **97**, 289–296.

Opdam, P., Verboom, J., and Pouwels, R. (2003). Landscape cohesion: An index for the conservation potential of landscapes for biodiversity. *Landsc. Ecol.* **18**, 113–126.

O'Riain, J.M., Jarvis, J.U.M., and Faulkes, C.G. (1996). A dispersive morph in the naked mole rat. *Nature* **380**, 619–621.

Orr, H.A. (1995). The population genetics of speciation: The evolution of hybrid incompatibilities. *Genetics* **139**, 1803–1813.

Orr, H.A., and Orr, L.H. (1996). Waiting for speciation: The effect of population subdivision on the waiting time to speciation. *Evolution* **50**, 1742–1749.

Orr, H.A., and Turelli, M. (2001). The evolution of postzygotic isolation: Accumulating Dobzhansky-Muller incompatibilities. *Evolution* **55**, 1085–1094.

Orzack, S.H., and Tuljapurkar, S. (2001). Reproductive effort in variable environments, or environmental variation is for the birds. *Ecology* **82**, 2659–2665.

Ostfeld, R.S. (1994). The fence effect reconsidered. *Oikos* **70**, 340–348.

Ouborg, N.J. (1993). Isolation, population size and extinction: The classical and metapopulation approaches applied to vascular plants along the Dutch Rhine-system. *Oikos* **66**, 298–308.

Ouborg, N.J., Biere, A., and Mudde, C.L. (2000). Inbreeding effects on resistance and transmission-related traits in the *Silene – Microbotryum* pathosystem. *Ecology* **81**, 520–531.

Ouborg, N.J., Piquot, Y., and van Groenendael, J.M. (1999). Population genetics, molecular markers and the study of dispersal in plants. *J. Ecol.* **87**, 551–568.

Ouborg, N.J., and van Treuren, R. (1994). The significance of genetic erosion in the process of extinction. IV. Inbreeding load and heterosis in relation to population size in the mint *Salvia pratensis*. *Evolution* **48**, 996–1008.

Ouborg, N.J., and van Treuren, R. (1995). Variation in fitness-related characters among small and large populations of *Salvia pratensis*. *J. Ecol.* **87**, 551–568.

Ovaskainen, O. (2001). The quasi-stationary distribution of the stochastic logistic model. *J. Appl. Prob.* **38**, 898–907.

Ovaskainen, O. (2002a). The effective size of a metapopulation living in a heterogeneous patch network. *Am. Nat.* **160**, 612–628.

Ovaskainen, O. (2002b). Long-term persistence of species and the SLOSS problem. *J. Theor. Biol.* **218**, 419–433.

Ovaskainen, O. (2003). Habitat destruction, habitat restoration and eigenvector-eigenvalue relations. *Math. Biosci.* **181**, 165–176.

Ovaskainen, O. (2004). Habitat-specific movement parameters estimated using mark-recapture data and a diffusion model. *Ecology* **85**, 242–257.

Ovaskainen, O., and Cornell, S. (2003). Biased movement at boundary and conditional occupancy times for diffusion processes. *Journal of Applied Probability* **40**, 557–580.

Ovaskainen, O., and Grenfell, B. (2003). Mathematical tools for planning effective intervention scenarios for sexually transmitted diseases. *Sexually Transmitted Diseases.* **30**, 388–394.

Ovaskainen, O., and Hanski, I. (2001). Spatially structured metapopulation models: Global and local assessment of metapopulation capacity. *Theor. Popul. Biol.* **60**, 281–302.

Ovaskainen, O., and Hanski, I. (2002). Transient dynamics in metapopulation response to perturbation. *Theor. Popul. Biol.* **61**, 285–295.

Ovaskainen, O., and Hanski, I. (2003a). How much does an individual habitat fragment contribute to metapopulation dynamics and persistence? *Theor. Pop. Biol.* **64**, 481–495.

Ovaskainen, O., and Hanski, I. (2003b). Extinction threshold in metapopulation models. *Annal. Zool. Fenn.* **40**, 81–97.

Ovaskainen, O., Sato, K., Bascompte, J., and Hanski, I. (2002). Metapopulation models for extinction threshold in spatially correlated landscapes. *J. Theor. Biol.* **215**, 95–108.

Owen, N.W., Kent, M., and Dale, M.P. (2001). Spatial and temporal variability in seed bank dynamics of machair sand dune plant communities, the Outer Hebrides, Scotland. *J. Biogeogr.* **28**, 565–588.

Pacala, S., Canham, C., and Silander, J., Jr. (1993). Forest models defined by field-measurements. 1. The design of a northeastern forest simulator. *Can. J. For. Res.* **23**, 1980–1988.

Pacala, S.W., and Rees, M. (1998). Field experiments that test alternative hypotheses explaining successional diversity. *Am. Nat.* **152**, 729–737.

Pacala, S.W., and Roughgarden, J. (1982). Spatial heterogeneity and interspecific competition. *Theor. Popul. Biol.* **21**, 92–113.

Packer, A., and Clay, K. (2000). Soil pathogens and spatial patterns of seedling mortality in a temperate forest. *Nature* **404**, 278–281.

Packer, C., and Pusey, A.E. (1997). The ecology of relationships. *In* "Behavioral Ecology" (J.R. Krebs and N.B. Davies, eds.), pp. 254–283. Blackwell, Oxford.

Paetkau, D., Calvert, W., Stirling, I., and Strobeck, C. (1995). Microsatellite analysis of population-structure in Canadian polar bears. *Mol. Ecol.* **4**, 347–354.

Paine, R.T., and Levin, S.A. (1981). Intertidal landscapes: Disturbances and the dynamics of pattern. *Ecol. Monogr.* **51**, 145–178.

Pakkala, T., Hanski, I., and Tomppo, E. (2002). Spatial ecology of the three-toed woodpecker in managed forest landscapes. *Silva Fennica* **36**, 279–288.

Palmqvist, E., and Lundberg, P. (1998). Population extinctions in correlated environments. *Oikos* **83**, 359–367.

Pannell, J. (1997a). The maintenance of gynodioecy and androdioecy in a metapopulation. *Evolution* **51**, 10–20.

Pannell, J. (1997b). Widespread functional androdioecy in *Mercurialis annua* L. (Euphorbiaceae). *Biol. J. Lin. Soc.* **61**, 95–116.

Pannell, J.R. (2000). A hypothesis for the evolution of androdioecy: The joint influence of reproductive assurance and local mate competition in a metapopulation. *Evol. Ecol.* **14**, 195–211.

Pannell, J.R. (2003). Coalescence in a metapopulation with recurrent local extinction and recolonization, *Evolution*

Pannell, J.R., and Barrett, S.C.H. (1998). Baker's law revisited: Reproductive assurance in a metapopulation. *Evolution* **52**, 657–668.

Pannell, J.R., and Charlesworth, B. (1999). Neutral genetic diversity in a metapopulation with recurrent local extinction and recolonization. *Evolution* **53**, 664–676.

Pannell, J.R., and Charlesworth, B. (2000). Effects of metapopulation processes on measures of genetic diversity. *Philos. Trans. R. Soc. Lond. B Biol. Sci.* **355**, 1851–1864.

Paradis, E. (1995). Survival, immigration and habitat quality in the Mediterranean pine vole. *J. Anim. Ecol.* **64**, 579–591.

Paradis, E., Baillie, S.R., Sutherland, W.J., and Gregory, R.D. (1999). Dispersal and spatial scale affect synchrony in spatial population dynamics. *Ecol. Lett.* **2**, 114–120.

Parham, P., and Ohta, T. (1996). Population biology of antigen presentation by MHC class I molecules. *Science* **272**, 67–74.

Park, A.W., Gubbins, S., and Gilligan, C.A. (2001). Invasion and persistence of plant parasites in a spatially structured host population. *Oikos* **94**, 162–174.

Park, T. (1948). Experimental studies of interspecific competition. I. Competition between populations of flour beetles *Tribolium confusm* Duval and *T. castaneum* Herbst. *Physiol. Zool.* **18**, 265–308.

Parker, G.A., Chubb, J.C., Roberts, G.N., Michaud, M., and Milinski, M. (2003). Optimal growth strategies of larval helminths in their intermediate hosts. *J. Evol. Biol.* **16**, 47–54.

Parmesan, C. (1996). Climate and species' range. *Nature* **382**, 765–766.

Parmesan, C., Ryrholm, N., Stefanescu, C., Hill, J.K., Thomas, C.D., Descimon, H., Huntley, B., Kaila, L., Kullberg, J., Tammaru, T., Tennent, W.J., Thomas, J.A., and Warren, M. (1999). Poleward shifts in geographical ranges of butterfly species associated with regional warming. *Nature* **399**, 579–583.

Parsons, P.A. (1975). The effect of temperature and humidity on the distribution of *Drosophila inornata* in Victoria, Australia. *Environ. Entomol.* **4**, 961–964.

Parvinen, K. (1999). Evolution of migration in a metapopulation. *Bull. Math. Biol.* **61**, 531–550.

Parvinen, K. (2002). Evolutionary branching of dispersal strategies in structured metapopulations. *J. Math. Biol.* **45**, 106–124.

Parvinen, K., Dieckmann, U., Gyllenberg, M., and Metz, J.A.J. (2003). Evolution of dispersal in metapopulations with local density dependence and demographic stochasticity. *J. Evol. Biol.* **16**, 143–153.

Pascual, M., and Levin, S.A. (1999). Spatial scaling in a benthic population model with density-dependent disturbance. *Theor. Popul. Biol.* **56**, 106–122.

Paul, N.D., and Ayres, P.G. (1986). The impact of a pathogen (*Puccinia lagenophorae*) on populations of groundsel (*Senecio vulgaris*) overwintering in the field. II. Reproduction. *J. Ecol.* **74**, 1085–1094.

Peacock, M.M., and Ray, C. (2001). Dispersal in pikas (*Ochotona princeps*): Combining genetic and demographic approaches to reveal spatial and temporal patterns. *In* "Dispersal" (J. Clobert, E. Danchin, A. Dhondt, and J.D. Nichols eds.), pp. 43–56. Oxford Univ. Press, Oxford.

Peacock, M.M., and Smith, A.T. (1997). The effect of habitat fragmentation on dispersal patterns, mating behavior, and genetic variation in a pika (*Ochotona princeps*) metapopulation. *Oecologia*, **112**, 524–533.

Pearcy, W.G. (1992). "Ocean Ecology of North Pacific Salmon." University of Washington Press, Seattle, WA.

Pearson, S.M., Turner, M.G., Gardner, R.H., and O'Neill, R.V. (1996). An organism-based perspective of habitat fragmentation. *In* "Biodiversity in Managed Landscapes: Theory and Practice" (R.C. Szaro and D.W. Johnston, eds.), pp. 77–95. Oxford Univ. Press, New York.

Pelletier, J.D. (1999). Species-area relation and self-similarity in a biogeographic model of speciation and extinction. *Phys. Rev. Lett.* **82**, 1983–1986.

Peltonen, A., and Hanski, I. (1991). Patterns of island occupancy explained by colonization and extinction rates in shrews. *Ecology* **72**, 1698–1708.

Peltonen, M., Liebhold, A.M., Bjornstad, O.N., and Williams, D.W. (2002). Spatial synchrony in forest insect outbreaks: Roles of regional stochasticity and dispersal. *Ecology* **83**, 3120–3129.

Pen, I. (2000). Reproductive effort in viscous populations. *Evolution* **54**, 293–297.

Perlin, M.H. (1996). Pathovars or *formae speciales* of *Microbotryum violaceum* differ in electrophoretic karyotype. *Int. J. Plant Sci.* **157**, 447–452.

Perlin, M.H., Hughes, C., Welch, J., Akkaraju, S., Steinecker, D., Kumar, A., Smith, B., Garr, S.S., Brown , S.A., and Andom, T. (1997). Molecular approaches to differentiate subpopulations or *formae speciales* of the fungal phytopathogen *Microbotryum violaceum. Int. J. Plant Sci.* **158**, 568–574.

Perrin, N., and Goudet, J. (2001). Inbreeding, kinship, and the evolution of natal dispersal. *In* "Dispersal" (J. Clobert, E. Danchin, A.A. Dhondt, and J.D. Nichols, eds.), pp. 123–142. Oxford Univ. Press, Oxford.

Perrin, N., and Malazov, V. (1999). Dispersal and inbreeding avoidance. *Am. Nat.* **154**, 282–292.

Perrin, N., and Malazov, V. (2000). Local competition, inbreeding, and the evolution of sex-biaised dispersal. *Am. Nat.* **155**, 116–127.

Perry, W.L., Lodge, D.M., Feder, J.L., and Ibarra, E.L. (2002). Importance of hybridization between indigenous and nonindigenous freshwater species: An overlooked threat to North American biodiversity. *Syst. Biol.* **51**, 255–275.

Peters, A.D., Halligan, D.L., Whitlock, M.C., Keightley, P.D. (2003). Dominance and overdominance of mildly deleterious induced mutations for fitness traits in *Caenorhabditis elegans. Genetics* **165**:589–599.

Peters, J.C., and Shaw, M.W. (1996). Effect of artificial exclusion and augmentation of fungal plant pathogens on a regenerating grassland. *New Phytol.* **134**, 295–307.

Petit, S., and Burel, F. (1998). Effects of landscape dynamics on the metapopulation of a ground beetle (Coleoptera: Carabidae) in a hedgerow network. *Agric. Ecosyst. Environ.* **69**, 243–252.

Petit, S., Moilanen, A., Hanski, I., and Baguette, M. (2001). Metapopulation dynamics of the bog fritillary butterfly: Movements between habitat patches. *Oikos* **92**, 491–500.

Pettifor, R.A., Perrins, C.M., and McCleery, R.H. (1988). Individual optimization of clutch size in great tits. *Nature* **336**, 160–162.

Petrie, M., Krupa, A. and Burke, T. (1999). Peacocks lek with relatives even in the absence of social and environmental cues. Nature, 40(9): 155–157.

Pfrender, M., Spitze, K., Hicks, J., Morgan, K., Latta, L., and Lynch, M. (2001). Lack of concordance between genetic diversity estimates at the molecular and quantitative-trait levels. *Conserv. Genet.* **1**, 263–269.

Phillips, P.D., Brash, T.E., Yasman, I., Subagyo, P., and van Gardingen, P.R. (2003). An individual-based spatially explicit tree growth model for forests in East Kalimantan (Indonesian Borneo). *Ecol. Model.* **159**, 1–26.

Pinel-Alloul, B., Niyonsenga, T., and Legendre, P., (1995). Spatial and environmental components of freshwater zooplankton structure. *Ecoscience* **2**, 1–19.

Piquot, Y., Petit, D., Valero, M., Cuguen, J., de Laguerie, P., and Vernet, P. (1998). Variation in sexual and asexual reproduction among young and old populations of the perennial macrophyte *Sparganium erectum. Oikos* **82**, 139–148.

Pither, J., and Taylor, P.D. (1998). An experimental assessment of landscape connectivity. *Oikos* **83**, 166–174.

Platt, W.J., and Weis, I.M. (1977). Resource partitioning and competition within a guild of fugitive prairie plants. *Am. Nat.* **111**, 479–513.

Plotnick, R.E., Gardner, R.H., and O'Neill, R.V. (1993). Lacunarity indices as measures of landscape texture. *Landsc. Ecol.* **8**, 201–211.

Pluzhnikov, A., and Donnelly, P. (1996). Optimal sequencing strategies for surveying molecular genetic diversity. *Genetics* **144**, 1247–1262.

Poethke, H.J., and Hovestadt, T. (2002). Evolution of density- and patch-size-dependent dispersal rates. *Proc. R. Soc. Lond. B* **269**, 637–645.

Poethke, H.J., Hovestadt, T., and Mitesser, O. (2003). Local extinction and the evolution of dispersal rates: Causes and correlations. *Am. Nat.* **161**, 631–640.

Pokki, J. (1981). Distribution, demography and dispersal of the field vole, *Microtus agrestis* (L.), in the Tvärminne Archipelago, Finland. *Acta Zool. Fenn.* **164**, 1–48.

Polasky, S.J., Camm, J., and Garber-Yonts, B. (2001). Selecting biological reserves cost-effectively: An application to terrestrial vertebrates in Oregon. *Land Econ.* 77, 68–78.

Policansky, D., and Magnuson, J.J. (1998). Genetics, metapopulations, and ecosystem management of fisheries. *Ecol. Appl.* **8**, S119–S123.

Pollard, A.J. (1980). Diversity of metal tolerances in *Plantago lanceolata* L. from the southeastern United States. *New Phytol.* **86**, 109–117.

Poon, A., and Otto, S.P. (2000). Compensating for our load of mutations: Freezing the mutational meltdown. *Evolution* **54**, 1467–1479.

Possingham, H.P. (1996). Decision theory and biodiversity management: How to manage a metapopulation. *In* "Frontiers of Population Ecology" (R.B., Floyd, A.W. Sheppard, and P., Wellings, eds.), pp. 391–398. CSIRO Publishing, Canberra, Australia.

Possingham, H.P. (1997). State-dependent decision analysis for conservation biology. *In* "The Ecological Basis of Conservation: Heterogeneity, Ecosystems and Biodiversity" (S.T.A. Pickett, R.S. Ostfeld, M., Shachak, and G.E., Likens, eds.), pp. 298–304. Chapman and Hall, New York.

Possingham, H.P., Andelman, S.J., Noon, B.R., Trombulak, S., and Pulliam, H.R. (2001). Making smart conservation decisions. *In* "Conservation Biology: Research Priorities for the Next Decade" (M.E., Soulé, and G.H., Orians, eds.), pp. 225–244. Island Press, Washington, D.C.

Possingham, H.P., Ball, I., and Andelman, S. (2000). Mathematical methods for reserve system design. *In* "Quantitative Methods for Conservation Biology" (S. Ferson and M. Burgman, eds.), pp. 291–306. Springer-Verlag, New York.

Possingham, H.P., and Davies, I. (1995). ALEX: A model for the viability of spatially structured populations. *Biol. Conserv.* **73**, 143–150.

Possingham, H., Day, J., Goldfinch, M., and Salzborn, F. (1993). *In* "Proceedings of the 12th Australian Operation Research Conference" (D.J., Sutton, C.E.M. Pearce, and E.A., Cousins, eds.), pp. 536–545. Adelaide University, Adelaide.

Possingham, H.P., and Noble, I.R. (1991). An evaluation of population viability analysis for assessing the risk of extinction. Research Consultancy for the Research Assessment Commission, Forest and Timber Inquiry, Canberra.

Powell, J.R., and Wistrand, H. (1978). The effect of heterogeneous environments and a competitor on genetic variation in *Drosophila*. *Am. Nat.* **112**, 935–947.

Pressey, R.L. (1994) *Ad hoc* reservations: Forward or backward steps in developing representative reserve systems? *Conserv. Biol.* **8**, 662–668.

Pressey, R.L. (1999). Systematic conservation planning for the real world. *Parks* **9**, 1–6.

Pressey, R.L., and Taffs, K.H. (2001). Scheduling conservation action in production landscapes: Priority areas in western New South Wales defined by irreplaceability and vulnerability to vegetation loss. *Biol. Conserv.* **100**, 155–376.

Pressey, R.L., Johnson, I.R., and Wilson, P.D. (1994). Shades of irreplaceability: Towards a measure of the contribution of sites to a reservation goal. *Biodiv. Conserv.* **3**, 242–62.

Pressey, R.L., and Nicholls, A.O. (1989). Efficiency in conservation evaluation: Scoring vs iterative approaches. *Biol. Conserv.* **50**, 199–218.

Pressey, R.L., Possingham, H.P., and Day, J.R. (1997). Effectiveness of alternative heuristic algorithms for identifying indicative minimum requirements for conservation reserves. *Biol. Conserv.* **80**, 207–219.

Pressey, R.L., Possingham, H.P., and Margules, C.R. (1996) Optimality in reserve selection algorithms: when does it matter and how much? *Biol. Conserv.* **76**, 259–267.

Pressey, R. and Taffs change to Pressey, R. L. and Taffs and reorder.

Price, G.R. (1970). Selection and covariance. *Nature* **227**, 520–521.

Price, P.W., and Cappuccino, N. (eds.) (1995). "Population Dynamics: New Approaches and Synthesis". Academic Press, San Diego.

Pritchard, J., Stephens, M., and Donnelly, P. (2000). Inference of population structure using multilocus genotype data. *Genetics* **155**, 945–959.

Prout, R. (1981). A note on the island model with sex dependent migration. *Theor. Appl. Genet.* **59**, 327–332.

Provine, W.B. (2001). "The Origins of Theoretical Population Genetics." University of Chicago Press, Chicago, IL.

Puget Sound Technical Recovery Team (PSTRT) (2001). Independent populations of chinook salmon in Puget Sound. Public review draft. April, 2001. Northwest Fisheries Science Center, NOAA, Seattle, WA.

Pulliam, H.R. (1988). Sources, sinks and population regulation. *Am. Nat.* **132**, 652–661.

Pulliam, H.R. (1996). Sources and sinks: Empirical evidence and population consequences. *In* "Population Dynamics in Ecological Space and Time" (O.E. Rhodes, Jr., R.K. Chester, and M.H. Smith, eds.), pp. 45–70. University of Chicago Press, Chicago.

Pulliam, H.R. (2000). On the relationship between niche and distribution. *Ecol. Lett.* **3**, 349–361.

Pulliam, H.R., and Danielson, B.J. (1991). Sources, sinks and habitat selection: A landscape perspective on population dynamics. *Am. Nat.* **137**, S50–S66.

Quinn, T.P. (1993). A review of homing and straying of wild and hatchery-produced salmon. *Fish. Res.* **18**, 29–44.

Quintana-Ascencio, P.F., Dolan, R.W., and Menges, E.S. (1998). *Hypericum cumulicola* demography in unoccupied and occupied Florida scrub patches with different time-since-fire. *J. Ecol.* **86**, 640–651.

Quintana-Ascencio, R.F., and Menges, E.S. (1996). Inferring metapopulation dynamics from patch-level incidence of Florida scrub plants. *Conserv. Biol.* **10**, 1210–1219.

Rabinowitz, D. (1981). Buried viable seeds in a North American tall-grass prairie: The resemblance of their abundance and composition to dispersing seeds. *Oikos* **36**, 191–195.

Raijmann, L.E.L., van Leeuwen, N.C., Kersten, R., Oostermeijer, J.G.B., den Nijs, J.C.M., and Menken, S.B.J. (1994). Genetic variation and outcrossing rate in relation to population size in *Gentiana pneumonanthe* L. *Conserv. Biol.* 1014–1025.

Ralls, K., Beissinger, S.R., and Cochrane, J.F. (2002). Guidelines for using population viability analysis in endangered-species management. *In* "Population Viability Analysis" (S.R. Beissinger and D.R. McCullough, eds.), pp. 521–550. University of Chicago Press, Chicago, IL.

Ralls, K., and Taylor, B.L. (1997). How viable is population viability analysis? *In* "The Ecological Basis of Conservation: Heterogeneity, Ecosystems, and Biodiversity." (S.T.A. Pickett, R.S. Ostfeld, M. Shachak, and G.E. Likens, eds.), pp. 228–235. Chapman and Hall, New York.

Rand, D.A. (1999). Correlation equations and pair approximations for spatial ecologies. *In* "Theoretical Ecology 2" (J. McGlade, ed.). Blackwell, Oxford.

Rand, D.A., Keeling, M.J., and Wilson, H.B. (1995). Invasion, stability and evolution to critically in spatially extended, artificial host-pathogen ecologies. *Proc. R. Soc. Lond. B* **259**, 55–63.

Rand, D.A., and Wilson, H.B. (1995). Using spatio-temporal chaos and intermediate-scale determinism to quantify spatially extended ecosystems. *Proc. R. Soc. Lond. Ser. B Biol. Sci.* **259**, 111–117.

Ranius, T. (2000). Minimum viable metapopulation size of a beetle, *Osmoderma eremita*, living in tree hollows. *Anim. Conserv.* **3**, 37–43.

Ranius,T., and Jansson, N. (2000). The influence of forest regrowth, original canopy cover and tree size on saproxylic beetles associated with old oaks. *Biol. Conserv.* **95**, 85–94.

Rannala, B., and Mountain, J.L. (1997). Detecting immigration by using multilocus genotypes. *Proc. Nat. Acad. Sci. USA* **94**, 9197–9201.

Ranta, E., Kaitala, V., and Lindström, J. (1997). Dynamics of canadian lynx populations in space and time. *Ecography* **20**, 454–460.

Ranta, E., Kaitala, V., and Lundberg, P. (1998). Population variability in space and time: The dynamics of synchronous population fluctuations. *Oikos* **83**, 376–382.

Raufaste, N., and Rousset, F. (2001). Are partial Mantel tests adequate? *Evolution* **55**, 1703–1705.

Ray, C. (2001). Maintaining genetic diversity despite local extinctions: Effect of population scale. *Biol. Conserv.* **100**, 3–14.

Ray, C., Gilpin, M.E., and Smith, A.T. (1991). The effect of conspecific attraction on metapopulation dynamics. *Biol. J. Linn. Soc.* **42**, 123–134.

Ray, N., Currat, M., and Excoffier, L. (2003). Intra-deme molecular diversity in spatially expanding populations. *Mol. Biol. Evol.* **20**.

Raybould, A.F., Clarke, R.T., Bond, J.M., Welters, R.E., and Gliddon, C.J. (2002). Inferring patterns of dispersal from allele ferquency data. *In* "Dispersal Ecology" (J.M. Bullock, R. Kenward, and R. Hails eds.), Blackwell Science, Oxford.

Rebelo, A.G., and Siegfried, W.R. (1990). Protection of fynbos vegetation: Ideal and real-world options. *Biol. Conserv.* **54**, 15–31.

Reboud, X., and Bell, G. (1998). Experimental evolution in *Chlamydomonas*. III. Evolution of specialist and generalist types in environments that vary in space and time. *Heredity* **78**, 507–514.

Reed, J.M., Mills, L.S., Dunning, J.B., Jr., Menges, E.S., McKelvey, K.S., Frye, R., Beissinger, S.R., Anstett, M.C., and Miller, P. (2002). Emerging issues in population viability analysis. *Conserv. Biol.* **16**, 7–19.

Rees, M. (1994). Delayed germination of seeds: A look at the effects of adult longevity, the timing of reproduction, and population age/stage structure. *Am. Nat.* **144**, 43–64.

Rees, M. (1996). Evolutionary ecology of seed dormancy and seed size. *Phil. Trans. R. Soc. B Biol. Sci.* **351**, 1299–1308.

Reid, W.V., and Miller, K.R. (1989). "Keeping Options Alive: The Scientifis Basis for Conserving Biodiversity". World Resources Institute, Washington, DC.

Reidys, C.M. (1997). Random induced subgraphs of generalized n-cubes. *Adv. App. Math.* **19**, 360–377.

Reidys, C.M., Stadler, P.F., and Schuster, P. (1997). Generic properties of combinatory maps: Neutral networks of RNA secondary structures. *Bull. Math. Biol.* **59**, 339–397.

Remes, V. (2000). How can maladaptive habitat choice generate source-sink population dynamics? *Oikos* **91**, 579–582.

Restrepo, C., Gomez, N., and Heredia, S. (1999). Anthropogenic edges, treefall gaps, and fruit-frugivore interactions in a neotropical montane forest. *Ecology* **80**, 668–685.

Ribble, D.O. (1992). Dispersal in a monogamous rodent *Peromyscus californicus*. *Ecology* **73**, 859–866.

Rice, W.R., and Salt, G.W. (1990). The evolution of reproductive isolation as a correlated character under sympatric conditions: Experimental evidence. *Evolution* **44**, 1140–1152.

Richards, C.M. (2000). Inbreeding depression and genetic rescue in a plant metapopulation. *Am. Nat.* **155**, 383–394.

Richter-Dyn, N., and Goel, N.S. (1972). On the extinction of a colonizing species. *Theor. Popul. Biol.* **3**, 406–433.

Ricketts, T. (2001). The matrix matters: Effective isolation in fragmented landscapes. *Am. Nat.* **158**, 87–99.

Rieman, B.E., and Dunham, J.B. (2000). Metapopulations and salmonids: A synthesis of life history patterns and empirical observations. *Ecol. Freshw. Fish* **9**, 51–64.

Rieman, B.E., and McIntyre, J.D. (1995). Occurrence of bull trout in naturally fragmented habitat patches of varied size. *Trans. Am. Fish. Soc.* **124**, 285–296.

Ritchie, M. (1997). Populations in a landscape context: Sources, sinks, and metapopulations. *In* "Wildlife and Landscape Ecology: Effects of Pattern and Scale" (J.A. Bissonette, ed.), pp. 160–184. Springer, New York.

Roach, J.L., Stapp, P., Van Horne, B., and Antolin, M.F. (2001). Genetic structure of a metapopulation of black-tailed prairie dogs. *J. Mammal.* **82**, 946–959.

Robinson, S.K., Thompson III, F.R., Donovan, T.M., Whitehead, D.R., and Faaborg, J. (1995). Regional forest fragmentation and the nesting success of migratory birds. *Science* **267**, 1987–1990.

Rodrigues, A.S., and Gaston, K.J. (2002) Optimisation in reserve selection procedures: Why not? *Biol. Conserv.* **107**, 123–129.

Rodrigues, A.S., Gregory, R.D., and Gaston, K.J. (2000) Robustness of reserve selection procedures under temporal species turnover. *Proc. R. Soc. Lond. B* **267**, 49–55.

Roff, D.A. (1975). Population stability and the evolution of dispersal in a heterogeneous environment. *Oecologia* **19**, 217–237.

Roff, D.A. (1992). "The Evolution of Life Histories." Chapman & Hall, New York.

Roff, D.A. (1994). Habitat persistence and the evolution of wing dimorphism in insects. *Am. Nat.* **144**, 772–798.

Roff, D.A., and Fairbairn, D.J. (2001). The genetic basis of dispersal and migration, and its consequences for the evolution of correlated traits. *In* "Dispersal" (J. Clobert, E. Danchin, A.A. Dhondt, and J.D. Nichols, eds.), pp. 191–202. Oxford Univ. Press, Oxford.

Roff, D.A., and Simons, A.M. (1997). The quantitative genetics of wing dimorphism under laboratory and 'field' conditions in the cricquet *Gryllus pennsylvanicus*. *Heredity* **78**, 235–240.

Roff, D.A., Stirling, G., and Fairbairn, D.J. (1997). The evolution of threshold traits: A quantitative genetic analysis of the physiology and life history correlates of wing dimorphism in the land cricket. *Evolution* **51**, 1910–1919.

Rohani, P., Earn, D.J.D., and Grenfell, B.T. (2000). Impact of immunisation on pertussis transmission in England and Wales. *Lancet* **355**, 285–286.

Rohani, P., Keeling, M.J., and Grenfell, B.T. (2002). The interplay between determinism and stochasticity in childhood diseases. *Am. Nat.* **159**, 469–481.

Rohani, P., May, R.M., and Hassell, M.P. (1996). Metapopulations and local stability: The effects of spatial structures. *J. Theor. Biol.* **181**, 97–109.

Roland, J., Keyghobadi, N, and Fownes, S. (2000). Alpine *Parnassius* butterfly dispersal: Effects of landscape and population size. *Ecology* **81**, 1642–1653.

Roland, J., and Taylor, P.D. (1997). Insect parasitoid species respond to forest structure at different spatial scales. *Nature* **386**, 710–713.

Ronce, O., Clobert, J., and Massot, M. (1998). Natal dispersal and senescence. *Proc. Natl. Acad. Sci. USA* **95**, 600–605.

Ronce, O., Gandon, S., and Rousset, F. (2000a). Kin selection and natal dispersal in an age-structured population. *Theor. Popul. Biol.* **58**, 143–159.

Ronce, O., and Kirkpatrick, M. (2001). When sources become sinks: Migrational meltdown in heterogeneous habitats. *Evolution* **55**, 1520–1531.

Ronce, O., and Olivieri, I. (1997). Evolution of reproductive effort in a metapopulation with local extinctions and ecological succession. *Am. Nat.* **150**, 220–249.

Ronce, O., Olivieri, I., Clobert, J., and Danchin, E. (2001). Perspectives on the study of dispersal evolution. *In* "Dispersal" (J. Clobert, E. Danchin, A.A. Dhondt, and J.D. Nichols, eds.), pp. 341–357. Oxford Univ. Press, Oxford.

Ronce, O., Perret, F., and Olivieri, I. (2000b). Evolutionarily stable dispersal rates do not always increase with local extinction rates. *Am. Nat.* **155**, 485–496.

Ronce, O., Perret, F., and Olivieri, I. (2000c). Landscape dynamics and evolution of colonizer syndromes: Interactions between reproductive effort and dispersal in a metapopulation. *Evol. Ecol.* **14**, 233–260.

Rood, J.P. (1986). Ecology and social evolution in the mongooses. *In* "Ecological Aspects of Social Evolution" (D.I. Rubenstein and R.W. Wrangham eds.), pp. 131–152. Princeton Univ. Press, Princeton, NJ.

Rosenberg, R., Nilsson, H.C., Hollertz, K. and Hellman, B (1997): Density-dependent migration in an *Amphiura filiformis* (Amphiuridae, Echinodermata) infaunal population. Marine Ecology Progress Series, 159: 121–131.

Rosenzweig, M.L. (1995). "Species Diversity in Space and Time." Cambridge Univ. Press, Cambridge.

Roughgarden, J.D. (1974). Population dynamics in a spatially varying environment: How population size tracks spatial variation in carrying capacity. *Am. Nat.* **108**, 649–664.

Roughgarden, J. (1977). Patchiness in the spatial distribution of a population caused by stochastic fluctuations in resources. *Oikos* **29**, 52–59.

Roughgarden, J. (1978). Influence of competition on patchiness in a random environment. *Theor. Pop. Biol.* **14**, 185–203.

Rousset, F. (1999a). Reproductive values vs sources and sinks. *Oikos* **86**, 591–596.

Rousset, F. (1999b). Genetic differentiation in populations with different classes of individuals. *Theor. Popul. Biol.* **55**, 297–308.

Rousset, F. (1999c). Genetic differentiation within and between two habitats. *Genetics*, **151**, 397–407.

Rousset, F. (2000). Genetic differentiation between individuals. *J. Evol. Biol.* **13**, 58–62.

Rousset, F. (2003). Effective size in simple metapopulation models. *Heredity,* 91: 107–111.

Rousset, F. (2004). "Genetic Structure and Selection in Subdivided Populations." Princeton Univ. Press, Princeton, NJ.

Rousset, F., and Gandon, S. (2002). Evolution of the distribution of dispersal distance under distance-dependent cost of dispersal. *J. Evol. Biol.* **15**, 515–523.

Rousset, F. and Ronce, P. (2004). Inclusive fitness for traits affecting metapopulation demography. *Theor. Popul. Biol.*, in press.

Roze, D., and Rousset, F. (2003). Diffusion approximations for selection and drift in subdivided populations: A straightforward method and examples involving dominance, selfing and local extinctions. Submitted for publication.

Ruckleshaus, M., Hartway, C., and Kareiva, P. (1997). Assessing the data requirements of spatially explicit dispersal models. *Conserv. Biol.* **11**, 1298–1306.

Ruckelshaus, M.H. (1996). Estimation of genetic neighbourhood parameters from pollen and seed dispersal in the marine angiosperm *Zostera marina L. Evolution* **50**, 856–864.

Rushton, S.P., Lurz, P.W.W., Fuller, R., and Garson, P.J. (1997). Modelling the distribution of the red and gray squirrel at the landscape scale: A combined GIS and population dynamics approach. *J. Appl. Ecol.* **34**, 1137–1154.

Ruxton, G. D. (1996). Dispersal and chaos in spatially structured models: An individual-level approach. *J. Anim. Ecol.* **65**, 161–169.

Ruxton, G., Gonzales-Andujar, J. L., and Perry, J. N. (1997). Mortality during dispersal stabilizes local population fluctuations. *J. Anim. Ecol.* **66**, 289–292.

Saccheri, I., Kuussaari, M., Kankare, M., Vikman, P., Fortelius, W., and Hanski, I. (1998). Inbreeding and extinction in a butterfly metapopulation. *Nature* **392**, 491–494.

Saccheri, I. J., and Brakefield, P. M. (2002). Rapid spread of immigrant genomes into inbred populations. *Proc. R. Soci. Lon. Ser. B Biol. Sci.* **269**, 1073–1078.

Sætersdal, M., Line, J.M., and Birks, H.J.B. (1993). How to maximize biological diversity in nature reserve selection: Vascular plants and breeding birds in deciduous woodlands, Western Norway. *Biol. Conserv.* **66**, 131–138.

Saether, B.-E., and Engen, S. (2002). Including uncertainties in population viability analysis using population prediction intervals. *In* "Population Viability Analysis". (S. R. Beissinger and D. R. McCullough, eds.), pp. 191–212. The University of Chicago Press, Chicago.

Saether, B.-E., Engen, S., Islam, A., McCleery, R., and Perrins, C. (1998a). Environmental stochasticity and extinction risk in a population of a small songbird, the great tit. *Am. Nat.* **151**, 441–450.

Saether, B.-E., Engen, S., and Lande, R. (1999). Finite metapopulation models with density-dependent migration and stochastic local dynamics. *Proc. R. Soc. Lond. B Biol. Sci.* **266**, 113–118.

Saether, B.-E., Engen, S., Swenson, J. E., Bakke, Ø., and Sandegren, F. (1998b). Assessing the viability of Scandinavian brown bear, *Ursus arctos*, populations: The effects of uncertain parameter estimates. *Oikos* **83**, 403–416.

Saether, B.-E., Ringsby, T. H., and Røskaft, E. (1996). Life history variation, population processes and priorities in species conservation: Towards a reunion of research paradigms. *Oikos* **77**, 217–226.

Sandell, M., Ågrell, J., Erling, S., and Nelson, J. (1990). Natal dispersal in relation to population density and sex ratio in the field vole, *Microtus agrestis. Oecologia* **83**, 145–149.

Sandell, M., Ågrell, J., Erlinge, S., and Nelson, J. (1991). Adult philopatry and dispersal in the field vole *Microtus agrestis. Oecologia* **86**, 153–158.

Sanz, L., and Bravo de la Parra, R. (1998). Variables aggregation in time varying discrete systems. *Acta Biotheor.* **46**, 273–297.

SAS Institute Inc. (1999). SAS/STAT User's Guide, Version 8. SAS Institute Inc., Cary, NC.

Sasaki, A. (1997). Clumped distribution by neighborhood competition. *J. Theor. Biol.* **186**, 415–430.

Satō, K., and Iwasa, Y., (2000). Pair approximations for lattice-based ecological models. In the Geometry of Ecological Interactions: Simplifying Spatial Complexity (U. Dieckman, R. Law, and J.A.J. Metz, eds.), pp. 341–358. Cambridge Univ. Press, Cambridge.

Satō, K., Matsuda, H., and Sasaki, A. (1994). Pathogen invasion and host extinction in lattice structured populations. *J. Math. Biol.* **32**, 251–268.

Saum, M., and Gavrilets, S. (2003). *In* "Complexity in Ecological Systems: On the Nature of Nature" (J. Drake, C. Zimmerman, S. Gavrilets, and T. Fukami, eds.), Columbia Univ. Press, New York.

Sawchik, J., Dufrene, M., Lebrun, P., Schtickzelle, N., and Baguette, M. (2002). Metapopulation dynamics of the bog fritillary butterfly: Modelling the effect of habitat fragmentation. *Acta Oecol.* **23**, 287–296.

Sawyer, S. (1977a). Asymptotic properties of the equilibrium probability of identity in a geographically structured population. *Adv. Appl. Prob.* **9**, 268–282.

Sawyer, S. (1977b). Rates of consolidation in a selectively neutral migration model. *Ann. Prob.* **5**, 486–493.

Sawyer, S. (1979). A limit theorem for patch size in a selectively-neutral migration model. *J. Appl. Prob.* **16**, 482–495.

Schaffer, W. M. (1985). Order and chaos in ecological systems. *Ecology* **66**, 93–106.

Schat, H., Vooijs, R., and Kuiper, E. (1996). Identical major gene loci for heavy metal tolerances that have independently evolved in different local populations and subspecies of *Silene vulgaris. Evolution* **50**, 1888–1895.

Scheiner, S.M., and Rey-Benayas, J.M. (1997). Placing empirical limits on metapopulation models for terrestrial plants. *Evol. Ecol.* **11**, 275–288.

Schenzle, D. (1984). An age-structured model of pre- and post-vaccination measles transmission. *I.M.A. J. Math. Appl. Med. Biol.* **1**, 169–191.

Schierup, M. H. (1998). The number of self-incompatibility alleles in a finite, subdivided population. *Genetics* **149**, 1153–1162.

Schierup, M. H., Vekemans, X., and Charlesworth, D. (2000). The effect of subdivision on variation at multi-allelic loci under balancing selection. *Genet. Res. Camb.* **76**, 51–62.

Schippers, P., Verboom, J., Knaapen, J. P., and van Apeldoorn, R. C. (1996). Dispersal and habitat connectivity in complex heterogeneous landscapes: An analysis with a GIS-based random walk model. *Ecography* **19**, 97–106.

Schlichting, C. D., and Pigliucci, M. (1998). "Phenotypic Evolution: A Reaction Norm Perspective." Sinauer Associates, Sunderland, MA.

Schluter, D. (1998). Ecological causes of speciation. *In* "Endless Forms: Species and Speciation" (D. J. Howard and S. H. Berlocher, eds.), pp. 114–129. Oxford Univ. Press, New York.

Schluter, D. (2000). "The Ecology of Adaptive Radiation." Oxford Univ. Press, Oxford.

Schluter, D. (2001). Ecology and the origin of species. *Trends Ecol. Evol.* **16**, 372–380.

Schmidt, K.P., and Levin, D.A. (1988). The comparative demography of reciprocally sown populations of *Phlox drummondii* Hood. I. Survivorships, fecundities, and finite rates of increase. *Evolution* **39**, 396–404.

Schmitt, J., and Gamble, S.E. (1990). The effect of distance from the parental site on offspring performance and inbreeding depression in *Impatiens capensis*: A test of the local adaptation hypothesis. *Evolution* **44**, 2022–2030.

Schmitt, T., and Seitz, A. (2002). Influence of habitat fragmentation on the genetic structure of *Polyommatus coridon* (Lepidoptera: Lycaenidae): Implications for conservation. *Biol. Conserv.* **107**, 291–297.

Schmitz, O.J., and Booth, G. (1997). Modelling food web complexity: The consequences of individual-based, spatially explicit behavioural ecology on trophic interactions. *Evol. Ecol.* **11**, 379–398.

Schnabel, A., Nason, J.D., and Hamrick, J.L. (1998). Understanding the population genetic structure of *Gleditsia triacanthos* L.: Seed dispersal and variation in female reproductive success. *Mol. Ecol.* **7**, 819–832.

Schoener, T., Clobert, J., Legendre, S., and Spiller, D. (2003). Life history models of extinction: a test with Island spiders. *Am. Nat.*

Schonewald-Cox, C.M., Chambers, S.M., MacBryde, B., and Thomas, L. (eds.) (1983). "Genetics and Conservation." Bejamin/Cummings, Menlo Park, CA.

Schöps, K., Emberson, R.M., and Wratten, S.D. (1998). Does host-plant exploitation influence the population dynamics of a rare weevil? *In* "Proceedings of the Ecology and Population Dynamics Section of the 20th Congress of Entomology" (B.J.F. Manly, J. Baumgörtner, and F. Brandlmayer, eds.) Florence, Italy, 25–31 August 1996.

Schott, G.W., and Hamburg, S.P. (1997). The seed rain and seed bank of an adjacent tallgrass prairie and old field. *Can. J. Bot.* **75**, 1–7.

Schrott, G.R., With, K.A., and King, A.W. (2003) Assessing extinction risk for migratory songbirds in dynamic landscapes. *Ecology.*

Schtickzelle, N., Le Boulenge, E., and Baguette, M. (2002). Metapopulation dynamics of the bog fritillary butterfly: Demographic processes in a patchy population. *Oikos* **97**, 349–360.

Schulte-Hostedde, A.I., Gibbs, H.L., and Millar, J.S. (2001). Microgeographic genetic structure in the yellow-pine chipmunk (*Tamias amoenus*). *Mol. Ecol.* **10**, 1625–1631.

Schultz, C.B. (1998). Dispersal behaviour and its implications for reserve design in a rare Oregon butterfly. *Conserv. Biol.* **12**, 284–292.

Schultz C.B., and Crone E.E. (2001). Edge-mediated dispersal behavior in a prairie butterfly. *Ecology* **82**, 1879–1892.

Schumaker, N.H. (1996). Using landscape indices to predict habitat connectivity. *Ecology* **77**, 1210–1225.

Schupp, E.W. (1995). Seed-seedling conflicts, habitat choice, and patterns of plant recruitment. *Am. J. Bot.* **82**, 399–409.

Schwartz, M., Tallmon, D.A., and Luikart, G. (1998). Review of DNA-based census and effective population size estimators. *Anim. Conserv.* **1**, 293–299.

Schwinning, S., and Parsons, A.J. (1996). A spatially explicit population model of stoloniferous N-fixing legumes in mixed pasture with grass. *J. Ecol.* **84**, 815–826.

Seger, J., and Brockmann, H.J. (1987). What Is Bet-Hedging? (P.H. Harvey and L. Partridge, eds.), pp. 182–211. Oxford Univ. Press.

Serneels, S., and Lambin, E.F. (2001). Proximate causes of land-use change in Narok District, Kenya: A special statistical model. *Agric. Ecosyst. Environ.* **85**, 65–81.

Settele, J. Henle, K., and Bender, C. (1996). Metapopulation und biotopverbund: Theorie und praxis am beispiel von tagfaltern und reptilien. *Zeitschr. Ökol. Natur.* **5**, 187–206.

Shaffer, M.L. (1981). Minimum population sizes for species conservation. *BioScience* **31**, 131–134.

Shahabuddin, G., and Terborgh, J.W. (1999). Frugivorous butterflies in Venezuelan forest fragments: Abundance, diversity and the effects of isolation. *J. Trop. Ecol.* **15**, 703–722.

Sharov, A.A., and Liebhold, A.M. (1998). Model of slowing the spread of the gypsy moth (Lepidoptera: Lymantriidae) with a barrier zone. *Ecol. Appl.* **8**, 1170–1179.

Shaw, F. H., Geyer, C. J., and Shaw, R. G. (2002). A comprehensive model of mutation affecting fitness and inferences for *Arabidopsis thaliana. Evolution* **56**, 453–463.

Shaw, R.G., Shaw, F.H., and Geyer, C.J. (2003). What fraction of mutations reduce fitness? A reply to Keightley and Lynch. *Evolution* **57**, 686–689.

Sheftel, B.I. (1989). Long-term and seasonla dynamics of shrews in central Siberia. *Ann. Zool. Fenn.* **26**, 357–370.

Shigesada, N., and Kawasaki, K. (1997). "Biological Invasions: Theory and Practice." Oxford Univ. Press, New York.

Shimizu, K., and Iwase, K. (1981). Uniformly minimum variance unbiased estimation in lognormal and related distributions. *Commun. Stat. A Theory Methods* **10**, 1127–1147.

Shrewsbury, J.F.D. (1970). "A History of Bubonic Plague in the British Isles." Cambridge Univ. Press, Cambridge.

Shugart, H. (1984). "A Theory of Forest Dynamics." Springer, New York.

Shurin, J.B. (2000). Dispersal limitation, invasion, resistance, and the structure of pond zooplankton communities. *Ecology* **81**, 3074–3086.

Shurin, J.B. (2001). Interactive effects of predation and dispersal on zooplankton communities. *Ecology* **82**, 3404–3416.

Shurin, J.B., and Allen, E.G. (2001). Effects of competition, predation and dispersal on local and regional species richness. *Am. Nat.* **158**, 624–637.

Shurin, J.B., Amarasekare, P., Chase, J.M., Holt, R.D., and Leibold, M.A. (2003). Alternate stable states and regional community structure. Submitted for publication.

Shurin, J.B., Havel, J.E., Leibold, M.A., and Pinel-Alloul, B. (2000). Local and regional zooplankton richness: A scale-independent test for saturation. *Ecology* **81**, 3062–3073.

Shuster, S.M. and Wade, M.J. (2003). "Mating Systems and Mating Strategies." Princeton University Press, Princeton, NJ.

Silvertown, J. (1987). "Introduction to Plant Population Ecology." Longman, Harlow.

Silvertown, J. (1991). Dorothy's dilemma and the unification of plant population biology. *Trends Ecol. Evol.* **4**, 24–26.

Silvertown, J., and Antonovics, J. (2001). "Integrating Ecology and Evolution in a Spatial Context." Blackwell, Oxford.

Silvertown, J., and Franco, M. (1993). Plant demography and habitat: A comparative approach. *Plant Spec. Biol.* **8**, 67–73.

Silvertown, J., Holtier, S., Johnson, J., and Dale, P. (1992). Cellular automaton models of interspecific competition for space: The effect of pattern on process. *J. Ecol.* **80**, 527–534.

Silvertown, J., Lines, C.E.M., and Dale, M.P. (1994). Spatial competition between grasses: Rates of mutual invasion between four species and the interaction with grazing. *J. Ecol.* **82**, 31–38.

Simberloff, D. (1988). The contribution of population and community biology to conservation science. *Annu. Rev. Ecol. Syst.* **19**, 473–512.

Simberloff, D. (1994). The ecology of extinction. *Acta Palaeontol. Polon.* **38**, 159–174.

Simberloff, D., and Cox, J. (1987). Consequences and costs of conservation corridors. *Conserv. Biol.* **1**, 63–71.

Simonsen, K.L., Churchill, G.A., and Aquadro, C.F. (1995). Properties of statistical tests of neutrality for DNA polymorphism data. *Genetics* **141**, 413–429.

Sinclair, A.R.E. (1989). Population regulation in animals. *In* "Ecological Concepts" (J.M. Cherrett, ed.), pp. 197–242. Blackwell, Oxford.

Sinervo, B., Calsbeek, R., and Clobert, J. (2003). Genetic and maternal determinants of dispersal in color morphs of side-blotched lizards. *Evolution.*

Sinervo, B., and Clobert, J. (2003). Morphs, dispersal behavior, genetic similarity, and the evolution of cooperation. *Science*, **300**: 1949–1951.

Sinervo, B., and Lively, C.M. (1996). The rock-scissors-paper game and the evolution of alternative males strategies. *Nature* **340**, 440–441.

Sinervo, B., Svensson, E., and Comendant, T. (2000). Density cycles and an offspring quantity and quality game driven by natural selection. *Nature* **406**, 985–988.

Singer, M.C. (1972). Complex components of habitat suitability within a butterfly colony. *Science* **176**, 75–77.

Singer, M.C., and Thomas, C.D. (1996). Evolutionary responses of a butterfly metapopulation to human and climate-caused environmental variation. *Am. Nat.* **148**, S9–S39.

Sjögren-Gulve, P., and Ebenhard, T. ((eds.)2000). The use of population viability analyses in conservation planning. *Ecol. Bull.* **48**.

Sjögren-Gulve, P., and Hanski, I. (2000). Metapopulation viability analysis using occupancy models. *Ecol. Bull.* **48**, 53–71.

Sjögren-Gulve, P., and Ray, C. (1996). Using logistic regression to model metapopulation dynamics: Large scale forestry extirpates the pool frog. *In* "Metapopulations and Wildlife Conservation." (D.L. McCulloch, ed.), pp. 111–138. Island Press, Washington, DC.

Skarpaas, O. (2003). "Plant Population Dynamics in Fragmented Landscapes." Ph.D. thesis, University of Oslo.

Skarpaas, O., and Stabbetorp, O.E. (2001). Diaspore ecology of *Mertensia maritima*: Effects of physical treatments and their relative timing on dispersal and germination. *Oikos* **95**, 374–382.

Skellam, J.G. (1951). Random dispersal in theoretical populations. *Biometrika* **38**, 196–218. Reprinted in "Foundations of Ecology: Classic Papers with Commentaries" (L.A. Real and J.H. Brown, eds.). University of Chicago Press, Chicago, 1991.

Slatkin, M. (1974). Competition and regional coexistence. *Ecology* **55**, 128–134.

Slatkin, M. (1977). Gene flow and genetic drift in a species subject to frequent local extinctions. *Theor. Popul. Biol.* **12**, 253–262.

Slatkin, M. (1978). Spatial patterns in the distributions of polygenic characters. *J. Theor. Biol.* **70**, 213–228.

Slatkin, M., (1981). Fixation probabilities and fixation times in a subdivided population. *Evolution* **35**, 477–488.

Slatkin, M. (1991). Inbreeding coefficients and coalescence times. *Genet. Res. Camb.* **58**, 167–175.

Slatkin, M. (1995). Epistatic selection opposed by immigration in multiple locus genetic systems. *J. Evol. Biol.* **8**, 623–633.

Slatkin, M., and Hudson, R.R. (1991). Pairwise comparisons of mitochondrial DNA sequences in stable and exponentially growing populations. *Genetics* **129**, 555–562.

Smith, A. (1974). The distribution and dispersal of pikas: Consequences of insular population structure. *Ecology* **55**, 1112–1119.

Smith, A. and Ivins, B. (1984). Spatial relationships and social organization in adult pikas: A facultatively monogamous mammal. *Z. Tierpsychol.* **66**, 289–308.

Smith, A.T., and Gilpin, M. (1997a). Spatially correlated dynamics in a pika metapopulation. *In* "Metapopulation Biology: Ecology, Genetics and Evolution" (I. Hanski and M.E. Gilpin eds.), pp. 407–428. Academic Press, San Diego.

Smith, D.L., Lucey, B., Waller, L.A., Childs, J.E., and Real, L.A. (2002). Predicting the spatial dynamics of rabies epidemics on heterogeneous landscapes. *Proc. Natl. Acad. Sci. USA* **99**, 3668–3672.

Smith, J.N.M., Taitt, M.J., Rogers, C.M., Arcese, P., Keller, L.F., Cassidy, A.L.E.V., and Hochachka, W.M. (1996). A metapopulation approach to the population biology of the song sparrow *Melospiza melodia. Ibis* **138**, 120–128.

Smith, S.E., Read, D.J. (1997b). "Mycorrhizal Symbioses." 2nd ed. Academic Press, London.

Smouse, P.E. (1978). Multiple locus problems in human-genetics. *Genetics* **88**, S93–S93.

Smouse, P.E., Long, J.C., and Sokal, R.R. (1986). Multiple-regression and correlation extensions of the Mantel test of matrix correspondence. *System. Zool.* **35**, 627–632.

Smouse, P.E., Spielman, R.S., and Park, M.H. (1982). Multiple-locus allocation of individuals to groups as a function of the genetic-variation within and differences among human-populations. *Am. Nat.* **119**, 445–463.

Smouse, P.E., Waples, R.S., and Tworek, J.A. (1990a). A genetic mixture analysis for use with incomplete source population-data. *Can. J. Fish. Aquat. Sci.* **47**, 620–634.

Smouse, P.E., Waples, R.S., and Tworek, J.A. (1990b). Genetic analysis of Chinook salmon taken at Bonneville Dam: A mixed fishery model for use with incomplete baseline data. *Can. J. Fish. Aquat. Sci.*

Snyder, R.E., and Chesson, P. (2003). Local dispersal can facilitate coexistence in the presence of long-lasting spatial heterogeneity. *Ecol. Lett.* **6**, 1–9.

Soons, M.B., Heil, G.W. (2002). Reduced colonisation capacity in fragmented populations of wind-dispersed grassland forbs. *J. Ecol.*

Sorci, G., Morand, S., and Hugot, J.P. (1997). Host-parasite coevolution: Comparative evidence for covariation of life history traits in primates and oxyurid parasites. *Proc. R. Soc. Lond. Ser. B Biol. Sci.* **264**, 285–289.

Soulé, M., and Wilcox, B.A. (1980). "Conservation Biology: An Evolutionary-Ecological Approach." Sinauer Sunderland, MA.

Soulé, M.E. (1986). "Conservation Biology." Sinauer, Sunderland, MA.

Sousa, W.P. (1984). The role of disturbance in natural communities. *Annu. Rev. Ecol. Syst.* **15**, 353–391.

South, A. (1999). Dispersal in spatially explicit population models. *Conserv. Biol.* **13**, 1039–1046.

Southwood, T.R.E. (1962). Migration of terrestrial arthropods in relation to habitat. *Biol. Rev.* **37**, 170–214.

Southwood, T.R.E. (1988). Tactics, strategies and templets. *Oikos* **52**, 3–18.

Spiegelhalter, D.J., Best, N.G., Carlin, B. R., and van der Linde, A. (2002). Bayesian measures of model complexity and fit. *J. R. Stat. Soc. Ser. B Statist. Methodol.* **64**, 583–616.

Srivastava, D.S. (1999). Using local-regional richness plots to test for species saturation: Pitfalls and potentials. *J. Anim. Ecol.* **68**, 1–17.

Stacey, P., Johnson, V., and Taper, M. (1997). Migration within metapopulations: The impact upon local population dynamics. *In* "Metapopulation Biology: Ecology, Genetics and Evolution" (I. Hanski and M.E. Gilpin eds.), pp. 267–291. Academic Press, San Diego.

Stamps, J.A. (1987). Conspecific as cues to territory quality: A preference of juvenile lizard (*Anolis aenus*) for previously used territory. *Am. Nat.* **129**, 629–642.

Stamps, J.A. (1988). Conspecific attraction and aggregation in territorial species. *Am. Nat.* **131**, 329–347.

Stamps, J.A. (1991). The effect of conspecifics on habitat selection in territorial species. *Behav. Ecol. Sociobiol.* **28**, 29–36.

Stamps J.A. (2001). Habitat selection by dispersers: integrating proximate and ultimate approaches. In "Dispersal" (J. Clobert, E. Danchin, A.A. Dhondt, and J.D. Nichols, eds), pp. 230–242. Oxford University Press, Oxford.

Stanton, M.L., and Galen, C. (1997). Life on the edge: Adaptation versus environmentally mediated gene flow in the snow buttercup, *Rannunculus adoneus. Am. Nat.* **150**, 143–178.

Stearns, S.C. (1992). "The Evolution of Life Histories." Oxford Univ. Press, New York.

Stearns, S.C. (1994). The evolutionary links between fixed and variable traits. *Acta Paleontol. Polon.* **38**, 215–232.

Stearns, S.C., Ackermann, M., Doebeli, M., and Kaiser, M. (2000). Experimental evolution of aging, growth, and reproduction in fruitflies. *Proc. Natl. Acad. Sci. USA* **97**, 3309–3313.

Stearns, S.C., and Sage, R.D. (1980). Maladaptation in a marginal population of the mosquito fish, *Gambusia affinis. Evolution* **34**, 65–75.

Steffan-Dewenter, I., Münzenberg, U. Bürger, C., Thies, C., and Tscharntke, T. (2002). Scale-dependent effects of landscape context on three pollinator guilds. *Ecology* **83**, 1421–1432.

Steffan-Dewenter, I., Münzenberg, U., and Tscharntke, T. (2001). Pollination, seed set and seed predation on a landscape scale. *Proc. R. Soc. Lond. B* **268**, 1685–1690.

Steiner, C.F., and Leibold, M.A. (2003). Cyclic assembly trajectories and the generation of scale-dependent productivity-diversity relationships. Submitted for publication.

Stenseth, N.C. (1983). Causes and consequences of dispersal in small mammals. *In* "The Ecology of Animal movement" (I.R. Swingland and P.J. Greenwood, eds.), pp. 63–101. Clarendon Press, Oxford.

Stenseth, N.C., and Lidicker, W.Z., Jr. (eds.) (1992). "Animal Dispersal: Small Mammal as a Model." Chapman & Hall, London.

Stephan, T. (1993). "Stochastische Modelle zur Extinktion von Populationen." Doctoral Thesis, University of Marburg, Germany.

Stephens, M. (2001). Inferences under the coalescent. *In* "Handbook of Statistical Genetics" (D.J. Balding, M.J. Bishop, and C. Cannings, eds.). Wiley, Chichester, England.

Stephens, P.A., and Sutherland, W.J. (1999). Consequences of the Allee effect for behaviour, ecology, and conservation. *Trends Ecol. Evol.* **14**, 401–405.

Stephens, P.A., Sutherland, W.J., and Freckleton, R.P. (1999). What is the Allee effect? *Oikos* **87**, 185–190.

Stewart, R.R., Noyce, T., and Possingham, H.P. (2003). The opportunity cost of *ad hoc* marine reserve decisions: An example from South Australia. *Mar. Ecol. Progr. Ser.*

Stewart, W.A., Dallas, J.F., Piertney, S.B., Marshall, F., Lambin, X., and Telfer, S. (1999). Metapopulation genetic structure in the water vole, *Arvicola terrestris*, in NE Scotland. *Biol. J. Linn. Soc.* **68**, 159–171.

Stoddart, D. (1970). Individual range, dispersion and dispersal in a population of water voles (*Arvicola terrestris* (L.)). *J. Anim. Ecol.* **39**, 403–425.

Storz, J.F., Ramakrishnan, U., and Alberts, S.C. (2001). Determinants of effective population size for loci with different modes of inheritance. *J. Hered.* **92**, 497–502.

Strachan, C., Strachan, R., and Jefferies, D. (2000). Preliminary report on the changes in the water vole population of Britain as shown by the national survey of 1989–1990 and 1996–1998. The Vincent Wildlife Trust, London.

Strickland, D. (1991). Juvenile dispersal in gray jays: Dominant brood member expels siblings from natal territory. *Can. J. Zool.* **69**, 2935–2945.

Sugg, D.W., and Chesser, R.K. (1994). Effective population sizes with multiple paternity. *Genetics* **137**, 1147–1155.

Sugg, D.W., Chesser, R.K., Dobson, F.S., and Hoogland, J.L. (1996). Population genetics meets behavioral ecology. *Trends Ecol. Evol.* **11**, 338–342.

Sumner, J., Rousset, F., Estoup, A., and Moritz, C. (2001). Neighbourhood' size, dispersal and density estimates in the prickly forest skink (*Gnypetoscincus queenslandiae*) using individual genetic and demographic methods. *Mol. Ecol.* **10**(8), 1917–1927.

Sutcliffe, O.L., and Thomas, C.D. (1996). Open corridors appear to facilitate dispersal by ringlet butterflies (*Aphantopus hyperantus*) between woodland clearings. *Conserv. Biol.* **10**, 1359–1365.

Sutcliffe, O.L., Thomas, C.D., and Peggie, D. (1997). Area-dependent migration by ringlet butterflies generates a mixture of patchy-population and metapopulation attributes. *Oecologia* **109**, 229–234.

Sutherland, G.D., Harestad, A.S., Price, K., and Lertzman, K.P. (2000). Scaling of natal dispersal distances in terrestrial birds and mammals. *Conserv. Ecol.* **4**, 16.

Swingland, I.R. (1983). Intraspecific differences in movement. *In* "The Ecology of Animal movement" (I.R. Swingland and P.J. Greenwood, eds.), pp. 102–115. Clarendon Press, Oxford.

Swinton, J., Harwood, J., Grenfell, B.T., and Gilligan, C.A. (1998). Persistence thresholds for phocine distemper virus infection in harbour seal *Phoca vitulina* metapopulations. *J. Anim. Ecol.* **67**, 54–68.

Szacki, J. (1999). Spatially structured populations: How much do they match the classic metapopulation concept? *Landsc. Ecol.* **14**, 369–379.

Tachida, H., and Iizuka, M. (1991). Fixation probability in spatially changing environments. *Genet. Res.* **58**, 243–251.

Tackenberg, O. (2003). Modelling long distance dispersal of plant diaspores by wind. *Ecol. Monogr.*

Tackenberg, O., Poschlod, P., and Bonn, S. (2003). Assessment of wind dispersal potential in plant species *Ecol. Monogr.*

Tainaka, K. (1988). Lattice model for the Lotka-Volterra system. *J. Phy. Soci. Jpn.* **57**, 2588–2590.

Tainaka, K. (1994). Vortices and strings in a model ecosystem. *Phys. Rev. E.* **50**, 3401–3409.

Tajima, F. (1983). Evolutionary relationship of DNA sequences in finite populations. *Genetics* **105**, 437–460.

Tajima, F. (1989). Statistical method for testing the neutral mutation hypothesis by DNA polymorphism. *Genetics* **123**, 585–595.

Tajima, F. (1993). Measurement of DNA polymorphism. *In* "Mechanisms of Molecular Evolution" (N. Takahata and A.G. Clark, eds.), pp. 37–60. Sinauer, Sunderland, MA.

Takenaka, Y., Matsuda, H., and Iwasa, Y. (1997). Competition and evolutionary stability of plants in a spatially structured habitat. *Res. Popul. Ecol.* **39**, 67–75.

Taneyhill, D.E. (2000). Metapopulation dynamics of multiple species: The geometry of competition in a fragmented habitat. *Ecol. Monogr.* **70**, 495–516.

Taper, M.L. (1990). Experimental character displacement in the adzuki bean weevil, *Callosobruchus chinensis*. *In* “Bruchids and Legumes: Economics, Ecology and Coevolution.” (K. Fujii, ed.), pp. 289–301. Kluwer Academic.

Tavaré, S. (1984). Lines-of-descent and genealogical processes, and their application in population genetic models. *Theor. Popul. Biol.* **26**, 119–164.

Tavaré, S., Balding, D.J., Griffiths, R.C., and Donnelly, P. (1997). Inferring coalescence times from DNA sequence data. *Genetics* **145**, 505–518.

Taylor, D.R., Aarssen, L.W., and Loehle, C. (1990). On the relationship between *r/K* selection and environmental carrying capacity: A new habitat templet for plant life history strategies. *Oikos* **58**, 239–250.

Taylor, P.D. (1988). An inclusive fitness model for dispersal of offspring. *J. Theor. Biol.* **130**, 363–378.

Taylor, P.D., Fahrig, L., Henein, K., and Merriam, G. (1993). Connectivity is a vital element of landscape structure. *Oikos* **68**, 571–573.

Taylor, P.D., and Irwin, A.J. (2000). Overlapping generations can promote altruistic behaviour. *Evolution* **54**, 1135–1141.

Teleky, S.B. (1980). Egg cannibalism in *Tribolium* as a model for interference competition. *Res. Popul. Ecol.* **21**, 217–227.

Telfer, S., Dallas, J.F., Aars, J., Piertney, S.B., Stewart, W.A., and Lambin X. (2003a). Population structure and genetic composition of fossorial water voles *Arvicola terrestris* on Scottish islands. *J. Zool. (Lond.)*, **259**, 23–29.

Telfer, S., Holt, A., Donaldson, R., and Lambin, X. (2001). Metapopulation processes and persistence in remnant water vole populations. *Oikos* **95**, 31–42.

Telfer, S., Piertney, S., Dallas, J., Stewart, W., Marshall, F., Gow, J., and Lambin, X. (2003b). Parentage assignment detects frequent and large-scale dispersal in water voles. *Mol. Ecol.*

Templeton, A.R. (1981). Mechanisms of speciation: A population genetic approach. *Annu. Rev. Ecol. System.* **12**, 23–48.

Ter Braak, C.J.F., and Etienne, R.S. (2003). Improved Bayesian analysis of metapopulation data with an application to a tree frog metapopulation. *Ecology* **84**, 231–241.

Ter Braak, C.J.F., Hanski, I., and Verboom, J. (1998). The incidence function approach to the modeling of metapopulation dynamics. *In* “Modeling Spatiotemporal Dynamics in Ecology.” (J. Bascompte and R.V. Solé, eds.), pp. 167–188. Springer-Verlag, Berlin.

Tewksbury, J.J., Levey, D.J., Haddad, N.M., Sargent, S., Orrock, J.L., Weldon, A., Danielsen, B.J., Brinkerhoff, J., Damschen, E.I., and Townsend, P. (2002). Corridors affect plants, animals, and their interactions in fragmented landscapes. *Proc. Nat. Acad. Sci. USA* **99**, 12923–12926.

Theodorou, K., and Couvet, D. (2002). Inbreeding depression and heterosis in a subdivided population: Influence of the mating system. *Genet. Res.* **80**, 107–116.

Thiede, D.A., and Augspurger, C.K. (1996). Intraspecific variation in seed dispersion of *Lepidium campestre* (brassicaceae). *Am. J. Bot.* **83**, 856–866.

Thies, C., and Tscharntke, T. (1999). Landscape structure and biological control in agroecosystems. *Science* **285**, 893–895.

Thomas, C.D. (1985a). Specializations and polyphagy of *Plebejus argus* (Lepidoptera: Lycaenidae) in North Wales. *Ecol. Entomol.* **10**, 325–340.

Thomas, C.D. (1985b). The status and conservation of the butterfly *Plebejus argus* (Lepidoptera: Lycaenidae) in North West Britain. *Biol. Conserv.* **33**, 29–51.

Thomas, C.D. (1991). Spatial and temporal variability in a butterfly population. *Oecologia (Berl.)* **87**, 577–580.

Thomas, C.D. (1994a). Extinction, colonization and metapopulations: Environmental tracking by rare species. *Conserv. Biol.* **8**, 373–378.

Thomas, C.D. (1994b). Local extinctions, colonizations and distributions: Habitat tracking by British butterflies. *In* "Individuals, Populations and Patterns in Ecology" (S.R. Leather, A.D. Watt, N.J. Mills, and K.F.A. Walters, eds.), pp. 319–336. Intercept Ltd., Andover.

Thomas, C.D. (1996). Essential ingredients of real metapopulations, exemplified by the butterfly *Plebejus argus*. *In* "Aspects of the Genesis and Maintenance of Biological Diversity" (M.E. Hochberg, J. Clobert, and R. Barbault, eds.), pp. 292–307. Oxford University Press, Oxford.

Thomas, C.D., Baguette, M., and Lewis, O.T. (2000). Butterfly movement and conservation in patchy landscapes. *In* "Behaviour and Conservation" (L.M.Gosling and W.J. Sutherland, eds.), pp. 85–104. Cambridge Univ. Press, Cambridge.

Thomas, C.D., Bodsworth, E.J., Wilson, R.J., Simmons, A.D., Davies, Z.G., Musche, M., and Conradt, L. (2001a). Ecological and evolutionary processes at expanding range margins. *Nature* **411**, 577–581.

Thomas, C.D., Glen, S. W.T., Lewis, O.T., Hill, J.K., and Blakeley, D.S. (1999a). Population differentiation and conservation of endemic races: The butterfly *Plebejus argus*. *Anim. Conserv.* **2**, 15–21.

Thomas, C.D., and Hanski, I. (1997). Butterfly metapopulations. *In* "Metapopulation Biology: Ecology, Genetics, and Evolution." (I. Hanski and M.E. Gilpin, eds.), pp. 359–386. Academic Press, San Diego.

Thomas, C.D., and Harrison, S. (1992). Spatial dynamics of a patchily-distributed butterfly species. *J. Anim. Ecol.* **61**, 437–446.

Thomas, C.D., Hill, J.K., and Lewis, O.T. (1998). Evolutionary consequences of habitat fragmentation in a localized butterfly. *J. Anim. Ecol.* **67**, 485–497.

Thomas, C.D., and Jones,T.M. (1993). Partial recovery of a skipper butterfly (*Hesperia comma*) from population refuges: Lessons for conservation in a fragmented landscape. *J. Anim. Ecol.* **62**, 472–481.

Thomas, C.D., and Kunin, W.E. (1999). The spatial structure of populations. *J. Anim. Ecol.* **68**, 647–657.

Thomas, C.D., and Singer, M.C. (1987). Variation in host preference affects movement patterns in a butterfly population. *Ecology* **68**, 1262–1267.

Thomas, C.D., Singer M.C., and Boughton D.A. (1996). Catastrophic extinction of population sources in a butterfly metapopulation. *Am. Nat.* **148**, 957–975.

Thomas, C.D., Thomas, J.A., and Warren M.S. (1992). Distributions of occupied and vacant butterfly habitats in fragmented landscapes. *Oecologia* **92**, 563–567.

Thomas, C.D., Wilson, R.J., and Lewis, O.T. (2002a). Short-term studies underestimate 30-generation changes in a butterfly metapopulation. *Proc. R. Soci. Lond. Seri. B* **269**, 563–569.

Thomas, F., Brown, S.P., Sukhdeo, M., and Renaud, F. (2002b). Understanding parasite strategies: A state-dependent approach? *Trends Parasitol.* **18**, 387–390.

Thomas, J.A. (1984). The conservation of butterflies in temperate countries: Past efforts and lessons for the future. *Symp. R. Entomol. Soc. Lond.* **11**, 333–353.

Thomas, J.A., Bourn, N.A.D., Clarke, R.T., Stewart, K.E., Simcox, D.J., Pearman, G.S., Curtis, R., and Goodger, B. (2001b). The quality and isolation of habitat patches both determine where butterflies persist in fragmented landscapes. *Proc. R. Soci. Lond. Seri. B* **268**, 1791–1796.

Thomas, J.A., Rose, R.J., Clarke, R.T., Thomas, C.D., and Webb, N.R. (1999b). Intraspecific variation in habitat availability among ectothermic animals near their climatic limits and their centres of range. *Funct. Ecol.* **13**(Suppl. 1), 55–64.

Thomas, J.A., Thomas, C.D., Simcox, D.J. and Clarke, R.T. (1986). The ecology and declining status of the silver-spotted skipper butterfly (*Hesperia comma*) in Britain. *J. Appl. Ecol.* **23**, 365–380.

Thompson, J.N. (1994). The Coevolutionary Process. University of Chicago Press, Chicago, IL.

Thompson, K., Bakker, J., and Bekker, R. (1997). "The Soil Seed Banks of North West Europe: Methodology, Density and Longevity. Cambridge Univ. Press, Cambridge.

Thompson, K., Bakker, J., Bekker, R.M., and Hodgson, J.G. (1998). Ecological correlates of seed persistence in soil in the north-west European flora. *J. Ecol.* **86**, 163–169.

Thompson, J.N., and Burdon, J.J. (1992). Gene-for-gene coevolution between plants and parasites. *Nature* **360**, 121–125.

Thrall, P.H., and Antonovics, J. (1995). Theoretical and empirical studies of metapopulations: Population and genetic dynamics of the *Silene-Ustilago* system. *Can. J. Bot.* **73** (Suppl.), 1249–1258.

Thrall, P.H., and Jarosz, A. M. (1994a). Host-pathogen dynamics in experimental populations of *Silene alba* and *Ustilago violacea*. I. Ecological and genetic determinants of disease spread. *J. Ecol.* **82**, 549–559.

Thrall, P.H., and Jarosz, A.M. (1994b). Host-pathogen dynamics in experimental populations of *Silene alba* and *Ustilago violacea*. II. Experimental tests of theoretical models. *J. Ecol.* **82**, 561–570.

Tikka, P.M., Högmander, H., and Koski, P.S. (2001). Road and railway verges serve as dispersal corridors for grassland plants. *Landsc. Ecol.* **16**, 659–666.

Tilman, D. (1980). Resources: A graphical-mechanistic approach to competition and predation. *Am. Nat.* **116**, 362–393.

Tilman, D. (1994). Competition and biodiversity in spatially structured habitats. *Ecology* **75**, 2–16.

Tilman, D. (1999). The ecological consequences of changes in biodiversity: A search for general principles. *Ecology* **80**, 1455–1474.

Tilman, D., Lehman, C.L., and Yin, C.J. (1997). Habitat destruction, dispersal, and deterministic extinction in competitive communities. *Am. Nat.* **149**, 407–435.

Tilman, D., May, R.M., Lehman, C.L., and Nowak, M.A. (1994). Habitat destruction and the extinction debt. *Nature* **371**, 65–66.

Tilman, D., and Pacala, S. (1993). The maintenance of species richness in plant communities. *In* "Species Diversity in Ecological Communities," (R. Ricklefs, and D. Schluter, eds.), pp. 13–25. University of Chicago Press, Chicago.

Tischendorf, L., and Fahrig, L. (2000a). How should we measure landscape connectivity? *Landsc. Ecol.* **15**, 633–641.

Tischendorf, L., and Fahrig, L. (2000b). On the usage and measurement of landscape connectivity. *Oikos* **90**, 7–19.

Tischendorf, L., and Fahrig, L. (2001). On the use of connectivity measures in spatial ecology: A reply. *Oikos* **95**, 152–155.

Travis, J.M.J., and Dytham, C. (1998). The evolution of dispersal in a metapopulation: A spatially explicit, individual-based model. *Proc. R. Soc. Lond. B Biol. Sci.* **265**, 17–23.

Travis, J.M.J., and Dytham, C. (1999). Habitat persistence, habitat availability and the evolution of dispersal. *Proc. R. Soc. Lond. B Biol. Sci.* **266**, 723–728.

Travis, J.M.J., Murrell, D.J., and Dytham, C. (1999). The evolution of density-dependent dispersal. *Proc. R. Soc. Lond. B Biol. Sci.* **266**, 1837–1842.

Trefilov, A., Berard, J., Krawczak, M., and Schmidtke, J. (2000). Natal dispersal in Rhesus Macaques is related to transporter gene promoter variation. *Behav. Genet.* **30**, 295–301.

Tregenza, T. (1995). Building on the ideal free distribution. *Adv. Ecol. Res.* **26**, 253–307.

Trzcinski, M.K., Fahrig, L., and Merriam, G. (1999). Independent effects of forest cover and fragmentation on forest breeding birds. *Ecol. Appl.* **9**, 586–593.

Tuljapurkar, S.D. (1982). Population dynamics in variable environments. III. Evolutionary dynamics of r-selection. *Theor. Popul. Biol.* **21**, 141–165.

Tuljapurkar, S.D. (1990). "Population Dynamics in Variable Environments." Springer-Verlag, New York.

Tuljapurkar, S., and Orzack, S.H. (1980). Population dynamics in variable environments. I. Long-run growth rates and extinction. *Theor. Popul. Biol.* **18**, 314–342.

Turchin, P. (1991). Translating foraging movements in heterogeneous environments into the spatial distribution of foragers. *Ecology* **72**, 1253–1266.

Turchin, P. (1995). Population regulation: Old arguments and a new synthesis. *In* "Population Dynamics: New Approaches and Synthesis." (P. Price and N. Cappuccino eds.), pp. 19–40. Academic Press, London.

Turchin, P. (1998). "Quantitative Analysis of Movement: Measuring and Modeling Population Redistribution in Animals and Plants." Sinauer Associates, Sunderland, MA.

Turchin, P. (2003). "Complex Population Dynamics." Princeton Univ. Press, Princeton, NJ.

Turelli, M., and Barton, N.H. (1994). Genetic and statistical analyses of strong selection on polygenic traits: What, me normal? *Genetics* **138**, 913–941.

Turelli, M., Barton, N.H., and Coyne, J.A. (2001). Theory and speciation. *Trends Ecol. Evol.* **16**, 330–390.

Turkington, R., and Harper, J.L. (1979). The growth distribution and neighbour relationships of *Trifolium repens* in a permanent pasture. IV. Fine-scale biotic differentiation. *J. Ecol.* **67**, 245–254.

Turnbull, L.A., Crawley, M.J., and Rees, M. (2000). Are plant populations seed-limited? A review of seed sowing experiments. *Oikos* **88**, 225–238.

Turner, M.G. (1989). Landscape ecology: The effect of pattern on process. *Annu. Rev. Ecol. System.* **20**, 171–197.

Turner, M.G., and Gardner, R.H. (1991). Quantitative methods in landscape ecology: An introduction. *In* "Quantitative Methods in Landscape Ecology" (M.G. Turner and R.H. Gardner, eds.), pp. 3–14. Springer-Verlag, New York.

Turner, M.G., Gardner, R.H., and O'Neill, R.V. (2001). "Landscape Ecology in Theory and Practice: Pattern and Process." Springer, New York.

Turner, M.G., Wu, Y.A., Wallace, L.L., Romme, W.H., and Brenkert, A. (1994). Simulating winter interactions among ungulates, vegetation, and fire in northern Yellowstone park. *Ecol. Appl.* **4**, 472–486.

Tyre, A.J., Possingham, H.P., and Niejalke, D.P. (2001). Detecting environmental impacts on metapopulations of mound spring invertebrates: Assessing an incidence function model. *Environ. Int.* **27**, 225–229.

Underhill, L.G. (1994). Optimal and suboptimal reserve selection algorithms. *Biol. Conserv.* **70**, 85–87.

Urban, D., Bonan, G., Smith, T., and Shugart, H. (1991). Spatial applications of gap models. *For. Ecol. Manage.* **42**, 95–110.

Urban, D., and Keitt, T. (2001). Landscape connectivity: A graph-theoretic perspective. *Ecology* **82**, 1205–1218.

Valderde, T., and Silvertown, J. (1998). Variation in the demography of a woodland understorey herb (*Primula vulgaris*) along the forest regeneration cycle: Projection matrix analysis. *J. Ecol.* **86**, 545–562.

Valentine, J. W. (1980). Determinants of diversity in higher taxonomic categories. *Paleobiology* **6**, 444–450.

van Baalen, M. (2000). Pair approximations for different spatial geometries. In the Geometry of Eological Interactions: Simplifying Spatial Complexity (U. Dieckman, R. Law, and J.A.J. Metz, eds.), pp. 359–387. Cambridge Univ. Press, Cambridge.

van Baalen, M., and Hochberg, M.E. (2001). Dispersal in antagonistic interactions. *In* "Dispersal" (J. Clobert, E. Danchin, A.A. Dhondt, and J.D. Nichols, eds.), pp. 299–310. Oxford Univ. Press, Oxford.

van Baalen, M., and Rand, D.A. (1998). The unit of selection in viscous populations and the evolution of altruism. *J. Theor. Biol.* **193**, 631–648.

van Baalen, M., and Sabelis, M.W. (1993). Coevolution of patch selection strategies of predator and prey and the consequences for ecological stability. *Am. Nat.* **142**, 646–670.

van der Heijden, M.G.A., Klironomos, J.N., Ursic, M., Moutoglis, P., Streitwolf-Engel, R., Boller, T., Wiemken, A., and Sanders, I.A. (1998). Mycorrhizal fungal diversity determines plant biodiversity, ecosystem variability and productivity. *Nature* **396**, 69–72.

Vandermeer, J., and Carvajal, R. (2001). Metapopulation dynamics and the quality of the matrix. *Am. Nat.* **158**, 211–220.

van der Meijden, E., Klinkhamer, P.G.L., de Jong, T.J., and van Wijk, C.A.M. (1992). Metapopulation dynamics of biennial plants: How to explore temporary habitats. *Acta Bot. Neerl.* **41**, 249–270

van der Putten, W.H., Van Dijk, C., and Peters, B.A.M. (1993). Plant-specific soil-borne diseases contribute to succession in foredune vegetation. *Nature* **362**, 53–55.

van Dorp, D., and Opdam, P. (1987). Effects of patch size, isolation and regional abundance on forest bird communities. *Landsc. Ecol.* **1**, 59–73.

van Dorp, D., Schippers, P., and van Groenendael, J.M. (1997). Migration rates of grassland plants along corridors in fragmented landscapes assessed with a cellular automation model. *Landsc. Ecol.* **12**, 39–50.

van Langevelde, F. (2000). Scale of habitat connectivity and colonization in fragmented nuthatch populations. *Ecography* **23**, 614–622.

van Nouhuys, S., and Hanski, I. (1999). Host diet affects extinctions and colonizations in a parasitoid metapopulation. *J. Anim. Ecol.* **68**, 1248–1258.

van Nouhuys, S., and Hanski, I. (2002). Colonization rates and distances of a host butterfly and two specific parasitoids in a fragmented landscape. *J. Anim. Ecol.* **71**, 630–650.

van Swaay, C.A.M., and Warren, M.S. (1999). "Red Data book of European butterflies (Rhopalocera)." Nature and Environment, No. 99, Council of Europe Publishing, Strasbourg.

van Tienderen, P.H. (1989). On the Morphology of *Plantago lanceolata*: Selection, Adaptation, Constraints. Ph.D. thesis, University of Utrecht, The Netherlands.

van Tienderen, P.H. (1997). Generalists, specialists, and the evolution of phenotypic plasticity in sympatric populations of distinct species. *Evolution* **51**, 1372–1380.

Van Treuren, R., Bijlsma, R., Ouborg, N.J., and Kwak, M.M. (1994). Relationships between plant density, outcrossing rates and seed set in natural and experimental populations of *Scabiosa columbaria*. *J. Evol. Biol.* **7**, 287–302.

Van Treuren, R., Bijlsma, R., Ouborg, N.J., and van Delden, W. (1993). The effects of population size and plant density on outcrossing rates in locally endangered *Salvia pratensis*. *Evolution* **47**, 1094–1104.

van Valen, L. (1971). Group selection and the evolution of dispersal. *Evolution* **25**, 591–598.

Vane-Wright, R.I., Humphries, C.J., and Williams, P.H. (1991). What to protect? Systematics and the agony of choice. *Biol. Conserv.* **55**, 235–254.

Vasek, F.C. (1980). Creosote buch: Long-lived clones in the Mohave Desert. *Am. J. Bot.* **67**, 246–255.

Vassiliadis, C., Valero, M., Saumitou-Laprade, P., and Godelle, B. (2000). A model for the evolution of high frequencies of males in an androdioecious plant based on a cross-compatibility advantage of males. *Heredity* **85**, 413–422.

Veldkamp, A., and Lambin, E.F. (2001) Predicting land-use change. *Agric. Ecosys. Environ.* **85**, 1–6.

Venable, D.L., and Brown, J.S. (1988). The selective interactions of dispersal, dormancy, and seed size as adaptations for reducing risk in variable environments. *Am. Nat.* **131**, 360–384.

Venable, D.L., and Burquez, A.M. (1989). Quantitative genetics of size, shape, life-history and fruit characteristics of the seed-heteromorphic composite *Heterosperma pinnatum*. I. Variation within and among populations. *Evolution* **43**, 113–124.

Verboom, J., Foppen, R., Chardon, P., Opdam, P., and Luttikhuizen, P. (2001). Introducing the key patch approach for habitat networks with persistent populations: An example for marshland birds. *Biol. Conserv.* **100**, 89–101.

Verboom, J., Lankester, K., and Metz, J.A.J. (1991a). Linking local and regional dynamics in stochastic metapopulation models. *Biol. J. Linn. Soc.* **42**, 39–55.

Verboom, J., Schotman, A., Opdam, P., and Metz, J.A.J. (1991b). European nuthatch metapopulations in a fragmented agricultural landscape. *Oikos* **61**, 149–156.

Verdonck, M.V. (1987). Adaptation to environmental heterogeneity in populations of *Drosophila melanogaster. Genet. Res. Camb.* **49**, 1–10.

Vergeer, P., Rengelink, R., Copal, A., and Ouborg, N.J. (2003). The interacting effects of genetic variation, habitat quality and population size on performance of *Succisa pratensis. J. Ecol.*

Verheyen, K., and Hermy, M. (2001). The relative importance of dispersal limitation of vascular plants in secondary forest succession in Muizen Forest, Belgium. *J. Ecol.* **89**, 829–840.

Via, S. (2001). *Trends Ecol. Evol.* **16**, 381–390.

Villard, M.-A., Trzcinski, M.K., and Merriam, G. (1999). Fragmentation effects on forest birds: Relative influence of woodland cover and configuration on landscape occupancy. *Conserv. Biol.* **13**, 774–783.

Vitalis, R., and Couvet, D. (2001a). Estimation of effective population size and migration rate from one- and two-locus identity measures. *Genetics* **157**, 911–925.

Vitalis, R., and Couvet, D. (2001b). Two-locus identity probabilities and identity disequilibrium in a partially selfing subdivided population. *Genet. Res. Camb.* **77**, 67–81.

Vos, C.C. (1999). "A Frog's-Eye View of the Landscape; Quantifying Connectivity for Fragmented Amphibian Populations." Ph.D. Thesis, Wageningen University, Wageningen, The Netherlands. Also published as IBN Scientific Contributions 18. DLO Institute for Forestry and Nature Research, Wageningen, The Netherlands.

Vos, C.C., Baveco, H., and Grashof-Borkdam, C.J. (2002). Corridors and species dispersal. *In* "Applying Landscape Ecology in Biological Conservation" (K.J. Gutzwiller, ed.), pp. 84–104. Springer, New York.

Vos, C.C., and Stumpel, A.H.P. (1996). Comparison of habitat-isolation parameters in relation to fragmented distribution patterns in the tree frog (*Hyla arborea*). *Landsc. Ecol.* **11**, 203–214.

Vos, C.C., ter Braak, C.J.F., and Nieuwenhuizen, W. (2000). Incidence function modelling and conservation of the tree frog *Hyla arborea* in the Netherlands. *Ecol. Bull.* **48**, 165–180.

Vos, C.C., Verboom, J., Opdam, P.F.M., and Ter Braak, C.J.F. (2001). Toward ecologically scaled landscape indices. *Am. Nat.* **183**, 24–41.

Waagepetersen, R., and Sorensen, D. (2001). A tutorial on reversible jump MCMC with a view toward applications in QTL-mapping. *Int. Stat. Rev.* **69**, 49–61.

Wade, M.J. (1978). A critical review of the models of group selection. *Q. Rev. Biol.* **53**, 101–114.

Wade, M.J. (1979). The primary characteristics of *Tribolium* populations group selected for increased and decreased population size. *Evolution* **33**, 749–764.

Wade, M.J. (1980). An experimental study of kin selection. *Evolution* **34**, 844–855.

Wade, M.J. (1982). The evolution of interference competition by individual, family, and group selection. *Proc. Natl. Acad. Sci. USA* **79**, 3575–3578.

Wade, M.J. (1985). The effects of genotypic interactions on evolution in structured populations. *In* "Proceedings of the XV International Congress of Genetics," pp. 283–290. IBH Publishing, New York.

Wade, M.J., (1990). Genotype-environment interaction for climate and competition in a natural population of flour beetles, *Tribolium castaneum*. *Evolution* **44**, 2004–2011.

Wade, M.J. (1996). Adaptation in subdivided populations: Kin selection and interdemic selection. *In* "Evolutionary Biology and Adaptation" (M.R. Rose, and G. Lauder, eds.), pp. 381–405. Sinauer Associates, Sunderland, MA.

Wade, M.J. (2000). Epistasis: Genetic constraint within populations and accelerant of divergence among them. "Epistasis and the Evolutionary Process." (J.B. Wolf, E.D. Brodie, and M.J. Wade, eds.), pp. 213–231. Oxford Univ. Press, Oxford.

Wade, P.R. (2001a). The conservation of exploited species in an uncertain world: Novel methods and the failure of traditional techniques. *In* "Conservation of Exploited Species" (J.D. Reynolds and G. Mace, eds.). Cambridge Univ. Press, Cambridge.

Wade, M.J. (2001b). Epistasis, complex traits, and rates of evolution. *Genetica* **112**, 59–69.

Wade, M.J. (2002). A gene's eye view of epistasis, selection, and speciation. *J. Evol. Biol.* **15**, 337–346.

Wade, M.J. (2003). Community genetics and species interactions: A commentary. *Ecology* **84**, 583–585.

Wade, M.J., and Goodnight, C.J. (1991). Wright's shifting balance theory: An experimental study. *Science* **253**, 1015–1018.

Wade, M.J., and Goodnight, C.J. (1998). The theories of Fisher and Wright in the context of metapopulations: When nature does many small experiments. *Evolution* **52**, 1537–1553.

Wade, M.J., and Griesemer, J.R. (1998). Populational heritability: Empirical studies of evolution in metapopulations. *Am. Nat.* **151**, 135–147.

Wade, M.J., and McCauley, D.E. (1980). Group selection: The phenotypic and genotypic differentiation of small populations. *Evolution* **34**, 799–812.

Wade, M.J., and McCauley, D.E. (1988). Extinction and colonization: Their effects on the genetic differentiation among populations. *Evolution* **42**, 995–1005.

Wagner, P.J. (1995). Testing evolutionary constraint hypothesis with early Paleozoic gastropods. *Paleobiology* **21**, 248–272.

Wahlberg, N., Klemetti, T., and Hanski, I. (2002a). Dynamic populations in a dynamic landscape: The metapopulation structure of the marsh fritillary butterfly. *Ecography* **25**, 224–232.

Wahlberg, N., Klemetti, T., Selonen, V., and Hanski, I. (2002b) Metapopulation structure and movements in five species of checkerspot butterflies. *Oecologia* **130**, 33–43.

Wahlberg, N., Moilanen, A., and Hanski, I. (1996). Predicting the occurrence of endangered species in fragmented landscapes. *Science* **273**, 1536–1538.

Wahlund, S. (1928). Zusammensetnung von populationen und korreletionserscheinungen vom standpunkt der vererbungslehre. *Hereditas* **11**, 65–105.

Wake, D.B. (1997). Incipient species formation in salamanders of the *Ensatina* complex. *Proc. Natl. Acad. Sci. USA* **94**, 7761–7767.

Wakeley, J. (1998). Segregating sites in Wright's island model. *Theor. Popul. Biol.* **53**, 166–175.

Wakeley, J. (1999). Non-equilibrium migration in human history. *Genetics* **153**, 1863–1871.

Wakeley, J. (2000). The effects of population subdivision on the genetic divergence of populations and species. *Evolution* **54**, 1092–1101.

Wakeley, J. (2001). The coalescent in an island model of population subdivision with variation among demes. *Theor. Popul. Biol.* **59**, 133–144.

Wakeley, J., and Aliacar, N. (2001),. Gene genealogies in a metapopulation. *Genetics* **159**, 893–905 (erratum to Fig. 2, Corrigenda (2002), *Genetics* **160**, 1263–1264).

Wakeley, J., and Hey, J. (1997). Estimating ancestral population parameters. *Genetics* **145**, 847–855.

Wakeley, J., Nielsen, R., Lui-Cordero, S.N., and Ardlie, K. (2001). The discovery of single nucleotide polymorphisms and inferences about human historical demography. *Am. J. Hum. Genet.* **69**, 1332–1347.

Waldmann, P. (1999). The effect of inbreeding and population hybridization on developmental instability in petals and leaves of the rare plant *Silene diclinis* (Caryophyllaceae). *Heredity* **83**, 138–144.

Walker, P.A.a.C., K.D. (1991) HABITAT: A procedure for modelling a disjoint environmental envelope for a plant or animal species. *Glob. Ecol. Biogeogr. Lett.* **1**, 108–118.

Walsh, J. J. 2003. Rates of adaptation in complex genetic systems. Ph.D. thesis, University of Sussex.

Walters, S. (2001). Landscape pattern and productivity effects on source-sink dynamics of deer populations. *Ecol. Model.* **143**, 17–32.

Wang, J., and Caballero, A. (1999). Developments in predicting the effective size of subdivided populations. *Heredity* **82**, 212–226.

Waples, R.S. (1991). Definition of "species" under the Endangered Species Act: Application to Pacific salmon. U.S. Department of Commerce, NOAA Tech. Memo. NMFS F/NWC-194.

Waples, R.S., and Smouse, P.E. (1990). Gametic disequilibrium analysis as a means of identifying mixtures of salmon populations. *Am. Fish. Soc. Symp.* **7**, 439–458.

Ware, D.M. (1995). A century and a half of change in the climate of the NE Pacific. *Fish. Oceanogr.* **4**, 267–277.

Warren, M.S. (1987). The ecology and conservation of the heath fritillary butterfly, *Mellicta athalia*. III. Population dynamics and the effect of habitat management. *J. Appl. Ecol.* **24**, 499–513.

Warren, M.S. (1992). Butterfly populations. *In* "The Ecology of Butterflies in Britain" (R.L.H. Dennis ed.), pp. 246–247. Oxford Univ. Press, Oxford.

Warren, M.S. (1994). The UK status and suspected metapopulation structure of a threatened European butterfly, the marsh fritillary *Eurodryas aurinia*. *Biol. Conserv.* **67**, 239–249.

Warren, P.H. (1996a). Dispersal and destruction in a multiple habitat system: An experimental approach using protist communities. *Oikos* **77**, 317–325.

Warren, P.H. (1996b). The effects of between-habitat dispersal rate on protist communities and metacommunities in microcosms at two spatial scales. *Oecologia* **105**, 132–140.

Warren, M.S., Hill, J.K., Thomas, J.A., Asher, J., Fox, R., Huntley, B., Roy, D.B., Telfer, M.G., Jeffcoate, S., Harding, P., Jeffcoate, G., Willis, S.G., Greatorex-Davies, J.N., Moss, D., and Thomas, C.D. (2001). Rapid responses of British butterflies to opposing forces of climate and habitat change. *Nature* **414**, 65–69.

Waser, N.M., Price, M.V., and Shaw, R.G. (2000). Outbreeding depression varies among cohorts of *Ipomopsis aggregata* planted in nature. *Evolution* **54**, 485–491.

Wasserman, S.S., and Futuyma, D.J. (1981). Evolution of host plant utilization in laboratory populations of the southern cowpea weevil, *Callosobruchus maculatus* Fabricius (Coleoptera: Bruchidae). *Evolution* **35**, 605–617.

Watkinson, A.R. (1985). On the abundance of plants along an environmental gradient. *J. Ecol.* **73**, 569–578.

Watkinson, A.R., Freckleton, R.P., and Forrester (2000). Population dynamics of *Vulpia ciliata*: Regional, metapopulation and local dynamics. *J. Ecol.* **88**, 1012–1029.

Watkinson, A.R., and Sutherland, W.J. (1995). Sources, sinks and pseudo-sinks. *J. Anim. Ecol.* **64**, 126–130.

Watt, A.S. (1947). Pattern and process in the plant community. *J. Ecol.* **35**, 1–22.

Watterson, G.A. (1975). On the number of segregating sites in genetical models without recombination. *Theor. Popul. Biol.* 7, 256–276.

Weddell, B.J. (1991). Distribution and movements of Columbian ground-squirrels (*Spermophilus-columbianus* (Ord) – are habitat patches like islands. *J. Biogeogr.* **18**, 385–394.

Weiss, G., and von Haeseler, A. (1998). Inference of population history using a likelihood approach. *Genetics* **149**, 1539–1546.

Weisser, W.W., Braendle, C., and Minoretti, N. (1999). Predator-induced morphological shift in the pea aphid. *Proc. R. Soc. Lond. B* **266**, 749–754.

Wells, H., Strauss, E.G., Rutter, M.A., and Wells, P.H. (1998). Mate location, population growth, and species extinction. *Biol. Conserv.* **86**, 317–324.

Wennergren, U., Ruckelshaus, M., and Kareiva, P. (1995). The promise and limitations of spatial models in conservation biology. *Oikos* **74**, 349–356.

Werner, F.E., Quinlan, J.A., Lough, R.G., and Lynch, D.R. (2001). Spatially explicit individual based modeling of marine populations: A review of the advances in the 1990s. *Sarsia* **86**, 411–421.

Westoby, M. (1998). A leaf-height-seed (LHS) plant ecology strategy scheme. *Plant Soil* **199**, 213–227.

Westphal, M.I., and Possingham, H.P. (2003). Applying a decision-theory framework to landscape planning for biodiversity: Follow-up to Watson et al. *Conserv. Biol.* **17**:327.

Whittaker, R.H. (1975). "Communities and Ecosystems," 2nd edn. Macmillan, New York.

Whittaker, R.H., and Levin, S.A. (1977). The role of mosaic phenomena in natural communities. *Theor. Popul. Biol.* **12**, 117–139.

Whitlock, M.C. (1992a). Nonequilibrium population structure in forked fungus beetles: Extinction, colonization, and the genetic variance among populations. *Am. Nat.* **139**, 952–970.

Whitlock, M.C. (1992b). Temporal fluctuations in demographic parameters and the genetic variance among populations. *Evolution* **46**, 608–615.

Whitlock, M.C. (1994). Fission and the genetic variance among populations: The changing demography of forked fungus beetle populations. *Am. Nat.* **143**, 820–829.

Whitlock, M.C. (2000). Fixation of new alleles and the extinction of small populations: Drift load, beneficial alleles, and sexual selection. *Evolution* **54**, 1855–1861.

Whitlock, M.C. (2002). Selection, load, and inbreeding depression in a large metapopulation. *Genetics* **160**, 1191–1202.

Whitlock, M.C. (2003). Fixation probability and time in a metapopulation. *Genetics*

Whitlock, M.C., and Barton, N.H. (1997). The effective size of a subdivided population. *Genetics* **146**, 427–441.

Whitlock, M.C., Ingvarsson, P.K., and Hatfield, T. (2000). Local drift load and the heterosis of interconnected populations. *Heredity* **84**, 452–457.

Whitlock, M.C., and McCauley, D.E. (1990). Some population genetic consequences of colony formation and extinction: Genetic correlations within founding groups. *Evolution* **44**, 1717–1724.

Whitlock, M.C., and McCauley, D.E. (1999). Indirect measures of gene flow and migration: F-ST not equal 1/(4Nm+1). *Heredity* **82**, 117–125.

Whitlock, M.C., and Phillips, P.C. (2000). The exquisite corpse: A shifting view of the shifting balance. *Trends Evol. Ecol.* **15**, 347–348.

Whitlock, M.C., Phillips, P.C., Moore, F.B.G., and Tonsor, S. (1995). Multiple fitness peaks and epistasis. *Annu. Rev. Ecol. Syst.* **26**, 601–629.

Whitlock, M.C., Phillips, P.C., and Wade, M.J. (1993). Gene interaction affects the additive genetic variance in subdivided populations with migration and extinction. *Evolution* **47**, 1758–1769.

Wiegand, T., Jeltsch, F., Hanski, I., and Grimm, V. (2003). Using pattern-oriented modeling for revealing hidden information: A key for reconciling ecological theory and application. *Oikos* **100**, 209–222.

Wiegand, T., Moloney, K.A., Naves, J., and Knauer, F. (1999). Finding the missing link between landscape structure and population dynamics: A spatially explicit perspective. *Am. Nat.* **154**, 605–627.

Wiegand, T., Naves, J., Stephan, T., and Fernandez, A. (1998). Assessing the risk of extinction for the brown bear (*Ursus arctos*) in the Cordillera Cantabrica, Spain. *Ecol. Monogr.* **68**, 539–570.

Wiens, J.A. (1989). Spatial scaling in ecology. *Funct. Ecol.* **3**, 385–397.

Wiens, J.A. (1996). Wildlife in patchy environments: Metapopulations, mosaics, and management. *In* "Metapopulations and Wildlife Conservation" (D.R. McCullough, ed.), pp. 53–84. Island Press, Washington, DC.

Wiens, J.A. (1997). Metapopulation dynamics and landscape ecology. *In* "Metapopulation Biology: Ecology, Genetics, and Evolution" (I. Hanski and M.E. Gilpin, eds.), pp. 43–62. Academic Press, San Diego.

Wiens, J.A. (2001): The landscape context of dispersal. In "Dispersal" (J. Clobert, E. Danchin, A.A. Dhondt, and J.D. Nichols, eds), pp. 96–109. Oxford University Press, Oxford.

Wiens, J.A. (2002). Central concepts and issues of landscape ecology. *In* "Applying Landscape Ecology in Biological Conservation" (K.J. Gutzwiller, ed.), pp. 3–21. Springer, New York.

Wiens, J.A., and Milne, B.T. (1989). Scaling of 'landscapes' in landscape ecology, or, landscape ecology from a beetle's perspective. *Landsc. Ecol.* **3**, 87–96.

Wiens, J.A., Schooley, R.L., and Weeks, R.D., Jr. (1997). Patchy landscapes and animal movements: Do beetles percolate? *Oikos* **78**, 257–264.

Wiens, J.A., Stenseth, N.C., Van Horne, B., and Ims, R.A. (1993). Ecological mechanisms and landscape ecology. *Oikos* **66**, 369–380.

Wilcove, D.S., Rothstein, D., Dubow, J., Phillips, A., and Losos, E. (1998). Assessing the relative importance of habitat destruction, alien species, pollution, overexploitation, and disease. *BioScience* **48**, 607–616.

Wilkinson-Herbots, H.M. (1998). Genealogy and subpopulation differentiation under various models of population structure. *J. Math. Biol.* **37**, 535–585.

Williams, D.W., and Liebhold, A.M. (2000). Spatial synchrony of spruce budworm outbreaks in eastern North America. *Ecology* **81**, 2753–2766.

Williams, G.C. (1966). "Adaptation and Natural Selection." Princeton Univ. Press, Princeton, NJ.

Williams, P. (2001) Complementarity. *In* "Encyclopedia of Biodiversity," pp. 813–829. Academic Press, London.

Williams, P.H., and Araújo M.B. (2000). Using probability of persistence to indentify important areas for biodiversity conservation. *Proc. R. Soc. Lon.* B, **267**:1959–1966.

Williams, P.H., and Araújo M.B. (2002). Apples, oranges, and probabilities: Integrating multiple factors into biodiversity conservation with consistency. *Env. Mod. Ass.*, 7, 139–151.

Williams, P.H., Burgess, N.D., and Rahbek, C. (2000). Flagship species, ecological complementarity, and conserving the diversity of mammals and birds in sub-Saharan Africa. *Anim. Conserv.* **3**, 249–260.

Williams, P.H., Margules, C.R., and Hilbert, D.W. (2002) Data requirements and data sources for biodiversity priority area selection. *J. Biosci.* **4**, 327–33.

Williams P.H. & Araújo M.B. (2002) Apples, oranges, and probabilities: Integrating multiple factors into biodiversity conservation with consistency. Environmental Modeling and Assessment, **7**: 139–151.

Wilson, D.S. (1980). "The Natural Selection of Populations and Communities." Benjamin/Cummings, Mento Park, CA.

Wilson, D.S. (1992). Complex interactions in metacommunities, with implications for biodiversity and higher levels of selection. *Ecology* **73**, 1984–2000.

Wilson, D.S., Pollock, G.B., and Dugatkin, L.A. (1992). Can altruism evolve in purely viscous populations. *Evol. Ecol.* **6**, 331–341.

Wilson, E.O. (1988). The current state of biological diversity. *In* "Biodiversity" (E.O. Wilson and F.M. Peter, eds.) National Academy Press, Washington, DC.

Wilson, E.O. (1989). Threats to biodiversity. *Sci. Am.* **261**, 60–66.

Wilson, G., and Rannala, B. (2003). Bayesian inference of recent migration rates using multilocus genotypes. *Genetics* **163**, 1177–1191.

Wilson, H.B. (2001). The evolution of dispersal from source to sink populations. *Evol. Ecol. Res.* **3**, 27–35.

Wilson, H.B., and Hassell, M.P. (1997). Host-parasitoid spatial models: The interplay of demographic stochasticity and dynamics. *Proc. R. Soc. Lond. B* **264**, 1189–1195.

Wilson, M.F., Rice, B.L., and Westoby, M. (1990). Seed dispersal spectra: A comparison of temperate plant communities. *J. Veg. Sci.* **1**, 547–562.

Wilson, R.J., Ellis, S.M., Baker, J.S., Lineham, M.E., Whitehead, R., and Thomas, C.D. (2002). Large-scale patterns of distribution and persistence at the range margins of a butterfly. *Ecology* **83**, 3357–3368.

Wilson, R.J., and Thomas, C.D. (2002). Dispersal and the spatial dynamics of butterfly populations. *In* "Dispersal Ecology" (J.M. Bullock, R.E. Kenward, and R.S. Hails, eds.), pp. 257–278. Blackwell Science, Oxford.

Wilson, W.G. (1998). Resolving discrepancies between deterministic population models and individual-based simulations. *Am. Nat.* **151**, 116–134.

Wilson, W.G., De Roos, A.M., and McCauley, E. (1993). Spatial instabilities within the diffusive Lotka-Volterra system: Individual-based simulation results. *Theor. Popul. Biol.* **43**, 91–127.

Wissmar, R.C., Smith, J.E., McIntosh, B.A., Li, H. W., Reeves, G.H., and Sedell, J.R. (1994). A history of use and disturbance in riverine basins of eastern Oregon and Washington (Early 1800s–1900s). *Northw. Sci.* **68**, 1–35.

With, K.A. (1994). Using fractal analysis to assess how species perceive landscape structure. *Landsc. Ecol.* **9**, 25–36.

With, K.A. (1997). The application of neutral landscape models in conservation biology. *Conserv. Biol.* **11**, 1069–1080.

With, K.A. (1999). Is landscape connectivity necessary and sufficient for wildlife management? *In* "Forest Fragmentation: Wildlife and Management Implications" (J.A. Rochelle, L.A. Lehmann, and J. Wisniewski, eds.), pp. 97–115. Brill, Leiden, The Netherlands.

With, K.A. (2002). Using percolation theory to assess landscape connectivity and effects of habitat fragmentation. *In* "Applying Landscape Ecology in Biological Conservation" (K.J. Gutzwiller, ed.), pp. 105–130. Springer-Verlag, New York.

With, K.A. (2003). Landscape conservation: A new paradigm for the conservation of biodiversity. *In* "Issues in Landscape Ecology" (J. A. Wiens and M. R. Moss, eds.). Cambridge Univ. Press, New York.

With, K.A., Cadaret, S.J., and Davis, C. (1999). Movement responses to patch structure in experimental fractal landscapes. *Ecology* **80**, 1340–1353.

With, K.A., and Crist, T.O. (1995). Critical thresholds in species' responses to landscape structure. *Ecology* **76**, 2446–2459.

With, K.A., and Crist, T.O. (1996). Translating across scales: Simulating species distributions as the aggregate response of individuals to heterogeneity. *Ecol. Model.* **93**, 125–137.

With, K.A., Gardner, R.H., and Turner, M.G. (1997). Landscape connectivity and population distributions in heterogeneous environments. *Oikos* **78**, 151–169.

With, K.A., and King, A.W. (1997). The use and misuse of neutral landscape models in ecology. *Oikos* **79**, 219–229.

With, K.A., and King, A.W. (1999a). Dispersal success on fractal landscapes: A consequence of lacunarity thresholds. *Landsc. Ecol.* **14**, 73–82.

With, K.A., and King, A.W. (1999b). Extinction thresholds for species in fractal landscapes. *Conserv. Biol.* **13**, 314–326.

With, K.A., and King, A.W. (2001). Analysis of landscapes sources and sinks: The effect of spatial pattern on avian demography. *Biol. Conserv.* **100**, 75–88.

Wiuf, C., and Hein, J. (1999). Recombination as a point process along sequences, *Theor. Popul. Biol.* **55**, 248–259.

Woiwood, I.P., Reynolds, D.R., and Thomas, C.D. (eds.) (2001). "Insect Movement: Mechanisms and Consequences." CAB Publication, Wallingford.

Wolf, D.E., Takebayashi, N., and Rieseberg, L.H. (2001). Predicting the risk of extinction through hybridization. *Conserv. Biol.* **15**, 1039–1053.

Wolf, J.B. (2000). Indirect genetic effects and gene interactions. *In* "Epistasis and the Evolutionary Process." (J.B. Wolf, E.D. Brodie, III, and M.J. Wade, eds.), pp. 158–176. Oxford Univ. Press, Oxford.

Wolf, J.B., Brodie, E.D., III, Cheverud, J.M., Moore, A.J., and Wade, M.J. (1998). Evolutionary consequences of indirect genetic effects. *Trends Ecol. Evol.* **13**, 64–69.

Wolf, J.B., Brodie, E.D., III, and Moore, A.J. (1999). Interacting phenotypes and the evolutionary process. II. Selection resulting from social interactions. *Am. Nat.* **153**, 254–266.

Wolf, J.B., Brodie, E.D., III, and Wade, M.J. (2003). Genotype-environment interaction and evolution when the environment contains genes. *In* "Phenotypic Plasticity: Functional and Conceptual Approaches." (T. DeWitt and S. Scheiner, eds.). Oxford Univ. Press, Oxford.

Wolf, J.B., and Wade, M.J. (2001). On the assignment of fitness to parents and offspring: Whose fitness is it and when does it matter? *J. Evol. Biol.* **14**, 347–358.

Wolff, J.O. (1992). Parents suppress reproduction and stimulate dispersal in opposite-sex juvenile white-fronted mice. *Nature* **359**, 409–410.

Woodbury, P.B., Beloin, R.M., Swaney, D.P., Gollands, B.E., and Weinstein, D.A. (2002). Using the ECLPSS software environment to build a spatially explicit component-based model of ozone effects on forest ecosystems. *Ecol. Model.* **150**, 211–238.

Woodroffe, R.B. (2000). Strategies for carnivore conservation: Lessons from contemporary extinctions. *In* "Carnivore Conservation," (J.L. Gittleman, R.K. Wayne, D.W. Macdonald, and S. Funk, eds.). Cambridge Univ. Press, Cambridge.

Wright, S. (1931). Evolution in Mendelian populations. *Genetics* **16**, 97–159.

Wright, S. (1932). The roles of mutation, inbreeding, crossbreeding and selection in evolution. *In* "Proceedings of the Sixth International Congress on Genetics," Vol. 1, pp. 356–366.

Wright, S. (1938). Size of population and breeding structure in relation to evolution. *Science* **87**, 430–431.

Wright, S. (1939). "Statistical Genetics in Relation to Evolution". Actualites scientifiques et industrielles 802. Exposes de biometrie et de la statistique biologique XIII. Hermann et Cie. Paris.

Wright, S. (1940). Breeding structure of populations in relation to speciation. *Am. Nat.* **74**, 232–248.

Wright, S. (1943). Isolation by distance. *Genetics* **28**, 114–138.

Wright, S. (1946). Isolation by distance under diverse systems of mating. *Genetics* **31**, 39–59.

Wright, S. (1951). The genetical structure of populations. *Ann. Eugen.* **15**, 323–354.

Wright, S. (1969). "The Theory of Gene Frequencies." The University of Chicago Press, Chicago, IL.

Wu, C.-I. (2001). The genic view of the process of speciation. *J. Evol. Biol.* **14**, 851–865.

Wu, C.-I., and Palopoli, M. F. (1994). Genetics of postmating reproductive isolation in animals. *Annu. Rev. Genet.* **27**, 283–308.

Yodzis, P. (1978). Competition for space and the structure of ecological communities. *In* "Lecture Notes in Biomathematics," Vol. 25. Springer-Verlag, Berlin.

Young, A., Boyle, T., and Brown, T. (1996). The population genetic consequences of habitat fragmentation for plants. *Trends Ecol. Evol.* **11**, 413–418.

Yu, D.W., Terborgh, J.W., and Potts, M.D. (1998). Can high tree species richness be explained by Hubbell's model? *Ecol. Lett.* **1**, 193–199.

Yu, D.W., and Wilson, H.B. (2001). The competition-colonization trade-off is dead; long live the competition-colonization trade-off. *Am. Nat.* **158**, 49–63.

Yu D.W, Wilson, H.B., and Pierce, N.E. (2001). An empirical model of species coexistence in a spatially structured environment. *Ecology* **82**, 1761–1771.

Zabel, J., and Tscharntke, T. (1998). Does fragmentation of *Urtica* habitats affect phytophagous and predatory insects differentially? *Oecologia* **116**, 419–425.

Zera, A.J., and Denno, R.F. (1997). Physiology and ecology of dispersal polymorphism in insects. *Annu. Rev. Ecol. Syst.* **42**, 207–231.

Zeyl, C., Mizesko, M., and de Visser, J. (2001). Mutational meltdown in laboratory yeast populations. *Evolution* **55**, 909–917.

Zhang D.Y., and Lin, K. (1997). The effects of competitive asymmetry on the rate of competitive displacement: How robust is Hubbell's community drift model? *J. Theor. Biol.* **188**, 361–367.

INDEX

C

Q

R

S

W